Dr. Ronald Clothier
Lecturer Environmental Studies
Northbrook College
? ?

Dutch Elm Disease v
viruses ? ?

The physiology of plant growth
and development

Consulting editor
Professor M. B. Wilkins
University of Nottingham

The physiology of plant growth and development

Edited by
Malcolm B. Wilkins
Professor of Plant Physiology
University of Nottingham

McGRAW-HILL · LONDON

New York · Toronto · Sydney · Mexico · Johannesburg · Panama

Published by
McGRAW-HILL Publishing Company Limited
MAIDENHEAD · BERKSHIRE · ENGLAND
07 094088 6

PRINTED AND BOUND IN GREAT BRITAIN

Preface

This book has been prepared for students taking advanced courses in plant physiology. For a number of years these students have had to rely heavily on review articles which are not always readily available and which, in most cases, are written for readers who already have a specialist knowledge of the subject. The level at which the topics are discussed in this book has been geared to meet the needs of students who have a sophisticated general appreciation of plant physiology and plant biochemistry, and who are now embarking upon advanced courses such as those found in, for example, the final year of undergraduate studies or the first year of graduate studies.

The number of papers being published in the scientific literature has now risen to a level which makes it virtually impossible for the individual scientist to keep abreast of developments in more than two or three fields. Furthermore, some areas of study have become so complex and specialized that it is extraordinarily difficult to evaluate, and place in proper perspective, a new development, unless one has an active research interest in that particular area. These factors have made it increasingly difficult, if not impossible, for one person to write an advanced textbook which deals authoritatively with a range of topics. A major aim of the present book is to overcome these difficulties by bringing together authors with a specialist knowledge in the fields chosen for coverage.

Several difficulties have to be faced in producing a multi-author textbook, the three principal ones being (a) changes in the literary style from one chapter to another, (b) variation in the depth of treatment of the subjects of the various chapters, and (c) continuity. The first of these difficulties must be accepted. Short of an editor rewriting each contribution there is no way in which uniformity of literary style can be attained. Such efforts would be quite useless and unwarranted since the clarity with which the authors have expressed themselves would be unlikely to be improved upon, and they would certainly not engender a happy relationship between the editor and the contributors! The second difficulty, variation in depth of treatment from chapter to chapter, is undoubtedly real, but it is questionable whether it is any greater in a multiple-author book than in a single-author book. In the former, it reflects the success, or otherwise, of the editor in giving the contributors an adequate number of guidelines on which to prepare their contributions. In the latter, it perhaps reflects the author's major fields of interest and the immense problems facing an individual in assessing research data in a wide range of topics. The third difficulty, continuity, is the most difficult to overcome. In an elementary textbook it is of critical importance since the reader's attention must be drawn to the inter-relationships between the various processes and phenomena under discussion. In an advanced textbook the problems of continuity are much less critical since the reader is assumed to have a general elementary knowledge of most of the topics discussed and of their general relationships to one another. In this book the arrangement of the chapters has been made with the aim of reducing to a minimum the problems of continuity, but in a multiple-author book they cannot be eliminated entirely. The Editor has resisted the temptation to write short connecting pieces between successive chapters because he does not believe that this is necessary in an advanced textbook. The Editor is convinced that in an advanced textbook all these difficulties are of minor significance when compared with the advantages of having each chapter written by a specialist, whose assessment of our understanding of a particular phenomenon is based on a thorough knowledge of both the fundamental problems involved and the experimental techniques which have been employed.

The Editor has been responsible for selecting both the topics and the contributors. The contributors have been responsible for the precise contents of their chapters and, within certain editorial guidelines, the way in which the topics have been approached. In selecting the topics to be included in this book the Editor was guided by what he felt ought to be included in an advanced course in whole-plant physiology. Some of the topics included in this book have been paid scant attention, or omitted altogether, in other books. Inevitably, some interesting topics have been omitted; an attempt to make the book encyclopaedic in nature would have defeated a second aim of the project: to produce a book within a certain price limit. Undoubtedly other editors would have selected, with equal justification, somewhat different topics, and omitted some of those included in this book. The physiology of plants cannot be studied in isolation; it must be studied in conjunction with biochemistry and the physical aspects of biology into which, of course, it merges imperceptibly. Chapters on various aspects of plant biochemistry have been deliberately omitted from this book because a number of excellent advanced textbooks on plant biochemistry are already available. A chapter on photosynthesis has been included because the processes involved in photon capture and electron transfer, on which Dr Kok has placed special emphasis, have a direct relevance to problems discussed in several other chapters, for example, those on other light-induced responses and on ion transport.

The Editor would like to thank his friends and fellow scientists who so readily accepted his invitation to contribute to this book. The care and precision with which they prepared their manuscripts in accordance with the rather complicated editorial instructions was greatly appreciated, and transformed the editing of this book from a task to a pleasure. In addition the Editor is particularly grateful to the editorial staff of the McGraw-Hill Publishing Company Ltd, for their help and advice during the production of this book. The Editor is also indebted to Dr T. K. Scott of Oberlin College, Ohio, who kindly read some of the chapters during his sabbatical year in Nottingham.

It has already been indicated that the aims of this project were to provide, at a reasonable price, an advanced textbook which gives an authoritative treatment of all the aspects of plant physiology selected for inclusion. Whether we have achieved these aims only time, and our critics, will tell.

MALCOLM B. WILKINS

Contributors

PROFESSOR L. J. AUDUS — Botany Department, University of London, Bedford College, Regents Park, London, N.W.1., U.K.

SIR NIGEL BALL, BT. — 4 Ennerdale Road, Kew Gardens, Surrey, U.K.

DR R. E. CLELAND — Botany Department, University of Washington, Seattle, Washington, 98105, U.S.A.

DR G. M. CURRY — Department of Biology, Tufts University, Medford, Massachusetts, 02155, U.S.A.

PROFESSOR J. DAINTY — School of Biological Sciences, University of East Anglia, University Plain, Norwich, NOR 24B, U.K.

DR J. E. FOX — Department of Botany, The University of Kansas, Lawrence, Kansas, 66045, U.S.A.

DR M. H. M. GOLDSMITH — Kline Biological Laboratories, Yale University, New Haven, Connecticut, 06520, U.S.A.

PROFESSOR O. V. S. HEATH — University of Reading, Horticultural Research Laboratories, Shinfield Grange, Shinfield, Berkshire, U.K.

DR W. S. HILLMAN — Biology Department, Brookhaven National Laboratory, Upton, New York, 11973, U.S.A.

DR BESSEL KOK — Bioscience Department, Research Institute for Advanced Studies, 1450 South Rolling Road, Baltimore, Maryland, 21227, U.S.A.

DR T. A. MANSFIELD — Department of Biology, Lancaster University, Lancaster, U.K.

PROFESSOR H. MOHR — Botanisches Institut der Universität 78 Freiburg i. Br., Schanzlestrasse 9–11, Germany.

DR I. D. J. PHILLIPS — Forestry Commission, Research Branch, Government Buildings, Bankhead Avenue, Edinburgh, 11, U.K.

DR H. W. SIEGELMAN	Biology Department, Brookhaven National Laboratory, Upton, New York, 11973, U.S.A.
PROFESSOR K. V. THIMANN	Division of Natural Sciences, University of California, Santa Cruz, California, 95060, U.S.A.
PROFESSOR P. F. WAREING	Botany Department, University College of Wales, Aberystwyth, Cardiganshire, U.K.
PROFESSOR M. B. WILKINS	Department of Physiology and Environmental Studies, University of Nottingham, School of Agricultural Sciences, Sutton Bonington, Loughborough, Leicestershire, U.K.
DR MARTIN H. ZIMMERMANN	Cabot Foundation of Harvard University, Harvard Forest, Petersham, Massachusetts, 01366, U.S.A.

Contents

Chapter 5   APICAL DOMINANCE   *by I. D. J. Phillips*

Chapter 6 GEOTROPISM *by L. J. Audus*

Chapter 7 PHOTOTROPISM *by George M. Curry*

Chapter 11   TRANSLOCATION OF NUTRIENTS
by Martin Zimmermann

Chapter 12   THE WATER RELATIONS OF PLANTS by J. Dainty

Chapter 13   THE IONIC RELATIONS OF PLANTS   *by J. Dainty*

Chapter 14   PHYTOCHROME   *by H. W. Siegelman*

Chapter 15   PHOTOMORPHOGENESIS   *by H. Mohr*

Chapter 16   PHOTOPERIODISM AND VERNALIZATION
       by W. S. Hillman

Chapter 17   GERMINATION AND DORMANCY
by P. F. Wareing

Chapter 18   CIRCADIAN RHYTHMS IN PLANTS
by Malcolm B. Wilkins

1. The auxins

Kenneth V. Thimann

Introduction

The fundamental concept that the growth of plants, and the interrelation between their parts, is controlled by hormones stems from the classical work of Charles and Francis Darwin. In Darwin's last book, *The Power of Movement in Plants* (1880),* a great many experiments are described bearing on the responses of plants to light and gravity. Among these was the critical finding that the extreme tip of a grass seedling controls the ability of the whole seedling to curve towards light. Covering the tip with an opaque cap made it insensitive (Fig. 1.1).

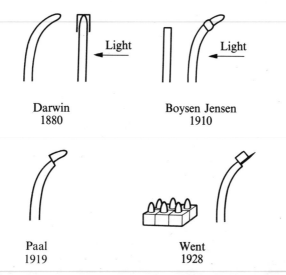

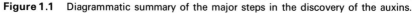

Figure 1.1   Diagrammatic summary of the major steps in the discovery of the auxins.

The coleoptile, a cylindrical sheath surrounding the first leaves of the grasses and cereals, was used in most of these early experiments. Many years later (1910), Boysen Jensen cut off the tip of the coleoptile and stuck it back on again with gelatin; the rest of the plant could still curve towards light, although if the tip had been completely removed it could not. It was this experiment that gave the first evidence that a material substance, rather than some ill-defined 'stimulus', was in control of growth. The explanation was given by Páal (1919),

* References for this first section may all be found in the book, *Phytohormones* (Went and Thimann).

2*

who did not use curvature towards light but cut off the tip and stuck it on a little to one side; then the plant curved even in darkness (Fig. 1.1). This shows, concluded Páal, that the tip secretes a substance which promotes the growth of the part below it. When the tip is *in situ*, of course, growth is promoted symmetrically. It follows that the curvature towards light, which is an unsymmetrical growth, must be due to an unsymmetrical distribution of this growth substance, a disproportionately large amount becoming present on the side of the seedling away from the light, causing that side to grow more and thus the plant to curve. The idea was generalized for all tropisms by Cholodny (1927) and supported also by straight growth measurements of intact and decapitated coleoptiles, especially by Söding (1925). Final proof was given by Went (1928) and Dolk (1930) in Holland. Went first found out how to determine the growth substance quantitatively, by placing the tips of numerous coleoptiles on a block of agar, then putting small cubes of the agar on one side of other coleoptiles whose tips had been removed (Fig. 1.1). If the times and conditions are carefully adhered to, the curvature is proportional to the number of tips used and to the length of time they have stayed on the agar. It is evident that we have to do with a substance which is continuously produced by the coleoptile tip, is readily diffusible and active in small quantities. Using this test, Went compared the amount of the substance diffusing out of the two sides of a coleoptile illuminated unilaterally, and showed that the shaded side yielded about twice as much as the illuminated side. Similarly, in a very extensive series of experiments Dolk showed that the lower side of a coleoptile tip which had been laid horizontally yielded about twice as much as the upper side. Thus in each case the redistribution of growth substance was enough to account for the redistribution of growth, that is, the curvature. In the case of geotropism, Dolk showed that the *total* growth did not change, so that the redistribution must be accomplished without loss.

These findings were soon broadened to apply to other seedlings and to mature plants of all types. No species-specificity was found. Thus a single group of auxins has become established as growth-controlling hormones throughout the vascular plants. Their role in the non-vascular plants is less clear.

In recent years it has become clear that total control of the plant is vested not in a single hormonal type—that of auxin—but is shared by several, specifically auxins, cytokinins, gibberellins, and ethylene, and this is further subject to modification by certain naturally occurring inhibitors, namely the phenols, flavonols, and abscisic acid. In this chapter we shall survey the auxins and some of their actions.

Chemical nature

Indole-acetic acid and its derivatives

The first studies of the chemical nature of auxin were directed to materials in which auxin activity is present in large amount, e.g., human urine and the culture media of several fungi. The isolation from human urine led to a peculiar false start, namely the claim for two closely-related acid substances, said to be derivatives of cyclopentene, and named auxin *a* and *b*. Neither of these has ever been isolated again, the chemical argument on which the formulae were based has been questioned, and the existence of these compounds is now no longer believed. Re-examination of urine (Kögl *et al.*, 1934) and examination of large-scale cultures of *Rhizopus suinus* (Thimann, 1935) and of yeast plasmolysate (Kögl and Kostermans, 1934) led to the isolation of a quite different substance, indole-3-acetic acid, I. Subsequently it was isolated from corn seeds (Haagen Smit *et al.*, 1942, 1946). It has now been identified, by chromatographic, chemical, and colorimetric methods, in a great many higher plants, including the *Avena* coleoptile, and has come to be recognized as the principal, or perhaps the only, true auxin of higher plants. It is found also in a number of fungi (see Gruen, 1959). The identification procedures include Rf's in different solvents, the colour reactions with Ehrlich's and more particularly Salkowski's reagent, the pK, molecular weight, and instability to hot acid, as well as the sensitivity to oxidation by peroxidases and the characteristics of its biological activity.

Supporting evidence for the widespread occurrence of Indole-3-acetic acid, abbreviated below as IAA, is also given by the isolation and identification of several of its derivatives, and the demonstration of their close relationship to the parent substance. These derivatives are the following:

(*a*) *Indole-3-acetaldehyde* (IAAld), II, was identified in etiolated seedlings, and later in other plant material, as a neutral substance readily converted to an auxin by soil, or, more specifically, by aldehyde dehydrogenase preparations (Larsen, 1944). Since this enzyme is almost universal in occurrence, the small and variable auxin effects which the aldehyde exerts (both in promoting growth of shoots and inhibiting growth of roots) are most probably due to conversion to the acid *in vivo*:

$$ICH_2CHO + H_2O - 2(H) \longrightarrow ICH_2COOH$$
$$(I = Indole\ nucleus)$$

(*b*) *Indole-3-pyruvic acid,* IPyA, III, was observed on chromatograms from corn seeds as a spot giving the same carmine colour as IAA with Salkowski reagent (Yamaki and Nakamura, 1952; Stowe and Thimann, 1954). It has been reported also in leaves and roots. It is converted to IAA by both spontaneous and enzymatic reactions, and also undergoes tautomeric change between keto and enol forms (Kaper and Veldstra, 1958; Schwarz, 1957; Kaper *et al.*, 1963). The solid is in the enol form, but alkaline pH favours the keto form and since this has quite a different UV spectrum and does not react with ammoniacal silver nitrate, the change was at first ascribed to breakdown of IPyA, and the

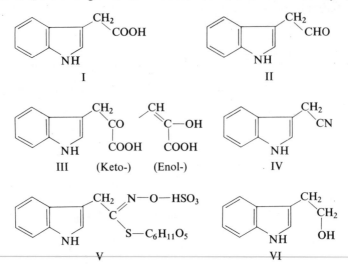

natural occurrence of IPyA doubted. However, the recent isolation of IPyA from corn, by a procedure involving no exposure to alkali, re-establishes its occurrence in plants (Winter, 1965).

Breakdown does occur as well, especially in alkaline pH, in light, and in air, and, of the series of seven or eight products resulting, only IAA and indole-aldehyde have been identified for certain; IAAld and indole-lactic acid are also probable, particularly in the reactions caused by *Agrobact. tumefaciens* (Kaper and Veldstra, 1958).

(c) **Indole-3-acetonitrile**, IAN, IV, was isolated from cabbage in 1952 (Jones *et al.*, 1952; Henbest *et al.*, 1953) and its presence has since been indicated by chromatography of 'neutral auxin' extracts from a number of plant materials. However, it is probable now that IAN does not occur free in plants, but only as a constituent of a thioglucoside named glucobrassicin, V. The IAN is liberated from V by the enzyme myrosinase, common in crucifers (Gmelin and Virtanen, 1961). If the hydrolysis is allowed to occur in tissue rich in ascorbic acid, an IAN-ascorbic complex, which was independently isolated under the name ascorbigen, is obtained (Prochazka and Sanda, 1958).

The most characteristic feature of the auxin activity of IAN is that it acts only on certain plants. In its discovery *Avena* coleoptiles were used for bio-assay, and its activity on them is even higher than that of IAA (Bentley and Housley, 1952). But on corn coleoptiles its activity is small and reaches a maximum over a very short concentration range, while on *Lupinus* hypocotyls or pea stems it is quite inactive (Thimann, 1954). This behaviour is due to the fact that it is converted to IAA by an enzyme called nitrilase, which occurs only in a few plant families, principally in the Cruciferae, Gramineae (though very weakly in some members), and Musaceae (Thimann and Mahadevan, 1964). Though it was discovered by this reaction, nitrilase hydrolyses many nitriles, both aromatic and aliphatic; the reaction requires two water molecules, apparently does not proceed via the expected amide as intermediate, and yields ammonia stoichiometrically, just like hydrolysis with hot alkali:

$$ICH_2CN + 2H_2O \longrightarrow ICH_2COO^- + NH_4^+$$

Its kinetics and the structural requirements for its action have been studied in some detail (Mahadevan and Thimann, 1964).

Since the activity of IAN on some plants is so high, it follows that hydrolysis must be rapid and virtually complete, as studies of the isolated enzyme confirm. However, activity *higher* than that of IAA calls for a further explanation, and this has been given by its more rapid entry than IAA into coleoptile tissue (Poole and Thimann, 1964). Unlike IAA, its uptake is also independent of pH.

(*d*) **Indole-3-ethanol**, IEtOH, VI, has been isolated in small yield from cucumber seedlings (Rayle and Purvis, 1967). It causes elongation in segments of cucumber hypocotyl and of wheat coleoptile, but not in those of the closely-related squash and oats respectively. It is likely, therefore, that its activity, like that of IAN, is due to enzymatic conversion to IAA.

In addition to the above four, indole-3-carboxylic acid, probably derived from IAA by oxidation, and ethyl-3-indoleacetate, probably formed from IAA in presence of an esterase, have been identified occasionally in plants. The former has no auxin activity; the latter, like the methyl ester, has good activity, perhaps due to hydrolysis by an esterase in the plant used for assay.

(*e*) **Bound auxin and auxin precursors.** It was early discovered that when some tissues are placed on agar they yield little or no 'diffusible' auxin, yet on extraction with ether considerable amounts of an auxin (apparently IAA or similar to IAA in general properties) are obtained. This gave rise to the concept of 'bound auxin', an idea which has been supported and broadened over the years, though little studied in detail until lately.

In 1940–2 it was found that the bound auxin of green tissues could not be extracted with ether when they had been dried, but was slowly liberated in presence of water (Thimann and Skoog, 1940, 1942). The liberation was greatly accelerated by proteolytic enzymes, especially chymotrypsin, which indicated that the auxin was bound to protein. Recently, after transport experiments with ^{14}C-IAA, it has been found that when all transport has ceased much isotope still remains in the coleoptile tissue, and is in an insoluble fraction; this too can be liberated with pure proteinases, but not by RNase or other enzymes. Since the liberated auxin was identified as unchanged IAA (Winter and Thimann, 1966), it may have been either an IAA-protein adsorption product, or else (less likely) an IAA peptide of large enough molecular weight to be insoluble. The earlier experiments make it probable that such products are of wide natural occurrence.

On the other hand, Avery and co-workers found that cereal seeds contain a bound auxin from which free auxin is liberated by alkaline hydrolysis, and they prepared from maize by extraction with aqueous acetone a low-molecular weight 'auxin precursor' (Berger and Avery, 1944). This substance has recently been restudied; it comprises a group of four compounds, two of which contained IAA and inositol (1:1), while the other two contain IAA, inositol, and arabinose (1:1:1) (Labarca *et al.*, 1966). One of the first two has been crystallized and identified as indole-3-acetyl-2-O-meso-inositol:

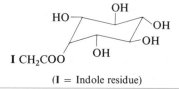

(I = Indole residue)

The other is formed from this on storage, and doubtless contains the IAA radical on the adjacent OH. When arabinose is present it is attached to the inositol to form a 'disaccharide'. These compounds yield IAA readily in very dilute alkali or on extraction with ether. They show some auxin activity but this may be due to their ready decomposition *in vivo* (Nicholls and Bandurski, 1966). A compound tentatively identified as indoleacetylarabinose has been isolated from tissue cultures; its relation to the above group is not yet clear (Shantz and Steward, 1957).

When IAA and NAA are fed to plant tissues they are rapidly converted to peptides with aspartic or glutamic acid (Andreae and Good, 1955; Andreae and van Ysselstein, 1956), and to glucosyl esters (Zenk, 1961). Coenzyme A participates in the first case:

$$ICH_2COOH + CoA + ATP \longrightarrow ICH_2COCoA + ADP + H_2O.$$

$$
ICH_2COCoA + \begin{array}{c} COOH \\ | \\ CH_2 \\ | \\ CHNH_2 \\ | \\ COOH \end{array} \longrightarrow \begin{array}{c} COOH \\ | \\ CH_2 \\ | \\ CHNHCOCH_2I \\ | \\ COOH \end{array} + CoA
$$

(I = Indole residue)

The peptides are rather stable and show no auxin activity, but the glucosyl esters are readily hydrolysed at pH 8 or in the ammoniacal solvents often used for chromatography. Indoleacetyl-aspartic acid is particularly characteristic of the Leguminosae, but members of all the families studied form both naphthyl-1-acetyl aspartic acid and glucosyl-naphthyl-1-acetate when NAA is supplied (Zenk, 1962). Of course, these last are storage or 'detoxification' products, but the materials described on pp. 7 and 8 represent additional naturally occurring forms of IAA. They may explain the frequent observations of 'water-soluble auxins' in the literature.

In recent years a California group has concentrated and partly purified an auxin from lemon seeds and other citrus fruit. This 'citrus auxin' fails to give indole reactions and separates from IAA on column chromatography (Khalifah *et al.*, 1966). Since it produces curvature in the standard *Avena* test, especially after purification, it seems possible that it may be a rather labile IAA-complex.

Thus, in spite of frequent suggestions to the contrary, IAA remains the only naturally occurring auxin whose existence is definitely established, and the wide occurrence as well of such a number of its derivatives makes it evident that IAA is very widely—perhaps universally—distributed throughout the plant kingdom.

Synthetic auxins

The simple structure of IAA led at once to the synthesis of related compounds and testing them for growth-promoting activity. The first to be found effective, indene-3- (VII), benzofurane-2- and 3- (VIII), and naphthalene-1 (IX) acetic acids, represented changes in the nucleus with maintenance of the acetic side-chain. The auxin activity of indole-3-butyric acid, X, showed that a longer side-chain could still confer activity, but it was not long before this was shown (with another series of homologues) to be due to oxidation at the position β to the carboxyl, to reduce the chain length to that of acetic acid (Synerholm and

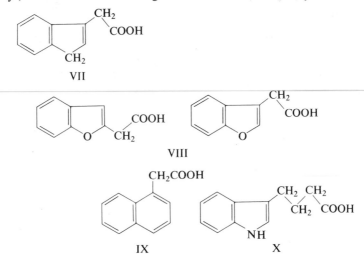

Zimmermann, 1947; Wain, 1953). Many years later it was confirmed, for compounds with several different rings and much longer chain-lengths, that *in vivo* the length becomes reduced systematically by two carbon atoms at a time, so that even-numbered chains eventually give rise to compounds with acetic side chains and thus show high auxin activity, while odd-numbered chains reduce to propionic and formic derivatives, which have much weaker, or zero, activity (Fawcett *et al.*, 1956).

With the discovery of the phenoxy compounds, of which 2,4-dichlorophenoxyacetic acid, '2,4-D', XI, is the most active, a new type of side chain was introduced, the last stage of whose β-oxidation was a phenol. Many such compounds have been shown to give rise in plant tissue to phenols, but here the ability of the oxidation system to cleave the side chain was found to vary with the plant and with the substitution on the ring; 2,4,5-trichlorophenoxybutyric acid, XII, and its higher homologues, were not attacked in peas, though they were in wheat, while the 2,4-dichloro series were β-oxidized vigorously in both (Wain and Wightman, 1954; Baloyannis *et al.*, 1965). Also the corresponding

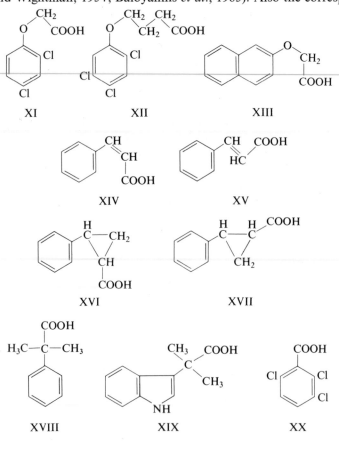

naphthalene derivatives show another interesting structural specificity, for naphthoxy-1-acetic acid is almost inactive, while naphthoxy-2-acetic acid (also called β-naphthoxyacetic acid, XIII) has activity comparable to that of IAA.

Another kind of structural specificity is shown by the cinnamic acids, for *cis*-cinnamic acid, XIV, has real, though low, activity, while the commoner *trans*-acid, XV, has none. Several substituted cinnamic acid derivatives show the same relationship, and the trimethylene analogues, XVI and XVII, represent a corresponding case, only XVI being active (see Veldstra, 1953).

Of a variety of rings studied, all those that are aromatic or contain at least one double bond were found to confer activity (with an acetic acid side chain), but saturated rings, or open chains, were always inactive. At first it was thought, therefore, that the requirements for activity were simply a ring, one or more double bonds in it, an acetic side chain (or one convertible to it) or some other type of acid group, situated at least one carbon atom away from the ring, and, finally, some particular spatial structure which either *maintained* the acid group near to the ring or *allowed* it to occupy such a position (Koepfli *et al.*, 1938). It was later thought that this spatial relationship meant that the axis of the carboxyl group was held perpendicular to the plane of the ring (Veldstra, 1944). It was also suggested that the auxin molecule has to attach to some specific substrate at two points, e.g., the double bond in the ring and the acid group. For this 'two-point attachment' a kinetic theory was even worked out (see Bonner, 1954). Since compounds like α,α-dimethyltoluic acid, XVIII, or indole-3-isobutyric acid, XIX, are inactive, a 'three-point attachment' has been suggested, one of the methylene H atoms furnishing the third point (Wain, 1953). These theories, which are noted also below in connection with cell enlargement, have been discussed in detail several times (see Veldstra, 1953; Audus, 1960). To complete the picture, however, a few additional points must be made:

(a) The concept of a minimum side-chain length is now untenable, because several substituted benzoic acids are auxins; 2,3,6-trichlorobenzoic, XX, is the most active in this group, but the 2,5- and 2,6-dichloro acids also show true activity. These compounds also disprove the suggestion, made at one time, that the position on the ring *ortho* to the side chain must be unsubstituted (Muir and Hansch, 1951).

(b) With the benzoic, phenylacetic, and phenoxy derivatives, a large number of differently substituted compounds has been examined, and it is clear that the type of substituent is highly critical. For example, halogen substitution, as in 2,4-D, confers activity, while hydroxyl substitution, either on the ring or in the side chain (e.g., mandelic acid) greatly decreases or even nullifies it. Similarly 2-chloro and 6-chloro-indoleacetic acids are considerably more active than IAA, but 5-hydroxy and 7-hydroxy-indoleacetic acids have only 3 per cent of the activity of IAA (Thimann, 1958). Methyl substitution generally confers activity, though less strongly than chlorine, and 2-methyl, 4-chloro-phenoxy-acetic acid ('methoxone,' XXI) is highly active. These changes of activity are not paralleled by changes in the dissociation constants of the acids. It was at first thought that the action of chlorine substitution was to confer lipoid-solubility (see Veldstra, 1953), but the following paragraph disposes of that idea.

(c) Within the phenoxy and benzoic groups the location of the halogens is a further critical factor. Of many pairs that could be mentioned the most striking are these; 2,4-D is the most active compound in the phenoxy series, while 3,5-D has zero activity; 2,4,5-trichlorophenoxy-acetic acid ('2,4,5-T,' XXII) has the second highest activity in the series while '2,4,6-T,' XXIII, is almost without activity. The inactivity of XXIII cannot be ascribed to steric

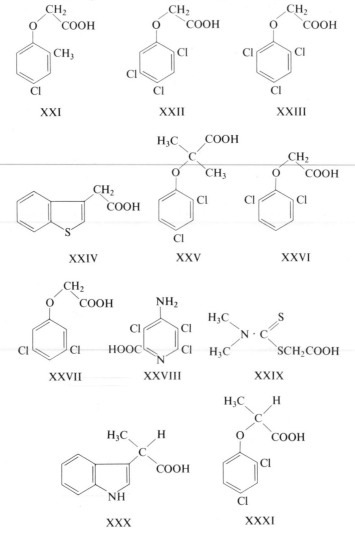

hindrance of the acid group by the two Cl atoms *ortho* to it, because the same would be true to an even greater degree in 2,3,6-trichlorobenzoic acid, XX (since its acid group is not held out on a side chain) yet this compound is about as active as IAA. Also (*in re* an earlier theory) the carboxyl must be even more out of the plane of the ring in XXIII than in XI. The influence of the location of substituents is thus more than merely steric; it must be electronic.

The latest proposal for the structural basis of auxin activity is that it depends on the presence of a fractional positive charge on the ring, situated at a distance of about 5·5 Å from the negative charge of the carboxyl (Porter and Thimann, 1965; Thimann, 1963). This is the condition that holds in indole-3, indene-3, benzofurane-3, and benzothiophene-3 (XXIV) acetic acids, in all of which the hetero atom (or the CH_2 of indene) tends to contribute an additional electron to the 5-membered ring, and thus carries an appreciable part of a positive charge. But the critically sited fractional + charge is much more widespread. In 2,4-D, the Cl atoms tend to withdraw electrons from the ring, leaving especially the position *meta* to them positive; this would be the 6-position, which is just 5·5 Å from the carboxyl oxygen when the side chain is in its mean position (see diagram). With the 6-position substituted, however, as in XXIII, no combination there can take place and hence 2,4,6-T is, as noted above, biologically inactive. In the case of '2,6-D' (XXVI) the 4-position is made positive by the two Cl atoms *meta* to it, and though in the mean position it is remote from the side chain, at one end of the chain's free motion it reaches 5·5 Å from the 4-position; hence the compound does have weak activity (about 3 per cent of that of IAA in the pea test). The case of '3,5-D' (XXVII) is interesting and critical because by all other theories it should be fully active (its sidechain orientation is as in 2,4-D, there is no steric hindrance, the position *ortho* to the side chain is free, and its lipophilic character is the same as that of 2,4-D). Yet it is totally inactive. In light of the theory, we see that the Cl atoms at the 3 and 5 position would place the fractional positive charge on the 1 position, which is already substituted; hence zero activity would be *expected*.

The activity of benzoic acids also finds a ready explanation on this basis; to be 5·5 Å from the negative charge the fractional positivity here must be located in the *4-position*. Correspondingly the active acids are the 2,6-dichloro-, the 2,3,6-trichloro- (XX), and to a lesser extent the 2,5-dichlorobenzoic acids. In contrast to the phenoxy series, 2,4-dichloro- and 2,4,5-trichlorobenzoic acids are inactive (because the 4-position, where the fractional + charge should lie, is not free). Indeed no benzoic acids with 4-position bearing anything larger than fluorine have auxin activity (Veldstra, 1956).

Many other compounds fit into this scheme. The relatively low activity of hydroxy derivatives is explained by the well-known tendency of the HO-group to feed electrons into the ring and thus oppose its positivity. Nitro groups, on the other hand, withdraw electrons and correspondingly 2-bromo-,3-nitro-benzoic acid and several other nitro compounds have real, though not very high, activity. However, when 2,5-dichlorobenzoic acid is further substituted with a 3-amino group, activity is greatly increased (Hicks *et al.*, 1964; Keitt and Baker, 1966). Since —NH_2 would be expected to donate electrons to the ring this behaviour is against the theory.

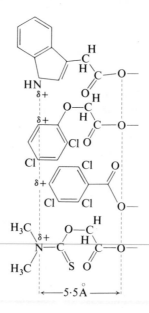

Striking confirmation of the theory has been provided by the recently discovered 2,3,5-trichloro-4-aminopicolinic acid, XXVIII. In this compound the three Cl atoms strongly withdraw electrons so that, while the pyridine N would normally be somewhat negative (e.g., pyridine-3-acetic acid has no auxin activity at all, and 2-pyridoxy acetic acid is very weak (Veldstra, 1956)), the resonance with the 4-amino group should be expected to give the latter a strong fractional positive charge, and this charge would be almost exactly 5·5 Å from the COO⁻ (Kefford and Caso, 1966). Indeed, even the two Cl atoms of 3,5-dichloro, 2-pyridoxy acetic acid are sufficient to give activity approaching that of IAA (Veldstra, 1956). In the case of compound XXVIII, this compound, commercially called 'Tordon', is being widely used as a synthetic auxin. In the pea test it has several times the activity of IAA, in the *Avena* straight growth test about one-third of that of IAA (data confirmed by the author).

Electron-withdrawing groups, like halogen atoms, even give some activity to phenols themselves; 2,6-dichlorophenol and 2,6-dibromophenol have real activity, and so do some related diphenols (Wain and Harper, 1967). However, in accordance with the above theory, the 4-position cannot be substituted without losing all activity.

Another striking agreement with the theory is provided by the small but real activity of thiocarbamates like XXIX (see Veldstra, 1956), the only known aliphatic auxins. Resonance here would make the molecule flat and give the nitrogen some positivity; the diagram on p. 14 shows this to lie just 5·5 Å from the COOH.

If this theory, or some close modification of it, is correct, the inference would be that the activity of an auxin is dependent on its combination (or close association) with a specific surface, whose configuration and charge distribution have to fit it precisely. This in turn helps to explain why, in addition to all the charge distribution and location factors discussed above, optical activity also enters in. Many of the compounds, e.g., 2′-substituted butyric and propionic acids such as XXX and XXXI, are optically active, and in every one of many cases studied it is the D-form which is active (Matell, 1953; also Smith and Wain, 1952). The L-forms are either of low activity or inactive and in several cases they even exert moderate antagonistic effects (Åberg, 1953). However, the nearest to a true anti-auxin appears to be 2,4-dichlorophenoxy isobutyric acid, XXV. Such compounds prevent the growth inhibition of roots which auxins induce, and usually somewhat promote root elongation by themselves (see below and Fig. 1.3). The mechanism of this action is not clear, especially now that it seems as if the growth inhibition of roots is due to a secondary effect, namely the production of ethylene in response to auxin treatment (see below). In any event the importance of optical activity in determining auxin activity certainly helps to support the concept of close and specific association with a charged and conformationally fixed surface. Further research will be needed to identify the molecule on which this surface is located and the way in which its combination with auxin causes it to become activated. Since the evidence summarized in the final section points to the action of auxin on an RNA, the spatial similarity between the indole and purine rings is obviously suggestive.

The functions of auxin

Cell enlargement

(*a*) *Definitions.* The most characteristic action of auxin is to promote cell enlargement. Indeed, the author has defined the auxins by this property: 'Auxins are organic substances which at low concentrations (<0.001 M) promote growth along the longitudinal axis, when applied to shoots of plants freed as far as practical from their own inherent growth-promoting substance.' The definition would now, unfortunately, hold for the gibberellins too, so that it needs amplification. The simplest way to do this is to add 'and inhibit the elongation of roots'.

Growth in plants is defined as *irreversible increase in volume*; it may or may not be accompanied by cell division.* It follows that a *growth* hormone must necessarily cause cell enlargement. In some tissues, like fruits and tubers, the enlargement is more or less equal in all directions—'isodiametric', in others like stems, coleoptiles, and roots, it is primarily in one direction—elongation. Since in the coleoptiles of the grasses all cell divisions are completed at an early age (before 10 mm length in *Avena*) subsequent growth is pure cell elongation, and is classically subject to auxin control.

(b) Bio-assays.   The numerous bio-assays for auxin mostly depend on cell elongation. In the *Avena curvature test* the plants are decapitated to remove the auxin source in the tip, left for 3 hours to use up much of the auxin in the tissue, and redecapitated (i.e., the topmost remaining 1 mm removed). The primary leaf is pulled loose but left partly in. Then the auxin, dissolved in 1·5 per cent agar, is applied as a 10 mm^3 block on the cut surface, against the leaf as support. The side of the coleoptile below the agar block now elongates more than the opposite side, and the resulting curvature reaches an optimum value in about 110 minutes. Thereafter, small curvatures decrease, large ones continue to increase. Usually therefore the curvatures are photographed at 90–110 minutes, and subsequently measured on the photograph. The curvature is linearly proportional to the IAA concentration up to about 0·15 p.p.m., then constant (Fig. 1.2A). Because the smallness of the agar blocks makes them dry out very fast, a room with high and controlled humidity is essential. For this and other reasons, many workers use tests carried out in solution.

In these, coleoptile tips are removed from seedlings 76 hours old at 25°C, and the adjacent subapical segment, commonly 10 mm long, is floated on the test solution. Sucrose 1–2 per cent, a weak buffer (e.g., 0·01 M pH 5·5 phosphate, citrate, or maleate) and sometimes $CoCl_2$ 3×10^{-5} M are added. Since almost all ions affect the growth, the use of buffers requires a caution which is often overlooked. Controls without auxin and a few calibrating concentrations of IAA are essential, since the sensitivity of the segments is critically dependent on the age and length of the plants used. The growth is linear for 12 hours or so, after a short lag, and continues more slowly up to 40 hours. The growth increment above that of the controls, measured after 20 hours, is approximately proportional to the log of the auxin concentration (Fig. 1.2B). With *Avena*, 1 mm of coleoptile growth at 25°C is brought about by $2·2 \times 10^{-5}$ mg of IAA.

Instead of coleoptiles, segments of mesocotyls can be used; they have the advantage of being considerably more sensitive to very low IAA concentrations (Fig. 1.2B). The *Avena* seedlings must be grown in total darkness, or the

---

* In seedlings heavily dosed with gamma-rays virtually all cell division ceases, yet growth continues at normal rates for several days, showing that cell division is not a limiting factor (Haber and Foard, 1964). It could be wished that biologists would not continue to speak of 'growth by cell division', for the division of cells does not of itself *cause* enlargement. Growth is always *by* cell enlargement.

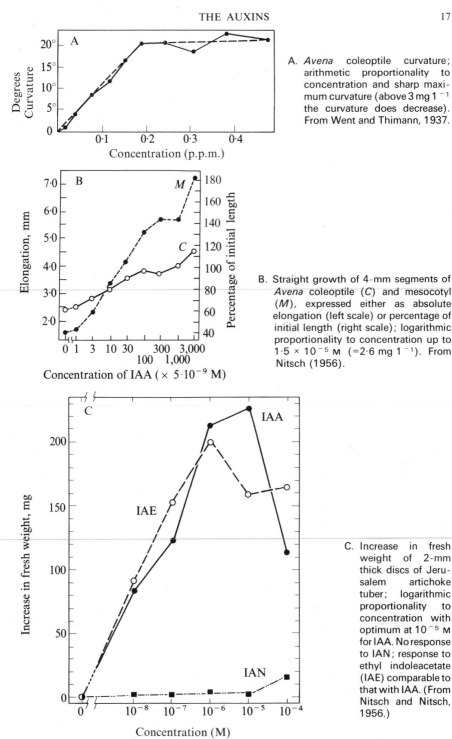

A. *Avena* coleoptile curvature; arithmetic proportionality to concentration and sharp maximum curvature (above 3 mg 1^{-1} the curvature does decrease). From Went and Thimann, 1937.

B. Straight growth of 4-mm segments of *Avena* coleoptile (*C*) and mesocotyl (*M*), expressed either as absolute elongation (left scale) or percentage of initial length (right scale); logarithmic proportionality to concentration up to 1.5×10^{-5} M (=2·6 mg 1^{-1}). From Nitsch (1956).

C. Increase in fresh weight of 2-mm thick discs of Jerusalem artichoke tuber; logarithmic proportionality to concentration with optimum at 10^{-5} M for IAA. No response to IAN; response to ethyl indoleacetate (IAE) comparable to that with IAA. (From Nitsch and Nitsch, 1956.)

Figure 1.2 Cell enlargement as a function of auxin concentration in three types of response.

mesocotyls will be too short, and the coleoptiles are cut off (just barely below the coleoptilar node) and the segments cut, in a weak green light; the elongation of the mesocotyl is controlled by the phytochrome system, being strongly inhibited in red light, and the inhibition more or less reversed by the far red around 730 nm. Presumably the operations might be carried out in orange light and the segments then exposed to far-red just before the growth period, when in the test solution, but this has not generally been done.

Stem segments from etiolated pea seedlings behave similarly. They are cut from just below the apical hook, using (in the case of Alaska peas) plants 7 days old at 25°C; the apical 10 or 20 mm of the then elongating fourth internode is usually used. Sucrose is not necessary, though it promotes growth a little; $CoCl_2$ is optimal at 3×10^{-4} M. Such pea stem segments stop growing in auxin solution in about 16 hours and are a little less sensitive to low auxin concentrations than coleoptile segments, but the final growth increment is again roughly proportional to the logarithm of the auxin concentration. Stems of other rapidly elongating seedlings can be used too, but they seldom are.

A more dramatic elongation assay is provided by similar pea stem segments slit lengthwise for 20 mm. In water the halves curve outwards about 120°; in auxin they curve first outward and then after 2–3 hours strongly inward, reaching up to 900° (more than two complete circles) in some concentrations of synthetic auxin solutions. The curvatures are usually measured (to the nearest 10°) after 22–26 hours (see Thimann and Schneider, 1938). Again the response is roughly linear to a logarithmic series of concentrations. The halves after slitting should be left in water for 2 hours, as this enables those most symmetrically slit to be selected, and most of the residual endogenous auxin to be used up. Slit coleoptiles can be used similarly. This test is the most widely responsive and hence particularly useful for studying synthetic compounds of low auxin activity. By contrast, the *Avena* curvature test with agar blocks is the least widely responsive, and many compounds of high activity in other tests give only very small curvatures.

The inhibition of root elongation is almost as characteristic as the promotion of shoot elongation, and has often been used as a bio-assay (see Åberg, 1957). Unfortunately, it now appears that the effect is indirect, auxins causing the root tissue to evolve ethylene, which is the direct cause of the inhibition (Chadwick and Burg, 1967). However, with very few exceptions, shoot-growth-promoting auxins always inhibit root growth. Tests with the roots growing on moist filter paper are erratic, and best results are obtained by supporting young seedlings, with roots only a few mm in length, above the test solution with the roots immersed. Wheat and cucumber (*Cucumis sativus*) have been the most used, but oats and many other seedlings are satisfactory. Elongation is usually measured after 24 hours at 25°C. The response is usually proportional to the log of the auxin concentration over part of the range, but not only do the slopes vary, but also a number of synthetic auxins give strongly divergent curve types, and some compounds even show promotion (above the controls) at specific low concentrations (see Fig. 1.3).

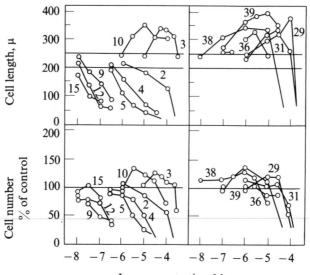

Figure 1.3 Elongation of wheat root epidermal cells, and calculated cell numbers, after 3 days in solutions of a series of synthetic auxins. Inhibition, roughly proportional to logarithm of concentration, by auxin-active substances; promotion, with optimum curve, by inactive or antagonistic substances.

Left: arylacetic acids:
  2 Phenylacetic
  3 Phenoxyacetic
  4 2-methylphenoxyacetic
  5 2,4-dimethylphenoxyacetic
  9 2,4-dichlorophenoxyacetic (2,4-D)
  10 3,5-dichlorophenoxyacetic
  12 2-methyl, 4-chloro-phenoxyacetic
  15 1-naphthaleneacetic (NAA)
Right; aryl-isobutyric acids:
  29 α-phenoxyisobutyric
  31 α-(2,4-dimethylphenoxy)-isobutyric
  36 α-(2,4-dichlorophenoxy)-isobutyric
  38 α-(2-methyl, 4,6-dichlorophenoxy)-isobutyric
  39 α-3-indoleisobutyric
Selected from Hansen, 1954.

Isodiametric cell enlargement is less suited for bio-assay. It is well shown by 1 mm slices of tubers of potato or artichoke (*Helianthus tuberosus*). These are usually washed or 'aged' for 24 hours at 25° before use, during which time the respiratory rate rises (some 300 per cent in potatoes), the respiration becomes largely insensitive to cyanide and to CO in the dark, proteins are synthesized, and the sensitivity to auxin is greatly increased (Hackett *et al.*, 1953). In potato the sensitivity to IAA is never very great, and best experiments are made with naphthalene-acetic acid. The sections must be well aerated, and grow best when supported at the surface of shallow layers of solution. Growth is followed by blotting (in a standardized manner) and weighing from time to time (Reinders, 1938; Hackett and Thimann, 1952). The potato sections in water grow slowly for about 4 days at 25°C, usually undergoing two cell divisions in the first cell layer beneath the surface. In auxin solutions they grow more rapidly, after a lag of several hours, and continue for at least 6 days; cell divisions do not occur. Thus in this tissue cell division and cell enlargement are more or less mutually exclusive. The artichoke sections show very little growth in water, but excellent growth in IAA or NAA, accompanied by active synthesis of both protein and RNA (Thimann and Loos, 1957; Setterfield, 1963). Such behaviour, incidentally, in which auxin induces a very large increase in growth rate and the controls grow minimally, provides almost ideal experimental material; the sections in optimal auxin grow 10 times as fast as those in water. The response is satisfactorily proportional to the log of the concentration of auxin, but unlike oats and wheat, the sections do not respond to indole-acetonitrile (Nitsch and Nitsch, 1956; see Fig. 1.2C). During the growth period their respiration rate is increased by as much as 400 per cent, this being the largest effect of auxin on respiration ever observed. For the first few days sections of both species absorb solutes vigorously from the solution, keeping the conductivity low; no carbohydrates or ions need be added, for this is storage tissue. Eventually, however, leakage of solutes begins, and infection follows rapidly, making it impossible to continue the experiments beyond 8–10 days.

(c) *The optimum curve.*   In all elongation experiments, the growth increment increases with increasing auxin concentration (or log of concentration) up to a maximum. At this point there is a plateau, but as the concentrations further increase, elongation decreases again. The resulting optimum curve resembles that given in a variety of physiological responses of both plants and animals, especially drug effects (cf. Thimann, 1956). In the case of auxins, toxicity, as shown by loss of turgor and leakage of solutes, usually supervenes at 25–50 mg l^{-1}, but this is a sensitive function of pH, and the absolute concentrations vary with the plant species, the time, the pH, the buffer, and the other ions. Elaborate interpretations of the optimum, in terms of the two points of attachment of the auxin to its substrate, have been developed, and the Michaelis–Menten treatment of enzyme substrate combination $(E + S \rightleftharpoons ES \rightarrow E + P)$ has been adapted to the elongation system (see Bonner 1954). This requires that growth rate be a hyperbolic function of auxin concentration, i.e., that the reciprocal of the growth rate (less that of auxin-free controls) should vary linearly with the reciprocal of auxin concentration, i.e., $1/V = K(1/[S])$, or for growth:

$$\frac{1}{V_{aux} - V_{contr}} = K\frac{1}{(Aux)}$$

(In this formulation auxin plays the role of the substrate, S.) When the elongation proceeds linearly with time the above relationship does hold satisfactorily (Bonner and Foster, 1956; Bonner, 1954).

However, the Michaelis–Menten system for enzymes can be considered to lead to two types of reactions, depending on the auxin (substrate) concentration:

$$E + S \rightleftharpoons ES \rightarrow \text{growth}, \quad \text{and} \quad E + 2S \rightarrow ES_2 \text{ (inactive)}.$$

It is to this that the optimum in the dose-response curve is ascribed, excess of auxin causing combination of two auxin molecules with the growth enzyme, and hence growth inhibition. Such behaviour would call for a steady decrease in the linear growth rate at auxin levels above the optimum, and here there is controversy. For most workers find that at high auxin levels growth rates are not constant, but start high and then rapidly decrease as secondary effects (loss of turgor, etc.) set in (Bennet-Clark, 1956; Marinos, 1957). Buffering can reduce and even minimize this effect, but low growth rates remaining linear with time at high auxin levels remain most elusive, and most workers feel the decreased growth at high auxin levels to be really a secondary effect. The whole concept of two-point attachment, which depends on this response, thus remains uncertain. It is notable that the rate of auxin uptake into the tissue, at least with 2,4-D as auxin, shows no optimum but continues to increase up to at least 100 mg l^{-1} (Bonner and Johnson, 1956).

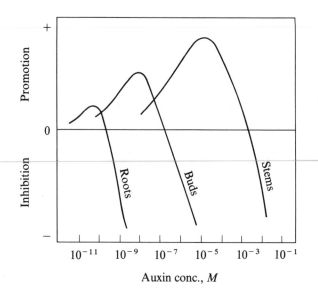

Figure 1.4 Schematic representation of the growth responses of roots, buds, and stems to a range of auxin concentrations, each organ having a promotive and an inhibitory range. (From Thimann, Redrawn by Leopold, 1964.)

The optimum curves, whatever their cause, are of interest for the difference in position of the optimum in different organs (see Fig. 1.4). This suggests a generalization for the control of growth and development by auxin, though recent work suggests strongly that the situation is more complex.

Cell division

Auxin causes cell division in some tissues but not in all, a fact which can best be interpreted as meaning that it is but one of several potentially limiting factors. Since cell division has to be preceded by DNA synthesis and chromosome doubling, and also involves organelle formation as well as karyokinesis proper, it need not be surprising that it is under multiple control.

The most apparent case of auxin control of cell division is in the cambium. First demonstrated by Snow (1935) in seedlings and Söding (1936) in woody plants, it has since been widely observed. Often the effect is transmitted only a few centimetres below the point of application of the auxin, but in the spring it can travel all the way down from near the apex almost to the base of the stem (Gouwentak and Maas, 1940). Evidently this represents the natural 'activation of cambium' which occurs in the spring, when the buds enlarge and cambium then begins to divide, first in the twigs and then all the way into the trunk. The wave of activation moves polarly downward, as auxin does, and at about the observed rate of auxin transport, i.e., one foot per day. The buds can be shown to produce auxin, and cambial scrapings are very rich in it (Söding, 1940); the number of cambial divisions (in apple twigs) correlates with the amount of auxin detectable (Avery et al., 1937).

In an herbaceous stem, when a transverse cut severs the xylem, new xylem differentiates from cortical cells to remake connection around the cut. This reaction does not involve cambium but is comparable in some respects. It starts above the cut and moves polarly downward; it is favoured by presence of the young leaves or the terminal bud, both of which are known auxin sources. Indeed, if IAA is applied to decapitated Coleus shoots, the number of xylem elements resulting is strictly proportional to the IAA applied (Jacobs, 1952; Jacobs and Morrow, 1957). Similarly, IAA applied to the upper surfaces of cultures of Syringa callus caused formation of xylem, and, if sugar was supplied, of phloem also (Rier, 1962; Wetmore and Rier, 1963). Phloem is also induced by IAA in Coleus stems whose phloem has been severed (LaMotte and Jacobs, 1963).

IAA applied to white pine seedlings causes formation of a characteristic wood type, the cells being rounded, thick-walled and reddish in colour—rotholz or redwood (Wershing and Bailey, 1942). Since this occurs on the underside of leaning trunks or lateral branches it has for a long time been ascribed to the compression of the lower side and called 'compression wood'. Evidently it results from the excess IAA which reaches the lower side under the influence of gravity (see chapter 6). The red colour is due to heavy lignification, which is promoted by IAA. The very light-coloured so-called 'tension wood' formed on the upper side of horizontal branches of dicotyledonous trees is apparently of the opposite type, being low in lignin; it is ascribed to the subnormal auxin level on the upper side (Thimann, 1964). This decreased auxin level on the upper side of horizontal *Populus* branches has been confirmed by using ^{14}C-IAA, but curiously, the cambium is not activated on the lower side (Leach and Wareing, 1967). It may be that an inhibitor is also displaced to the lower side. Furthermore, there is evidence that the complete phenomenon of *normal* wood formation, including cambial division, transversal enlargement of the daughter cells, and typical lignification, requires interaction of auxin with gibberellin and cytokinin (see Wareing *et al.*, 1964).

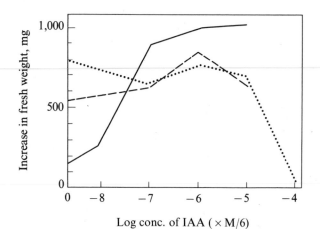

Figure 1.5　Response of three types of *Scorzonera* tissue to indole-acetic acid. Solid line, normal tissue; dashed line, accoutumé tissue; dotted line, tissue from crown gall. (From data of Gautheret, 1950.)

The whole development of plant tissue cultures since the late nineteen thirties, which has been so valuable for the study of morphogenesis, depends on the action of auxin on cell division. For isolated fragments of willow cambium or carrot tissue placed on a nutrient medium undergo a few cell divisions and then die, whereas in presence of 10^{-6} M IAA they divide vigorously and yield lasting cultures (see Gautheret, 1943, for review). The action of synthetic auxins on tissue cultures is varied, some promoting mainly cell division while others promote mainly cell enlargement (Steward and Shantz, 1956). With tobacco pith, cell division begins as soon as IAA is added, but a cytokinin is needed for continuous growth; apparently IAA induces DNA synthesis and mitosis, while the cytokinin also participates in mitosis but is needed for karyokinesis (Patau *et al.*, 1957).

Many tissue cultures show the curious and unexplained phenomenon of 'adaptation'. After growth for several transfers on an auxin-containing medium they become friable and semi-transparent and are now able to grow without added IAA; they have begun to synthesize their own auxin (see Gautheret, 1955). Indeed their growth is now inhibited by IAA at all but minimal concentrations. The auxin relations of these 'accoutumé' tissues now resemble those of cultures of crown gall tumours; these cultures require no added auxin for growth and are even more strongly inhibited by auxin (Gautheret, 1950) (see Fig. 1.5). However, the conclusions to be drawn should not be over-simplified, for the conversion to crown gall due to *Agrobact. tumefaciens* appears to involve more than just the synthesis of auxin. Besides, in the spruce (*Picea glauca*) tumour the change has taken place in the opposite sense, for the normal spruce tissue, which grows slowly, is not affected by auxin, while it is the tumour tissue which is dependent on auxin (de Török and Thimann, 1961).

Root formation

A process in which cell division plays an important part is the formation of roots. Van der Lek's evidence (1925; 1934) that active buds on cuttings promote root formation below them led at an early stage to the idea of a downward-moving root-forming hormone, and this was borne out later by the demonstrated activity of extracts of leaves and seeds (Bouillenne and Went, 1933). A quantitative assay was soon developed, based on the formation of roots at the base of stems of etiolated pea seedlings which had been treated apically with the test solution and then kept two weeks in sucrose solution. This led to the identification of the root-forming hormone with auxin (see Went and Thimann, 1937). In some cases nitrogen nutrition, as nitrate, ammonium or amino acids, may also promote rooting (Bouillenne, 1949, aminoacids; Gregory and Samantarai, 1950, asparagine; Thimann and Poutasse, 1941, nitrate; van Overbeek and Gregory, ammonium, arginine) but auxin still appears to be the major limiting factor. Ever since the late 'thirties indeed, synthetic auxins have been used by nurserymen to promote rooting of cuttings. For convenience they are applied at the base, and produce roots close to the zone of application. The response depends widely on species, and is also modified by age of the plant from which the cuttings are taken, time of year, temperature of the rooting bed, etc. Extensive tables of rooting results have been published (Avery and Johnson, 1947; Thimann and Behnke, 1950; Audus, 1960). Although there is often a small improvement in the rooting if the base of the cutting is extensively wounded, there is as yet no basis for invoking a wound hormone, or indeed any hormonal action other than that of auxin, in the process. After the root initial has appeared, however, thiamine and other compounds needed for root elongation may play a part.

Less uniquely under auxin control is the formation of lateral roots on the main root. Treatment of seedling roots with auxin will usually initiate a crop of laterals, but a second treatment adds no additional group (Torrey, 1950). This suggests that a factor other than auxin is necessary and has been used up. Since a continuously growing root-tip, after the auxin treatment, produces laterals again slowly and steadily, the limiting factor may be formed in the tip (Pecket, 1957); however, there is evidence also of its formation in cotyledons. In isolated root segments several vitamins promote lateral root formation (Torrey, 1956). All in all, this appears to be a process in which auxin and one or more other internal factors interact.

Bud inhibition and apical dominance

The inhibition of the growth of lateral buds on a dicotyledonous stem was historically the first growth inhibition to be proved to be due to auxin, and subsequently a number of other phenomena, such as the formation of short shoots in gymnosperms, or of tillers in the grasses, have been found to be cognate with it. Now that the inhibition of root elongation has come to seem like an indirect effect (mediated by ethylene), bud inhibition stands out all the more clearly as one of the two primary auxin-induced inhibitions. However, the inhibition of bud growth cannot usefully be discussed apart from its release, which is brought about by cytokinins; for this reason the whole matter will be postponed to chapter 5, under Apical Dominance.

Auxins and the abscission of leaves and fruits

The second clear-cut inhibiting action of auxin is exerted on the abscission process. The role of the leaf-blade in this was brought to light by Küster in 1916, who showed that *Coleus* petioles soon fall off if their leaf blades are removed, while as little as 100 mm^2 of leaf blade would delay abscission for many days. Later, Laibach (1933) and Mai (1934) showed that an orchid pollinium (known to be a source of auxin) similarly delayed the abscission of debladed petioles, and shortly afterwards LaRue (1935) obtained comparable results, with both *Coleus* and *Ricinus*, by applying pure IAA. Comparative studies by Myers (1940) and Gardner and Cooper (1943) showed that many auxins other than IAA were effective, while compounds without auxin activity had no action on abscission. The process depends on formation of several rows of small cells at the base of the petiole, and the subsequent separation of their walls (see, e.g., Wetmore and Jacobs, 1953).

Essentially the same action is exerted on the pedicels of many fruits. This effect has proved most important in fruit growing, and auxin sprays have for many years been used to prevent pre-harvest drop of apples (see the reviews of Vyvyan, 1949, and Luckwill, 1959). There are promising effects on oranges also, especially with 2,4-D and 2,4,5-T as auxins (Randhawa and Sharma, 1967).

Measurements of the amount of auxin present in extracts of the fruits usually show a general correlation between the fluctuations in auxin content and the periods of fruit retention on the tree (Luckwill, 1959; see Fig. 1.6), though with grapes the correlation is poor (Nitsch *et al.*, 1960). In tomatoes, both flowers and fruits fall readily, and this is prevented by naphthalene-acetic acid (Reinders and Bing, 1948). In *Nicotiana*, removal of the leaves accelerates abscission of

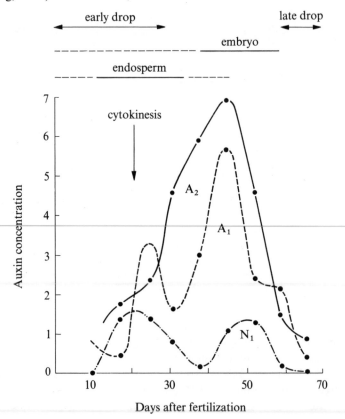

Figure 1.6   Changes in concentration of three auxins (A_1, A_2, and N_1) in the fruit of black-currant (*Ribes nigrum*) during development. Growth phases shown above. (From Luckwill, 1959; drawn from data of S.T.C. Wright.)

the flowers, and IAA substituted for the leaves delays abscission just as well (Yager, 1960a). Thus the inhibiting action of auxin on formation of abscission layers is evidently general.

    Among complications is the fact that under some circumstances auxins can accelerate the abscission. This has been brought out by the following widely-used assay technique. In *Phaseolus* there are abscission layers at the base of each leaflet as well as at the base of the petiole, an arrangement which makes possible a simplified system consisting only of part of the petiole and the leafstalk of the attached leaflet. On deblading, these 'explants' show abscission at the base of the leafstalk in a few days, and, as might be expected, it is delayed by applying IAA to the distal end (i.e., in place of the leaf-blade). But if IAA is applied to the proximal end (i.e., to the petiole base) abscission is accelerated (Addicott and Lynch, 1951; Biggs and Leopold, 1958). A similar system in *Coleus*, using the node with a short stretch of internode and petiole attached, has been worked up in detail as a bio-assay for auxins, etc. (Gorter, 1957). Abscission of *Lupinus* flowers is similarly promoted by applying IAA or NAA to a point basal to them (van Steveninck, 1957; 1958).

Perhaps because of the production of auxin by leaves which are not yet senescent, it seems that leaves at the nodes adjacent to the one under test (on intact plants) can also promote abscission. The effect is not large, but is clearly present (cf., e.g., Rossetter and Jacobs, 1953). The suggestion was made that it might be due to ethylene produced by the leaves, but no firm evidence supports this.

A second complication is due to the presence in many fruits and leaves of an abscission-accelerating substance. Abscisic Acid, or Dormin, was isolated from *young* cotton fruits and is only moderately active in promoting abscission; its ability to induce dormancy is far more notable. Nevertheless it does accelerate abscission in the *Phaseolus* system mentioned above and it seems likely to be of common occurrence in plants. A neutral substance in lupin pods which inhibits coleoptile elongation also strongly promotes abscission (van Steveninck, 1959); this may well be the same or a related compound. Ethylene drastically promotes abscission, and its formation is induced by high auxin levels. Several amino acids also promote leaf abscission (see Rubinstein and Leopold, 1964), and ascorbic acid somewhat promotes the fall of some fruits. With all these factors interacting, the abscission process as a whole must represent a balance between the powerfully inhibiting effect of auxin and the more modest promoting influences of several other substances. Ethylene would certainly be included among the latter group, and indeed the abscission it produces is definitely antagonized by IAA (Abeles and Holm, 1966).

Nevertheless, auxin itself does appear to exert two opposing kinds of action—inhibition and promotion or acceleration—and the basis for this has elicited a succession of theories. At first an auxin gradient was thought to be the controlling factor, abscission occurring when the auxin level below the abscission zone was higher than that above it (Addicott *et al.*, 1955). Subsequently evidence was adduced that the action of auxin simply followed the classical optimum curve, low concentrations promoting and high concentrations inhibiting (Biggs and Leopold, 1958). Rubinstein and Leopold (1963, 1964) believed these effects could be separated in time, the first phase lasting six hours or so and promoted by auxin, the second longer-lasting and inhibited by auxin. Attempts to involve enzymes that might function to separate cell walls, such as pectin methyl esterase, have shown only a very slight influence of auxin (mainly at unphysiologically

3

high concentrations) and no sign of the required optimum type of curve (Yager, 1960b; Albersheim and Killias, 1962). Lately some evidence for the participation of protein and RNA synthesis has been brought up (see, e.g., Lewis and Bakshi, 1966), and Scott and Leopold (1966) have likened abscission to senescence in that a gradual mobilization of materials appears to take place; soluble protein, total RNA, and chlorophyll decrease in the distal part and either remain constant or increase in the proximal part, so that a mass migration of organic materials takes place out of the part distal to the abscission zone.

It seems that before abscission as a whole can be understood, it might be well to concentrate on the primary problem of how auxin inhibits formation of the abscission zone.

The destruction of indole-acetic acid

A compound so widely effective as IAA must of necessity be subject to rapid removal in the tissue, otherwise pathological overgrowths would be common. The rapid and effective polar transport system (see chapter 4) removes auxin from the region where it is produced to other parts of the plant. Besides this, however, auxin destruction is by oxidation. The complex and sometimes conflicting literature on this oxidation has been well reviewed (Ray, 1958) and work on the enzymes has recently been fully covered (Mahadevan, 1963), so that only brief attention need be given here. There are two basic processes, an enzymatic (dark) reaction and a photo-oxidation.

The oxidizing enzyme as found in leaves and in certain fungi is usually a peroxidase acting as an oxidase. Only around $\frac{1}{10}$ mole of H_2O_2 per mole of IAA is needed, and peroxidation in the strict sense (i.e., by H_2O_2 in the absence of O_2) does not occur. One mole of O_2 is taken up, and one mole of CO_2 evolved from the carboxyl group. The reaction proceeds via an intermediate whose UV spectrum has been deduced but whose exact nature is not known (Ray, 1956), and it probably involves a free radical mechanism. The most probable product is 3-methylene-oxindole (Hinman et al., 1961). The enzyme system from higher plants requires manganese, and is partly independent of added H_2O_2 because it appears to carry a peroxide-producing system along with it (either as impurity or as a bound flavin moiety). The enzyme from the coffee leaf-spot fungus, Omphalia flavida, which has been highly purified, contains no manganese and yet does not require it, though added manganese does modify its action somewhat. It requires H_2O_2 and is inhibited or delayed by catalase. Both enzymes are accelerated by monophenols such as p-hydroxybenzoic acid or 2,4-dichlorophenol (often an impurity in 2,4-D); they are inhibited by ortho-diphenols such as catechol and pyrogallol. Flavonols act in the same way: kaempferol promotes and quercetin inhibits (Furuya et al., 1962). Since phenolic substances occur widely in plants they must have some importance as modifiers of auxin destruction. More remarkable is the renewed significance which now attaches to phenoloxidases like tyrosinase, for these enzymes specifically have the power of inserting a second hydroxyl into phenols (e.g., converting tyrosine to dihydroxy-phenylalanine) and thus can alter an oxidation accelerator into an inhibitor. The actions of the phenols have the result that in small quantities (ca. 2×10^{-6} M) they greatly modify growth tests and bio-assays, and the CO_2 evolved is readily detectable with ^{14}C-IAA (Tomaszewski and Thimann, 1966).

Peroxidases are mixtures of isozymes (McCune, 1961) and thus probably differ somewhat from plant to plant. Incubation of plant tissues with auxins (IAA or other) often increases the peroxidase activity, but, since auxins which are not oxidized act as well as IAA, the change is evidently not a true induction of enzyme activity but a result of generally enhanced syntheses (Pilet and Galston, 1955). One isozyme is even repressed by IAA (Ockerse et al., 1966) and in tobacco tissue the whole system is similarly repressed (Kerstetter and Keitt, 1966).

If cytochrome and cytochrome oxidase are present the oxidation route differs and indole-3-aldehyde is the main product (Stutz, 1958).

The photoreaction is quite different; riboflavin, eosin, and other fluorescent dyes are the catalysts. Again one mole of O_2 is taken up per mole IAA destroyed. Very large light dosages—in millions of ergs cm^{-2}—are required, and the process seems to have no bearing on phototropism, which in the *Avena* coleoptile gives good curvatures with 2 to 15 ergs cm^{-2}. The products are multiple, and may include indole-glycolic and indole-carboxylic acids, as well as 3-methylene-oxindole and its hydration product, 3-methyl, 3-hydroxy-oxindole (Stell et al., 1965). These compounds are weakly inhibitory to growth, but less so on plants than on bacteria.

Evidence as to the mode of action of auxin

The above discussion of the molecular structure which an auxin requires for its activity leads to the concept of a specific charge distribution on a surface, at which the auxin molecule becomes oriented. In some way the placement of the auxin molecule there activates that surface, or, in terms of the current popular allosteric concept, may activate *another part* of the surface which thus acquires enzymatic activity. The older, cruder statement was that auxin acts as a coenzyme; this in its simple form now appears unlikely, although the difference is one of degree and not of principle. What sort of an enzyme could it be that is thus activated?

Obviously many enzymes function in the growth process, but we are only concerned with those that constitute limiting factors. Evidence as to a number of these has been obtained, over the years, from inhibitor studies. Since the simplest and most studied growth system, uncomplicated by cell division, is the *Avena* coleoptile, work with this system will be kept in the foreground.

(a) *Sulphhydryl enzymes.*   All sulphhydryl reagents studied inhibit the elongation of oat coleoptile segments, as well as of other growing organs so far as they
have been studied. The compounds include iodoacetate, phenyl mercuric salts,
parachloromercuribenzoate, arsenite (Thimann and Bonner, 1950), cobalt salts,
coumarin and protoanemonin (Thimann and Bonner, 1949a), maleic hydrazide,
and N-ethylmaleimide (van Overbeek *et al.*, 1955) (see Fig. 1.7 for examples).
The growth inhibition takes place at concentrations which have little or no effect
on oxygen consumption, and is therefore not exerted indirectly via a respiratory
system, but more directly on some component of growth. The inhibition is not
reversible, once it has set in, but in some cases can be prevented by simultaneous
administration of another reagent. Thus potassium malate and citrate prevent
the iodoacetate inhibition, though they do not affect that due to the mercury
compounds. The effect is partly due to the K^+ ion (Thimann and Schneider,
1938; Cooil, 1952). The inhibition by high cobalt concentrations is similarly
prevented by BAL (Thimann, 1958). Many of the reagents cause marked growth
promotion when at sub-inhibiting concentrations, an effect which has not been
satisfactorily explained. It may be that they inhibit anaerobic reactions which
metabolize uselessly certain substrates that could have supported growth (see
Slocum and Little, 1957).

(b) *Reagents for the cytochrome system.*   The earliest such observations were
with cyanide (Bonner, 1936), with which the concentrations effective on growth
and on respiration (of coleoptile segments) were very similar. A much more
specific cytochrome reagent is carbon monoxide; this inhibits elongation of
coleoptile and pea stem segments, and isodiametric enlargement of tuber slices.
The inhibitions are more or less completely reversed by light (Hackett and
Schneiderman, 1953; Hackett *et al.*, 1953). The clear-cut light reversal is specific
evidence that cytochrome oxidase controls cell enlargement. After incubation
for 1–2 days, potato slices show a curious phenomenon; their respiration rate
increases by 200–300 per cent and at the same time becomes less sensitive, or
nearly insensitive, to HCN and to CO in the dark. Evidently a second oxidative
system becomes formed; this has been variously thought to involve an autoxi-
dizable cytochrome, a flavoprotein, or a non-haem iron compound. A *de novo*
enzyme synthesis, controlled by a messenger-type RNA with a half-life of 2
hours, is involved (Click and Hackett, 1963). At any rate, the change in limiting
enzyme for respiration is not paralleled by a similar change in the limiting

---

Figure 1.7   Inhibition of elongation by metabolic poisons.
Top: pea stem segments in arsenite (A) and fluoride (B); arithmetical plot. (From Christiansen
and Thimann, 1950).
Middle: (C) pea stem segments in arsenite; logarithmic plot. (From Thimann and Bonner,
1949b.)
Bottom: (D) pea stem segments in arsenate, with (white circles) and without (black circles)
sucrose. (E) pea root segments in arsenate, all with sucrose. (From Audus, 1952.) Growth
rates shown are after 24 hours in solutions.

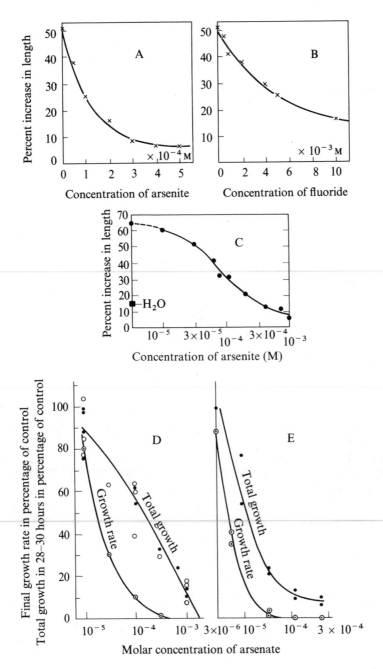

Figure 1.7

system for growth, for this remains sensitive to HCN and CO. Thus again the systems active on respiration as a whole are clearly separable from those controlling growth. The differences in effective concentration of several inhibitors for respiration and for growth are summarized in Table 1.1.

Table 1.1   Relation between the inhibition of growth and of respiration

When growth of:	is inhibited 50% by:	which requires a concentration of ($\times$ 10^{-3}M)	Respiration is inhibited by (%)
Pea stems	Iodoacetate	0·6	26
	Arsenite	0·1	13
	Fluoride	5	ca. 0
Oat coleoptiles	Iodoacetate	0·04	10
	Fluoride	2·5	9
	Arsenate	0·02	0
Potato discs	Dinitrophenol (at pH 4·5)	0·02	ca. 0
	Carbon monoxide	5CO:1O_2	25

(*c*) *Fluoroacetate.*   This substance, which intervenes in the conversion of citrate to isocitrate by being converted to fluorocitrate, again inhibits both elongation and isodiametric growth (Thimann and Bonner, 1950). Its effect can be partially prevented by acetate. Evidently, therefore, Krebs cycle oxidations participate in the growth process.

(*d*) *Phosphorylation inhibitors.*   Arsenate, fluoride, and dinitrophenol all inhibit growth strongly, and at concentrations which exert much smaller effects on overall respiration (Bonner, 1950; Hackett and Thimann, 1953). The inhibition by fluoride has a complex dependence on concentration, indicating that several processes participate; in general, however, as with other reagents in solution, inhibition of growth is about linearly proportional to the log of inhibitor concentration (Fig. 1.8). These inhibitions are not clearly prevented by any reagent, though added phosphate has some effect. The endogenously-controlled elongation of root segments is even more sensitive to arsenate than the auxin-induced enlargement of shoot tissues (Audus, 1952) (see Fig. 1.7).

The role of protein and nucleic acid synthesis

All the above evidence shows that a variety of energy-yielding oxidations furnish ATP for the growth process—the Krebs cycle, one or more –SH enzymes, phosphate transfers, and the cytochrome system. Correspondingly, growth does not occur in nitrogen. Some chemical evidence on the reactions involved supplements these growth studies, for the inhibition of growth of pea-stem segments is accompanied by little increase in soluble sugars, but a marked accumulation of fats. Growth is accompanied by decreases in free amino acids and increases in asparagine and in protein, and these increases are prevented when growth is inhibited by the reagents described above (Christiansen and Thimann, 1950). It follows that the substrates for the new cell materials formed in growth of pea-stem tissue are not primarily sugars, but mainly fats and amino acids, and that protein synthesis, along with asparagine formation, is an important constituent process. The last conclusion is supported by the observations that canavanine, an inhibitor of arginine metabolism, hydroxyproline, which may interfere with proline metabolism (Bonner, 1949), and parafluorophenylalanine,

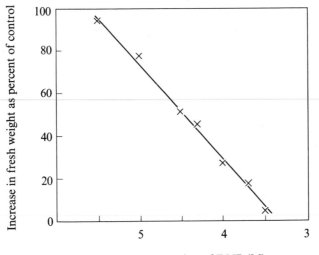

Figure 1.8   Inhibition of growth of artichoke discs in IAA by 2,4-dinitrophenol; log-linear relationship. (From Hackett, unpub.)

a clear-cut inhibitor of phenylalanine and tyrosine metabolism (Noodén and Thimann, 1963) all inhibit elongation, the last-named acting on several types of growing tissues. The proteins which are synthesized as part of the growth process must only be certain specific ones, for neither in coleoptiles (Boroughs and Bonner, 1953) nor in wheat roots (Burström, 1951, 1953) is there any overall increase in protein during elongation.

A fresh turn has been given to the investigations by recent evidence implicating the nucleic acids in growth. Actinomycin D, which specifically inhibits the DNA-controlled formation of messenger-RNA, puromycin, which inhibits the action of the messenger on ribosomes, and chloramphenicol, which inhibits protein synthesis at the ribosome level, all inhibit growth, and at very low concentrations (Noodén and Thimann, 1963, 1965; Key, 1964; Key and Ingle,

1964). All three can be shown, *in vivo*, to inhibit the incorporation of ^{14}C-amino acids into protein, in parallel with the inhibition of growth (see Fig. 1.9). Earlier, Masuda (1959) reported that ribonuclease treatment prevented IAA-induced elongation of coleoptile segments. A suggestive observation, too, was that both IAA and kinetin increase total RNA synthesis, both in tuber tissue (Setterfield, 1963) and elsewhere (Silberger and Skoog, 1953).

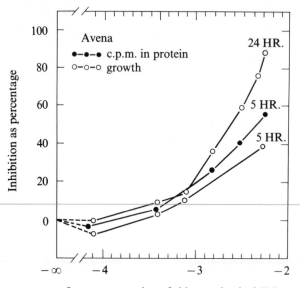

Figure 1.9 Comparison of the percentage inhibition of elongation and incorporation of ^{14}C-leucine into protein of oat coleoptile sections in the presence of IAA. (From Noodén and Thimann, 1963.)

Fractionation of the nucleic acids of growing soybean hypocotyls has brought out new evidence on the types of RNA involved (Key and Ingle, 1964). Unlike other purine antagonists, 5-fluorouracil does not inhibit growth of these segments, but it does inhibit RNA formation; however, the inhibition affects mainly the synthesis of those fractions that are eluted from the MAK column at low salt concentrations, e.g., ribosomal RNA formation was inhibited 90 per cent and that of soluble RNA 75 per cent. The fraction least eluted by salt was in- hibited only 10 per cent, and this is one whose properties resemble those of messenger-RNA (see Fig. 1.10). Thus when growth is not inhibited this is the fraction whose formation is also not inhibited, and thus is likely to be the one associated with growth (Ingle *et al.*, 1965; Lin *et al.*, 1966). These results, taken

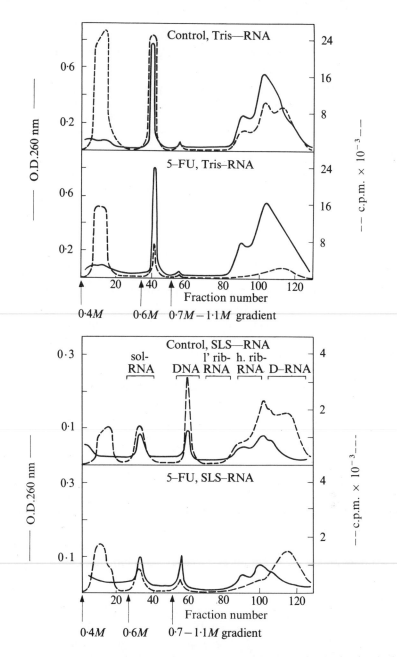

Figure 1.10 Fractionation of nucleic acids from soybean hypocotyl tissue after incubation with ^{32}P phosphate for 7 hours.
Top figure: Tris-RNA, mainly soluble and ribosomal RNA; *Bottom figure:* SLS-RNA, mainly DNA and DNA-like RNA ('D-RNA'). Solid lines, optical density; dashed lines, radioactivity. *Above:* control; *below:* with 5-fluoro-uracil. Note failure to inhibit labelling of the 'D-RNA', parallel with failure to inhibit elongation. (From Key and Ingle, 1964.)

together with other findings, can best be interpreted on the following scheme:

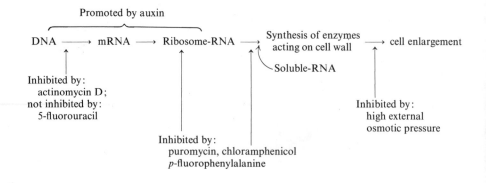

It would mean that auxin functions by somehow activating a messenger-type RNA and thus induces the synthesis of specific enzymes. These would bring about the insertion of new materials into the cell wall, causing its extension. Enlargement of the cell then follows, but only if there is sufficient water-entering tendency ('water potential', 'suction pressure', or 'DPD'). Inert osmotica, like mannitol, carbowax, or certain salts, inhibit growth at this last stage by denying water uptake. When elongation is inhibited by calcium salts, IAA still increases the rate of cell wall formation (Baker and Ray, 1965), though with other growth inhibitors this is not the case. Usually, in coleoptiles and in tuber tissues, growth resumes when the osmoticum is removed. There is no reason any longer to implicate a metabolic component in the water entry process itself.

A few points remain to be added. Auxin action is very rapid. Increase in growth rate can be observed within less than 15 minutes, with a sensitive registration system (Ray and Ruesink, 1962). Decrease in the growth rate of roots is almost as rapid, but this is apparently an indirect effect due to the auxin-induced production of ethylene which is the immediate growth inhibitor (Chadwick and Burg, 1967). Increase in the rate of respiration generally appears within about half an hour. Increase in the rate of protoplasmic streaming is even more rapid, appearing in 2 minutes (Sweeney and Thimann, 1938). This acceleration of streaming parallels the acceleration of growth in many ways: dependence on sugar supply and on oxygen, sensitivity to inhibitors, interaction with malic acid. Thus both the complex chemical behaviour and the kinetics are, in the broad, those of enzymatic reactions and not of mechanical changes like plastic flow, which has often been envisaged as a limiting factor in cell wall extension. Also the auxin-induced growth has an instantaneous response to raised temperature, and a reasonably high Q_{10} (Ray and Ruesink, 1962). The connections between accelerated streaming, increased respiration, and cell-wall modification remain obscure, and it may be that more than one enzyme system is simultaneously stimulated in the manner shown above. To explain the multiple responses, Armstrong (1966) invokes an auxin action on soluble RNA, regulating total RNA synthesis.

The lethal effect of high auxin concentrations has not been explained. Since high gibberellin concentrations, though sometimes causing tremendous growth, have no such toxic effect, it is obviously not (as at first thought) a *result* of excessive growth. Its understanding will have to await a complete elucidation of the mode of auxin action.

As noted at the outset, many plant growth responses result from the interaction of auxin with other hormones. These are discussed in other chapters. But this fact means that the full understanding of the role of hormones in the growth, integration and organ formation of plants is still a long way off.

Bibliography

Selected further reading

Books

Audus, L. J., 1960. *Plant Growth Substances*. Leonard Hill Ltd., London; Interscience Publ. Inc., New York.

Leopold, A. C., 1955. *Auxins and Plant Growth*. Univ. of Calif. Press, Berkeley & Los Angeles, Calif.

Pilet, P. E., 1961. *Les Phytohormones de Croissance*. Masson et Cie, Paris.

Söding, H., 1952. *Die Wuchsstofflehre*. Georg Thieme Verlag, Stuttgart.

Went, F. W. and K. V. Thimann, 1937. *Phytohormones*. MacMillan, N.Y.

Reviews

Åberg, B., 1957. Auxin relations in roots. *Ann. Revs. Plant Physiol.* **8**, 153–180.

Bentley, J. A., 1958. The naturally occurring auxins and inhibitors. *Ann. Revs. Plant Physiol.* **9**, 47–80.

Fawcett, C. H., 1961. Indole auxins. *Ann. Revs. Plant Physiol.* **12**, 345–368 (226 refs.)

Gruen, H., 1959. Auxins and fungi. *Ann. Revs. Plant Physiol.* **10**, 405–440 (267 refs.)

Larsen, P., 1951. Formation, occurrence and inactivation of growth substances. *Ann. Revs. Plant Physiol.* **2**, 169–198.

Leopold, A. C., 1964. Plant growth hormones. In *The Hormones*, Vol. IV, ed. G. Pincus, K. V. Thimann, and E. B. Astwood, Academic Press, New York, pp. 1–66.

Luckwill, L. C., 1959. Fruit growth in relation to internal and external chemical stimuli. In *Cell, Organism & Milieu*, XVIIth Growth Symp., ed. D. Rudnick, Ronald Press, New York, pp. 223–251.

Mahadevan, S., 1963. Enzymes involved in synthesis and breakdown of indoleacetic acid. In *Modern Methods of Plant Analysis*, Vol. VII. Springer, Berlin.

Ray, P. M., 1958. Destruction of auxin. *Ann. Revs. Plant Physiol.* **9**, 81–118.

Rubinstein, B. and A. C. Leopold, 1964. The nature of leaf abscission. *Quart. Rev. Biol.* **39**, 356–372.

Thimann, K. V., 1951. The Synthetic Auxins. Relation between structure and activity. In *Plant Growth Substances*, ed. F. Skoog. Univ. of Wisconsin Press, Madison.

Veldstra, H., 1953. The relation of chemical structure to biological activity in growth substances. *Ann. Revs. Plant Physiol.* **4**, 151–198 (190 refs.)

Papers cited in text

Abeles, F. B. and R. E. Holm, 1966. Enhancement of RNA synthesis, protein synthesis and abscission by ethylene. *Plant Physiol.* **41**, 1337–1342.

Åberg, B., 1953. On optically active plant growth regulators. *Ann. Agric. Coll. Sweden*, **20**, 241–295.

Addicott, F. T. and R. S. Lynch, 1951. Acceleration and retardation of abscission by indoleacetic acid. *Science*, **114**, 688–689.

Addicott, F. T., R. S. Lynch, and H. R. Carns, 1955. Auxin gradient theory of abscission regulation. *Science*, **121**, 644–645.

Albersheim, P. and U. Killias, 1962. Studies relating to the purification and properties of Pectin trans-eliminase. *Arch. Biochem. Biophys.* **97**, 107–115.

Andreae, W. A. and N. E. Good, 1955. The formation of Indoleacetyl-aspartic acid in pea seedlings. *Plant Physiol.* **30**, 380–382.

Andreae, W. A. and M. W. H. van Ysselstein, 1956. Studies on 3-Indoleacetic acid metabolism III. The uptake of IAA by pea epicotyls and its conversion to 3-Indole-acetyl-aspartic acid. *Plant Physiol.* **31**, 235–240.

Audus, L. J., 1952. The time factor in studies of growth inhibition in excised organ sections. *J. Exp. Bot.* **3**, 375–386.

Avery, G. S., P. R. Burkholder and H. B. Creighton, 1937. Production and distribution of growth hormone in shoots of Aesculus and Malus, and its probable role in stimulating cambial activity. *Am. J. Bot.* **24**, 51–58.

Avery, G. S. and Johnson, 1947. *Hormones and Horticulture.* McGraw-Hill, New York.

Baker, D. B. and P. M. Ray, 1965. Relation between effects of auxin on cell wall synthesis and cell elongation. *Plant Physiol.* **40**, 360–368.

Baloyannis, P. G., R. S. Smith, and R. L. Wain, 1965. Studies on plant growth regulating substances. XX. The metabolism of γ(2,4,5-trichlorophenoxy)-butyric acid in wheat and pea stem tissues. *Ann. Appl. Biol.* **55**, 261–265.

Bennet-Clark, T. A., 1956. The kinetics of auxin-induced growth. In *Chem. and Mode of Action of Plant Growth Substances*, ed. R. L. Wain and F. Wightman. Butterworths, London, pp. 310–312.

Bentley, J. A. and S. Housley, 1952. Studies on plant growth hormones. I. Biological activities of 3-Indolyl-acetaldehyde and 3-Indolylacetonitrile. *J. Exp. Bot.* **31**, 393–405.

Berger, J. and G. S. Avery, Jr. 1944. Isolation of auxin precursor and an auxin (indole-acetic acid) from maize. Chemical and physiological properties of maize auxin precursor. *Am. J. Bot.* **31**, 199–203.

Biggs, R. H. and A. C. Leopold, 1958. The two-phase action of auxin on abscission. *Am. J. Bot.* **45**, 547–551.

Bonner, J., 1936. The growth and respiration of the Avena coleoptile. *J. Gen. Physiol.* **20**, 1–11.

Bonner, J., 1949. Limiting factors and growth inhibitors in the growth of the Avena coleoptile. *Am. J. Bot.* **36**, 323–332.

Bonner, J., 1950. Arsenate as a selective inhibitor of growth substance action. *Plant Physiol.* **25**, 181–184.

Bonner, J., 1954. The hormonal control of plant growth. *Harvey Lectures*, series 48. Academic Press, New York.

Bonner, J. and R. J. Foster, 1956. The kinetics of auxin-induced growth. In *Chem. and Mode of Action of Plant Growth Substs.*, ed. R. L. Wain and F. Wightman. Butterworths, London, pp. 245–309.

Bonner, J. and M. P. Johnson, 1956. The uptake of auxin by plant tissue. *Physiol. Plantarum*, **9**, 102–118.

Boroughs, H. and J. Bonner, 1953. Effects of indoleacetic acid on metabolic pathways. *Arch. Biochem. Biophys.* **46**, 279–290.

Bouillenne, R., 1950. La Rhizogenèse. *L'Année Biol.* **26**, 597–628. (Proc. Strasbourg Symp., 1949).

Bouillenne, R. and F. W. Went, 1933. Recherches expérimentales sur la néoformation des racines dans les plantules et les boutures des plantes supérieures. *Ann. jard. bot. Buitenzorg*, **43**, 25–202.

Burström, H., 1951. Studies on growth and metabolism of roots. V. Cell elongation and dry matter content. *Physiol. Plantarum*, **4**, 199–207.

Burström, H., 1953. Studies on growth and metabolism of roots. IX. Cell elongation and water absorption. *Physiol. Plantarum*, **6**, 262–276.

Chadwick, A. V. and S. P. Burg, 1967. An explanation of the inhibition of root growth caused by indole-3-acetic acid. *Plant Physiol.* **42**, 415–420.

Christiansen, G. S. and K. V. Thimann, 1950. The metabolism of stem tissue during growth and its inhibition. II. Respiration and ether-soluble material. *Arch. Biochem. Biophys.* **26**, 248–259. III. Nitrogen Metabolism. Ibid. **28**, 117–129.

Click, R. E. and D. P. Hackett, 1963. The role of protein and nucleic acid synthesis in the development of respiration of potato tuber slices. *Proc. Nat. Acad. Sci.* (U.S.) **50**, 243–250.

Cooil, B. J., 1952. Relationships of certain nutrients, metabolites and inhibitors to growth in the Avena coleoptile. *Plant Physiol.* **27**, 49–69.

de Török, D. and K. V. Thimann, 1961. The auxin requirement and the effect of an auxin antagonist on tumorous and normal tissues of *Picea glauca*. *Physiol. Plantarum*, **14**, 543–547.

Fawcett, C. H., H. F. Taylor, R. L. Wain, and F. Wightman, 1956. The degradation of certain phenoxy acids, amides and nitriles within plant tissues. In *The Chem. and Mode of Action of Plant Growth Substs.*, ed. R. L. Wain and F. Wightman. Butterworths, London, pp. 187–194.

Furuya, M., A. W. Galston, and B. B. Stowe, 1962. Isolation from peas of co-factors and inhibitors of indolyl-3-acetic acid oxidase. *Nature, Lond.* **193**, 456–457.

Gardner, F. E. and W. C. Cooper, 1943. Effectiveness of growth substances in delaying abscission of Coleus petioles. *Botan. Gaz.* **105**, 80–89.

Gautheret, R-J., 1942. Recherches sur le développement de tissus végétaux cultivés *in vitro*. *Rev. cyt. cytophysiol. vég.* **6**, 84–180.

Gautheret, R-J., 1950. La culture des tissus végétaux et les phenomènes d'histogenèse. *L'Anneé Biol.* **26**, 719–744 (Proc. Strasbourg Symp., 1949).

Gautheret, R-J., 1953. Recherches anatomiques sur la Culture des Tissus de Rhizomes de Topinambour et d'Hybrides de Soleil et de Topinambour. *Rev. gén. Botan.* **60**, 129–219 (with 8 plates).

Gautheret, R-J., 1955. Sur la variabilité des propriétés physiologiques des cultures de tissus végétaux. *Rev. gén. Botan.* **62**, 1–107.

Gmelin, R. and A. I. Virtanen, 1961. Glucobrassicin, precursor of the thiocyanate ion, Indolylacetonitrile and Ascorbigen, in *Brassica oleracea* and related spp. *Ann. Acad. Sci. Fennicae*, Ser. A II, 107–121.

Gorter, C. J., 1957. Abscission as a Bio-assay for the determination of plant growth regulators. *Physiol. Plantarum*, **10**, 858–868.

Gouwentak, C. A. and A. L. Maas, 1940. Kambiumtätigkeit und Wuchsstoff. II. *Med. Landbouw-hoogeschool Wageningen*, **44**, 3–16.

Gregory, F. G. and B. Samantarai, 1950. Factors concerned in the rooting responses of isolated leaves. *J. Exp. Bot.* **1**, 159–193.

Haagen Smit, A. J., W. B. Dandliker, S. H. Wittwer, and A. E. Murneek, 1946. Isolation of 3-Indoleacetic acid from immature corn kernels. *Am. J. Bot.* **33**, 118–120.

Haagen Smit, A. J., W. D. Leach, and W. R. Bergren, 1942. The estimation, isolation and identification of auxins in plant materials. *Am. J. Bot.* **29**, 500–506.

Haber, A. H. and D. E. Foard, 1964. Further studies of gamma-irradiated wheat and their relevance to use of mitotic inhibition for developmental studies. *Am. J. Bot.* **51**, 151–159.

Hackett, D. P. and H. A. Schneiderman, 1953. Terminal oxidases and growth in plant tissues. I. The terminal oxidase mediating growth in Avena coleoptile and Pisum stem sections. *Arch. Biochem. Biophys.* **47**, 190–204.

Hackett, D. P., H. A. Schneiderman, and K. V. Thimann, 1953. Terminal oxidases and growth in plant tissues. II. The terminal oxidase mediating water uptake by potato tissue. *Arch. Biochem. Biophys.* **47**, 205–213.

Hackett, D. P. and K. V. Thimann, 1952. The effect of auxin on growth and respiration of artichoke tissue. *Proc. Nat. Acad. Sci.* 770–775.

Hackett, D. P. and K. V. Thimann, 1953. The nature of the auxin-induced water uptake by potato tissue. II. The relation between respiration and water absorption. *Am. J. Bot.* **40**, 183–188.

Hansen, B. A. M., 1954. The physiological classification of 'shoot auxins' and 'root auxins'. I and II. *Botan. Notiser (Lund)* **3**, 230–268, 318–325.

Henbest, H. B., E. R. H. Jones, and G. F. Smith, 1953. Isolation of a new plant growth hormone, 3-Indolylacetonitrile. *J. Chem. Soc.* 3796–3801.

Hicks, G. S., G. Setterfield, and F. Wightman, 1964. Effects of synthetic auxins on cell division and cell expansion in Jerusalem artichoke tissue. *Proc. Canad. Soc. Pl. Physiol.* **5**, 18–21.

Hinman, R. L., C. Bauman, and J. Lang, 1961. Conversion of indole-3-acetic acid to 3-methylene-oxindole in presence of peroxidase. *Biochem. Biophys. Res. Comms.* **5**, 250–254.

Ingle, J., J. L. Key, and R. E. Holm, 1965. The demonstration and characterization of a DNA-like RNA in excised plant tissue. *J. Mol. Biol.* **11**, 730–746.

Jacobs, W. P., 1952. The role of auxin in differentiation of xylem around a wound. *Am. J. Bot.* **39**, 301–309.

Jacobs, W. P. and I. B. Morrow, 1957. A quantitative study of xylem development in the vegetative shoot apex of Coleus. *Am. J. Bot.* **44**, 823–842.

Jones, E. R. H., H. B. Henbest, G. F. Smith, and J. A. Bentley, 1952. 3-Indolyl-aceto-nitrile, a naturally occurring plant growth hormone, *Nature, Lond.* **169**, 455–486.

Kaper, J. M., O. Gebhard, C. J. van den Burg, and H. Veldstra, 1963. Studies on Indolepyruvic acid. I. Synthesis and Paper Chromatography. II. Ultraviolet Spectrophotometry. *Arch. Biochem. Biophys.* **103**, 469–474, 475–487.

Kaper, J. M. and H. Veldstra, 1958. On the metabolism of tryptophan by *Agrobacterium tumefaciens*. *Biochim. Biophys. Acta* **30**, 401–420.

Kefford, N. P. and O. H. Caso, 1965. A potent auxin with unique chemical structure; 4-amino-3,5,6-trichloropicolinic acid. *Botan. Gaz.* **127**, 159–163.

Kefford, N. P. and P. L. Goldacre, 1961. The changing concept of auxin. *Am. J. Bot.* **48**, 643–650.

Keitt, G. W., Jr. and R. A. Baker, 1966. Auxin activity of substituted benzoic acids and their effect on auxin transport. I. Growth assay by tobacco pith tissue culture. *Plant Physiol.* **41**, 1561–1569.

Kennedy, R. W. and J. L. Farrar, 1965. Induction of tension wood with the anti-auxin 2,3,5-triiodobenzoic acid. *Nature, Lond.* **208**, 406–407.

Kerstetter, R. E. and G. W. Keitt, 1966. Direct assay of IAA decarboxylation rate in excised tobacco pith; Relation to aging. *Plant Physiol.* **41**, 903–906.

Key, J. L. and J. Ingle, 1964. Requirement for the synthesis of DNA-like RNA for growth of excised plant tissue. *Proc. Nat. Acad. Sci.* **52**, 1382–1388.

Khalifah, R. A., N. L. Lewis, and C. W. Coggins, 1966. Differentiation between IAA and the Citrus Auxin by column chromatography. *Plant Physiol.* **41**, 208–210.

Koepfli, J. B., K. V. Thimann, and F. W. Went, 1938. Phytohormones: Structure and physiological activity. I. *J. Biol. Chem.* **122**, 763–780.

Kögl, F., A. J. Haagen Smit, and H. Erxleben, 1934. Über ein neues Auxin ('Heteroauxin') aus Harn. XI Mitt. *Zeit. physiol. Chem.* **228**, 104–112.

Kögl, F. and D. R. F. G. Kostermans, 1934. Heteroauxin als Stoffwechselprodukt niederer pflanzlicher Organismen. Isolierung aus Hefe. XIII Mitt. *Zeit. physiol. Chem.* **228**, 113–121.

Labarca, C., R. S. Bandurski, and P. B. Nicholls, 1966. Separation and characterization of four low-molecular-weight bound auxins. *Ann. Report MSU/AEC Plant Res. Lab.*, East Lansing, Michigan. 38–40.

Laibach, F., 1933. Wuchsstoffversuche mit lebenden Orchid-pollinien. *Ber. deut. botan. Ges.* **51**, 336–340.

LaMotte, C. E. and W. P. Jacobs, 1963. A role of auxin in phloem regeneration in Coleus internodes. *Developmental Biol.*, **8**, 80–98.

Larsen, P., 1944. 3-Indoleacetaldehyde as a growth hormone in higher plants. *Dansk. Bot. Arkiv*, **11**, 11–132.

LaRue, C. D., 1935. The role of auxin in the development of intumescences on poplar leaves; in the production of cell outgrowths in the tunnels of leaf-miners; in the leaf-fall of *Coleus. Am. J. Bot.* **22**, 908.

LaRue, C. D., 1936. The effect of auxin on the abscission of petioles. *Proc. Nat. Acad. Sci.* **22**, 254–259.

Leach, R. W. A. and P. F. Wareing, 1967. Distribution of auxin in horizontal woody stems in relation to gravimorphism. *Nature, Lond.* **214**, 1025–1027.

Lewis, L. N. and J. C. Bakshi, 1966. Protein synthesis in leaf abscission. *Plant. Physiol.* **41**, Suppl., Proc. Ann. Mtgs., iv.

Lin, C. Y., J. L. Key, and C. E. Bracker, 1966. Association of D-RNA with polyribosomes in the soybean root. *Plant Physiol.* **41**, 976–982.

Luckwill, L. C., 1959. Fruit growth in relation to internal and external chemical stimuli. In *Cell, Organism and Milieu*, Growth Symp. XVII, ed. D. Rudnick. Ronald Press, New York.

Mahadevan, S. and K. V. Thimann, 1964. Nitrilase II. Substrate specificity and possible mode of action. *Arch. Biochem. Biophys.* **107**, 72–68.

Mai, G., 1934. Korrelationsuntersuchungen an entspreiteten Blattstielen mittels lebender Orchideenpollinien als Wuchsstoffquelle. *Jahrb. wiss. Bot.* **79**, 681–713.

Marinos, N. G., 1957. Responses of Avena coleoptile sections to high concentrations of auxin. *Austral. J. Biol. Sci.* **10**, 147–163.

Masuda, Y., 1959. Role of cellular ribonucleic acid in the growth response of *Avena* coleoptile to auxin. *Physiol. Plantarum*, **12**, 324–335.

Matell, M., 1953. *Stereochemical studies on plant growth substances*. Diss., Almquist and Wiksells, Uppsala.

McCune, D. C., 1961. Multiple peroxidases in corn. *Ann. N.Y. Acad. Sci.* **94**, 723–730.

Muir, R. M. and C. Hansch, 1951. The relationship of structure and plant growth activity of substituted benzoic and phenoxyacetic acids. *Plant Physiol.* **26**, 369–374.

Myers, R. M., 1940. Effect of growth substances on the absciss layer in leaves of Coleus. *Botan. Gaz.* **102**, 323–338.

Nicholls, P. B. and R. S. Bandurski, 1966. Further analysis of the IAA-inositol Complex 'B–2'. *Ann. Report MSU/AEC Plant Res. Lab.*, East Lansing, Michigan, 40–43.

Nitsch, J. P. 1956. Methods for the investigation of natural auxins and growth inhibitors. In *The Chem. and Mode of Action of Plant Growth Substances*, ed. R. L. Wain and F. Wightman. Butterworths, London, pp. 3–31.

Nitsch, J. P. and C. Nitsch, 1956. Auxin-dependent growth of excised *Helianthus tuberosus* tissues. *Am. J. Bot.* **43**, 839–851.

Nitsch, J. P., C. Pratt, C. Nitsch, and N. J. Shaulis, 1960. Natural growth substances in Concord and Concord Seedless grapes in relation to berry development. *Am. J. Bot.* **47**, 566–576.

Noodén, L. and K. V. Thimann, 1963. Evidence for a requirement for protein synthesis for auxin-induced cell enlargement. *Proc. Nat. Acad. Sci.* (U.S.) **50**, 194–200.

Noodén, L. and K. V. Thimann, 1965. Inhibition of protein synthesis and of auxin-induced growth by Chloramphenicol. *Plant Physiol.* **40**, 193–201.

Ockerse, R., B. Z. Siegel, and A. W. Galston, 1966. Hormone-induced repression of a peroxidase isozyme in plant tissue. *Science*, **151**, 452–453.

Overbeek, J. van, R. Blondeau, and V. Horne, 1955. Maleimides as auxin antagonists. *Am. J. Bot.* **42**, 205–213.

Patau, K., N. K. Das, and F. Skoog, 1957. Induction of DNA Synthesis by Kinetin and IAA in excised tobacco pith tissue. *Physiol. Plantarum*, **10**, 949–966.

Pecket, R. C., 1957. The initiation and development of lateral meristems in the pea root. I. The effect of young and of mature tissue. *J. Exp. Bot.* **8**, 172–180. II. The effect of indole-3-acetic acid. *Ibid.*, 181–194.

Pilet, P. E. and A. W. Galston, 1955. Auxin destruction, peroxidase activity and peroxide genesis in the roots of *Lens culinaris*. *Physiol. Plantarum*, **8**, 888–898.

Poole, R. J. and K. V. Thimann, 1964. Uptake of Indole-3-acetic acid and Indole-3-acetonitrile by *Avena* coleoptile sections. *Plant Physiol.* **39**, 98–103.

Porter, W. L. and K. V. Thimann, 1965. Molecular requirements for auxin action I. Halogenated indoles and indoleacetic acid. *Phytochem.* **4**, 229–243.

Prochazka, Z. and V. Sanda, 1958. Combined ascorbic acid. XII. Isolation of pure Ascorbigen and other Indole derivatives from cabbage. *Chem. Listy*, **52**, 2378–2387 (*Chem. Abstr.* **53**, 8255; **54**, 10008).

Randhawa, G. S. and B. B. Sharma, 1967. Studies on fruit set and fruit drop in sweet orange. *Int. Symp. on Plant Growth Substances*, Calcutta Univ., ed. S. M. Sircar; Abstr. 3.24.

Ray, P. M., 1956. The destruction of indoleacetic acid. II. Spectrophotometric study of the enzymatic reaction. *Arch. Biochem. Biophys.* **64**, 193–216.

Ray, P. M. and A. L. Ruesink, 1962. Kinetic experiments on the nature of the growth mechanism in oat coleoptile cell walls. *Developm. Biol.* **4**, 377–397.

Rayle, D. L. and W. K. Purves, 1967. Isolation and identification of Indole-3-Ethanol (Tryptophol) from Cucumber Seedlings. *Plant Physiol.* **42**, 520–524.

Reinders, D. E., 1942. Intake of water by parenchymatic tissue. *Rec. trav. bot. Néerl.* **39**, 1–140.

Reinders-Gouwentak, C. A. and F. Bing, 1948. Action de l'acide α-naphthyl-acétique contre la chute des fleurs et des fruits de la tomate et son influence sur la couche séparatrice des pédicelles. *Proc. Kon. Akad. Wetensch.*, Amsterdam, **51**, 1183–1192.

Rier, J. P., Jr., 1962. The differentiation of vascular tissues in callus of angiosperms. *Am. J. Bot.* **49**, 663–667.

Rossetter, F. N. and W. P. Jacobs, 1953. Studies on abscission: The stimulating role of nearby leaves. *Am. J. Bot.* **40**, 276–280.

Rubinstein, B. and A. C. Leopold, 1963. Analysis of the auxin control of bean leaf abscission. *Plant Physiol.* **38**, 262–267.

Schwarz, K. 1957. Espectros U.V. dos produtos de decomposiçao espontanea de Acido Indolpiruvico. *Arq. Inst. Biol. São Paulo*, **24** (6), 81–91.

Scott, P. C. and A. C. Leopold, 1966. Abscission as a mobilization phenomenon. *Plant Physiol.* **41**, 826–830.

Setterfield, G., 1963. Growth regulation in excised slices of Jerusalem artichoke tuber tissue. *XVIIth Symp. Soc. Exp. Biol.*, ed. G. E. Fogg. Univ. Press, Cambridge, pp. 98–126.

Shantz, E. M. and F. C. Steward, 1957. The growth stimulating substances in extracts of immature corn grain. *Plant Phys.* **32**, suppl. viii.

Silberger, J. Jr. and F. Skoog, 1953. Changes induced by IAA in nucleic acid contents and growth of tobacco pith tissue. *Science*, **118**, 443–444.

Slocum, D. H. and J. E. Little, 1957. Growth stimulation of Avena coleoptiles. *Plant Physiol.* **32**, 192–196.

Smith, M. S. and R. L. Wain, 1952. The plant growth regulating activity of dextro- and laevo-α(2-naphthoxy)propionic acid. *Proc. Roy. Soc. B*, **139**, 118–127.

Snow, R., 1935. Activation of cambial growth by pure hormones. *New Phytol.* **34**, 347–360.

Söding, H., 1936. Über den Einfluss von Wuchsstoff auf das Dickenwachstum der Bäume. *Ber. deut. bot. Ges.* **54**, 291–304.

Söding, H., 1940. Weitere Untersuchungen über die Wuchsstoffregulation der Kambium-tätigkeit. *Z. Bot.*, **36**, 113–141.

Stell, C. C., T. T. Fukuyama, and H. S. Moyed, 1965. Inhibitory oxidation products of IAA: Mechanism of action and route of detoxification. *J. Biol. Chem.* **240**, 2612–2618.

Steward, F. C. and E. M. Shantz, 1956. The chemical induction of growth in plant tissue cultures. In *The Chem. and Mode of Action of Plant Growth Substs.*, ed. R. L. Wain and F. Wightman. Butterworths, London, pp. 165–186.

Stowe, B. B. and K. V. Thimann, 1954. The paper chromatography of indole compounds and some indole-containing auxins of plant tissues. *Arch. Biochem. Biophys.* **51**, 499–516.

Stutz, R. E. 1958. Enzymatic formation of Indole-3-carboxaldehyde from Indole-3-acetic acid. *Plant Physiol.* **33**, 207–212.

Sweeney, B. M. and K. V. Thimann, 1938. The effect of auxins on protoplasmic streaming. II. *J. Gen. Physiol.* **21**, 439–461.

Synerholm, M, and P. W. Zimmerman, 1947. Preparation of a series of ω-(2,4-dichloro-phenoxy)-aliphatic acids and some related compounds with a consideration of their biochemical role as plant growth regulators. *Contrib. Boyce Thompson Inst.* **14**, 369–382.

Thimann, K. V., 1935. On the plant growth hormone produced by Rhizopus suinus. *J. Biol. Chem.* **109**, 279–291.

Thimann, K. V., 1954. The physiology of growth in plant tissues. *Am. Scientist*, **42**, 589–606.

Thimann, K. V., 1956. Promotion and inhibition; twin themes of physiology. *Am. Naturalist*, **90**, 145–162.

Thimann, K. V., 1956. Studies on the growth and inhibition of isolated plant parts. V. The effects of cobalt and other metals. *Am. J. Bot.* **43**, 241–250.

Thimann, K. V., 1958. Auxin activity of some indole derivatives. *Plant Physiol.* **33**, 311–321.

Thimann, K. V., 1963. Plant growth substances; past, present and future. *Ann. Revs. Plant Physiol.* **14**, 1–18.

Thimann, K. V., 1964. In *Formation of Wood in Forest Trees*, ed. M. H. Zimmermann. Academic Press, New York, pp. 452–454.

Thimann, K. V. and J. Behnke, 1950. *The use of auxins in the rooting of woody cuttings* (with supplementary tables). Petersham, Mass., Harvard Forest, Cabot Foundn. Pub. No. 1.

Thimann, K. V. and W. D. Bonner, Jr. 1949a. Inhibition of plant growth by protoane-monin and coumarin and its prevention by BAL. *Proc. Nat. Acad. Sci.* **35**, 272–276.

Thimann, K. V. and W. D. Bonner, Jr., 1949b. Experiments on the growth and inhibition of isolated plant parts. II. The action of several enzyme inhibitors on the growth of the Avena coleoptile and on Pisum internodes. *Am. J. Bot.* **36**, 214–222.

Thimann, K. V. and W. D. Bonner, Jr., 1950. Experiments on the growth and inhibition of isolated plant parts. III. Action of some inhibitors of pyruvate metabolism. *Am. J. Bot.* **37**, 66–75.

Thimann, K. V. and Loos, G. M., 1957. Protein synthesis during water uptake by tuber tissue. *Plant Physiol.* **32**, 274–279.

Thimann, K. V. and S. Mahadevan, 1964. Nitrilase. I. Occurrence, preparation, and general properties of the enzyme. *Arch. Biochem. Biophys.* **105**, 133–144.

Thimann, K. V. and E. F. Poutasse, 1941. Factors affecting root formation of *Phaseolus vulgaris. Plant Physiol.* **16**, 585–598.

Thimann, K. V. and C. L. Schneider, 1938. Differential growth in plant tissues. I and II. *Am. J. Bot.* **25**, 627–641; **26**, 792–799.

Thimann, K. V. and C. L. Schneider, 1938. The role of salts, hydrogen ion concentration and agar in the response of Avena coleoptile to auxins. *Am. J. Bot.* **25**, 270–280.

Thimann, K. V. and F. Skoog, 1940, 1942. The extraction of auxin from plant tissues. I and II. *Am. J. Bot.* **27**, 951–960; **29**, 598–606.

Tomaszewski, M. and K. V. Thimann, 1966. Interactions of phenolic acids, metallic ions and chelating agents on auxin-induced growth. *Plant Physiol.* **41**, 1443–1454.

Torrey, J. G., 1950. The induction of lateral roots by indoleacetic acid and root decapitation. *Am. J. Bot.* **37**, 257–264.

Torrey, J. G., 1956. Chemical factors limiting lateral root formation in isolated pea roots. *Physiol. Plantarum*, **9**, 370–387.

van der Lek, H. A. A., 1925. *Over de Wortelvorming van houtige stekken.* Diss. Utrecht.

van der Lek, H. A. A., 1934. Over den invloed der knoppen op de Wortelvorming der stekken. *Med. Landbouw-hoogeschool Wageningen*, **38**, 1–95.

van Overbeek, J. and L. E. Gregory, 1945. A physiological separation of two factors necessary for the formation of roots on cuttings. *Am. J. Bot.* **32**, 336–341.

Van Steveninck, R. F. M., 1957–58. Factors affecting the abscission of reproductive organs in yellow lupins. I and II. *J. Exp. Bot.* **8**, 373–381; **9**, 366–372.

Van Steveninck, R. F. M., 1959. III. Endogenous growth substances in virus infected and healthy plants and their effect on abscission. *J. Exp. Bot.* **10**, 367–376.

Veldstra, H., 1944. Researches on plant growth substances, IV and V. Relation between chemical structure and physiological activity. *Enzymologia*, **II**, 97–136, 137–163.

Vyvyan, M. C., 1949. The use of growth substances to control the shedding of fruit. *Ann. Appl. Biol.* **36**, 553–558.

Wain, R. L., 1953. Plant Growth Substances. *Roy. Inst. Chem. Lectures, Monogr. and Reports.* No. 2.

Wain, R. L. and D. B. Harper, 1967. New phenolic plant growth-regulating compounds. *Nature, Lond.* **213**, 1155–1156.

Wain, R. L. and F. Wightman, 1954. The growth-regulating activity of certain ω-substituted alkyl carboxylic acids in relation to their β-oxidation within the plant. *Proc. Roy. Soc.* **B 142**, 525–536.

Wareing, P. F., C. E. A. Hanney, and J. Digby, 1964. In *Formation of Wood in Forest Trees*, ed. M. H. Zimmermann, Academic Press, New York, pp. 323–344.

Wershing, H. F. and I. W. Bailey, 1942. Seedlings as experimental material in the study of " Redwood " in conifers. *J. Forestry*, **40**, 411–414.

Wetmore, R. H. and W. P. Jacobs, 1953. Studies on abscission: the inhibiting effect of auxin. *Am. J. Bot.* **40**, 272–276.

Wetmore, R. H. and J. P. Rier, 1963. Experimental induction of vascular tissues in callus of Angiosperms. *Am. J. Bot.* **50**, 418–430.

Winter, A. H., 1964. Evidence for the occurrence of indolepyruvic acid *in vivo*. *Arch. Biochem. Biophys.* **106**, 131–137.

Winter, A. and K. V. Thimann, 1966. Bound Indoleacetic acid in *Avena* coleoptiles. *Plant Physiol.* **41**, 335–342.

Wright, S. T. C., 1956. Studies of fruit development in relation to plant hormones. III. Auxins in relation to fruit morphogenesis and fruit drop in the black currant, *Ribes nigrum. J. Hort. Sci.* **31**, 196–211.

Yager, R. E., 1960a. Some effects of leaves and IAA upon floral abscission in Nicotiana tabacum. *Bot. Gaz.* **121**, 244–249.

Yager, R. E., 1960b. Possible role of pectic enzymes in abscission. *Plant Physiol.* **35**, 157–162.

Yamaki, T. and Nakamura, 1952. Formation of indoleacetic acid in maize embryo. *Sci. Papers Coll. Gen. Educ., Univ. Tokyo*, **2**, 81–98.

Zenk, M. H., 1962. Aufnahme und Stoffwechsel von *-Naphthyl-essigsäure durch Erbsenepicotyle. *Planta*, **58**, 75–94.

2. The gibberellins

Robert E. Cleland

Introduction

For two decades following its discovery in 1927, auxin was the only natural growth regulator known to the western world. At first many physiologists attempted to explain the control of all aspects of plant growth in terms of auxin, but as time progressed it became increasingly apparent that other natural growth regulators must exist. For instance, the difference in growth rate between tall and dwarf plants could not be explained simply in terms of auxin. Ironically, a second natural growth regulator had actually been discovered in Japan in the year preceding the discovery of auxin. (A summary of the early history of gibberellin research can be found in Stowe *et al.*, 1961.) The young physiologist Kurosawa had been attempting to determine why infection of rice seedlings with the fungus *Fusarium heterosporum* (also called *Gibberella fujikuroi*) results in an excess longitudinal growth of the stems. In 1926 he showed that the fungus must produce a growth-promoting substance since a culture broth in which the fungus had grown was capable of causing this growth promotion even after the fungus had been removed. Shortly thereafter, the active substance was isolated and named gibberellin. In 1938, Yabuta and Sumiki succeeded in crystallizing gibberellin but the crystals proved to be a mixture of substances which could not be separated with the techniques then available. Thus the chemical nature of gibberellin remained unknown. During the 'thirties and 'forties, however, the Japanese had tested the biological activity of gibberellin and had shown that it affected a variety of developmental processes in addition to stem elongation.

This pioneering work was all published in Japanese with the result that it remained unknown to the western world until English translations were made available after the Second World War. Physiologists then realized the potential

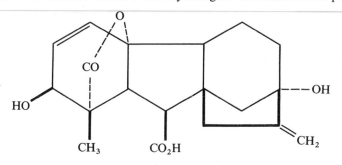

Figure 2.1 Structure of gibberellic acid (GA$_3$).

importance of gibberellin and set to work to determine its identity and its effects on plants. Brian and co-workers had the good fortune to obtain a *Fusarium* culture which produced large amounts of a single gibberellin, and in 1955 they isolated gibberellic acid. Within a few years Cross *et al.* (1961) had worked out its structure (Fig. 2.1). Meanwhile, physiologists had demonstrated that the gibberellins were capable of causing amazing changes in the growth of plants. For instance, microgram quantities of a gibberellin could convert dwarf beans into pole beans (Fig. 2.2) or cause *Hyoscyamus niger* plants to flower. With the demonstration by Phinney *et al.* in 1957 that gibberellins are present in higher plants, it became apparent that plants possess a second group of growth regulators which might play an important role in the control of growth and development. Work on the role of the gibberellins then began in earnest.

Responses of plants to gibberellins

The initial experiments had suggested that gibberellin, like auxin, might be capable of influencing a variety of aspects of growth and development. To test this, gibberellic acid was added to a wide range of intact plants under a variety of conditions and the resulting effects were carefully measured and recorded. The results indicated that the gibberellins were, indeed, a most active group of growth regulators. An indication of the range of activity of the gibberellins can be seen by describing briefly some of the most dramatic of the responses of plants to them.

Plant growth and its response to gibberellin

One of the most pronounced effects of the gibberellins is to modify plant growth. Indeed, we have already seen that it was the ability of gibberellin to enhance the elongation of rice stems that led to its discovery. The effects of the gibberellins on growth are diverse, however, and differ from organ to organ and from plant to plant. This is not unexpected since growth itself is a complex phenomenon. For instance, the organs of the plant differ in regard to the location of the growth and to the way in which the growth occurs. Furthermore, since growth can occur in more than one way, what may appear to be identical changes in the overall growth of an organ may have resulted in entirely different ways. Therefore it is not sufficient to simply describe the effects of gibberellin on the growth of the whole organ; the effects must be further identified as to the location of the effect and the type of process that is changed. In order to do this we must first remind ourselves just what growth is and where it occurs in each organ (for a detailed discussion of the location of growth in a plant see Esau, 1953).

There are many ways in which growth can be defined. We shall use the term to indicate an increase in volume or in length of a plant or plant part. Growth

Figure 2.2 Effect of gibberellin on the growth of dwarf bean plants. Application of 20 μg of gibberellin has been made to plant on the right. (From Wittwer and Bukovac, 1957.)

meristem organizes form

can only come about through an increase in volume of the individual cells. When a cell expands, it may result in a cell with a larger volume or it may divide to produce two cells, each of which can then enlarge or divide again. Two processes, then, contribute to growth of the plant as a whole; cell division, which leads to an increase in the number of cells, and cell enlargement. In most organs both processes tend to occur in concentrated, distinctly localized regions. A region of cell division is known as a meristem, while the region of cell enlargement is usually known as the elongation zone.

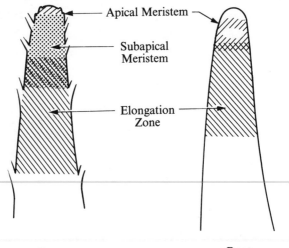

Figure 2.3   Location of growth in stems and roots. All stems contain a small apical meristem. In many cases this is followed by a large subapical meristem which overlaps the elongation zone. Roots possess only the apical meristem which is larger and overlaps the elongation zone.

Meristems may possess a second function in addition to the production of new cells. This is to provide the form for the organ. Whenever cells simply divide at random with no regulation as to the time or the direction of the division, they result in the formation of an unorganized callus. In a meristem, however, division is regulated so that the order of the organ is maintained and the form of the organ is assured. For instance, leaves are formed in precise positions on the stem and protoxylem cells are produced in the proper position to connect with the already differentiated xylem only because the divisions in the apical meristem of the stem are organized and regulated.

In stems, roots, and grass leaves the processes of cell division and cell elongation are occurring simultaneously but tend to be spatially separated (Fig. 2.3). Such an arrangement is characteristic of indeterminate growth. If one traverses a root from its apex to its junction with the stem, one encounters just behind the root cap a meristem which provides both the form and the new cells for the root. Next comes an elongation zone in which these new cells undergo elongation. Finally, the cells differentiate, cease elongating, and join the mature cells which make up the bulk of the root. The stem has a similar arrangement except that the meristem is divided into two parts. All stems possess an apical meristem which determines the form of the stem and provides some new cells. In addition, some, but not all stems have a second meristem below this, called the subapical or intercalary meristem, which produces great numbers of new cells for the developing stem. The length of the stem is largely determined by the activity of this subapical meristem; when this meristem is inactive a dwarf plant is formed. The cells which are produced by these two meristems then undergo elongation and finally, differentiation. In the grass leaf, the same growth pattern prevails with a meristem located at the base of the leaf, an elongation zone distal to it and finally, the bulk of the leaf composed of mature, non-growing cells.

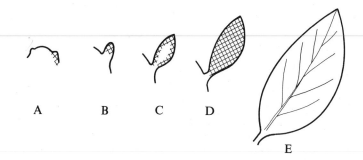

Figure 2.4 Sequence of growth in a dicot leaf. After initiation (A), growth occurs first in an apical meristem (B), then in lateral meristems (C) and finally throughout the blade (D). This is followed by a period of extensive cell enlargement to produce a fully expanded leaf (E).

A different pattern of growth is found in dicot leaves, flowers, and fruits. Here, cell division and enlargement tend to be separated in time rather than space. This is known as determinate growth. For example, in the dicot leaf, growth is, at first, primarily due to division in an apical meristem and later to divisions in lateral and rib meristems (Fig. 2.4). Following this, there is a period in which cell enlargement occurs throughout the entire leaf and which leads to the expanded, mature leaf.

With this background, we can now return to a description of the effects of gibberellin on the growth of intact plants. Certainly one of the most dramatic of its effects is its ability to increase the growth rate of dwarf stems. For example, addition of gibberellin to a cabbage plant converts the 'head' or dwarf stem into a stem which is 6–8 feet tall. An extreme case of dwarfism occurs in the 'rosette' plants such as sugar beet where, in the rosette form, the stem is an extreme dwarf. The stem can be made to undergo rapid growth or 'bolting' if the plant is subjected to the proper set of environmental conditions, or, in the absence of these conditions, if the plant is treated with gibberellin (Lang, 1956). In contrast to the situation with dwarf plants, treatment of caulescent or non-dwarf plants with gibberellin results in little or no stimulation of growth.

In only a few cases has there been any attempt to determine the nature of the growth promotion in dwarf stems. Sachs *et al.* (1959) clearly demonstrated that the induction of bolting in rosette plants is due to an activation of cell division in the normally inactive subapical meristem. The growth, in this case, is a result of the increase in the number of cells in the stem. In other cases, such as dwarf peas (Arney and Mancinelli, 1966) gibberellin stimulates both cell division in the subapical meristem and cell expansion, but the major effect is still the induction of cell division.

A second way in which gibberellin influences stem growth is by its ability to reverse some of the inhibitory effects of light. When dark-grown seedlings are exposed to dim red light a reduction in cell elongation takes place. This inhibition can be completely reversed in plants such as the Alaska pea (Lockhart, 1956) by treatment of the plant with gibberellin.

In other organs the effects of gibberellins on growth are less pronounced. However, significant increases in the rates of both cell division and cell enlargement can be induced by gibberellin in grass (Cleland, 1964) and dicot leaves. Roots, on the other hand, rarely respond to gibberellin to any really significant extent.

A special aspect of growth is dormancy. In deciduous plants the apical buds become dormant each autumn and growth resumes only when the dormancy is broken by cold treatment or by the proper light-dark cycles. In some plants, such as birch and sycamore (Eagles and Wareing, 1964), gibberellin is capable of substituting for the cold or light treatments, with the result that the growth capacity of the bud is restored. Seeds, too, often show dormancy. In certain varieties of lettuce and tobacco the seeds will fail to germinate unless they are exposed to red light. Treatment of these seeds with gibberellin in the dark will also induce germination.

A final growth effect of gibberellin is its ability to cause parthenocarpic growth of fruits. Normally, fertilization is required before fruit growth is initiated, but in some cases the fruit can develop in the absence of fertilization; this process is known as parthenocarpy. Among the plants in which parthenocarpic fruit growth can be induced by gibberellin are tomatoes, apples, and pears (Crane, 1964).

Developmental responses of intact plants to added gibberellin

The activity of gibberellin is not limited to a promotion of growth as the development of the plant can also be modified by gibberellin. This change in development may manifest itself through an altered pattern of cell division with the result that different organs are formed, or it may result in a change in the enzymes

(Substitute for light?)

which are produced by individual cells. One of the most significant developmental effects is the ability of gibberellin to cause certain plants to flower, i.e., to cause the conversion of vegetative apices into floral apices. Particularly receptive to gibberellin are the rosette plants which normally flower only after exposure to long days and, in some cases, a pretreatment with cold. In some rosette plants gibberellin will substitute for both the cold and the long-day treatments while in others it will substitute for only one (Lang and Reinhard, 1961). Not all plants can be made to flower with gibberellin, as caulescent plants and plants which require short-day conditions for flowering are insensitive to gibberellin. A second effect of gibberellin on flowering is its ability to modify the sex of flowers. In plants such as hops and cucumber (Galun, 1959) treatment of the developing flowers with gibberellin leads to shift towards maleness (stamens) in the flowers. Similarly, the formation of male sexual organs (antheridia) can be induced in many fern gametophytes by gibberellin.

A number of biochemical reactions in the plant have now been shown to be influenced by added gibberellin. The most thoroughly studied of these is the *de novo* production of α-amylase in the aleurone layers of cereal grains. A more thorough discussion of the role of gibberellin in this process will be presented later. Likewise, the wall-bound invertase of artichoke tubers is almost doubled by treatment with gibberellin (Edelman and Hall, 1964).

It should be apparent from this abbreviated list that application of gibberellin can produce a wide variety of changes in intact plants. It can be expected from this that the natural gibberellins will play a significant role in the control of plant growth and development. In order to understand the role, we must now discuss our knowledge concerning the natural gibberellins.

The natural gibberellins

What is a gibberellin?

It was apparent from the time when the first attempts were made to obtain a pure gibberellin that more than one compound possessed 'gibberellin-like' activity. This raised the important question: what is a gibberellin? Up to now we have been using the term without defining it.

There are two ways to characterize any biological compound; by its structure and by its physiological action. Both of these factors are incorporated into the definition of a gibberellin. When the original gibberellin mixtures were finally resolved into a series of physiologically active compounds, Cross *et al.* (1961) showed, by chemical analysis that they all possessed a common feature, the gibbane ring skeleton (Fig. 2.5). This requirement for a gibbane ring skeleton was then accepted as one property which a compound must possess in order to be called a gibberellin. But this could not be the sole factor which defined a gibberellin, for there are many compounds possessing a gibbane ring skeleton which are totally inactive in producing the gibberellin-like physiological effects. It is necessary, then, to have as a second part of the definition that the compound must be capable of producing the same physiological responses as gibberellic acid, i.e., must be active in specific gibberellin bio-assays.

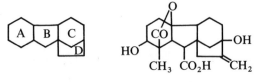

Gibbane skeleton GA_1

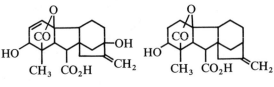

GA_3 GA_4

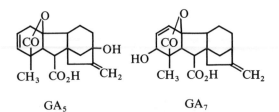

GA_5 GA_7

Figure 2.5 Comparison of the structures of five gibberellins. All possess the gibbane skeleton, but differ in the presence or absence of a hydroxyl at the junction of rings C and D, and the presence or absence of a hydroxyl and double bond on ring A.

A good bio-assay must be specific, easy to use and responsive to minute amounts of the substance to be tested. It is not easy to find bio-assays which meet these requirements. We have already seen that gibberellin can elicit a variety of physiological responses, but only a few are suitable for a bio-assay. Some responses, such as fruit growth or leaf expansion, are non-specific in that they can be induced by other substances in addition to the gibberellins. Other responses, such as flowering or bud dormancy are too difficult to use in that they require too much material or make use of material which is hard to obtain. The two responses which are most widely used as assays for gibberellins are the reversal of stem dwarfism and the induction of α-amylase in barley endosperm.

A variety of different plants has been used in the stem dwarfism assay, but the most commonly used at present are the dwarf corn and the dwarf pea. To illustrate how a gibberellin bio-assay is performed, the dwarf pea bio-assay of Hayashi and Rappaport (1966) will be described. Seeds of the variety Progress No. 9 are soaked for 6–8 hours and allowed to germinate in moist vermiculite in the dark. After 4 days, the seedlings are given red light. On the fifth day a 10 μl drop of 50 per cent methanol containing the material to be tested is added to the apex of each plant. After an additional five days the length of the epicotyl is measured and the lengths of the plants treated with the unknown material are compared with plants which received only methanol. A greater growth in the plants treated with the unknown material indicates that this material possesses gibberellin-like activity. The amount of activity can be determined by comparing the amount of increased growth with that caused by known amounts of GA_3. In order to check these results one can then employ one of the other common gibberellin bio-assays such as those listed in Table 2.1.

Table 2.1 Some commonly-used bio-assays for gibberellins.

Bio-assay	Response assayed	Assay period days	Minimum detectable GA_3—mμg	Reference
Dwarf corn	Length of 1st leaf sheath	10	10	Phinney, 1957
Dwarf pea	Length of epicotyl	5	1	Hayashi and Rappaport, 1966
Lettuce hypocotyl	Length of hypocotyl	5	100	Frankland and Wareing, 1960
Avena leaf	Length of leaf section	3	1	Bentley-Mowat, 1966
Barley endosperm	Production of α-amylase	1	0·2	Jones and Varner, 1967
Rumex leaf	Delay in loss of chlorophyll	4–5	0·2	Whyte and Luckwill, 1966

A gibberellin, then, is a compound which is active in gibberellin bio-assays and possesses a gibbane ring skeleton. For many substances which are isolated from plants it is easy to demonstrate the former, but almost impossible to demonstrate the latter for lack of sufficient material for chemical analysis. Then there are compounds, such as kaurene, which are active in some of the assays but do not have a gibbane ring. Compounds such as these are called 'gibberellin-like' rather than gibberellins.

The diversity of the natural gibberellins

Armed with suitable bio-assays, scientists have set out to study the natural gibberellins. One of the first questions that was asked was how many gibberellins exist in nature. Careful analysis of the products formed by the fungus *Fusarium* resulted in the identification of six different gibberellins (Cross *et al.*, 1961). Meanwhile three additional ones had been isolated from bean seeds (MacMillan *et al.*, 1961). Rather than differentiate these nine gibberellins by means of trivial or chemical names, the wise decision was made to designate them as GA with a subscript 1 to 9. Thus gibberellic acid became GA_3 while the gibberellins from *Fusarium* were designated GA_1, GA_2, GA_3, GA_4, GA_7, and GA_9, and the gibberellins originally isolated from higher plants were labelled GA_5, GA_6, and GA_8. As additional gibberellins are discovered they are assigned numbers; the list now exceeds twenty (see Paleg, 1965, for structures of GA_{1-13}).

A glance at the structure of the gibberellins should be sufficient to indicate their similarities. The five gibberellins shown in Fig 2.5 all possess the gibbane ring system, the lactone ring attached to ring A and the carboxyl group on ring B. They differ only in the presence or absence of a hydroxyl and a double bond in ring A or a hydroxyl at the juncture between rings C and D. The other gibberellins differ from these in similar ways.

In addition to these gibberellins which are all acidic in character, many plants also contain neutral gibberellins. The structures of these neutral gibberellins are still in doubt, but the fact that they can be converted into acidic gibberellins suggests that they are simply acidic gibberellins whose carboxyl group is masked by esterification. Evidence to support this has been obtained by Hashimoto and Rappaport (1966) when they demonstrated that radioactive GA_1, fed to bean seeds, is rapidly converted into a neutral gibberellin from which the original GA_1 can be recovered.

The gibberellins are all active in one or more of the standard bio-assays, but one should not think of them as having identical physiological activities (Brian *et al.*, 1964). A comparison of their effectiveness in a series of assays is shown in

Table 2.2 Comparison of the effectiveness of six gibberellins in a series of gibberellin bio-assays. (After Brian, 1966.)

Bio-assay	Effectiveness of gibberellin in bio-assay*					
	A_1	A_3	A_4	A_5	A_7	A_8
Dwarf corn, d-1 †	+ + +	+ + +	+ + +	+ + +	+ + +	0
Bean stem †	+	+ + +	0	+ + +	+ + +	0
Dwarf pea stem ‡	+ + +	+ + +		0		
Cucumber hypocotyl †	+ +	+ +	+ + +	+	+ + +	0
Myosotis flowering §	+ +	0	0	0	+ + +	0
Silene flowering §	0	0	0	0	+ + +	

* Activity is indicated from highly active (+ + +) to inactive (0). Blank means not tested.
† Brian *et al.*, 1964.
‡ Kende and Lang, 1964.
§ Michniewicz and Lang, 1962.

Table 2.2. Note that some processes, such as the growth of bean stems, respond to only certain gibberellins while other processes, such as flowering in *Myosotis*, respond to a different set of gibberellins. This raises the possibility that each of the gibberellins may be maximally effective on a certain set of processes while being inactive or only slightly active on others. Furthermore, it is essential to realize that the common bio-assays are not universally sensitive to the gibberellins. In general they select for those gibberellins which are most active in the promotion of stem growth. Should other gibberellins be isolated they might easily be ignored if they are inactive in stem growth but active on some other process such as sex determination or flowering. A similar situation exists with the steroids in animals, where no single bio-assay will detect all of them. For example, if the only bio-assays for steroids selected for male hormone activity, the number of steroids would appear to be limited. It should be apparent that there exists a need for additional gibberellin bio-assays which make use of some physiological response other than stem growth. With a proper spectrum of assays we might well find that the natural gibberellins number in the hundreds rather than the tens.

Distribution and synthesis of gibberellins in plants

The gibberellins are widely distributed in nature. Every Angiosperm and Gymnosperm which has been carefully examined has been found to possess at least one gibberellin. In addition, gibberellins have been detected in ferns, mosses, algae, fungi, and bacteria. It would appear, however, that they are restricted to only certain species of fungi and bacteria.

Determining which gibberellins are present in plants has not been an easy job. The difficulties in obtaining sufficient pure material has made chemical analysis impossible in most cases. Instead, the techniques of chromatography and IR spectroscopy have made it possible to show that GA_1 and GA_{3-9} all exist in higher plants. As yet, GA_2 has only been isolated from *Fusarium*. In addition there appear to be at least several other gibberellins whose identity is still unknown because they are different from any of the already characterized gibberellins.

No plant has yet been found which contains all of the gibberellins, but most plants possess at least several of them. For instance, *Echinocystis* seeds have yielded GA_1, GA_3, GA_4, GA_7, and one as yet unidentified gibberellin (Elson *et al.*, 1964), while *Phaseolus* seeds were found to have GA_1, GA_5, GA_6, and GA_8 (MacMillan *et al.*, 1961). The occurrence of a variety of gibberellins in the same plant should be contrasted with the situation with auxins and cytokinins where one principal one seems to predominate in most plants (see chapters 1 and 3).

The gibberellins can exist in more than one state within the plant. As has already been mentioned, in some plants a portion of the gibberellins are found

4

in a neutral, presumedly esterified form. A careful analysis of the rise and fall of these neutral gibberellins during the maturation of bean seeds has led Hashimoto and Rappaport (1966) to suggest that they act as a storage reservoir of gibberellin from which active, acidic gibberellins can be withdrawn when needed. In addition, bound forms of gibberellin exist. For instance, McComb (1961) showed that gibberellin can be released from a bound form by the proteolytic enzyme ficin.

All of the organs of the higher plant contain gibberellins, but the gibberellin concentration is by no means constant throughout the plant. The highest levels have been found in seeds, with exceptional levels occurring in the liquid endosperm of some seeds. For instance, a concentration of 470 μg GA_3 equivalents/g fresh weight has been recorded for *Echinocystis* endosperm (Corcoran and Phinney, 1962). The levels in the vegetative regions are more in the order of 1–10 μg GA_3 equivalents/g fresh weight. Young leaves are rich in gibberellins in comparison with the older leaves and mature stems (Radley, 1958). At first there was some doubt as to whether gibberellins occur in roots, as extractions of roots yielded little gibberellin. But in 1964 two groups (e.g., Phillips and Jones, 1964) demonstrated that the bleeding sap from roots of plants such as sunflower or *Lupinus alba* are particularly rich in gibberellins. The roots apparently produce the gibberellin which is then transported to the rest of the plant via the xylem. In general, then, the gibberellins are concentrated in the most rapidly growing and developing regions of the plant, as might be expected for substances which are involved in the regulation of the plant's growth and development.

The gibberellin molecule is somewhat unique in structure among the naturally occurring compounds. As a result the pathway of its biosynthesis was of more than usual interest to the biochemists. In 1958 Birch *et al.* conclusively demonstrated that the gibberellins were formed through the normal isoprenoid pathway of terpene biosynthesis, starting with acetate. The remaining details have now been largely filled in as a result of studies with both *Fusarium* and *Echinocystis* and are shown in Fig. 2.6. Acetate is first converted to mevalonate and then to isopentyl pyrophosphate. Four units of this (5 compound) are, in turn, condensed to form geranylgeranyl pyrophosphate, a C_{20} diterpene (Graebe *et al.*, 1965). Closure of the rings produces (−)kaurene, a four-ringed compound, which subsequently undergoes a rearrangement in ring B to produce the gibbane ring system and the free carboxyl group. The first identifiable gibberellin to appear in *Fusarium* is GA_4 (Geissman *et al.*, 1966). This can be converted to either GA_7 or GA_1, both of which appear to act as precursors for GA_3. The origin of the remaining gibberellins is not yet known.

The sites of gibberellin synthesis are not known for certain as it has not been possible to locate the biosynthetic enzymes with the exception of the *Echinocystis* endosperm. Experiments in which the gibberellin content of a plant is determined after excision of a particular region (Jones and Phillips, 1966) have suggested that it is the regions which normally have the highest levels of gibberellins which are the sites of synthesis. These include the apices of stem and root, the young leaves and the embryos and endosperm of developing seeds. It must be admitted, however, that our present evidence is far from satisfactory. The recently developed technique of collecting diffusible gibberellin from an excised

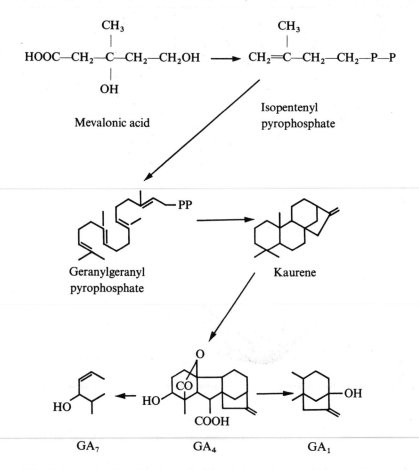

Figure 2.6 Pathway of gibberellin biosynthesis as deduced from experiments with *Fusarium moniliforme*.

portion of the plant (Jones and Phillips, 1964) may provide us with the necessary information. If the amount of gibberellin which diffuses from a region is greater than the amount that can be extracted at the original time of excision, then it is fairly certain that gibberellin synthesis has taken place in that region.

Although much of the gibberellin seems to remain close to the sites of synthesis, some movement can occur in the plant. This was clearly shown by

Zweig *et al.* (1961) who applied labelled gibberellin to the cotyledons of young bean seedlings and found that the gibberellin could soon be detected in both the shoot apex and the root. Apparently gibberellin can travel in both directions from the point of application; thus, unlike auxin, its movement is non-polar. In addition to the transport of applied gibberellin, there is also a significant movement of endogenous gibberellins in the plant. The bleeding sap of the root, which gives an indication of the composition of the transpiration stream, is rich in gibberellin in certain plants such as the sunflower (Phillips and Jones, 1964). This gibberellin is apparently produced in the roots and transported to the stem, but its function there is unknown. Likewise, a movement of gibberellin from the embryo to the aleurone layer is an essential step in the germination of cereal seeds.

Gibberellin can move in several different ways in the plant. Jones and Phillips (1964) have shown that certain of the gibberellins (e.g., GA_1) have the ability to move by diffusion in the plant, although other gibberellins (e.g., GA_5) apparently are unable to move in this fashion. Export of gibberellin from the root occurs in the xylem. Movement of gibberellin in the phloem is indicated by its presence in sieve-tube sap (Kluge *et al.*, 1964). Apparently gibberellins move about the plant in a manner which is similar to other organic metabolites such as the amino acids and carbohydrates.

In conclusion, it is now apparent that each higher plant contains a series of gibberellins, which differ in biological activity as well as in structure. They are concentrated in the regions where the development of the plant is taking place. They are presumed to be synthesized in these regions although they may also have arrived there by transport from some other part of the plant.

The role of endogenous gibberellin in the plant

Techniques for studying the role of endogenous gibberellins

The presence of natural gibberellins in the plant, combined with our knowledge of the physiological effects of added gibberellins, should suggest that the endogenous gibberellins must play some role in the regulation of the plant's development. But what role do they play? This is not an easy question to answer. The problem is that healthy, rapidly growing plants must be assumed already to possess optimal levels of the growth-regulating substances. Addition of gibberellin will result in super-optimal levels of gibberellin and the response of plants to super-optimal levels may be quite different than their responses to optimal gibberellin levels. For instance, the growth of grass leaves can be promoted by high levels of added gibberellin, but there is no evidence that gibberellin is needed for the normal growth of the grass leaf (Cleland, 1964). Moreover, the added gibberellin will be identical, at best, to only one of the natural gibberellins. The fact that a process fails to respond to added gibberellin cannot be used as evidence that the process does not require gibberellin, since it may be a different gibberellin which regulates this process. For instance, addition of gibberellic acid (GA_3) fails to cause *Silene* to flower (Michniewicz and Lang, 1962) but the fact that flowering in this plant can be induced by GA_7 suggests that gibberellins are involved in the flowering process. It should be obvious that other techniques must be used in order to assess the role of endogenous gibberellins.

If a process involves a distinct change in the developmental pattern of the plant (e.g., the change from the vegetative to floral state), one can attempt to correlate changes in amounts or types of gibberellin with the onset of the change. This has proved to be a useful technique, but it must always be kept in mind that this type of analysis will not reveal whether a change in the gibberellins is the cause or the result of the developmental change. If the change in the gibberellins precedes the developmental change one can be more certain that the gibberellins are regulating this process.

A second approach is to obtain plants or plant parts in which the gibberellin level is reduced, and to show that this reduction results in an inhibition of the process under study. In addition, one must show that added gibberellin will then restore the activity. Plants or plant parts with a reduced gibberellin content can be obtained in several ways. The easiest method, in theory, is to excise a portion of the plant, such as a stem or root segment, and remove it from the gibberellin supply. Of course, the excised segment must be incapable of synthesizing gibberellin itself. This technique has been of great value in studies involving auxins where much has been learned through the use of section tests, but it has been less useful with the gibberellins since regions of the plant which respond to gibberellin also appear to produce it in most cases. Callus tissues have also proved useful in the studies with auxins and cytokinins, but the general lack of response of calluses to gibberellins has made them of little use in gibberellin research.

The second method is to utilize plants in which there is a genetic block in the pathway of gibberellin synthesis. These plants can often be recognized by their 'dwarf' growth habit. Care must be used in selecting the plants since dwarfism can be caused by other factors such as the presence of inhibitors or a lack of ability to respond to gibberellins. Before using a particular dwarf, one must determine the level of gibberellin and show that this is considerably less than in the tall variety of the same plant.

The third method is to make use of plants in which the gibberellin level has been reduced by treatment with one of the 'growth retardants'. This term is used to describe a group of chemicals which have the common property of causing dwarfism when applied to plants (Cathey, 1964). For example, when tall varieties of chrysanthemum or poinsettia are treated with a retardant they revert to the dwarf form. Best known of these compounds are CCC, Phosfon D, Amo-1618, and B-995. The structures of these compounds are shown in Fig. 2.7. It can be seen that they differ in their structure and that none of them is closely related to gibberellin. However, most of the physiological effects of the retardants can be reversed by gibberellin. Because of this, the growth retardants have often been called 'antigibberellins', although there is no evidence that they can act as competitive inhibitors of gibberellin in the way that the antiauxins compete with auxin. The use of the term 'antigibberellin' is unfortunate and now seems to be disappearing.

Three of the growth retardants, CCC, Phosfon D, and Amo-1618, have now been shown to be potent inhibitors of one or more steps in the biosynthesis of gibberellin (e.g., Dennis *et al.*, 1965). As a result, treatment with these retardants

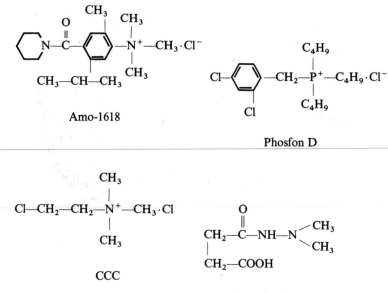

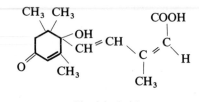

Figure 2.7 Structures of five growth inhibiting compounds. Amo-1618, Phosfon D, CCC, and B995 are synthetic compounds which cause dwarfism in some plants and are known as 'growth retardants'. The first three can act as inhibitors of gibberellin biosynthesis. Abscisic acid is a naturally occurring growth inhibitor which interacts with gibberellin.

can lead to a reduction in the endogenous gibberellin level. For instance, Baldev et al. (1965) found that the gibberellin content of developing pea seeds was reduced by over 85 per cent by 50 mg 1^{-1} of Amo-1618. Caution must be used in interpreting the results of studies with the retardants, since it now seems clear that they can act in more than one way. This is indicated by the fact that only certain of the effects of the retardants can be reversed by added gibberellin (Cathey, 1964). Many plants fail to show any response to the growth retardants. In these cases the retardants may be failing to penetrate, may be detoxified after entry, or may be unable to interfere with gibberellin biosynthesis. Before one can conclude from the insensitivity of a process to the retardants that the process is not regulated by gibberellin, one must first demonstrate that the retardant has actually lowered the endogenous gibberellin level.

The techniques, then, are available for studying the role of endogenous gibberellins in the control of the plant's growth and development. An attempt will now be made briefly to summarize the present state of our knowledge concerning this subject.

The role of endogenous gibberellin

Let us first consider the role of gibberellin in the growth of the various plant organs. It will be recalled that the stem has three regions which contribute to its longitudinal growth: the apical meristem, the subapical meristem and the elongation zone. The first of these appears to function independently of the presence or absence of gibberellin. This is indicated by such facts as that the mitotic activity of this region is the same in dwarf and tall varieties of Samolus parviflorus (Sachs et al., 1959) and that treatment of plants such as chrysanthemum with the growth retardants CCC and Amo-1618 does not affect the apical meristem even though it causes the plant to be dwarfed (Sachs et al., 1960).

The activity of the apical meristem is not, however, completely divorced from gibberellin. Under certain environmental conditions, the apical buds of many perennials become dormant and the mitotic activity ceases. This dormancy can be reversed in plants such as the peach by addition of gibberellin (Donaho and Walker, 1957). Furthermore, the natural breaking of dormancy by cold is usually accompanied by an increase in gibberellins (Eagles and Wareing, 1964). This might be taken as evidence that the activity of this region is regulated by gibberellin. However, the picture is not that simple. Associated with the decrease in gibberellins at the onset of dormancy there is usually an accumulation in the bud of one or more growth inhibitors. Perhaps the best studied of these is dormin (Fig. 2.7) which has been implicated in the dormancy of sycamore and birch buds (Eagles and Wareing, 1964). Although the exact mode of action of dormin is yet to be discovered, its ability to effectively block growth in a variety of systems is now well established (El-Antably et al., 1967). In many systems, added gibberellin can counteract the effects of dormin. This suggests that the role of gibberellin in the apical meristem may simply be to protect it from the inhibiting effects of endogenous growth inhibitors such as dormin.

A more direct role of gibberellin is found in the subapical meristem, where the mitotic activity is largely regulated by the gibberellin level. This was first indicated by the studies of Sachs *et al.* (1959) on *Hyoscyamus niger*. This plant in its vegetative phase assumes the rosette growth form, i.e., the stem is severely dwarfed due to a lack of any activity in its subapical meristem. When gibberellin is applied, the subapical meristem becomes active, cell division in the longitudinal direction increases markedly and the stem begins to grow (Fig. 2.8).

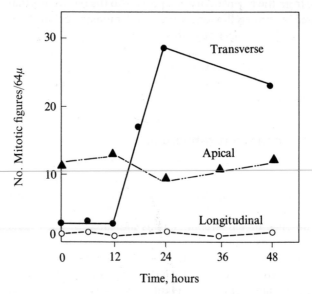

Figure 2.8   Effect of gibberellin on cell division in stems of *Hyoscyamus niger*. Plants treated with 25 μg GA$_3$ and numbers of divisions determined in apical meristem and in transverse and longitudinal directions in the subapical meristem. Curves for controls not shown as values remained at zero hour level. (Adapted from Sachs *et al.*, 1959.)

This is known as 'bolting'. Normally bolting only occurs when the plant changes from the vegetative to the flowering phase. Analysis of the gibberellin level during this transition showed that this bolting is correlated with a marked increase in the level of one of the gibberellins (Lang, 1960).

Further evidence concerning the relationship between gibberellin and the subapical meristem has come from the studies of Sachs *et al.* (1960) on the effects of growth retardants on plants which normally possess an active subapical meristem. For example, treatment of tall varieties of chrysanthemum with Amo-1618 resulted in cessation of cell division in the subapical meristem and a dwarfing of the stem. The activity could be restored by subsequent addition of gibberellin.

It is tempting to assume that gibberellin directly affects the mitotic processes in the subapical meristem, but such a conclusion is premature. When one cell produces two cells of identical size, both cell division and cell enlargement must take place. We do not know which of these processes must come first. Haber (1964) has suggested that cell division in the subapical meristem only occurs when the volume of a cell exceeds a certain limit. If this is true, stimulation of cell enlargement by gibberellin would result in an increase in cell number but no increase in the final size of the individual cells, i.e., cell division would be stimulated by the gibberellin-induced enlargement rather than by the gibberellin itself. At present there is simply not enough evidence to determine whether the primary effect of gibberellin is on cell division or cell enlargement, but it is now clear from the correlation between the gibberellin level and the activity of the subapical meristem that it is gibberellin which is the principal agent which regulates this region.

The relationship between gibberellin and cell elongation in the stem is less clear. One of the reasons for this is that despite the large number of cases where it has been shown that gibberellin controls stem growth, in only a handful of cases has the effect of gibberellin on cell elongation been actually checked. In some plants such as *Hyoscyamus* (Sachs *et al.*, 1959) the final cell length is unaffected by gibberellin while in other dwarfs such as the Meteor pea (Arney and Mancinelli, 1966) the final cell length is increased to some extent by gibberellin. Perhaps the best evidence for a role of gibberellin in the control of cell elongation comes from studies on the inhibition of stem growth by red light. Treatment of dark-grown seedlings with red light results in an inhibition of both cell elongation and cell division. Lockhart (1956) showed that with Alaska pea and Pinto bean gibberellin could completely reverse the light-induced inhibition. Furthermore, the decrease in wall extensibility which was caused by red light and which is presumedly the cause of the inhibition of cell elongation was also reversed by gibberellin (Lockhart, 1960). It is apparent that at least in these plants the presence of gibberellin results in an enhancement of cell elongation.

On the other hand, there is good reason to believe that the role of gibberellin in the control of cell elongation is a minor one. Most excised stem sections, where the gibberellin content might be expected to be low, respond to only a limited extent to added gibberellin (Brian and Hemming, 1958). When one compares the response of stem sections to auxin and gibberellin, one can only conclude that auxin is the agent which exerts the principal control of cell elongation in the stem with a relatively minor role being played by gibberellin.

One other aspect of the interaction between light and gibberellin needs to be mentioned at this point. In certain varieties of plants, such as the Progress No. 9 variety of pea, the growth rate of dark-grown plants is normal, but illumination of the seedlings with red light causes a severe inhibition of the growth rate; these plants only show the dwarf character when exposed to red light. Analysis of these plants by Kende and Lang (1964) indicated that the light caused no real difference in either the amounts or types of gibberellins present in the stem. The sensitivity of the plants to the various gibberellins was, however, greatly altered by light. The light treatment resulted in a loss of ability to respond to GA_5 or to the major gibberellin that is present in the pea stems, while the ability of the plant to respond to GA_1 was unaffected (Table 2.3). One must keep in mind,

Table 2.3   Effect of red light on the ability of dwarf peas (Progress No. 9) to respond to the different gibberellins. Peas dwarfed with red light or Amo-1618 and treated with the two gibberellins present in dwarf peas, GA_1 and GA_5, at 0·1 µg/plant. (Kende and Lang, 1964.)

Type of dwarf	Height of pea stem, mm		
	0	GA_1	GA_5
Dwarfed by red light	37	120	54
Dwarfed by Amo-1618	44	105	105

then, that regulation of gibberellin-sensitive processes can occur not only by a change in the amounts and types of the gibberellins but also by a change in the ability of the plant to respond to the gibberellins.

In conclusion, the three growth regions of the stem differ in the degree to which their activity is normally regulated by gibberellin; from the subapical meristem whose activity is primarily controlled by the gibberellin level, to the elongation zone where gibberellin probably plays a secondary role in the regulation, to the apical meristem whose activity is essentially independent of gibberellin.

In contrast, the evidence to suggest that gibberellin is involved in the regulation of root growth is scarce. There are some reports of a stimulation of root growth by gibberellin (e.g., Mertz, 1966), but in many cases added gibberellin was without effect. In addition, the general failure of dwarf varieties or growth-retardant treated plants to show reduced root growth indicates that gibberellin is unlikely to be a major factor in the control of growth in this organ. This may be due to the lack of a subapical meristem in the root, since it is this region of the stem which is most markedly affected by gibberellin.

Our understanding of the control of leaf expansion is complicated by the fact that so many factors seem to play a part in this process. For example, the expansion of discs cut from etiolated bean leaves can be enhanced by cytokinins, red light, and cobalt in addition to gibberellin (Humphries and Wheeler, 1963). It must be noted, however, that in none of these cases does the expansion of the discs even begin to equal the expansion which would occur in the intact leaf when exposed to light. Thus experiments involving excised leaf discs may be of little use in determining the role of gibberellin in leaf growth. The best indication that gibberellin may be involved in the regulation of leaf expansion is the demonstration of Wheeler (1960) that the gibberellin level of bean leaves rises sharply at the time that light-induced expansion begins. On the other hand, in plants in which the stem growth is inhibited by growth retardants or by dwarf genes, and in which the gibberellin level is presumedly reduced, leaf expansion is usually unaffected (Cathey, 1964). Thus it seems doubtful that gibberellin is required for leaf expansion although, when present in elevated levels, it may be able to stimulate leaf growth.

The evidence for a role of the gibberellins in the control of fruit growth continues to mount. It is now well established that unfertilized flowers of many plants such as the tomato and certain varieties of apple can be made to set normal-looking, but seedless fruits when treated with gibberellin (Crane, 1964). Bradley and Crane (1962) showed that parthenocarpic peaches produced by gibberellin treatment were similar to the normal peaches in size and in the ratio of cell number to cell size. In addition, a strong correlation has been shown in normal fruits between the gibberellin content at various stages and the growth rate of the fruit (Jackson and Coombe, 1966). Following fertilization, synthesis of gibberellin takes place in the endosperm and embryo. This gibberellin, in turn, is required in order to allow growth of the fruit to proceed.

The gibberellins are important in the regulation and integration of development as well as growth processes in the plant. One of the most interesting of these processes is the initiation of flowering. In many plants the conversion of vegetative apices to floral apices only occurs when the plant has been subjected to a particular set of environmental conditions, such as specific regimes of light and dark or a protracted period of cold (see chapter 16). The first indication that gibberellin might be involved in this transition came from the demonstrations of Lang (1956, 1957) that added gibberellin could substitute for the proper environmental conditions in some plants. For example, *Hyoscyamus niger*, which normally will remain vegetative indefinitely if kept under short days, can be made to flower by an application of gibberellin. Likewise, the Early French Forcing variety of carrot requires a protracted cold treatment before it will flower, but can be made to flower without the cold period if treated with gibberellin (Fig. 2.9). In addition, analysis of the gibberellins during the induction of flowering has shown that this transition is accompanied by an increase in one or more of the gibberellins (Lang, 1960).

The process of flowering is believed to be mediated by the hormone, florigen, which is formed in the leaves under the proper environmental conditions and then migrates to the apex which it causes to change from the vegetative to the floral condition (see chapter 16). It is tempting to believe that florigen is a gibberellin, but the evidence would seem to be against this. Only certain plants can be induced to flower by added gibberellin (Lang and Reinhard, 1961). Most of these are rosette plants which require long days, a protracted cold period (vernalization) or both for flowering. The failure of the known gibberellins to

Figure 2.9   Ability of gibberellin to replace the cold requirement for the flowering of carrot. Early French Forcing variety of carrot grown without cold treatment (left), with an 8-week cold treatment (right) or with 10 μg gibberellin daily in place of the cold treatment (middle). (From Lang, 1957.)

cause flowering in short-day plants indicates that none of these gibberellins can be florigen, since there is ample evidence that florigen is the same in long-day and short-day plants (Lang, 1965). This is certainly not conclusive evidence since florigen may simply be an as yet unknown gibberellin and the plants which flower in response to added GA_3 may do so by converting the GA_3 to the flowering gibberellin. Such a flowering gibberellin might easily be missed if it had little or no ability to induce stem elongation.

On the other hand, it should be noted that one of the first steps in the flowering of rosette plants is the bolting of the stem. The action of the added gibberellin may simply be to activate the subapical meristem and thus produce bolting, which in turn, allows floral initiation to take place. Further evidence will be needed before an unequivocal statement can be made concerning the role of gibberellin in flowering.

A second developmental process in which gibberellin plays an essential part is the germination of cereal grains. Nourishment for the embryo during the first few days of germination comes from the endosperm by degradation and solubilization of the starch and proteins of this region. In order for this solubilization to occur in the endosperm, the embryo must be present. If the embryo is excised, the amylolytic activity fails to develop in the endosperm, but it can be induced by a diffusate from the embryo or by purified gibberellins (Paleg, 1960). The finding that the diffusate from the embryo is a rich source of gibberellin provides the final evidence that it is a diffusion of gibberellin from the embryo to the endosperm which initiates the solubilization processes (Yomo and Jinoma, 1966).

In conclusion, it is now apparent that gibberellin influences the growth and development of a plant through exerting its effects on a variety of different processes. Certainly, the most dramatic of the known roles of gibberellin are its regulation of the activity of the subapical meristem and the induction of flowering in certain plants. It must be remembered, though, that we are not yet in a position to assess the role in the plant of any gibberellin which is relatively inactive in the promotion of stem elongation since the growth-promoting gibberellins are the only ones which are now available and for which adequate methods for study are available.

The mode of action of gibberellin

One cannot claim to understand the role of gibberellin in the plant unless one knows the mechanisms by which gibberellin exerts its effects. Consequently, one of the goals of plant physiologists has been to determine the mode or modes of action of the gibberellins.

It has already been shown that gibberellin influences processes such as stem growth and fruit growth which are also affected by auxin. At first this led to the suggestion that gibberellins might simply be a form of auxin, but this possibility had to be abandoned when it became clear that the biological activities of the two are, in fact, quite different. A comparison of the developmental effects of auxin and gibberellin is given in Table 2.4. The inability of gibberellin to influence root initiation or cause *Avena* curvature, and the inability of auxin to induce α-amylase in cereal endosperm or cause dwarf stems to grow, is a good indication of the fact that these two groups of regulators are not interchangeable. Likewise, in certain processes such as the determination of sexuality in flowers, the two hormones have antagonistic effects. Clearly, the gibberellins are a distinct group of growth regulators.

Table 2.4   Comparison of some of the biological effects of gibberellins and auxins.

Response	Effect of:	
	Gibberellin	Auxin
Dwarf pea stem growth, sections	n.e.	promotes
Dwarf pea stem growth, intact	promotes	n.e.
Cucumber hypocotyl growth, intact	promotes	promotes
Root growth	n.e.	inhibits
Parthenocarpic fruit growth, tomato	promotes	promotes
Cell division, tobacco pith	n.e.	promotes
Root initiation	n.e.	promotes
Flower initiation, long-day plants	promotes	n.e.
Seed germination, Grand Rapids lettuce	promotes	n.e.
Sex expression, cucumbers	promotes maleness	promotes femaleness

n.e.; no effect or only weak response.

Nevertheless, the number of situations in which auxins and gibberellins produce the same types of response (e.g., fruit set) or are both required for the same process (e.g., stem growth) suggests that the actions of the two substances may be connected. The question, then, is the nature of this connection.

One possibility that has received wide attention is that gibberellin exerts its physiological effects by altering the auxin status of the tissue. If this alteration results in an increase in the endogenous auxin level, then any process which is sensitive to the auxin concentration will, in turn, be affected. In other words, the actual morphogenesis-regulating substance will be auxin; the effect of gibberellin on the process will be indirect.

For processes in which gibberellin acts in this manner, it should be possible to show that the auxin level in the affected tissue increases upon treatment with gibberellin. This has now been demonstrated for a variety of plants and for several different gibberellin responses. For example, Kuraishi and Muir (1963) found that treatment of rosette *Hyoscyamus* plants caused a forty-fold increase in auxin level. Furthermore, this increase in auxin began just prior to the commencement of stem growth. In bean shoots an increase in auxin can be detected within four hours after addition of gibberellin (Nitsch and Nitsch, 1959). Similarly, increases in diffusible or extractable auxin have been shown to be induced by gibberellin in the shoots of sunflower, *Centaurea*, winter wheat, corn, cabbage, and sugar beets, in *Parthenocissus* tissue culture, and in parthenocarpic rose and tomato fruits. In some cases it appears that it is the auxin which is already present which increases in amount, but in other cases, such as *Centaurea* stems and rose fruits, the increase is reported to be due to the appearance of new, non-IAA auxins (Kuraishi and Muir, 1964a).

For processes in which gibberellin acts through an effect on auxin metabolism, it should also be possible to show that the effectiveness of gibberellin is blocked whenever auxin-action is inhibited by means of an antiauxin. To date, only a few of the gibberellin-mediated processes have been subjected to such a test. The necessity for active auxin in the gibberellin-induced elongation of *Avena* leaf sections was shown in this manner (Cleland, 1964). Similarly, the presence of antiauxins inhibits the gibberellin-induced elongation in *Ipomea* petioles and intact rice coleoptiles (Kefford, 1962).

Finally, if gibberellin acts through auxin it should be possible to obtain the same response by treatment with either auxin *or* gibberellin. For most gibberellin-sensitive processes it has not been possible to replace the gibberellin with added auxin. Possible reasons for these failures will be discussed later. In two cases, however, such a replacement has been possible. Unfertilized flowers will not normally set fruit, but addition of gibberellin will allow normal-sized fruits to develop in tomato, cucumber, and apple plants (Crane, 1964). This same response can be obtained by application of an auxin instead of the gibberellin. Similarly, the elongation of cucumber hypocotyls is stimulated by the addition of either gibberellin or auxin (Katsumi *et al.*, 1965), although there is some doubt as to whether the same type of growth is induced by both hormones. However, in these two cases it is clear that gibberellin is not needed as long as sufficient auxin is present.

It is indeed likely that at least some of the morphogenetic effects of gibberellin are due to an increased endogenous auxin level. Since it is now clear that the auxin level does increase in many tissues in the presence of gibberellin, one must expect that any process which is sensitive to auxin will show at least some response to an increase in gibberellin. One such process may be cell elongation in stems. This type of gibberellin action would make it easy to understand why gibberellin has less effect on excised stem segments than it does on the intact plant (Brian and Hemming, 1958). In intact plants gibberellin can stimulate auxin production in the apex while in sections, where little or no auxin synthesis occurs, gibberellin will be without effect. A second process in which such a mode of gibberellin action seems likely is the growth of fruits such as the tomato in which added auxin and gibberellin produce identical effects.

The manner in which gibberellin brings about the increase in auxin level has been a matter of some controversy. It could result from a decrease in the rate of auxin destruction or an increase in the rate of auxin production. Evidence for an effect of gibberellin on both processes has now been obtained. For instance, treatment of dwarf peas and dwarf corn with gibberellin results in a decrease in the levels of extractable IAA-oxidase and peroxidase and a change in the pattern of the peroxidases (Galston and McCune, 1961). As a result, both the types and levels of these enzymes more closely resemble those in the tall varieties. The manner in which gibberellin exerts these effects is still uncertain. The isolated IAA-oxidase is insensitive to gibberellin. The amount of the enzyme may be changed in response to gibberellin, but it seems more likely that the control is exerted at the level of an enzyme inhibitor. IAA-oxidase is particularly sensitive to inhibition by a variety of naturally occurring diphenols. Kögl and Elema (1960) have shown that treatment of peas with gibberellin results in the increase in these inhibitors with the result that the extracted oxidase has a reduced activity. It should be noted, however, that it is still uncertain whether or not IAA-oxidase is a particularly effective enzyme in intact cells; the changes in IAA-oxidase level are of little importance unless it can be conclusively shown that the level of this enzyme does regulate the endogenous auxin level.

A second possibility is that gibberellin stimulates the production of auxin. Evidence to support this idea, which was originally proposed by Brian and Hemming in 1958, was obtained by Sastry and Muir (1963) when they showed that the level of diffusible auxin that could be obtained from unfertilized tomato ovaries and from apices of dwarf pea stems was markedly increased by gibberellin. Subsequently, they demonstrated (1965) that *Avena* coleoptile tips would convert tryptophan to IAA more rapidly if the tips were treated with gibberellin. As yet we do not know what step in the pathway is enhanced by gibberellin; this must wait until we have a clearer idea of just how this conversion takes place.

It would appear that an increase in endogenous auxin levels follows gibberellin treatment because of both a decreased rate of destruction and an increased rate of synthesis. The increased level of auxin, in turn, produces a series of morphogenetic changes in the plant.

Not all gibberellin responses can be explained in this manner, however. It is increasingly clear that at least certain responses are independent of the presence or absence of auxin. This is most clearly shown in the stimulation of α-amylase production in barley endosperm. Here auxin clearly will not replace gibberellin in inducing the response (Paleg, 1960). Furthermore, the gibberellin response occurs even when the action of the endogenous auxin is prevented by an anti-auxin (Cleland and McCombs, 1965). There is no evidence for any effect of auxin on this process.

For many other processes the evidence is less clearcut but still points to a similar conclusion; gibberellin acts independently of auxin. An example of this is the elongation of oat leaf section, which is stimulated by gibberellin. This increased elongation requires the presence of endogenous auxin, as indicated by the fact that an antiauxin will prevent this response and that added auxin can then restore the sensitivity to gibberellin (Cleland, 1964). But the fact that added auxin alone cannot replace gibberellin, even though it can replace the endogenous auxin, indicates that the gibberellin is not acting through an increase in the endogenous auxin level. In this case, gibberellin must act *with* auxin rather than *through* auxin to exert its response.

The inability of auxin to replace gibberellin has been noted in a variety of gibberellin responses such as the activation of the subapical meristem or the breaking of dormancy. This lack of sensitivity to added auxin is usually explained as being due to an inability of applied auxin to reach the active sites in the plant. As evidence for this, Kuraishi and Muir (1964b) have cited the fact that gibberellin but not auxin promotes elongation in intact dwarf pea stems while auxin but not gibberellin causes some elongation when applied to decapitated stems. They interpret this as indicating that gibberellin causes an increased auxin production by the apex which, in turn, causes the elongation. They have ignored, however, the fact that the site of the gibberellin-induced elongation occurs in the region *above* the point of decapitation, while the auxin-effect is entirely below this point. In general, there is a lack of any direct evidence to support the idea that added auxin is inactive because of failure to reach the active sites. It is more probable that in processes such as the activation of the subapical meristem, flowering in long-day plants and sex expression that gibberellin acts in some as yet unknown manner rather than through any effect on the auxin status of the plant.

A possible mode of action which has been receiving increasing attention is that gibberellin acts at the gene level to cause de-repression of specific genes. The activated genes would, in turn, through production of new enzymes bring about the observed morphogenetic changes. The most extensive evidence to support such a mode of action comes from the studies of Varner et al. (1965) on the induction of α-amylase in barley endosperm. In the absence of gibberellin, the aleurone cells of the barley endosperm contain only traces of α-amylase. Treatment of the aleurone cells with gibberellin or with a natural factor from the embryo, which is believed to be a gibberellin, results in a spectacular increase in the level of α-amylase after a 7–8 hour lag (Fig. 2.10). Evidence that the enzyme has been synthesized *de novo* rather than just released from an inactive form has come in several ways. Not only is the appearance blocked by inhibitors of protein synthesis, but the enzyme is labelled throughout when the incubation is carried out in the presence of labelled amino acids (Varner and Ram Chandra, 1964). Furthermore, the appearance of α-amylase requires the synthesis of new RNA, presumably the messenger-RNA which codes for the α-amylase protein. The timing of events leading to α-amylase production has been elucidated by Varner et al. (1965) by use of actinomycin D, an inhibitor of RNA synthesis. The period when the synthesis of α-amylase can be blocked by actinomycin D is the first 7–8 hours after treatment with gibberellin; before any increase in α-amylase can be detected. Thereafter the process is insensitive to actinomycin D. Apparently all of the necessary m-RNA has been formed in this initial period and the synthesis of α-amylase can then proceed, using the m-RNA as the template for the protein synthesis.

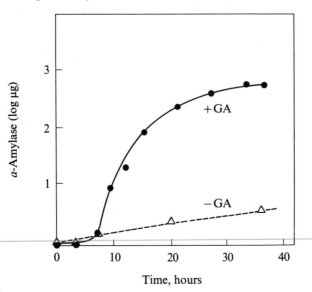

Figure 2.10   Effect of gibberellin on α-amylase production by barley endosperm. Ten endosperm halves were incubated in buffer with 10^{-6} M GA_3 where indicated. (After Varner *et al.*, 1965.)

Clearly, gibberellin has caused the de-repression of the gene for α-amylase in this system, but it is premature to conclude that it has done so by acting directly at the gene level. For one thing, α-amylase is not the only enzyme whose level is affected by gibberellin. A number of other enzymes, such as proteinases and β-amylase, also increased markedly following addition of gibberellin, but since in these cases sizable amounts of enzyme were present prior to the gibberellin treatment it is less likely that gibberellin acted as a de-repressor of these genes. Gibberellin may have affected these enzymes in an indirect manner, however, by de-repressing some as yet unidentified gene whose product, in turn, stimulates the production of these other enzymes. Furthermore, gibberellin is required for continued α-amylase production even after m-RNA synthesis has been completed. Further studies on the aleurone system, especially with cell-free systems, may provide us with the answers as to ability of gibberellin to act at the gene level.

While the work with the barley endosperm system was in progress, a different sort of system was providing additional evidence for a role of gibberellin at the gene level. The hormone ecdysone, which promotes the moulting of insects, is believed to act as a specific gene de-repressor. This is based on the fact that cytological observation of the insect chromosomes after ecdysone treatment showed that certain specific regions of its chromosome would form puffs (Clever and Karlson, 1960). Puffing of a gene has been shown to be associated with the initiation of RNA synthesis at that site, and it has now been shown that ecdysone causes the formation of specific messenger-RNAs which lead to formation of proteins needed in moulting (Karlson, 1965). The fact that other steroid hormones do not cause moulting indicates that de-repression of the moulting genes requires specific activators.

In 1963, Carlisle *et al.* made the surprising discovery that gibberellin possessed low but definite activity in a locust moulting assay. Furthermore, ecdysone was shown to cause some growth in the dwarf pea assay. The simplest explanation of these results is that gibberellin and ecdysone are similar enough in their modes of action to be able to substitute for each other to some extent. Then, if ecdysone is acting at the gene level, gibberellin must also be acting in such a manner. Many additional details are, however, needed in order to be certain of this. For instance, it must be shown that gibberellin, itself, is active in the moulting assay and is not just activating ecdysone or being converted into it.

The idea that gibberellins act at the gene level is especially attractive because of the diversity of gibberellins in plants. If a substance is to bring about a morphogenetic change by gene activation, it must be fairly specific for certain genes; it cannot cause change if it causes a general activation of all genes. Such specificity means that a variety of distinct substances must be present, each of which is maximally effective on certain genes. Of the known growth hormones, only the gibberellins possess the necessary diversity. In addition, their similarity to the steroids in structure and biosynthetic pathway suggests that the gibberellins might have similar functions in cells. Although the functions of the steroids is far from settled, there is an increasing amount of evidence to indicate that many of the steroids do act as specific gene de-repressors.

Bibliography

Selected further reading

Brian, P. W., 1966. The gibberellins as hormones. *Int. Rev. Cytol.* **19**, 229–266. (The latest review on gibberellins.)

Paleg, L. G., 1965. The physiological effects of gibberellins. *Ann. Rev. Plant Physiol.* **16**, 291–322.

Leopold, A. C., 1964. Gibberellins, chapter 7, in *Plant Growth and Development.* McGraw-Hill Co., New York. (A short, concise treatment of the gibberellins.)

Phinney, B. O. and C. A. West, 1961. Gibberellins and plant growth. *Encyl. Plant Physiol.* **14**, 1085–1227. (A comprehensive treatment of gibberellins as it was known up to 1960.)

Stowe, B. B. and T. Yamaki, 1959. Gibberellin: stimulants of plant growth. *Science*, **129**, 807–816. (A readable account of the early history of gibberellin research.)

Stodola, F. H., 1957. *Source Book on Gibberellin* (1827–1957). Agr. Res. Serv., USDA, Peoria, Ill, 560 pp. (Abstracts of all of the early gibberellin papers. Includes an English translation of Kurosawa's original paper.)

Papers cited in text

Arney, S. E. and P. Mancinelli, 1966. The basic action of gibberellic acid in elongation of 'Meteor' pea stems. *New Phytol.* **65**, 161–175.

Baldev, B., A. Lang, and A. O. Agatep, 1965. Gibberellin production in pea seeds developing in excised pods: effect of growth retardant Amo-1618. *Science*, **147**, 155–157.

Bentley-Mowat, J. A., 1966. Activity of gibberellins A_1 to A_9 in the *Avena* first-leaf bio-assay, and location after chromatography. *Ann. Bot.* **30**, 165–171.

Birch, A. J., R. W. Rickards, and H. Smith, 1958. Biosynthesis of gibberellic acid. *Proc. Chem. Soc.* (London), pp. 192–193.

Bradley, M. V. and J. C. Crane, 1962. Cell division and enlargement in mesocarp parenchyma of gibberellin-induced parthenocarpic peaches. *Bot. Gaz.* **123**, 243–246.

Brian, P. W. and H. G. Hemming, 1958. Complementary action of gibberellic acid and auxins in pea internode extension. *Ann. Bot.* N.S. **22**, 1–17.

Brian, P. W., H. G. Hemming, and D. Lowe, 1964. Comparative potency of nine gibberellins. *Ann. Bot.* N.S. **28**, 369–389.

Brian, P. W., G. W. Elson, H. G. Hemming, and M. Radley, 1955. The growth-promoting properties of gibberellic acid, a metabolic product of the fungus *Gibberella fujikuroi*. *J. Sci. Food Agr.* **5**, 602–612.

Carlisle, D. B., D. J. Osborne, P. E. Ellis, and J. E. Moorhouse, 1963. Reciprocal effects of insect and plant-growth substances. *Nature, Lond.* **200**, 1230.

Cathey, H. G., 1964. Physiology of growth retarding chemicals. *Ann. Rev. Plant Physiol.* **15**, 271–302.

Cleland, R., 1964. The role of endogenous auxin in the elongation of *Avena* leaf sections. *Physiol. Plantarum*, **17**, 126–135.

Cleland, R. and N. McCombs, 1965. Gibberellic acid: action in barley endosperm does not require endogenous auxin. *Science*, **150**, 497–498.

Clever, U. and P. Karlson, 1960. Induktion von Puff-Veränderungen in den Speicheldrüsen-chromosomen von *Chironomus tentans* durch Ecdyson. *Exp. Cell Res.* **20**, 623–626.

Corcoran, M. R. and B. O. Phinney, 1962. Changes in amounts of gibberellin-like substances in developing seed of *Echinocystis, Lupinus* and *Phaseolus. Physiol. Plantarum*, 252–262.

Crane, J. C., 1964. Growth substances in fruit setting and development. *Ann. Rev. Plant Physiol.* **15**, 303–326.

Cross, B. E., J. F. Grove, P. McClosky, J. MacMillan, J. S. Moffatt, and T. P. C. Mulholland, 1961. The structures of fungal gibberellins. *Adv. Chem. Ser.* **28**, 3–17.

Dennis, D. T., C. D. Upper, and C. A. West, 1965. An enzymatic site of inhibition of gibberellin biosynthesis by Amo-1618 and other plant growth retardants. *Plant Physiol.* **40**, 948–952.

Donoho, C. W. and D. W. Walker, 1957. Effect of gibberellic acid on breaking of the rest period in Elberta peach. *Science*, **126**, 1178–1179.

Eagles, C. F. and P. F. Wareing, 1964. The role of growth substances in the regulation of bud dormancy. *Physiol. Plantarum*, **17**, 697–709.

Edelman, J. and M. A. Hall, 1964. Effect of growth hormone on the development of invertase associated with cell walls. *Nature, Lond.* **201**, 296–297.

El-Antably, H. M. N., P. F. Wareing, and J. Hillman. 1967. Some physiological responses to *d,l*, Abscisin (Dormin). *Planta*, **73**, 74–90.

Elson, G. W., D. F. Jones, J. MacMillan, and P. J. Suter, 1964. Plant hormones. IV. Identification of the gibberellins of *Echinocystis macrocarpa* Greene by thin-layer chromatography. *Phytochem.* **3**, 93–101.

Esau, K., 1953. *Plant Anatomy*. John Wiley & Sons, Inc., New York, 735 pp.

Fletcher, R. A. and D. J. Osborne, 1966. Gibberellin, as a regulator of protein and ribonucleic acid synthesis during senescence in leaf cells of *Taraxacum officinale*. *Can. J. Bot.* **44**, 739–746.

Frankland, B., and P. F. Wareing, 1960. Effect of gibberellic acid on hypocotyl growth of lettuce seedlings. *Nature, Lond.* **185**, 255–256.

Galun, E., 1959. Effects of gibberellic acid and naphthaleneacetic acid on sex expression and some morphological characters in the cucumber plant. *Phyton* (Buenos Aires) **13**, 1–8.

Galston, A. W. and D. C. McCune, 1961. An analysis of gibberellin-auxin interaction and its possible metabolic basis. In *Plant Growth Regulation*, ed. R. M. Klein. Iowa State Press, Ames, pp. 611–624.

Geismann, T. A., A. J. Verbiscar, B. O. Phinney, and G. Cragg, 1966. Studies on the biosynthesis of gibberellins from (−)-kaurenoic acid cultures of *Gibberella fujikuroi*. *Phytochem.* **5**, 933–947.

Graebe, J., D. T. Dennis, C. D. Upper, and C. A. West, 1965. Biosynthesis of gibberellin. *J. Biol. Chem.* **240**, 1847–1854.

Haber, A., 1964. Interpretations concerning cell division and growth. In *Regulateurs Naturels de la Croissance Végétale*, ed. J. P. Nitsch. Ed. Cent. Nat. Recherche Scient., Paris, pp. 493–503.

Hashimoto, T. and L. Rappaport, 1966. Variations in endogenous gibberellins in developing bean seeds. II. Changes induced in acidic and neutral fractions by GA_1. *Plant Physiol.* **41**, 629–632.

Hayashi, F. and L. Rappaport, 1966. Isolation, crystallization and partial identification of potato factor II from potato tubers. *Plant Physiol.* **41**, 53–58.

Humphries, E. C. and A. W. Wheeler, 1963. The physiology of leaf growth. *Ann. Rev. Plant Physiol.* **14**, 385–410.

Jackson, D. I. and B. G. Coombe, 1966. Gibberellin-like substances in the developing apricot fruit. *Science*, **154**, 277–278.

Jones, R. L. and I. D. J. Phillips, 1964. Agar-diffusion technique for estimating gibberellin production by plant organs. *Nature, Lond.* **204**, 497–499.

Jones, R. L. and I. D. J. Phillips, 1966. Organs of gibberellin synthesis in light-grown sunflower plants. *Plant Physiol.* **41**, 1381–1386.

Jones, R. L. and J. E. Varner, 1967. The bio-assay of gibberellin. *Planta*, **72**, 155–161.

Karlson, P., 1965. Biochemical studies of Ecdysone control of chromosomal activity. *J. Cell. Comp. Physiol.* **66**, Suppl. 69–75.

Katsumi, M., B. O. Phinney, and W. K. Purves, 1965. The role of gibberellin and auxin in cucumber hypocotyl growth. *Physiol. Plantarum*, **18**, 462–473.

Kefford, N. P., 1962. Auxin-gibberellin interactions in rice coleoptile elongation. *Plant Physiol.* **37**, 380–386.

Kende, H. and A. Lang, 1964. Gibberellins and light inhibition of stem growth in peas. *Plant Physiol.* **39**, 435–440.

Kluge, M., E. Reinhard, and H. Ziegler, 1964. Gibberellinaktivität von Siebröhrensäften. *Naturwiss.* **51**, 145–146.

Kögl, F. and J. Elema, 1960. Wirkungsbeziehungen zwischen Indol-3-essigsäure und Gibberellinsäure. *Naturwiss.* **47**, 90.

Kuraishi, S. and R. M. Muir, 1963. Diffusible auxin increase in a rosette plant treated with gibberellin. *Naturwiss.* **50**, 337–338.

Kuraishi, S. and R. M. Muir, 1964a. The relationship of gibberellin and auxin in plant growth. *Plant & Cell Physiol.* **5**, 61–69.

Kuraishi, S. and R. M. Muir, 1964b. The mechanism of gibberellin action in the dwarf pea. *Plant & Cell Physiol.* **5**, 259–271.

Kurosawa, E., 1926. Experimental studies on the secretion of *Fusarium heterosporum* on rice plants. *J. Nat. Hist. Soc. Formosa*, **16**, 213–227.

Lang, A., 1956. Induction of flower formation in biennial *Hyoscyamus* by treatment with gibberellin. *Naturwiss.* **12**, 284–285.

Lang, A., 1957. Effect of gibberellin upon flower formation. *Proc. Nat. Acad. Sci.* **43**, 709–717.

Lang, A., 1960. Gibberellin-like substances in photoinduced and vegetative *Hyoscyamus* plants. *Planta*, **54**, 498–504.

Lang, A., 1965. Physiology of flower initiation. *Encyl. Plant Physiol.* **15**(1), 1380–1536.

Lang, A. and E. Reinhard, 1961. Gibberellins and flower formation. *Adv. Chem. Ser.* **28**, 71–79.

Lockhart, J. A., 1956. Reversal of the light inhibition of pea stem growth by gibberellins. *Proc. Nat. Acad. Sci., U.S.* **42**, 841–848.

Lockhart, J. A., 1960. Intracellular mechanism of growth inhibition by radiant energy. *Plant Physiol.* **35**, 129–135.

MacMillan, J., J. C. Seaton, and P. J. Suter, 1961. Isolation and structures of gibberellins from higher plants. *Adv. Chem. Ser.* **28**, 18–25.

McComb, A. J., 1961. Bound gibberellin in mature runner bean seeds. *Nature, Lond.* **192**, 575–576.

Mertz, D., 1966. Hormonal control of root growth. *Plant & Cell Physiol.* **7**, 125–135.

Michniewicz, M. and A. Lang, 1962. Effect of nine different gibberellins on stem elongation and flower formation in cold requiring and photoperiodic plants grown under non-inductive conditions. *Planta*, **58**, 549–563.

Nitsch, J. P. and C. Nitsch, 1959. Modification du metabolisme des auxines par l'acide gibberellique. *Bull. Soc. France Physiol. Veg.* **5**, 20–23.

Paleg, L., 1960. Physiological effects of gibberellic acid. I. On the carbohydrate metabolism and amylase activity of the barley endosperm. *Plant Physiol.* **35**, 293–299.

Paleg, L., 1965. Physiological effects of gibberellins. *Ann. Rev. Plant Physiol.* **16**, 291–322.

Phillips, I. D. J. and R. L. Jones, 1964. Gibberellin-like activity in bleeding-sap of root systems of *Helianthus annuus* detected by a new dwarf pea epicotyl assay and other methods. *Planta*, **63**, 269–278.

Phinney, B. O., 1957. Growth response of single-gene dwarf mutants in maize to gibberellic acid. *Proc. Nat. Acad. Sci.* **42**, 185–189.

Phinney, B. O., C. A. West, M. Ritzel, and P. M. Neely, 1957. Evidence for 'gibberellin-like' substances from flowering plants. *Proc. Nat. Acad. Sci., U.S.* **43**, 398–404.

Radley, M., 1958. The distribution of substances similar to gibberellic acid in higher plants. *Ann. Bot. N.S.* **22**, 297–307.

Sachs, R. M., C. F. Bretz, and A. Lang, 1959. Shoot histogenesis: the early effects of gibberellin upon stem elongation in two rosette plants. *Am. J. Bot.* **46**, 376–384.

Sachs, R. M., A. Lang, C. F. Bretz, and J. Roach, 1960. Shoot histogenesis: subapical meristematic activity in a caulescent plant and the action of gibberellic acid and Amo-1618. *Am. J. Bot.* **47**, 260–266.

Sastry, K. K. S. and R. M. Muir, 1963. Gibberellin: effect on diffusible auxin in fruit development. *Science*, **140**, 494–495.

Sastry, K. K. S. and R. M. Muir, 1965. Effects of gibberellic acid on utilization of auxin precursors by apical segments of the *Avena* coleoptile. *Plant Physiol.* **40**, 294–298.

Stowe, B. B., F. H. Stodola, T. Hayashi, and P. W. Brian, 1961. The early history of gibberellin research. In *Plant Growth Regulation*, ed. R. M. Klein. Iowa State Press, Ames, Iowa, pp. 465–471.

Varner, J. and G. Ram Chandra, 1964. Hormonal control of enzyme synthesis in barley endosperm. *Proc. Nat. Acad. Sci., U.S.* **52**, 100–106.

Varner, J., G. Ram Chamdra, and M. J. Chrispeels, 1965. Gibberellic-acid controlled synthesis of α-amylase in barley endosperm. *J. Cell. Comp. Physiol.* **66**, Suppl. 55–68.

Wheeler, A. W., 1960. Changes in the leaf-growth substances in cotyledons and primary leaves during the growth of dwarf bean seedlings. *J. Exp. Bot.* **11**, 222–229.

Whyte, P., and L. C. Luckwill, 1966. A sensitive bio-assay for gibberellins based on retardation of leaf senescence in *Rumex obtusifolius. Nature, Lond.* **210**, 1360.

Wittwer, S. H. and M. J. Bukovac, 1957. Gibberellins: new chemicals for crop production. *Mich. State Univ. Agr. Expt. Sta. Quart. Bull.* **39**, 469–494.

Yomo, H. and H. Jinoma, 1966. Production of gibberellin-like substances in the embryo of barley during germination. *Planta*, **71**, 113–118.

Zweig, G., S. Yamaguchi, and G. W. Mason, 1961. Translocation of C^{14}-gibberellin in red kidney bean, normal corn and dwarf corn. *Adv. Chem. Ser.* **28**, 112–134.

REFERENCES

3. The cytokinins

J. Eugene Fox

Introduction

Discovery of kinetin

In recent years many of us interested in growth and its regulation have become excited by the implications of a group of plant growth substances known as the cytokinins. Although discovered only in the 1950's, the cytokinins have had a decided impact on plant physiology and biochemistry because of the increasing evidence that they play a key role in the life of higher plants. Studies on the function of these substances within the plant and their interaction with auxins and gibberellins are really just beginning and very likely will form an important part of the efforts of plant scientists in the near future.

The discovery of the cytokinins was a direct outcome of tissue culture studies carried on for a number of years in the laboratory of Professor Folke Skoog at the University of Wisconsin. Skoog, for long a student of plant growth, was particularly interested in the chemical control of differentiation and as a result of this interest began in 1941 to work with a new research tool, plant tissues or organs cultured heterotrophically *in vitro**. Skoog first used the tobacco hybrid tissue isolated by Philip White (1939), but this was lost during the war and in 1947 he initiated the culturing of stem internode segments of *Nicotiana tabacum* cv. Wisconsin No. 38.

Unlike White's hybrid tobacco tissue which could grow on a simple mineral medium containing sucrose and a few vitamins, the stem segments required other components. In particular Skoog and Tsui (1948) reported that small pieces of stem segments showed no measurable growth unless supplied with an auxin such as indole-acetic acid, in which case enlargement and proliferation was noted, especially in the pith. In contrast to this situation, Jablonski and Skoog (1954) demonstrated that if the pith tissue were freed from vascular and cortical elements of the stem, it responded to the same auxin-containing medium by '. . . an enormous cell enlargement entirely unaccompanied by cell division'. Vascular tissue placed in contact with the pith sections would restore cell division. This observation was instrumental in the discovery of the cytokinins since it led to a search for pure substances which could induce cell division in a manner similar to that of the unknown substance in vascular tissue.

The first attempt to isolate active material was made with aqueous extracts from tobacco stems which gave positive but inconsistent results (Jablonski and

* The first successful plant tissue cultures had just been reported independently and nearly simultaneously by Gautheret (1939) Nobecourt (1939) and White (1939).

Skoog, 1954). Because of this poor reproducibility the investigators turned to other sources and found that coconut milk and malt extract (Jablonski and Skoog, 1954), yeast extract, and autoclaved DNA (Miller et al., 1955a) were highly active. From autoclaved herring sperm DNA the first crystals of a cell-division-inducing substance were isolated by Carlos Miller in December of 1954.

The pure product proved to be active at astonishingly great dilutions; it induced cell division in tobacco pith at concentrations as low as 1 part per billion when auxin was also present in the medium (Miller et al., 1956). Because of its specific effect on cytokinesis the newly isolated product was named *kinetin*; later as similar molecules were shown to have much the same biological activity, the generic name *kinin* was proposed. The latter, however, was an unfortunate choice since animal physiologists had previously used this term for a class of polypeptides having specific physiological properties quite unlike those of kinetin (Lewis, 1960). Because of the ensuing confusion a number of other names such as *kinetenoid* (Burström, 1961), *phytokinin* (Dedolph et al., 1963), and phytocytomine (Pilet, 1965), have been used to designate kinetin-like materials; but the term *cytokinin* proposed by Letham (1963) has met with the widest, though by no means universal acceptance.

Kinetin was found to contain only carbon, hydrogen, nitrogen, and oxygen in the ratio $C_{10}H_9N_5O$ and to absorb strongly in the ultraviolet with a single band near 268 mμ (Miller et al., 1955a). These findings together with the fact that it had been isolated from DNA indicated a purine derivative and in fact the empirical formula of kinetin is that of a dehydrated deoxyadenosine. Acid hydrolysis of kinetin yielded adenine (Miller et al., 1955b) which left only a C_5H_5O moiety to be accounted for. It was logical to assume that the remainder was a dehydration product of deoxyribose and since the infrared spectrum of kinetin indicated an ether linkage, Miller and his colleagues (1956) proposed that '. . . the group comprised a furan nucleus plus one extra carbon present either as a side methyl on the furan ring or as a methylene radical between the furan and purine nuclei'. Since a Kuhn–Roth test showed no carbon-methyl groups (Miller et al., 1955b) the latter possibility was deemed the more likely,

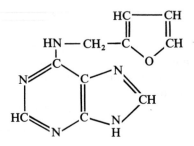

Figure 3.1   Structure of kinetin.

and it was assumed that kinetin was an adenine bearing a furfuryl substituent at the 9 position (Strong, 1956), a perfectly reasonable proposition as purines are combined in this manner with sugars in DNA and many other substances of natural origin. Its reasonableness notwithstanding, the assumption was shown to be false since kinetin was readily precipitated with silver under slightly acid conditions (Miller et al., 1955b), indicative of a free 9 position. Numerous other bits of evidence led to the conclusion that the pentose derivative had migrated to the 6 position of the adenine ring during autoclaving of the DNA and the structure shown in Fig. 3.1 was proposed (Miller et al., 1955b). This detective story was brought to an intellectually satisfying conclusion with the finding that synthetic material of the structure shown in Fig. 3.1 had the same chemical and biological properties as kinetin isolated from DNA.

Analogues of kinetin: the cytokinins

Shortly after the discovery of kinetin, the Wisconsin group synthesized a large number of compounds and tested them for kinetin-like properties. It quickly became apparent that kinetin was not unique but that many other compounds stimulated cell division in plant tissues. In general this work (summarized by Strong, 1958) demonstrated that the side chain of kinetin could be replaced with a wide variety of substituents yielding compounds with activities in the tobacco test equal to or greater than kinetin itself. However alterations in the adenine moiety either greatly reduced activity or inactivated the compound entirely. Later workers using various tests for kinetin-like activity have by and large confirmed these findings with a few exceptions.

Substitutions in the side chain

In Fig. 3.2 are shown some side chain components which confer high cytokinin activity on adenine. Substitution of a benzene nucleus for the furan gives a compound somewhat more effective than kinetin (Miller, 1961). Likewise the thiophene, napthalene (Strong, 1958), thenyl, pyridyl (Kuraishi, 1959), and several other ring systems substituted here result in highly effective compounds. The ring may be attached directly to the amino group as in N^6-phenylamino-purine without significant loss in activity (Strong, 1958). Substituting the ortho position of the benzene ring with methyl, amino, hydroxyl, or chloro groups, e.g., N^6-(2-chlorobenzyl)aminopurine, results in compounds more effective than N^6-benzylaminopurine, while the same substituents at the meta or para positions reduce activity (Kuraishi, 1959). Even straight-chain aliphatic groups confer cytokinin activity on adenine so that compounds such as N^6-n-hexyl-aminopurine (Table 3.1) are quite effective (Strong, 1958). While the n-amyl-, iso-amyl-, and n-hexyl-side chains show excellent stimulatory ability, activity falls off in the *n*-butyl-, *n*-propyl-, *n*-ethyl-, and *n*-heptyl-analogues and it was originally reported that the methyl and *n*-octyl derivatives were completely inactive (Strong, 1958). It was later shown by Kuraishi (1958), however, that both substances had kinetin-like activity, a result confirmed with N^6-methylaminopurine by Miller (1962) and Klämbt, Thies, and Skoog (1966).

R = N^6-substituted adenine

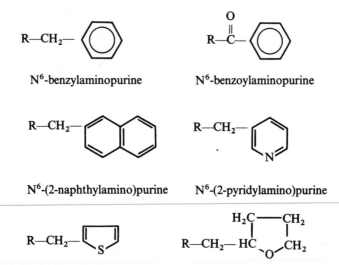

N⁶-benzylaminopurine   N⁶-benzoylaminopurine

N⁶-(2-naphthylamino)purine   N⁶-(2-pyridylamino)purine

N⁶-(2-thenylamino)purine   N⁶-tetrahydrofurfurylaminopurine

Figure 3.2   Some chemical moieties which confer high cytokinin activity on adenine when substituted at the N^6 position.

One of the most active substances yet synthesized is N^6-(γ,γ-dimethylallylamino) purine (Table 3.1) also known as N^6-isopentenyladenine which was first shown to be a cytokinin by Beauchesne and Goutarel (1963) and is reportedly (Rogozinska *et al.*, 1964) about ten times as active as kinetin. That the presence of a double bond in the side chain should boost the activity to this degree is surprising indeed and demonstrates our lack of knowledge concerning the relationship between side chain structure and activity. It is clear, however, that the side chain must be relatively non-polar since hydrazino and succinyl derivatives are without activity (Strong, 1958).

Substitutions in the purine ring

Very few analogues of kinetin which have alterations in the purine moiety show any demonstrable cytokinin activity and none show greater activity than kinetin.

Table 3.1 Alkylated adenines and cytokinin activity: influence of chain length and structure.

Side chain attached at the N^6 position of adenine	Compounds listed in relative order of activity
Structure	Name

1.
$$\text{R—CH}_2\overset{\displaystyle H}{\diagdown}\ \ \overset{\displaystyle CH_2OH}{\diagup}$$
H   CH₂OH
  \   /
   C=C
  /   \
R—CH₂   CH₃
    N^6-(*Trans*-γ-hydroxymethyl-γ-methylallyl) adenine (Zeatin)

2.
H   CH₃
  \   /
   C=C
  /   \
R—CH₂   CH₃
    N^6-(Δ^2-isopentenyl) adenine

3. R—CH₂—CH₂—CH₂—CH₂—CH₃ N^6-Pentyladenine

4. R—CH₂—CH₂—CH₂—CH₃ N^6-Butyladenine

5. R—CH₂—CH=CH₂ N^6-Allyladenine

6. R—CH₂—CH₂—CH₂—CH₂—CH₂—CH₃ N^6-Hexyladenine

7. R—CH₂—CH₂—CH₃ N^6-Propyladenine

8.
   CH₃
  /
R—CH₃ N^6-Dimethyladenine

9. R—CH₃ N^6-Methyladenine

10. R—H   Adenine

Data from Strong, 1958, and Skoog *et al.*, 1967.

Thus furfurylguanine (2-furfurylamino-6-hydroxypurine) is completely inactive as are benzylguanine and benzylcytosine (2-hydroxy-4-benzylaminopyrimidine) (van Eyk and Veldstra 1966). Another interesting series of analogues are the 4-substituted pyrazolo-(3-4-d)-pyrimidines which are completely inactive but differ from highly active forms only in the interchange of the carbon and nitrogen atoms at positions 7 and 8 in the purine ring (Strong, 1958).

Even minor substitutions on the adenine nucleus of cytokinins usually reduce their effectiveness. There is evidence to suggest that the 1 position must be free for cytokinin activity since a methyl group at the 1 position of N^6-benzylamino-purine removes practically all of the activity (Leonard *et al.*, 1966). The 2 position is somewhat less sensitive as both 2-methyl- and 2-amino-N^6-furfuryl-aminopurine have kinetin-like activity although less than the unsubstituted N^6-furfurylaminopurine (Strong, 1958). Similar activity is noted with the 2-fluoro analogue (Leonard *et al.*, 1961) while surprisingly a large number of

2-hydroxy analogues of active cytokinins were without effect (Leonard et al., 1962). Isomers of active cytokinins in which the side chain is on the 3 position are themselves inactive (Cave, et al., 1962).

Little information is available concerning substitutions at other loci on the purine ring except for the 9 position. Here it would appear that certain alterations can be made without greatly reducing activity. Thus both the riboside (Strong, 1958) and the ribotide (Von Saltza, 1958) of kinetin are approximately equal to the free base in activity. Similarly the ribosides of the methyl, dimethyl, and benzyl analogues are at least as effective as their respective free bases.* On the other hand, the replacement of the hydrogen at the 9 position of N^6-benzylaminopurine with a 3-hydroxypropyl group reduces activity about tenfold (Fox and Chen, 1968) although a similar substitution with a tetrahydropyran ring is said to yield a good 'kinin' (Weaver, van Overbeek, and Pool, 1965). Replacement of the amino group at the 6 position with a sulphur atom results in a compound, 6-benzylmercaptopurine which has very low but definite activity in the tobacco pith test (Strong, 1958).

Miscellaneous, non-purine compounds with cytokinin activity

In Fig. 3.3 are shown a number of substances reported to have kinetin-like activity in one test system or another. Among these are urea derivatives such as N,N'-diphenylurea which has weak and delayed kinetin-like effects in the tobacco pith system (Strong, 1958) and N-3-chlorophenyl-N'-phenylurea said to be—along with several other related compounds—even more effective (Bruce, Zwar, and Kefford, 1965). Similar in structure to the purines and clearly active in inducing cell divisions in plant tissues is 8-azakinetin (Miller, 1961). Similarly benzimidazole duplicates certain effects of kinetin (Wang et al., 1961). Perhaps the substance whose kinetin-like activity is most surprising is 2-benzthiazolyloxyacetic acid which Steward and Shantz (1956) claim to be nearly as effective as kinetin.

What is a cytokinin?

With such a wide variety of active substances, we are justified in asking at this point if there are any rules which relate structure to activity. To answer that question we must first have a definition of kinetin-like activity. The only definition yet offered is that of Skoog, Strong, and Miller (1965), who state that cytokinins are chemicals which, regardless of their other activities, promote cytokinesis in cells of various plant origins. In their opinion, 'At present rigid proof of cytokinin activity is limited entirely to 6-substituted purine derivatives.' In the absence of detailed knowledge regarding the metabolism and mode of action of cytokinins such a statement is probably premature.

* Fox, unpublished data.

Nevertheless, it was pointed out some time ago by Miller (1961) that at least one non-purine with cytokinin activity, N,N'-diphenylurea might well be used by the cell as a precursor of a kinetin-type compound. If for example the urea portion of the molecule were to supply the carbon for position 6 of a purine, then a molecule with a phenylamino group already attached, an effective cytokinin, would result. It is not difficult to imagine similar possibilities for nearly every other compound shown in Fig. 3.3 although there is absolutely no experimental evidence on these points. Consequently with regard to a rule relating

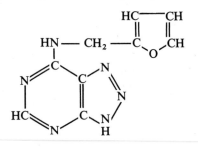

8-azakinetin

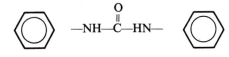

N,N'-diphenylurea

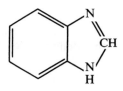

Benzimidazole

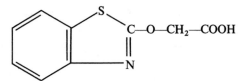

2-benzthiazolyloxyacetic acid

Figure 3.3   Miscellaneous non-purine compounds reported to have cytokinin activity.

structure to activity we can probably do little better than to amend slightly that of Strong (1958) to read: the structure needed for cell division stimulating activity is that of adenine bearing on the amino group in the 6 position, a side chain of a non-polar character. It may be instructive to put it in another way similar to that of Kuraishi (1959): Cytokinins are substances composed of one hydrophilic group of high specificity (adenine) and one lipophilic group without specificity.

Naturally occurring cytokinins

It was pointed out not long after their discovery that kinetin and its analogues are either synthetic or artifacts of DNA degradation and it has been held in some quarters that the cytokinins have, therefore, no natural function in plant growth. In fact as recently as 1966 at least two authors felt there was no proof for the natural occurrence of substances with kinetin-like activity but that '. . . it is still undecided whether we really do have physiologically functioning compounds . . . or are merely concerned with an interesting chapter on phytopharmacology' (van Eyk and Veldstra, 1966).

In such statements, however, a long line of evidence dating from the early part of this century is ignored, for the idea that cell division in plants might be controlled by specific substances is not a new one but goes back at least to Wiesner (1892). The first experimental evidence for such substances was described by Haberlandt (1913) who found that phloem diffusates could induce cell division in potato parenchyma and later (1921) that cell division induced by wounding could be prevented by rinsing of the wounded surface and restored by application of crushed tissue to the wound surface in several plant tissues. Haberlandt clearly saw the significance of this work since he used the term 'wundhormone'. Although a substance termed *traumatic acid* was later isolated by Bonner and English (1938) and identified as a fatty acid ($C_{12}H_{20}O_4$), it is inactive in the tests for cytokinins and of doubtful physiological significance. Apparently the bean assay (Wehnelt, 1927) on which its isolation was based is not highly specific; even distilled water gives a positive response in some cases (Strong, 1958).

In recent years a large variety of extracts and juices of plant origin have shown cell division activity in one test or another. Prominent among these is the endosperm of the coconut, *Cocos nucifera* (referred to as coconut milk in its liquid state) which van Overbeek et al. (1941) found to contain factors stimulatory to the growth of embryos in tissue culture and which has subsequently been widely used as a means of initiating growth in many plant tissues *in vitro*. Notable among the many attempts to isolate and identify the active ingredients of coconut endosperm are those of F. C. Steward and his colleagues at Cornell University beginning in the late 'forties (Caplin and Steward, 1948) and continuing up to the present time, of the Wisconsin group (Mauney *et al.*, 1952), of Kuraishi and Okumura (1961), and of Shaw and Srivastava (1964).

Coconut milk and kinetin can be used interchangeably, if an auxin is also supplied in the medium, to support the growth of many plant tissues—e.g., secondary phloem from the root of *Daucus carota* (Halperin and Wetherell, 1964), mesocarp from the fruit of *Prunus Persica* (Sommer *et al.*, 1962)—and there can no longer be any question but that substances with kinetin-like activity are present in coconut endosperm. In fact a number of purine-type substances have been isolated from coconut endosperm, and these had kinetin-like activity in delaying leaf senescence although they were not tested in assays involving cell division; in the opinion of the authors (Shaw and Srivastava, 1964): 'In view of their general resemblance to kinetin and to some of the substances from coconuts first described by Steward's group, it would be surprising if some of the coconut substances . . . were not, in fact, kinins.' This view is, however, at variance with that of Steward who has repeatedly emphasized that 'no single substance unlocks the door of cell division', but that the growth-promoting effect of complex nutrients such as coconut milk is the result of a subtle and involved interplay of many substances. (See for example Steward and Ram, 1961.) There can be no doubt that Steward's data show just such a complicated situation in the carrot root phloem tissue which he uses as an assay for the active ingredients in coconut milk. Obviously this tissue does in fact require a number of substances for *in vitro* growth supplied by coconut endosperm, some of which, such as *myo-inositol*, have been identified (Pollard *et al.*, 1961).

It seems pointless, however, to refer to the growth-promoting ability of myo-inositol in this situation as a 'kinin' activity as has been done in at least one instance (Leopold, 1964). Such thinking leads to the rather frightening chaos in which the vagaries of metabolism of the particular assay tissue determine to which chemical or group of chemicals the investigator will assign the term, *cell division regulator* (or cytokinin if you will). The strain of tobacco pith tissue used in the original isolation of kinetin provides a particularly illustrative example of the semantic difficulties to which one might succumb. The initial isolates required an auxin such as indole-acetic acid and a cytokinin* in addition to a standard medium containing a carbohydrate such as sucrose, minerals, and a few vitamins. In recent years, however, substrains have been derived from the original isolates which have lost their requirements either for the auxin or the cytokinin or both (Fox, 1962, 1963, and unpublished data). What are we now to say about the substrain for example which has an exogenous requirement only for an auxin in addition to the basal medium? Shall we consider auxin as having cell division regulatory activity? If so, then we must face the even stickier problem of deciding what to do about the substrain of tobacco pith which has

---

* Much has been made of the fact that cytokinins are ineffective in the absence of auxins but it may be just as meaningful to say that cytokinins are ineffective in the absence of iron as most assuredly no growth of the tissue takes place if the latter is left out of the medium.

spontaneously gained the ability to grow on a simple basal medium.* No one would deny that the interactions among the various minerals, sucrose, and the vitamins in regulating the growth of the tissue are both complex and delicate. If we follow this chain of thought to its logically absurd conclusion we might wish to designate sucrose as a cytokinin (witness the fact that the addition of a small amount of sucrose to a culture formerly deprived of carbohydrates induces a remarkable amount of cell division). It is difficult to disagree with Bottomly *et al.* (1963) who conclude that rigorous proof that any substance is a specific cell-division inducer must await elucidation of its mechanism of action. Quite clearly, though, the ultimate criterion for a cell-division-regulating substance is whether or not it is utilized physiologically for this purpose in the intact plant, i.e., is the limiting factor in those cells whose machinery for division is genetically shut down. It is not inconceivable that plants may use a variety of methods to control their various growth processes, and that a substance limiting for cell division and thus a regulator in one instance need not be in another. The semantic difficulties involved in this discussion are by themselves sufficient to be depressing.

Zeatin

In addition to coconut milk, cytokinin activity has been detected in tomato juice (Nitsch and Nitsch, 1961), in extracts of the flowers and fruits of several varieties of apple (*Pyrus malus* L.), in the fruits of quince (*Cydonia oblonga* Mill.), Pear (*Pyrus communis* L.), plum (*Prunus cerasifera* Ehrh. cult. Nigra), and tomato (*Lycopersicon esculentum* cult. Grosse Lisse), as well as from cambial tissues in the stems of *Pinys radiata*, *Eucalyptus regnans*, and *Nicotiana tabacum* cv. Wisconsin Havana 38 by Goldacre and Bottomly (1959) and Bottomly *et al.* (1963). Excellent cytokinin activity is likewise displayed in extracts of the immature fruits of *Zea mays*, *Aesculus noerlitzensis*, *Juglans* sp., and *Musa* sp. and the female gametophyte of *Ginkgo biloba* (Steward and Shantz, 1959, Beauchesne, 1961). A natural, highly active, kinetin-like substance is present in the fruitlets, embryo, and endosperm of peach (*Prunus persica*) (Lavee, 1963, Powell and Pratt, 1964). Pea seedlings (Zwar and Skoog, 1963), sunflower root exudate (Kende, 1964), tobacco tumour tissue (Braun, 1956), and an autonomous strain of tobacco pith tissue which spontaneously lost its cytokinin requirement (Fox, 1962) are other sources of natural cytokinins among a rapidly increasing list.

The wide array of cytokinin sources is good circumstantial evidence for the natural occurrence of such compounds although in the work described above little more than preliminary evidence is available as to the nature of the active

---

* Such changes in growth requirements are extremely common in plant and animal tissues in culture and have been described by a number of terms, common among which is *habituated* (Gautheret, 1955).

ingredients. In 1961, however, Carlos Miller described a number of the properties of a partially purified substance from immature corn grains with high cytokinin activity and concluded that the factor was a substituted purine. Later evidence (Miller, 1962) indicated that the corn factor was adenine substituted in the 6 amino group with an unsaturated side chain having at least one –OH component.

This work was followed up by D. S. Letham (1963) who obtained the factor in crystalline form, reported further on its properties and named the material *zeatin*. Its structure (Letham *et al.*, 1964) and synthesis (Shaw and Wilson, 1964) showed it to be another analogue of kinetin (Fig. 3.4) and confirmed Miller's view as to the nature of the side chain. Although some doubt has been expressed

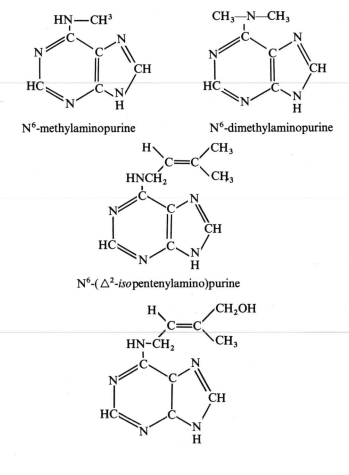

N^6-methylaminopurine N^6-dimethylaminopurine

N^6-(Δ^2-*iso*pentenylamino)purine

Zeatin

Figure 3.4   Some naturally occurring cytokinins.

as to the natural occurrence of zeatin (Kefford, 1963), convincing evidence is available that it is not an artifact of extraction but occurs naturally not only as the free base but also as its nucleoside and nucleotide (Miller, 1965, Letham, 1966) and as an integral part of corn RNA (Hall *et al.*, 1966). The partial characterization of factors chromatographically and spectroscopically identical with zeatin from plum fruitlets (Letham, 1964) and immature sunflower fruits (Miller and Witham, 1964) make it likely that this cytokinin is widely distributed in the plant kingdom.

Other naturally occurring cytokinins

Although clearly an important plant cytokinin, zeatin is not the first natural product shown to have cytokinin activity. This distinction should go to N^6-methylaminopurine, a compound widespread in the RNA's of plants (Dunn, Smith, and Spahr, 1960) first shown to have cell-division stimulating activity by Miller (1962) although its kinetin-like effect in a radish leaf disc test dependent on cell enlargement had been previously noted (Kuraishi, 1959). Even though described by some workers as practically inactive in the tobacco test (Rogozinska *et al.*, 1965) and therefore discounted as a cytokinin having real physiological significance, N^6-methylaminopurine can be shown in fact to have a pronounced kinetin-like effect on Wisconsin No. 38 tobacco pith tissue.*

It should be noted that the relationship of the activity of an exogenously supplied substance to its importance within the cell may be very deceiving. In addition to questions of solubility and penetration, one must consider the possibility that substances such as 6-methylaminopurine which are structurally very similar to important, naturally occurring compounds such as adenine may antagonize some of the functions of the latter when supplied exogenously at relatively high levels. The potentiation of N^6-methylaminopurine's cytokinin effect by adenine supplied at the same time (Miller, 1962) is in fact good evidence for an interaction of this sort. Miller (1962) visualized 'a critical methylation of adenine already properly located within the cell' a remarkably prescient statement in view of the announcement a few weeks later of the discovery of enzymes which perform an *in situ* methylation of bases in polymerized nucleic acids (Fleissner and Borek, 1962) leading to the formation of such substituted adenines as N^6-methylaminopurine and N^6,N^6-dimethylaminopurine. The latter substance incidentally is likewise a cytokinin (Kuraishi, 1959).

The most recent additions to the list of cytokinins which occur naturally is N^6-(γ,γ-dimethylallyl amino) purine, whose activity had been previously demonstrated with synthetic material (Beauchesne and Goutarel, 1963), and which has now been isolated from serine transfer-RNA of yeast (Hall *et al.*, 1966, and Zachau *et al.*, 1966) and the *cis*-isomer of zeatin isolated from sweet corn (Hall *et al.*, 1967). The natural occurrence of the former has subsequently

* Fox, unpublished data.

been confirmed with its isolation from *Corynebacterium fascians* (Helgeson and Leonard, 1966). This cytokinin together with zeatin constitute by far the most active cytokinins yet discovered having effects at concentrations at least an order of magnitude less than any synthetic substance.

Biological activity

Mitosis and cell division

The property most characteristically associated with the cytokinins is their stimulation of cell division in plant tissue cultures. The discovery and isolation of kinetin depended on this property and the bulk of the studies on cytokinins emphasize this effect. It is a very common experience to find that non-meristematic tissues isolated from higher plants require for *in vitro* growth a cytokinin or its naturally occurring equivalent in the form of coconut milk, tomato juice, etc., and often an auxin as well.

There are of course plant tissues in which presumably other, as yet unidentified, substances are limiting so that these cannot be grown on defined media. In like manner there are numerous examples of tissues which grow perfectly well on a simple medium in the absence of cytokinins, auxins, or both. This behaviour is especially characteristic of tissues from meristematic regions and of tumour tissues, particularly those resulting from the activities of the bacterium *Agrobacterium tumefacians*. In one of the most thoroughly studied of such systems, *Vinca rosea* L., Braun and Wood (1962) have demonstrated that the conversion of a normal *Vinca* cell which has an absolute exogenous cytokinin requirement to a tumour cell involves a specific activation of the cytokinin synthesizing system in the plant cell through some activity of the bacterium. A natural conclusion of this study is that the cytokinin synthesizing system is operative in meristematic cells but is turned off through gene repression during the course of differentiation of this tissue into a non-dividing one such as pith. In a detailed study of mitosis and cell division in cultured tobacco pith tissue Das *et al.* (1956) demonstrated that with IAA by itself some mitoses occurred in the tissue resulting primarily in binucleate cells since only a few of the mitoses were followed by cell division; kinetin in combination with IAA, however, induced many mitoses virtually all of which were followed by cytokinesis. In the same vein, Guttman (1956) presented data which were interpreted to mean that kinetin acted sometime during the interphase of mitosis in root cells of onion, *Allium cepa*, to trigger the subsequent prophase.

A somewhat similar result was obtained by Torrey (1961) with intact roots of the garden pea, *Pisum sativum* cv. Alaska. Here cytokinins brought into division mature cells in the root cortex which normally may undergo chromosome doubling but rarely mitosis or cell division. Haber and Luippold (1960) observed a specific initiation of mitosis in the radicle of ungerminated seeds of lettuce,

Lactuca sativa L. cv. New York and considered kinetin as '. . . a true cell division factor in nongerminated lettuce seed'. Cytokinins are likewise reported to increase the rates of cell division in several micro-organisms including a protozoan *Paramecium caudatum* (Guttman and Back, 1960), a phytoflagellate *Polytoma uvella* (Moewus, 1959) and bacteria such as *Escherichia coli* (Kennell, 1959) and a number of others (Maruzzella and Garner, 1963). Of particular interest is the observation (Quinn *et al.*, 1963) that kinetin can replace the requirement for yeast extract in the culture of certain strains of *Clostridium thermocellum*. In addition there is some evidence that kinetin can induce a high rate of multiplication in the cells of higher animals as well (Ogawa *et al.*, 1957).

Certain of the other biological effects displayed by the cytokinins likewise have cell division as an underlying motif, and cytokinins are widely referred to as regulators of cell division. Nevertheless until it is demonstrated that the intact plant makes use of naturally occurring cytokinins to regulate cell division in the normal course of events, such an assignation really cannot be made. We do well to keep in mind that cell division may be only one, admittedly dramatic response to a more fundamental activity of the cytokinins.

Cell enlargement

Not long after the first observations of kinetin-stimulated cell division, three groups of workers reported nearly simultaneously that cell enlargement in discs of etiolated leaves was markedly increased in the presence of kinetin and certain of its analogues active in cell division (Miller, 1956, Kuraishi and Okumura, 1956; Scott and Liverman, 1956, Table 3.2). This effect could not be duplicated

Table 3.2   Effect of kinetin on expansion of etiolated bean discs during a 48-hour growth period.

Kinetin concentration M	Increase in diameter mm
0	$1 \cdot 05 \pm 0 \cdot 04$*
5×10^{-6}	$1 \cdot 73 \pm 0 \cdot 06$
5×10^{-5}	$2 \cdot 48 \pm 0 \cdot 03$
5×10^{-4}	$2 \cdot 96 \pm 0 \cdot 07$

* Standard error. Ten discs per treatment initially 5·5 mm in diameter from leaves of Burpee Dwarf Stringless Greenpod beans grown 7 days in the dark. Data from Miller, 1956.

with auxin and was observed even in light-grown, fully mature leaves which had ceased expanding (Kuraishi and Okumura, 1956). Since this work, evidence for a cytokinin role in cell expansion has been increasing. Arora *et al.* (1959) reported that cortical cells of tobacco roots enlarge up to four times their normal size in the presence of kinetin, and Glasziou (1957) demonstrated that under certain conditions low concentrations of kinetin brought about an increase in cell elongation of tobacco pith. Although kinetin inhibits elongation of pea stem sections (Brian and Hemming, 1957) it appears to favour a lateral extension of cells causing an increase in the diameter of the sections (Katsumi, 1962). This observation might be related to that of de Ropp (1956) who found that kinetin treatment of sunflower hypocotyls caused an increase in fresh and dry weight more than double that of the controls although no elongation occurred. The kinetin effect on leaf growth has since been extended to a number of species (Kuraishi, 1959), and although this cytokinin effect is by no means universal, its generality seems clear.

Morphogenesis

One of the truly dramatic responses to cytokinins is the formation of organs which takes place under appropriate conditions in a variety of tissue cultures. The antecedents of this phenomenon were established by Skoog and Tsui (1948) and Miller and Skoog (1953) who noted that in short term cultures of tobacco stem segments the addition of 5 µg l^{-1} of IAA to the culture medium prevented the formation of buds which spontaneously appeared on the control medium. This effect of IAA could be counterbalanced, however, by adenine at 75 mg l^{-1}, i.e., at a ratio of about 15,000 adenine molecules to 1 of IAA. With this background it was not surprising to find that kinetin exhibited a similar interaction with IAA. What has been striking, however, is the delicate and quantitative interaction between the auxins and the cytokinins in the control of morphogenesis.

In the tobacco pith system (Skoog and Miller, 1957) the situation in general is that at certain balanced levels such as 2 mg l^{-1} IAA and 0·02 mg l^{-1} kinetin the pith grows as an amorphous, undifferentiated callus. Increasing the ratio of kinetin with respect to auxin—either by lowering the auxin concentration or raising the kinetin level in the medium—results in the formation of buds which may grow into shoots and ultimately into complete tobacco plants under the right conditions. In the reverse situation, where the kinetin to auxin ratio is lowered, roots appear on the pith. This beautiful example of the exogenous control of differentiation has implications of considerable importance to the field of morphogenesis and its control in plants.

Numerous other examples of cytokinin stimulation of bud formation exist in the literature. Thus Plummer and Leopold (1957) demonstrated a decided increase of budding in leaf cuttings of *Saintpaulia ionantha* after a 24-hour immersion of the petioles in a kinetin solution. Schraudolf and Reinert (1959) observed in discs of *Begonia rex* leaves a dramatic kinetin effect. In controls, roots followed by shoots appeared at the basal end of main veins while in treated material new shoots appeared around the entire circumference of the disc although root growth was suppressed. A positive effect of kinetin on the regeneration of shoots from cultured root segments has been reported in *Isatis tinctoria* (Danckwardt-Lillieström, 1957) and *Convolvulus arvensis* (Torrey, 1958).

Heide (1965) found that certain other cytokinins are many times more effective than kinetin itself at inducing bud formation and further that specific species or cultivars of *Begonia* which normally do not form adventitious buds on leaves or do so only rarely, with cytokinin treatment exhibited profuse bud formation. Stimulation of bud formation has also been observed in non-angiosperms. For example in the moss, *Tortella caespitosa*, cytokinins increase the number of buds from which gametophores develop (Gorton *et al.*, 1957) and in *Pohlia nutans*, kinetin over a wide range of concentrations stimulated budding in protonemata.

That the relationship between the cytokinins and bud induction is not a simple one was brought out by work with leaf squares of *Peperomia sandersii* (Harris and Hart, 1964). Here bud formation was entirely dependent upon the prior appearance of a root and seemingly the only influence of cytokinins was to inhibit bud generation presumably by inhibiting root initiation. In isolated instances, however, exceedingly high levels of kinetin resulted in bud formation in the absence of roots. Although the general bud initiating effect of the cytokinins is well grounded in experimental evidence, many tissue cultures just as with some intact systems are refractory to cytokinin treatment and in fact do not form buds under any conditions. For this reason Miller (1962) has cautioned against regarding the cytokinins as specific bud-formers; obviously other, as yet unrecognized factors can be limiting.

In nearly every case reported, exogenous application of the cytokinins is deleterious to the initiation and elongation of the main axis of roots (see for example Miller *et al.*, 1955; Skinner and Shive, 1955; de Ropp, 1956; Schraudolf and Reinert, 1959; Harris and Hart, 1964). That this should be true is a curious circumstance inasmuch as there is no longer any doubt but that cytokinins are produced by roots; perhaps exogenous applications result in supra-optimal levels. A rather fascinating aspect of the situation is that in certain cases cytokinins stimulate lateral root formation at the same concentrations which inhibit the growth of the main axis (Skinner and Shive, 1955, Fries, 1960). Furthermore, although inhibiting elongation of the primary root, cytokinins strikingly promote an increase in its diameter (Skoog and Miller, 1957) due to their stimulation, along with auxins, of root cambial activity (Loomis and Torrey, 1964).

Miller (1962) pointed out the need for clearly differentiating between root initiation and subsequent elongation since there exists some evidence that cytokinins in conjunction with other substances stimulate the former (Skoog and Miller, 1957). Again an unusually complex situation is suggested by the work of Bachelard and Stowe (1963) who noted that kinetin applied to the base of cuttings of *Acer rubrum* inhibited root formation but sprayed on the leaves had a promoting effect. The data in this study, however, do not rule out the possibility that we are only dealing with a concentration effect, the amount transported from the leaves to the base being the point in question. A striking example of root initiation induced by cytokinin treatment was shown in a study with the water fern *Marsilea drummondii* (Allsop and Szweykowski, 1960). Kinetin at 0.01 mg l^{-1} induced copious root formation from the adaxial surface of leaves from sporelings cultured aseptically in a liquid medium. The result is particularly interesting since similar root formation had not been previously observed on *Marsilea* leaves even with auxin treatment. A clear-cut role for the cytokinins in root growth has not yet emerged.

In addition to root and shoot morphogenesis cytokinins have been implicated in a large variety of systems which involve differentiation in one way or another. Among a rapidly increasing list are reports which indicate a specific role for these growth substances in the maturing of proplastids into plastids (Stetler and Laetsch, 1965), in the differentiation of tracheids through activation of lignin biosynthesis (Bergmann, 1964), in the induction of parthenocarpy in certain fruits (Weaver *et al.*, 1965, Crane, 1965) and in the induction of flowers in a plant which normally requires a cold treatment and long days for this process (Michniewicz and Kamienska, 1964). A measure of our ignorance about the role of cytokinins in plants is that there have been no unifying concepts offered to bring together and explain these and a host of other morphogenetic responses to cytokinin treatment.

Dormancy

Promotion of seed germination is an effect of the cytokinins announced within a few months of their discovery. In lettuce, *Lactuca sativa*, cytokinins can substitute for the red light requirement in breaking seed dormancy (Miller, 1956; Haber and Tolbert, 1957; Evanari *et al.*, 1958), as well as synergize with light in promoting germination (Fig. 3.5). Similar responses to kinetin or its analogues are exhibited by seeds of tobacco (Kuraishi, 1959), white clover, and carpet grass (Skinner *et al.*, 1958), and the upper, dormant seed in the bur of *Xanthium pensylvanicum* (Khan, 1966). Furthermore, kinetin in combination with red light reversed the effect of certain naturally occurring inhibitors of seed germination such as coumarin and xanthatin (Khan and Tolbert, 1965).

This particular activity of the cytokinins is probably only another aspect of their stimulatory effect on cell enlargement, as indicated by the work of Haber and Luippold (1960) on gamma-irradiated lettuce seeds. In this material mitosis was delayed more than cell expansion, and it was possible to show that kinetin-induced germination preceded mitotic activity and was apparently the result of root cell enlargement. The latter point is in dispute, however, as Ikuma and Thimann (1963) believe that the site of cytokinin action on seed germination is in the cotyledon. Cytokinin induced cotyledon expansion results in rupture of the seed coat presumably initiating the train of processes leading to germination. The regulation of seed germination by cytokinins may have important ecological implications. For example the seeds of the parasitic plant *Striga asiatica* (witch-weed) germinate in response to root secretions of the host plant or to kinetin (Worsham *et al.*, 1959). Since root exudates have been shown to have kinetin-like activity in several assay systems (Kende, 1965), a role for cytokinins in seed germination under natural conditions seems likely.

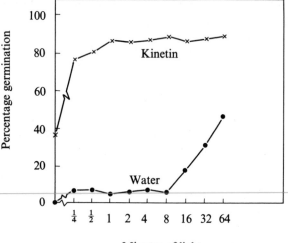

Figure 3.5   Promotion of lettuce seed germination by red light in the presence or absence of kinetin. Light exposures at an intensity of 98 erg cm^{-2} s^{-1} and for the indicated times. 5×10^{-5} M kinetin was used. From Miller, 1958 with the permission of the author and the publisher.

Cytokinins likewise break dormancy in other plant organs including the resting buds of Hydrocharis (Kurz and Kummerow, 1957) and *Vitis vinifera* (Weaver, 1963). The dormancy of *Lemna minor* is similarly interrupted by cytokinin treatment (Deysson, 1959). The physiological basis for this effect of the cytokinins is not known.

Apical dominance

In the plates illustrating Skoog and Miller's now classic paper concerning the hormonal regulation of morphogenesis (1957) one sees that kinetin-induced buds appear not to inhibit one another but to grow in this tissue culture system as though normal apical dominance were inoperative. That cytokinins do in fact counteract the usual dominance of the apical bud was later confirmed in intact plants (Wickson and Thimann, 1958, Sachs and Thimann, 1964). The underlying cause of this effect is not completely clear but is apparently related to the action of cytokinins on vascular tissue differentiation. Sorokin and Thimann (1964) demonstrated that the kinetin-induced release of lateral buds from inhibition by the apical bud coincided with the growth and juncture of xylem elements differentiating basipetally from the bud and acropetally from within the stem internode. This establishment of a vascular connection, normally inhibited by auxin elaborated by the apical bud, allowed further growth and development because of the increased water and solute supply. The interaction between cytokinins and other growth regulators in this system is quite involved (see for example Davies et al., 1965), and not yet well worked out.

An interesting relationship between the cytokinin effect on apical dominance and fasciation of plants caused by the micro-organism *Corynebacterium fascians* was described by Samuels (1961). The symptoms of this disease—loss of apical dominance with the subsequent appearance of a 'witches broom' of growing shoots— could be duplicated entirely by kinetin treatment, suggesting that natural cytokinins were involved in the etiology of the disease. This idea was later confirmed with the extraction from the bacterium of materials which had cytokinin activity in various tests (Klämbt et al., 1966; Thimann and Sachs, 1966), and the identification of active substances as N^6-methylaminopurine and N^6-(γ,γ-dimethylallylamino) purine (Helgeson and Leonard, 1966). A widespread role for naturally occurring cytokinins in host-parasite relationships has been visualized (Thimann and Sachs, 1966).

Delay of senescence: the Richmond–Lang effect

In 1957 a curious and significant activity of the cytokinins in senescence retardation was described by Richmond and Lang. In detached leaves of *Xanthium* kinetin was able to postpone for a number of days the disappearance of chlorophyll and the degradation of proteins which normally accompanies the ageing process of leaves: 'In one experiment the treated leaves were still fully green after a period of 20 days, while the controls were completely yellow and were dying at the tip and margins' (Richmond and Lang, 1957). This phenomenon has stimulated an increasing amount of interest and research in recent years, and of particular importance is that of K. Möthes, his students and colleagues. Möthes has long been interested in ageing processes in leaves, and described in 1928 the role of actively growing centres in the plant in hastening the hydrolysis of components in older leaves. The Richmond–Lang effect suggested that cytokinins may be involved under natural conditions in ageing phenomena, and this idea was given considerable impetus with the finding that kinetin acted as a mobilizing agent, directing the movement of numerous substances to areas of the plant treated with this growth substance (Möthes and Engelbrecht, 1959, Möthes et al., 1959). Especially impressive are radio-autographs of leaves sprayed in one spot with kinetin and supplied with a ^{14}C labelled amino acid either through the petiole or on the surface of the leaf at a site remote from the kinetin-treated area; radioactive material may be seen moving through the main veins and congregating in precisely the area sprayed with kinetin (see for example Möthes, 1960). Cytokinins have the ability to attract in this manner a large number of substances including auxins (Möthes and Engelbrecht, 1961), and to prevent the movement of leaf components out of the treated area (Möthes, 1960).

Much more is involved, however, than a simple mobilizing of the flow of nutrients. Osborne (1962) showed that isolated discs of *Xanthium* leaves were similarly retarded in their senescence by a kinetin treatment even though there was no adjacent untreated area with a reservoir of nutrients on which to draw. Her observation that kinetin treatment maintained (in fact augmented) the ratio of RNA or protein to DNA suggests that a critical action of kinetin in senescence might be the maintenance of the protein synthesizing machinery perhaps by regulating RNA synthesis (Osborne, 1962).

Almost identical results and conclusions had been previously presented by Wollgiehn (1961) and later by Sugiura *et al.* (1962), Shaw *et al.* (1965), Srivastava and Ware (1965), and many others. Support for a role of the cytokinins in RNA metabolism is therefore widespread although other hypotheses to explain cytokinin-induced delay of senescence have been offered. Among alternative possibilities is that of Wittwer and his colleagues who suggested that increased longevity may be a consequence of a decrease in respiration observed in certain vegetables sprayed with the cytokinin, benzyladenine (Dedolph *et al.*, 1961, 1962). This effect of benzyladenine was thought to occur through its inhibition of glycolytic kinases (Tuli *et al.*, 1964). More recent work, however, which demonstrates that a respiration decrease is not a necessary concomitant of chemically induced delay of senescence (Halevy *et al.*, 1966), makes this idea of doubtful significance.

An interesting set of experiments by Leopold and Kawase (1964) indicates that within the intact plant the mobilizing effect of cytokinins which accompanies their delay of senescence may actually induce senescence in other parts of the plant. In young bean (*Phaseolus vulgaris*) plants, primary leaves aged, as measured by chlorophyll loss, much faster when an opposing primary leaf was treated with benzyladenine than in untreated controls. This observation suggests a correlative action of the cytokinins which might account for the senescence inducing effect of fruits and stem apices on leaves. Natural cytokinins induce the flow of nutrients to these sites thus preventing an adequate supply from reaching given leaves (Leopold and Kawase, 1962). Probably related to this phenomenon are the observations that cytokinins both retard and accelerate abscission of leaf petioles depending upon the site of application of the exogenous growth substance and its concentration (Osborne and Moss, 1963, Chatterjee and Leopold, 1964).

The mechanism by which cytokinins induce nutrient flow to treated areas, or to those plant parts in which natural cytokinins are synthesized, is as yet unknown but work by Müller and Leopold (1966) has an important bearing on this question. They concluded that kinetin-induced transport of ^{32}P takes place in the phloem since it occurred along the axis of vascular bundles independently of water flow in the xylem and was blocked by steam-killed zones or by metabolic inhibitors. Furthermore they presented evidence which supports the concept that mass flow in the phloem is stimulated by kinetin-induced 'mobilizing centres' which act as suction pumps through changes in osmotic potential. The latter is presumably brought about through an increase in the volume of mature cells, a well-known effect of cytokinins on many tissues.

Mobility within the plant

Much of what we know about the action of exogenously applied cytokinins leads to the conclusion that these substances are very restricted in their movement within the plant, if indeed they move at all from the original site of application (Thimann, 1963). Studies dealing with the mobilizing effect of cytokinins have been particularly illustrative. Thus, Möthes (1960), Gunning and Barkley (1963), and others have shown that amino acids, phosphates, and various other substances accumulate in plant tissue directly under or very close to the site of application of a cytokinin.

Similarly a study dealing with the effect of kinetin in overcoming apical dominance (Sachs and Thimann, 1964) leads one to conclude that cytokinins are practically immobile in the plant. Here, a stimulation of lateral bud growth was observed only when kinetin was placed directly on the bud. Application to the stem or to the trifid bract, even within 2 mm of the bud tip, was without effect.

On the other hand Kaminek (1965) found that when the morphologically basal end of 39-mm stem sections of *Pisum sativum* were immersed into a kinetin solution, buds near the apex swelled and were stimulated to grow to a significant degree over the control. Because of the very real possibility that kinetin was simply carried along in the transpiration stream by the xylem, this report is of doubtful significance. Chvojka *et al.* (1961) injected radioactive benzyladenine into the petioles of apple leaves and concluded '. . . that it is localized in the vicinity of the point of application, predominantly in the bark and in the phloem'. However, a small amount of movement was detected, predominantly in an apex to base direction (basipetal), by means of autoradiography. Whether the cytokinin itself was involved or a radioactive metabolite later isolated from the petiole was not clear.

More recently a series of papers by Osborne and colleagues not only suggest that cytokinins are indeed mobile within some plants, but also support the idea that movement is in a predominantly polar, basipetal, direction. Although the amounts transported are very small—roughly 2 per cent of the radioactivity supplied is transported in 24 hours basipetally through 5·4 mm petiole segments of *Phaseolus vulgaris* (Osborne and Black, 1964)—polarity of movement is an apparently consistent feature of the experiments (Osborne and McCready, 1965; Black and Osborne, 1965).

Data indicating polarity of movement are so far limited to *Phaseolus vulgaris* and, perhaps, *Pyrus malus*. Although a similar situation has been claimed for epicotyls of *Pisum sativum* cv. Alaska (Osborne and McCready, 1965), the amounts translocated are so small as to make the conclusion highly questionable. By contrast Fox and Weis (1964) failed to find evidence for polar transport in petioles of *Phaseolus vulgaris*, coleoptiles of *Avena sativa*, epicotyls of *Pisum sativum*, or petioles of *Coleus blumei*, either in the presence or absence of added IAA. Similarly polar movement could not be detected in petioles of *Xanthium pennsylvanicum* (Osborne and McCready, 1965) nor in stems or petioles of *Gossypium hirsutum* (Lagerstedt and Langston, 1966).

The discrepancy in results may be linked to a phenomenon described by Seth *et al.* (1966) who observed that IAA simultaneously supplied with kinetin greatly increased the mobility of the latter in stems of *Phaseolus vulgaris*. In the absence of IAA, kinetin was relatively immobile. Similar effects of auxin on the basipetal movement of benzyladenine in *Phaseolus* petioles had been previously noted by Osborne and Black (1964) and by Fox and Weis (1965) for movement in both directions in *Avena coleoptiles*. Furthermore the results of Lagerstedt and Langston (1966) lend themselves to an explanation involving the internal auxin physiology. These workers were unable to confirm Osborne and Black's results when using lower concentrations of benzyladenine or young petioles. However, in the presence of added IAA, polarity of kinetin transport was detected. In older petioles much greater movement of kinetin occurred in both directions although about twice as much moved in a basipetal manner; the addition of IAA in this experiment abolished polarity. The work of Guern (1966), Guern and Hugon (1966), and Guern and Sadorge (1967) likewise bears on this point. They have shown that the application of the cytokinin N^6-benzyladenine to an intact plant results in its metabolism to a nucleoside, the transport of which depends upon the site of injection with respect to actively growing centres in the plant. Under certain circumstances an acropetally polarized transport could be demonstrated.

Obviously we are dealing with a complex, highly interrelated system. It is probable that polar transport is real and demonstrable in some plants by the manipulation of certain experimental variables. The essential ingredient may well be a physiological state in which substantial amounts of IAA are being polarly transported. In such a case the polar transport of cytokinins as well as other substances may be set up as a kind of secondary phenomenon; the polar transport of 'an unidentified metabolite of adenine' in the bean petiole system (Black and Osborne, 1965) may be an example of just this sort. Whether or not the basipetal transport of cytokinins plays a significant role in plant biology cannot be answered with present data. It seems not to be a general phenomenon, however, and whether it occurs under physiological conditions is a moot point. The concentrations used in experiments in which polar transport has been observed are highly lethal to plant tissues *in vitro*, being roughly 500 times greater than optimum levels for callus growth, and artifacts induced by such unphysiological levels cannot be ruled out. There is increasing reason to believe, however, that the natural cytokinins are in fact mobile, perhaps formed in the roots and moved to other parts of the plant in the transpiration stream (Loeffler and van Overbeek, 1964). It would be well to keep an open mind on this point for the future.

Miscellaneous effects

The increasing interest in the cytokinins has been reflected in a steadily accelerating flow of papers describing a wide variety of biological systems on which cytokinins have an effect of one sort or another. A large part of these do not conveniently lend themselves to ready classification but seem to stand apart as isolated phenomena. No doubt many of the biological effects of the cytokinins, perhaps all, have a common basis, but it is a measure of our ignorance concerning this group of plant growth substances that the lines of connection are not readily apparent. The list of such cytokinin activities is now too large to include here, but the following are among those which at the moment appear to have potential significance.

One of the more interesting activities of the cytokinins is their partial prevention of the toxic effects of certain phytopathogens. For example kinetin reduces the number and size of lesions caused by tomato spotted-wilt virus (Selman, 1964) and antagonizes the toxic effect of *Pseudomonas tabaci* which causes 'wildfire' disease of tobacco (Lovrekovich and Farkas, 1963). In the latter case the organism produces a toxin, a structural analogue of methionine, which damages RNA metabolism of the host leading to a breakdown in protein synthesis (Lovrekovich *et al.*, 1964). The protection provided by cytokinins in this system may well be related to their senescence inhibiting capacity which seems to involve a maintenance of protein synthesis.

Other cytokinin influences of special interest involve effects on the activities of certain enzymes, on the metabolism of various substances, and on the respiration of a number of tissues and organs. An important paper by Steinhart *et al.* (1964) which has not been followed up demonstrates that kinetin markedly increases the rate of synthesis of tyramine methylpherase in barley roots, while having no effect on four other enzymes shown to be present. Similarly, cytokinin treatment resulted in an increase in aminoacyl-s-RNA synthetase activity in leaf discs (Anderson and Rowan, 1966), and a decrease in the activities of ribonuclease (Srivastava and Ware, 1965), and enzymes of the hexose monophosphate shunt (Scott *et al.*, 1964).

Cytokinins are reported to reduce alkaloid content in *Datura meteloides* (Ambrose and Sciuchetti, 1962), to activate thiamine biosynthesis in tobacco callus tissue (Digby and Skoog, 1966), to regulate in conjunction with auxin the levels of scopoletin and scopolin in tobacco tissue cultures (Skoog and Montaldi, 1961), and to greatly increase anthocyanin production in cultured petals of *Impatiens balsamina* (Klein and Hagen, 1961), (to name only a few in a very long list of naturally occurring substances whose metabolism is influenced by cytokinins). Such diverse manifestations of cytokinin activity are no doubt secondary phenomena, appearing in response to some, as yet unknown, more basic role of these growth substances.

Depending upon the circumstances, cytokinins have been reported to stimulate respiration (Glasziou, 1957; Gilbart and Dedolph, 1965; Ballantyne, 1966), or to inhibit it (Katsumi, 1963; Bergmann, 1963; Dedolph *et al.*, 1961). To measure respiration rates in various plant tissues before and after cytokinin treatment is probably an exercise in futility in the absence of information about the level of endogenous, natural cytokinins and the requirement of the tissues for these substances. For similar reasons one wonders about the significance of studies in which cytokinins sprayed or otherwise applied to intact plants or plant parts are shown subsequently to affect the metabolism of given substances.

Bio-assays for cytokinins

Cell division tests

Although specific problems or applications may have a bearing, one normally wishes to achieve with any bio-assay the best compromise possible between sensitivity, specificity, speed, linearity of response, applicability over a wide concentration range, and ease of manipulation. Among the assays for cytokinins, those which satisfy the relevant factors to the greatest degree are based on the induction of cell division in cytokinin-requiring tissue cultures. Although many tissues will serve the purpose, the best known and most widely used are tobacco pith tissue used in the original isolation of kinetin but described in detail more recently by Murashige and Skoog (1962) and soybean callus tissue (Miller, 1963). Of the two, the latter offers the greater advantages. Although all tissue culture systems suffer because of their relatively slow response, experienced observers can detect the presence of highly active compounds in the soybean test in less than 48 hours by visual observation. Normally the tissues are grown for a period of 2 to 4 weeks on media into which the test-substance has been incorporated, after which fresh and dry weights are compared with suitable controls.

Both tissues are extremely sensitive, responding with significant dry weight increases in the case of soybean to as little as 2×10^{-8} M kinetin, and highly specific (although it should be pointed out that this statement involves some circular reasoning, considering that cytokinins are defined in terms of this assay). The soybean system is much the more quantitative in response, exhibiting a linear increase in fresh weight in proportion to the logarithm of the kinetin concentration over a several hundred-fold range (Fig. 3.6). The major dis-

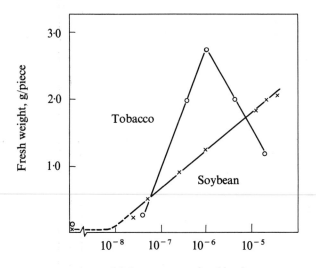

Molar concentration kinetin

Figure 3.6 The influence of kinetin concentration on the growth of soybean and tobacco. The tissues were grown for 21 days on identical media containing 5 mg 1 $^{-1}$ IAA. Data for soybean from Miller, 1963, for tobacco from Fox, unpublished.

advantage of the tobacco system is that its response is linear over a relatively narrow concentration range. For the bio-assay of many crude preparations, however, tobacco tissue is superior since it is a good deal less sensitive to the presence of toxic or inhibitory substances commonly encountered in tissue extracts.

Chlorophyll preservation tests

Bio-assays based on the Richmond–Lang (1957) effect are coming to be used more and more primarily because of their speed and convenience. Osborne and McCalla (1961) described an assay in which discs from *Xanthium pennsylvanicum* leaves aged in dim light are placed on filter paper wetted with the test solution and kept in the dark for 48 hours. Chlorophyll is then extracted and measured photometrically. A linear relationship is claimed between the amount of chlorophyll retained and the log of the concentration (in the case of kinetin from 0·1 to 10 mg l $^{-1}$).

An even more convenient test based on the same response makes use of detached blades of wheat or oat seedlings; apparently this test has been developed independently by a number of workers (but see Gunning and Barkley, 1963, for a description of the activity of kinetin in this system). The assay is especially valuable where the amount of test material is limited since normally a single drop of roughly 0·01 ml is placed on each leaf blade for testing. However, the sensitivity of tests depending upon chlorophyll preservation is a good deal less than the tissue culture assay. An even more serious objection is that the assay is not particularly specific, as auxins, sucrose, and other substances have pronounced activity under certain conditions.

Cell enlargement tests

Kuraishi (1959) made use of the ability of cytokinins to promote cell enlargement in leaf discs for a study of kinetin analogues and described a particularly responsive variety of *Raphanus sativus*. In this test immature foliage leaves aged for 24 hours in the dark are used as a source of discs punched out with a cork borer. The discs are floated on solutions to be tested for 18 hours at 28° in the light and then weighed after blotting off excess liquid. The test has the advantage of speed and may be almost as sensitive as the soybean test but is probably affected by compounds not normally considered to be cytokinins. Leaf discs from dark-grown bean seedlings have been used in a similar manner (Scott and Liverman, 1956; Miller, 1956), but may be less specific than the radish test and seemingly offer no advantages.

Germination tests

The promotive effect on seed germination of cytokinins described earlier in this chapter offers obvious bio-assay possibilities, and in fact Skinner *et al.* (1956), have used lettuce seeds for this purpose. Seeds which have been soaked in the test solution in the dark at 25° for a period of 8 hours are tested for their germination during a subsequent 24 to 60 hour period in the dark at 30°. A seed germination test would not be the method of choice for most applications since they appear to require high concentrations of cytokinins for effect and are influenced by a variety of other factors including red light, gibberellin, and other compounds (Miller, 1963).

Differentiation tests

Bio-assays for cytokinins which depend upon the induction of differentiation have been used very little but their neglect may not be justified since certain of these offer a number of advantages. In the moss, *Tortella caespitosa* for example, protonema respond to cytokinins by an increase in bud formation, responding in at least one instance to as little as 0·5 µg l^{-1} of kinetin (Gorton *et al.*, 1957).

Since greater increases were observed up to at least 5 mg l^{-1} this bio-assay would seem to rival soybean callus with respect both to sensitivity and range of response and is at least as fast. Because the assay method has not been widely used, its specificity is not clearly established. Nevertheless the method holds considerable promise for further exploitation.

Other mosses, e.g., *Pohlia nutans* (Mitra and Allsop, 1959) appear to yield very similar results. One might also take advantage of the cytokinin-auxin interaction on the formation of buds in tobacco pith tissue (described earlier) in the development of a bio-assay as suggested by Miller (1963). The principal drawback to such an assay is that the cytokinin effect is antagonized by auxins even at very low levels. This, together with the probable difficulty in quantifying results, as well as the length of time required to observe results, no doubt explain why the assay has not met with great favour.

Biological role and mode of action

Despite nearly forty years of study and the accumulation of voluminous amounts of information concerning nearly every aspect of the biology of the auxins, there are those who believe that we are very little closer to discovering their function in plant growth. Should we then expect more rapid results with the cytokinins which, compared to the auxins, are of recent vintage and little studied? The answer surprisingly enough may be yes. The application of tools and techniques which have been enormously fruitful in the molecular biology of micro-organisms to problems in plant growth is already yielding information of major significance both for the cytokinins and other plant growth regulators as well.

Effect on nucleic acid metabolism

For substances such as the cytokinins, auxins, and gibberellins which are involved in practically every facet of the life of higher plants, it is only logical to argue that their diverse and apparently unrelated biological effects are explainable only in terms of a master reaction which governs all cellular activities. Consequently it comes as no surprise to find that cytokinins have been evaluated extensively for their role in nucleic acid metabolism beginning with the study of Guttman (1957) who found that kinetin treatment of onion roots caused a quick increase in the amount of RNA in the nuclei. Similar results have now been described by many investigators including Olszewska (1959), Wollgiehn (1965), and Osborne (1962). Jensen *et al.* (1964) reported that kinetin caused a doubling of the RNA levels within 30 minutes in onion root tip cells although it reduced DNA (Table 3.3). By contrast kinetin alone induces a rapid increase in the DNA of tobacco pith cells although the effect can be duplicated by auxin (Patau *et al.*, 1957).

Cytokinins greatly increase the incorporation of ^{32}P into nucleic acids of detached oat leaves (Gunning and Barkeley, 1963), stimulate preferentially the synthesis of heavy RNA in barley leaves (Srivastava, 1965) but increase the incorporation of orotic acid into all classes of RNA (Burdett and Wareing,

Table 3.3   Effect of kinetin on RNA content of onion root cells.

Length of time immersed in kinetin solution*	RNA/cell
h	μμg
0	38
0·5	75
3·0	57
12·0	48
24·0	25
48·0	17

* The kinetin solution contained 1 p.p.m. kinetin in double distilled water. The O control contained double distilled water only. Measurements made on cells between 1·4 and 1·7 mm from root tip. Data from Jensen et al., 1964.

1966) and the incorporation of adenine into leaf RNA (Möthes, 1964). These and many other reports make it clear that cytokinins profoundly influence the nucleic acid metabolism of higher plants in some as yet unknown way.

Cytokinins and protein synthesis

Most of the studies showing a cytokinin-stimulated increase of nucleic acid metabolism in like manner report an effect on protein synthesis (e.g., Wollgiehn, 1961). That cytokinins play a role in this process was of course pointed out in the original paper by Richmond and Lang (1957) and strongly suggested by the work of Möthes and his colleagues. Gunning and Barkeley (1963) have reported data which indicate that cytokinin-directed transport of amino acids is a result of their utilization in protein synthesis. Even more direct evidence is available as Osborne (1962) demonstrated a kinetin stimulated increase in protein synthesis in leaf discs and felt that a primary action of kinetin might be to regulate RNA synthesis with a resulting regulation of one of the steps leading to protein synthesis.

Metabolism of the cytokinins

In order to discover the part played by the cytokinins in the stimulation of nucleic acid and protein synthesis, we need to know what happens to these substances in the plant. Unfortunately only limited information is available on this point and much work remains to be done. Cytokinins are oxidized by mammalian xanthine oxidase to the corresponding 2-mono and 2,8-dihydroxy derivatives and perhaps other compounds (Bergmann and Kwietny, 1958; Henderson *et al.*, 1962). Since some of the derivatives inhibited further action of the enzyme, a mechanism sparing the catabolism of normally occurring purines was postulated (Henderson *et al.*, 1962).

There is some information to indicate that benzyladenine is converted into unknown glycosides by members of the plant family Cruciferae (Loeffler and van Overbeek, 1964). However, the most extensive study of benzyladenine metabolism is that of McCalla *et al.* (1962), who reported that senescing cocklebur (*Xanthium pennsylvanicum*) and bean (*Phaseolus*) leaves convert this compound into a number of low molecular weight substances which include adenylic, guanylic, and inosinic acids, benzyladenosine and probably benzyladenylic acid as well as urea, allantoin, and an unidentified ureide. Further they reported that metabolites of the purine moiety appear in RNA although benzyladenine itself is not incorporated to any significant extent. The extensive degradation of cytokinins by plant tissues and the resulting scavenging of the purine ring for a variety of purposes including RNA formation has subsequently been confirmed by Wollgiehn (1965) and Fox (1966).

Incorporation into RNA

Fox (1964, 1965, 1966) presented evidence which indicates that plant tissues requiring cytokinins incorporate at least two of these growth substances, 6-benzyladenine and 6-methyladenine, into their RNA. This finding has subsequently been confirmed in at least two other laboratories (Wollgiehn, 1965; Srivastava, 1966) for 6-furfuryladenine. The incorporation in soybean tissue cultures has been almost exclusively into the sRNA fraction (Fox, 1966) with some indication that only certain subfractions of this RNA class contained the growth substance (Fox, 1967).

These findings suggest the obvious possibility that cytokinins exert their biological effects through incorporation into polynucleotides which play key roles in protein synthesis. This idea has been greatly strengthened recently with the isolation of the extremely active cytokinin, N^6-(Δ^2-isopentenylamino) purine, as a naturally occurring substance from yeast RNA (Hall *et al.*, 1966) and from serine-specific transfer RNA from yeast (Biemann *et al.*, 1966). Likewise zeatin and its *cis*-isomer are now known to exist naturally as an integral part of the RNA of *Zea mays* (Hall *et al.*, 1966, 1967). Of course the cytokinins

6-methylaminopurine and 6,6-dimethylaminopurine have been known to exist in yeast and wheat germ RNA (Littlefield and Dunn, 1958), and in tRNA from tobacco leaves (Bergquist and Matthews, 1962), for several years.

If in fact the activity of the cytokinins is dependent upon their presence in tRNA, we are still hard put to account for their specific role there. However, the exciting discovery by Zachau and his co-workers (Zachau et al., 1966) that the cytokinin N^6-(Δ^2-isopentenylamino) purine is localized in yeast serine transfer RNA adjacent to the anti-codon provides us with an important clue. There is increasing evidence that purines alkylated at the 6 position lie next to the anticodon in several tRNAs from a variety of organisms. For example, N^6-dimethylaminopurine is adjacent to the presumed anticodon in tyrosine transfer RNA from yeast (Madison, et al., 1966) and a general model of the anticodon loop involving an alkylated purine in this manner has been proposed (Fuller and Hodgson, 1967).

It is difficult to believe that the localization of cytokinins in the anticodon loop of certain tRNAs is merely a fortuitous circumstance; one is led to speculate that these molecules have important regulatory activity in the tRNA. Fuller and Hodgson (1967) point out that the modified purine is at a position which could prevent a wrong set of three nucleotides in the anticodon being recognized by the codon. There is in fact some evidence which suggests that the presence of the cytokinin is necessary for codon-anticodon interaction between messenger RNA and transfer RNA. Chemical modification of only the cytokinin residues in yeast serine transfer RNA (possible because of the extreme chemical reactivity of the side chain of N^6-(Δ^2-isopentenylamino) purine had no effect on the acceptance of the amino acid but interfered with the binding of the charged tRNA to the ribosome-messenger RNA complex (Fittler and Hall, 1966).

One ought not to close one's eyes to the possibility that the incorporation of cytokinins into RNA is an artifact which has nothing to do with their biological role. Adenine, the presumably critical and specific portion of cytokinin molecules, is after all an extremely ubiquitous compound in cellular metabolism playing many important parts other than its obvious one as a constitutent of nucleic acids. An action of the cytokinins in the soluble, non-polynucleotide fraction of the cell, enhancing or inhibiting some function of adenine (Miller, 1961), is not ruled out.

Bibliography

Papers cited in text

Allsop, A. and Alicja Szweykowska, 1960. Foliar abnormalities, including repeated branching and root formation, induced by kinetin in attached leaves of *Marsilea*. *Nature, Lond.* **186**, 813–814.

Ambrose, D. G. and L. A. Sciuchetti, 1962. Influence of kinetin and gibberellic acid on the growth and alkaloid patterns in *Datura meteloides*. *J. Pharm. Sciences*, **51**, 934–938.

Anderson, J. W. and K. S. Rowan, 1966. Activity of aminoacyl-transfer-ribonucleic acid synthetases in tobacco-leaf tissue in relation to senescence and to the action of 6-furfurylaminopurine. *Biochem. J.* **101**, 15–18.

Arora, N., F. Skoog, and O. N. Allen, 1959. Kinetin-induced pseudonodules on tobacco roots. *Am. J. Bot.* **46**, 610–613.

Ballantyne, D. J., 1966. Respiration of floral tissue of the daffodil (*Narcissus pseudo-narcissus* Linn.) treated with benzyladenine and auxin. *Can. J. Bot.* **44**, 117–119.

Beauchesne, G., 1961. Separation des substances de croissance d'extrait de maiz immature in Plant Growth Regulation (*Fourth International conference on Plant Growth Regulation*). Iowa State University Press, Ames, Iowa, pp. 667–674.

Beauchesne, G. and R. Goutarel, 1963. Activité de certaines purines substituees sur le développement des cultures de tissus de möelle de Tabac en présence d'acide indolyl-acétique. *Physiol. Plant*, **16**, 630–635.

Bergmann, F. and Hanna Kwietny, 1958. Oxidation of kinetin by mammalian xanthine oxidase. *Biochem. Biophys. Acta.* **28**, 100–103.

Bergmann, L., 1963. Die Wirkung von 6-furfurylaminopurin auf die Atmung von Zellsuspensionen von *N. tabacum*. *Z. f. Naturf.* **18b**, 942–946.

Bergmann, L., 1964. Der Einfluss von Kinetin auf die Ligninbildung und die Differenzierung in Gewebekulturen von *N. tabacum*. *Planta*, **62**, 221–254.

Bergquist, P. L. and R. E. F. Matthews, 1962. Occurrence and distribution of methylated purines in the ribonucleic acids of subcellular fractions. *Biochem. J.* **85**, 305–313.

Biemann, K., S. Tsunakawa, J. Sonnenbichler, H. Feldmann, D. Dütting, and H. G. Zachau, 1966. Struktur eines ungewöhnlichen Nucleosids aus serin-spezifischer Transfer-Ribonucleinsäure. *Angew. Chem.* **78**, 600–601.

Black, M. Kay and Daphne J. Osborne, 1965. Polarity of transport of benzyladenine, adenine, and indol-3-acetic acid in petiole segments of *Phaseolus vulgaris*. *Plant Physiol.* **40**, 676–680.

Bonner, J. and J. English, 1938. A chemical and physiological study of traumatin, a plant wound hormone. *Plant Physiol.* **13**, 331–348.

Bottomley, W., N. P. Kefford, J. A. Zwar, and P. L. Goldacre, 1963. Kinin activity from plant extracts I. Biological assay and sources of activity. *Austral. J. Biol. Sci.* **16**, 395–406.

Braun, A. C., 1956. The activation of two growth-substance systems accompanying the conversion of normal to tumor cells in crown gall. *Cancer Res.* **16**, 53–56.

Brian, P. W. and H. G. Hemming, 1957. Effects of gibberellic acid and kinetin on growth of pea stem sections. *Naturwiss.* **44**, 594.

Burdett, A. N. and P. F. Wareing, 1966. The effect of kinetin on the incorporation of labelled orotate into various fractions of ribonucleic acid of excised radish leaf discs. *Planta*, **17**, 20–26.

Bürstrom, H., 1961. Other biogenous growth promoters. In *Handbuch der Pflanzenphysiologie*. Springer-Verlag, Berlin, Vol. XIV, pp. 1156–1161.

Caplin, S. and F. Steward, 1948. Effect of coconut milk on the growth of explants from carrot root. *Science*, **108**, 655–657.

Cave, A., J. A. Deyrup, R. Goutarel, N. J. Leonard, and X. G. Monseur, 1962. Identite' de la triacanthine, de la togholamine et de la chidlovine. *Ann. Pharm. Franc.* **20**, 285–292.

Chatterjee, S. R. and A. C. Leopold, 1964. Kinetin and gibberellin actions on abscission processes. *Plant physiol.* **39**, 334–337.

Chvojka, J., K. Veres, and J. Kozel, 1961. The effect of kinins on the growth of apple-tree buds and on incorporation of radioactive phosphate. *Biol. Plantar.* **3**, 140–147 (in Russian).

Crane, J. C. and J. van Overbeek, 1965. Kinin-induced parthenocarpy in the fig, *Ficus carica* L. *Science*, **147**, 1468–1469.

Danckwardt-Lillieström, C., 1957. Kinetin induced shoot formation from isolated roots of *Isatis tinctoria*. *Physiol. Plant.* **10**, 794–797.

Das, N. K., K. Patau, and F. Skoog, 1956. Initiation of mitosis and cell division by kinetin and indoleacetic acid in excised tobacco pith tissue. *Physiol. Plant.* **9**, 640–651.

Davies, C. R., A. K. Seth, and P. F. Wareing, 1965. Auxin and kinetin interaction in apical dominance. *Science*, **151**, 468–469.

Dedolph, R. R., S. H. Wittwer, and D. C. MacLean, 1963. Letter to the editor. *Sci. Amer.* **208**, 13.

Dedolph, R. R., S. H. Wittwer, and V. Tuli, 1961. Senescence inhibition and respiration. *Science*, **134**, 1075.

Dedolph, R. R., S. H. Wittwer, V. Tuli, and D. Gilbart, 1962. Effect of N^6-benzylaminopurine on respiration and storage behavior of broccoli (*Brassica oleracea* var. *Italica*). *Plant Physiol.* **37**, 509–512.

de Ropp, R. S., 1956. Kinetin and auxin activity. *Plant Physiol.* **31**, 253–254.

Deysson, M. G., 1959. Action de la kinétine et de la thiokinétine sur la croissance de la lentille d'eau (*Lemna minor* L.). *Compt. Rend.* **248**, 841–843.

Digby, J. and F. Skoog, 1966. Cytokinin activation of thiamine biosynthesis in tobacco callus cultures. *Plant Physiol.* **41**, 647–652.

Evanari, M., G. Neuman, S. Blumenthal-Goldschmidt, A. M. Mayer, and A. Poljakoff-Mayber, 1958. The influence of gibberellic acid and kinetin on germination and seedling growth of lettuce. *Bull. Res. Counc. Israel*, **60**, 65–72.

Fittler, F. and R. Hall, 1966. Selective modification of yeast seryl-tRNA and its effect on the acceptance and binding functions. *Biochem. Biophys. Res. Comm.* **25**, 441–446.

Fleissner, E. and E. Borek, 1962. A new enzyme of RNA Synthesis: RNA methylase. *Proc. Nat. Acad. Sci. U.S.* **48**, 1199–1203.

Fox, J. E., 1962. A kinetin-like substance in habituated tobacco tissue. *Plant Physiol.* **37**, (suppl.) xxxv.

Fox, J. E., 1963. Growth factor requirements and chromosome number in tobacco tissue cultures. *Physiol. Plantarum*, **16**, 793–803.

Fox, J. E., 1964. Incorporation of kinins into the RNA of plant tissue cultures. *Plant Physiol.* **39**, xxxi.

Fox, J. E., 1965. Characterization of labelled RNA from tissues grown on the kinin, N^6-benzyladenine-benzyl-^{14}C. *Plant Physiol.* **40**, LXXVii.

Fox, J. E., 1966. Incorporation of a kinin, N^6-benzyladenine into soluble RNA. *Plant Physiol.* **41**, 75–82.

Fox, J. E. and C. Chen, 1967. Characterization of labeled ribonucleic acid from tissue grown on ^{14}C-containing cytokinins. *J. Biol. Chem.* **242**, 4490–4494.

Fox, J. E. and C. Chen, 1968. Cytokinin incorporation into RNA and its possible role in plant growth. In *Biochemistry and Physiology of Plant Growth Substances*. The Runge Press, Ottawa.

Fox, J. E. and J. S. Weis, 1964. Transport of the kinin, N^6-benzyladenine: non-polar or polar? *Nature, Lond.* **206**, 678–679.

Fuller, W. and A. Hodgson. 1967. Conformation of the anticodon loop in tRNA. *Nature, Lond.* **215**, 817–821.

Gautheret, R., 1939. Sur la possibilité de realizer la culture indefinie des tissus de tubercules de carrote. *Compt. Rend. Acad. Sci.* Paris, **208**, 118–120.

Gilbart, D. A. and R. R. Dedolph, 1965. Phytokinin effects on respiration and photosynthesis in Roses and broccoli. *Proc. Am. Soc. Hort. Sci.* **86**, 774–778.

Glasziou, K. T., 1957. Respiration and levels of phosphate esters during kinetin-induced cell division in tobacco pith sections, *Nature, Lond.* **179**, 1083–1084.

Goldacre, P. L. and W. Bottomley, 1959. A kinin in apple fruitlets. *Nature, Lond.* **184**, 555–556.

Gorton, B. S., C. G. Skinner, and R. E. Eakin, 1957. Activity of some 6-(substituted) purines on the development of the moss *Tortella caespitosa*. *Arch. Biochem. Biophys.* **66**, 493–496.

Guern, J., 1966. Sur les relations entre metabolisme et transport de la 6-benzylamino-purine-α-^{14}C injectée dans les juenes plantes étiplées de *Cicer arictinum* L. *Comp. Rend. Acad. Sci.* Paris, **262**, 2340–2343.

Guern, J. and Éliane Hugon, 1966. Remarques á propos du transport de la 6-benzyl-aminopurine marquée au ^{14}C dans les juenes plantes éticlées de *Cieer arictinum* L. *Comp. Rend. Acad. Sci.* Paris, **262**, 2226–2229.

Guern, Noémi and Paulette Sadorge, 1967. Polarité du transport de la 6-benzylamino-purine dans de jeunes plantes étiolées de *Cicer arictinum* L. *Comp. Rend. Acad. Sci.* Paris, **264**, 2106–2109.

Gunning, B. and W. Barkley, 1963. Kinin-induced directed transport and senescence in detached oat leaves. *Nature, Lond.* **199**, 262–265.

Guttman, Ruth, 1956. Effects of kinetin on cell division with special reference to initiation and duration of mitosis. *Chromosoma*, **8**, 341–350.

Guttman, Ruth, 1957. Alterations in nuclear ribonucleic acid metabolism induced by kinetin. *J. Biophys. Biochem. Cytol.* **3**, 129–131.

Guttman, Ruth and A. Back, 1960. Effect of kinetin on *Paramecium caudatum* under varying culture conditions. *Science*, **131**, 986–987.

Haber, A. H. and H. J. Luippold, 1960. Separation of mechanisms initiating cell division and cell expansion in lettuce seed germination. *Plant Physiol.* **35**, 168–173.

Haber, A. H. and N. E. Tolbert, 1957. Interactions of kinetin and adenine in the germina-tion of lettuce seed. *Plant Physiol.* **32**, (suppl.) xlix.

Haberlandt, G., 1913. Zur Physiologie der Zellteilung. *Sitz. Ber. K. Preuss. Akad. Wiss* **1913**, 318–345.

Haberlandt, G., 1921. Wundhormone als Erreger von Zellteilungen. *Beitr. Allg. Bot.* **2**, 1–53.

Halevy, A. H., D. R. Dilley, and S. H. Wittwer, 1966. Senescence inhibition and respira-tion induced by growth retardants and N^6-benzyladenine. *Plant Physiol.* **41**, 1085–1089.

Hall, R. H., Lilla Csonka, H. David, and B. McLennen, Cytokinins in the soluble RNA of plant tissues. *Science*, **156**, 69–71.

Hall, R. H., M. J. Robins, and F. Fittler, 1966. *Amer. Chem. Soc.* (abstracts of New York meetings), p. 130.

Hall, R. H., M. J. Robins, L. Stasiuk, R. Thedford, 1966. Isolation of N^6-(γ,γ-Dimethyl-allyl) adenoisine from soluble ribonucleic acid. *J. Am. Chem. Soc.* **88**, 2614–2615.

Harris, G. P. and Ennid M. H. Hart, 1964. Regeneration from leaf squares of *Peperomia sandersii* A. DC: a relationship between rooting and budding. *Ann. Bot.* N.S. **28**, 509–526.

Heide, O. M., 1965. Interaction of temperature, auxins and kinins in the regeneration ability of *Begonia* leaf cuttings. *Physiol. Plantarum*, **18**, 891–920.

Helgeson, J. P. and N. J. Leonard, 1966. Cytokinins: identification of compounds isolated from *Corynebacterium facsians*. *Proc. Nat. Acad. Sci. U.S.* **56**, 60–63.

Henderson, T. H., C. G. Skinner, and R. E. Eakin, 1962. Kinetin and kinetin analogues as substrates and inhibitors of xanthine oxidase. *Plant Physiol.* **37**, 552–555.

Ikuma, H. and K. V. Thimann, 1963. Action of kinetin on photosensitive germination of lettuce seed as compared with that of gibberellic acid. *Plant Cell Physiol.* **4**, 113–128.

Jablonski, J. R. and F. Skoog, 1954. Cell enlargement and cell division in excised tobacco pith tissue. *Physiol. Plant.* **7**, 16–24.

Jensen, W. A., E. G. Pollock, P. Healey, and Mary Ashton, 1964. Kinetin and the nucleic acid content of onion root tips. *Exp. Cell Res.* **33**, 523–530.

Kaminek, M., 1965. Acropetal transport of kinetin in pea stem sections. *Biol. Plantar.* **7**, 394–396.

Katsumi, M., 1962. Physiological effects of kinetin. Effect on the thickening of etiolated pea stem sections. *Physiol. Plant.* **15**, 115–121.

Katsumi, M., 1963. Physiological effects of kinetin. Effect of kinetin on elongation, water uptake and oxygen uptake of etiolated pea stem sections. *Physiol. Plant.* **16**, 66–72.

Kefford, N. P., 1963. Natural Plant Growth regulators. Rept. on 5th Intern. Conf. on Plant Growth Regulation, Gif-sur-Yvette, France July 15–20, 1963, *Science*, **142**, 1495–1505.

Kende, H., 1964. Preservation of chlorophyll in leaf sections by substances obtained from root exudate. *Science*, **145**, 1066–1067.

Kennell, D., 1959. *The effects of IAA and kinetin on the growth of some microorganisms.* Doctoral thesis, University of California, Berkeley.

Khan, A. A., 1966. Breaking of dormancy in *Xanthium* seeds by kinetin mediated by light and DNA-dependent RNA synthesis. *Physiol. Plant.* **19**, 869–874.

Khan, A. A. and N. E. Tolbert, 1965. Reversal of inhibitors of seed germination by red light plus kinetin. *Physiol. Plant.* **18**, 41–43.

Klämbt, D., Thies, and F. Skoog, 1966. Isolation of cytokinins from *Cornyebacterium fascians. Proc. Nat. Acad. Sci. U.S.* **56**, 52–59.

Klein, A. O. and C. W. Hagen Jr. 1961. Anthocyanin production in detached petals of *Impatiens balsamina* L. *Plant Physiol.* **36**, 1–9.

Kuraishi, S., 1959. Effect of kinetin analogs on leaf growth. *Scientific Papers of the College of General Education University of Tokyo*, **9**, 67–104.

Kuraishi, S. and F. S. Okumura, 1956. The effect of kinetin on leaf growth. *Bot. Mag. Tokyo*, **69**, 817–818.

Kuraishi, S. and F. Okumura, 1961. A new green-leaf growth stimulating factor phyllo-cocosine, from coconut milk. *Nature, Lond.* **189**, 148–149.

Kurz, Luise and J. Kummerow, 1957. Brechung der Ruheperiode von *Hydrocharis* durch Kinetin. *J. Naturwiss.* **44**, 121.

Lagerstedt, H. B. and R. G. Langston, 1966. Transport of kinetin-8-^{14}C in petioles. *Physiol. Plantarum*, **19**, 734–740.

Lavee, S., 1963. Natural kinin in peach fruitlets. *Science*, **142**, 583–584.

Leonard, E. O., W. H. Orme-Johnson, T. R. McMurtray, C. G. Skinner and W. Shive, 1962. 2-hydroxy derivatives of some biologically active 6–(substituted) purines. *Arch. Biochem. Biophys.* **99**, 16–24.

Leonard, E. O., C. G. Skinner, and W. Shive, 1961. Synthesis of 2-fluoro-6-benzylamino-purine and other purine derivatives. *Arch. Biochem. Biophys.* **92**, 33–37.

Leonard, N., S. Achmatowicz, R. Loeppky, K. Carraway, W. Grimm, Alicja Szwey-kowska, H. Hamzi, and F. Skoog, 1966. Development of cytokinin activity by re-arrangement of 1-substituted adenines to 6-substituted aminopurines: inactivation by N^6, 1-cyclization. *Proc. Nat. Acad. Sci. U.S.* **56**, 709–716.

Leopold, A. C. and M. Kawase, 1964. Benzyladenine effects on bean leaf growth and senescence. *Am. J. Bot.* **51**, 294–298.

Letham, D. S., 1963. Regulators of cell division in plant tissues I. Inhibitors and stimu-lants of cell division in developing fruits: their properties and activity in relation to the cell division period. *New Zeal. J. Bot.* **1**, 336–350.

Letham, D. S., 1963. Zeatin, a factor inducing cell division from *Zea mays. Life Sci.* **8**, 569–573.

Letham, D. S., 1964. Isolation of a kinin from plum fruitlets and other tissues. In

Regulateurs Naturels de la Croissance Vegetale. (Collogue International Gif-sur-Yvette, (1963). Centre National de la Recherche Scientifique, Paris, p. 109.

Letham, D. S., 1966. Purification and probable identity of a new cytokinin in sweet corn extracts. *Life Sci.* **5**, 551–554.

Letham, D. S., J. S. Shannon, and I. R. McDonald, 1964. The structure of zeatin, a factor inducing cell division. *Proc. Chem. Soc.* (London) **1964**, 230–231.

Lewis, G. P., 1960. Active polypeptides derived from plasma proteins. *Physiol. Rev.* **40**, 647–676.

Littlefield, J. W. and D. B. Dunn, 1958. The occurrence and distribution of thymine and three methylated adenine bases in ribonucleic acids from several sources. *Biochem. J.* **70**, 642–651.

Loeffler, J. E. and J. van Overbeek, 1964. Kinin activity in coconut milk. In *Regulateurs naturels de la Croissance Vegetale* (Collogue International, Gif-sur-Yvette, 1965). Centre National de la Recherche Scientifique, Paris, pp. 77–82.

Loomis, R. S. and J. G. Torrey, 1964. Chemical control of vascular cambium initiation in isolated radish roots. *Proc. Nat. Acad. Sci. U.S.* **52**, 3–11.

Lovrekovich, L. and G. L. Farkes, 1963. Kinetin as an antagonist of the toxic effect of *Pseudomonas tabaci. Nature, Lond.* **198**, 710.

Lovrekovich, L., Z. Klement, and G. L. Farkas, 1964. Toxic effect of *Pseudomonas tabaci* on RNA metabolism in tobacco and its counteraction by kinetin. *Science*, **145**, 165.

Madison, J., G. Everett, and H. Kung, 1966. Nucleotide sequence of a yeast tyrosine transfer RNA. *Science*, **153**, 531–534.

Maruzzella, J. C. and J. G. Garner, 1963. Effect of kinetin on bacteria. *Nature, Lond.* **200**, 385.

Mauney, J., W. Hillman, C. Miller, F. Skoog, R. Clayton, and F. Strong, 1952. Bioassay, purification, and properties of a growth factor from coconut. *Physiol. Plant.* **5**, 485–497.

McCalla, D. R., D. J. Moore, and Daphne Osborne, 1962. The metabolism of a kinin, benzyladenine. *Biochem. Biophys. Acta*, **55**, 522–528.

Michniewicz, M. and A. Kamieńska, 1964. Flower formation induced by kinetin and vitamin E treatment in cold-requiring plant (*Cichorium intybus* L.) grown under non-inductive conditions. *Naturwiss.* **51**, 295–296.

Miller, C. O., 1956. Similarity of some kinetin and red light effects. *Plant Physiol.* **31**, 318–319.

Miller, C. O., 1958. The relationship of the kinetin and red light promotions of lettuce seed germination. *Plant Physiol.* **33**, 115–117.

Miller, C. O., 1961. A kinetin-like compound in maize. *Proc. Nat. Acad. Sci. U.S.* **47**, 170–174.

Miller, C., 1961. Kinetin and related compounds in plant growth. *Ann. Rev. Plant Physiol.* **12**, 395–408.

Miller, C., 1962. Interaction of 6-methylaminopurine and adenine in division of soybean cells. *Nature, Lond.* **194**, 787–788.

Miller, C. O., 1962. Additional information on the kinetin-like factor in maize. *Plant Physiol.* **37**, (suppl.) xxxv.

Miller, C. O., 1963. Kinetin and kinetin-like compounds. *Modern Methods of Plant Analysis*, **6**, 194–202.

Miller, C. O., 1965. Evidence for the natural occurrence of zeatin and derivatives: compounds from maize which promote cell division. *Proc. Nat. Acad. Sci. U.S.* **54**, 1052–1058.

Miller, C. O. and F. Skoog, 1953. Chemical control of bud formation in tobacco stem segments. *Am. J. Bot.* **40**, 768–773.

Miller, C. O., F. Skoog, F. S. Okumura, M. H. Von Saltza, and F. M. Strong, 1955b. Structure and synthesis of kinetin. *J. Am. Chem. Soc.* 77, 2662.

Miller, C. O., F. Skoog, F. S. Okumura, M. H. Von Saltza, and F. M. Strong, 1956. Isolation, structure and synthesis of kinetin, a substance promoting cell division. *J. Am. Chem. Soc.* 78, 1375–1380.

Miller, C. O., F. Skoog, M. H. Von Saltza, and F. M. Strong, 1955. Kinetin, a cell division factor from deoxyribonucleic acid. *J. Am. Chem. Soc.* 77, 1329.

Miller, C. O. and F. H. Witham, 1964. A kinetin-like factor from maize and other sources. In *Regulateurs Naturels de la Croissance Vegetale* (Collogue International Gif-sur-Yvette, 1963). Centre National de la Recherche Scientifique, Paris, p. I (erratum).

Mitra, G. C. and A. Allsopp, 1959. III. The effects of various physiologically active substances on the development of the protonema and bud formation in *Pohlia nutans* (Hedw.) Lindb. *Phytomorph.* 9, 64–71.

Möthes, K., 1928. Pflanzenphysiologische Untersuchungen über Alkaloide. I. Das Nikotin in der Tabakpflanze. *Planta*, 5, 563–615.

Möthes, K., 1960. Über des Altern der Blätter und Möglichkeit ihrer Wiederverjüngung. *Naturwiss.* 47, 337–350.

Möthes, K., 1964. The role of kinetin in plant regulation. In *Regulateurs naturels de la Croissance Vegetale* (Collogue International, Gif-sur-Yvette, 1963). Centre National de la Recherche Scientifique, Paris, pp. 131–140.

Möthes, K. and L. Engelbrecht, 1959. Kinetin und das Problem der Akkumulation löslicher Stickstoff-Verbindungen. *Monatsber. Deutsch. Akad. Wiss.* Berlin 1, 367–375.

Möthes, K., L. Engelbrecht, and O. Kulajewa, 1959. Über die Wirkung des Kinetins auf stickstoff-verteilung und Eiweisssynthese in isolierten Blättern. *Flora*, 147, 445–465.

Murashige, T. and F. Skoog, 1962. A revised medium for rapid growth and bio-assays with tobacco tissue cultures. *Physiol. Plant.* 15, 473–497.

Nobécourt, P., 1939. Sur la pérennité et l'augmentation de volume des cultures de tissus végétaux. *Compt. Rend. Soc. Biol.*, Paris, 130, 1270.

Ogawa, Y., Y. Abe, and K. Fujioka, 1957. Effects of kinetin on division of Yoshida sarcoma cells. *Nature, Lond.* 180, 983–986.

Olszewska, M. J., 1959. Étude autoradiographique de l'influence de la kinétine sur l'incorporation d'adénine dans les cellules du méristème radiculaire d'*Allium cepa*. *Exp. Cell Res.* 16, 193–201.

Osborne, Daphne J., 1962. Effect of kinetin on protein and nucleic acid metabolism in *Xanthium* leaves during senescence. *Plant Physiol.* 37, 595–602.

Osborne, Daphne J. and M. Kay Black, 1964. Polar transport of a kinin, benzyladenine. *Nature, Lond.* 201, 97.

Osborne, Daphne J. and D. R. McCalla, 1961. Rapid bio-assay for kinetin and kinins using senescing leaf tissue. *Plant Physiol.* 36, 219–221.

Osborne, Daphne J. and C. C. McCready, 1965. Transport of the kinin, N^6-benzyl-adenine: non-polar or polar? *Nature, Lond.* 206, 679–680.

Osborne, Daphne J. and Susan E. Moss, 1963. Effect of kinetin on senescence and abscission in explants of *Phaseolus vulgaris*. *Nature, Lond.* 200, 1299–1301.

Overbeek, J. van, M. Conklin, and A. Blakeslee, 1941. Factors in coconut milk essential for growth and development of *Datura* embryos. *Science*, 94, 350–351.

Patau, K., N. K. Das, and F. Skoog, 1957. Induction of DNA synthesis by kinetin and indoleacetic acid in excised tobacco pith tissue. *Physiol. Plant.* 10, 949–966.

Pilet, P. E., 1965. Sur la terminologie des kinines. *Revue Générale de Botanique.* 71, 250–253.

Plummer, T. H. and A. C. Leopold, 1957. Chemical treatment for bud formation in Saintpaulia. Proc. Am. Soc. Hort. Sci. **70**, 442–444.

Pollard, J., E. Shantz, and F. Steward, 1961. Hexitols in coconut milk: their role in nurture of dividing cells. Plant Physiol. **36**, 492–501.

Powell, L. E. and Charlotte Pratt, 1964. Kinins in the embryo and endosperm of Prunus persica. Nature, Lond. **204**, 602–603.

Quinn, L. Y., R. P. Oates, and T. S. Beers, 1963. Support of cellulose digestion by Clostridium thermocellum in a kinetin-supplemented basal medium. J. Bacteriol. **86**, 1359.

Richmond, A. and A. Lang, 1957. Effect of kinetin on protein content and survival of detached Xanthium leaves. Science, **125**, 650–651.

Rogozinska, Janina, J. Helgeson, and F. Skoog, 1964. Tests for kinetin-like growth promoting activities of triacanthine and its isomer 6–(γ,γ-Dimethylallylamino)-purine. Physiol. Plant. **17**, 165–176.

Rogozinska, Janina, J. Helgeson, F. Skoog, S. Lipton, and F. Strong, 1965: Partial purification of a cell-division factor from peas. Plant Physiol. **40**, 469–476.

Sachs, T. and K. V. Thimann, 1964. Release of lateral buds from apical dominance. Nature, Lond. **201**, 939–940.

Samuels, R. M., 1961. Bacterial-induced fasciation in Pisum sativum var. Alaska. Ph.D Thesis Indiana University.

Schraudolf, H. and J. Reinert, 1959. Interaction of plant growth regulators in regeneration processes. Nature, Lond. **184**, 465–466.

Scott, K. J., J. Daly, and H. H. Smith, 1964. Effects of indoleacetic acid and kinetin on activities of enzymes of the hexose monophosphate shunt in tissue cultures of Nicotiana. Plant Physiol. **39**, 709–711.

Scott, R. A. and J. L. Liverman, 1956. Promotion of leaf expansion by kinetin and benzylaminopurine. Plant Physiol. **31**, 321–322.

Selman, I. W., 1964. The effect of kinetin on infection of petunia and tomato spotted-wilt virus. Ann. Appl. Biol. **53**, 67–76.

Seth, A. K., C. R. Davies, and P. F. Wareing, 1966. Auxin effects on the mobility of kinetin in the plant. Science, **151**, 587–588.

Shantz, E. M. and F. C. Steward, 1955. The identification of compound A from coconut· milk as 1, 3-diphenylurea. J. Am. Chem. Soc. **77**, 6351–6353.

Shaw, G. and D. V. Wilson, 1964. A synthesis of zeatin. Proc. Chem. Soc. (London) **1964**, 231.

Shaw, M., P. K. Bhattacharya, and W. A. Quick, 1965. Chlorophyll, protein, and nucleic acid levels in detached, senescing wheat leaves. Can. J. Bot. **43**, 739–746.

Shaw, M. and B. Srivastava, 1964. Purine-like substances from coconut endosperm and their effect on senescence in excised cereal leaves. Plant Physiol. **39**, 528–532.

Skinner, C. G., J. R. Claybrook, F. D. Talbert, and W. Shive, 1956. Stimulation of seed germination by 6-(substituted) Thiopurines. Arch. Biochem. Biophys. **65**, 567–569.

Skinner, C. G. and W. Shive, 1955. Synthesis of some 6-(substituted)-aminopurines. J. Am. Chem. Soc. **77**, 6692–6693.

Skinner, G., F. D. Talbert, and W. Shive, 1958. Effect of 6-(substituted) purines and gibberellin on the rate of seed germination. Plant Physiol. **33**, 190–194.

Skoog, F., H. Hamzi, Alicia Szweykowska, N. Leonard, K. Carraway, T. Fujii, J. Helegson, and R. Loeppky, 1967. Cytokinins: structure/activity relationships. Phytochem. **6**, 1169–1192.

Skoog, F. and E. Montaldi, 1961. Auxin-kinetin interaction regulating the scopoletin and scopolin levels in tobacco tissue cultures. Proc. Natl. Acad. Sci. U.S. **47**, 36–39.

Skoog, F. and C. Tsui, 1948. Chemical control of growth and bud formation in tobacco stem segments and callus cultured in vitro. Am. J. Bot. **35**, 782–787.

Sommer, N. F., Muriel V. Bradley, and M. T. Creasy, 1962. Peach mesocarp explant enlargement and callus production *in vitro*. *Science*, **136**, 264–265.

Sorokin, Helen and K. V. Thimann, 1964. The histological basis for inhibition of axillary buds in *Pisum sativum* and the effects of auxins and kinetin on xylem development. *Protoplasma*, **59**, 326–350.

Srivastava, B. I. S., 1965. Effect of kinetin on the ecteola cellulose elution profile and other properties of RNA from the excised first seedling leaves of barley. *Arch. Biochem. Biophys.* **110**, 97–103.

Srivastava, B. I. S., 1966. Metabolism of kinetin in excised barley leaves and in tobacco pith tissue culture. *Plant Physiol.* **41**, IXii–Ixiii.

Srivastava, B. I. S. and G. Ware, 1965. The effect of kinetin on nucleic acids and nucleases of excised barley leaves. *Plant Physiol.* **40**, 62–64.

Steinhart, Carol, J. D. Mann, and S. H. Mudd, 1964. Alkaloids and plant metabolism. VII. The kinetin-produced elevation in tyramine methylpherase levels. *Plant Physiol.* **39**, 1030–1038.

Stetler, D. A. and W. M. Laetsch, 1965. Kinetin-induced chloroplast maturation in cultures of tobacco tissue. *Science*, **149**, 1387–1388.

Steward, F. and H. Mohan Ram, 1961. Determining factors in cell growth: some implications for morphogenesis in plants. *Advances in Morphogenesis*, **1**, 189–265.

Steward, F. and E. Shantz, 1956. In *The Chemistry and Mode of Action of Plant Growth Substances* (Proc. Symp. Univ. London, 1955), eds Wain and F. Wightman. Butterworths Scientific Pub., London, pp. 168–186.

Steward, F. C. and E. M. Shantz, 1959. The chemical regulation of growth (some substances and extracts which induce growth and morphogenesis). *Ann. Rev. Plant Physiol.* **10**, 379–404.

Strong, F. M., 1958. *Topics in Microbial Chemistry*. John Wiley and Sons Inc., New York, pp. 98–157.

Sugiura, M., K. Umemura and V. Oota, 1962. The effect of kinetin on protein level of tobacco leaf discs. *Physiol. Plant.* **15**, 457–464.

Thimann, K., 1963. Plant growth substances; past, present and future. *Ann. Rev. Plant Physiol.* **14**, 1–18.

Thimann, K. V. and T. Sachs, 1966. The role of cytokinins in the 'fasciation' disease caused by *Corynebacterium fascians*. *Am. J. Bot.* **53**, 731–739.

Torrey, J. G., 1958. Endogenous bud and root formation by isolated roots of *Convolvulus* grown *in vitro*. *Plant Physiol.* **33**, 258–263.

Tuli, V., D. R. Dilley, and S. H. Wittwer, 1964. N^6-Benzyladenine: inhibitor of respiratory kinases. *Science*, **146**, 1477–1479.

van Eyck, J. and H. Veldstra, 1966. A comparative investigation of kinetin (6-Furfurylaminopurine) and some similarly substituted purines and pyrimidines with *Lemna minor* (L.). *Phytochemistry*, **5**, 457–462.

Von Saltza, M. H., 1958. *The synthesis and properties of the nucleotide of kinetin and of some kinetin analogs*. Doctoral thesis. University of Wisconsin.

Weaver, R. J., 1963. Use of kinin in breaking rest in buds of *Vitis vinifera*. *Nature, Lond.* **198**, 207–208.

Weaver, R. J., J. van Overbeek, R. M. Pool, 1965. Induction of fruit set in *Vitis vinifera* L. by a kinin. *Nature, Lond.* **206**, 952–953.

Wehnelt, B., 1927. Untersuchungen über das Wundhormon der Pflanzen. *Jahrb. Wiss. Bot.* **66**, 778–813.

White, P., 1939. Potentially unlimited growth of excised plant callus in an artificial nutrient. *Am. J. Bot.* **26**, 59–64.

Wickson, Margaret and K. V. Thimann, 1958. The antagonism of auxin and kinetin in apical dominance. *Physiol. Plant.* **11**, 62–74.

Wiesner, J., 1892. *Die Elementarstruktur und das Wachstum der lebenden Substanz*. A. Hölder, Vienna.

Wollgiehn, R., 1961. Untersuchungen über den Einfluss des Kinetins auf den Nuclein-säure und Proteinstoffwechsel isolierter Blätter. *Flora*, **151**, 411–437.

Wollgiehn, R., 1965. Kinetin and Nucleinsäurestoffwechsel. *Flora*, **156**, 291–302.

Zachau, H., D. Dütting, and H. Feldman, 1966. Nucleotide sequence of two serine-specific transfer ribonucleic acids. *Angew. Chem.* internation. Edit. **5**, 422.

Zwar, J. A. and F. Skoog, 1963. Promotion of cell division by extracts from pea seedlings. *Austral. J. Biol. Sci.* **16**, 129–139.

4. Transport of plant growth regulators

Mary Helen M. Goldsmith

Introduction

The endogenous hormones in intact plants must move from one or several locations where they are synthesized to other regions where they direct growth, differentiation, and morphogenesis. Thus, elucidating the direction, pattern, and mechanism of movement of hormones is important in the study of the hormonal control of growth of plants. Much of what we know about the transport of auxin, cytokinin, and gibberellin has been learned by studying the distribution of hormone exogenously applied to either isolated sections of various organs or to different organs of intact or nearly intact plants.

In 1928, Went devised the system that is still used to study the movement of hormones, particularly auxins, in isolated segments of plant tissue. He applied an agar donor block containing the hormone to one cut surface of an isolated section and an agar receiving block initially devoid of any growth substance to the opposite end of the section to collect any hormone that moved from the donor through the tissue (Fig. 4.1). The distinctive property of auxin movement

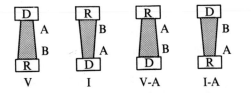

Figure 4.1 The experimental system. D, donor block; R, receiver block; A, morphological apical end of section; B, morphological basal end of section. (V), (I), basipetal movement in vertical (V) and inverted (I) sections; (V-A), (I-A), acropetal movement in vertical (V-A) and inverted (I-A) sections (From Little and Goldsmith, 1967).

in sections of shoots is that much more auxin moves basipetally than acropetally (Went, 1928; van der Weij, 1932). This polarity has been repeatedly confirmed in a variety of tissues and a number of species (Went and Thimann, 1937; Jacobs, 1961; Leopold, 1963; McCready, 1966). Besides the major naturally occurring auxin, IAA,* the only synthetic auxins tested, 2,4-D, NAA, and IBA

---

* Abbreviations: IAA, indole-3-acetic acid; BA, benzyladenine; 2,4-D, 2,4-dichlorophenoxyacetic acid; 2,4-DNP, 2,4-dinitrophenol; GA, gibberellic acid; IBA, indole-3-n-butyric acid; NPA, N-1-naphthylphthalamic acid; PCMB, p-chloromercuribenzoic acid; TIBA 2,3,5-triiodobenzoic acid.

also move polarly in this system (McCready, 1963; Gorter and Veen, 1966; Leopold and Lam, 1961). By contrast with the auxins, the movement of cytokinins and gibberellins has been studied relatively infrequently. Although polarity of movement of the cytokinins benzyladenine and kinetin has been demonstrated in a very limited number of cases (Black and Osborne, 1965; Lagerstedt and Langston, 1966), the factors determining whether movement is polar or not are unknown (Fox and Weis, 1965; Osborne and McCready, 1965). The available data on movement of gibberellins in isolated sections are even more limited than for cytokinins, but in all cases so far tested, gibberellic acid moves non-polarly (Kato, 1958; Clor, 1967).

Polarity of auxin movement exists in isolated segments of the non-vascular tissue of coleoptiles, and *Coleus* pith (Jacobs and McCready, 1967) and also in the veins of tobacco leaves (Avery, 1935), so that both parenchymatous tissue and vascular tissue may be capable of polar transport. It is unlikely, however, that the normal mechanisms of transport in either xylem or phloem can be operating in the short (5 to 20 mm) sections used in Went's system. This view is confirmed by velocities of polar movement of IAA ranging from about 5 up to a maximum of about 20 mm h^{-1} (van der Weij, 1932; Went and White, 1939; Hertel and Leopold, 1963b). These values are rather low for normal vascular transport.

On the other hand, if growth regulators are applied to intact plants and are able to penetrate the surface to which they are applied, they generally reach the vascular tissue and are rapidly distributed at velocities of 50 mm h^{-1} or more (Skoog, 1938; Crafts, 1961; McComb, 1964; McCready, 1966). This movement is non-polar, being both upward and downward in the plant, and the hormones usually accumulate in regions of rapid growth and expansion, as for example in the young leaves. Under conditions favouring transpiration, the applied hormones move in the xylem; under conditions favourable for translocation of sugar, they also appear to move in the phloem. Thus whereas any of the hormones appears able to move in the vascular system, auxin alone consistently moves polarly in isolated sections.

Which of these systems is most likely then to yield reliable information about the movement of endogenous hormones in intact plants? The predominantly basipetal movement of auxin in isolated sections agrees with what is known indirectly of the movement of endogenous auxin. The endogenous auxin is produced in the young and growing tissues of the shoot or in the tips of the coleoptile and then migrates basipetally (Went, 1928; Skoog, 1938; Oserkowsky, 1942; Jacobs, 1952; Scott and Briggs, 1960). The predominantly basipetal migration of geo- and phototropic responses (Went and Thimann, 1937), cambial activation (Snow, 1935 and 1937), and vascular regeneration about a wound (Jacobs, 1967), as well as the familiar dominance of apical over lower lateral shoots (Thimann and Skoog, 1934), indicate that these processes depend on auxin moving predominantly basipetally in the shoot. On the other hand, the situation with endogenous cytokinins and gibberellins appears to be rather different from auxin. Apparently cytokinins and gibberellins are supplied to the shoot not only from the young leaves (Jones and Phillips, 1966) but also to a considerable extent from the roots, and are translocated upwards in the xylem (Kende, 1965; Phillips and Jones, 1964; Carr, Reid, and Skene, 1964). Endogenous auxin is probably polarly transported in intact plants and thus the isolated section is an advantageous system in which to analyse this transport. Endogenous cytokinins and gibberellins, however, may move predominantly in the vascular system, and thus for these hormones, experiments on their circulation in more nearly intact plants is more likely to be meaningful.

The movement of auxins in isolated sections

Definitions

In experiments with isolated sections, one can readily measure (a) uptake, (b) distribution along the section, and (c) movement into receivers. *Uptake* is routinely measured by the decrease of auxin in the donor in a given period. If no loss of auxin occurs, uptake also equals the total auxin recovered in tissue plus receiver. Since the system is a complex one, there is no assurance that the uptake from the donor is equivalent to uptake by a polar transport system. At present no method is available to determine how much of the uptake might be due to other processes besides polar transport; e.g., diffusion, utilization, etc.

Commonly the term *transport* is simply synonymous with movement of auxin. Transport is usually measured, however, as the amount of auxin appearing in the receiver. If receivers are apical, acropetal transport is measured; if they are basal, basipetal transport is measured (Fig. 4.1). Collecting auxin in agar receiver blocks does not apparently introduce any significant artifacts; the amount transported basipetally through sections of oat coleoptiles is the same whether the receiver is a block of agar or more of the section (Goldsmith and Thimann, 1962). The appearance of auxin in the receiver gives no indication of the mechanism involved in moving from donor to receiver. Some movement by diffusion is inevitable, but whether diffusing auxin actually reaches the receiver during a given experimental period will depend on the concentration gradient, the time of transport, the length of section, and the area of the paths available for diffusion. This chapter will present the evidence that basipetal transport in organs of shoots possesses a component that is distinct from diffusion in being polar, metabolically dependent, and molecularly specific.

In the broad sense, *metabolic transport* refers to any transport dependent on metabolism and is not meant as a synonym for active transport. According to the generally accepted definition *active transport* is limited to situations where the movement is independent of the passive physical forces such as solvent drag or thermodynamic potential gradient (Rosenberg, 1954; Dainty, 1962). For a neutral substance, the thermodynamic potential gradient depends only on the concentration gradient, but for a charged species, gradients of both electrical and chemical potential must be considered. Thermodynamically, metabolic transport may be either active or passive. Active transports require metabolic energy to perform the work of moving a substance up a potential gradient, but in the living system, physically passive transports may also use metabolic energy; for example, to maintain the structural integrity of the transport system, or to drive protoplasmic streaming, or to provide energy of activation for passage across membranes. Thus properties such as (a) specificity, (b) temperature dependence, and (c) sensitivity to metabolic inhibitors may indicate that auxin transport is a metabolically dependent process but cannot distinguish whether this transport is active or passive.

It is not known whether auxin is transported as a charged species which could be influenced by the gradient of electrical potential that exists longitudinally in plant organs. If auxin transport is active, basipetal transport into receivers will occur in the absence of both concentration and electrical gradients. A definitive experiment to test this possibility has not been performed. In experiments where the electrical gradient was not recorded, auxin continued to move to receivers even when the receiver's concentration was higher than the donor's (Table 4.1)

Table 4.1   Movement of auxin against a concentration gradient (after van der Weij, 1934).

Original donor	Original receiver	After transport for 250 min*		Receiver
		Donor	Receiver	
				(Δ)
100†	0	33	60	60
100	100	35	159	59
100	200	33	255	55
100	300	36	352	52

* By 1 mm coleoptile sections.
† 100 = 12·6° ± 0·8 *Avena* curvature bio-assay.
This experiment shows that uptake (calculated from column 3) and transport to receivers (last column) is independent of the concentration in the receivers.

(van der Weij, 1932 and 1934; McCready, 1963). In other experiments in which the concentrations in donor and receiver were approximately equal by the end of 1·5 hours of transport, the amount of transport was not affected by applying external voltages to the section that either increased or reversed the inherent longitudinal potential of several millivolts by as much as 40 to 60 mV (Clark,

1937). Furthermore, a potential difference of scores of millivolts exists across each cell membrane in the section (Etherton and Higinbotham, 1960), and this fact also diminishes the likelihood that the longitudinal potential gradient can be an effective force in auxin movement. Auxin would have to move alternately up and down these potential jumps in passing through each cell in the file, and thus it is unlikely that an extracellular potential of much smaller magnitude per unit length could drive the movement of auxin. The evidence, therefore, suggests, but does not prove, that basipetal auxin transport is independent of the longitudinal potential and when occurring against a concentration gradient is an active transport in the strict thermodynamic sense.

The fact that auxin can continue to move into receivers against a concentration gradient also has a bearing on the suggestion that transport is polar because growth establishes a sink for auxin (Zaerr and Mitchell, 1967). This explanation is plausible only if concentration gradients are a prerequisite for polar transport.

Cellular versus symplastic path of movement

Two rather different paths of movement in the section can be visualized. In the first case, auxin would pass through the cell membrane of the first cell of the section, move through the cytoplasm, and pass across the cell membrane into extracellular space before entering the next cell. In the second case, auxin would simply enter the cytoplasm of the first cell and migrate to successive cells via the cytoplasmic connections of the symplast, becoming extracellular only as it leaves the basal cell to enter the receiver. Plasmodesmata have been considered to be likely channels of transport and communication between cells; however, the complexity of their structure in electron micrographs argues that they are by no means simple cytoplasmic bridges (O'Brien and Thimann, 1967). No experimental evidence is available as to whether or not plasmodesma function in auxin transport. In fact, little experimental evidence of any kind is available on their function. Arisz (1958) has shown that during transport of chloride in the water plant, *Vallisneria*, this ion does not leak to the bathing medium and therefore Cl^- probably travels from cell to cell within plasmodesma. A new approach to determining whether plasmodesma are pathways for transport is suggested by the discovery of connections of low electrical resistance between certain animal cells (e.g., Potter, Furshpan, and Lennox, 1966). The electrical approach should also be feasible with plant cells, and in *Nitella* has already shown that the electrical coupling through the nodal cells which are penetrated by many plasmodesma is sufficient for an action potential in an internodal cell to trigger a potential in an adjacent internodal cell (Spanswick and Costerton, 1967). If plasmodesma of higher plant cells can be shown to pass electric currents which are indicative of passive movements of ions between cells this would enhance the likelihood that they may also serve as channels for movement of auxin and other molecules.

Detection of transported auxin

With optimum donor concentrations the amount of auxin arriving in receivers at the base of sections of corn and oat coleoptiles 6–7 mm long is small, ca. 2×10^{-10} g h^{-1} mm^{-2}; therefore, the receivers must be assayed by an appropriately sensitive method. The *Avena* curvature test (Went and Thimann, 1937) has proved the most satisfactory assay in experiments with endogenous or unlabelled auxin. The availability of IAA and synthetic auxins labelled with ^{14}C at high specific activity has greatly facilitated and stimulated research on auxin movement. With auxins having a specific activity of 5–20 c mole^{-1} and a sensitive counting system to detect radioactivity, experiments on single sections are feasible.

Verification that movement of radioactivity reflects movement of auxin

The importance of showing that the movement of radioactivity reflects the movement of auxin cannot be over-emphasized. The initial donor should be free of radioactive contaminants. Carboxyl labelled IAA has been widely used on the premise that since the initial step in oxidative breakdown is decarboxylation, the only radioactive product would be ^{14}CO$_2$, which would be lost rapidly from the blocks or tissue. Independent evidence that movement of ^{14}C represents movement of IAA is necessary, however, because dark or photosynthetic fixation of ^{14}CO$_2$ could give rise to mobile products.

Several methods have been used to examine the transported radioactivity:

(a) Goldsmith and Thimann (1962) verified that the ^{14}C which has moved through sections of oat coleoptiles has the same auxin activity as the starting material by showing that the biological activity/c.p.m. ratio is the same in donor and receiver blocks.

(b) Pilet (1965), in six-hour experiments with sections of *Lens culinaris* stems, obtained quantitatively identical results with both carboxyl and ring labelled IAA. This result indicates that the *Lens* tissue does not decarboxylate IAA and suggests that IAA may be the only radioactive substance in receivers.

(c) The most widely used method has been to extract the radioactivity quantitatively from separate parts of the system (donor, section, receiver) and examine the extract chromatographically with several different developing solvents. Results are now available in various tissues with ^{14}C-IAA, 2,4-D, and NAA and in general appear gratifyingly simple. Ninety per cent or more of the radioactivity in the receivers cochromatographs with either cold auxin or a fresh sample of ^{14}C auxin (Hertel and Leopold, 1963b; Naqvi and Gordon, 1965; McCready, 1963; Gorter and Veen, 1966). Other radioactive substances may appear in the tissue during such experiments, but their absence in receivers indicates that transport is relatively specific for auxin.

The effect of concentration on auxin transport

Evidence for the existence of a relatively specific, metabolically dependent transport of auxin has been sought in experiments on the relation between transport and concentration and also in the experiments on the velocity, specificity, polarity, and inhibition of auxin transport which will be considered in succeeding sections of this chapter.

If the kinetics of auxin transport should prove analogous to the behaviour of an enzymatic reaction, this would constitute support for a specific transport site. Just as the rate of an enzymatic reaction at a given enzyme concentration

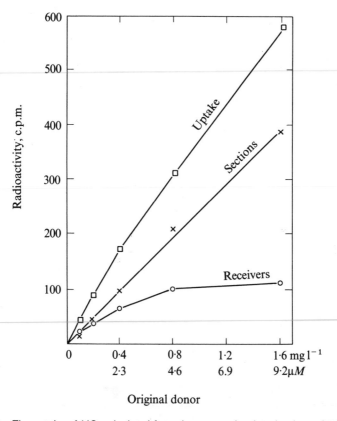

Original donor

Figure 4.2   The uptake of ^{14}C, calculated from the amount leaving the donor (□), and the amount in the section (×), and receiver (○) as a function of concentration of IAA in the donor. Uptake and amount in the section increase more with concentration than transport to the receiver. About 90 per cent of the uptake is accounted for in the sections and receivers. Sections of *Avena* coleoptiles 7 mm long; time of transport 3 hours, but qualitatively similar results were obtained for shorter times. (Adapted from Goldsmith and Thimann, 1962.)

saturates as the substrate concentration is raised, transport into the receiver would be a hyperbolic function of auxin uptake if combination with a specific site limits the rate of auxin transport. The influence of donor concentration on auxin transport has been carefully investigated in relatively few cases, and the amount of auxin reaching receivers fails to increase linearly with donor concentration (Fig. 4.2) (Goldsmith and Thimann, 1962; Hertel and Leopold, 1963b; Gillespie and Thimann, 1963; Scott and Jacobs, 1963; Veen, 1967). In the case of oat coleoptiles, however, the relationship does not appear to be hyperbolic, and furthermore, a Lineweaver–Burk plot is not linear. Finally, simple enzyme kinetics cannot account for the fact that the relationship between concentration and transport depends on the length of the section (Goldsmith and Thimann, 1962).

Although basipetal transport to receivers is not proportional to concentration, uptake is (Fig. 4.2); therefore, the higher the concentration, the smaller the proportion of uptake that reaches the receiver. We can validly compare the proportion of the IAA taken up that is transported in different tissues only if we take into account that the fraction of uptake reaching the receiver depends on length of section and period of transport as well as concentration. At concentrations where transport is approximately proportional to concentration, transport varies from a few per cent of uptake as in bean petioles to as much as 70 per cent of uptake in corn coleoptiles.

Immobile auxin

Direct evidence exists for irreversible immobilization of IAA during transport (Goldsmith and Thimann, 1962). If the donor with ^{14}C-IAA is replaced with either an agar block with no auxin or one containing unlabelled auxin at the same concentration, the rate of appearance of ^{14}C in the basal receiver decreases rapidly, and some radioactivity always remains immobile in the tissue. The proportion of immobile activity increases with time of uptake and concentration of IAA. After one hour with 0·46 μM IAA in the donor, only 15 per cent of the uptake is immobile, but after two hours with donors at a 20-fold higher concentration, 70 per cent is immobile.

The appearance of immobile activity correlates well with the gradual decline in rate of transport as a function of donor concentration. A similar correlation between immobilization and transport emerges from a recent study of the transport of ^{14}C-NAA in Coleus (Veen, 1966; 1967). This suggests that the apparent saturation of transport as the concentration is raised is the result of immobilization and transport competing for auxin. Such irreversible removal of auxin during its transport should produce a logarithmic decline in radioactivity with distance from the source (Horwitz, 1958), as is the case for IAA (Goldsmith and Thimann, 1962). Immobilization could also account for the rapid decrease in transport flux at high concentrations (Thornton and Thimann, 1967).

Localization and identity of immobile radioactivity

The immobile ^{14}C might represent (a) decarboxylation of auxin and refixation of $^{14}CO_2$, (b) retention of auxin in a tight compartment of the cell (e.g., vacuole) remote from the transport system, or (c) chemical or physical combination of the labelled auxin with other substances.

Ether extracts all the immobile activity from sections of oat coleoptiles (Winter and Thimann, 1966), and chromatography confirms that it is IAA. A crude fractionation by centrifugation at $240g$ shows that during an export period the radioactivity of the supernatant fraction containing the cell sap and small organelles falls nearly 20-fold to a very low value, but that the radioactivity of the tissue fraction does not change. Clearly the mobile activity is present in the supernatant while the immobile activity is associated with the sediment of cell walls and debris.

Immobile radioactivity has also been localized by microautoradiograms of histological sections of *Coleus* pith (Veen, 1966). After transporting ^{14}C-NAA for 24 hours, radioactivity is localized in the peripheral cytoplasm adjacent to the cell walls. Thus both the autoradiographic and fractionation experiments indicate that immobile auxin is not within the vacuole.

In the fractionation experiments, the radioactivity can be extracted from the sediment by 10 per cent urea, trypsin, and chymotrypsin, but not water, phosphate buffer, or ribonuclease. Ether and urea would not be expected to break covalent bonds, and thus the immobile IAA seems to be linked by hydrogen bonds to protein (Winter and Thimann, 1966).

The situation with other tissues is more complicated. Immobile radioactivity cannot be totally extracted from etiolated pea epicotyls by ether (Winter and Thimann, 1966). Indoleacetylaspartate as well as IAA are present in the ether extract, and subsequent extraction with other solvents yields unidentified radioactive substances. Acetonitrile extracts of *Coleus* after transport of NAA yield five substances besides NAA (Veen, 1966).

Studies on pea epicotyls indicate that endogenous auxin is also immobilized. Scott and Briggs (1960, 1962) collected auxin in agar receiver blocks from intact epicotyls of light-grown 9-day-old peas that had been cut off at various distances from the apex. They compared the distribution of this mobile auxin with that of the total auxin extracted by ether. Throughout the top internode, the concentration of extractable auxin is constant, but the fraction of this auxin that is mobile decreases linearly with distance from the apex.

In pea epicotyls, the regions of greatest growth are those in which auxin disappears from the transport system without a corresponding decrease in the total auxin of the tissue (Scott and Briggs, 1960). This was also found in a developing rachis of a fern (Steeves and Briggs, 1960). Since immobile auxin is almost certainly unavailable for growth (Went, 1942; Winter and Thimann, 1966), growth may be closely associated with the process of conversion of auxin from a mobile to an immobile form (Scott and Briggs, 1960).

The velocity of auxin movement

Van der Weij (1932) found that the amount of auxin in a basal receiver increases nearly linearly with time. Assuming that the intercept of this transport on the time axis represented the time for auxin molecules to traverse the section, he divided the distance travelled by the time required for passage (Fig. 4.3). Under

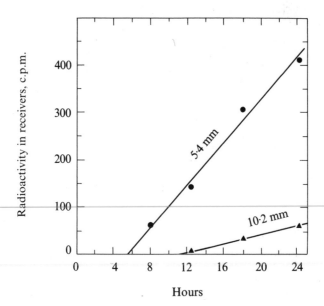

Figure 4.3 Determination of velocity of basipetal movement of 2,4-D in bean petioles by the intercept method. The extrapolated intercept with the time axis is 5·5 and 11·6 hours for sections 5·4 and 10·2 mm long respectively; the velocities 0·98 and 0·88 mm hr^{-1} respectively. This verifies the independence of the calculated velocity on the length of the section. (From McCready, 1963.)

the fairly limited conditions so far tested (Went and White, 1939; McCready, 1963; McCready and Jacobs, 1963a; Pilet, 1965), this quotient is independent of (a) distance (i.e., the length of the section), and (b) the concentration of the donor, and can be thought of as a velocity. The fact that the transport of IAA has a particular velocity as would a particle or molecule carried along in a moving stream indicates that the rate-limiting step in auxin movement does not possess the kinetics of a diffusion process.

The velocity obtained for IAA by this method varies from 5 to 20 mm h^{-1}. Presumably auxin moves in the cytoplasm where protoplasmic streaming would participate in the transport. The minimum velocity of protoplasmic streaming in oat coleoptiles, 36 mm h^{-1} (Thimann and Sweeney, 1937) is thus more than sufficient to account for the velocity of auxin movement. Presumably if proto-plasmic streaming participates in transport, the time consumed in moving from one cell to the next must account for the considerable differences between the velocity of streaming and transport. Furthermore, synthetic auxins and a cytokinin move with rather different velocities from IAA (Table 4.2). This fact indicates that the rate limiting step has specificity.

The slope of the time course is also of interest (Fig. 4.3). This gives the auxin flux, the amount of auxin moved divided by the time, and is often referred to as the *capacity* of transport.

Over the years, van der Weij's method has been routinely employed to measure the velocity of auxin movement, and the possible limitations of this method have often been overlooked. With the advent of radio-isotopes, the sensitivity with which auxin could be detected increased, and the flux into receivers was often found to change with time (de la Fuente and Leopold, 1966; Thornton and Thimann, 1967). One suggestion has been to use the intercept on the time axis of the linear portion of the time course to compute the velocity (McCready, 1966); however, by this method, one measures the time for the steady state to be achieved rather than the time for a particular group of mole-cules to traverse the section. Put another way, we simply do not know exactly when the molecules appearing in the receiver during any specific time interval actually left the donor.

Pulse experiments alleviate the difficulty of knowing how much of the measured time is required to traverse the section and how much to establish the steady state concentration, because the same group of auxin molecules can be detected at increasingly distant points within the tissue. In these experiments, the donor containing ^{14}C auxin is supplied for only a few minutes and the movement of the ^{14}C pulse down the section is followed after the radioactive donor is removed. The shape of the pulse changes as it migrates down the coleoptile; nevertheless, its peak can be identified (Fig. 4.4). The pulse travels down corn coleoptiles at 12 to 15 mm h^{-1} (Goldsmith, 1967a), which agrees with the velocity determined for this tissue by the classical method (Hertel and Leopold, 1963b). So far this method has been successfully applied only to move-ment of IAA in sections of corn coleoptiles.

The velocity of auxin transport in decapitated but otherwise intact oat and corn coleoptiles has also been measured electrically (Newman, 1963). Within a minute after IAA is applied to the decapitated stump, a series of damped oscil-lations in electric potential migrate down the coleoptile. These oscillations have a wavelength of about 5 mm and a maximum amplitude of 10 to 16 mV measured over a 2 mm interval. The electrical disturbance moves downward at about 14 mm h^{-1} in *Avena* and is apparently associated with the transport of auxin. Evidence supporting this view is (a) the electrical disturbance is not propagated upward when auxin is supplied to the base of the coleoptiles; (b) a similar disturbance occurs on a unilateral illumination of intact coleoptiles; (c) the

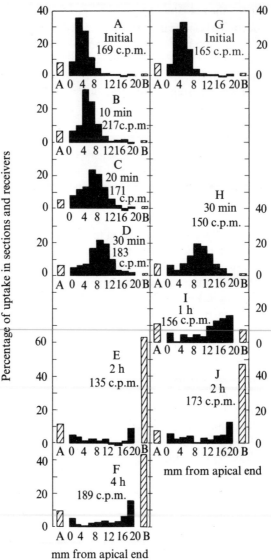

mm from apical end

Figure 4.4 Movement of a pulse of ^{14}C-IAA down aerobic sections of corn coleoptiles. Experiment 1, A–F; experiment 2, G–J. Donors containing ^{14}C-IAA supplied for 15 minutes. Fifteen minutes after removal of donors (A, G), the peak of the radioactive pulse is located several millimetres below the cut surface of the section. In the next hour the peak migrates downward at 12–15 mm hr^{-1} (B–D, H, I). Note the spreading out of the pulse during basipetal movement. Eventually the major part of the ^{14}C appears in the basal receiver. The fall in content of the receiver between 2 and 4 hours indicates some acropetal uptake. The histograms show the percentage of the total uptake in each 2 mm piece of the section and in apical (A) and basal (B) receivers. The apical receiver was present only during the first 15 minutes after the ^{14}C donor was removed. (From Goldsmith, 1967a.)

threshold concentration for evoking the disturbance is about 0·1 μM, or about the same as the threshold for the stimulation of growth which depends on transport; (d) a disturbance is not produced by 2,4-D which moves much slower than IAA (Table 4.2).

Table 4.2   Some typical velocities of basipetal movement determined by the intercept method

Tissue	Substance	Approximate velocity (mm h^{-1})	Reference
Zea coleoptile	IAA	14	Hertel and Leopold, 1963b
Zea root	IAA	6	Hertel and Leopold, 1963b
Avena coleoptile	IAA	8–15	Van der Weij, 1932; Went and White, 1939
	NAA	3·7	Went and White, 1939
	IBA	6·6	Went and White, 1939
	Anthracene acetic acid	5·3	Went and White, 1939
Helianthus epicotyl	IAA	7·5	Leopold and Lam, 1961
	NAA	6·7	Leopold and Lam, 1961
	IBA	3·2	Leopold and Lam, 1961
Phaseolus epicotyl	IAA	5·7	McCready and Jacobs, 1963a
	2,4-D	0·6–1·0	McCready, 1963
	BA (a kinin)	0·7	Black and Osborne, 1965
Arachis gynophore (an intercalary meristem)	IAA	10	Jacobs, 1961

These three methods, van der Weij's, Goldsmith's and Newman's represent the best available for measuring the velocity of auxin transport. The latter two yield velocities of IAA movement consistent with those obtained by the classical procedure.

Inhibition of auxin movement

Studies with inhibitors also support the view that the basipetal transport has a specific metabolic component. Unfortunately, however, the results obtained with inhibitors are frequently open to more than one interpretation, and most of the data are of a qualitative nature. In general, the inhibitors fit into one of three categories: (a) other auxins and structurally related molecules that are assumed to compete with IAA for a specific transport site, (b) 2,4-DNP, lack of carbohydrate, and respiratory inhibitors that reduce the supply of metabolic energy, (c) exposure to ethylene or lack of auxin that apparently leads to progressive deterioration of transport.

Auxins and structurally related substances
A number of auxins and related substances including IAA itself, 2,4-D and other substituted phenyoxyacetic acids, TIBA and other substituted benzoic acids, and NPA inhibit the basipetal movement of IAA (Zwar and Rijven, 1956; Niedergang-Kamien and Leopold, 1959; Hertel and Leopold, 1963b; Keitt and Baker, 1966). Evidence for the view that these substances compete for sites on the transport system is not strong. It would be useful to know if the inhibition is competitive,* decreasing with increasing IAA concentration. Variation of inhibition with IAA concentration has only been studied in corn coleoptiles (Christie and Leopold, 1965b). The inhibition of entry and exit of IAA in the presence of the weak auxin NPA and the strong one, 2,4-D, is apparently non-competitive, and in this respect, their effects are not separable from DNP, an uncoupler of oxidative phosphorylation or PCMB, an inactivator of sulphydryl enzymes.

TIBA is known to be a weak auxin (Thimann and Bonner, 1948) and to combine with sulphydryl groups (Niedergang-Kamien and Leopold, 1957). The first indirect evidence that it inhibits auxin transport was that both apical dominance and elongation are reduced below a ring of TIBA (Kuse, 1953, 1954). TIBA also affects the polarity of callus formation (Niedergang-Kamien and Skoog, 1956). If cylinders of tobacco stem are planted on a medium of moderate auxin concentration, callus forms at their base; but in the presence of TIBA callus is nearly uniformly distributed. Furthermore, the effect of TIBA on auxin movement in such cylinders is two-fold: (a) endogenous auxin no longer accumulates at the base of the cylinders, and (b) the amount of auxin extracted by ether from the cylinders decreases about three-fold. Since callus formation requires auxin, the decreased polarity in the distribution of endogenous auxin may be causally related to the decreased polarity of callus formation.

Not only TIBA but other inactivators of sulphydryl groups inhibit auxin transport (Niedergang-Kamien and Leopold, 1957). Since these substances are generally more effective in reducing transport and growth of isolated sections than in inhibiting respiration (Thimann and Bonner, 1949; Niedergang-Kamien and Leopold, 1957; but cf., McCready, 1966), Leopold and his co-workers suggest that TIBA affects IAA transport directly perhaps by combining with a hypothetical transport site. Alternatively, TIBA might inhibit transport indirectly by stimulating reactions that reduce the amount of auxin in the tissue available for transport. This latter suggestion receives support from the discovery that the presence of TIBA in either donor or receiver leads to an increase in the immobilization of auxin (Winter, 1967). Certain results of TIBA treatment including the promotion of uptake (Hertel and Leopold, 1963b), the decrease in extractable auxin (Niedergang-Kamien and Skoog, 1956), and more movement of ^{14}C-IAA through a TIBA than a plain lanolin ring (Vardar, 1959) are also consistent with an indirect action of TIBA on transport.

* Non-competitive inhibition does not eliminate the possibility that the inhibitor is combining with a transport site, but competitive inhibition would favour this hypothesis.

Respiratory inhibitors

The maximum rate of transport is achieved when respiration is fully functioning. Transport is inhibited by cyanide, 2,4-DNP (duBuy and Olson, 1940; Niedergang-Kamien and Leopold, 1957), iodoacetate, azide (Reiff and Guttenberg, 1961), reduced oxygen tension (Gregory and Hancock, 1955; Goldsmith, 1966a; Wilkins and Martin, 1967), and depletion of endogenous carbohydrates (Guttenberg and Zetsch, 1956).

Reduced oxygen tensions have been most frequently employed to investigate the relationships between the movement of auxin and aerobic metabolism. The results from various experiments are qualitatively similar: when nitrogen is substituted for air, less auxin is recovered in the basal receivers. In most cases, however, it is impossible to decide whether failure to achieve full inhibition occurred because conditions were not sufficiently anaerobic or because some transport persists anaerobically.

The basipetal transport to receivers in *Avena* coleoptiles is nearly independent of oxygen concentration between 21 and about 5 per cent oxygen and is only 50 per cent inhibited at about 2 per cent oxygen (Goldsmith, 1968b). This fact points up the difficulty of distinguishing experimentally between a residual aerobic transport and an anaerobic transport. Equilibrating bulky tissues or

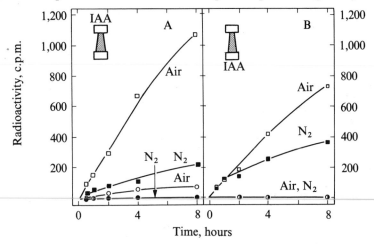

Figure 4.5 Effect of anaerobic conditions on basipetal (A) and acropetal (B) movement of IAA in 10 mm sections of oat coleoptiles. Only for basipetal movement through aerobic sections (open symbols) does significant activity appear in receivers (circles) during 8 hours. Anaerobic conditions (solid symbols) reduce the uptake (squares) immediately at the apical end of the tissue but not for nearly two hours at the base. Absolute rates of acropetal and basipetal uptake are not directly comparable as the experiments were done at different times and not corrected for the differences in specific activity or cross-sectional area of the tissue. Uptake from 10 μM ^{14}C. IAA. The insets show the direction of movement. (Adapted from Goldsmith, 1966a.)

□—uptake in air   ○—movement into receiver in air
■—uptake in N_2 ●—movement into receiver in N_2.

coleoptiles that have a large central hollow space with an anaerobic atmosphere may require considerable time, especially if the cut surfaces are covered with agar blocks. For this reason a wise precaution is to remove air from tissues and agar blocks by evacuating and releasing to nitrogen. Evacuation and release to atmospheric pressure does not affect the transport in aerobic tissues (Goldsmith, 1966a).

Using these techniques, apical uptake (2 hours) by *Avena* coleoptiles is reduced 75–85 per cent (Goldsmith, 1966a; Wilkins and Martin, 1967), (Fig. 4.5A), and about 75 per cent in corn coleoptiles (Wilkins and Martin, 1967). No activity penetrates to receivers of either 5 or 10 mm *Avena* sections (Fig. 4.5A); activity at the base of 5 mm corn sections is reduced 45 per cent, and none is found at the base of 10 mm sections. Furthermore, within 10 minutes after transfer from air to nitrogen, movement of a pulse of auxin within corn coleoptiles falls from about 15 mm h^{-1} to 1–2 mm h^{-1}. This inhibition is completely reversed by return to air (Goldsmith, 1967a, b). These pulse experiments clearly demonstrate inhibition of transport; uptake of IAA could not be affected because it is completed before the transfer to N_2. In summary, under anaerobic conditions, polarity disappears in oats, and movement is consistent with diffusion (Goldsmith, 1966a), whereas in corn coleoptiles some polar transport is apparently maintained by anaerobic metabolism (Goldsmith, 1967a; Wilkins and Martin, 1967; Wilkins and Whyte, 1968).

Indirect inhibition by ethylene or decreased auxin supply

If the plants are exposed to ethylene for 15 hours prior to the transport test (Morgan and Gausman, 1966) transport into receivers is inhibited by about 90 per cent; however, if the gas is only present during the 3-hour transport period, transport is not affected (Burg and Burg, 1966; Abeles, 1966). About 5 hours in ethylene are required to inhibit the transport system, and this accounts for its ineffectiveness on isolated sections. By this time transport has also ceased in isolated control sections. The progressive inhibition exerted by ethylene suggests it is not a direct inhibitor but rather affects the maintenance of the transport system.

Increased auxin concentrations induce the formation of ethylene (Burg and Burg, 1966). If ethylene in turn reduces auxin transport, this would lower the endogenous auxin concentration in the tissue at sites removed from the place of synthesis. This type of interaction between auxin and ethylene could provide feedback regulation of auxin levels in the tissue.

A supply of endogenous or applied auxin is necessary to maintain a fully functional transport system. Transport in sections from hypocotyls of sunflower seedlings falls 40 per cent after decapitation, but returns to above normal after lateral buds at the cotyledonary node start to grow (Leopold and Lam, 1962). Substitution of NAA for the intact apex prevents the decline in transport after decapitation. Similarly pea epicotyls whose endogenous auxin has been depleted for 12 hours do not move auxin until given a massive dose of 0·1 per cent IAA in lanolin (Scott and Briggs, 1962).

Attempts to identify a metabolically dependent step

Interest has centred recently on locating the metabolically controlled step in auxin transport. This is not an easy task because inhibiting any step in the sequence: uptake, movement through the section, and exit to the receiver, may reduce flux through preceding as well as subsequent steps. As a general rule, basipetal transport into the receiver is more effectively reduced if an inhibitor is applied to the basal than to the apical end of a section (Christie and Leopold, 1965a, b; Winter, 1967). Christie and Leopold have attempted to account for this result using a modified experimental system to separate cellular uptake from exit. Just as auxin began arriving in the receiver, they applied various inhibitors to the base of the section for 5 to 10 minutes and found a pronounced decrease in the amount of auxin collected in the receiving solution. On the other hand, application of the same inhibitors at the apical end for a similar time did not appreciably alter entry into cells. Therefore they suggest that entry into cells is a passive process whereas exit from the base of the cell is the metabolic step. This interpretation, however, is not beyond criticism. (a) By restricting the choices to cellular uptake or exit, passage across the cell is arbitrarily dropped from consideration as a possible site of inhibition. (b) In these experiments, the apical cells where uptake is measured and the basal ones where exit is determined are 5 mm apart. Coming from different parts of the plant, these cells may have different sensitivities to inhibitors. (c) In addition the apical and basal cells are under different conditions; most notably the concentration of mobile auxin in apical cells is 10 to 100 times that in basal ones. Thus differences attributed to different sensitivities of cellular uptake and exit might be due either to differences in sensitivity of transport in apical and basal cells or differences in auxin concentration. (d) The distinction between cellular uptake and exit on metabolic grounds might be more apparent than real and depend simply on the length of the section. This is because the contribution of diffusion to the total flux falls rapidly with distance from the donor block. In conclusion, appropriately controlled experiments are needed to verify the suggestion that cellular uptake is passive while exit is metabolic. These two processes should be compared on similar cells transporting similar auxin concentrations, and the sensitivity of exit to receivers should be compared on sections of different lengths.

The polarity of auxin transport

The easiest and most unequivocal way to demonstrate that a specific transport is involved in auxin movement is to show that this transport is polar. Of course, in comparing basipetal and acropetal transport, transport should be expressed per unit cross-sectional area of section at the donor end. The degree of polarity of transport can be expressed as the ratio of basipetal to acropetal transport, but with isolated sections this ratio has no fundamental significance since it depends on the experimental conditions. The degree of polarity of isolated sections is a function of length of section, period of transport, concentration, temperature, orientation, and physiological state of the tissue, for each of these variables affects acropetal and basipetal transport differently.

Much remains to be learned about both the molecular specificities and the tissues capable of polar movement. A single report exists that the ions PO_4^{3-}, Br^-, and Na^+ do not move polarly in coleoptiles (Went, 1939); and although there is a recent report that sucrose and mannitol move polarly in isolated bean hypocotyls, the polarity was not consistently observed in all the experiments (Zaerr and Mitchell, 1967). The polarity of movement of small organic molecules common to plant cells—organic acids, amides, amino acids, and sugars— remains to be investigated.

Auxin moves polarly in internodes, petioles, hypocotyls, coleoptiles, leaves, and roots (Went and Thimann, 1937; Jacobs, 1961; Leopold, 1963). In roots, polarity may be the reverse of that in the shoot, acropetal movement being greater than basipetal (McCready, 1966). Auxin moves polarly in isolated segments containing only parenchyma, but other cells may also be capable of polar transport (Jacobs and McCready, 1967; Goldsmith, 1968a).

A basic question raised by polarity is whether the mechanism of acropetal and basipetal transport is the same or different. Both probably have physical components; the question is whether or not acropetal transport has a significant metabolic component. Experiments with transport of IAA in coleoptiles and 2,4,5-trichlorophenoxyacetic acid in bean petioles give clear results. Polarity disappears in narcotized (van der Weij, 1934) or anaerobic sections of oat coleoptiles (Goldsmith, 1966a; Wilkins and Martin, 1967) as well as in sections of bean petioles treated in both donors and receivers with the inhibitors, naphthylphthalamic acid, fluorenolcarboxylic acid, TIBA, or sodium azide (McCready, 1968). In all these cases, the inhibitory treatments greatly reduce basipetal transport without much effect on acropetal movement. This result implies that for oat coleoptiles and bean petioles the mechanism of auxin movement must be different in the two directions and that acropetal movement lacks a significant metabolic component. The effect of an increase in temperature from 18 to 25°C on transport through 5·4 mm sections of bean petioles is also consistent with this conclusion. The temperature increment has a negligible effect on the amount of acropetal transport of either IAA or 2,4-D, whereas basipetal transport is increased (McCready, 1968).

Evidence for interaction between basipetal transport and acropetal movement

In a section where acropetal movement is predominantly diffusion and basipetal movement a metabolic transport, polarity would be increased if basipetal transport recycled acropetally moving auxin. Direct evidence that recycling occurs comes from experiments with oat coleoptiles (Goldsmith, 1966b). The sections are preloaded anaerobically from donors at their bases and then suspended without basal receivers after removing the donors. In anaerobic sections, movement continues acropetally down the existing concentration gradient, but this movement is reversed on transferring these sections to air because the basipetal transport is reinstated. In similar experiments with apical donors, acropetal movement into apical receiver (after removal of the donor) increases when oxygen is removed. In both experiments, net acropetal movement either within the section or from section into receiver occurs in the direction of the prevailing concentration gradient.

Views on the mechanism of transport

Evidence obtained primarily from only two tissues, oat coleoptiles and bean petioles, indicates that auxin is transported polarly in isolated 5–10 mm sections because basipetal movement involves a specific metabolic transport. This polar transport recycles basally some of the acropetally diffusing auxin in each cell. The effects of anaerobic conditions (Goldsmith, 1966a; Wilkins and Martin, 1967) as well as various inhibitors and promotors (McCready, 1968), length of section, duration of transport, concentration of donor (McCready, 1963; McCready and Jacobs, 1963a) and temperature (McCready, 1968) are all consistent with the view that basipetal transport in 5 to 10 mm sections is primarily metabolic but acropetal movement is diffusion.

Close examination of some experimental data indicates this is perhaps an oversimplified view of a complex system. For example, initially acropetal uptake is not much affected anaerobically; but after several hours, more auxin has been taken up by aerobic than anaerobic sections (Fig. 4.5B). This is probably because immobilization and protoplasmic streaming, which help maintain a high concentration gradient from donor to tissue, are both aerobic processes (Goldsmith, 1966a; Eymers and Bottelier, 1937). Despite this inhibition of uptake, however, the fraction of uptake that moves acropetally out of the basal millimetre of the section is greater in N_2 (Fig. 4.6D) than air (Fig. 4.6C). This is expected since auxin moving acropetally in an anaerobic section would be subject to neither metabolic basipetal transport nor immobilization. Basically acropetal movement may be diffusion, but this diffusion is most surely modified by one or more metabolic processes, such as protoplasmic streaming, immobilization, and basipetal transport. This situation could accommodate a wide range of variation in the effects of a particular treatment on acropetal movement (e.g., TIBA, see below).

A divergent view is that metabolic transport exists in both directions. De la Fuente and Leopold (1966) have elegant data for four different tissues showing a logarithmic increase of polarity with length of section. This relationship leads

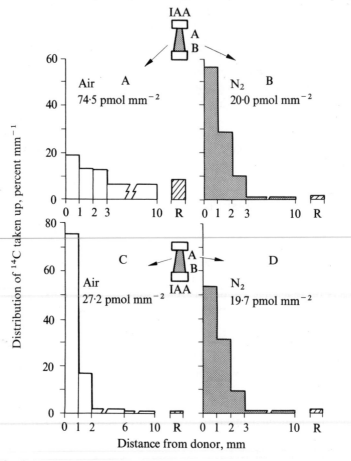

Figure 4.6 Comparison of basipetal (A, B) and acropetal (C, D) movement of IAA in sections of oat coleoptiles. For aerobic sections, the amount of uptake, distribution in the section, and amount reaching the receiver (R) all reflect greater basipetal (A) than acropetal (C) movement. Aerobic apical uptake is about 3 times basal and the gradient of ^{14}C in the section is much shallower for basipetal than acropetal movement. These differences disappear anaerobically indicating loss of polarity. With acropetal movement (C, D) anaerobic conditions inhibit uptake at the basal end of the section, but a larger fraction moves acropetally. The uptake (pmoles mm^{-2}) during 4 hours at 10 μM ^{14}C-IAA is given for each experiment. Note that the percentages indicated in the bar graphs must be multiplied by the length to obtain the total percentage in any portion of the section; e.g., in A the first millimetre of the section accounts for 19 per cent of the uptake, while the region 3–10 mm contains 6·5 per cent mm^{-1} or 45 per cent of the total uptake. The insets show the direction of movement. (Adapted from Goldsmith, 1966a.)

to the suggestion that a small difference in the polarity of transport through an individual cell may be amplified many times by passing through a longitudinal file of similar cells (Leopold and Hall, 1966). For example, corn coleoptiles are noteworthy for a high degree of polarity; the basipetal movement into receivers of a 4 mm section is 54 times the acropetal movement. As these sections are only 31 cells in length, a calculated difference of only 3 per cent in the acropetal and basipetal movement through an individual cell is sufficient to produce the observed polar ratio (de la Fuente and Leopold, 1966).

Some evidence has been presented in support of a metabolic acropetal transport. (a) The time for transport through 2 to 4 mm sections in several tissues appears similar in both directions (de la Fuente and Leopold, 1966). (b) TIBA, a powerful inhibitor of basipetal transport inhibits a fraction of acropetal movement (Leopold, 1963; Keitt and Baker, 1967).

The similar time of arrival in either direction through short sections is not so much evidence for metabolic transport in both directions as for a similar mechanism of transport in both directions. Although the basipetal transport of auxin has the properties of a stream, and the time required for auxin to pass basipetally through a section is proportional to the length of the section, this is not so for acropetal transport. When the length of a 3 mm section of coleoptile is doubled, much more than twice the time is required for acropetal movement (Went and White, 1939). With longer sections, de la Fuente and Leopold (1966) also obtained different arrival times for basipetal and acropetal movement. Particularly with the highly polar 6 mm sections of sunflower hypocotyl, auxin arrives at the base in less than one hour, whereas none arrives at the apex of either 4 or 6 mm sections after as much as 3·5 hours of transport. In other words although basipetal transport has a velocity, acropetal transport, like diffusion, does not. Thus with short sections, which have a small polar ratio, diffusion to receivers may well predominate in both directions, but may become insignificant with longer sections.

Secondly, inhibition by TIBA is not sufficient proof of metabolic transport. Since TIBA increases immobilization of auxin (Winter, 1967), diffusion as well as metabolic transport might be reduced. In some cases TIBA treatment is not sufficient to abolish polarity (Leopold, 1963), but when it does its inhibition of acropetal movement is usually small compared to basipetal movement (Keitt and Baker, 1967). On the other hand, TIBA abolishes polarity in bean petioles while promoting acropetal movement (McCready, 1968).

Recycling of acropetally diffusing auxin presents a reasonable alternative to metabolic transport in both directions. If each cell of the section re-exports basally some of its acropetally moving auxin, this would give rise to an exponential relation between length of section and polarity of transport.

The acropetal movement has been carefully examined in only a relatively few tissues. Even with bean petioles and coleoptiles, the possibility has still not been excluded that a small fraction of acropetal movement is metabolic. While diffusion is the most likely explanation for the major portion of the acropetal movement in these two tissues, our view may broaden as more tissues are examined under more conditions.

Effect of light

Depending on the nutritional status of the tissue, light either promotes or inhibits basipetal movement. Etiolated coleoptiles (Naqvi and Gordon, 1967; Thornton and Thimann, 1967) and green sunflower plants (Lam and Leopold, 1964) are not nutritionally limited, and here light apparently inhibits basipetal transport. On the other hand, after 40 to 50 hours in darkness, stem tips of sunflower, tomato, and tobacco no longer transport endogenous auxin to receivers even though the amount of auxin extracted from these tips with acidic ether increases. Clearly in this case, transport fails although synthesis of auxin continues. Similarly, sections cut from sunflowers after several days in darkness fail to transport exogenously applied auxin (Guttenberg and Zetsche, 1956), and etiolated pea epicotyls move auxin less readily than light-grown plants (Scott and Briggs, 1963). Spraying darkened sunflower plants several times daily with sucrose solutions restores auxin transport in both the apices and isolated sections to normal levels. In this case, light is promoting transport by supplying photosynthetic products.

In one case the effect of light on both acropetal and basipetal movement has been examined. Irradiation of green sections of pea epicotyls during uptake reduces polarity by promoting acropetal more than basipetal movement (Thimann and Wardlaw, 1963).

Effect of inversion

Although inversion of the section may not appreciably affect basipetal movement or polarity in some cases (van der Weij, 1932; Skoog, 1938; Pilet, 1965), this is not always the case (Naqvi and Gordon, 1966; Little and Goldsmith, 1967). During inversion of oat coleoptiles for 8 hours, basipetal transport is inhibited 66 per cent, but acropetal movement is promoted (Fig. 4.1). After about 4 hours, the fluxes into apical and basal receivers of inverted sections are about equal indicating that inversion can lead to loss of polarity. The longitudinal growth of the main axis decreases when its orientation is altered from the vertical (Anker, 1960; Little and Goldsmith, 1967). This decrease in growth may be the result of the decreased basipetal transport of auxin.

Longer term experiments are possible with cuttings that are inverted and rooted at the apical end. Although during 9 weeks, the original basipetal transport in inverted cuttings of *Tagetes* did not change, after 5 weeks, the acropetal transport rose to over 60 per cent of the basipetal (Went, 1941).

Table 4.3 Changes in auxin movement during development.

Tissue	Developmental change	Transport period (h)	Length of section (mm)	Basipetal movement	Acropetal movement	Change in degree of polarity (B/A) during development	Reference
Petioles *Phaseolus*	elongation from 10–75 mm	4	5·4	no change	increases	230–1·7	McCready and Jacobs, 1963b
Coleus	from leaf 2–8	up to 8	5·0	decreases (ca. 25 fold)	undetectable	(not determinable)	Jacobs, 1967
Hypocotyl *Phaseolus*	elongation from 15–70 mm	3	5·0	increases (ca. 4–5 fold)	undetectable	(not determinable)	Jacobs, 1950
Internodes *Coleus*	vegetative–flowering apex	3	5·0	decreases	increases	3–1·3	Naqvi and Gordon, 1965
	2nd to 5th	5	5·0–7·0	increases 1·2–3 fold depending on concentration	decreases to undetectable	3–(not determinable)	Jacobs, 1967

Effect of physiological state and maturity of the tissue

In a number of tissues, the degree of polarity changes during development (Table 4.3). Such changes may be the result of alteration in the amount of either acropetal or basipetal movement.

In the development of bean petioles acropetal movement increases without any change in basipetal movement (Table 4.3). In this tissue, the pith breaks down so that the mature petiole is hollow, and it seems likely that these morphological changes might lead to increased acropetal diffusion rather than increased metabolic transport (McCready and Jacobs, 1967).

Relation of IAA movement to tropisms in coleoptiles

This discussion will be limited to coleoptiles because extensive experimentation with them has led to a relatively clear understanding of the relationships between auxin movement and tropisms in these organs. Geotropic or phototropic stimulation results in lateral movement of endogenous auxin in the stimulated coleoptile (Went and Thimann, 1937; Briggs et al., 1957; Gillespie and Briggs, 1961). Lateral movement has also been unequivocally demonstrated with ^{14}C-IAA by two techniques. In the first method, a donor is applied to the apical surface of a longitudinally split coleoptile and the radioactivity moving laterally collected in a receiver applied to the longitudinal cut surface (Hertel and Leopold, 1963a; Gillespie and Thimann, 1963). In the second method, the donor is applied asymmetrically to sections; and after a period of transport, the section is split longitudinally, and the distribution of auxin in each half of the tissue determined for vertical controls and horizontal sections (Goldsmith and Wilkins, 1964). With both methods, lateral movement is 2–3 times greater in horizontal than vertical sections. Furthermore, lateral movement in horizontal sections is polar: as much as five times more lateral movement occurs downward as upward.

In still other experiments, ^{14}C-IAA applied symmetrically to corn coleoptiles given either geo- or phototropic stimuli emerges asymmetrically into receiver blocks at the basal end of the tissue. In each case the side that would grow faster in executing the tropic curvature delivered more auxin to the receiver. Since recovery of the ^{14}C applied is nearly complete, such asymmetries are also consistent with lateral movement (Gillespie and Thimann, 1963; Pickard and Thimann, 1964).

Early failures to detect lateral movement with ^{14}C-IAA can be attributed to the experimental design (Ching and Fang, 1958; Reisener, 1958; Reisener and

Simon, 1960). Since at that time experimenters were unaware of the existence of immobile auxin, they supplied the ^{14}C-IAA before instead of during the tropic stimulation, and determined the distribution of the total ^{14}C in the tissue rather than the transported ^{14}C. Under these conditions, most of the ^{14}C-IAA in the tissue is likely to be immobile, obscuring any redistribution of mobile auxin.

As with the longitudinal movement of auxin, potential gradients are not implicated in lateral auxin movement. Recent evidence indicates that the transverse electric potentials which develop after tropic stimulation are the result, rather than the cause, of the auxin asymmetry (Grahm, 1964; Wilkins and Woodcock, 1965).

Although both lateral and longitudinal movement of auxin are polar, suggesting the basic mechanism of movement in the two systems might be similar, they react differently to the hormone ethylene. In peas, lateral movement is immediately inhibited by ethylene, but longitudinal movement is only affected by prolonged treatment (Burg and Burg, 1966). In coleoptiles, geotropism and lateral movement are insensitive, but as with peas, longitudinal transport is inhibited by long exposures to ethylene.

Significance of polar transport

Polar transport arouses interest not only because the process is intrinsically interesting but because it might be related to the morphological polarity of plants and could be a controlling factor in growth and differentiation.

If polar transport is causally related to morphological polarity it should occur in developing embryos and be present from the time of germination onward. The time during development at which polar transport makes an appearance has not received much attention. In 2–3 day old bean hypocotyls, polar transport of auxin apparently does not occur; equal amounts of endogenous auxin are collected in apical and basal receivers, and apparently auxin does not move through a 5 mm section from donor to receiver in 3 hours. Basipetal transport of applied auxin becomes detectable with sections from 5 day old seedlings, and polar transport continues to increase during the next three days (Jacobs, 1950). The important point is that the morphological polarity of the embryo and seedling precede the development of polar transport. Thus polar transport is more likely to be a physiological manifestation of the inherent longitudinal polarity of the plant than a cause of this polarity.

The elegant investigations on the control of vascular differentiation in *Coleus* are one of the most complete descriptions available of the interaction between auxin transport and differentiation. The endogenous auxin of *Coleus* is produced predominantly by the young leaves (Jacobs, 1952) and has been identified by chromatography and bio-assay as IAA (Scott and Jacobs, 1964). Auxin limits the regeneration of both xylem tracheids and phloem sieve tubes around a wound severing a vascular strand in the stem (Jacobs, 1952 and 1956; Thompson and Jacobs, 1966). When different numbers of young leaves above a severed vascular strand are removed, the number of xylem strands that regenerate is proportional to the amount of auxin produced by the remaining leaves (Jacobs, 1952 and 1956). Furthermore, the reduction in xylem regeneration when all the leaves above the wound are removed can be completely overcome if the amount of auxin they supply is replaced by applying IAA (about 10 μM) to the debladed petiole of one of the leaves (Fig. 4.7).

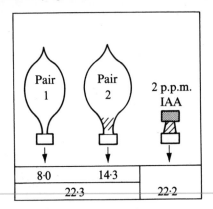

Auxin from Coleus leaves (units h^{-1})

Figure 4.7 Diagram to show the method of determining the auxin production of the first and second pair (from apex) of expanded leaves in *Coleus* and the exogenous concentration necessary to exactly replace the natural supply. The average amount of auxin obtained from leaf pair 1 plus 2 equalled the amount of basipetal transport by a petiole section (shaded portion) from one of the second pair of leaves given 2 mg l^{-1} IAA in the donor. One unit is equivalent to 10^{-11} g of IAA. (Reprinted from 'Internal factors controlling cell differentiation in the flowering plants' in *The American Naturalist* by W. P. Jacobs, **90**, 163–169, 1956, by permission of the University of Chicago Press. © 1956 University of Chicago.)

Normal xylem differentiation also appears to be controlled by auxin (Jacobs and Morrow, 1957). The rate of auxin production varies from leaf to leaf and correlates extremely well with the rate of production of xylem cells in the petiole.

If all the leaves are left above a wound in the second internode (that below the second pair of expanded leaves from the apex) about three times as much xylem regenerates in a week as when only leaves below the wound are present (Jacobs, 1956). The occurrence of acropetal regeneration in *Coleus* without any leaves above the wound is evidence for a physiologically effective acropetal transport of auxin.

Transport and vascular regeneration have been compared in isolated second and fifth internodes (Thompson and Jacobs, 1966; Jacobs, 1967). The second internode is rapidly elongating while the fifth internode has stopped increasing in length and cambial activity has just started. Both transport and regeneration can occur acropetally in the second, whereas both appear strictly polar in the fifth internode (Thompson and Jacobs, 1966). Furthermore transport is a somewhat different function of concentration in the two internodes, and regeneration reflects these differences (Fig. 4.8) (Jacobs, 1967). In internode 5, both

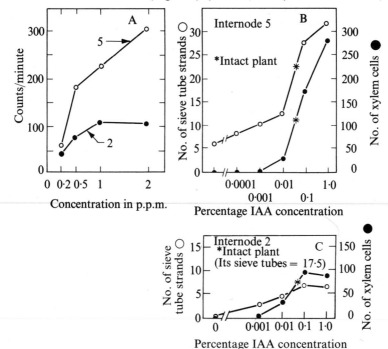

Figure 4.8 Comparison of the effect of donor concentration on the amount of basipetal auxin transport (A) and vascular regeneration (B, C) in internode 2 and 5 of *Coleus*. A—Basipetal transport through 7 mm sections to agar receivers during eight hours. B, C—Effect of IAA applied apically in lanolin to isolated second (C) and fifth (B) internodes on xylem and sieve tube regeneration around a wound. The amount of regeneration around a similar wound in an intact plant is indicated by the asterisks. (From W. P. Jacobs, 1967 in Annals of *The New York Academy of Sciences*, **144**, 102–117.)

transport and vascular regeneration continue to increase with increasing concentration. In internode 2, on the other hand, both transport and regeneration level off at high concentrations of auxin. This correlation is consistent with control of regeneration by auxin. The amount of auxin reaching the regenerating cell depends both on the amount supplied—either endogenously or exogenously —and the ability of the stem to transport it to the differentiating cells.

Experiments with *Coleus* have also shown that physiologically effective amounts of auxin are transported over the sorts of distances required in the intact plant (Thompson, 1966). Removing all the leaves and buds of *Coleus* plants completely eliminated xylem regeneration around a wound in the fifth internode and significantly reduced sieve tube regeneration, but application of

IAA to the decapitated shoot 3·5 internodes above the wound largely restored vascular regeneration (Thompson, 1966).

Movement of gibberellins and cytokinins

Less is known as to whether growth substances other than auxin move in the plant by (a) diffusion, (b) movement in xylem, (c) movement in the phloem, or (d) polar transport. The effect of concentration, distance, and duration of transport, as well as metabolic inhibitors on the movement of these hormones has not been systematically investigated. Furthermore, it is well to remember that auxins applied to shoots and roots may move in both xylem and phloem (Skoog, 1938), whereas endogenous auxin moves by polar transport. Thus the transport in the vascular tissue of gibberellic acid (McComb, 1964) and of kinetin (Lagerstedt and Langston, 1967) applied to intact or semi-intact plants does not prove that the endogenous hormones are transported in this manner. Successful studies of the movement of endogenous cytokinins and gibberellins depend on knowing where in the plant synthesis of these hormones occurs.

Gibberellins

Gibberellins labelled with ^{14}C are apparently quite stable in the plant (McComb, 1964), and the rapid distribution of radioactivity to shoot and roots shown in radioautographs when ^{14}C gibberellins are applied to leaves is believed to reflect the distribution of gibberellins (Zweig et al., 1961; McComb, 1964). Gibberellic acid applied to a leaflet of a pea plant moves acropetally at 50 mm h^{-1}. The movement is probably in the phloem because the pattern of GA movement is reminiscent of other organic compounds; GA did not move into leaves opposite the one to which it was applied, it immediately moved out of mature leaves into immature ones, and it remained localized in an expanding leaf (McComb, 1964).

The movement of GA in isolated sections from etiolated pea epicotyls is not polar (Kato, 1958; Clor, 1967); however, sections of pea epicotyls are capable of much more uptake and movement of GA than are pieces of potato tuber (Clor, 1967).

Movement of endogenous gibberellin also appears to be non-polar. In peas and sunflowers, endogenous gibberellins synthesized in the stem tips (Lockhart, 1957; Jones and Phillips, 1966), probably in the young leaves, move basipetally as indicated by their effect on the growth of young internodes. Acropetal movement of endogenous gibberellin is apparent in sunflower where the growth of the fourth internode is reduced to half by excising the pair of leaves immediately below this internode (Jones and Phillips, 1966). In sunflowers, the region of maximum gibberellin synthesis correlates well with the amount of internode elongation. Apparently the gibberellin produced in the leaves influences the growth of immediately adjacent internodes, and the distances that gibberellin must be transported need not be great (Jones and Phillips, 1966).

In sunflowers, gibberellins are also synthesized in the apical 3 to 4 mm of the roots (Jones and Phillips, 1966) and have also been detected in the bleeding sap of sunflowers (Phillips and Jones, 1964) lupin, pea, and balsam plants (Carr, Reid, and Skene, 1964). This sap, which is low in sugars (Carr, Reid, and Skene, 1964) and can be collected above a region of the stem girdled by steam, is exuded from the xylem when a plant is cut off at ground level (Bollard, 1960). The presence of gibberellins in bleeding sap suggests that they are supplied by the roots via the transpiration stream to the stem and mature leaves. In all likelihood, they can then enter the phloem and move to immature portions of the shoot.

Cytokinins

Until recently the movement of cytokinins within plants was thought to be restricted. For example, direct application of kinetin to inhibited lateral buds of pea seedlings breaks apical dominance, whereas application as little as 2 mm from the inhibited bud is ineffective (Sachs and Thimann, 1964). Moreover, ^{14}C kinetin does not migrate readily in bean seedlings (Seth, et al., 1966).

In certain respects, the movement of the cytokinin, BA, in 5·4 mm sections of bean petioles appears similar to IAA (Black and Osborne, 1965). After 24 hours, about 5 per cent of the uptake of either hormone is recovered in basal receivers, but the degree of polarity is somewhat more for auxin (10 to 20) than for BA (about 7).

Unlike IAA, the movement of BA is not notably polar early in the experiment; but the polarity increases after about 10 hours, especially if IAA, which enhances basipetal but not acropetal movement of BA, is also supplied.

Although these results seem convincing, another laboratory could find no polarity of BA movement in similar experiments with a variety of tissues including petioles. This difference in results is presently inexplicable (Fox and Weis, 1965; Osborne and McCready, 1965).

Evidence is accumulating that, as with gibberellins, considerable amounts of endogenous cytokinins may also be synthesized in the roots and exported to the shoot via the xylem. The formation of adventitious roots on petioles delays senescence of detached leaves in the same way as applying cytokinins to their laminae. This suggests that roots might supply cytokinins to leaves (Möthes and Engelbrecht, 1963). This suggestion is further supported by the discovery of cytokinin activity in bleeding sap of several plants (Kende, 1965; Loeffler and van Overbeek, 1964). In sunflowers, the concentration of cytokinins in the exudate from the xylem remains high over a period of several days, suggesting that cytokinins must be continuously produced in the roots and translocated in the xylem to the shoot.

Hormonal Interaction in polar transport

Recent experiments indicate that different growth substances can interact to promote each other's movement. The basipetal movement of auxin in decapitated seedlings or isolated sections is enhanced by kinetin (Davies *et al.*, 1966; McCready, *et al.*, 1965) or GA (Jacobs and Case, 1965; Pilet, 1965), and similarly, the basipetal movement of kinetin in decapitated bean seedlings (Seth *et al.*, 1966) is promoted by IAA. IAA is particularly effective in enhancing the movement of BA in isolated petioles that have been aged for 18 hours without any growth substances (Black and Osborne, 1965).

In view of the promotion of movement of one growth substance by another, it is not surprising that in studies of apical dominance auxin supplied with either GA (Jacobs and Case, 1965) or kinetin (Davies *et al.*, 1966) is a more effective substitute for the intact apex than auxin alone. Nevertheless, the increased apical dominance in the presence of a second hormone might not be just a result of enhanced auxin transport. Cytokinins are especially noted for causing movement of amino acids, ions, and other substances toward their site of application, and similarly the movement of metabolites to the decapitated stump, and hence away from the laterals, is greater when both kinetin and auxin are present at the apex (Davies *et al.*, 1966).

Concluding remarks

Among plant hormones, the transport of auxin has been the most studied. Endogenous auxin as well as auxin applied to isolated sections is transported polarly. In shoot tissue the polar basipetal transport is metabolically dependent and may even be active transport in the strict sense of operating against an electrochemical potential gradient. The analysis of auxin transport is complicated by the presence of an aerobic reaction which apparently binds auxin to proteins so it is unavailable for transport. This immobilization, which becomes increasingly prominent at longer times and higher concentrations, probably accounts for the failure of transport to increase linearly with concentration. The velocity of transport varies for different auxins and is always less than protoplasmic streaming, so that the specific and rate limiting step may be located during passage from cell to cell rather than passage across a cell.

Available evidence suggests that the mechanisms of acropetal and basipetal transport are different. Under a variety of conditions, the amount of acropetal movement varies inversely with basipetal movement. In at least two tissues, oat coleoptiles and bean petioles, acropetal transport is largely non-metabolic and appears to be diffusion. The metabolic basipetal transport enhances polarity by recycling acropetally moving auxin. In young *Coleus* internodes, acropetally moving auxin is physiologically active in vascular regeneration.

Further understanding of the mechanism of polar transport now depends on exploring aspects such as molecular specificity, the tissues capable of polar transport, the path of transport on the cellular level, the chemical form of auxin during transport, and the role played by protoplasmic streaming, plasmodesma, and cellular membranes.

The situation with two other growth hormones, cytokinins and gibberellins is even less clear than for auxins. At times both of these substances apparently circulate in the vascular tissues of the plant, and it may be that the primary transport of these hormones is different from auxin. Neither endogenous gibberellins nor gibberellins applied to isolated sections have shown any indication of polar movement, and although cytokinins may show polar transport in sections, the conditions necessary for this transport are not understood.

Bibliography

Selected further reading

Goldsmith, M. H. M., 1968a. The transport of auxin. *Ann. Rev. Plant Physiol.* **19**, 347–360.

Jacobs, W. P., 1961. The polar movement of auxin in the shoots of higher plants: its occurrence and physiological significance. In *Plant Growth Regulation.* The Iowa State University Press, Ames, Iowa, pp. 397–409.

Leopold, A. C., 1963. The polarity of auxin transport. In *Meristems and Differentiation.* Brookhaven Symposia in Biology, No. 16, pp. 218–233.

McCready, C. C., 1966. Translocation of growth regulators. *Ann. Rev. Plant Physiol.* **17**, 283–294.

Went, F. W. and K. V. Thimann, 1937. *Phytohormones.* The Macmillan Company, New York.

Papers cited in text

Abeles, F. B., 1966. Effect of ethylene on auxin transport. *Plant Physiol.* **41**, 946–948.

Anker, L., 1960. On a geo-growth reaction of the *Avena* coleoptile. *Acta Bot. Neerl.* **9**, 411–415.

Arisz, W. H., 1958. Influence of inhibitors on the uptake and the transport of chloride ions in leaves of *Vallisneria spiralis. Acta Bot. Neerl.* **7**, 1–32.

Avery, G. S., Jr., 1935. Differential distribution of a phytohormone in the developing leaf of *Nicotiana*, and its relation to polarized growth. *Bull. Torrey Bot. Club*, **62**, 313–330.

Black, M. K. and D. J. Osborne, 1965. Polarity of transport of benzyladenine, adenine and indole-3-acetic acid in petiole segments of *Phaseolus vulgaris. Plant Physiol.* **40**, 676–680.

Bollard, E. G., 1960. Transport in the xylem. *Ann. Rev. Plant Physiol.* **11**, 141–166.

Briggs, W. R., R. D. Tocher, and J. F. Wilson, 1957. Phototropic auxin redistribution in corn coleoptiles. *Science*, **126**, 210–212.

Burg, S. P. and E. A. Burg, 1966. The interaction between auxin and ethylene and its role in plant growth. *Proc. Nat. Acad. Sci. U.S.* **55**, 262–269.

Carr, D. J., D. M. Reid, and K. G. M. Skene, 1964. The supply of gibberellins from the root to the shoot. *Planta*, **63**, 382–392.

Ching, T. M. and S. C. Fang, 1958. The redistribution of radioactivity in geotropically stimulated plants pretreated with radioactive indoleacetic acid. *Physiol. Plantarum,* **11**, 722–727.

Christie, A. E. and A. C. Leopold, 1965a. On the manner of triiodobenzoic acid inhibition of auxin transport. *Plant Cell Physiol.* **6**, 337–345.

Christie, A. E. and A. C. Leopold, 1965b. Entry and exit of indoleacetic acid in corn coleoptiles. *Plant Cell Physiol.* **6**, 453–465.

Clark, W. G., 1937. Polar transport of auxin and electrical polarity in coleoptiles of *Avena. Plant Physiol.* **12**, 737–754.

Clor, M. A., 1967. Translocation of tritium-labeled gibberellic acid in pea stem segments or potato tuber cylinders. *Nature, Lond.* **214**, 1263–1264.

Crafts, A. S., 1961. *Translocation in plants.* Holt, Rinehart and Winston, New York, 182 pp.

Dainty, J., 1962. Ion transport and electrical potentials in plant cells. *Ann. Rev. Plant Physiol.* **13**, 379–402.

Davies, C. R., A. K. Seth, and P. F. Wareing, 1966. Auxin and kinetin interaction in apical dominance. *Science,* **151**, 468–469.

de la Fuente, R. K. and A. C. Leopold, 1966. Kinetics of polar auxin transport. *Plant Physiol.* **41**, 1481–1484.

duBuy, H. G. and R. A. Olson, 1940. The relation between respiration, protoplasmic streaming and auxin transport in the *Avena* coleoptile, using a polarographic microrespirometer. *Am. J. Bot.* **27**, 401–413.

Etherton, B. and N. Higinbotham, 1960. Transmembrane potential measurements of cells of higher plants as related to salt uptake. *Science,* **131**, 409–410.

Eymers, J. G. and H. P. Bottelier, 1937. Protoplasmic movement in the *Avena* coleoptile as related to oxygen pressure and age. *Proc. Kon. Akad. Wetens.* Amsterdam. **40**, 589–595.

Fox, J. E. and J. S. Weis, 1965. Transport of the kinin, N_6 benzyladenine: Nonpolar or polar? *Nature, Lond.* **206**, 678–679.

Gillespie, B. and W. R. Briggs, 1961. Mediation of geotropic response by lateral transport of auxin. *Plant Physiol.* **36**, 364–368.

Gillespie, B. and K. V. Thimann, 1963. Transport and distribution of auxin during tropistic response. I. The lateral migration of auxin in geotropism. *Plant Physiol.* **38**, 214–225.

Goldsmith, M. H. M., 1966a. Movement of indoleacetic acid in coleoptiles of *Avena sativa* L. II. Suspension of polarity by total inhibition of the basipetal transport. *Plant Physiol.* **41**, 15–27.

Goldsmith, M. H. M., 1966b. Maintenance of polarity of auxin movement by basipetal transport. *Plant Physiol.* **41**, 749–754.

Goldsmith, M. H. M., 1967a. Movement of pulses of labeled auxin in corn coleoptiles. *Plant Physiol.* **42**, 258–263.

Goldsmith, M. H. M., 1967b. Separation of transit of auxin from uptake: average velocity and reversible inhibition by anaerobic conditions. *Science,* **156**, 661–663.

Goldsmith, M. H. M., 1968b. Comparison of aerobic and anaerobic movement of indole-3-acetic acid in coleoptiles. In *Biochemistry and Physiology of Plant Growth Substances,* Proc. 6th Intern. Conf. Plant Growth Substances, Rung Press, Ottawa: in press.

Goldsmith, M. H. M. and K. V. Thimann, 1962. Some characteristics of movement of indoleacetic acid in coleoptiles of *Avena.* I. Uptake, destruction, immobilization, and distribution of IAA during basipetal translocation. *Plant Physiol.* **37**, 492–505.

Goldsmith, M. H. M. and M. B. Wilkins, 1964. Movement of auxin in coleoptiles of *Zea mays* L. during geotropic stimulation. *Plant Physiol.* **39**, 151–162.

Gorter, Chr. J. and H. Veen, 1966. Auxin transport in explants of *Coleus. Plant Physiol.* **41**, 83–86.

Grahm, L., 1964. Measurements of geoelectric and auxin-induced potentials in coleoptiles with a fine vibrating electrode technique. *Physiol. Plantarum*, **17**, 231–261.

Gregory, F. G. and C. R. Hancock, 1955. The rate of transport of natural auxin in woody shoots. *Ann. Bot.* N.S. **19**, 451–465.

Guttenberg, H. von and K. Zetsche, 1956. Der Einfluss des Lichtes auf die Auxinbildung und den Auxintransport. *Planta*, **48**, 99–134.

Hertel, R. and A. C. Leopold, 1963a. Auxin relations in geotropism of corn coleoptiles. *Naturwiss.* **50**, 695–696.

Hertel, R. and A. C. Leopold, 1963b. Versuche zur Analyse des Auxintransports in der Koleoptile von *Zea mays* L. *Planta*, **59**, 535–562.

Horwitz, L., 1958. Some simplified mathematical treatments of translocation in plants. *Plant Physiol.* **33**, 81–93.

Jacobs, W. P., 1950. Auxin-transport in the hypocotyl of *Phaseolus vulgaris* L. *Am. J. Bot.* **37**, 248–254.

Jacobs, W. P., 1952. The role of auxin in differentiation of xylem around a wound. *Am. J. Bot.* **39**, 301–309.

Jacobs, W. P., 1956. Internal factors controlling cell differentiation in the flowering plants. *Am. Nat.* **90**, 163–169.

Jacobs, W. P., 1967. Comparison of the movement and vascular differentiation effects of the endogenous auxin and of phenoxyacetic weedkillers in stems and petioles of *Coleus* and *Phaseolus*. *Ann. N.Y. Acad. Sci.* **144**, 102–117.

Jacobs, W. P. and D. B. Case, 1965. Auxin transport, gibberellin and apical dominance. *Science*, **148**, 1729–1731.

Jacobs, W. P. and C. C. McCready, 1967. Polar transport of growth-regulators in pith and vascular tissues of *Coleus* stems. *Am. J. Bot.* **54**, 1035–1040.

Jacobs, W. P. and I. B. Morrow, 1957. A quantitative study of xylem development in the vegetative shoot apex of *Coleus*. *Am. J. Bot.* **44**, 823–842.

Jones, R. L. and I. D. J. Phillips, 1966. Organs of gibberellin synthesis in light-grown sunflower plants. *Plant Physiol.* **41**, 1381–1386.

Kato, J., 1958. Nonpolar transport of gibberellin through pea stem and a method for its determination. *Science*, **128**, 1008–1009.

Keitt, G. W., Jr. and R. A. Baker, 1966. Auxin activity of substituted benzoic acids and their effect on polar auxin transport. *Plant Physiol.* **41**, 1561–1569.

Keitt, G. W., Jr. and R. A. Baker, 1967. Acropetal movement of auxin: Dependence on temperature. *Science*, **156**, 1380–1381.

Kende, H., 1965. Kinetin-like factors in the root exudate of sunflowers. *Proc. Nat. Acad. Sci. U.S.* **53**, 1302–1307.

Kuse, G., 1953. Effect of 2,3,5-triiodobenzoic acid on the growth of lateral bud and on tropism of petiole. *Mem. Coll. Sci. Univ. Kyoto* B, **20**, 207–215.

Kuse, G., 1954. Bud inhibition and correlative growth of petiole in sweet potato stem. *Mem. Coll. Sci. Univ. Kyoto* B, **21**, 107–114.

Lagerstedt, H. B. and R. G. Langston, 1966. Transport of kinetin-8-^{14}C in petioles. *Physiol. Plantarum*, **19**, 734–740.

Lagerstedt, H. B. and R. G. Langston, 1967. Translocation of radioactive kinetin. *Plant Physiol.* **42**, 611–622.

Lam, S. L. and A. C. Leopold, 1964. Effect of light on auxin transport. *Plant Physiol.* suppl. **39**, xxxviii.

Leopold, A. C. and O. F. Hall, 1966. Mathematical model of polar auxin transport. *Plant Physiol.* **41**, 1476–1480.

Leopold, A. C. and S. L. Lam, 1961. Polar transport of three auxins. In *Plant Growth Regulators*. Iowa State University Press. Ames, Iowa.

Leopold, A. C. and S. L. Lam, 1962. The auxin transport gradient. *Physiol. Plantarum,* **15,** 631–638.

Little, C. H. A. and M. H. M. Goldsmith, 1967. Effect of inversion on growth and movement of indole-3-acetic acid in coleoptiles. *Plant Physiol.* **42,** 1239–1245.

Lockhart, J. A., 1957. Studies on the organ of production of the natural gibberellin factor in higher plants. *Plant Physiol.* **32,** 204–207.

Loeffler, J. E. and J. van Overbeek, 1964. Kinin activity in coconut milk. In *Regulateurs Naturels de la Croissance Vegetale.* Proc. 5th Intern. Conf. Plant Growth Substances, Editions du Centre National de la Recherche Scientifique, Paris, pp. 77–82.

McComb, A. J., 1964. The stability and movement of gibberellic acid in pea seedlings. *Ann. Bot.* **28,** 669–687.

McCready, C. C., 1963. Movement of growth regulators in plants. I. Polar transport of 2,4-dichlorophenoxyacetic acid in segments from the petioles of *Phaseolus vulgaris.* *New Phytologist,* **62,** 3–18.

McCready, C. C., 1968. The polarity of auxin movement in segments excised from petioles of *Phaseolus vulgaris* L. In *Biochemistry and Physiology of Plant Growth Substances,* Proc. 6th Intern. Conf. Plant Growth Substances, Rung Press, Ottawa, in press.

McCready, C. C. and W. P. Jacobs, 1963a. Movement of growth regulators in plants. II. Polar transport of radioactivity from indoleacetic acid-(^{14}C) and 2,4-dichlorophenoxyacetic acid-(^{14}C) in petioles of *Phaseolus vulgaris. New Phytologist,* **62,** 19–34.

McCready, C. C. and W. P. Jacobs, 1963b. Movement of growth regulators in plants. IV. Relationships between age, growth and polar transport in petioles of *Phaseolus vulgaris. New Phytologist,* **62,** 360–366.

McCready, C. C. and W. P. Jacobs, 1967. Movement of growth regulators in plants. V. A further note on the relationship between polar transport and growth. *New Phytologist,* **66,** 485–488.

McCready, C. C., D. J. Osborne and M. K. Black, 1965. Promotion by kinetin of the polar transport of two auxins. *Nature, Lond.* **208,** 1065–1067.

Morgan, P. W. and H. W. Gausman, 1966. Effects of ethylene on auxin transport. *Plant Physiol,* **41,** 45–52.

Mothes, K. and L. Engelbrecht, 1963. On the activity of a kinetin-like root factor. *Life Sci.* **2,** 852–857.

Naqvi, S. M. and S. A. Gordon, 1965. Auxin transport in flowering and vegetative shoots of *Coleus blumei* Benth. *Plant Physiol.* **40,** 116–118.

Naqvi, S. M. and S. A. Gordon, 1966. Auxin transport in *Zea mays* L. coleoptiles. I. Influence of gravity on the transport of indoleacetic acid-2-^{14}C. *Plant Physiol.* **41,** 1113–1118.

Naqvi, S. M. and S. A. Gordon, 1967. Auxin transport in *Zea mays* coleoptiles. II. Influence of light on the transport of indoleacetic acid-2-^{14}C. *Plant Physiol.* **42,** 138–143.

Newman, I. A., 1963. Electric potential and auxin translocation in *Avena. Aust. J. Biol. Sci.* **16,** 629–646.

Niedergang-Kamien, E. and A. C. Leopold, 1957. Inhibitors of polar auxin transport. *Physiol. Plantarum,* **10,** 29–38.

Niedergang-Kamien, E. and A. C. Leopold, 1959. The inhibition of transport of indoleacetic acid by phenoxyacetic acids. *Physiol. Plantarum,* **12,** 776–785.

Niedergang-Kamien, E. and F. Skoog, 1956. Studies on polarity and auxin transport in plants. I. Modification of polarity and auxin transport by triiodobenzoic acid. *Physiol. Plantarum,* **9,** 60–73.

O'Brien, T. P. and K. V. Thimann, 1967. Observations on the fine structure of the oat coleoptile. VI. The parenchymal cells of the apex. *Protoplasma,* **63,** 417–442.

Osborne, D. J. and C. C. McCready, 1965. Transport of the kinin, N_6 benzyladenine: Nonpolar or polar? *Nature, Lond.* **206**, 679–680.

Oserkowsky, J., 1942. Polar and apolar transport of auxin in woody stems. *Am. J. Bot.* **29**, 858–866.

Phillips, I. D. J. and R. L. Jones, 1964. Gibberellin-like activity in bleeding sap of root systems of *Helianthus annuus* detected by a new dwarf pea epicotyl assay and other methods. *Planta*, **63**, 269–278.

Pickard, B. G. and K. V. Thimann, 1964. Transport and distribution of auxin during tropistic response. II. The lateral migration of auxin in phototropism of coleoptiles. *Plant Physiol.* **39**, 341–350.

Pilet, P. E., 1965. Polar transport of radioactivity from ^{14}C-labelled-β-indolylacetic acid in stems of *Lens culinaris. Physiol. Plantarum*, **18**, 687–702.

Potter, D. D., E. J. Furshpan, and E. S. Lennox, 1966. Connections between cells of the developing squid as revealed by electrophysiological methods. *Proc. Nat. Acad. Sci. U.S.* **55**, 328–335.

Reiff, B. und H. v. Guttenberg, 1961. Der polare Wuchsstofftransport von *Helianthus annuus* in seiner Abhangigkeit von Alter, Quellungszustand und Kohlenhydratversorgung des Gewebes. *Flora*, **151**, 44–72.

Reisener, H. J., 1958. Untersuchungen über den Phototropismus der Hafer-Koleoptile. *Z. Bot.* **46**, 474–505.

Reisener, H. J. and H. Simon, 1960. Weitere Geotropismusversuche mit radioaktive β-Indolylessigsaure. *Z. Bot.* **48**, 66–70.

Rosenberg, T., 1954. The concept and definition of active transport. *Symp. Soc. Exp. Biol.* **8**, Active transport and secretion, pp. 27–41.

Sachs, T. and K. V. Thimann, 1964. Release of lateral buds from apical dominance. *Nature, Lond.* **201**, 939–940.

Scott, T. K. and W. R. Briggs, 1960. Auxin relationships in the Alaska pea (*Pisum sativum*). *Am. J. Bot.* **47**, 492–499.

Scott, T. K. and W. R. Briggs, 1962. Recovery of native and applied auxin from the light-grown 'Alaska' pea seedling. *Am. J. Bot.* **49**, 1056–1063.

Scott, T. K. and W. R. Briggs, 1963. Recovery of native and applied auxin from the dark-grown 'Alaska' pea seedling. *Am. J. Bot.* **50**, 652–657.

Scott, T. K. and W. P. Jacobs, 1963. Auxin in *Coleus* stems: limitation of transport at higher concentrations. *Science*, **139**, 589–590.

Scott, T. K. and W. P. Jacobs, 1964. Critical assessment of techniques for identifying the physiologically significant auxins in plants. In *Regulateurs Naturels de la Croissance Vegetale*, Proc. 5th Intern. Conf. Plant Growth Substances, Gif s/Yvette, 1963, Editions du Centre National de la Recherche Scientifique, Paris, 457–474.

Seth, A. K., C. R. Davies and P. F. Wareing, 1966. Auxin effects on the mobility of kinetin in the plant. *Science*, **151**, 587–588.

Skoog, F., 1938. Absorption and translocation of auxin. *Am. J. Bot.* **25**, 361–372.

Snow, R., 1935. Activation of cambial growth by pure hormones. *New Phytologist*, **34**, 347–360.

Snow, R., 1937. On the nature of correlative inhibition. *New Phytologist*, **36**, 283–300.

Spanswick, R. M. and J. W. F. Costerton, 1967. Plasmodesmata in *Nitella translucens*: Structure and electrical resistance. *Cell Sci.* **2**, 451–464.

Steeves, T. A. and W. R. Briggs, 1960. Morphogenetic studies on *Osmunda cinnamomea* L. The auxin relationships of expanding fronds. *J. Exp. Bot.* **11**, 45–67.

Thimann, K. V. and W. D. Bonner, 1948. The action of triiodobenzoic acid on growth. *Plant Physiol.* **23**, 158–161.

Thimann, K. V. and W. D. Bonner, 1949. Experiments on the growth and inhibition of isolated plant parts. II. The action of several enzyme inhibitors on the growth of the *Avena* coleoptiles and on *Pisum* internodes. *Am. J. Bot.* **36**, 214–221.

Thimann, K. V. and F. Skoog, 1934. On the inhibition of bud development and other functions of growth substance in *Vicia faba*. *Proc. Roy. Soc.* B **114**, 317–339.

Thimann, K. V. and B. M. Sweeney, 1937. The effect of auxins upon protoplasmic streaming. *J. Gen. Physiol.* **21**, 123–135.

Thimann, K. V. and I. F. Wardlaw, 1963. The effect of light on the uptake and transport of indoleacetic acid in the green stem of the pea. *Physiol. Plantarum*, **16**, 368–377.

Thompson, N. P., 1966. Vascular regeneration and long distance transport of indole-3-acetic acid in *Coleus* stems. *Plant Physiol.* **41**, 1106–1112.

Thompson, N. P. and W. P. Jacobs, 1966. Polarity of IAA effect on sieve-tube and xylem regeneration in *Coleus* and tomato stems. *Plant Physiol.* **41**, 673–682.

Thornton, R. M. and K. V. Thimann, 1967. Transient effects of light on auxin transport in the *Avena* coleoptile. *Plant Physiol.* **42**, 247–257.

Vardar, Y., 1959. Some confirmatory experiments performed with IAA-C^{14} concerning the effect of TIBA upon auxin transport. *Rev. Fac. Sci. Univ. Istanbul*, Ser. B **24**, 133–145.

Veen, H., 1966. Transport, immobilization and localization of naphthylacetic acid-l-^{14}C in *Coleus* explants. *Acta Botan. Neerl.* **15**, 419–433.

Veen, H., 1967. On the relation between auxin transport and auxin metabolism in explants of *Coleus*. *Planta*, **73**, 281–295.

Weij, H. G. van der, 1932. Der Mechanismus des Wuchsstofftransportes. *Rec. Trav. Botan. Neerl.* **29**, 380–496.

Weij, H. G. van der, 1934. Der Mechanismus des Wuchsstofftransportes II. *Rec. Trav. Botan. Neerl.* **31**, 810–857.

Went, F. W., 1928. Wuchsstoff und Wachstum. *Rec. Trav. Botan. Neerl.* **25**, 1–116.

Went, F. W., 1939. Transport of inorganic ions in polar plant tissues. *Plant Physiol.* **14**, 365–369.

Went, F. W., 1941. Polarity of auxin transport in inverted *Tagetes* cuttings. *Bot. Gaz.* **103**, 386–390.

Went, F. W., 1942. Growth, auxin and tropisms in decapitated *Avena* coleoptiles. *Plant Physiol.* **17**, 236–249.

Went, F. W. and R. White, 1939. Experiments on the transport of auxin. *Bot. Gaz.* **100**, 465–484.

Wilkins, M. B. and M. Martin, 1967. The dependence of basipetal polar transport of auxin upon aerobic metabolism. *Plant Physiol.* **42**, 831–839.

Wilkins, M. B. and P. Whyte, 1968. In *Biochemistry and Physiology of Plant Growth Substances*. Proc. 6th Intern. Conf. Plant Growth Substances, Rung Press, Ottawa, in press.

Wilkins, M. B. and A. E. R. Woodcock, 1965. Origin of the geoelectric effect in plants. *Nature Lond.* **208**, 990–992.

Winter, A., 1967. The promotion of the immobilization of auxin in *Avena* coleoptiles by triiodobenzoic acid, *Physiol. Plantarum*, **20**, 330–336.

Winter, A. and K. V. Thimann, 1966. Bound indoleacetic acid in *Avena* coleoptiles. *Plant Physiol.* **41**, 335–342.

Zaerr, J. B. and J. M. Mitchell, 1967. Polar transport related to mobilization of plant constituents. *Plant Physiol.* **42**, 863–874.

Zwar, J. A. and A. H. G. C. Rijven, 1956. Inhibition of transport of indole-3-acetic acid in the etiolated hypocotyl of *Phaseolus vulgaris* L. *Aust. J. Biol. Sci.* **9**, 528–538.

Zweig, G., S. Yamaguchi and G. W. Mason, 1961. Translocation of C^{14} gibberellin in red kidney bean, normal corn and dwarf corn. *Adv. Chem.* **28**, 122–134.

5. Apical dominance

I. D. J. Phillips

Introduction

The various cells, tissues, and organs which together constitute a multicellular plant do not exist independently of one another, but rather their activities are interrelated; that is, all the components of the plant body form one integrated system. These interrelationships are termed *correlations*, and where the growth and development of various parts of a plant are subject to reciprocal influences then the term *growth correlations* is used to describe the phenomena. The best studied example of a growth correlation in plants is that known as *apical dominance* in the shoot. Apical dominance is manifest in at least three ways: (a) by complete or almost complete inhibition of growth in the axillary, or lateral, buds by the presence of the apical bud, (b) by inhibition of the growth of one shoot by the presence of another dominant shoot, and (c) in effects of the apical part of the shoot upon orientation and development of lateral organs such as branches, leaves, rhizomes, and stolons. The *degree* of apical dominance in a shoot is determined by genetic and environmental factors and is also greatly influenced by the physiological age of the plant.

Thus, axillary buds are usually subject to *correlative inhibition* by the apical bud. This phenomenon is of widespread occurrence in the plant kingdom, having been observed in some algae, bryophytes, pteridophytes, gymnosperms, and angiosperms. The capacity of the apical bud to inhibit lateral buds is seen not only in aerial shoots, but also in rhizomes, tubers, bulbs, and corms. The root apex, too, exerts dominance in that its presence inhibits lateral root initiation, but this problem has received somewhat less study than has that of apical dominance in the shoot and is less well defined at the present time.

This chapter is almost exclusively an account of the results of studies of apical dominance in shoots of higher plants. However, it is reasonable to assume that the mechanism of apical dominance in shoots may be at least basically similar to other examples of growth correlations in plants, and when completely understood may aid in interpreting these.

Most experiments designed to elucidate the mechanism of apical dominance have been performed using herbaceous dicotyledonous species, and rather less attention has been paid to woody species. It would appear, nevertheless, that the situation is similar in both herbaceous and woody plants, but that the expression of apical dominance is modified in many woody plants by the intervention of a dormant period in each year's growth cycle. Because of this the subject of apical dominance in woody species is considered briefly under a separate heading in the following account, although experimental results obtained with woody plants are drawn upon elsewhere if they appear to have general significance.

Early studies of the physiological basis of correlative inhibition and the 'nutritive' theory

Most early investigators favoured the view that competition for nutrients between the apical and lateral bud meristems is the determining factor in apical dominance. Thus, since the apical meristem is already present in the embryo and proceeds to grow at germination, it will continue to command a supply of necessary nutritive substances to the detriment of the later-formed lateral buds. In other words, it was thought that the apical bud constitutes a metabolic sink to which substances move along concentration gradients and, by being 'first away', it will continue to enjoy a preferential supply of nutrients relative to the younger axillary buds so long as it is in an active state of growth.

This concept of the mechanism of apical dominance came to be known as the *nutritive theory*. Essentially, therefore, this theory held that apical dominance is but one example of quantitative correlation or compensation of growth, in that every plant forms more primordial organs (e.g., shoots, buds, seeds, leaves) than it is able to support to maturity and that there is a 'struggle for existence' between different organs in a plant. The general phenomenological evidence in favour of such a view was comprehensively summarized by Goebel (1900), together with a certain amount of experimentally derived data showing that the size to which leaves grow, 'can be much greater when each single leaf has a plentiful supply of food-material than when a fixed quantity of food-material is distributed over a number of leaves' (Goebel, 1900, pp. 209–210). This relationship between leaf size and nutrient availability was inferred from experiments where excision of some leaves resulted in compensatory growth of remaining leaves.

The earliest detailed experimental work which supported Goebel's view that competition for nutritive materials was the basis of apical dominance, was performed by Loeb, working primarily with the tropical plant, *Bryophyllum calycinum*. This plant possesses definitely located buds, one in the axil of each leaf and one in each of the notches of the lamina margin. As long as the leaf is attached to the intact plant none of these buds grow out, but detachment of a leaf from the stem and maintenance of humid conditions around that leaf allows the outgrowth of buds from the leaf marginal notches. Loeb (1918) demonstrated that the mass of the shoots which an isolated leaf produced from its notches was a function of the mass of the leaf. Sister leaves (i.e., opposite members of a leaf pair) of equal size, when isolated from the stem, produced equal masses of shoots in a similar time under similar conditions, even if the number of shoots produced differed greatly in the two leaves. The relationship between mass of regenerated tissue and mass of parent organ was further demonstrated in experiments where portions were cut out of isolated leaves while their isolated sister leaves were left intact. This showed that the mass of shoots produced by the two sets of sister leaves varied approximately in proportion with the mass of the leaves (Loeb, 1918). A similar relationship was found between axillary shoot development on an isolated nodal stem segment, and shoot development from a leaf left attached to the stem. The mass of the stem increased (due to axillary bud growth) by an amount equal to the reduction in shoot mass development from the attached leaf as compared with a completely detached sister leaf (Loeb, 1915; 1917a). In summary, Loeb concluded from these experiments that 'equal masses of tissue of either isolated leaves or portions of stem, always regenerated equal masses of regenerated roots and shoots'. Although Loeb was subsequently to alter his views on the matter (see below), at the time he clearly considered that it was the availability of nutritive substances in an isolated organ which determined the capacity for regenerative bud growth from that organ, and that competition took place for these nutrients between outgrowing buds. Extrapolation of his data to the condition in the whole plant led him to propose that leaf margin buds in *Bryophyllum calycinum* do not grow out unless the leaf is excised from the parent plant because of a drain of nutrients out of the leaf to the apical bud, together with a similar preferential movement of substances from the roots to the stem apex.

Experiments on isolated nodal segments of *Scrophularia nodosa* were also interpreted by Dostál (1926) in terms of nutrient competition. Dostál noted that an isolated nodal segment exhibited slow outgrowth of the previously correlatively inhibited axillary buds from the axil of each of the pair of attached leaves, but that removal of one of the pair of sister leaves resulted in much more vigorous growth of the bud on the side of the stem from which the leaf had been excised. The effect of removal of one leaf was therefore to upset the 'balance' between the buds, so that one grew faster than the other. The more vigorously growing bud then strongly inhibited the growth of the opposite bud to such an extent that the development of the latter could be completely suppressed. Thus Dostál concluded that the leaf inhibits the bud which it subtends by extracting nutrients and water from the stem which would otherwise be available to the bud. Removal of the leaf would therefore, on his reasoning, increase the nutrient and water supply to the previously subtended bud which would consequently grow out more rapidly, and gradually become a more effective sink for these substances than the slower growing opposite bud, thereby increasing the starvation of the latter.

In essentials, that was the experimental evidence up to the mid-nineteen-twenties which supported the earlier conjectures of Goebel (1900) and led to the establishment of the nutritive theory. However, other research, some contemporary with that already discussed, and other pieces of work carried out later on, yielded results which pointed to the existence in plants of diffusible correlative inhibitors.

Diffusible substances concerned in correlative inhibition of lateral shoots and buds

As early as 1904, Errera envisaged apical dominance as being due to 'internal secretions' or, as we might say today, hormones. Also, notwithstanding his previous conclusions, Loeb did a number of experiments the results of which he interpreted in terms of a diffusible lateral bud inhibitor. In one experiment Loeb (1917b) placed horizontally a completely defoliated stem segment of *Bryophyllum calycinum* containing six or more nodes, in such a manner that a line connecting the morphologically most apical pair of buds was vertical. Both buds of the apical node then commenced growth, but only the physically uppermost of the pair continued to grow, while growth of the lower bud soon stopped altogether or was considerably retarded. None of the buds at other nodes grew out. If the physically uppermost of the two apical buds was removed then the physically lower apical bud grew out, and also one or both of the buds at the next morphologically lower node. If the lower apical bud was removed then only the uppermost apical bud grew out, and the morphologically basal buds remained inhibited. Loeb concluded that when a bud grows out it produces and transmits inhibitory substances towards the base of the stem. At the same time Loeb recognized that these experiments had not excluded the possibility that since the apical bud is the first to grow out then it could perhaps consume all the 'shoot-forming material', which may have a 'tendency to rise' to the highest point in the stem. Consequently Loeb (1917b) performed a similar experiment with a horizontal stem segment with all buds left intact, but in this case leaving the apical pair of delaminated petioles attached to the stem. All other laminae and petioles were removed. The petioles senesced in a week or so, but until this occurred they prevented or retarded the growth of the apical buds in their axils. However, the buds at the next node did start growing. As soon as the petioles fell off, the axillary buds at the apical end of the stem segment began to grow out and soon retarded or stopped the growth of the morphologically lower buds. This observation was taken by Loeb to demonstrate unequivocally that a 'growing bud sends out substances toward the base of the stem which directly inhibit growth of the other buds'.

Both Loeb and Dostál appeared to vacillate somewhat in interpreting their experimental data either in terms of the original nutritive theory or by invoking the existence of diffusible bud-inhibiting principles. As mentioned above, Dostál in 1926 favoured the nutritive theory, whereas earlier (Dostál 1909) he had speculated that leaves were a source of a substance inhibitory to lateral bud outgrowth. The fact of these workers taking different stands at different times is understandable in view of their confusing data. Even in the light of recent research, current workers in this field are inclined, as we shall see later, to favour a theory of apical dominance which contains elements of the old nutritive theory and the involvement of diffusible lateral bud growth suppressors.

The best early evidence for the existence of a special diffusible bud-inhibiting substance came from the work of Snow (1925). Snow (1925) revealed that the inhibiting influence of the apical bud could pass across a water-gap between adjacent stem tissues and suppress the outgrowth of axillary buds on a decapitated piece of stem. This was demonstrated with *Phaseolus* seedlings by means of 'grafting' experiments of the type illustrated in Fig. 5.1A. In these experiments Snow ensured that no graft union formed between the adjacent cut stem surfaces, but that the only connection between the adjacent stem tissues was by way of a water-filled gap. Snow realized that the simple nutritive theory was not adequate to explain this result, and proposed that correlative inhibition of axillary buds by the apical bud is achieved by the agency of a diffusible substance originating at the growing apex.

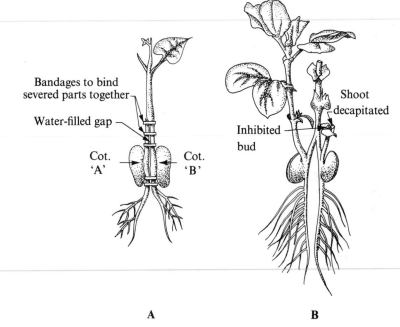

A   B

Figure 5.1

A. Diagrammatic representation of an experiment by Snow with *Phaseolus vulgaris* seedlings, which demonstrated that correlative inhibition was able to pass across a water-filled gap and prevent the outgrowth of the axillary bud of cotyledon 'A'.

B. Illustrates another of Snow's experiments, using 'two-shoot' pea seedlings, which showed that correlative inhibition could move both basipetally and acropetally. In the plant shown, the axillary bud on the decapitated shoot remained inhibited as long as the other actively growing shoot was present. Decapitation of the second shoot resulted in axillary bud outgrowth from both shoots. (Fig. A after Snow, 1925. Fig. B after Snow, 1937.)

Further evidence that the simple nutritive theory was not adequate was provided by Harvey (1920) and Snow (1925), who both showed that girdling a zone of the stem of *Phaseolus* or *Vicia faba* with a steam-jet resulted in outgrowth of axillary buds situated below the girdled region whilst the growth of the main stem apex continued unimpaired. Such girdling experiments have been repeated with similar results, which are clearly in conflict with the nutritive theory of apical dominance in that the growing main apex presumably still consumes nutrients and yet the lower lateral buds are released from inhibition by the presence of a girdled zone above the point of their insertion on the stem. Snow (1925) drew the obvious conclusion that the diffusible correlative inhibitor coming from the apical bud could not pass the girdled region.

It was not until auxin was recognized as a chemically discrete substance occurring naturally in plant tissues, and which appeared to be indole-3-acetic acid (IAA), that further progress was made in elucidating the nature of the inhibitory influence transmitted from the apical to lateral buds. Thus it was found by Thimann and Skoog (1933; 1934), that repeated applications of agar blocks containing an auxin preparation derived from cultures of the fungus *Rhizopus suinus* (probably, principally IAA) to the cut surface of an internode after excision of the apical bud from *V. faba* plants (i.e., 'decapitated' plants), resulted in maintenance of lateral bud inhibition just as if the growing apical bud were present. Exogenous auxin thus simulated the effect of the apical bud with respect to correlative inhibition of axillary buds. This finding immediately suggested that it was auxin diffusing from the apical bud which was responsible for apical dominance, and Thimann and Skoog (1934) were able to establish that the apical buds of green *V. faba* plants did in fact secrete auxin to lower parts of the stem. A similar situation had earlier been shown to exist in the etiolated oat coleoptile where the tip is the source of auxin necessary for the elongation of the lower regions of that organ. Thimann and Skoog (1934) also noted that exogenous auxin could cause elongation of the decapitated stem in *V. faba*, particularly in darkness, and that the amount of lateral bud inhibition produced in their experiments depended on the concentration of auxin applied to the cut stem surface. These results could possibly be interpreted as indicating that an application of exogenous auxin to a decapitated plant stimulated growth in the treated internode, thus producing a metabolic sink in the stem which caused continued 'starvation' of the lateral buds. However, Thimann and Skoog (1934) recorded that the amount of auxin required for stem elongation was smaller than that required for bud inhibition, and these workers suggested that *apically synthesized auxin does indeed reach the lateral buds* and in some way prevents their own production of auxin. This possibility was substantiated by the fact that lateral buds were found not to produce auxin when subject to correlative inhibition, but following release from inhibition in decapitated plants they produced considerable quantities of auxin (Thimann and Skoog, 1934).

At about the same time that Thimann and Skoog (*loc. cit.*) demonstrated that exogenous auxin could replace the apical bud in correlative inhibition of axillary buds, similar reports were made by Laibach (1933) and Müller (1935) who used lanolin pastes containing urine or orchid pollinia on a number of different plant species. The urine and orchid pollinia were known at that time to be rich sources of auxin. In contrast to Thimann and Skoog (1934), both Müller (1935) and Laibach (1933) ascribed the lateral bud-inhibiting effect of exogenous auxin application to the local growth-stimulating effect of the hormone. That is, they too noted that auxin-treated stem stumps in decapitated plants became swollen or elongated, or both, and in their view lateral bud inhibition was only a secondary result of this growth which drained nutrients away from the lateral buds. Because of these interpretations, Thimann (1937) made careful measurements of stem and bud dry weights in decapitated *Pisum sativum* (c.v. 'Alaska') plants which had or had not received applications of lanolin pastes containing 4 per cent IAA. His results showed that inhibition of lateral buds by IAA was not accompanied by any compensating increase of growth elsewhere in the plant and involved a real decrease in total dry weight. In the same experiments (Thimann, 1937) the effects of applications of IAA to lateral buds of decapitated pea plants and also to the roots of pea plants were determined. Direct applications of very low IAA concentrations to lateral buds slightly increased their dry weight but not elongation; higher IAA concentrations similarly applied were as inhibitory to bud growth as the same concentrations applied to the stem stump. Roots bathed in IAA solutions showed either slightly stimulated elongation when the hormone concentration was very low, or more generally inhibited elongation without a corresponding increase in diameter, reflecting, as with the buds, a real decrease in dry weight. As a result of these experiments and earlier ones which demonstrated that normal outgrowth of buds is associated with the capacity of the buds to synthesize auxin (Thimann and Skoog, 1934; Zimmermann, 1936), and on the assumption that all plant growth requires the presence of auxin, Thimann (1937) proposed that the responses of buds, roots, and stems to auxin all show an optimum type of dose-response curve but with different optima for different organs (see Fig. 1.4 of Chapter 1). Thus Thimann envisaged that buds, although requiring auxin for their growth, are inhibited by a concentration of auxin which would be optimal for the elongation of the main stem. These observations and conclusions of Thimann provided the basis for what came to be known as the '*direct*' *theory* of auxin action in apical dominance. This theory therefore assumed that auxin enters lateral buds from the stem at concentrations similar to those pertaining in the main shoot axis.

The present status of the 'direct' theory

Even before Thimann and Skoog (1934) and Thimann (1937) formulated the 'direct' theory of auxin inhibition in correlative phenomena, work by Snow and others had yielded results which raised apparently insurmountable difficulties for it. Mogk (1913) and Snow (1931) noted that not only buds but also the elongation of quite long growing shoots can be inhibited by other growing shoots, which on the basis of the 'direct' theory would involve acropetal movement of auxin up the whole length of the inhibited shoot. Also, Snow (1931) showed that in 'two-shoot' pea or bean seedlings (produced by excision of the epicotyl resulting in more or less equal outgrowth of both cotyledonary buds) removal of the young actively growing leaves from one of the two shoots caused severe growth inhibition in the defoliated shoot, whereas if the other shoot was removed then the defoliated shoot would grow on unchecked. Snow (1931; 1932) pointed out that removal of the young leaves, which are principal centres of auxin production, must have diminished the supply of auxin to the defoliated stem. Thus it was difficult to see how any auxin coming from the non-defoliated shoot could be the cause of inhibition of elongation in the defoliated shoot. In fact, from further experiments (Snow, 1937) it appeared that auxin emanating from the young leaves 'protects' a shoot from inhibition by another actively growing shoot on the same plant. It was found (Snow, 1937) that a lanolin paste containing IAA could substitute for the young leaves in providing protection against inhibition by the non-defoliated shoot. It has also been shown that the stem of an inhibited pea shoot contains little diffusible auxin, whereas a growing shoot does (Le Fanu, 1936; Scott and Briggs, 1960), yet on the basis of the 'direct' theory, if applicable to shoots as well as buds, one might expect that if inhibited shoots were inhibited by auxin passing into them then the auxin would be recoverable from them by agar-diffusion methods.

Another difficulty associated with the 'direct' theory is, as has already been touched upon above, the *upward* or acropetal transmission of the correlative inhibitory influence which can be observed in many types of experiment and in a number of different species, whether this inhibition is derived from sources within the plant or from exogenous applications of auxin to lower parts of the stem (Le Fanu, 1936; Snow, 1936; Champagnat, 1955). As auxin translocation is known to be polar in a predominantly basipetal direction (see chapter 4 in this book) then it might appear impossible to ascribe such an upward spread of inhibitory influence to the acropetal movement of auxin itself. In an attempt to remove this difficulty it has been suggested (Thimann and Skoog, 1934) that some auxin can travel upwards with the transpiration stream in the xylem vessels. However, by means of an ingenious experiment with pea seedlings Snow (1937) found that the inhibitory influence could travel acropetally even against the direction of the transpiration stream. Then, using elegant surgical procedures Snow (1937) further demonstrated that the inhibiting influence could travel where auxin does not appear able to travel. Most of these experiments of Snow involved the use of 'two-shoot' pea and bean seedlings subjected to varying degrees of dissection designed to increase the difficulty of auxin (endogenous or exogenous) movement from its known source to the positions of inhibited buds (see Fig. 5.1B). As a check that auxin did not pass through the regions of the

plant between the auxin source and the inhibited buds it was observed by histological examination that no activation of the cambium had occurred in these
regions. It had earlier been found that auxin during its passage through a stem
induces division of the cambium leading to the formation of secondary vascular
tissues (Snow, 1935; Gouwentak, 1936). Thus Snow (1937) derived the theory
that auxin does not, so far as apical dominance is concerned, act in such a direct
manner as proposed by Thimann and Skoog (1934) and Thimann (1937), but
rather that auxin has an 'indirect' influence by stimulating in the stem through
which it travels the production of a secondary inhibiting influence which can
travel in a non-polar manner in the plant (the possible existence of a special
correlative inhibitor other than auxin is discussed later). As a corollary of this
hypothesis, Snow suggested that auxin present in a stem 'protects' that stem
from inhibition by the auxin-induced inhibitory principle, but that auxin-
deficient tissues (such as axillary buds, or partially defoliated shoots), are not
resistant to the effects of this inhibitory influence. Other workers have obtained
results which demonstrate that the presence of auxin in a growing shoot can
'protect' that shoot from inhibition by other more vigorous shoots on the same
plant. For example, Libbert (1954a) removed and replaced the apical bud from
Pisum sativum plants, with a layer of agar in the junction. This resulted in release
of an upper lateral bud from correlative inhibition (in itself, as pointed out by
Audus (1959), a surprising result in view of the earlier experiments of Snow
demonstrating the diffusible property of the correlative inhibitor). After a period
of growth of the lateral bud, the apical bud established tissue union with the stem
stump and either the lateral shoot maintained or the apical shoot reassumed
dominance. Application of IAA directly to the apex of the apical bud of the
inhibited main shoot in such plants resulted in renewed growth of that shoot,
whether it was previously inhibited by an apical bud on the dominant shoot or
by IAA applied to the stump of a decapitated domain shoot. In his experiments
Libbert found that the auxin had to be applied only to the apex of the apical bud,
and that if it were allowed to come into contact with the young internodes then
it did not protect the grafted apex from inhibition by the growing lateral shoot.
It will be noted that in all these examples of auxin-protection of shoots from
inhibition by other dominant shoots the experimental subjects were plants with
two *elongated* shoots. Where attempts have been made to release small unelon-
gated axillary *buds* from correlative inhibition by direct applications of auxin to
them (e.g., Thimann, 1937), the result has been, if anything, increased inhibition.
It has been suggested (Libbert, 1954a) that the reason such buds cannot be
released by direct auxin applications is that it is not technically possible to apply
the hormone specifically to the apex of a small axillary bud. However, Sachs
and Thimann (1967) showed that auxin, whether applied to the apex or the
stem of an inhibited but elongated shoot of a 'two-shoot' pea seedling, was
effective in protecting that shoot against inhibition by the dominant shoot. Sachs
and Thimann (1967) suggested that the reason for this discrepancy between

their findings and those of Libbert (1954a) is that the latter used etiolated seedlings in which auxin treatment of the stem results in abnormal swelling and inhibits elongation, whereas they experimented with green plants. Sachs and Thimann (1967) also showed that auxin had no significant influence on large dominant shoots, indicating that in such shoots the level of auxin is non-limiting. It appears therefore that auxin availability limits the extension of elongated lateral shoots which are still subject to some correlative inhibition, but that there is another component of apical dominance which determines the initial outgrowth of axillary buds. The possibility that this other component involves cytokinins is discussed later.

Primarily as a result of the availability of isotopically labelled IAA, we have today a somewhat clearer picture of the capacities of various plant tissues to transport auxins in one direction or another. In general we are now aware that the transport of IAA can be both basipetal and acropetal in any tissue but that basipetal transport predominates (see chapter 4).

Thus, the objection to the 'direct' theory on the grounds of strict basipetal polar auxin transport is less soundly based than it appeared to be at the time of Snow's contributions. Also, it has been specifically shown (Wickson and Thimann, 1960) that supply of ^{14}C − labelled IAA to a nodal segment obtained from a pea plant which had been decapitated three days earlier, resulted in acropetal movement of radioactivity into the outgrowing axillary bud. Further, a linear relationship was found between the inhibition of bud growth thus re-established and the content of isotope in the bud tissues (Fig. 5.2). This relationship suggested that all the radioactive material in the bud was in fact inhibitory, although it was not shown unequivocally that the radioactivity was confined to free molecules of IAA. Nevertheless, Wickson and Thimann (1960) did establish in their experiments that ^{14}C-IAA retained biological activity after passage through excised sections of pea stems.

A reasonable line of investigation into establishing whether or not auxin levels in lateral buds subjected to correlative inhibition are supra-optimal for their growth would seem to be one where auxin is extracted from such buds, and from buds released from inhibition by excision of the apical bud. If the 'direct' theory is to be upheld it would be expected that following release from inhibition there would be a measurable *fall* in auxin concentration within the bud tissues. In fact, very few such studies have been made, understandably in view of the technical difficulty of obtaining sufficient bud tissues for auxin estimations. However, what information is at hand suggests that in at least *Lupinus* and pea seedlings, inhibited lateral buds contain *less* extractable auxin than buds released from inhibition by decapitation (Ferman, 1938; Van Overbeek, 1938). If these findings are of general significance they argue very strongly against the concept of direct inhibition of lateral buds by auxin. It has to be borne in mind, however, that neither Ferman nor Van Overbeek (*loc. cit.*) had available effective means of fractionating their bud extracts, so that their determinations of

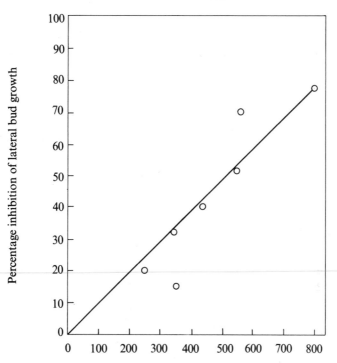

Radioactivity (^{14}C) in lateral buds, counts per minute/mg dry weight

Figure 5.2   Correlation between total radioactivity present in pea lateral buds and degree of inhibition of their growth, following treatment with ^{14}C-labelled IAA at different concentrations and for different times. (After Wickson and Thimann, 1960.)

total auxin activity probably represented the product of both auxin and any growth-inhibitors present in the buds.

There is other more recent evidence that auxin concentrations in inhibited buds may be sub- rather than supra-optimal. It has been found that lateral buds released from correlative inhibition by direct application of a cytokinin to them (Sachs and Thimann, 1964; 1967; Panigrahi and Audus, 1966) continue growth only if auxin also is supplied directly to the outgrowing buds (Sachs and Thimann, 1967). This observation, together with the findings that low concentrations of auxin when applied to the cut stump of decapitated plants can actually promote lateral bud outgrowth and that higher concentrations are necessary to effect bud inhibition (Ferman, 1938; Meinl and Guttenberg, 1954; Šebánek, 1966a; 1967) strongly suggests that the 'direct' theory of auxin inhibition of lateral buds is no longer tenable.

Cytokinins and correlative inhibition

Mention has already been made that direct application of a cytokinin (see chapter 3) to an axillary bud can bring about release of that bud from correlative inhibition induced by the apical bud or by exogenous auxin applied to the cut surface of the stem in a decapitated plant. However, the first indication that cytokinins might be involved in apical dominance phenomena came from studies of the effect of kinetin upon inhibited lateral buds on isolated pea stem segments (Wickson and Thimann, 1958). These workers obtained inspiration for their work from the observation that the numerous shoot-buds which are initiated in a single tobacco-pith tissue culture in the presence of kinetin (Skoog and Miller, 1957) proceed to elongate side by side without any evidence of mutual interference. Thus Wickson and Thimann considered the possibility that the lack of correlative inhibition in these multi-shoot cultures was due to the free availability of cytokinin, in the form of kinetin, in the nutrient medium which supported their growth. These workers used excised, partially etiolated, nodal stem segments floating on a sugar solution in light. The presence of IAA in the sugar solution resulted in inhibition of bud growth, and kinetin was found to counteract this inhibition. The auxin: kinetin ratio determined the completeness of the counteraction. Similarly, using both green and etiolated pea shoot cuttings standing upright in kinetin solutions they revealed that cytokinin so supplied would also release buds inhibited by the intact apical bud; however, probably because of the limited transport of kinetin, only the lowest buds were able to grow out. In a somewhat similar manner, von Maltzahn (1959) demonstrated that application of kinetin in agar-gel to the top of a decapitated gametophore in *Splachnum ampullaceum* antagonized the inhibitory effect on lateral buds of a simultaneous application of IAA.

The next important step in establishing the involvement of cytokinins in apical dominance was the demonstration that lateral buds could be released from correlative inhibition in intact whole plants by application of kinetin directly to the buds themselves. This was first shown in *Pisum sativum, Coleus, Scabiosa, Helianthus annuus*, and *Helianthus tuberosus* by Sachs and Thimann (1964), and later in *Vicia faba* by Panigrahi and Audus (1966). However, direct applications of kinetin to lateral buds, even if made repeatedly, gave only a shortlived release from correlative inhibition by the apical bud, for dominance was always reimposed when they had elongated by one or two centimetres. It therefore appeared that cytokinin may be concerned in the initial release from dominance but that subsequent growth was dependent on other factors. The possibility that gibberellin is the 'missing factor' was explored by Sachs and Thimann (1964) and by Panigrahi and Audus (1966), but they found that addition of gibberellic acid (GA_3) gave only a slight additive effect to that of kinetin. It was earlier observed (Wickson and Thimann, 1958) that with isolated nodal stem segments also, GA_3 gave promotion of bud elongation only when the bud was not subject to correlative inhibition. However, it has subsequently been clearly demonstrated that buds released from apical dominance by direct kinetin applications can be made to elongate more or less normally by careful application of auxin (IAA) to the apices of the partially expanded buds. However, the leaves of buds treated with both kinetin and auxin remained small, and no treatment has yet been found (apart from decapitation) which results in their normal expansion (Sachs and Thimann, 1967). The parallel between the discovery that auxin application to a cytokinin-released lateral bud can at least reduce residual apical dominance, and earlier work (discussed above) showing that auxin can 'protect' an elongated *shoot* against correlative inhibition by a dominant shoot, is of course very striking. In fact it was concluded by Sachs and Thimann (1967) that growing shoots are relatively insensitive to correlative inhibition because they synthesize both auxin which may antagonize the inhibitory effect on internode elongation, and cytokinins which may be necessary for the apex itself to develop. In general terms, therefore, these authors have taken the view that axillary buds are inhibited because they are unable to synthesize cytokinins (perhaps because of the inhibiting effect of auxin originating outside the bud itself), but that auxin originating in the bud apical region actually stimulates cytokinin synthesis within the bud which is a necessary prerequisite for continued bud growth. It has, in this connection, been found that addition of kinetin to an inhibited shoot of a pea plant can prevent its senescence and cause renewed growth (Sachs, 1966).

Although based on good experimental data, the hypothesis put forward by Sachs and Thimann (1967) is nevertheless still highly speculative. Other interpretations are available, particularly if one considers the possible implication of cytokinins derived from sources (such as the root system) outside the bud, and the effects of hormones on metabolite translocation. Such an alternative possibility is considered below within the section dealing with hormone-directed transport.

Gibberellins and correlative inhibition

Early studies of the physiological effects of gibberellins in plants revealed that application of a gibberellin to the cut stem surface in decapitated plants did not result in inhibition but rather a promotion of axillary bud outgrowth, as compared with buds on untreated decapitated plants (Kato, 1953; Brian *et al.*, 1955), thus contrasting with the inhibiting effect of an auxin application in such a system. Similarly, it has been reported by some workers that gibberellin treatment of an intact shoot can promote lateral shoot growth (Marth *et al.*, 1956), although much more usually it is found that treatment of an intact shoot with a gibberellin results in promotion of main stem elongation but with a simultaneous inhibition of growth in the laterals (Lona and Bocchi, 1956; Brian *et al.*, 1959; Stoddart, 1959; Bradley and Crane, 1960; Bruinsma and Patil, 1963). In general it appears that gibberellin treatment promotes axillary bud or shoot growth only when correlative inhibition has already been abolished or decreased by some other means, and represents the usual elongation response of a growing shoot to exogenous gibberellin (Brian and Hemming, 1957; Wickson and Thimann, 1958; Brian *et al.*, 1959; Sachs and Thimann, 1964; Panigrahi and Audus, 1966). However, Šebánek (1965) found that release of cotyledonary buds in pea from correlative inhibition resulted in a measurable rise in gibberellin content of the buds within forty-eight hours, which preceded visible morphological development of the buds. Growing apical buds are known to be sites of active gibberellin synthesis (Jones and Phillips, 1966). Even in those cases where gibberellin treatment of an intact shoot led to enhanced lateral bud growth, this occurred only after growth in the main shoot had subsided following its initial burst of elongation after the gibberellin application (Marth *et al.*, 1956), suggesting that the lowered vigour of the main shoot at that time was reflected in a diminished capacity of that shoot to dominate the laterals.

There is no doubt that exogenous gibberellin can *increase* apical dominance in intact plants (references cited in previous paragraph). It is, however, not so clear whether application of gibberellin increases bud inhibition in plants whose axillary buds are in a state of correlative inhibition due to the presence of exogenous auxin rather than the apical bud. In some cases antagonism between the effects of added gibberellin and auxin in correlative bud inhibition has been recorded (Kato, 1958; Wickson and Thimann, 1958; Nakamura, 1965) and in others synergism (Jacobs and Case, 1965; Scott *et al.*, 1967). Reports other than those of Jacobs and Case (1965) and Scott *et al.* (1967) that gibberellin can participate with auxin in maintenance of apical dominance are somewhat more ambiguous in their experimental bases. The finding that GA_3 will induce normal negative geotropic reaction of upper lateral branches in decapitated *Cupressus arizonica* seedlings following the inhibition of this reaction by various growth retardants (Pharis *et al.*, 1965), can be tentatively interpreted as being due to no more than the replacement with exogenous GA_3 of endogenous gibberellin required for cambial activity (Wareing, 1958) in the lower side of the branch. Where GA_3 applications to growth retardant-treated, decapitated soybean (*Glycine max*) and redwood (*Sequoia sempervirens*) seedlings restored the normal ability of upper lateral buds to achieve dominance over lower buds (Ruddat and Pharis, 1966), what possibly occurred was that the GA_3 restored vigorous growth to the upper outgrowing lateral shoots followed by a typical basipetal spread of correlative inhibition from such actively growing shoots. These interpretations are based upon the available evidence that growth-retardants of the type used by Pharis and co-workers (Amo-1618, B-995) can inhibit gibberellin biosynthesis in fungal cultures (Kende *et al.*, 1963; Ninnemann *et al.*, 1964;

Ruddat *et al.*, 1965) and in higher plant tissues (Baldev *et al.*, 1965; Dennis *et al.*, 1965; Harada and Lang, 1965; Zeevart, 1966; Jones and Phillips, 1967).

Nevertheless, there is a hard core of information showing that gibberellin treatment of *intact* plants often causes an increase in apical dominance (Lona and Bocchi, 1956; Brian *et al.*, 1959; Stoddart, 1959; Bradley and Crane, 1960; Bruinsma and Patil, 1963; Nakamura, 1965). In general it appears that exogenous gibberellin enhances the growth of vigorously growing shoots, but with a concomitant increase of correlative inhibition, and it is possible that part of the reason for this is gibberellin-enhanced auxin levels. It has been frequently observed that gibberellin treatment of intact shoots leads to increased auxin levels in stems, leaves and apical buds (Hayashi and Murakami, 1953; Phillips *et al.*, 1959; Pilet and Wurgler, 1958; Brian, 1959; Galston and McCune, 1961; Kuraishi and Muir, 1962; 1963; 1964a, b; Halevey, 1963; Sastry and Muir, 1965). But, even if increased auxin synthesis or decreased auxin destruction, in gibberellin-treated intact shoots is not the means by which apical dominance is increased by exogenous gibberellin, it is a generally applicable rule that the more vigorous the growth of a shoot (and it is usually increased by gibberellin treatment) then the greater is the ability of that shoot to exert dominance over other less vigorous shoots on the same plant. In other words, reported examples of gibberellin-enhancement of apical dominance can all be traced back to the general stimulatory effect of gibberellins on growth, except perhaps for the reports that GA_3 can enhance the inhibitory effect of IAA in decapitated seedlings (Jacobs and Case, 1965; Scott *et al.*, 1967).

Role of endogenous growth inhibitors in correlative inhibition

Results of various experiments, most of which we have considered earlier in this account, led to the suggestion that there exists a special correlative growth-inhibitory hormone, the formation of which may be induced by auxin as it moves through stem tissues (Snow, 1929; 1937; 1938; 1939a; 1940). Following reports in the literature that it was possible to extract substances from plant tissues which were very inhibitory to coleoptile growth (Stewart *et al.*, 1939; Stewart, 1939), it was found by Snow (1939b) that pea leaves and leafy lateral shoots contained inhibitor, whereas growing apical buds and stems did not. More recent work in this field has made use of chromatographic techniques to separate such inhibitors from other constituents in extracts, but it is difficult to be certain that endogenous growth inhibitors play a part in correlative inhibition. Libbert (1964) has taken the results of many of his experiments as evidence for the existence of a special correlative growth-inhibitor, the formation of which is induced by auxin. Some of these results have, however, been subjected to criticism on grounds of the techniques used (Audus, 1959). Further, other workers have obtained results conflicting with those of Libbert, although the latter has attempted to resolve these apparent contradictions (see Libbert, 1964).

Following reports of a correlation between levels of endogenous ' β-inhibitor' (later shown to be abscisic acid, see chapter 17) and depth of innate-dormancy in terminal buds of sycamore (*Acer pseudoplatanus*) (Phillips and Wareing, 1958a, b; 1959), it was found by Dörffling that, in sycamore and pea, there was a similar relationship between endogenous β-inhibitor and correlatively-imposed lateral bud inhibition. Thus, Dörffling (1963a, b; 1964; 1965; 1966) measured a fall in extractable β-inhibitor in lateral buds following their release from apical dominance by decapitation and defoliation of the shoot, and was able to prevent the outgrowth of such released buds by application of β-inhibitor directly to the buds. Dörffling envisages correlative inhibition of buds involving both auxins and β-inhibitor, but has also found that a number of growth-inhibitors may be involved and that GA_3 can partially overcome the inhibitory influence of these on lateral bud growth (Dörffling, 1966).

Potato tubers have been found to contain a neutral substance which inhibits the growth of lateral buds when supplied at a 'physiological concentration' (Goodwin and Cansfield, 1967). This inhibitor was present only in tubers bearing actively growing dominant sprouts, and it disappeared from tuber tissue within 24 hours of removing the growing sprouts. Consequently, Goodwin and Cansfield (1967) have suggested that this inhibitor is the controlling factor in potato tuber apical dominance.

Hormone-directed transport and the nutrient-diversion theory

There has been a revival of interest in recent years in the possibility that the supply of metabolites to axillary buds is a determining factor in apical dominance. However, the modern concept which has developed is somewhat different from that of the earlier Nutritive Theory. The latter held that nutrients flowed from point to point in the plant solely in response to concentration gradients whereas currently the view is, in general terms, that patterns of nutrient movement are determined by physiological effects of growth hormones.

Many studies have demonstrated that both inorganic and organic nutrients are indeed translocated predominantly to regions of active growth, such as the shoot apical bud, root tips, developing seeds and fruits, and the cambium (Williams, 1955). But, of course, such observations do not establish a causal relationship between lack of supply of nutrients to, and inhibited growth of, lateral buds. The first indication that the level of nutrients present in a plant can determine the extent of lateral bud growth was provided by Gregory and Veale (1957), who found that in flax (*Linum usitatissimum*) correlative inhibition of lateral buds was much more evident at low than at high levels of nitrogen supply and concluded that growth of lateral buds in flax is inhibited by a lack of nutrient such as nitrogen. Similarly, McIntyre (1964; 1965) reported that in the rhizome of *Agropyron repens* apical dominance was complete with a low nitrogen supply but that axillary buds grew out with higher nitrogen levels. Also Troughton (1967) observed that tillering in *Lolium perenne* was greatly increased by a favourable NPK supply.

The first suggestions that auxin influences metabolite translocation were made by Went (1936; 1939), who obtained evidence that a high auxin concentration in the apical bud in some unspecified manner induced a flow of nutrients and other growth factors towards the shoot tip. This *'nutrient-diversion' theory* was immediately subjected to strong criticism by some other workers, particularly Snow (see Snow, 1937; 1940), although direct evidence was obtained at that time by Mitchell and Martin (1937) and Stuart (1938) that nitrogenous substances and carbohydrates accumulated in auxin-treated regions of stems in bean plants. However, these latter workers allowed several days to elapse between the time of auxin application and analysis of the tissues for inorganic and organic constituents, and it was probable that the observed effects were due to the creation of a new growth centre at the place of auxin treatment with consequent flow of metabolites to the new 'sink'. It was not until the 'sixties that further evidence was adduced implicating auxin in translocation. It was then found that ^{14}C-assimilates and ^{14}C-sucrose accumulated in auxin-treated stumps of decapitated potato and pea plants within a few hours of the time of auxin application. This indicated that the accumulation was perhaps independent of a local growth-stimulating effect of auxin, although it did not completely discount that possibility (Booth *et al.*, 1962). Similar results were obtained by Nakamura (1965) and by Šebánek (1965; 1966a, b; 1967) for ^{32}P accumulation in pea internodes following supply of $H_3{}^{32}PO_4$ or $KH_2{}^{32}PO_4$ to the roots, although in these experiments at least twenty-four hours elapsed between the time of IAA and ^{32}P applications and analysis of distribution of radioactivity.

Davies and Wareing (1965) supplied ^{32}P-labelled sodium orthophosphate to the basal region of pea stems or poplar (*Populus robusta*) twigs and, as expected, found that ^{32}P moved to and accumulated in decapitated stem stumps treated with either IAA or Naphthoxyacetic acid (NOAA), but did so only weakly into untreated stem stumps. The accumulation of ^{32}P in response to auxin occurred within a few hours of the time of simultaneous auxin and ^{32}P application, confirming the earlier results of Booth *et al.* (1962). Further, by decapitating to low levels in the stem it was found that ^{32}P accumulated in old non-growing internodes in response to auxin application. These results of Davies and Wareing (1965) indicated therefore that auxin does not necessarily have to induce local growth at its point of application before ^{32}P mobilization is effected, and led these workers to suggest that auxin has some rather more direct effect upon the translocation system. Other experiments by Davies and Wareing (1965) established that acropetal transport of ^{32}P took place primarily via the phloem, and that a ring of tri-iodobenzoic acid (TIBA) in lanolin around the stem between auxin and ^{32}P sources greatly reduced ^{32}P accumulation in the auxin-treated stem stump; this effect of TIBA has been confirmed by Šebánek (1966b). The blocking effect of TIBA upon basipetal movement of IAA is well established (Niedergang-Kamien and Skoog, 1956; Panigrahi and Audus, 1966). Thus in the experiments carried out by Davies and Wareing (1965) and by Šebánek (1966b), it can be presumed that the immediate ability of IAA to induce a local growth reaction in the stem stump was not impaired by the presence of a TIBA ring below, yet it was found that ^{32}P movement to the stump was prevented. On the assumption that TIBA did not block movement of solutes in the phloem Davies and Wareing (1965) favoured the conclusion that auxin does not stimulate nutrient transport through a local effect in the stem stump, but rather that in some unknown manner basipetal transport of auxin affects solute transport along the whole length of the translocation system. Results obtained by Libbert (1959) showing that movement of sucrose in phloem was not impeded by a TIBA ring around a stem lends support to this conclusion. Surprisingly, Davies and Wareing found that both indole-3-acetonitrile (IAN) and 2,4-dichlorophenoxyacetic acid (2,4-D) were inactive in inducing accumulation of ^{32}P in stem stumps of pea, and yet 2,4-D appears to show polar basipetal transport in much the same way as IAA (McCready, 1963). However, it is very likely that any particular auxin will have contrasting effects on nutrient transport in different systems, for IAA has been found not to influence metabolite transport in some leaves (Mothes, 1960; Gunning and Barkley, 1963), whereas 2,4-D often does (Osborne and Hallaway, 1959.) Similarly, although cytokinin (Mothes and Englebrecht, 1961) and gibberellin (Davies, 1963) have been observed to influence metabolite movement in leaves, it was found by Davies and Wareing (1965) that neither kinetin nor gibberellic acid (GA$_3$) when applied alone was able to induce significant accumulation of ^{32}P in stem stumps of decapitated pea plants. However, simultaneous treatment of a stem stump with IAA and

Table 5.1 The effects of exogenous hormone applications to the cut surface of decapitated internodes of *Phaseolus vulgaris* on the accumulation of ^{32}P in the subjacent tissues. (After Seth and Wareing, 1967.)

Treatment	Mean count rate (counts/minute)	Standard error ($\pm$)
Lanolin control	23·6	4·7
GA$_3$	84·8	16·9
Kinetin	58·1	17·8
IAA	453·4*	143·1
IAA + GA$_3$	849·6*	278·6
IAA + Kinetin	889·4*	255·4
IAA + GA$_3$ + Kinetin	1,943·1*	312·4

* Significantly different from lanolin control at 1 per cent probability.

GA_3, or IAA and kinetin has been found to result in greater accumulation of ^{32}P in the stump than with IAA alone (Seth and Wareing, 1964; Šebánek, 1966b; Wareing and Seth, 1967), and application of IAA, GA_3 and kinetin at the same time to a stem stump caused even greater ^{32}P accumulation (Davies, Seth, and Wareing, 1966; Seth and Wareing, 1967), (see Table 5.1). At first sight it might be said that these synergisms between auxin, cytokinin, and gibberellin indicate that it is the local growth-stimulatory effect of the hormones in the stem stump which is responsible for ^{32}P accumulation, but the observations that neither GA_3 nor kinetin induced ^{32}P accumulation (Davies and Wareing, 1965; Seth and Wareing, 1967) suggest otherwise, as both these hormones would be likely to stimulate local growth. An alternative explanation for synergism between added auxin and other hormones in ^{32}P accumulation is offered by the findings that auxin uptake and/or basipetal translocation in decapitated stems was significantly increased by the simultaneous addition of either kinetin or GA_3, and that such simultaneous hormone applications can also increase lateral bud inhibition relative to the effect of IAA alone (Davies, Seth and Wareing, 1966; Jacobs and Case, 1965; Pilet, 1965; McCready et al., 1965). Thus, it is possible that the effect of GA_3 or kinetin when supplied along with IAA is to increase the level of IAA in the stem, which in turn increases the inhibition of lateral buds.

There is, therefore, a body of evidence which is best interpreted in terms of the nutrient-diversion theory involving the concept of hormone-directed metabolite transport. Unfortunately there is also information available which cannot be so easily fitted to the nutrient-diversion theory. Thus, Goodwin and Cansfield (1967) found that correlatively-inhibited, non-dormant, lateral buds in potato tubers were not induced to grow by direct introduction of nutrient solutions containing cytokinin, auxin, and gibberellin to the inhibited buds, and yet the same nutrient solutions would support the growth of similar but isolated buds. These authors therefore concluded that absence of growth in potato tuber lateral buds cannot be ascribed to a shortage of common nutrients or growth substances. Also, Panigrahi and Audus (1966) found that following application of uracil-2-^{14}C to cotyledons of Vicia faba seedlings, there were no differences in the distribution of radioactivity in the stem in relation to decapitation or IAA-treatment of the stem stump. However, the concentration of uracil-2-^{14}C in axillary buds increased rapidly following decapitation, accumulation therefore preceding bud growth. Application of IAA to the stem stump caused both reduced bud outgrowth and reduced uracil-2-^{14}C accumulation in the bud. Panigrahi and Audus (1966) consequently concluded that diversion of nutrient flow towards regions of high auxin concentration plays no significant part in correlative inhibition. But it seems reasonable to suggest that although uracil accumulation in lateral buds released from apical dominance largely preceded growth in Panigrahi and Audus' experiments, this may have been no more than a reflection of necessary RNA synthesis in the very early stages of bud growth.

Panigrahi and Audus (1966) did not study incorporation of uracil-2-^{14}C into RNA or other cellular constituents. The lack of uracil-2-^{14}C accumulation in stem stumps supplied with IAA suggests that exogenous auxin did not stimulate RNA synthesis and growth in the stem tissues, a possibility which if true supports the data and conclusions of Davies and Wareing (1965). An apparently more serious deficiency in the nutrient diversion theory is suggested by the findings of Šebánek (1966a; 1967) who observed that in decapitated pea seedlings treatment of the stem stump with IAA in lanolin resulted in axillary bud inhibition only if the auxin concentration was greater than 0·25 per cent, and yet such higher concentrations promoted ^{32}P accumulation in the stump only slightly more than a concentration of 0·007 per cent IAA which actually *promoted* lateral bud outgrowth. Šebánek concluded that higher IAA concentrations inhibit the growth of axillary buds not by attracting nutrients, but because of a 'toxic effect' in the buds.

The question raised at this point by the results of Šebánek (1966a; 1967) as to whether axillary buds in decapitated plants can be inhibited by auxin at concentrations which are similar to those pertaining in the intact plant ('physiological concentrations') is difficult to answer with any degree of certainty. As early as 1934 Thimann and Skoog recorded that ten times more auxin than could be obtained by diffusion from an apical bud had to be applied to the top of a decapitated *Vicia faba* plant in order to see inhibition of lateral shoot growth. This result was confirmed for *Coleus* by Jacobs et al. (1959), but Libbert (1964) found that in pea a 'physiological concentration' of applied IAA did inhibit axillary buds. It might be significant here that in *Vicia faba* and *Coleus* there is less natural apical dominance than in pea, so that Libbert (1964) was measuring the effects of auxin on the growth of axillary buds whereas Jacobs et al. (1959) and Thimann and Skoog (1934) were concerned with the growth of fairly well developed lateral shoots. It will be remembered from earlier considerations in this chapter that direct applications of auxin have very different effects on axillary buds and lateral shoots.

If hormone-directed transport of some sort is a major component of the mechanism of apical dominance, one possibility worthy of consideration here is, whether apically synthesized auxin influences the distribution of endogenous cytokinins in the shoot. This is of particular interest due to the instances of release of buds from correlative inhibition by kinetin (references cited earlier), and evidence that the root system is a major source of cytokinins for the shoot (Loeffler and van Overbeek, 1964; Kende, 1964; Carr and Burrows, 1966). It is known also that auxin stimulates basipetal cytokinin transport in stem tissues (Seth, Davies, and Wareing, 1966; Osborne and Black, 1964; Black and Osborne, 1965; Lagerstedt and Langston, 1966) although it is much less certain that basipetally moving auxin can induce acropetal cytokinin transport. In fact there are a number of reports that cytokinin transport in isolated stem or petiole segments is predominantly basipetal (Osborne and Black, 1964; Black and Osborne, 1965; Lagerstedt and Langston, 1966) but on the other hand in the intact shoot cytokinin movement appears to be predominantly acropetal (Kamínek, 1965; Pilet, 1968). It is tempting to suggest therefore that it is the availability of cytokinins, perhaps derived from the roots, which determines the outgrowth of axillary buds and that apically synthesized auxin in the shoot 'directs' the flow of cytokinins to the stem apex.

In concluding this section we may consider possible ways in which the various examples of hormone-directed metabolite transport come about. There is the possibility that a metabolic sink is created at the point of auxin application, and there has been the suggestion that auxin has an effect along the whole path of metabolite transport, both of which have been considered above in some detail. Another possible mechanism involves the vascular connections between axillary bud and main stem. It was suggested by van Overbeek (1938) that high auxin concentrations in the stem interfere with the transporting qualities of the phloem of the vascular trace to the bud, but there has been no experimental evidence to support this idea. Histological studies in flax by Gregory and Veale (1957) revealed a correlation between speed of axillary bud growth following decapitation or treatment with nitrogen, and the amount of vascular tissue present at the base of the bud. Detailed studies on pea plants by Sorokin and Thimann (1964) demonstrated that the establishment of contact between the xylem differentiating basipetally in the bud and the xylem differentiating acropetally in the traces leading upwards to it coincided with the visible release from correlative inhibition in the bud, and that IAA inhibited the development of a xylem connection and bud growth. In *Vicia faba* it was found by Panigrahi and Audus (1966) that the total volume of phloem tissue at the base of a lateral bud increased rapidly after decapitation and that this increase could be prevented by application of IAA to the stem stump. It would appear then that the presence of basipetally moving auxin in the stem inhibits the establishment of a full vascular connection between axillary bud and stem, and it might be that it is this particular effect of apically produced auxin which is responsible for a limitation in the flow of growth-promoting substances (nutrients included) to the bud and consequent inhibition of bud growth. Sorokin and Thimann (1964) also observed that kinetin greatly hastened the contact between the two vascular strands and suggested that the effectiveness of kinetin in releasing axillary buds from correlative inhibition is due to its effect on xylem differentiation.

Correlative bud inhibition in woody species

It is probable that the basic mechanism of apical dominance is similar in herbaceous and woody species, but there has been some confusion in the past due to attempts to explain the adult forms of trees and shrubs in terms of apical dominance as it is understood in herbaceous plants. Thus, for example,

Kozlowski (1964) characterized trees with an excurrent branching habit (most conifers and a few angiosperms) as having strong apical dominance, and decurrent trees (most angiosperms) as having a weak apical dominance. However, as recently pointed out by Brown *et al.* (1967) the very reverse is really the case, for in decurrent species almost all lateral buds on the current year's twigs are completely inhibited and it is only in the spring after a period of dormancy that one or more of the uppermost lateral buds are released from apical dominance. Conversely, in excurrent forms, such as cone-shaped conifers with a well-defined trunk, many of the lateral buds on the current year's shoots elongate to varying degrees reflecting the existence of only weak apical dominance. Because of these features, Brown *et al.* (1967) have suggested that the term apical dominance is misleading when applied to tree crowns and that its use should be restricted to the pattern of bud inhibition on currently elongating shoots, and have proposed another term, *apical control*, as being more suitable to describe the physiological basis of tree form.

Other factors than apical dominance are therefore involved in the determination of shoot habit in woody plants. Attention has been drawn by Champagnat (1965) to the parts played by basal branch outgrowth ('basitonie') and gravimorphic influences (Wareing and Nasr, 1958; 1961; Smith and Wareing, 1964a, b; 1966) on the outgrowth of physically upper ('epitonie') or lower ('hypotonie') axillary buds on plagiotropic branches, in the establishment of typical shapes of shrubs and trees. The relative propensities of buds at different morphological positions ('basal' or 'apical' buds) to grow has also been revealed in experiments with herbaceous species such as flax (Gregory and Veale 1957) and pea (Nakamura, 1965). A further factor involved in the establishment of overall form of the shoot in woody perennials is that even in strongly excurrent species such as some conifers there is a gradual loss of 'apical control' with increasing tree size. This process has been called 'ageing' by Moorby and Wareing (1963) and leads to the loss of a clear leading shoot, and the formation of a rounded crown with no effective further increase in height.

It seems clear therefore, that studies of apical dominance in trees and shrubs should relate strictly to processes going on in actively growing first-year shoots. To date there have been a number of such investigations which in general reveal that the mechanism of correlative inhibition of axillary buds is basically similar in woody and herbaceous species (see, Kozlowski, 1964; Champagnat, 1961; 1965; Brown *et al.* 1967).

Orientation of rhizomes, stolons, branches, and leaves in relation to apical dominance

Although it is clear that light intensity, and perhaps photoperiod, are important factors in the control of plagiotropism or diageotropism in rhizomes and stolons

(Bennet-Clark and Ball, 1951; Palmer, 1956) there is evidence that apical dominance is also involved. It was shown by Palmer (1955) that positive geotropic responses of rhizomes in *Agropyron repens* occur in response to a transmitted stimulus from the aerial parts of the plant, the induction of which is produced by high light intensity. Similarly, Palmer (1956) showed that in several tropical grass species a light stimulus of the shoot evoked positive geotropism in the stolons, even if the latter were kept in darkness, which counterbalanced the stolons' inherent negative geotropism resulting in a plagiotropic growth habit. Also, removal of a plagiotropic rhizome from the parent plant often results in the upturning of the rhizome tip (negative geotropism) (Wareing, 1964) although on the other hand diageotropic rhizomes usually persist in horizontal growth even when severed from the parent plant (Bennet-Clark and Ball, 1951). Stolon development in potato appears to be at least partially under the control of the physiological activities of the stem apex, in that plagiotropism in the stolons can be changed to negative geotropism by decapitation and removal of axillary buds from the shoot (Booth, 1959). It appears also that endogenous growth hormones participate in the control of rhizome and stolon orientation, for treatment of potato plants with auxin and/or gibberellin can have profound effects upon stolon behaviour and these hormonal effects are related to apical dominance (Booth, 1959; 1963). Similarly, cytokinins appear to be involved in the same phenomena for Kumar (1966) found that treatment of plagiotropic potato stolon tips with kinetin or 6-benzylamino-purine resulted in upturning of the stolons and development of leaves, and that simultaneous treatment of decapitated plants with auxin, gibberellin and cytokinin resulted in the formation of horizontally-growing leafy lateral shoots instead of stolons. It is likely therefore, that rhizome and stolon behaviour is influenced by apical dominance and that growth hormones serve as correlation factors in the process.

In somewhat the same way it has been shown that the plagiotropic habit of leaves and lateral branches is influenced by apical dominance. Decapitation of the shoot commonly results in an upward (hyponastic, see chapter 8) movement of leaves and lateral branches, and application of an auxin to the stem stump can either maintain or increase (i.e., induce epinasty) the normal angle of orientation of such lateral organs depending upon the concentration applied (Snow, 1945; 1947; Leike and Guttenberg, 1961; Verner, 1938; 1955; Preston and Barlow, 1950; Jankiewicz *et al.* 1961; Palmer and Phillips, 1963).

Environmental effects in apical dominance phenomena

Relatively little attention has been paid to the effects of environment upon branching in the shoot, although it is recognized that effects of the environment in all classes of apical dominance phenomena are likely to be mediated through influences on both the hormonal and nutritional status of the plant.

Inorganic nutrients.   As has already been mentioned above, an ample supply of nitrogen can stimulate the growth of axillary buds and shoots. Nitrogen has also been reported to increase leaf epinasty (Palmer and Phillips, 1963). Less evidence is available on the effects of other minerals in apical dominance. Gardner (1942) recorded increased tillering in wheat under conditions of calcium and phosphorus deficiency, whereas Nakamura (1965) demonstrated slightly suppressed branching in pea when the level of calcium or magnesium was low, and reduced tillering was found under phosphorus-deficient conditions in barley (Brenchley, 1929) and rice (Takahashi *et al.*, 1957). Similarly, there are conflicting reports on the effects of potassium; Nakamura (1965) found no effect of potassium on bud growth in pea, but Hartt (1934) observed inhibition of tillering in sugar-cane by potassium deficiency and Takahashi *et al.* (1957) recorded that tiller production in rice was increased with reduced potassium availability. Also, epinasty in leaves has been found to be greater under conditions of potassium deficiency (Eckerson, 1931).

Light and temperature effects.   Generally, apical dominance is increased when plants are grown at low light intensity (Ooizumi and Nishiiri, 1956; Gregory and Veale, 1957; Takeuchi, 1957; Ryle, 1961; Nakamura, 1965), although Gregory and Veale (1957) found that in flax apical dominance with restricted nitrogen supply was independent of light intensity, indicating that photoassimilate availability is of less importance than inorganic nutrient supply. Nevertheless, Ballard and Wildman (1964) found that mitotic activity in sunflower cotyledonary buds was similarly stimulated by either excision of the apical bud or by applications of sucrose to the buds in intact plants.

The effects of light intensity on orientation of leaves and lateral branches have received some attention and it appears that high light favours epinasty and low light hyponasty (Snow, 1945; Bottelier, 1954).

Although the influence of light quality in apical dominance phenomena does not appear to have been investigated, except for the finding that red light enhances leaf epinasty (Meijer, 1957), there are a number of recorded cases where daylength had marked effects on the magnitude of apical dominance. Usually, short days (SD) promote axillary bud and branch growth and long days (LD) enhance apical dominance (Garner and Allard, 1923; Doroshenko, 1927; Yoshii, 1927; Kondo, 1934; Chinoy and Nanda, 1951; Kitamura and Kubo, 1952; Piringer and Cathey, 1960; Nakamura, 1965). However, complex interactions occur between photoperiod and temperature in apical dominance (Reath and Wittwer, 1952; Nakamura, 1965). The effects of temperature, and diurnal fluctuations of this parameter, have been most rigorously studied in pea by Nakamura (1965).

It is difficult to interpret the effects of environmental changes on axillary bud growth, since it has been frequently observed that buds at different morphological positions exhibit opposite growth responses to changes in the same environmental factor. This is particularly true for photoperiod and temperature (Gregory and Veale, 1957; Nakamura, 1965), and it appears likely that this is at least partially attributable to interactions between vegetative growth and reproductive development. This suggestion is supported by experiments which showed effects of seed-chilling on both branching and flowering behaviour in plants grown from such seeds (Shinohara and Sano, 1957; Nakamura, 1965).

Clearly more work is required to clarify relationships between components of the environment and apical dominance. Even nitrogen, whose supply appears often to limit axillary bud growth, may have only an indirect influence by virtue of effects on the hormonal or carbohydrate status of the plant. There is, for example, some evidence that mineral supply can influence auxin synthesis in the shoot tip (Avery et al., 1937) and carbohydrate balance (see Allsopp, 1965). Also, there are many recorded instances of temperature and daylength affecting levels of auxins and gibberellins in plants (see Nitsch, 1963). In the same way, where synthetic growth retardants such as Cycocel, Amo-1618, Phosfon-D and B-995 have been found to reduce apical dominance (Oota, 1962; Pharis et al., 1965; Ruddat and Pharis, 1966) it is likely that these compounds simulated short days in their effects on endogenous gibberellin levels.

Other environmental effects. Gravity can often play an important role in the control of leaf and branch orientation (Lyon, 1963a, b; Palmer, 1964) and correlative inhibition of axillary buds (Wareing and Nasr, 1958; 1961; Smith and Wareing, 1964a, b; 1966). The results of Wareing and co-workers were obtained during studies of 'gravimorphism' (Wareing and Nasr, 1958) involving the training of shoots of woody plants in various ways, which in general showed that moving a normally dominant shoot from its previous vertical to a horizontal position, or completely looping the main stem, resulted in a reduction of apical dominance. This led Wareing and Nasr (1961) to postulate that nutrients are diverted to the highest upwardly directed meristem, and that other things being equal, proximity of laterals to the roots appears to confer an advantage. This suggestion is very reminiscent of that of Loeb (1917), discussed earlier in this account, who proposed that there exists 'shoot-forming material' which has a natural 'tendency to rise' to the highest point in the stem.

Finally, anaerobic conditions in the soil around roots can enhance the effect of the apical bud in inducing epinastic curvatures in leaves (Kramer, 1951; Jackson, 1956; Phillips, 1964a, b), and here again there is evidence that this environmental effect is mediated through endogenous auxins and gibberellins (Phillips, 1964a, b; Phillips and Jones, 1964).

Summary

The best studied aspect of apical dominance is *correlative inhibition* of axillary buds or shoots by the apical bud.

The 'nutritive' theory

The earliest ideas of the physiological basis of correlative inhibition were expressed in the '*nutritive*' *theory*. This held that the apical bud constitutes a metabolic sink to which most available nutrients flow solely in response to their concentration gradients. Thus, according to this view, since the apical meristem is already present in the embryo it will, at germination, proceed to grow before later-formed axillary bud meristems and so monopolize available nutrients. In other words, there is a 'struggle for existence' between meristems and the oldest meristem will always enjoy a better supply of necessary nutrients than younger meristems. A number of workers obtained experimental results which were interpreted in terms of the nutritive theory (particularly Goebel, Loeb, and Dostál). These experiments revealed that the extent to which leaves, or whole regenerating shoots, grow is closely related to the availability of nutritive substances. It was therefore assumed by these early investigators that lack of axillary bud growth is simply a consequence of a drain of all available nutrients to the apical bud.

Diffusible correlative factors

Some early experiments (including some by Loeb and Dostál) yielded results which pointed to the existence of special diffusible substances concerned in correlative bud inhibition. Thus, experiments with isolated stems of *Bryophyllum calycinum* (Loeb, 1917b) showed that a growing bud transmitted some principle inhibitory to the growth of other buds, and that this inhibitory principle moved basipetally in the stem. It was later shown by Snow (1925) that this inhibitor could pass through a water-filled gap between adjacent stem tissues. Also, Harvey (1920) and Snow (1925) found that axillary buds could be released from correlative inhibition by killing a region of the stem above the buds, and yet at the same time the main apex was still growing and consuming nutrients. This argued very strongly against the simple nutritive theory.

The 'direct' and 'indirect' theories of auxin inhibition in apical dominance

It was reported by Thimann and Skoog (1933; 1934) that an application of auxin to the cut stem-surface in decapitated plants simulated the apical bud with respect to correlative inhibition of axillary buds. This, together with knowledge that the apical bud is the principal site of auxin synthesis in the shoot, led to the suggestion that it is auxin diffusing from the apical bud which is responsible for correlative bud inhibition. Some workers, particularly Thimann and Skoog (1934) and Thimann (1937) favoured the view that apically synthesized auxin enters axillary buds and directly inhibits their growth due to a greater sensitivity of lateral buds to auxin. This concept of apical dominance was contained in the *'direct' theory of auxin inhibition*, and was most vigorously promulgated by Thimann (1937). A number of other workers, particularly Snow, obtained results which led to the suggestion that auxin does *not* act in such a direct manner as that proposed by Thimann, but rather that apically secreted auxin has an *indirect action* in bringing about correlative inhibition of buds and shoots. This idea came to be known as the *'indirect' theory of auxin inhibition.* Much of Snow's work involved elegant surgical procedures which demonstrated the ability of the correlative inhibitory influence to move both acropetally and basipetally for some considerable distance in the plant. As auxin moves predominantly in a basipetal direction, Snow pointed out that correlative inhibition could not be achieved if auxin itself were the inhibitory principle. But, modern research with isotopically labelled auxin has revealed that auxin can move acropetally as well as basipetally, though not so readily, and that apically-applied auxin can move down the stem and up into axillary buds (Wickson and Thimann, 1960). Even so, there is a considerable body of evidence available which shows that the auxin concentration in correlatively inhibited buds may be sub-optimal rather than supra-optimal, and this certainly does not support the 'direct' theory.

Growth-inhibitors and apical dominance

Hormones other than auxin may also be involved in apical dominance phenomena. As early as 1937, Snow suggested that there exists a special correlative growth-inhibitor, and that the formation of this substance is enhanced in stem tissues by the presence of auxin. As a corollary of this suggestion, Snow (1937) proposed that auxin 'protects' a stem from inhibition by the inhibitor, but that auxin-deficient tissues (such as axillary buds) are not resistant to the growth-inhibitor. In more recent years the existence of endogenous growth-inhibitors in plants has been well established, and a number of workers have shown that correlatively inhibited lateral buds contain higher concentrations of inhibitors than do buds released from apical dominance (Libbert, 1964; Dörffling 1963a, b; 1964; 1965; 1966; Goodwin and Cansfield, 1967). Despite this fact, it is clear that much work is required before growth-inhibitors can be regarded as playing a role in correlative bud inhibition.

Cytokinins and apical dominance

Cytokinins do, however, appear to play a part in apical dominance. This has been revealed by experiments which have shown that axillary buds can be released from correlative inhibition by treatment with kinetin (Wickson and Thimann, 1958; von Maltzahn, 1959; Sachs and Thimann, 1964; 1967; Panigrahi and Audus, 1966). In all cases of bud release by kinetin, however, it was found that growth continued for only a short time, after which apical dominance was reimposed. Only when auxin was supplied directly to outgrowing kinetin-treated buds did growth continue (Sachs and Thimann, 1967). Consequently it appears that sufficient cytokinin and auxin have to be present before a bud is able to grow. We know from other research that cytokinins are exported from the root to the shoot, and it is possible that the availability of these endogenous cytokinins to axillary buds limits their initial outgrowth.

Gibberellins and apical dominance

Treatment of intact plants with a gibberellin can often lead to enhanced apical dominance. It has been suggested by several workers (Pharis et al., 1965; Jacobs and Case, 1965; Ruddat and Pharis, 1966; Scott et al., 1967) that gibberellin participates with auxin in maintaining lateral buds in an inhibited state. Whether this is so is certainly not at all clear at the present time. Gibberellin treatment commonly results in an increased growth rate in the dominant shoot of a plant, and it seems that it is only as a consequence of this more vigorous growth that the dominant shoot achieves even greater dominance over weaker buds or shoots. The reason for this is obscure, but may involve increased nutrient drain to the dominant shoot, and/or increased auxin synthesis in the dominant shoot.

Unfortunately, there are a few reports in the literature concerning gibberellin and apical dominance that are difficult to interpret. These are that gibberellic acid (GA_3) increases the inhibitory effect of IAA on axillary buds, when both hormones are supplied to the upper cut end of the stem in a decapitated plant (Jacobs and Case, 1965; Scott et al., 1967). The difficulty in understanding these latter reports is increased by observations that GA_3 antagonizes the inhibitory effect of IAA in correlative inhibition (Kato, 1958; Wickson and Thimann, 1958; Nakamura, 1965). Thus, we cannot yet be certain that gibberellins play any direct role in apical dominance.

The nutrient-diversion theory

Apart from the 'nutritive', 'direct', and 'indirect' theories of apical dominance which we have already summarized, there is a fourth concept known as the 'nutrient-diversion' theory. This was first formulated by Went (1936; 1939), and suggests that nutrients flow preferentially towards regions of high auxin concentration. Thus, the nutrient-diversion theory differs from the earlier nutritive theory in that the latter held that nutrients moved in the plant solely in response to concentration gradients, whereas the nutrient-diversion theory suggests that it is some physiological effect of auxin, other than that of promoting growth, which directs metabolite translocation.

Many studies with isotopes such as ^{32}P and ^{14}C-sucrose have demonstrated that nutrients do indeed move to and accumulate in regions of high auxin concentration. In fact, if not only auxin but also gibberellin and/or cytokinin are applied to the cut end of a stem, then even greater nutrient accumulation occurs in the hormone-treated region. This general phenomenon is known as *hormone-directed transport*, but how it comes about is not certain. Possibilities which can be seen are:

(a) That there is stimulated growth at the point of hormone application which creates a metabolic sink (such an interpretation cannot be excluded, but appears unlikely in view of certain experimental results considered in detail).

(b) That auxin has an effect along the whole path of metabolite transport (this possibility also cannot be excluded, for the results of Davies and Wareing (1965) cannot be easily interpreted in any other way).

(c) That auxin moving down the stem inhibits the development of vascular connections between axillary buds and the main stele, and that it is because of lack of good vascular connections that axillary buds are inhibited due to a resulting impedence in supply of nutrients. (The correlation between vascular connection development and axillary bud growth has been established in histological studies by Gregory and Veale (1957), Sorokin and Thimann (1964) and Panigrahi and Audus (1966).)

There seems little doubt that hormone-directed transport, whatever its basis, plays an important part in correlative bud inhibition. We have in this account suggested that it is the availability of cytokinins, perhaps derived originally from the roots, which determines the outgrowth of axillary buds and that apically synthesized auxin in the shoot 'directs' the flow of cytokinins to the apical bud.

Correlative bud inhibition in woody species

There has been, in the past, some confusion due to attempts to explain the adult forms of trees and shrubs in terms of apical dominance. It is now clear that the mechanism of correlative inhibition is similar in herbaceous and woody species, but that in the latter the final form of the shoot is determined not only by apical dominance, but also by certain other influences such as 'gravimorphism' and 'ageing'. 'Apical control' is a recently suggested term to describe the physiological basis of tree form.

Orientation of rhizomes, stolons, branches, and leaves in relation to apical dominance

The apical bud has influence on the orientation and development of lateral organs. In most cases this effect of the stem apex appears to be produced by the agency of endogenous auxins, gibberellins and cytokinins.

Environmental effects in apical dominance phenomena

Inorganic nutrients, particularly nitrogen, when in plentiful supply often lower the degree of apical dominance in the shoot. To a lesser extent light intensity, photoperiod, temperature, gravity, and soil aeration all influence apical dominance.

Much more work is needed to clarify relationships between components of the environment and apical dominance, but at present it appears that environmental influences are mediated through effects on endogenous hormone levels and distribution within the plant.

Bibliography

Selected further reading

Audus, L. J., 1959. Correlations. *J. Linn. Soc. (Bot.)*, **56**, 177–187.
Booth, A., 1959. Some factors concerned in the growth of stolons in potato. *J. Linn. Soc. (Bot.)*, **56**, 166–169.
Brown, C. L., R. G. McAlpine and P. P. Kormanik, 1967. Apical dominance in woody plants: A reappraisal. *Am. J. Bot.* **54**, 153–162.
Champagnat, P., 1961. Dominance apicale. Tropismes, épinastie. In *Encyc. Plant Physiol.* XIV, ed. W. Ruhland. Springer-Verlag, Berlin, pp. 872–908.
Champagnat, P., 1965. Physologie de la croissance et l'inhibition des bourgeons: Dominance apicale et phenomènes analogues. In *Encyc. Plant Physiol.* XV/I, ed. W. Ruhland. Springer-Verlag, Berlin, pp. 1106–1164.
Gregory, F. G. and J. A. Veale, 1957. A reassessment of the problem of apical dominance. *Symp. Soc. Exp. Biol.* **XI**, 1–20.
Loeb, J. 1924. *Regeneration from a physicochemical viewpoint*. McGraw-Hill, New York.
Snow, R., 1937. On the nature of correlative inhibition. *New Phytol.* **36**, 283–300.

Papers cited in text

Allsopp, A., 1965. The significance for development of water supply, osmotic relations and nutrition. In *Encyc. Plant Physiol.* **XV/I**, ed. W. Ruhland. Springer-Verlag, Berlin, pp. 504–552.
Audus, L. J., 1959. Correlations. *J. Linn. Soc. (Bot.)*, **56**, 177–187.
Avery, G. S., P. R. Burkholder, and H. B. Creighton, 1937. Nutrient deficiencies and growth hormone concentration in Helianthus and Nicotiana. *Am. J. Bot.* **24**, 553–557.
Baldev, B., A. Lang, and A. O. Agatep, 1965. Gibberellin production in pea seeds developing in excised pods. Effect of growth retardant AMO-1618. *Science*, **147**, 155–157.

Ballard, L. A. T. and S. G. Wildman, 1964. Induction of mitosis in excised and attached dormant buds of sunflower (*Helianthus annuus*). *Austr. J. Biol. Sci.* **17**, 36–43.

Bennet-Clark, T. A. and N. G. Ball, 1951. The diageotropic behaviour of rhizomes. *J. Exp. Bot.* **5**, 169–203.

Black, M. K. and D. J. Osborne, 1965. Polarity of transport of benzyladenine, adenine and indole-3-acetic acid in petiole segments of *Phaseolus vulgaris*. *Plant Physiol.* **40**, 676–680.

Bodlaender, K. B. A., 1963. Influence of temperature, radiation and photoperiod on development and yield. In *The Growth of the Potato*, ed. J. D. Ivins and F. L. Milthorpe, Butterworths, London, pp. 199–210.

Booth, A., 1959. Some factors concerned in the growth of stolons in potato. *J. Linn. Soc.* **56**, 166–169.

Booth, A., 1963. The role of growth substances in the development of stolons. In *The Growth of the Potato*, ed. J. D. Ivins and F. L. Milthorpe. Butterworths, London, pp. 99–113.

Booth, A., J. Moorby, C. R. Davies, H. Jones, and P. F. Wareing, 1962. Effects of indolyl-3-acetic acid on the movement of nutrients within plants. *Nature, Lond.* **194**, 204–205.

Bottelier, H. P., 1954. *Ageratum Houstonianum* Mill. as test object for growth substances. I. The relation between epinastic curving of the leaf petioles and the concentration of indole-acetic acid and the effect of light thereon. *Ann. Bogoriensis*, **1**, 185–200.

Bradley, V. and J. C. Crane, 1960. Gibberellin induced inhibition of bud development in some species of Prunus. *Science*, **131**, 825–826.

Brenchley, W. E., 1929. The phosphate requirement of barley at different periods of growth. *Ann. Bot.* (Lond.), **43**, 89–100.

Brian, P. W., 1959. Effects of gibberellins on plant growth and development. *Biol. Rev.* **34**, 37–84.

Brian, P. W., H. G. Hemming and M. Radley, 1955. A physiological comparison of gibberellic acid with some auxins. *Physiol. Plant.* **8**, 899–912.

Brian, P. W. and H. G. Hemming, 1957. The effect of maleic hydrazide on the growth response of plants to gibberellic acid. *Ann. Appl. Biol.* **45**, 489–497.

Brian, P. W., H. G. Hemming, and D. Lowe, 1959. The effect of gibberellic acid on shoot growth of cupid sweet peas. *Physiol. Plant.* **12**, 15–29.

Brown, C. L., R. G. McAlpine, and P. P. Kormanik, 1967. Apical dominance in woody plants: A reappraisal. *Am. J. Bot.* **54**, 153–162.

Bruinsma, J. and S. S. Patil, 1963. The effects of 3-indoleacetic acid, gibberellic acid and vitamin E on flower initiation in unvernalized Petkus winter rye plants. *Naturwiss.* **50**, 505.

Bünsow, R., 1961a. Zur physiologie der achsengestaltung bei Kalanchoë Blossfeldiana I. Die Hauptachse. *Planta*, **57**, 71–87.

Bünsow, R., 1961b. Zur physiologie der achsengestaltung bei Kalanchoë Blossfeldiana II. Die Seitenachsen. *Planta*, **57**, 88–110.

Carr, D. J. and W. J. Burrows, 1966. Evidence of the presence in xylem sap of substances with kinetin-like activity. *Life Sci.* **5**, 2061–2077.

Champagnat, P., 1955. Les correlations entre fenilles et bourgeons de la pousse herbacee du Lilas. *Rev. Gen. Bot.* **62**, 325–372.

Champagnat, P., 1965. Physiologie de la croissance et l'inhibition des bourgeons: Dominance apicale et phenomenes analogues, In *Encyc. Plant Physiol.* **XV/I**, ed. W. Ruhland. Springer-Verlag, Berlin, pp. 1106–1164.

Chinoy, J. J. and K. K. Nanda, 1951. Effect of vernalization and photoperiodic treatments on growth and development of crop plants. *Physiol. Plant.* **4**, 427–436.

Davies, C. R., 1963. *Studies on nutrient transport in relation to indolyl-acetic acid and other plant hormones.* Ph.D. Thesis, University of Wales.

Davies, C. R., A. K. Seth, and P. F. Wareing, 1966. Auxin and kinetin interaction in apical dominance. *Science*, **151**, 468–469.

Davies, C. R. and P. F. Wareing, 1965. Auxin-directed transport of radiophosphorus in stems. *Planta* (Berl.), **65**, 139–156.

Dennis, D. T., C. D. Upper, and C. A. West, 1965. An enzymic site of inhibition of gibberellin biosynthesis by AMO-1618 and other plant growth retardants. *Plant Physiol.* **40**, 948–952.

Dörffling, K., 1963a. Die bedeutung von Inhibitor β für die korrelative hemmung und für die winterruhe der knospen von *Acer pseudoplatanus*. *Planta* (Berl.) **59**, 346–350.

Dörffling, K., 1963b. Über das wuchsstoffe-heminstoffsystem von *Acer pseudoplatanus* L. I. Der jahresgang der wuchs- und hemmstoffe in knospen blättern und im kambium. *Planta* (Berl.) **60**, 390–412.

Dörffling, K., 1964. Über das wuchsstoffe-hemmstoffsystem von *Acer pseudoplatanus* L. II. Die bedeutung von 'Inhibitor β' für die korrelative knospenhemmung und für die regulation der kambium tätigkeit. *Planta* (Berl.), **60**, 413–433.

Dörffling, K., 1965. Die rolle von natürlichen wachstumsregulatoren bei der korrelativen knospenhemmung (Versuche mit *Pisum sativum*). *Ber. Deut. Botan. Ges.* **78**, 122–128.

Dörffling, K., 1966. Weitere untersuchungen über korrelative knospenhemmung. Anwendung zweier biotests mit knospen bei der papierchromatographischen untersuchung von extrakten aus *Pisum* Pflanzen. *Planta* (Berl.), **70**, 257–274.

Doroshenko, A., 1927. Photoperiodism of some cultivated crops with relation to their geographical origins. *Bull. Appl. Bot. Gen. Plant Breeding*, **17**, 167–220.

Dostál, R., 1909. Die korrelationsbeziehung zwischen dem Blatt und seiner Axillarknospe. *Ber. Deut. Bot. Ges.* **27**, 547–554.

Dostál, R., 1926. Über die wachstumsregulierende Wirkung des Laubblattes. *Act. Soc. Sci. Nat. Moravicae*, **3**, 83–209.

Eckerson, S. H., 1931. Influence of phosphorus deficiency on metabolism of the tomato (*Lycopersicum esculentum*). *Contr. Boyce Thompson Inst.* **3**, 197–217.

Errera, L., 1904. Conflits de préséance et excitations inhibitoires chez les vegetaux. *Bull. Soc. Roy. Bot.* Belgique, **42**, 27–43.

Ferman, J. H. G., 1938. The rôle of auxin in the correlative inhibition of the development of lateral buds and shoots. *Rec. Trav. Bot. Néerl.* **35**, 177–287.

Galston, A. W. and D. C. McCune, 1961. An analysis of gibberellin-auxin interaction and its possible metabolic basis. In *Plant Growth Regulation*, ed. R. M. Klein. Iowa State University Press. pp. 611–625.

Gardner, J. L., 1942. Studies in tillering. *Ecology*, **23**, 162–174.

Garner, W. W. and H. A. Allard, 1923. Further studies in photoperiodism. *J. Agr. Res.* **23**, 871–920.

Goebel, K., 1900. *Organography of plants especially of the Archegoniatae and Spermaphyta*. Part I. General Organography. Clarendon Press, Oxford.

Goodwin, P. B. and P. E. Cansfield, 1967. The control of branch growth on potato tubers III. The basis of correlative inhibition. *J. Exp. Bot.* **18**, 297–307.

Gouwentak, C. A., 1936. Kambiumtätigkeit und Wuchsstoff I. *Meded. Landb Hoogesch. Wagenginen*, **40**, pt. 3.

Gregory, F. G. and J. A. Veale, 1957. A reassessment of the problem of apical dominance. *Symp. Soc. Exp. Biol.* **XI**, 1–20.

Gunckel, J. W., K. V. Thimann, and R. H. Wetmore, 1949. Studies of development in long shoots and short shoots of *Ginkgo biloba* IV. Growth habit, shoot expression and the mechanism of its control. *Am. J. Bot.* **36**, 309–316.

Gunning, B. E. S. and W. K. Barkley, 1963. Kinin-induced directed transport and senescence in detached oat-leaves. *Nature (Lond.)* **199**, 262–265.

Halevy, A. H., 1963. Interaction of growth-retarding compounds and gibberellin on indoleacetic acid oxidase and peroxidase of cucumber seedlings. *Plant Physiol.* **38**, 731–737.

Harada, H. and A. Lang, 1965. Effect of some 2-chloroethyl trimethylammonium chloride analogs and other growth retardants on gibberellin biosynthesis in *Fusarium moniliforme. Plant Physiol.* **40**, 176–183.

Hartt, C. E., 1934. Some effects of potassium upon the growth of sugar cane and upon the absorption and migration of ash constituents. *Plant Physiol.* **9**, 399–452.

Harvey, E. N., 1920. An experiment on regulation in plants. *Am. Nat.* **54**, 362–367.

Hyashi, T., and Y. Murakami, 1953. The biochemistry of the bakanai fungus. The physiological action of gibberellin. The effect of gibberellin on the straight growth of etiolated pea epicotyl sections. *J. Agr. Chem. Soc. Japan*, **27**, 675.

Jackson, W. P., 1956. The relative importance of factors causing injury to shoots of flooded tomato plants. *Am. J. Bot.* **43**, 637–639.

Jacobs, W. P. and B. Bullwinkel, 1953. Compensatory growth in *Coleus* shoots. *Am. J. Bot.* **40**, 385–392.

Jacobs, W. P. and D. B. Case, 1965. Auxin transport, gibberellin, and apical dominance. *Science*, **148**, 1729–1731.

Jacobs, W. P., J. Danielson, V. Hurst, and P. Adams, 1959. What substance normally controls a given biological process? II. The relation of auxin to apical dominance. *Devel. Biol.* **1**, 534–554.

Jankiewicz, L. S., H. Szpunar, R. Baranska, R. Rumplowa, and K. Fiutowska, 1961. The use of auxin to widen crotch angles in young apple trees. *Acta Agrobot.* **10**, 151.

Jones, R. L. and I. D. J. Phillips, 1966. Organs of gibberellin synthesis in light-grown sunflower plants. *Plant Physiol.* **41**, 1381–1386.

Jones, R. L. and I. D. J. Phillips, 1967. Effect of CCC on the gibberellin content of excised sunflower organs. *Planta* (Berl.), **72**, 53–59.

Kamínek, M., 1965. Acropetal transport of kinetin in pea sections. *Biol. Plant.* (Praha), **7**, 394–396.

Kato, J. 1953. Studies on the physiological effect of gibberellin: I. On the differential activity between gibberellin and auxin. *Mem. Coll. Sci. Kyoto.* Ser. B. **20**, 189–193.

Kato, J. 1958. Studies on the physiological effect of gibberellin: II. On the interaction of gibberellin with auxins and growth inhibitors. *Physiol. Plant.* **11**, 10–15.

Kende, H., 1964. Preservation of chlorophyll in leaf sections by substances obtained from root exudate. *Science*, **145**, 1066–1067.

Kende, H., H. Ninnemann, and A. Lang, 1963. Inhibition of gibberellin biosynthesis in *Fusarium moniliforme* by AMO-1618, and CCC. *Naturwiss.* **50**, 599–600.

Kitamura, S. and M. Kubo, 1952. Influence of daylength on peas. *Rep. Congr. Assoc. Hort. Japan*, **22**.

Kondo, M., 1934. Untersuchungen über 'photoperiodismus' der reispflanzen II. *Ohara Inst. Landw. Forsch.* (Berl.) **6**, 307–322.

Kozlowski, T. T., 1964. Shoot growth in woody plants. *Bot. Rev.* **30**, 335–392.

Kramer, P. J., 1951. Causes of injury to plants resulting from flooding of the soil. *Plant Physiol.* **45**, 722–736.

Kumar, D., 1966. *The physiology of stolon development, tuberization and dormancy in the potato.* Ph.D. Thesis, University of Wales.

Kuraishi, S. and R. M. Muir, 1962. Increase in diffusible auxin after treatment with gibberellin. *Science*, **137**, 760–761.

Kuraishi, S. and R. M. Muir, 1963. Paper chromatographic study of diffusible auxin: *Plant Physiol.* **39**, 23–28.

Kuraishi, S. and R. M. Muir, 1964a. The relationship of gibberellin and auxin in plant growth. *Plant and Cell Physiol.* **5**, 61–68.

Kuraishi, S. and R. M. Muir, 1964b. The mechanism of gibberellin action in the dwarf pea. *Plant and Cell Physiol.* **5**, 259–271.

Lagerstedt, H. B. and R. G. Langston, 1966. Transport of kinetin-8-^{14}C in petioles. *Physiol. Plant.* **19**, 734–740.

Laibach, F., 1933. Wuchstoffversuche mit lebenden Orchideenpollinen. *Ber. Deut. Bot. Ges.* **51**, 336–340.

Le Fanu, B., 1936. Auxin and correlative inhibition. *New Phytol.* **35**, 205–220.

Leike, H. and H. von Guttenberg, 1961. Über die Bedeutung des Auxins für die epinastiche Bewegung plagiotroper Seitensprosse. *Naturwiss.* **48**, 604–605.

Libbert, E., 1954a. Zur Frage nach der Natur der korrelativen Hemmung. *Flora*, **141**, 269–297.

Libbert, E. 1954b. Das Zusammenwirkung von Wuchs-und Hemmstoffen bei der korrelativen knospenhemmung. *Planta* (Berl.), **44**, 286–318.

Libbert, E., 1959. Trijodobenzolsäure und die stoffleitung bei höhren pflanzen. *Planta* (Berl.), **53**, 612–627.

Libbert, E., 1964. Significance and mechanism of action of natural inhibitors. In *Régulateurs Naturels de la Croissance Végétale*. Éditions du Centre National de la Recherche Scientifique, Paris, pp. 387–405.

Loeb, J., 1915. Rules and mechanism of inhibition and correlation in the regeneration of *Bryophyllum calycinum*. *Bot. Gaz.* **60**, 249–276.

Loeb, J., 1917a. The chemical basis of regeneration and geotropism. *Science*, **46**, 115–118.

Loeb, J., 1917b. The chemical basis of axial polarity in regeneration. *Science*, **46**, 547–551.

Loeb, J. 1918. Chemical Basis of Correlation I. Production of equal masses of shoots by equal masses of sister leaves in *Bryophyllum calycinum*. *Bot. Gaz.* **65**, 150–174.

Loeffler, J. E., and J. van Overbeek, 1964. Kinin activity in coconut milk. In *Régulateurs Naturels de la Croissance Végétale*. Édition du Centre National de la Recherche Scientifique (Paris), pp. 77–82.

Lona, F. and A. Bocchi, 1956. Caratteristiche d'accrescimento e sviluppo di alcune rasse invernali di cereali trattati con acido gibberellico. *Riv. intern. Agr.* **6**, 44–47.

Lyon, C. J., 1963a. Auxin factor in branch epinasty. *Plant Physiol.* **38**, 145–152.

Lyon, C. J., 1963b. Auxin transport in leaf epinasty. *Plant Physiol.* **38**, 567–574.

Maltzahn, K. E. von, 1959. Interaction between kinetin and indoleacetic acid in the control of bud reactivation in *Splachnum ampullaceum* (L). Hedw. *Nature* (*Lond.*) **183**, 60–61.

Marth, P. C., W. V. Andia, and J. W. Mitchel, 1956. Effects of gibberellic acid on growth and development of plants of various genera and species. *Bot. Gaz.* **118**, 106–111.

McIntyre, G. I., 1964. Mechanism of apical dominance in plants. *Nature* (*Lond*), **203**, 1190–1191.

McIntyre, G. I., 1965. Some effects of the nitrogen supply on the growth and development of *Agropyron repens*. *Weed Res.* **5**, 1–12.

McCready, C. C., 1963. Movement of growth regulators in plants. I. Polar transport of 2,4-dichlorophenoxyacetic acid in segments from the petioles of *Phaseolus vulgaris*. *New Phytol.* **62**, 3–18.

McCready, C. C., Osborne, D. J. and M. K. Black, 1965. Promotion by kinetin of the polar transport of two auxins. *Nature* (*Lond.*), **208**, 1065–1067.

Meijer, G., 1957. The influence of light quality on the flowering response of Salvia occidentalis. *Acta. Bot. Neerl.* **6**, 395–406.

Meinl, G. and H. von Guttenberg, 1954. Über Förderung und Hemming der Entwicklung von Axillarsprossen durch Wirkstoffe. *Planta*, **44**, 121–135.

Mitchell, J. W. and W. E. Martin, 1937. Effect of indoleacetic acid on growth and chemical composition of etiolated bean plants. *Bot. Gaz.* **99**, 171–183.

Mogk, W., 1913. Untersuchungen über Korrelationen von Knospen und Sprossen. *Arch. EntwMech. Org.* **38**, 584.

Moorby, J. and P. F. Wareing, 1963. Ageing in woody plants. *Ann. Bot.* (Lond.). **27**, 291–308.

Mothes, K., 1960. Über das altern der blätter und die möglichkeit ihrer wiederverjüngung. *Naturwiss.* **47**, 337–351.

Mothes, K. and L. Engelbrecht, 1961. Kinetin-induced directed transport of substances in excised leaves in the dark. *Phytochem.* **1**, 58–62.

Müller, A. M., 1935. Über den Einfluss von Wuchsstoff auf das Austreiben der Seitenknospen und auf die Wurzelbildung. *Jahrb. wiss. Bot.* **81**, 497–540.

Nakamura, E., 1965. *Studies on the branching in Pisum sativum L.* Special Report of the Laboratory of Horticulture, Shiga Agricultural College (Japan).

Niedergang-Kamien, E. and F. Skoog, 1956. Studies on polarity and auxin transport, I. Modification of polarity and auxin transport by tri-iodobenzoic acid. *Physiol. Plant.* **9**, 60–73.

Ninnemann, H., H. Kende, J. A. D. Zeevaart, and A. Lang. The plant growth retardant CCC as inhibitor of gibberellin biosynthesis in *Fusarium moniliforme*. *Planta* (Berl.), **61**, 229–235.

Nitsch, J. P., 1963. The mediation of climatic effects through endogenous regulating substances. In *Environmental Control of Plant Growth*, ed. L. T. Evans. Academic Press, N.Y. and London, pp. 175–214.

Ooizumi, H. and K. Nishiiri, 1956. Effects of shading during the early part of growing period of soybean plants on their growth and their nitrogen and carbohydrate contents. *Proc. Crop Sci. Soc. Japan*, **24**, 188.

Osborne, D. J. and M. K. Black, 1964. Polar transport of a kinin, benzyladenine, *Nature* (*Lond.*), **201**, 97.

Osborne, D. J. and M. Hallaway, 1959. The role of auxins in the control of leaf senescence: Some effects of local applications of 2,4-D on carbon and nitrogen metabolism. In *Plant Growth Regulation*. IVth Intern. Congr. on Plant Growth Regulation. Iowa State Univ. Press, pp. 329–338.

Overbeek, J. van, 1938. Auxin distribution in seedlings and its bearing on the problem of bud inhibition. *Bot. Gaz.* **100**, 133–166.

Palmer, J. H., 1955. *An investigation into the behaviour of the rhizome of* Agropyron repens *Beauv., with special reference to the effect of light*. Ph.D. Thesis, University of Sheffield.

Palmer, J. H., 1956. The nature of the growth response to sunlight shown by certain stoloniferous and prostrate tropical plants. *New Phytol.* **55**, 346–355.

Palmer, J. H., 1964. Comparative study of the effects of applied indoleacetic acid and horizontal orientation of the primary shoot, upon internode extension and petiole orientation in *Helianthus annuus*, and the modifying influence of gibberellic acid. *Planta*, **61**, 283–297.

Palmer, J. H. and I. D. J. Phillips, 1963. The effect of the terminal bud, indoleacetic acid, and nitrogen supply on the growth and orientation of the petiole of *Helianthus annuus*. *Physiol. Plant.* **16**, 572–584.

Panigrahi, B. M. and L. J. Audus, 1966. Apical dominance in *Vicia faba*. *Ann. Bot.* N.S. **30**, 457–473.

Pharis, R. P., M. Ruddat, C. Phillips, and E. Heftmann, 1965. Gibberellin, growth retardants, and apical dominance in Arizona cypress. *Naturwiss.* **52**, 88–89.

Phillips, I. D. J., 1964a. Root-shoot hormone relations. I. The importance of an aerated root system in the regulation of growth hormone levels in the shoot of *Helianthus annuus*. *Ann. Bot.* N.S. **28**, 17–35.

Phillips, I. D. J., 1964b. Root-shoot hormone relations. II. Changes in endogenous auxin concentration produced by flooding of the root system in *Helianthus annuus*. *Ann. Bot.* N.S. **28**, 37–45.

Phillips, I. D. J. and R. L. Jones, 1964. Gibberellin-like activity in bleeding sap of root systems of *Helianthus annuus* detected by a new dwarf pea epicotyl assay and other methods. *Planta* (Berl.), **63**, 269–278.

Phillips, I. D. J., A. J. Vlitos, and H. Cutler, 1959. The influence of gibberellic acid upon the endogenous growth substances of the Alaska pea. *Contrib. Boyce Thompson Inst.* **20**, 111–120.

Phillips, I. D. J. and P. F. Wareing, 1958a. Studies in dormancy of sycamore. I. Seasonal changes in the growth substance content of the shoot. *J. Exp. Bot.* **9**, 350–364.

Phillips, I. D. J. and P. F. Wareing, 1958b. Effect of photoperiodic conditions on the level of growth inhibitors in *Acer pseudoplatanus*. *Naturwiss.* **45**, 317.

Phillips, I. D. J. and P. F. Wareing, 1959. Studies in dormancy of sycamore. II. The effect of daylength on the natural growth-inhibitor content of the shoot. *J. Exp. Bot.* **10**, 504–514.

Pilet, P. E., 1965. Action of gibberellic acid on auxin transport. *Nature, Lond.* **208**, 1344.

Pilet, P. E., 1968. *In vivo* and *in vitro* auxin and kinin transport. In *Biochemistry and Physiology of Plant Growth Substances*, ed. G. Setterfield and F. Wightman. Runge Press, Ottawa.

Pilet, P. E. and W. Wurgler, 1958. Action des gibberellines sur la croissance et l'activité auxines-oxydasique du *Trifolium ochroleucum* Hudson. *Bull Soc. Botan. Suisse*, **68**, 54–63.

Piringer, A. A. and H. M. Cathey, 1960. Effect of photoperiod, kind of supplemental light and temperature on the growth and flowering of Petunia plants. *Proc. Am. Soc. Hort. Sci.* **76**, 649–660.

Preston, A. P. and H. W. B. Barlow, 1950. The use of growth substances to widen crotch angles. *Rep. East Malling Res. Sta.* No. 76.

Reath, A. N. and S. H. Wittwer, 1952. The effect of temperature and photoperiod on the development of pea varieties. *Proc. Am. Soc. Hort. Sci.* **60**, 301–308.

Ruddat, M., E. Heftmann, and A. Lang, 1965. Chemical evidence for the mode of action of AMO-1618, a plant growth retardant. *Naturwiss.* **52**, 267.

Ruddat, M. and R. P. Pharis, 1966. Participation of gibberellin in the control of apical dominance in soybean and redwood. *Planta* (Berl.), **71**, 222–228.

Ryle, G. J. A., 1961. Effects of light intensity on reproduction in S48 timothy (*Phleum pratense* L.). *Nature, Lond.* **191**, 196–197.

Sachs, T., 1966. Senescence of inhibited shoots of peas and apical dominance. *Ann. Bot.* N.S. **30**, 447–456.

Sachs, T. and K. V. Thimann, 1964. Release of lateral buds from apical dominance. *Nature, Lond.* **201**, 939–940.

Sachs, T. and K. V. Thimann, 1967. The role of auxins and cytokinins in the release of buds from dominance. *Am. J. Bot.* **54**, 136–144.

Sastry, K. S. K. and R. M. Muir, 1965. Effects of gibberellic acid on utilization of auxin precursors by apical segments of the Avena coleoptile. *Plant Physiol.* **40**, 294–298.

Scott, T. K. and W. R. Briggs, 1960. Auxin relationships in the Alaska pea (*Pisum sativum*). *Am. J. Bot.* **47**, 492–499.

Scott, T. K., D. B. Case, and W. P. Jacobs, 1967. Auxin-gibberellin interaction in apical dominance. *Plant Physiol.* **42**, 1329–1333.

Šebánek, J., 1965. The interaction of endogenous gibberellins in correlation with cotyledons and axillary buds in the pea (*Pisum sativum* L.). *Biol. Plant.* (Praha) 7, 194–198.

Šebánek, J., 1966a. Investigation of apical dominance in pea seedlings by means of ^{32}P transport. *Acta Univ. Agr.* (Brno) A 4, 587–596.

Šebánek, J., 1966b. Interaction of indoleacetic acid with synthetic and native growth regulators during transfer of ^{32}P into epicotyls of etiolated pea seedlings. *Biol. Plant.* (Praha) **8**, 213–219.

Šebánek, J., 1967. Einfluss von pasten mit verschiedenen indolylessigsäure-konzentrationen auf den transport von ^{32}P in dekapitierten erbsenepikotylen. *Planta* (Berl.) **75**, 283–285.

Seth, A. K., C. R. Davies, and P. F. Wareing, 1966. Auxin effects on the mobility of kinetin in the plant. *Science*, **151**, 587–588.

Seth, A. K. and P. F. Wareing, 1964. Interaction between auxins gibberellins and kinins in hormone-directed transport. *Life Sci.* **3**, 1483–1486.

Seth, A. K. and P. F. Wareing, 1967. Hormone-directed transport of metabolites and its possible role in plant senescence. *J. Exp. Bot.* **18**, 65–77.

Shinohara, S. and M. Sano, 1957. Studies on the vernalization of leguminous plants. I. On the effect of vernalization on germinating seeds of various varieties of broad beans (*Vicia faba*). *Bull. Shizuoka Agr. Expt. Sta.* **2**, 43–55.

Skoog, F. and C. O. Miller, 1957. Chemical regulation of growth and organ formation in plant tissues cultured *in vitro*. *Symp. Soc. Exp. Biol.* **11**, 118–130.

Smith, H. and P. F. Wareing, 1964a. Gravimorphism in trees 2. The effect of gravity on bud-break in Osier Willow. *Ann. Bot.* N.S. **28**, 283–295.

Smith, H. and P. F. Wareing, 1964b. Gravimorphism in trees 3. The possible implication of a root factor in the growth and dominance relationships of the shoots. *Ann. Bot.* N.S. **28**, 297–309.

Smith, H. and P. F. Wareing, 1966. Apical dominance and the effect of gravity on nutrient distribution. *Planta* (Berl.) **70**, 87–94.

Snow, R., 1925. The correlative inhibition of the growth of axillary buds. *Ann. Bot.*, Lond. **39**, 841–859.

Snow, R., 1929. The transmission of inhibition through dead stretches of stem. *Ann. Bot.*, Lond. **43**, 261–267.

Snow, R., 1931. Experiments on growth and inhibition. II. New phenomena of inhibition. *Proc. Roy. Soc.* **B 108**, 209–223.

Snow, R., 1932. Experiments on growth and inhibition. III. Inhibition and growth promotion. *Proc. Roy. Soc.* **B 111**, 86–105.

Snow, R., 1935. Activation of cambial growth by pure hormones. *New Phytol.* **34**, 347–360.

Snow, R., 1936. Upward effects of auxin in coleoptiles and stems. *New Phytol.* **35**, 292–304.

Snow, R., 1937. On the nature correlative inhibition. *New Phytol.* **36**, 283–300.

Snow, R., 1938. On the upward inhibiting effect of auxin in shoots. *New Phytol.* **37**, 173–185.

Snow, R., 1939a. A second factor involved in inhibition by auxin in shoots. *New Phytol.* **38**, 210–223.

Snow, R., 1939b. An inhibitor of growth extracted from pea leaves. *Nature, Lond.* **144**, 906.

Snow, R., 1940. A hormone for correlative inhibition. *New Phytol.* **39**, 177–184.

Snow, R., 1945. Plagiotropism and Correlative inhibition. *New Phytol.* **44**, 110–117.

Snow, R., 1947. Further experiments on plagiotropism and correlative inhibition. *New Phytol.* **46**, 254–257.

Sorokin, H. P. and K. V. Thimann, 1964. The histological basis for inhibition of axillary buds in *Pisum sativum* and the effects of auxins and kinetin on xylem development. *Protoplasma*, **59**, 326–350.

Stewart, W. S., 1939. A plant growth inhibitor and plant growth inhibition. *Bot. Gaz.* **101**, 91–108.

Stewart, W. S., W. Bergren, and C. E. Redemann, 1939. A plant growth inhibitor. *Science*, N.S. **89**, 185–186.

Stoddart, J. L., 1959. The effects of gibberellic acid on growth habit and heading in late-flowering red clover (*Trifolium pratense* L.) *J. Agr. Sci.* **52**, 161–167.

Stuart, N. W., 1938. Nitrogen and carbohydrate of kidney bean cuttings as affected by treatment with indoleacetic acid. *Bot. Gaz.* **100**, 298–311.

Takahashi, N., H. Honda, S. Takagi, and T. Honda, 1957. Studies on the mechanism of the tiller development in rice plant. I. The tiller development of the rice plants planted under the element deficient condition. *Rep. Inst. Agr. Res.* (Tohoku Univ., Japan), **8**, 71–105.

Takeuchi, S., 1957. On the influence of shading the stem base of rice seedlings upon the emergence of tillers. *Proc. Crop. Sci. Soc. Japan*, **25**, 228.

Thimann, K. V., 1937. On the nature of inhibitions caused by auxin. *Am. J. Bot.* **24**, 407–412.

Thimann, K. V. and F. Skoog, 1933. Studies on the growth hormone of plants. III. The inhibiting action of the growth substance on bud development. *Proc. Nat. Acad. Sci.* **19**, 714–716.

Thimann, K. V. and F. Skoog, 1934. On the inhibition of bud development and other functions of growth substance in *Vicia faba*. *Proc. Roy. Soc.* **B. 114**, 317–339.

Troughton, A., 1967. The effect of mineral nutrition on the distribution of growth in *Lolium perenne*. *Ann. Bot.* N.S. **31**, 447–454.

Verner, L., 1938. The effect of a plant growth substance on crotch angles in young apple trees. *Proc. Am. Soc. Hort. Sci.* **36**, 415–422.

Verner, L., 1955. Hormone relations in the growth and training of apple trees. *Res. Bull. Idaho Agric. Exp. Sta.* No. 28.

Wareing, P. F., 1958. Interaction between indole acetic acid and gibberellic acid in cambial activity. *Nature, Lond.* **181**, 1744–1745.

Wareing, P. F., 1964. The developmental physiology of rhizomatous and creeping plants. *Proc. 7th. British Weed Control Conference*, **3**, 1020–1030.

Wareing, P. F. and T. A. A. Nasr, 1958. Gravimorphism in trees. Effects of gravity on growth, apical dominance and flowering in fruit-trees. *Nature, Lond.* **182**, 379–381.

Wareing, P. F. and T. A. A. Nasr, 1961. Gravimorphism in trees. I. Effects of gravity on growth and apical dominance in fruit trees. *Ann. Bot.* N.S. **25**, 321–340.

Wareing, P. F. and A. K. Seth, 1967. Ageing and senescence in the whole plant. *Symp. Soc. Exp. Biol.* **21**, 543–558.

Went, F. W., 1936. Algemeine Betrachtungen über das Auxin-Problem. *Biol. Zbl.* **56**, 449–463.

Went, F. W., 1939. Some experiments on bud growth. *Am. J. Bot.* **26**, 109–117.

Wickson, M. E. and K. V. Thimann, 1958. The antagonism of auxin and kinetin in apical dominance. *Physiol. Plant.* **11**, 62–74.

Wickson, M. E. and K. V. Thimann, 1960. The antagonism of auxin and kinetin in apical dominance. II. The transport of IAA in pea stems in relation to apical dominance. *Physiol. Plant.* **13**, 539–554.

Williams, R. F., 1955. Redistribution of mineral elements during development. *Ann. Rev. Plant Physiol.* **6**, 25–42.

Yoshii, Y., 1927. Some preliminary studies of the influence upon plants of the relative length of day and night. *Sci. Rep. Tohoku Imp. Univ. Ser. Biol.* **2**, 143–158.

Zeevaart, J. A. D., 1966. Reduction of gibberellin content of *Pharbitis* seeds by CCC and after-effects in the progeny. *Plant Physiol.* **41**, 856–862.

Zimmerman, W. A., 1936. Untersuchungen über die räumliche und zeitliche verteilung des wuchsstoffes bei Baumen. *Z. Bot.* **30**, 209–252.

6. Geotropism

L. J. Audus

Introduction

The term geotropism is applied to phenomena in which parts of multicellular plants assume orientations at angles specifically related to the direction of the plumb-line. Thus most plant organs attain stable equilibrium at a certain angle to the gravity force vector and any enforced departure therefrom will cause them to bend back into what might be called their 'preferred' orientations. The curvatures arise by the differential expansion of the two sides of the organ concerned; almost always these are permanent, non-reversible growth expansions, although sometimes (e.g., in the pulvini of *Phaseolus vulgaris*) reversible volume changes induced by change in hydrostatic pressure (turgor) of the cell contents may underlie the responses.

The first scientific demonstration of geotropic phenomena goes back to Dodart (1703). They were recognized to be related to gravity by Austruc (1709), who suggested that the nutrient plant 'juices', because of their density, would tend to move predominantly in the lower halves of horizontally disposed organs, thereby favouring growth and inducing upward curvatures. But a century had to pass before it was demonstrated that gravity was indeed the stimulus concerned. Gravity acts by causing the acceleration of mass and in 1806, Knight pointed out that if it were responsible for the curvatures then centrifugal force, which also causes mass acceleration, should be similarly effective. To check this he attached germinating seeds of *Phaseolus* to the rim of a wheel which could be rotated about a vertical axis to give a centrifugal force equivalent to 3–5 times gravity acting on the seedlings. As they grew the roots curved away from the axis of rotation while the shoots curved towards it, i.e., in the direction of and contrary to the centrifugal force respectively.

Frank (1868) was the first to realize and establish that geotropic curvatures were directly connected with growth and he coined the word *geotropism* to describe the phenomena. This was an unfortunate choice, since it implied that only the earth's field would bring about a response. A better description would have been gravitropism (L. *gravitas*—heavy), or, as suggested by Larsen (1962a) barytropism (Gk. *barytés*—weight).

Types of response

Geotropic responses are classified according to the equilibrium orientation of the organ relative to the direction of the gravitational pull. The central axes of most plants align themselves parallel to this direction and are said to be *ortho-geotropic* (see Fig. 6.1). If their growth is in the direction of the gravity vector

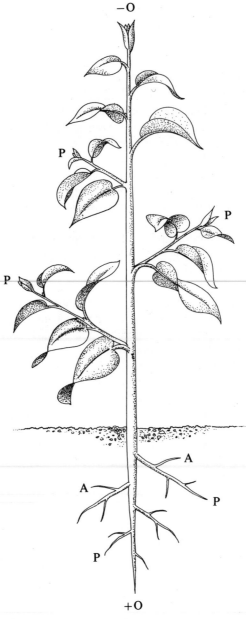

Figure 6.1   The geotropic responses of plant organs.
A. Shoot and root system of branched dicotyledonous plant.
(For explanation see under **D** and **E**, page 208).

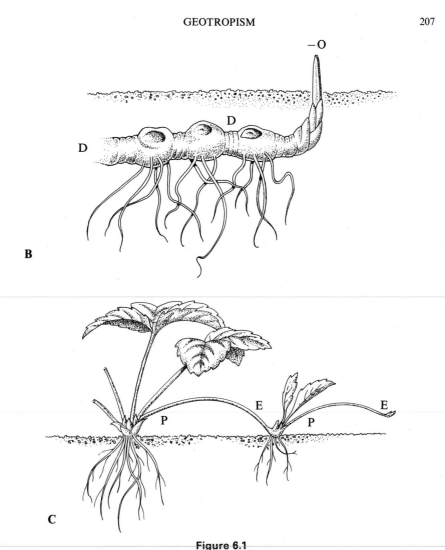

Figure 6.1
B. Rhizome of *Polygonatum multiflorum*. **C**. Runners of *Fragaria vesca*.
(For explanation see under **D** and **E**, page 208)

(e.g., primary roots) they are positively geotropic; if they grow in the opposed direction they are negatively geotropic (e.g., main stems). A lateral organ usually finds equilibrium at an angle to the gravity vector characteristic of its stage of development and the conditions under which it is growing. Such behaviour is said to be *plagiogeotropic* (see Fig. 6.1). A special case of plagiogeotropism is shown by many roots and stems (rhizomes) which grow horizontally, i.e., strictly at right angles to the gravity vector. Their behaviour is given a special name—*diageotropism*.

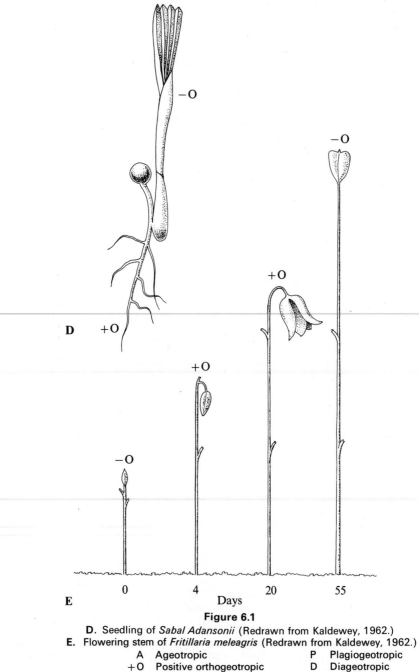

Figure 6.1
D. Seedling of *Sabal Adansonii* (Redrawn from Kaldewey, 1962.)
E. Flowering stem of *Fritillaria meleagris* (Redrawn from Kaldewey, 1962.)

A	Ageotropic	P	Plagiogeotropic
+O	Positive orthogeotropic	D	Diageotropic
−O	Negative orthogeotropic	E	Epinastic

The majority of orthogeotropic organs are radially symmetrical and response is due simply to a difference in the growth rates of those sides which happen to become upper and lower, the difference being maintained until the organ becomes vertical. The situation is more complicated in plagiogeotropic organs, which often have a bilateral symmetry, i.e., are *dorsiventral* with either structurally or biochemically distinct upper and lower sides. This dorsiventrality is the basis of certain nastic movement (see chapter 8), which may contribute to plagiogeotropic behaviour. Thus *epinasty* is the innate tendency for the dorsal (normally uppermost—or *adaxial*) surface to grow more rapidly than the ventral (normally lowermost—*abaxial*) surface; *hyponasty* is the reverse of this. There is experimental evidence to suggest, at least in some plants, that plagiogeotropic behaviour is the outcome of two opposing stimuli, one orthogeotropic and the other epinastic or hyponastic. For example, an organ which is negatively plagiogeotropic would be negatively orthogeotropic and epinastic, the equilibrium angle being attained where these two tendencies neutralized each other. The arguments for this interpretation cannot be pursued here but the interested reader is referred to extensive discussion in Rawitscher (1932) and Kaldewey (1962).

Other gravity-controlled movements are also linked with the internal asymmetry of organs. The leaves of many plants are at geotropic equilibrium when the plane of the lamina is perpendicular to the gravity vector with the adaxial side uppermost. Angular displacements will be followed by a twisting of the leaf stalk (petiole) or even of the supporting branch so as to bring the lamina back into the original (horizontal) plane. This phenomenon is known as *geostrophism*. For a recent review the reader is referred to Snow (1962).

What follows in the rest of this chapter is mainly concerned with orthogeotropism of main axes. There is a good reason for this; the responses of such organs are relatively uncomplicated and easy to study and research on mechanisms has, therefore, been confined almost exclusively to them.

Modifications of response by internal and external factors

Internal factors

The geotropic behaviour of an organ is not fixed and immutable; it can be changed by a range of circumstances. Thus one organ may modify the geotropic response of another, the best example being the influence of the apex of the main (orthogeotropic) axis on the lateral (plagiogeotropic) branches, a situation akin to apical dominance (see chapter 5). If the apical portion of the main axis of some monopodial trees (e.g., *Pinus strobus*) is removed, the neighbouring lateral branches will abandon their oblique growth and become vertical (negatively geotropic). Similarly, the plagiogeotropic rhizomes of potato turn erect and develop leaves when main shoots have their apical and axillary buds removed (Booth, 1959).

Then there are changes which accompany normal development processes such as production of orthogeotropic shoots from diageotropic rhizomes. Some organs, particularly flower and fruit stalks of many herbaceous plants (e.g., *Papaver, Fritillaria, Tussilago*), show complete reversals of response between bud and mature fruit stages. This is illustrated in Fig. 6.1E in *Fritillaria* where the changes are from a negative (flower bud) to a positive (partially opened flower) and finally to a negative (mature fruit) orthogeotropic response. The underlying causes of these changes are beyond the scope of this chapter; details should be sought in Kaldewey (1962).

External factors

As might be expected environmental factors can modify geotropic responses. Temperature changes, which by themselves induce organ movements (Thermonasty—see chapter 8), can modify geotropic responses qualitatively, e.g., the negatively orthogeotropic stems of *Lamium* become plagiogeotropic at temperatures just above freezing (Lidforss, 1903). Quantitative effects of temperature are many and important in that they throw light on mechanisms; they will be discussed at relevant points in subsequent sections.

One of the most important environmental factors is light, which produces qualitative and quantitative changes in geotropic response. Taking first the qualitative effects we find for example that some plagiogeotropic organs may become orthogeotropic when lighting conditions are changed. Thus the rhizomes of *Aegopodium podagraria* are diageotropic in darkness, but become positively plagiogeotropic even after a short exposure to red light (Bennet-Clark and Ball, 1951). On the other hand stolons of the grasses *Cynodon dactylon*, *Stenotaphrum secundatum*, and *Paspalum vaginatum* are diageotropic in the light but become negatively geotropic if the parent plant is darkened (Palmer, 1956). The most striking example of change is shown by the stamens of *Hosta caerulea* which are positively geotropic in darkness but become negatively geotropic when illuminated (Pilet, 1950).

As regards quantitative effects of light both the speed and the magnitude of the response can be changed, generally in an adverse manner. Thus in *Phycomyces* sporangiophore both the presentation time and latent time (see later section) are increased by illumination (Pilet, 1956) and the rate of curvature may be increased or decreased depending on the light intensity (Dennison, 1964). Perhaps the most striking results have come from studies on the wavelength dependence of such effects. Thus in *Avena* coleoptiles, irradiation with red light (661 nm) increased subsequent geotropic curvature response, a maximum effect being obtained with a dose of 7×10^3 erg cm^{-2}. Blue light (479 nm) suppressed the response progressively with increasing doses up to about

3×10^3 erg cm^{-2} and then the effect declined at higher doses (Blaauw, 1961). These effects were shown to arise from acceleration of the growth rate of the convex side of the coleoptile by red light and of the concave side by blue light.

The full complexity of the situation in *Avena* coleoptiles has been revealed by the work of Wilkins (1965) who has shown that the red light (660 nm) effect can be induced only once; a second exposure has no effect. It is one of the most sensitive photo-responses known in plants, being evoked by as little as 0·6 erg cm^{-2} of light energy. Wilkins also found that far-red light had no direct effect on geotropic sensitivity but that it would reverse the stimulating effects of red light. In the closely related *Zea* the geotropic response of coleoptiles is decreased by red light and the effect reversed by far-red light (Wilkins and Goldsmith, 1964). Thus the phytochrome system (see chapter 14), apparently participates in this phenomenon. How it does so is not known. Judging from the fact that it is an essential component of many photomorphogenic systems, it would seem that its action is most likely to be via the hormone effector of the response sequence rather than on the perception mechanism (see later sections).

Certain chemicals in the environment can markedly modify responses. In *Aegopodium*, for example, enrichment of the surrounding atmosphere with carbon dioxide will make the normally diageotropic rhizome negatively plagiogeotropic. Even mechanical stimuli can modify responses: for example main roots, re-orientating themselves from a horizontal position, will eventually reach the vertically downward direction (orthogeotropic) only if they grow in sharp sand; in moist air or in glass micro-beads (ballotini) they remain positively plagiogeotropic (Bennet-Clark, Younis, and Esnault, 1959).

The nature of the response

It was Sachs who first demonstrated in 1874 that the zone of geotropic curvature in primary roots coincides with the zone of rapid extension. His classical experiment using indian ink spots placed at points equidistant along the main axis has long been a basic school exercise. Since then this correlation has been repeatedly confirmed for most organs. Extension in these curving zones is the result of growth by cell elongation, a situation beautifully illustrated by the observations of Arslan and Bennet-Clark (1960) in the geotropically sensitive nodes of *Triticum* and *Bromus* and shown in Fig. 6.2B. Claims that in onion roots geotropic curvature is caused by different rates of cell division on the two sides (Wagner, 1936) prove, on statistical analysis of the data, to be unfounded.

Naturally, one would expect that the curvatures would be accompanied in the extending cells by *differential* changes in physical and chemical properties normally associated with extension growth. This is indeed so. For example Heyn (1934) a pioneer in these studies, demonstrated that in the hypocotyls of *Lupinus* seedlings, geotropic stimulation made cell walls on the lower side of the organ more plastic than those on the upper side. This is illustrated in Fig. 6.2A. Hypocotyls were mechanically bent to an angle of 35° either upwards or downwards and held there for 5 minutes. This induced a permanent curvature due, it was assumed, to the plastic extension of the cell walls on the convex side. The ratio of wall plasticity (lower/upper) determined from the respective residual curvatures was $22{\cdot}9/12{\cdot}5 = 1{\cdot}83$. This has subsequently been verified by Brauner and Brauner (1962), using a direct mechanical stretching of the upper and lower halves of split segments of hypocotyl.

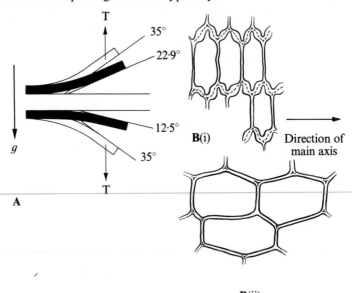

Figure 6.2
A. Diagram showing the effect of gravity on the plastic extensibility of the cell walls of upper and lower sides of *Lupinus* hypocotyl. Hollow outlines represent curvatures under mechanical force T. Solid outlines represent residual permanent curvatures. *g* represents direction of gravity vector. (Drawn from data of Heyn, 1934.)
B. Cells from the extension parenchyma of wheat stem nodes from the upper (i) and lower (ii) sides after geotropic curvature. Note the short thick periclinal walls of the upper side; on the lower side these have stretched and thinned as the cells extended during curvature. (Redrawn from Arslan and Bennet-Clark, 1960.)

Correlated with these changes in wall properties are changes in the water relations of the extending cells. As early as 1924 Ursprung and Blum had demonstrated that, although the molar concentration (osmotic pressure 9 atmospheres) of solutes in the cell vacuoles of geotropically curved *Vicia faba* roots was the same on upper and lower sides, yet the suction force of cells on the convex surface was 8 atmospheres while that on the concave surface was only 0·5 atmospheres. These differences must, therefore, be due to differences in wall pressure which were calculated to be 1 atmosphere on the convex surface and 8·5 atmospheres on the concave side; they obviously reflect the changes in wall properties already described. The development of these suction force changes in *Helianthus* hypocotyls (Brauner and Brauner, 1962) occurred as early as $2\frac{1}{2}$ minutes after the commencement of stimulation and was regarded as one of its primary effects. In spite of the observations of Ursprung and Blum, it looks as if some increase in osmotically active solutes in cell vacuoles may contribute to these suction force changes. Consistent evidence from several workers show that in *Helianthus* the sugar content of cells on the under side of curving hypocotyls exceeds that on the upper side (see survey in Brauner, 1962). Similar observations have been made for the nodes of *Triticum* (Arslan and Bennet-Clark, 1960). Increased growth requires increased metabolic energy supply and it is not surprising therefore that the convex sides of geotropically curving organs have a more intense respiration rate than the concave sides (see Brauner, 1962).

Other changes, not so easily explained as part of the general syndrome of a growth differential, include the redistribution of cations. In horizontal *Helianthus* seedlings there is an apparent movement of potassium to the lower side, the concentration rising to 5 per cent in excess of that on the upper side (Bode, 1959). Arslan-Çerim (1966) demonstrated that ^{45}Ca accumulated on the upper side but this did not take place under nitrogen. It is interesting that the potassium ion can stimulate extension growth, whereas the calcium ion can inhibit it (Cooil, 1952; Cooil and Bonner, 1957); the meaning of these ion redistributions and their relationship to the growth differential remain problems for the future.

Methods of study

As with other physiological phenomena, the study of geotropism requires techniques drawn from the whole of physical and chemical science. Some of these will be mentioned later when particular facets of the overall problem are being discussed. However, there are aspects of experimentation that are specific to geotropism and these require to be outlined at this stage.

In any study of stimulation physiology a basic requirement is the establishment of the quantitative relationships of stimulus to response; this means that the stimulus must be under precise control. For phototropism (see chapter 7) this is relatively simple; dark rooms and modern light sources are more than adequate to give us all the control we need. But, apart from installing our laboratory in an orbiting satellite (a possibility however, which may not be far distant), we cannot alter the force of gravity to any practical extent. We can, however, eliminate its unilateral action by rotating the organ slowly about its main axis which is orientated horizontally. In this way all sides receive successively and continuously the same stimulus and long experience has shown that provided the speed of rotation is carefully chosen, the organ usually continues to grow straight along its axis. The curvature-inducing effects of gravity are thus eliminated. The technique was pioneered by Sachs (1882a), who called the apparatus a *klinostat*. The general principle is illustrated in Fig. 6.3. Details of construction, operation and modern refinements should be sought in the article by Larsen (1962a). The main use of the klinostat is to study, under conditions of ineffective omnilateral stimulus, the results of exposure to a unilateral gravitational (or centrifugal) force, operating for a set period of time. In this way the quantity of stimulus (force × time) can be controlled and its relationship to response accurately followed. Curvatures, which are observed preferably during the course of klinostat rotation, must be recorded photographically, in present day research by high-speed synchronized flash illumination at phototropically ineffective wavelengths.(See chapter 7).

In practice of course such rotation on a klinostat does not remove all curvatures; innate randomness in growth causing equally random curvatures, although usually small, may thereby be uncovered (see analysis by Johnsson, 1966) and for any precise measurement, particularly of responses to threshold stimuli, this necessitates the study of large samples. Furthermore, innate growth potential gradients in dorsiventral organs (epinasty or hyponasty) may also be disclosed by klinostat rotations, and this illustrates a second use of the instrument, the unravelling of the complexities of orientation control in such organs. Rotation of plant organs with their axes *not parallel* to the horizontal axes of

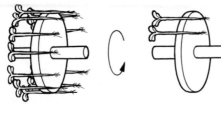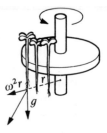

KLINOSTAT	**CENTRIFUGE**	
Slow rotation	Fast rotation	
Gravity omnilateral	Gravity omnilateral	Gravity along main axis
Centrifugal force negligible	Centrifugal force high	
		Resultant force as in diagram

Figure 6.3 Diagrams showing the principles of klinostat and centrifuge techniques in studies of geotropism.

rotation often gives rise to geotropic curvatures, particular in dorsiventral organs but even in radially symmetrical ones. Full expositions of the phenomena will be found in Rawitscher (1932) and Larsen (1962a).

To vary the intensity of the stimulus we must have recourse to centrifugal force. The apparatus used has the same principle as the klinostat; the difference lies in the speed of rotation which must be high enough to develop a force to which the organ is sensitive. Control of this speed (usually measured in radians per second (ω)) gives equal control of the force ($m\omega^2 r$) where r is the distance of the organ from the axis of rotation and m is the mass which is being accelerated. The axis of rotation may be horizontal, in which case the gravity vector is omnidirectional with regard to the organs, or it may be vertical, in which case gravitational and centrifugal forces act in quadrature and the resultant vector can be determined from the parallelogram of forces (Fig. 6.3). Such an arrangement can be of great value in the study of the effect of both direction and intensity of the resultant force (see next section).

The general laws and their implications

Parameters of response and their measurement

Since the earliest quantitative studies it has been known that geotropic response is a 'threshold' phenomenon; stimuli which fail to reach a certain level, specific to the particular organ, evoke no response. The minimum exposure necessary to induce a just detectable response is known as the *presentation time* for the given force. Since otherwise identical organs can differ widely in their sensitivities, the determination of this parameter lacks precision and is complicated by the use of different techniques which measure slightly different things. The most popular method has been to expose samples to given stimuli for different periods of time and to determine for each period the proportion of the organs in the sample subsequently showing a detectable curvature; the time period causing 50 per cent to curve is taken as the presentation time. This gives a fair estimate of the average sensitivity of the sample. Recent investigators have preferred to use more sensitive methods involving the determination of mean curvature rates for a period after stimulation. By plotting curvature rate against period of stimulation, the threshold for zero rate can be determined by backward extrapolation. This gives values which are considerably lower than those determined by the first technique, since they are based on the responses of the most sensitive individuals in the sample. An idea of the large differences found can be gleaned from work with cress (*Lepidium sativum*) roots. By the first method Zimmerman (1927) recorded a presentation time of 10 minutes; by the latter method a time of 0·5 minutes was obtained (Larsen, 1957). Such a situation is a little disturbing but consistent employment of either technique can yield equally informative data.

After the reception of a stimulus above the threshold value a further period has to elapse before curvature starts. The time taken for the first perceptible curvature to appear after the inception of the stimulus is known as the *reaction time* (*latent time* of some workers). The rate of curvature is initially constant but gradually declines as the new equilibrium orientation is approached. Organs rotated on a klinostat after stimulation will first curve and then eventually straighten completely by processes not yet understood.

From such parameters of response and their relationships to the intensity and nature of the stimulus emerge the general laws of geotropic response, which are our main guide to the basic mechanisms involved.

The presentation time

Presentation time with normal gravitational stimuli varies very widely from organ to organ and from species to species and is also very sensitive to environmental conditions, particularly temperature. The first type of variation is illustrated in Table 6.1 by examples selected from a wide range of species (determined by the first method described on p. 215). The variations do not seem to be related in any way to morphological or systematic characteristics.

Table 6.1   Presentation times (normal gravitational stimuli).

Species	Organ	Time in minutes	Reference
Artemisia absinthium	Radicle	0·5	Larsen (1957)
Osmunda regalis	Mature frond	0·5–1·0	Hawker (1932a)
Avena sativa	Coleoptile	2·5–5·0	Rutgers (1912) Pekelharing (1909)
Helianthus annuus	Hypocotyl	5–10	Hawker (1932a)
Lepidium sativum	Radicle	10	Zimmermann (1922)
Beta vulgaris	Hypocotyl	10–15	Hawker (1932a)
Picea pungeus	Hypocotyl	22–33	Hawker (1932a)
Asplenium bulbiferum	Young fronds	30	Waight (1923)

Presentation times are minimal between temperatures of 20° and 30°C, but above and below this range they increase dramatically. The sharp rise above 35°C is due to heat damage; that below 10–15°C is reversible and its high negative temperature coefficient (Q_{10}) reaches values of 3·25–3·7 in epicotyls of *Lathyrus odoratus* (Hawker, 1933), 2·6 in coleoptiles of *Avena sativa* (Rutgers, 1912), 3·75 in epicotyls of *Vicia faba* (Bach, 1907) and 2·25 in primary roots of *Lupinus albus* (Czapek, 1898).

The presentation time is closely dependent on the intensity of the stimulus. Observations with the centrifuge at the beginning of the century (Pekelharing,

1909, Rutten-Pekelharing, 1910) showed that the product of the presentation time (T) and the stimulus intensity was a constant over a very wide range of values of g. Thus for *Avena* coleoptiles the mean product Tg was 2·988 ($\pm$0·006) $\times 10^5$ cm sec^{-1} for 31 values of force ranging from 0·08g to 58·4g. For *Lepidium sativum* roots the product was 3·424 ($\pm$0·051) $\times 10^5$ cm sec^{-1} for seven values ranging from 0·28g to 12·8g. This was established as the *reciprocity rule*, which has subsequently formed the basis of all calculations of stimulus 'quantity', reckoned in terms of $g \times$ time of exposure, irrespective of the proportions of these two components. In view of its fundamental importance it is surprising that this 'rule' has been subjected to little further experimental testing, either to extend it to other organs or to establish the full range of values of g and T over which it is valid.

There seems to be a lower limit of g below which no response can be evoked even with exposure of long duration. The work of Czapek (1895) on various centrifuged organs indicated this to be between 5×10^{-4} and $10^{-3}g$. Lyon (1961), using the centrifugal force generated by vibrating wires obtained even lower values of $4 \times 10^{-5}g$ for *Zea* roots. The use of modern highly sophisticated klinostats suggests, however, that a more realistic threshold would be between 10^{-3} and $10^{-2}g$. (Gordon and Shen-Miller, 1966).

Curvatures can be obtained from a series of very short intermittent stimuli, each alone well below the threshold for response. Fitting (1905) first demonstrated in several organs that the threshold stimulus ($g \times$ presentation time) could be effective if given in two doses (each $g \times \frac{1}{2}$ presentation time) separated by a 'relaxation' period in the vertical (normal equilibrium) position of no stimulation, provided that that period did not exceed a certain maximum, characteristic of the organ. For example in *Vicia faba* epicotyls a relaxation period of five times a single (subthreshold) stimulus allowed complete summation of the two stimuli. Longer relaxation periods necessitated prolonging the total exposure times before curvatures were obtained. This suggests that the conditions induced in the organ by subthreshold (or indeed any) stimuli persist virtually unchanged for periods characteristic of the organ and then slowly revert to those of the unstimulated state. The duration of this changed state (i.e., the relaxation period) is longer on a klinostat than in the normal vertical orientation (Zielinski, 1911). But this picture, of successive stimuli of equal duration producing equal changes in the gravity-detecting system, is too simple since subsequent studies (Gunther-Massias, 1928; Bunning and Glatzle, 1948) have shown that summated subthreshold stimuli can induce responses which exceed those caused by a single continuous stimulus of the same 'quantity'. This is shown in Fig. 6.4A, from data on *Lepidium sativum* roots for relaxation period/stimulation period ratios between 1 and 4. This situation could be explained if the changes induced in the gravity-detecting system did not proceed linearly with time but were most rapid immediately after the start of the stimulation.

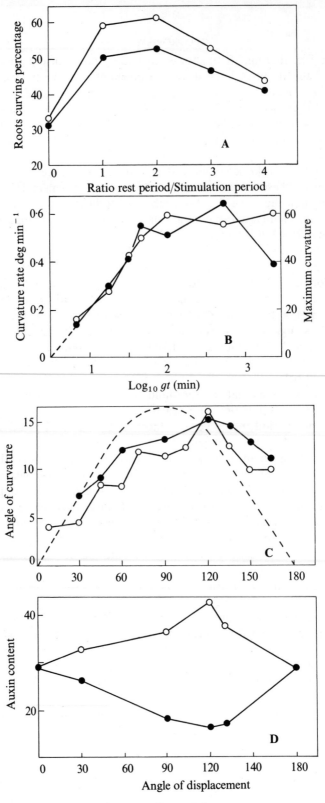

Figure 6.4

The response above the threshold

It is strange that there are very few direct observations on the relationships between response and 'quantity' of stimulus above these threshold values. What
there are strongly suggest that over certain limits the relationship follows the
well-known Weber–Fechner law of animal sensory physiology where the response increases arithmetically with a geometrical increase in the stimulus
intensity:

$$\Delta R = \frac{K.\Delta S}{S}$$

where R = response; S = stimulus, and K is a constant. Such a log/linear relationship is well illustrated by Fig. 6.4B, taken from data of *Pisum* roots
(Lundegårdh, 1926). Above a certain stimulus the relationship no longer holds
and response reaches a maximum or may even decline with further increase in
stimulus.

One very important stimulus factor still remains to be considered; this is the
angle of displacement from the normal equilibrium orientation. It would be
expected that maximum stimulation would be achieved when the axis of the
organ makes a right angle with the direction of the gravity vector. This was
appreciated by Sachs (1882b), who suggested that at other angles the effective
force would be the component of the net force transverse to the main axis, i.e.,
$g \sin \alpha$ where α is the angle of displacement from the vertical ($0°$ representing
the vertically downward direction). This came to be known as the *sine rule*.
It was soon demonstrated to be approximately true, yet data over the intervening years show quite consistent departures from it. The optimum angular
displacement for maximum stimulus is not $90°$, but falls between $120°$ and $135°$

Figure 6.4
A. Graphs from data of Gunther–Massias (1929) showing the effects of successive subthreshold stimuli in roots of *Lepidium*.
——●—— 2 stimuli of 1 minute duration each
——○—— 4 stimuli of $\frac{1}{2}$ minute duration each
(N.B. The value zero on the abscissa corresponds to one single two-minute stimulation.)
B. Graphs calculated from data of Lundegårdh (1926) showing the relationship between
geotropic response and stimulus intensity in *Pisum* roots. The seedlings were centrifuged
at various values of g on a klinostat.
——●—— Initial rates of curvature, determined by regression analysis
——○—— Maximum curvature obtained
C. Graphs showing the departure from the sine rule in geotropic curvature responses.
——●—— *Lepidium* root (from Larsen, 1962b)
——○—— *Vicia* root (Ahrens, 1933/4)
– – – – – – Sine rule
D. Graphs showing the departure from the sine rule as illustrated by auxin concentration
difference between upper and lower sides of *Vicia faba* root tips (from Amlong, 1939).
——○—— Lower side   ——●—— Upper side
Auxin concentration expressed in angle of curvature of test plants (*Helianthus* seedlings).

(see Fig. 6.4C) for both threshold and supra-threshold stimuli in coleoptiles, roots, and grass nodes (see Audus, 1964) and even in a related morphogenic phenomenon, the induction of tension wood in willow (Robards, 1966). Even hormone concentration differences (see later section) seem to be similarly related to displacement angle (Amlong, 1939) (see Fig. 6.4D).

Regions of perception

The simplest way to locate the sensitive regions of plants would be to apply the stimulus to various restricted areas and compare the resulting responses. This is the method used in studies of phototropism. Gravity, however, cannot be similarly 'screened' from any part of an organ and the methods available for the analysis of sensitivity distribution involves either excision of parts of the organ concerned or what might be called differential centrifugation.

Thus it has been known since Ciesielski (1872) that removal of the extreme tips of roots prevented their geotropic response. Similarly with coleoptiles removal of progressively larger portions of the apex causes a corresponding progressive depression of sensitivity (increase in presentation time) (Dolk, 1936). With epigeal dicotyledon seedlings, e.g., *Helianthus* (Brauner and Hager, 1958) removal of cotyledons and apical bud also completely prevents response. Such observations might be advanced as proof that the gravity-perceiving apparatus resides in the apices of these organs, were it not for the fact that the hormones controlling cell extension are produced also in these apices. Without the hormones there can be no growth and hence no response (see chapter 1). Attempts have been made to resolve this by supplying auxin to the decapitated and stimulated organ. In this way the extension zone also of *Avena* coleoptiles and *Helianthus* hypocotyl have been shown to have some sensitivity (Anker, 1959, Brauner and Hager, 1958).

The situation in roots has been more confusing. Thus it has been claimed that decapitated and stimulated root stumps of *Vicia faba* when placed vertically would curve if unstimulated tips were stuck back in position with agar (Amlong, 1939). This would imply some perception by the stump and release of response by the hormone coming from the re-attached tip. On the other hand, other workers have been unable to confirm response in 're-tipped' roots (Younis, 1954). The first clear-cut resolution of this problem came with the discovery that in some species (e.g., *Zea mays*) root caps can be dissected whole from the root without damage to the meristem (Juniper, Groves, Landau-Schachar, and Audus, 1966). Such decapped roots grow at an unchanged rate but are completely insensitive to gravity; sensitivity is regained as the root cap is regenerated by the cap meristem. In *Zea* at least the root cap would seem to be the sole centre of perception.

Another approach to this problem is due to Piccard and uses a special centrifuge technique (for details see Rawitscher, 1932, pp. 302–4; Larsen 1962a;

pp. 66–67), by which the distribution of sensitivity can be very roughly mapped. In roots maximum sensitivity is in the apical 1–2 mm (i.e., the root cap). In the *Avena* coleoptile it is in the apical 3–4 mm. In hypocotyls and epicotyls it is even more widely spread, the zone of high sensitivity stretching from the apical bud for a distance of 20 or more millimetres down the main axis.

Geotropism as a catenary process

Geotropism can thus be regarded as a catenary process, a chain of reactions causally linked in sequence. This overall process can be represented diagrammatically as follows:

Perception		*Transformation*		*Transmission*		*Response to*
Physical action	$\longrightarrow$	*of Information*	$\longrightarrow$	*of Information*	$\longrightarrow$	*Information*
of gravity						
stimulus						
(Susception)		(Reception)		(Hormone Transport)		(Reaction)

Presentation Time	Latent Phase	Curvature

The first step is the immediate action of gravity on some component mass of the system causing therein a physical change. This is regarded as only the first stage of perception by some workers and has been called *susception*. The second step is the transformation of that physical information (structural changes) into biochemical information (hormone changes). This has been called *reception*, in the writer's opinion a misnomer. The third step is the transmission of the information from the site of perception to the region of response (e.g., from root apex to extending zone). This step is reasonably certain to be mainly one of differential transport or supply of growth-substances. The final step is the response of the reacting zone to this differential concentration of growth factor.

In the remainder of this chapter the first three steps will be further analysed, firstly the mechanisms proposed for perception, secondly the evidence for the participation of hormones in the transmission phase, and lastly the possible mechanisms of the transformation phase which links the two.

The reactions of the perception phase

Theories of perception

Gravity can affect a plant organ only by causing movement (mass acceleration) of one or more of its component parts. Early ideas, that the organ as a whole might be the movable mass and that strains set up therein by sagging under its own weight might be the cause of the growth curvature were disproved at an early date by the argument that non-strained roots (i.e., those supported horizontally along their whole length) showed normal responses and that curving organs would overcome counteracting force of many times their own weight (see Christiansen, 1917; Giltay, 1910). The primary action of gravity must, therefore, be the displacement of something inside the sensitive cell.

The starch statolith theory

Cells in certain regions of plants possess starch grains that sediment under gravity, in contrast to storage starch which shows no such movement. It was Berthold in 1886 who first suggested a connection between this sedimentation and gravity perception in plants, but it was not until 1900 that Haberlandt and Nemeć made independent surveys of the occurrence of starch *statoliths*, as they are called, and based a theory of gravity perception upon their findings. The main strength of the theory is that statolith-containing cells (*statocysts* or

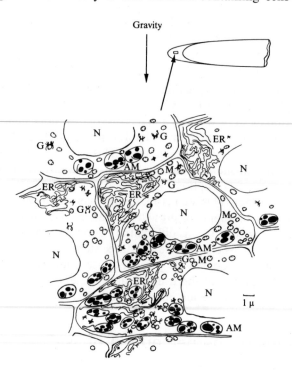

Figure 6.5

A. Drawing of a composite electron micrograph of cells from the central region of the root caps of *Vicia faba* showing the redistribution of amyloplasts under gravity. (From Griffiths and Audus, 1964.)

AM	Amyloplasts	G	Golgi body
N	Nucleus	ER	Endoplasmic reticulum
M	Mitochondrion		

statocytes) are almost invariably found in and confined to the geotropically-sensitive zones of plants. Thus they are characteristic of root cap cells, of coleoptile tips, and of the endodermis and starch sheaths that surround the vascular bundles of hypocotyls, epicotyls, young foliage leaves, etc. A drawing of such statocyte cells from the root cap of *Vicia faba* is shown in Fig. 6.5A. When the

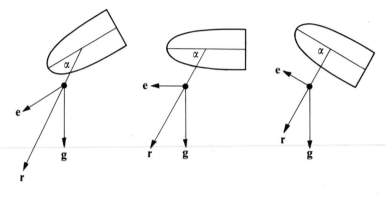

Figure 6.5

B. Diagrams illustrating the micropendulum statolith theory of Larsen (1959, 1962b).

      g   gravity force vector
      e   electrostatic force vector
      r   resultant vector and direction of micropendulum axis
      α angle of deflection of pendulum

organ is in the normal equilibrium orientation these statoliths rest on the lowermost wall of the statocyte (distal wall in root tips, proximal wall in shoot cells). Angular displacement of the organ will cause a sedimentation of the statoliths to the lateral walls where, according to the theory, mechanical pressure on the sensitive protoplasm lining these lateral walls will initiate the chain of physiological processes culminating in geotropic curvature. For example, compression of membranes could disturb their electrical states causing the development of electrical potentials which could themselves be transmitted or generate biochemical changes (Zollikofer, 1921). The most easily available description of the original theory and its supporting observations is by Haberlandt himself (1928).

The microsome theory

Owing to criticism of the starch statolith theory other sedimenting particles were sought by later workers. One was the 'microsome' of Cholodny (1922).

These were not the microsomes of current biology but particles visible in the light microscope, which we today would identify with either mitochondria or Golgi bodies. These 'microsomes' were supposed to be electrically charged and would, on sedimentation, bring about an associated migration of physiologically active ions such as K^+ and Ca^{2+}. This in turn would cause differential permeability changes in the plasmalemma on the two sides of the cell, leading ultimately to differential growth of these two sides. This theory had to fail because it did not recognize that perception and growth response do not normally take place in the same cell. Other objections will be discussed later.

The geoelectric theory

With the advent of the auxins and the Cholodny–Went theory of tropisms (see later) the time was ripe for a theory linking perception to lateral hormone movement. Gravity had to induce some lateral 'polarization' causing preferential movement of auxin from the site of its synthesis in the tip, to the lowermost side of the stimulated organ. Such a polarization was found in the geoelectric effect, a phenomenon first observed by Bose (1907) who showed that in horizontal or orthogeotropic organs a difference in electrostatic potential developed between upper and lower sides. These results were confirmed by Brauner (1927, 1928), who claimed that within a few minutes of displacement the under sides of hypocotyls, roots, and even pieces of undifferentiated apple and potato parenchyma became about 10 mV positive to the upper sides. Since auxin is a weak acid and is almost completely dissociated at the normal pH (5–6) of the cell, the growth-active anion would be attracted to the positive (under) side of the organ, there stimulating (in the case of the shoot) or inhibiting (in the case of the root) cell extension and causing curvatures.

The micropendulum theory

The fact that the departure from the strict sine rule had never been properly explained on previous theories led Larsen (1959, 1962b) to propose a gravity sensor (statolith) whose displacement would not be exactly 'in phase' with the angular displacement of the organ but would lag behind it, showing its maximum when the organ displacement reached 120°–130°. The sensor was visualized as a submicroscopic asymmetric particle (possibly a macromolecule) not free to fall in the cell but hinged at one end and capable therefore of swinging like a pendulum towards the vertical when the cell is turned. The magnitude of the stimulation would be proportional to the angular displacement of this sensor from its normal orientation along the main axis and would, therefore be maximal at 90°. In order to explain the 'angular lag' behind the gravity vector Larsen assumed the free ends (bobs) of his micropendulums to be electrically charged so that the potential gradient, which is known to exist along the main axis of many plant organs, would produce a torque tending to return the pendulums towards their normal position parallel to the main axis.

The final direction of the pendulum would be determined by the resultant of these two forces (see Fig. 6.5B). No suggestions as to the identity of these charged micropendulum sensors were put forward.

Critique of theories of perception and current ideas

Geoelectric effect. We will start our critique with the geoelectric effect which in student textbooks has long been the favoured explanation of hormone migration. This theory has not been without its critics; botanists have been sceptical of the efficiency and selectivity of the geoelectric effect as a prime-mover in auxin migration; physical chemists have suggested that it could be a property of the electrodes used and not a physiological response of the plants studied.

To resolve this last issue Grahm and Hertz (1962) devised an electrometer employing a vibrating reed electrode which avoids the complications of contact with the organ at the point studied. With *Avena* coleoptiles they were able to confirm the existence of a geoelectric effect of the same order of magnitude as that shown by Brauner, but their results differed in one fundamental respect; whereas Brauner's effect reached its maximum within a minute or two of organ displacement (Fig. 6.6A) Grahm and Hertz's effect did not appear for about 15 minutes, which corresponds almost exactly to the latent period for curvature response in this organ (Fig. 6.6B). Furthermore, both geoelectric and curvature responses are prevented by anaerobic conditions. The conclusion naturally drawn from such observations is that the geoelectric effect is a symptom of the *response* phase and does not occur until long after the perception processes are complete. It has a temperature coefficient close to that of respiration, suggesting that it may have its origins in some metabolic process. (Grahm and Hertz, 1964). In *Zea* coleoptiles and *Helianthus* hypocotyls, Woodcock and Wilkins (1969), using electrodes similar to those of Brauner finally demonstrated that his immediate rise in potential is a response of the electrodes superposed on the true geoelectric effect which follows later (Fig. 6.6C).

The development of the geoelectric effects seems to be directly linked to lateral differences in auxin concentration (see later), since it has been shown (Grahm 1964; Wilkins and Woodcock, 1965) that unilateral application of indole-3-acetic acid to the apical end of vertical decapitated coleoptiles can induce a potential difference of the same order in the growing zone below, the positive side being that with the highest auxin concentration. Furthermore, short-circuiting the geoelectric potential by filling the internal cavity of the *Avena* coleoptile with solutions of various electrolytes caused changes in geotropic response which ran closely parallel with, and were therefore probably indirectly caused by, direct effects on extension growth·(Anker and Biessels, 1964). We must, therefore, conclude that the geoelectric effect has nothing whatever to do with gravity perception processes.

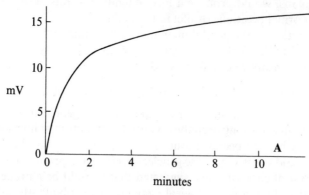

Figure 6.6

A. Graph of geoelectric effects in grass node showing immediate rise in potential at inception of stimulus (zero time). (From Brauner, 1956.)

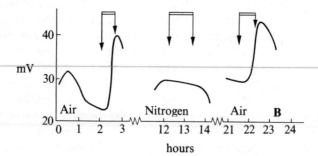

B. Graph of geoelectric effect in *Zea mays* coleoptiles showing the 15–20 minute lag after inception of stimulus and the absence of response in nitrogen. (From Grahm and Hertz, 1962.) Over the intervals shown by the pairs of vertical arrows the coleoptiles were in a horizontal position.

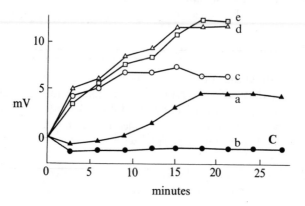

Theoretical considerations of statolith size. It is clear that if gravity is perceived by plant cells as a result of the displacement of some intracellular particle or particles of a density different from that of their immediate environment, then an appreciable movement (sedimentation) must take place during the presentation time. Consequently there have been attempts to calculate the theoretical rate of sedimentation of particles of various sizes and densities in the cytoplasm, using Stokes law, which governs such movement in a viscous medium. Although many approximations have to be made as to the physical properties of cell components, reasonably accurate estimations of sedimentation rates can be made. Thus, Audus (1962) has calculated that only a starch grain is large enough and dense enough to traverse a distance of half a cell width (10 μ) in an average presentation time (3 minutes). If smaller distances are involved (e.g., the particle's own diameter) then a much smaller body (e.g., a mitochondrion) could cover them in a reasonably short time (e.g., 6 minutes). Using a slightly different approach Gordon (1963) has calculated that any organelle smaller than a mitochondrion could not be a gravity sensor in plants. Thus theories based on statoliths of macromolecular dimensions (e.g., Larsen's micropendulums) must be ruled out.

Since the above calculations allow the possibility that particles of mitochondrial dimensions could be statoliths, Griffiths and Audus (1964) made direct observations of organelle distributions in the root-cap cells of *Vicia faba* under the electron microscope and analysed statistically the changes induced by angular displacement of the growing root (see Fig. 6.5A). Both mitochondria and Golgi bodies showed little change in distribution except as an indirect result of amyloplast* sedimentation. There was no evidence that they were directly affected by gravity over the stimulation period of two hours. On the other hand the endoplasmic reticulum (E.R.) which normally lies along the cell walls parallel to the long axis of the root, tended to accumulate in the upper half of the stimulated root cap cell; this could also be an indirect result of amyloplast movement.

---

* It should be noted that starch grains do not occur 'naked' in the cell but are enclosed, either singly or in clusters, inside protoplasmic membranes (see Fig. 6.5A). The whole constitutes an amyloplast, which sediments under gravity in the statocyte cell.

Figure 6.6

C. Graphs showing that the type of geoelectric effect shown in A is probably a composite one, being the resultant of a true geoelectric effect with a lag and a spurious immediate electrode effect. Data for hypocotyls of *Helianthus*.
- (a) ——▲—— Intact plants; flowing liquid electrode
- (b) ——●—— Decapitated control; flowing liquid electrode
- (c) ——○—— Decapitated control; Brauner type electrode
- (d) ——△—— (a) − (b) + (c) calculated
- (e) ——□—— Intact plant; Brauner type electrode

(From Woodcock and Wilkins, 1969)

Amyloplast statoliths.   Since all other theories so far mentioned seem to be ruled out, it remains for us to review briefly the evidence for and against the starch (amyloplast) statolith theory.

Firstly the correlation between the possession of geotropic sensitivity and the occurrence of amyloplast statoliths is almost perfect in the very wide range of plants studied over the last half century. Even plants which do not make storage starch (e.g., onion) have amyloplast statoliths in root caps and bundle sheaths. Very few exceptions have been reported; for example the aerial roots of the orchid *Laelia anceps* (Tischler, 1905) and the perianth of *Clivia nobilis* (Linsbauer, 1907) are supposed to be sensitive to gravity but have no starch statoliths. Modern confirmation of these observations is needed. Undisputed exceptions are the fungi where the gravity sensors may be other heavy particles, such as the highly-refractive bodies ('Glanzkörper') in the sensitive rhizoids of *Chara* (Buder, 1961; Sievers, 1965). In the development of higher plants acquisition of geotropic sensitivity coincides very closely with the development of a statenchyma in the organ concerned (Hawker, 1932a).

Direct measurements of the rates of amyloplast sedimentation in the cell also support the theory. In the much-studied *Avena* coleoptile the amyloplasts sediment across the statocyte in just under the presentation time (Zaepffel, 1923). In the seedling stem of *Lathyrus odoratus* an amyloplast moves one quarter the cell diameter in the presentation time (Hawker, 1933), which would mean that approximately one quarter of the cell's complement of amyloplasts, moving into contact with a lateral wall, would impose the threshold stimulus. Moreover, the effects of temperature on the rate of attainment of the stimulus threshold (reciprocal of presentation time) and on the observed rate of statolith movement are virtually identical and are closely similar to temperature-viscosity relationships in *Amoeba* protoplasm (Heilbrunn, 1930).

Even the departures from the strict sine rule (p. 219) can be explained in terms of amyloplast behaviour. Using plastic scale models of statocyte cells in which the cytoplasm was represented by a very viscous paraffin and the amyloplasts by particles of polythene, Audus (1964) has shown that the maximum number of these particles come to lie in contact with the lowermost wall of the cell, not when that wall is turned (from vertical) into a horizontal position (90° displacement) but when the displacement angle is about 120°.

Many attempts have been made to check the theory by manipulations whereby the amyloplasts could be manoeuvred into selected regions of the statocyte cell by special sequences of organ orientation, followed by observations of curvature response. The manipulations have been many and ingenious but only one or two of the most important can be described here. Thus Zimmerman (1927) demonstrated that inversion of an organ, subsequent to stimulation at 90° displacement, would greatly augment the subsequent curvature. As Fig. 6.7 shows, this can be most easily explained in terms of amyloplast movement. On the other hand von Ubisch (1928) claimed that inversion *before*

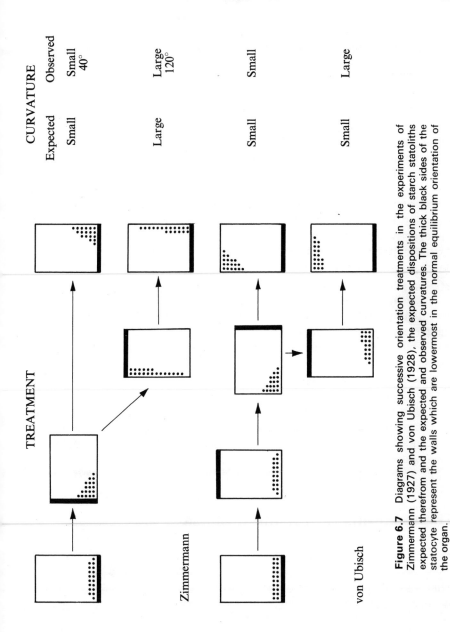

Figure 6.7 Diagrams showing successive orientation treatments in the experiments of Zimmermann (1927) and von Ubisch (1928), the expected dispositions of starch statoliths expected therefrom and the expected and observed curvatures. The thick black sides of the statocyte represent the walls which are lowermost in the normal equilibrium orientation of the organ.

stimulation does not alter the Zimmerman effect, whereas, on the statolith theory it should prevent it (see Fig. 6.7). Zimmerman's phenomenon has been confirmed several times, von Ubisch's modifications have not.

Another approach has been to try to induce the disappearance of statolith starch by exposure to more or less extreme physiological conditions before stimulation. Surprisingly, statolith starch, in contrast to storage starch, is extremely difficult to remove from statocyte cells. Long periods of starvation or exposure to cold or to heat have succeeded with some plants and there has been a correlated loss of geotropic sensitivity. The main objection to such techniques has been the lack of a convincing precision in the results and the abnormal state of the experimental plants after the treatment. The discovery that gibberellic acid would induce amylolytic enzymes in cereal grain led Pickard and Thimann (1966) to try this approach; they claimed that young wheat coleoptiles incubated in a mixture of gibberellic acid and kinetin lost their statolith starch but retained the capacity to curve geotropically. This is the first observation that has seriously threatened the very imposing edifice of circumstantial evidence supporting the starch statolith theory. The next step must be its confirmation and extension to a wide variety of organs.

On the basis of these results, Pickard and Thimann have abandoned theories of particulate statoliths and have returned to an original idea of Czapek (1898) that a cell might detect the weight of its own cytoplasm pressing on the membranes bounding the lowermost cell wall. This weight would be equivalent to a pressure about 1 dyn cm^{-2}, a force which, by itself, might modify molecular orientation in, and hence biochemical changes associated with those membranes. However, such membranes are pressed into contact with *all* cell walls in turgid plant cells by the hydrostatic (turgor) pressure of the cell contents amounting to at least 10^6 dyn cm^{-2}, and it seems unlikely that a small difference in hydrostatic pressure of one part in a million or less could be the basis of gravi-perception (Audus, 1962).

The problem still remains unresolved.

The reactions of the latent phase

The Cholodny–Went theory

Almost as soon as the existence of growth-controlling hormones, the auxins, had been recognized, these compounds were linked to tropisms in the now classical Cholodny–Went theory (see Went and Thimann, 1937, p. 157). According to this theory gravity induces a lateral polarization of the organ (for example, a transverse geoelectric potential gradient). Thus the normal symmetrical and strictly polar flow of auxin from the tip to the extension zone (see chapter 4) would be disturbed and there would be a lateral migration of the auxin, which would accumulate on the lower side of the organ, and be depleted on the upper side. In view of the greatly differing sensitivities of roots and shoots to auxins, and on the assumption that natural auxin levels were suboptimal for the growth of shoots and supraoptimal for that of roots, the negative and positive curvature responses of shoots and roots respectively could be explained (see Fig. 6.8A).

Evidence in support of the theory was first obtained for shoots by Dolk (1930, 1936) who invented a method of collecting auxin by diffusion from upper and lower sides of coleoptile tips into small blocks of agar, separated by a slip of mica (see Fig. 6.8B). He found about 60 per cent more auxin (measured by the *Avena* curvature test) in the lower than in the upper block. In the next few years similar redistribution of auxin under gravity was demonstrated, by both diffusion and extraction techniques, in coleoptiles of *Zea*, in hypocotyls of *Helianthus*, *Phaseolus*, and *Lupinus*, in epicotyls of *Vicia*, and in the nodes of a variety of grasses. The mechanism of this redistribution in coleoptiles was thought by Dolk to be a more effective transport of auxin on the lower side and this was supported by direct observations on horizontal coleoptile segments, using a modification of the auxin transport technique of van der Weij (1932) (see Fig. 6.8C). However, there are suggestions that accelerated auxin production on the lower side of the tip might be the cause of the unequal distribution. Thus shoots

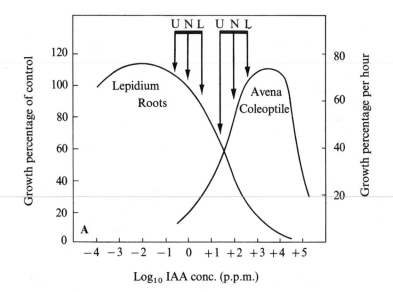

Figure 6.8

A. Diagram illustrating the principle of the Cholodny–Went theory of geotropism. The growth response from data of Moewus (1949) (*Lepidium sativum* roots) and of Bonner (1949) (*Avena* coleoptile segments). The arrows show the possible auxin concentrations and corresponding growth rates in normal vertical (N) organs and on the upper (U) and lower (L) side of geotropically stimulated organs.

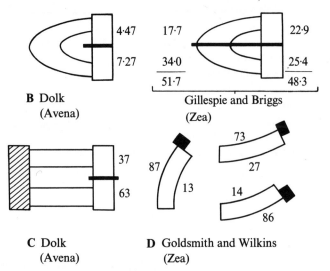

B Dolk
(Avena)

Gillespie and Briggs
(Zea)

C Dolk
(Avena)

D Goldsmith and Wilkins
(Zea)

Figure 6.8

B. Relative amounts of diffusible auxin collected from upper and lower halves of stimulated tips of *Avena* and *Zea* coleoptiles. (Data of Dolk, 1936, and Gillespie and Briggs, 1961.) Results with *Zea* show that with intact tips (left-hand column) twice as much auxin is collected from the lower side, as in the *Avena* experiments of Dolk; whereas with completely split tips (right-hand column) equal amounts are collected from both sides. Total amounts collected are the same for both treatments indicating a movement of auxin from top to bottom side of the intact tips.

C. Differential auxin transport in horizontal segments of *Avena* coleoptile. The shaded area represents the 'donor' block containing auxin. The amounts transported through the two sides are recorded by the right-hand (receiver) blocks. (Data of Dolk, 1936).

D. Diagrams showing the effect of gravity in the migration of radioactive IAA across a *Zea* coleoptile segment. The numbers by the segments show the relative amounts of auxin on the two sides (as percentage of total). The diagram also illustrates the effects of treatments on curvatures (not accurately drawn). (Data of Goldsmith and Wilkins, 1964.)

seem to grow faster and roots slower when continuously stimulated omnilaterally on a klinostat (see Larsen, 1962a, p.57); this points to an augmented auxin production. There have been claims that such rotation increases the auxin (indole-3-acetic acid) content of *Helianthus* hypocotyls by activating the enzymes synthesizing tryptophan (the supposed precursor of auxin) (Bara, 1962, 1965).

The situation for *Zea* coleoptiles was made reasonably clear by Gillespie and Briggs (1961) who showed that horizontal tips, completely split (see Fig. 6.8B), gave off equal amounts of auxin from upper and lower halves and that the amount of auxin collected from whole tips did not depend on orientation. With the advent of isotope techniques Reisener (1957) soaked excised *Avena* coleoptiles, normally orientated, for three hours in a solution of radioactive IAA. They were then stimulated geotropically and the upper and lower halves separately assayed for ^{14}C content. No differences could be detected, suggesting no lateral migration of IAA under gravity. The misleading nature of these results was demonstrated by Gillespie and Thimann (1963), who repeated Dolk's classical experiments using ^{14}C-labelled IAA. Like Reisener they could demonstrate no significant differences in the auxin contents of upper and lower sides of *Avena* coleoptile segments but the amounts transported through the lower half exceeded those through the upper half by 50 per cent. The differences in the small amounts of auxin being transported on the two sides were swamped in the tissue analyses by the great amounts bound inside the cells.

But this leaves unanswered the question whether gravity acts merely by promoting longitudinal transport on the lower side of the coleoptile or whether it changes the polarity of the tissue and causes a lateral movement of auxin from the upper to the lower side. Such a lateral movement was shown by Goldsmith and Wilkins (1964) by asymmetric application of radioactive IAA to the distal ends of 15 mm coleoptile segments of *Zea* (Fig. 6.8D). In vertical segments about 13 per cent of the absorbed auxin moved to the side opposite to that to which it was applied. In horizontal segments this movement was promoted (to 27 per cent) if the untreated side was lowermost whereas the movement was similar to the vertical control (14 per cent) if it was uppermost. There thus seems little doubt that a lateral movement of auxin is induced by gravity in coleoptile segments.

In roots the situation is still confused. Early support for the Cholodny–Went theory was obtained by Hawker (1932b) who collected auxin from upper and lower halves of stimulated root tips of *Vicia faba* by diffusion into agar and applied the agar blocks asymmetrically to the root stumps, using the curvature so obtained to assay the 'auxin' content. She demonstrated more 'auxin' from the lower half of the tip. This was confirmed by Boysen-Jensen (1933). The work of Cholodny (1924) and Keeble, Nelson, and Snow (1931) had indicated that the auxin mediating geotropic response was the same for roots and shoots, since coleoptile tips stuck symmetrically on root stumps would re-establish geotropic sensitivity just as effectively as the root tips themselves. The presence of auxins in root tips has now been firmly established but there is still some doubt whether these include indole-3-acetic acid. Thus in *V. faba*, the main auxins appear to be substances different from IAA (Lahiri and Audus, 1960; Burnett, Audus, and Zinsmeister, 1965). Very dilute solutions (one part in 10^{11}) of IAA promoted both the growth and the curvature of pea roots (Audus and Brownbridge, 1957) a situation which cannot be explained if this is also the endogenous auxin, present in supraoptimal concentrations and being redistributed under the action of gravity. The situation is further complicated by the fact that, unlike the situation in shoots, ^{14}C-labelled IAA is not transported in a polar manner from the tip to the extension zone in roots but tends to accumulate in the meristem (Yeomans and Audus, 1964). Furthermore, during geotropic curvature the normal growth rate of the roots is reduced and this was explained in terms of a stimulated production of an auxin in the tip (Audus and Brownbridge, 1957), a situation later confirmed by extraction and assay (Audus and Lahiri, 1961).

Summing up it would seem that an auxin, possibly not IAA, is produced in the meristem of the root; the root cap does not produce it since de-capping has no effect on root growth (Juniper *et al.*, 1966). Gravity, by some mechanism operating in the root cap, induces more of this auxin to be produced on the lower side of the meristem from where differential movement to the extending cells would result in differential growth and curvature. These theories cannot be checked until the auxins of roots have been identified and their movements under gravitational stimulation studied.

In grass nodes also the available evidence suggests that augmented auxin production on the lower side is responsible for the curvatures. This has been shown by Schmitz (1933) in *Triticum vulgare, Secale cereale, Lolium perenne, Holcus mollis*, and *Setaria verticillata*. An increase in auxin production was reported by van Overbeek, Olivo, and de Vasquez (1945) in growth rings of sugar-cane nodes but this did not occur until the fourth day after stimulation and no differences were to be observed between upper and lower sides. The situation calls for further study.

One type of indirect attack on the role of auxin transport in geotropic responses has employed synthetic antagonists of the polar movement of auxin. Results to date have been disappointing, partly because of inconsistency of responses and partly because of uncertainty as to the exact operation of the compounds used. The most promising substance is N-l-naphthylphthalamic acid (NPA) (Morgan and Söding, 1958). It renders both roots and shoots unresponsive to geotropic stimulation (Mentzer and Netien, 1950), indicating that in both, auxin transport is an essential part of the response mechanism. However NPA sometimes shows direct growth-regulating activity (Morgan and Söding, 1958; Hoffman and Smith, 1949) and could therefore modify the curvatures directly. Until we know much more about the full range of physiological actions of such compounds it could be very misleading to use them as definitive tools to check on the Cholodny–Went theory.

Another check of the theory has been made by studying the behaviour of decapitated organs. For example, excised *Avena* coleoptiles when their tips are removed, are insensitive to gravity but will curve if they are immersed in auxin solutions (Anker, 1954, 1956). It was found that maximum curvature is attained at an auxin concentration corresponding to the maximum slope of the growth response/auxin concentration curve for the same material (see Fig. 6.9), i.e., where the coleoptile was most sensitive to small differences in auxin concentration. Since the applied auxins enter the coleoptiles almost entirely at the cut ends, Anker interpreted these results in terms of a preferential movement of the auxin to the lower side of the coleoptile. However, the results could equally well be interpreted in terms of a *uniform* distribution of auxin but a heightened sensitivity of the cells of the lower side. This can be easily appreciated from Fig. 6.9,

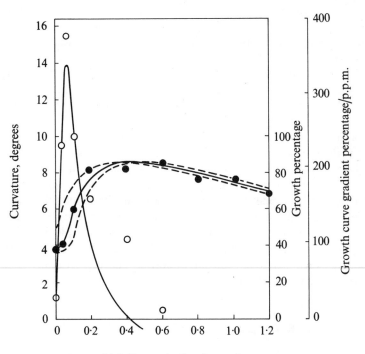

Figure 6.9 Graphs showing growth and geotropic curvature of *Avena* coleoptile segments in IAA solutions of various concentrations.

● Extension growth   ○ Geotropic curvature

The gradient of the extension growth curve has been fitted as closely as possible to the geotropic curvature points. The dotted lines show the results of shifting the extension growth curve laterally to reproduce the effects of small changes in cell sensitivity, the left-hand curve representing the lower side of the coleoptile and the right-hand curve the upper side. (Redrawn and adapted from the data of Anker, 1956.)

where shifting the growth/concentration curves laterally along the auxin axis will produce growth differences between upper and lower sides which also closely parallel the observed curvatures.

The co-factor theory

This suggestion of a change in auxin sensitivity of the two sides brings us to a theory involving growth hormone co-factors first proposed by Gradman (1930). By studying the rate of extension of the upper and lower halves of split hypocotyls of *Lupinus* placed horizontally, Gradman concluded that only lower sides of organs responded to gravity with changed growth rates, which were caused by the co-operation of two growth factors, neither of which acting alone had any effect. Substance A (presumably auxin) was produced by the apex and moved thence to the extension zone; substance B was induced to form locally in the *lower* half of the organ by the action of gravity. Similar explanations were made by Brauner and Hager (1958) of observations on the *Helianthus* hypocotyl. This organ can be stimulated (induced) by horizontal exposure at 4°C but will not respond since growth is stopped under these low-temperature conditions. Subsequent return to 20°C will allow curvature to take place. Such induction (co-factor production) at low temperatures can be 'remembered' (co-factor storage) for periods of up to 12 hours both when intact hypocotyls are kept vertical at 4°C and until decapitated plants are supplied with IAA after similar waiting periods. Brauner (1966) has also shown that in such induced hypocotyls subsequently split into two, the lower half is much more sensitive to auxin than the upper half. This increased sensitivity he thinks is also due to the presence of more local auxin co-factor. But there are some surprising features of Brauner and Hager's induction results, for example the ability of hypocotyls to remember and respond correctly in sequence to two successive and opposing stimuli, that call for corroboration and further study. The co-factor theory still remains an interesting speculation.

The linkage of perception to hormone distribution

This aspect of the mechanism is one characterized by pure speculation. In the days when electrophoretic migration of the auxin anion under the geoelectric potential difference was accepted as the link between perception and response little further thought was given to the matter. Now we know that this potential difference is the *result* of auxin movement, we must look for other 'prime movers' in auxin redistribution. Unfortunately, we cannot yet be certain by what means differences in auxin concentrations occur, whether by transverse polar transport, by differential longitudinal transport, by differential synthesis or by differential destruction (as suggested by observations on the distribution of IAA-oxidase in geotropically stimulated pea roots (Konings, 1965)).

Whether the amyloplast or the whole protoplast is the gravity sensor, it seems likely that mechanical pressure on membranes lining the cell walls is the immediate cause of the changes in auxin behaviour. If transverse auxin movement across the organ is involved then the pressure might be causing a polarization of the lateral membranes and consequently a one-way (downward) flow of the auxin from cell to cell. Such a flow would need to be metabolically powered since it would be against a concentration gradient. Hertel and Leopold (1963), regard such one-way movement as a secretion process. If, however, the stimulus causes differences in longitudinal transport or auxin metabolism or co-factor production, then the effects on the membranes would have to depend on whether the sensing cell were in the upper or lower half of the organ, i.e., whether the pressure was being exerted at the cell wall furthest away from or nearest to the central axis of the organ. Thus while pressure on the inner membranes (upper half of organ) could be ineffective, pressure on the outer ones (lower half of organ) could facilitate a longitudinal transport process, or accelerate auxin synthesis, or suppress auxin destruction, or induce co-factor synthesis, etc. This implies that membranes (e.g., the endoplasmic reticulum) may be closely associated either with auxin transport or with auxin metabolism or with both; so far we have no relevant experimental data and these ideas are mere pointers to future studies.

Bibliography

Selected further reading

Anker, L., 1942. Ortho-geotropism in shoots and coleoptiles. In *Encyclopedia of Plant Physiology, XVII/2 Physiology of Movements*, Springer, Berlin, Göttingen, Heidelberg, pp. 103–152.

Audus, L. J., 1962. The mechanism of the perception of gravity by plants. *Symp. Soc. exp. Biol.*, **16**, 197–226.

Banbury, G. H., 1962. Geotropism of lower plants. In *Encyclopedia of Plant Physiology XVII/2 Physiology of Movements*, Springer, Berlin, Göttingen, Heidelberg, pp. 344–377.

Brauner, L., 1954. Tropisms and nastic movements. *A. Rev. Pl. Physiol.*, **5**, 163–182.

Brauner, L., 1962. Primäreffekte der Schwerkraft bei der geotropischen Reaktion. In *Encyclopedia of Plant Physiology XVII/2 Physiology of Movements*, Springer, Berlin, Göttingen, Heidelberg, pp. 74–102.

Haupt, W., 1965. Perception of environmental stimuli, orientating growth and movement in lower plants. *A. Rev. Pl. Physiol.* **16**, 267–290.

Kaldewey, H., 1962. Plagio- und Diageotropismus der Sprosse und Blätter, einschliesslich Epinastie, Hyponastie, Entfaltungsbewegungen. In *Encyclopedia of Plant Physiology, XVII/2 Physiology of Movements*. Springer, Berlin, Göttingen, Heidelberg, pp. 200–321.

Larsen, P., 1962a. Geotropism. An Introduction. In *Encyclopedia of Plant Physiology, XVII/2 Physiology of Movements*. Springer, Berlin, Göttingen, Heidelberg, pp. 34–73.

Larsen, P., 1962b. Orthogeotropism in roots. In *Encyclopedia of Plant Physiology XVII/2 Physiology of Movements*. Springer, Berlin, Göttingen, Heidelberg, pp. 153–199.

Rawitscher, F., 1932. *Der Geotropismus der Pflanzen*. Fischer, Jena, pp. 420.

Rufelt, H., 1961. Geotropism in roots and shoots. *A. Rev. Pl. Physiol.* **12**, 409–430.

Rufelt, H., 1962. Plagiogeotropism in Roots. In *Encyclopedia of Plant Physiology, XVII/2 Physiology of Movements*, Springer, Berlin, Göttingen, Heidelberg, pp. 322–343.

Snow, R., 1962. Geostrophism. In *Encyclopedia of Plant Physiology, XVII/2 Physiology of Movements*, Springer, Berlin, Göttingen, Heidelberg, pp. 378–389.

Wilkins, M., 1966. Geotropism. *A. Rev. Pl. Physiol.* **17**, 379–408.

Papers cited in text

Ahrens, A., 1933/4. Untersuchungen über die optimale Reizlage bei der geotropischen Reaktion. *Z. Bot.* **26**, 561–596.

Amlong, H. U., 1939. Untersuchungen über Wirkung und Wanderung des Wuchsstoffes in der Wurzel. *Jb. wiss. Bot.* **88**, 421–469.

Anker, L., 1954. A comparative study on the recovery of the geotropic response of decapitated *Avena* coleoptiles by indoleacetic acid, indoleacetonitrile and naphthylacetic acid. *Proc. K. ned. Akad. Wet.* Ser. C. **57**, 304–316.

Anker, L., 1956. The auxin concentration rule for the geotropism of *Avena* coleoptiles. *Acta bot. neerl.* **5**, 335–341.

Anker, L., and H. W. A. Biessels, 1964. Effects of electrolytes on growth and geotropism of the *Avena* coleoptile. *Proc. K. ned. Akad. Wet.* Ser. C. **67**, 320.

Arslan-Çerim, N., 1966. The redistribution of radioactivity in geotropically stimulated hypocotyls of *Helianthus annuus* pretreated with radioactive calcium. *J. exp. Bot.* **17**, 235–240.

Arslan, N. and T. A. Bennet-Clark, 1960. Geotropic behaviour of grass nodes. *J. exp. Bot.* **11**, 1–12.

Audus, L. J., 1964. Geotropism and the modified sine rule: an interpretation based on the amyloplast statolith theory. *Physiologia Pl.* **17**, 737–745.

Audus, L. J., and M. E. Brownbridge, 1957. Studies on the geotropism of roots. I. Growth-rate distribution during response and the effects of applied auxins. *J. exp. Bot.* **8**, 105–124.

Audus, L. J. and A. N. Lahiri, 1961. Studies on the geotropism of roots. III. Effects of geotropic stimulation on growth substance concentrations in *Vicia faba* root tips. *J. exp. Bot.* **12**, 75–84.

Austruc, J., 1709. Conjecture sur le redressement des plantes inclinées à l'horizon. *Mem. Acad. r. Sci. Paris*, **1708**, 463–470.

Boch, H., 1907. Über die Abhangigkeit der geotropischen Präsentations- und Reaktionszeit von verschiedenen Aussenbedingungen. *Jb. wiss. Bot.* **44**, 57–123.

Bara, M., 1962. L'action de la gravitation sur le niveau auxinique des hypocotyles de l'*Helianthus annus*. *Physiologia Pl.* **15**, 725–8.

Bara, M., 1965. L'effet du clinostate sur la synthèse du tryptophan chez l'*Helianthus annuus*. *Physiologia Pl.* **18**, 1037.

Bennet-Clark, T. A. and N. G. Ball, 1951. The diageotropic behaviour of rhizomes. *J. exp. Bot.* **2**, 169–203.

Bennet-Clark, T. A., A. F. Younis, and R. Esnault. Geotropic behaviour of roots. *J. exp. Bot.* **10**, 69–86.

Berthold, G. 1886. *Der Geotropismus der Pflanzen*. Fischer, Jena, p. 308.

Blaauw, O. H., 1961. The influence of blue, red and far red light on geotropism and growth of the *Avena* coleoptile. *Acta bot. neerl.* **10**, 397–450.

Bode, H. R., 1959. Über den Einfluss der Geoinduktion auf die Kationenverteilung im Hypocotyl von *Helianthus annuus*. *Planta*, **54**, 15–33.

Bonner, J., 1949. Limiting factors and growth inhibitors in the growth of the *Avena* coleoptile. *Am. J. Bot.* **36**, 323–332.

Booth, A., 1959. Some factors concerned in the growth of stolons in potato. *J. Linn. Soc.* **16**, 166–169.

Bose, J. C., 1907. *Comparative Electrophysiology*, Longmans, Green, London.

Boysen-Jensen, P., 1933. Die Bedeutung des Wuchsstoffes für das Wachstum und die geotropische Krümmung der Wurzeln von *Vicia faba*. *Planta*, **20**, 688–698.

Brauner, L., 1927. Untersuchungen über das geolektrische Phänomen I. *Jb. wiss. Bot.* **66**, 781.

Brauner, L. 1928. Untersuchungen über das geoelektrische Phänomen II. *Jb. wiss Bot.* **68**, 711.

Brauner, L., 1942. New experiments on the geo-electric effect in membranes. *Instanb. Üniv. Fen. Fak. Mecm.* Ser. B. **7**, 46–102.

Brauner, L., 1956. Über die Primäreffekt der Schwerkraft beim Geotropismus der Pflanzen. *Naturw. Rdsch.* Stuttg. **12**, 466–470.

Brauner, L., 1966. Versuche zur Analyse der geotropischen Perzeption V. Über den Einfluss des Schwerefeldes auf die Auxinempfindlichkeit von *Helianthus*—hypokotylen. *Planta*, **69**, 299–318.

Brauner, L. and E. Appel, 1960. Zum Problem der Wuchsstoff-Querverschiebung bei der geotropischen Induktion. *Planta*, **55**, 226–234.

Brauner, L. and A. Böck, 1963. Versuche zur Analyse der geotropischen Perzeption IV. Untersuchungen über Auswirkung der Dekapitierung auf den Wuchsstoffgehalt, das Langenwachstum und die geotropische Krümmungsfähigkeit von *Helianthus* Hypokotylen. *Planta*, **60**, 109–130.

Brauner L. and M. Brauner, 1961. Versuche zur Analyse der geotropischen Perzeption. III. Über den Einfluss des Schwerefeldes auf die Dehnbarkeit der Zellwand und den osmotischen Wert des Zellsaftes. *Planta*, **58**, 301–325.

Brauner, L. and A. Hager, 1958. Versuche zur Analyse der geotropischen Perzeption, I. *Planta*, **51**, 115–147.

Buder, J., 1961. Geotropismus der Characeenrhizoide. *Ber. dt. bot. Ges.* **74**, 14–23.

Bünning, E. and D. Glatzle, 1948. Über die geotropische Erregung. *Planta*, **36**, 199–202.

Burnett, D., L. J. Audus, and H. D. Zinsmeister, 1965. Growth substances in the roots of *Vicia faba* II. *Phytochemistry*, **4**, 891–904.

Cholodny, N., 1922. Zur Theorie des Geotropismus. *Beih. bot. Zbl.* **39**, 222–230.

Cholodny, N., 1924. Über die hormonale Wirkung der Organspitze bei der geotropischen Krümmung. *Ber. dt. bot. Ges.* **42**, 356.

Christiansen, M., 1917. Bibliographie des Geotropismus. *Mitt. Inst. allg. Bot. Hamburg*, **2**, 1–118.

Ciesielski, T., 1872. Untersuchung über die Abwärtskrümmung der Wurzel. *Beitr. Biol. Pfl.* **1**, 30.

Cooil, B. J., 1952. Relationship of certain nutrients, metabolites and inhibitors to growth in the *Avena* coleoptile. *Pl. Physiol.* Lancaster, **27**, 49–69.

Cooil, B. J. and J. Bonner, 1957. The nature of growth inhibition by calcium in the *Avena* coleoptile. *Planta*, **48**, 696–723.

Czapek, F., 1895. Untersuchungen über Geotropismus. *Jb. wiss. Bot.* **27**, 243–339.

Czapek, F., 1898. Weitere Beitrage zur Kenntnis der geotropischen Reizbewegungen. *Jb. wiss. Bot.* **32**, 175–308.

Dennison, D. S., 1964. The effect of light on the geotropic responses of *Phycomyces* sporangiophores. *J. gen. Physiol.* **47**, 651–655.

Dodart, D., 1703. Sur l'affectation de la perpendiculaire, remarkable dans toutes les tiges, dans plusieurs racines, et autant qu'il est possible dans toutes les branches des plantes. *Mem. Acad. r. Sci. Paris*, **1700**, 47–63.

Dolk, H. E., 1930. *Geotropie en Groeistoff*. Doctoral Dissertation, Univ. Utrecht, pp. 129.

Dolk, H. E., 1936. Geotropism and the growth substance. *Recl. Trav. bot. néerl.* **33**, 509–585.

Frank, A. R., 1868. *Beiträge zur Pflanzenphysiologie. I. Ueber die durch Schwerkraft verursachte Bewegung von Pflanzentheilen.* W. Engelmann, Leipzig. pp. 167.

Fitting, H., 1905. Untersuchungen über den geotropischen Reizvorgang. *Jb. wiss. Bot.* **41**, 221–398.

Gillespie, B. and W. R. Briggs, 1961. Mediation of geotropic response by lateral transport of auxin. *Pl. Physiol.* Lancaster, **36**, 364–367.

Gillespie, B. and K. V. Thimann, 1963. Transport and distribution of auxin during tropistic response. I. The lateral migration of auxin in geotropism. *Pl. Physiol.* Lancaster, **38**, 214–225.

Giltay, E., 1910. Einige Betrachtungen und Versuche über Grundfragen beim Geotropismus der Wurzeln. *Z. Bot.* **2**, 305–331.

Goldsmith, M. H. M. and M. B. Wilkins, 1964. Movement of auxin in coleoptiles of *Zea mays* L. during geotropic stimulation *Pl. Physiol.* Lancaster, **39**, 151–162.

Gordon, S. A., 1963. Gravity and plant development: Bases for experiment. *Space Biology.* Proc. 24th Biol. Colloq. Oreg. State Univ. Press, pp. 75–105.

Gordon, S. A. and J. Shen-Miller, 1966. On the thresholds of gravitational force perception by plants. In *Life Sciences and Space Research*, IV, ed. A. H. Brown and Florkin. Spartan Books, Washington, D.C.

Gradmann, H., 1930. Die tropischen Krümmung als Auswirkungen eines gestörten Gleichgewichts. *Jb. wiss. Bot.* **72**, 513–610.

Grahm, L., 1964. Measurement of geoelectric and auxin-induced potentials in coleoptiles with a refined vibrating electrode technique. *Physiologia Pl.* **17**, 231–261.

Grahm, L. and C. H. Hertz, 1962. Measurements of the geoelectric effect in coleoptiles by a new technique. *Physiologia Pl.* **15**, 96–114.

Grahm, L. and C. H. Hertz, 1964. Measurement of the geoelectric effect in coleoptiles. *Physiologia Pl.* **17**, 186–201.

Griffiths, H. J. and L. J. Audus, 1964. Organelle distribution in the statocyte cells of the root-tip of *Vicia faba* in relation to geotropic simulation. *New Phytol.* **63**, 319–333.

Gunther-Massias, M., 1929. Über die Gultigkeit des Reizmengengesetzes bei der Summation unterschwelliger Reizung. *Z. Bot.* **21**, 129–172.

Haberlandt, G., 1900. Über die Perzeption des geotropischen Reizes. *Ber. dt. bot. Ges.* **18**, 261.

Haberlandt, G., 1928. *Physiological Plant Anatomy.* MacMillan, London, pp. 777.

Hawker, L. E., 1932a. A quantitative study of the geotropism of seedlings with special reference to the nature and development of their statolith apparatus. *Ann. Bot.* **46**, 121–157.

Hawker, L. E., 1932b. Experiments on the perception of gravity by roots. *New Phytol.* **31**, 321–328.

Hawker, L. E., 1933. The effect of temperature on the geotropism of seedlings of Lathyrus odoratus. *Ann. Bot.* **47**, 503–515.

Heyn, A. N. J., 1934. Weitere Untersuchungen über den Mechanismus der Zellstrekung und die Eigenschaften der Zellmembran. III. Die Änderung der Plastizität der Zellwand bei verschiedenen Organen. *Jb. wiss. Bot.* **79**, 753–789.

Heilbrunn, L. V., 1930. Protoplasmic viscosity of *Amoeba* at different temperatures. *Protoplasma*, **8**, 58–64.

Hertel, R. and A. C. Leopold, 1963. Versuche zur Analyse des Auxintransports in der Koleoptile von *Zea mays* L. *Planta*, **59**, 535–562.

Hoffmann, O. L. and A. E. Smith, 1940. A new group of growth regulators. *Science*, **109**, 588.

Huisingha, V., 1964. Influences of light on growth, geotropism and guttation of *Avena* seedlings grown in total darkness. *Acta bot. neerl*, **13**, 445–487.

Johnsson, A., 1966. Spontaneous movements in plants studied as a random walk process. *Physiologia Pl.* **19**, 1125–1137.

Juniper, B. E., S. Groves, B. Landau-Schachar, and L. J. Audus, 1966. Root cap and the perception of gravity. *Nature, Lond.* **209**, 93–94.

Keeble, F., M. G. Nelson, and R. Snow, 1931. The integration of plant behaviour. IV. Geotropism and growth substances. *Proc. R. Soc.* **B**. **108**, 537–545.

Knight, T. A., 1806. On the direction of the radicle and germen during the vegetation of seeds. *Phil. Trans. R. Soc.* 99–108.

Konings, H., 1965. On the indoleacetic acid converting enzyme of pea roots and its relation to geotropism, straight growth and cell wall properties. *Acta bot. neerl.* **13**, 566–622.

Lahiri, A. N. and L. J. Audus, 1960. Growth substances in the roots of *Vicia faba*. *J. exp. Bot.* **11**, 341–350.

Larsen, P., 1957. The development of geotropic and spontaneous curvatures in roots. *Physiologia Pl.* **10**, 127–163.

Larsen, P., 1959. The physical phase of gravitational stimulation. *Abstr. 9th Internat. bot. Congr. Montreal*, **2**, 216.

Lidforss, B., 1903. Über den Geotropismus einiger Frühjahrpflanzen. *Jb. wiss Bot.* **38**, 343–376.

Linsbauer, K. 1907. Über Wachstum and Geotropismus der Aroideen-Luftwurzel. *Flora*, **97**, 267–298.

Lundegårdh, H., 1926. Reizphysiologische Probleme. *Planta*, **2**, 152–240.

Lyon, C. J., 1961. Measurement of geotropic sensitivity of seedlings. *Science*, **133**, 194–195.

Moewus, F., 1949. Der Kresswurzeltest, ein neuer quantitativer Wuchsstofftest. *Biol. Zbl.* **68**, 118–140.

Morgan, D. G. and H. Söding, 1958. Über die Wirkungsweise von Phthalsäuremono-α-naphthylamid (PNA) auf das Wachstum der Haferkoleoptile. *Planta*, **52**, 235–249.

Nemeć, B., 1900. Über die Art der Wahrnehmung des Schwerkraftreizes bei den Pflanzen. *Jb. wiss. Bot.* **36**, 80.

Palmer, J. H., 1956. The nature of the growth response to sunlight shown by certain stoloniferous and prostrate tropical plants. *New Phytol.* **55**, 346–355.

Pekelharing, E. J., 1909. Onderzoekingen omtrent de betrekking tussen praesentatietijd en groote van de prickkel bij geotropische krommingen. *Proc. K. ned. Akad. Wet.* **18**, 11.

Pickard, B. G. and K. V. Thimann, 1966. Geotropic response of wheat coleoptiles in absence of amyloplast starch. *J. gen. Physiol.* **49**, 1065–1086.

Pilet, P. E., 1950. Nouvelle contribution à l'étude du géotropisme des étaimes *d'Hosta caerulea* Tratt. *Ber. Schweiz. bot. Ges.* **60**, 5–13.

Pilet, P. E., 1956. Sur l'apogéotropisme de *Phycomyces*. *Experientia*, **12**, 148–149.

Reisener, H. J., 1957. Versuche zum Geotropismus mit radioaktiver β-Indolessigsäure. *Naturwiss*, **44**, 120.

Robards, A. W., 1966. The application of the modified sine rule to tension wood production and eccentric growth in the stem of crack willow (*Salix fragilis*, L.). *Ann. Bot.* **30**, 513–522.

Rutgers, A. A. L., 1912. The influence of temperature on the geotropic presentation-time. *Recl. Trav. bot. néerl.*, **9**, 1–123.

Sachs, J., 1874. Über das Wachstum der Haupt- und Nebenwurzeln. *Arb. bot. Inst. Würzburg.* **1**, 385–474.

Sachs, J., 1882a. Über Ausschliessung der geotropischen und heliotropischen Krümmung während des Wachsens. *Arb. bot. Inst. Würzburg.* **2**, 209–225.

Sachs, J., 1882b. Über orthotrope und plagiotrope Pflanzenteile. *Arb. bot. Inst. Würzburg.* **2**, 226–284.

Schmitz, H., 1933. Über Wuchstoff und Geotropismus bei Gräsern. *Planta*, **19**, 614–635.

Sievers, A. 1965. Elektronenmikroskopsiche Untersuchungen zur geotropischen Reaktion. *Pflanzenphys.* **53**, 193–213.

Tischer, G., 1905. Über das Vorkommen von Statolithen bei wenig oder gar nicht geotropischen Wurzeln. *Flora*, **94**, 1–67.

van Overbeek, J., D. Olivo, and E. M. S. de Vasquez, 1945. Rapid extraction method for free auxin and its application in geotropic reactions of bean and sugar cane. *Bot. Gaz.* **106**, 440–451.

von Ubisch, G., 1928. Betrachtungen und Versuche zur Statolithenhypothese. *Biol. Zbl.* **48**, 172–190.

Wagner, N., 1936. Über der Mitosenverteilung in Wurzelspitzen bei geotropischen Krümmungen. *Planta*, **25**, 751–773.

Waight, F. M. O., 1923. On the presentation time and latent time for reaction to gravity in fronds of *Asplenium bulbiferun*. *Ann. Bot.* **37**, 56–61.

van der Weij, H. G., 1932. Der Mechanismus des Wuchsstofftransportes. *Recl. Trav. bot néerl.* **29**, 379–496.

Went, F. W. and K. V. Thimann, 1937. *Phytohormones*. MacMillan, New York, pp. 294.

Wilkins, M. B., 1965. Red light and the geotropic response of the *Avena* coleoptile. *Pl. Physiol*. Lancaster, **40**, 24–34.

Wilkins, M. B. and M. H. Goldsmith, 1964. The effects of red, far-red and blue light on the geotropic response of coleoptiles of *Zea mays*. *J. exp. Bot*. **15**, 600–615.

Wilkins, M. B., and A. E. R. Woodcock, 1965. Origin of the geoelectric effect in plants. *Nature, Lond*. **208**, 990–992.

Woodcock, A. E. R., and M. B. Wilkins., 1969. The geoelectric effect in plant shoots. I. The characteristics of the effect *J. exp. Bot*. **20**, 156–169.

Yeomans, L. M. and L. J. Audus, 1964. Auxin transport in roots; *Vicia faba*. *Nature, Lond*. **204**, 559–562.

Younis, A. F., 1954. Experiments on the growth and geotropism of roots. *J. exp. Bot*. **8**, 357–372.

Zaepffel, E., 1923. Contribution a l'étude du géotropisme *Ann. Sci. nat. bot*. 2nd. Ser. **5**, 97–192.

Zielinski, F., 1911. Über die gegenseitige Abhängigkeit geotropischer Reizmoment. *Z. Bot*. **3**, 81–101.

Zimmermann, W., 1927. Beitrage zur Kenntnis der Georeaktionen. I. Geotonische Längskraftwirkungen auf orthotrope Hauptwurzeln. *Jb. wiss. Bot*. **66**, 631–677.

Zollikofer, C., 1921. Über den Einfluss des Schwerereizes auf das Wachstum des Koleoptile von *Avena sativa*. *Recl. Trav. bot. néerl*. **18**, 237–321.

7. Phototropism

George M. Curry

Introduction

When the distribution of light falling on many growing plant organs becomes uneven, the organs develop curvatures in such a way that they are brought once again into a position in which they are evenly illuminated. In general, the movement is the result of a change in the pattern of growth, induced by a change in the spatial distribution of light falling on the plant. Such movements, in which the plane of curvature is determined by the spatial relationship of the organ and the light stimulus, are called 'phototropism'.

Phenomena of this kind have been known for a long time. They are particularly obvious in young, rapidly growing shoots where the long axis may be twisted 90° or more after the direction of the incident light is similarly changed. Geotropism, in most cases, also operates to determine spatial orientation, and there are many situations where light direction is of little consequence to the plant. However, from the simplest and most direct observations there is no doubt that light can dramatically affect the direction of growth. Studies of such phenomena were a central part of Charles Darwin's work, so elegantly described in his *Power of Movement in Plants* in 1880. His studies represent a major turning point in our understanding of phototropism. The object of this chapter is to outline what is known about mechanisms underlying the process and to point out those areas which must be further explored before we can unravel the process. Attention will be focused on one system, the *Avena* coleoptile, and a single scheme, the Cholodny–Went theory. For broader and more detailed treatments of the subject, the reader is referred to the many recent reviews.

As far as phototropism is concerned, Darwin's major contribution was the clear-cut demonstration that, after reception of the light stimulus in the apical part of grass seedlings, some 'influence' is transmitted to the lower regions, and it is in these regions that the bulk of the response occurs in the form of differential growth. Darwin's conclusions were drawn from simple but penetrating studies in which, by various ingenious techniques, parts of the coleoptile were shielded from the light and the magnitude and position of the bending response subsequently determined. In brief, he found that when only the apical 6 mm were unilaterally illuminated for a few hours, large curvature developed in the region below the illuminated zone, whereas when the 6 mm apex was shielded and the lower region exposed, little or no curvature ensued. The conclusion mentioned above, that some 'influence' is transmitted from the apex to lower regions, is inescapable, and this Darwin properly stressed. The further conclusion, that the light-sensitive system is restricted to the apex, is one which most investigators have subsequently assumed, but it is by no means as inescapable as the first. Here it is interesting to note that Darwin did not observe a *total* lack of curvature in all cases when tips were shielded, and he himself was much more cautious about reaching for the second conclusion. This point is of major significance when it comes to the question of photorecepter identity.

Much of Darwin's work on phototropism was carried out with grass coleoptiles, either *Phalaris* or *Avena*; the latter continues to be the favoured object for study today. There are several reasons for this, among which are the facts that coleoptiles are relatively simple structures, they are very sensitive to light, they respond relatively quickly, and they can readily be grown and prepared for experiments. Most of Darwin's studies were concerned with the effects of prolonged unilateral illumination, using either a paraffin lamp or daylight. In this connection, Darwin made another important observation, largely overlooked subsequently, to the effect that the magnitude of the response is singularly unaffected by the light intensity. When coleoptiles are unilaterally and continuously exposed at various distances from a lamp, curvatures are very similar in most cases, except at the intensity extremes.

Interestingly enough, the bulk of later investigations were carried out using short exposures, and, at the present time, there are serious difficulties in attempting to relate facts and theories about short-exposure experiments with those based on continuous, prolonged illumination, presumably the more 'natural' situation. The main contributor to the short-exposure aspect of phototropism in the early days was Blaauw (1914, 1915). He did extensive work on light-growth reactions and phototropism in *Avena* coleoptiles and in sporangiophores of the fungus *Phycomyces*. The light-growth reaction is a growth rate change induced by a *temporal* change in light intensity. In the first place, Blaauw established that short exposures do lead to a graded response, both for the light-growth reaction and, in the case of asymmetrical illumination, for phototropism. Thus, the magnitude of the response was shown to depend on the total light dosage, intensity times time ($I\,t$), at least over a limited range of dosages. Furthermore, because of the close parallel between the dosage-dependency of the light-growth reaction and the magnitude of phototropic curvature, Blaauw made the broad conclusion that phototropism is essentially a matter of localized, asymmetrical light-growth reactions, caused by an asymmetrical distribution of light in the plant. It is noteworthy that this conclusion largely overlooks Darwin's key point, that the light-receptive and the responsive zones are spatially separate. Even though Blaauw's postulate can be successfully modified to take this into account, we are still left with the problem of understanding the mechanism of the light-growth reaction.

Subsequent work added more details to the so-called 'dose-response' curve, and this aspect of the analysis continues at present, as will be discussed below.

During and after Blaauw's period, and stemming directly from Darwin's hypo-thesis, the role of hormones, particularly indole-3-acetic acid (IAA), in phototropism began to be revealed. In the late 'twenties and early 'thirties this culminated in the 'Cholodny–Went theory' of phototropism, which in its sim-plest form suggested that unilateral light leads to lateral displacement of growth hormone in the plant, thus causing curvature. The main foundation for this theory was Went's demonstration that, after prolonged unilateral illumination, more auxin than normal (i.e., than in dark controls) diffused out of the half of the coleoptile on the side away from the light, while less than normal came out of the side toward the light. This further led to the notion that unilateral light somehow induced a 'lateral transport' mechanism, leading to the asymmetry of auxin distribution.

Other lines of evidence suggest additional or alternative explanations to account for asymmetrical auxin distribution. The simplest of these, arising from the finding of Galston and Baker (1949) that IAA is destroyed *in vitro* when illuminated in the presence of riboflavin, is the notion that unilateral light de-stroys IAA asymmetrically. But present evidence is that little, if any, IAA is actually destroyed during phototropism. In experiments which confirmed Went's earlier findings, Briggs (1963) found that light does not decrease the total yield of auxin obtained in agar receiving blocks. Pickard and Thimann (1964), using radioactive IAA, demonstrated that no auxin is destroyed in corn coleoptile sections unilaterally exposed to 500 metre-candle white light for 3 hours (but see chapter 4). These workers also confirmed the asymmetrical dis-tribution of auxin after unilateral illumination with 100 metre-candle seconds of white light. The ratio of IAA-C^{14} radioactivity transported into agar re-ceivers under the lighted and shaded halves was 25:75; nearly the same ratio was measured in the tissue. If IAA photolysis is ruled out, a further alternative is that unilateral light may asymmetrically change the synthesis or consumption of IAA, but this idea also runs counter to the evidence cited above.

Therefore, the Cholodny–Went theory remains at the centre of present ideas concerning phototropism in higher plants. We are left with the following main questions:

(a) What is the nature of the photoreceptive system?
(b) What physical and chemical events immediately follow the primary light absorption event?
(c) How does the light receiving system interact with the auxin transport system to cause asymmetrical distribution?
(d) How does asymmetrically distributed auxin cause asymmetrical growth, hence curvature?

The questions concerning auxin transport and auxin control of growth are the subjects of chapters 1 and 4 of this book; I shall therefore confine my atten-tion to the first two questions, and part of the third, in the remainder of this chapter. At this point it should be admitted that although we have a few facts

Phytochrome — plant pigment energized by light to function as enzyme to iniate germinati growth stem leaf [handwritten margin notes]

to go on in dealing with the first question, we have practically nothing but specu-lations beyond this point.

The nature of the photoreceptor system

There are several lines of evidence which bear on the nature of the photorecep-tor. Among these are the following: action spectrum studies; the distribution of photosensitivity and of pigments in sensitive plants; dose-response con-siderations, and the possibility of more than one receptive system; intercon-nections between receptor systems for phototropism and other light-controlled processes such as photosynthesis and photomorphogenesis.

Action spectrum studies

In theory, at least, if one can obtain a detailed action spectrum (relative quantum effectiveness in inducing the response as a function of the wavelength of the quantum), one has a major criterion as to the nature of the photoreceptor, since the action spectrum may closely parallel the absorption spectrum of the receptor material. Classically, this is certainly the case for the chlorophylls in photosynthesis, for protochlorophyll conversion, for DNA in ultraviolet muta-genesis, and for visual pigments in vision. On recent evidence it seems also to be the case for the phytochromes in photomorphogenesis (see chapter 14). This parallelism is frequently found in spite of the presence of other pigments or materials not directly involved in the particular phenomenon under study. In fact, this frequent lack of distortion in the presence of masking pigments is often very puzzling.

The spectral sensitivity of phototropism has been studied for a number of years, and the techniques have recently been quite highly refined. With few exceptions the earlier work demonstrated that blue light was phototropically active, while green, yellow, and red light were not. On this note, it is interesting that Darwin, while watching plant position through a microscope, used light from a paraffin lamp filtered through a solution of bichromate of potassium 'which does not induce heliotropism'. This kind of observation led to the notion that the photoreceptor system must involve a yellow pigment, because only a yellow pigment would absorb blue light and not (at the same time) green or red light. This argument excludes the participation of chlorophyll, unless one can suppose that absorption of a blue quantum by chlorophyll can lead to photo-tropism, while absorption of a red quantum by the same system has no photo-tropic effect. This proposition, indeed, needs further testing.

A detailed action spectrum is shown in Fig. 7.1. It exhibits two peaks, or maxima, in the blue, centred near 445 and 475 nm, and a shoulder in the 425 nm region. Furthermore, there is another maximum, about half the size of the main peak, near 370 nm. This spectrum was obtained by determining the relative numbers of quanta required to obtain a given curvature response, restricting attention to the first-positive range of the dose-response curve (see below). Shropshire and Withrow (1958) found a very similar action spectrum except that the size of the 370 nm peak depended on the response 'aimed' for; to put it another way, the slope of the dose-response curve was wavelength dependent. This was not confirmed by Curry and Gruen (1961) who found that the dose-response curves at 365, 405, and 436 nm did not have significantly different slopes. Action spectra very similar to the above were obtained by Curry and Gruen (1959) and by Delbrück and Shropshire (1960) for the phototropism of *Phycomyces* sporangiophores, and by Delbrück and Shropshire for the light-growth reaction of *Phycomyces*.

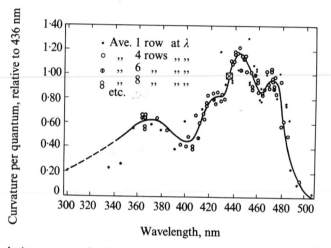

Figure 7.1 Action spectrum for the first-positive response of *Avena* coleoptiles, deduced from the pooled results of three large series of experiments.

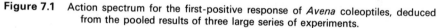

An earlier action spectrum obtained by Johnston (1934) should be mentioned here. He used a balancing technique, whereby *Avena* seedlings were illuminated by a standard (white) light from one side and simultaneously by a monochromatic test light from the other side. By adjusting the intensity of the test light until no curvature resulted after long exposure (null-point), Johnston was able to gauge the relative effectiveness of various test wavelengths. His tests encompassed only the blue region, and, although less detailed than the later spectra, the main features were the same. The important point here is that his system involved continuous illumination; reference will be made to this later. A similar 'balancing' system was used by Curry and Gruen in the *Phycomyces* work mentioned above.

All of these spectra support the idea that the photoreceptor is a yellow pigment; furthermore, their details narrow down the possibilities to a few specific candidates, known to be present in the responsive plants. At present, the two possibilities most seriously considered are a carotenoid or a flavin pigment. Typical absorption spectra for these are shown in Fig. 7.2. Direct comparison with the action spectrum indicates that the carotenoid matches exceedingly well in the 400–500 nm region, but not in the 370 nm region. The flavin does not fit the blue region as well as the carotenoid, but its large absorption peak near 370 nm offers better possibilities for explaining the action spectrum peak here than does the carotenoid. This issue remains unresolved, and it is doubtful that further

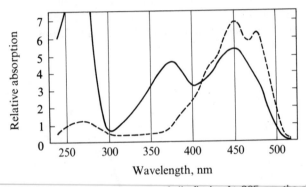

Figure 7.2   Solid line: absorption spectrum of riboflavin. At 265 nm the riboflavin curve rises to 13·4 on this relative scale. Broken line: absorption spectrum of β-carotene in hexane.

action spectrum work alone will resolve it. Several suggestions have been made to interpret the discrepancies, but they remain suggestions. Thus, it has been argued that a carotenoid may be the primary acceptor and that absorption of UV by a flavin, followed by blue fluorescence, might account for UV effectiveness. Briggs (1964) has, in fact, isolated a material with appropriate fluorescent properties in a buffer extract of oat coleoptiles. Alternatively, it has been suggested that a carotenoid having a large cis-peak in the 360 nm region would fit the curve, but such a pigment has not yet been detected in *Avena*. Along similar lines, Asomaning and Galston (1961) have suggested that a special flavin pigment, in a particular physical situation, might have appropriate absorption characteristics in the 400–500 nm region. This is based on the fact that flavins may indeed show multi-peaked or highly modified absorption characteristics in special solvent systems. Here again, however, the existence of such a system in the plant has not been demonstrated. Arguments based on 'masking' effects of one pigment on another have been invoked, both pro-carotenoid and pro-flavin, perhaps equally unconvincingly. A final suggestion is that *both* flavin and carotenoids are involved as photoreceptors in phototropism, or even that some as yet unknown class of pigments is really at the heart of the matter. All of this simply emphasizes the fact that the action spectrum alone does not provide an adequate basis for decision.

coleoptile protective sheath surrounding terminal bud of embryo plants

The distribution of photosensitivity and of pigments

It might be a simple matter to decide on the identity of the photoreceptor if there were a clear-cut coincidence between pigment location and the location of photosensitivity, as is the case for rhodopsin and visual sensitivity. Unfortunately, this is not the case in phototropism; neither the suspected pigments nor photosensitivity are localized in any very obvious way. There are, however, a few suggestive leads.

Ever since Darwin's observations, attention has been focused on the coleoptile apex as the site of photosensitivity. This was reinforced in the 'twenties by the observations of Lange (1927), who found that the apical millimetre of the coleoptile is more than 500 times as sensitive as a 1-mm zone 10 mm from the apex. In more detail, Lange's results indicated that greatest sensitivity is located in the 50 μ region extending from 50 to 100 μ below the extreme apex. Sensitivity falls off rapidly below 200 μ. Applying a geometrical correction to take into account the smaller 'cross-sectional' surface presented to the light by the extreme apex, Lange concluded that the extreme apical 50 μ region is, in fact, the most sensitive.

When one carefully examines the evidence leading to these conclusions, one or two serious questions arise. In the first place, the criterion for 'sensitivity' is open to question. (This is a perpetual problem in phototropism studies, as will be apparent in further discussion.) In these studies what was determined was the amount of unilateral light required to cause visible curvature in 50 per cent or more of the cases tested, the light being applied to narrow zones delimited by various shielding techniques. This is roughly equivalent to the 'threshold' approach frequently used by earlier workers. We know now that phototropism is not an 'all or none' phenomenon, and it seems quite certain that very small responses were either overlooked or ignored in these studies. In the second place, even if we ignore diffraction around the shield (see p. 263) light scattering in the tissue would probably lead to illumination of regions well outside the 50 μ limits. This would be especially true for illumination of the extreme apex, and it weakens the argument that highest sensitivity resides here.

It is noteworthy that Darwin did *not* find highest sensitivity in the extreme apex. For *Phalaris* he reported that the apex, for a length of 0·04 or 0·05 inch, has 'only a slight power of causing the lower part to bend'. Furthermore, he noted that exclusion of light from the apex for a length of 0·1 inch does not have a 'strong influence' on the curvature of the lower part. However, exclusion of light from the zone between 0·15 and 0·2 inch below the apex 'plainly prevents' curvature. Darwin's findings with *Avena* were similar to the above, but not as well defined.

The widespread assumption that phototropic sensitivity in coleoptiles is largely confined to the extreme apex must therefore be questioned. The evidence actually indicates that maximum sensitivity is located in a region approximately 100 μ below the apex.

Taken together, the action spectrum data and the high sensitivity of the tip lead one to look for a yellow pigment in the coleoptile apex. Bünning (1937) demonstrated the presence of a carotenoid in *Avena* coleoptiles using the Molisch microchemical method. He noted highest pigment concentrations just below the tip, but none in the extreme apex. These observations have recently been confirmed by Sorokin and Thimann (1960). Hexane extracts of coleoptile tips reveal the presence of at least three different carotenoids. The absorption spectrum of such an extract is strikingly similar to the action spectrum, except in the near UV (Fig. 7.3). Flavins are also present in *Avena* coleoptiles; Asoman-

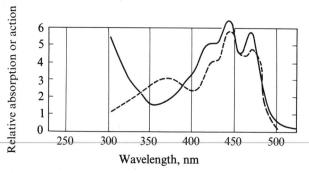

Figure 7.3   Solid line: absorption spectrum of a direct hexane extract of 500 coleoptile tips. Broken line: action spectrum.

ing and Galston (1961) found that coleoptile tips contain twice as much flavin (and three to five times as much carotenoid) as the bases. The absolute amounts of flavins (FMN) in the tips were found to be $56 \cdot 0\ \mu g\ g^{-1}$ fr. wt; total carotenoids were $3 \cdot 5\ \mu g\ g^{-1}$ fr. wt.

Dose-response relationships

As mentioned earlier, most growing plants, when continuously exposed to a unilateral white light, change their spatial configuration to some extent. The change may require several hours, or it may be almost complete in less than an hour. In *Phycomyces* sporangiophores, bending begins two or three minutes after the onset of unilateral illumination, and curvatures approaching 90° may be attained in twenty or thirty minutes. In *Avena* coleoptiles, bending may become apparent ten to fifteen minutes after the beginning of continuous stimulation, and 90° curvatures may be reached in less than two hours. The exact behaviour of a particular plant depends on a large number of factors: its background or history, its growth rate at the onset of illumination, the quality and intensity of the stimulus, and any other environmental factor affecting growth, including light other than the stimulus light. Precise analysis requires control of all of the factors affecting growth, and this, of course, can only be approached. One of the major problems in phototropism studies is the difficulty in controlling the total light régime for the experimental object, both prior to and during the experiment. Historically, red light has been used for preparations and observations, with the notion that it is equivalent to darkness in so far as phototropism is concerned; this is definitely not the case as will be discussed later (p. 261).

Blaauw's work initiated a whole series of studies on the relationship between light dosage and curvature response. This turns out to be a very complicated relationship, and it can well be argued that the complications introduced by such studies have led us farther away from, rather than closer to, an understanding of the fundamental mechanism underlying phototropism. It is important to recognize, however, that we cannot pretend to have an understanding of the process until we can adequately account for such complications. In fact, the curious nature of the dose-response curve seems to offer unique posssibilities for getting at basic mechanisms and for testing hypotheses. Zimmerman (1962) has recently worked out a complete thesis on this subject, and Briggs (1964) has also explored it in great detail, both with corn and oat coleoptiles. The major aspects of the dose-response relationships will be outlined below. In order to avoid confusion some of the minor complications will be ignored at this point.

Given a ten-second exposure to a unilateral blue (436 nm) light at an intensity of 2 erg cm^{-2} s^{-1}, a dark-adapted *Avena* coleoptile will begin to bend after 15 or 20 minutes, and 100 minutes after stimulation a curvature of 15 to 20 degrees will have been attained. This curvature begins to appear in the apical 3 mm and slowly migrates toward the base. It reaches its maximum development in approximately 100 minutes, after which the coleoptile begins to straighten (except on a klinostat, in which case development may continue for 6 hours or more). If less light energy is applied, smaller curvatures are seen after 100 minutes; with greater amounts of energy, larger curvatures result, but a point is reached beyond which further energy increments do not increase the response. The maximum response following a short stimulus is in the neighbourhood of 30° curvature. The response does not 'plateau' with increasing energy; rather, it begins to fall off quite rapidly, and for total dosages near 1000 erg cm^{-2} the plant may not be curved at all 100 minutes later. At still higher dosages, the coleoptile will typically exhibit small, but definite, negative curvatures; the apical region bends away from the light stimulus. This type of response also disappears as dosage is increased, and beyond this point neither positive nor negative curvatures are obtained unless the stimulus duration exceeds 3 or 4 minutes. If it does, positive curvatures extending well toward the base develop in the next 100 minutes. The magnitude of these curvatures depends on the *length* of the light exposure, not on the intensity of the light. The dose-response curves obtained using two monochromatic light sources (365 and 436 nm) are shown in Fig. 7.4; Fig. 7.5 shows a typical set of double shadowgraphs covering a 10,000-fold range of dosages. The oscillations in the dose-response curve gave rise to the terms 'first-positive', 'first-negative', and 'second-positive' curvatures to characterize the three different parts of the curve. du Buy and Nuernbergk (1934) reported data indicating another minimum beyond the second-positive maximum, and even a 'third-positive' range beyond this. It is difficult to assess the significance of such data based on very long, high intensity exposures, and most workers since have been content to deal with the complexities up to and including the second-positive range.

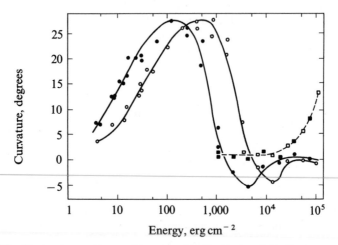

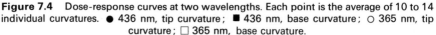

Figure 7.4   Dose-response curves at two wavelengths. Each point is the average of 10 to 14 individual curvatures. ● 436 nm, tip curvature; ■ 436 nm, base curvature; ○ 365 nm, tip curvature; □ 365 nm, base curvature.

The same dose-response curve is obtained, at least up to negative curvatures, when a variety of combinations of stimulus duration and intensity is used. Thus, 0·01 second at an intensity of 2000 erg cm^{-2} s^{-1} gives the same response as 10 seconds at 2 erg cm^{-2} s^{-1}. The Bunsen–Roscoe reciprocity law, which states that the photochemical effect of light should remain the same if the *product* of intensity and duration is the same, appears to be valid over a considerable range. Clearly (and significantly), it is not valid for second-positive curvatures; in fact, long exposures at low intensities give large 'second-positive' curvatures, even though the total dosage applied falls within the first negative range. To restate the case, the Bunsen–Roscoe law seems to be valid providing stimulus duration is short (less than four minutes); for longer-lasting stimuli, length of exposure becomes the determining factor, rather than total energy applied.

The fact that phototropism 'obeys' the reciprocity law, at least in part, implies that some relatively straight-forward photochemical event determines response magnitude. This raises another feature of the dose-response curve.

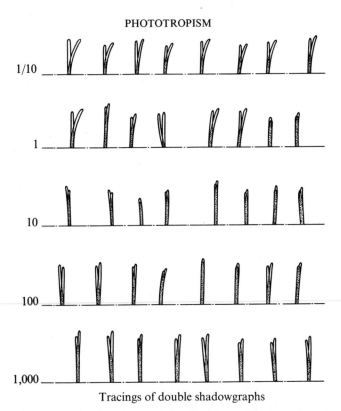

Tracings of double shadowgraphs

Figure 7.5 Curvatures of coleoptiles receiving fixed intensities for various time intervals. The source of light (high intensity 436 nm) was on the right. The intensity was *ca.* 1500 erg cm^{-2}s^{-1} at the right end of each row, and *ca.* 500 erg cm^{-2}s^{-1} at the left end. Exposure times indicated are seconds. Shadowgraphs taken immediately after application of the phototropic stimulus and 100 minutes later. Note: typical first-positive curvatures in frame 1; negative in left side of frame 3; second-positive in frame 5.

It will be noted that the rising part of the first-positive range of the curve is linear with the logarithm of light energy, except at its extremes. This is, of course, characteristic of stimulus-response curves in many photobiological systems. A simple interpretation of this is the idea that light acts by converting a photosensitive molecular species from some 'active' form to another form, the rate of conversion being proportional to the concentration of the 'active' form and light intensity (Curry, 1957). Schematically, this may be represented as follows:

$$A \xrightarrow[kI]{h\nu} B \quad \text{and} \quad \frac{dA}{dt} = -kIA$$

If one assumes that the curvature response depends on the amount of A changed by a particular light dosage, integration of the first-order differential equation leads to the expression: $C = C \max(1 - e^{-kIt})$ where C max is the maximum curvature obtained as light dosage (It) is increased. This relatively simple equation fits the rising arm of the first-positive range very well; it does not account for the descending arm. However, if we further assume that photoconversion occurs not only in the illuminated (front) flank of the coleoptile but also in the side away from the light (back), and if curvature depends on the difference in amount of conversion in the two sides, then the equation becomes:

$$C = C\max (e^{-kI_b t} - e^{-kI_f t})$$

where I_b and I_f are the light intensities on the back and front halves, respectively. The ratio I_b/I_f describes the light gradient or 'Lichtabfall' across the coleoptile. This equation predicts the descending arm of the first-positive range ($C \to 0$ as $It \to \infty$), but does not agree with the experimental curve in detail. It yields a symmetrical curve, rising and descending, and does not account for the rapid decline actually observed. The equation incorporates the idea of a light 'saturation' effect proposed earlier by many workers to account for declining responses with increased dosage in *Avena*, *Phycomyces*, and other phototropic systems.

The above interpretation also completely fails to account for first-negative and second-positive curvatures. Zimmerman and Briggs (1963a, 1963b) have developed much more complicated models and greatly extended their analysis. They include hypotheses about pigment regeneration, both by dark reactions and light-driven reactions, and they further suggest that three distinct systems (Systems I, II, and III) are involved in the total dose-response picture. Briggs' schematic summary of the total model is shown in Fig. 7.6. In Briggs' view, system-I curvatures are essentially equivalent to first-positive, system-II to first-negative, and system-III to the second-positive. The systems overlap in the distribution of the pigments involved, in their kinetics, and in the ways in which they affect the lateral transport of IAA. Briggs (1964) elegantly marshals the evidence from hormone collection studies that IAA destruction (or a change in the rate of IAA synthesis) is not fundamentally involved in any of the systems. Rather, systems I and III seem to be connected to lateral transport of hormone away from the light source, leading to positive curvatures, while system II may promote transport of IAA toward the light, giving rise to negative curvatures. The evidence on the last point is skimpy and needs to be amplified.

The extensive analysis of Briggs and Zimmerman goes far to pull together the multitude of complexities surrounding the dose-response phenomena. It also provides a very useful framework for testing hypotheses about phototropism in general. Its chief drawback is the fact that it is an exceedingly complicated model. Why should there be three or more different systems underlying phototropism? Does it make sense that a plant should have a special system for negative curvatures? Similarly, would one expect plants to have a special system designed for response to short stimuli, and a different system for continuous or prolonged stimuli? Or is it more reasonable to suppose that the complicated dose-response relationship is simply a consequence of the way the basic system

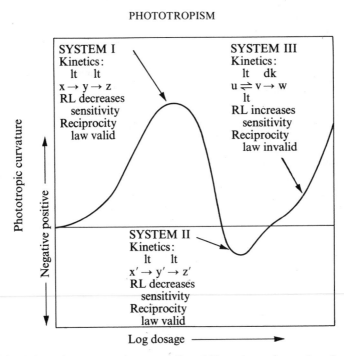

Figure 7.6 Schematic summary of systems I, II, and III curvature of oat coleoptiles. A phototropic dosage-response curve obtained with high intensity monochromatic light. Definitions: x, x′, u (primary photoreceptor molecules); y, y′, w (phototropically active forms); z, z′ (light-inactivated forms); v (light-activated form of u; may be photochemically reconverted to u or decay thermochemically to w); lt (light reaction); dk (dark reaction); RL (red light). (From Briggs, 1964)

is experimentally probed, especially in regard to abrupt 'on and off' exposures to light?

A more concrete objection to the three systems hypothesis comes from consideration of action spectra. There is no solid evidence that three different action spectra, and hence three different photoreceptors, are involved in phototropism. On the contrary, there is a considerable body of evidence that all parts of the dose-response relationship have the same spectral sensitivity, centred in the blue and cutting-off above 550 nm. It is true that we have very little evidence to go on for the first-negative and second-positive ranges, but responses in these ranges are certainly not obtained with wavelengths longer than 550 nm. It can be noted, also, that the dose-response curves for 365 and 436 nm run parallel *through* the first-negative range (Fig. 7.4). With respect to the second-positive, Johnston's action spectrum (mentioned on p. 249) was obtained using continuous illumination; on a total energy and time basis this spectrum should apply to the second-positive range. Its close similarity to the action spectrum for the first-positive supports the idea of a single photoreceptor. We need more firm evidence on this point; the problem is that, beyond the first positive range, it is very difficult to isolate specific parts of the dose-response curve to analyse their spectral sensitivity meaningfully.

Tip and base responses

Early in this century it was noted that phototropism in *Avena* has two morphologically distinct components, called tip and base responses. The high sensitivity of the tip compared to the base has already been noted. Went (1925) observed that the kinetics of curvature are different for tip and base illumination; curvatures begin more quickly after exposure of basal zones. Schrank (1954) noted that ultraviolet light induces curvatures which first appear near the base. Curry *et al.* (1956) concluded that UV-induced curvatures are much the same as those induced by 4 minutes or more exposure to intense blue light (second-positive). The main features of such 'base curvatures' are the following:

(a) They require more total light energy than do tip curvatures.
(b) They begin to appear earlier (10–15 minutes) than do tip curvatures (20–30 minutes).
(c) The curvatures start to appear along the entire length of the coleoptile, extending through the first node, whereas tip curvatures typically start near the apex and migrate basipetally.
(d) Illumination of apex *or* base with sufficient energy, especially with UV, gives base curvatures, whereas 'pure' tip curvatures are caused only by low-energy illumination of the apical region. Small amounts of UV applied exclusively to the tip give curvatures mostly confined to the apical half, but it is difficult to categorize these as 'tip' or 'base' curvatures.

UV-induced base curvatures are dose-dependent up to a 15-degree maximum. The latter fact permitted the determination of an action spectrum (Fig. 7.7). The spectrum is very similar to the absorption spectrum of IAA, except that it is shifted about 12 nm toward the visible region. This finding prompted the suggestion that UV base curvatures may arise from a direct photo-inactivation of IAA or a chemically bound form of IAA. (It may be noted here that the lack of a pronounced peak in the 260–270 nm region argues against participation of a flavin.) UV does destroy IAA in solution and in agar blocks (Curry, 1957), but 1000 times as much energy is required for this process as for the induction of base curvatures. A lateral transport mechanism is not ruled out; here again we need more information about the distribution of growth and of IAA during the response.

As in phototropism, Went (1925) found two types of light-growth reactions in *Avena*: the first, given only by illumination of the tip and at low energy levels, is a slowly initiated, long-lasting depression of the growth rate (tip-response); the second results from the illumination at higher energy levels of either the tip or the base, and it is a quickly initiated, transient depression (base response). Went postulated that a one-sided tip response could lead to first-

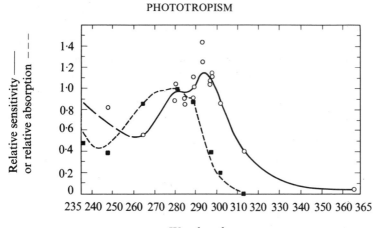

Figure 7.7 Action spectrum for the base curvature (circles and solid line). The ordinates give the reciprocal of the number of quanta needed for 8·6° curvature, relative to that at 289 nm. The dotted curve is the absorption spectrum of IAA, the optical density at 280 nm having been taken as 1·0. The black squares show the action spectrum for photo-oxidation of IAA solutions in vitro, the value at 280 nm being taken as 1·0.

positive curvatures, while second-positive might be attributable to an asymmetrical base response. Van Dillewijn (1927) confirmed the existence of these two different light-growth reactions and extended their analysis (Fig. 7.8). An important feature documented by him is the fact that the tip response disappears with increasing dosage; the base response does not. However, in the base response the growth depression is followed by an accelerating phase, so that total growth over a two-hour period is in many cases just what it would have been in the absence of a stimulus. A similar 'compensation' effect occurs in the light-growth reaction of *Phycomyces* (Castle, 1966b). From light-growth reaction data, and assuming unilateral illumination, van Dillewijn was able to account for the complete dose-response curve for phototropism, including the first-negative range.

The close parallel between light-growth reactions and phototropism in *Avena* cannot be doubted. This does not tell us much more about underlying mechanisms for either, but it does help to eliminate some of the possibilities. For one thing, since total growth is often the same in the presence or absence of a light stimulus, theories based on auxin destruction, or on substantial changes in auxin synthesis, are not applicable. Secondly, theories based on assumptions about light induced changes in cell-sensitivity to auxin do not fit the light-growth reaction data; in any case, such theories are out of line with the demonstrable separation between light sensitive and responsive zones of the organs. The Cholodny–Went theory, based on lateral redistribution, remains intact for phototropism and can be adjusted to account simultaneously for the light-growth reactions.

10*

An interesting point concerning UV effects in *Phycomyces* may be brought in here. Curry and Gruen (1957) discovered that unilateral short wavelength UV causes rapid bending away from the light source (negative curvatures). The action spectrum for this phenomenon is in general agreement with that for UV base curvatures in *Avena*. Symmetrical UV illumination gives a positive light-growth reaction, as does symmetrical illumination with blue light. Positive

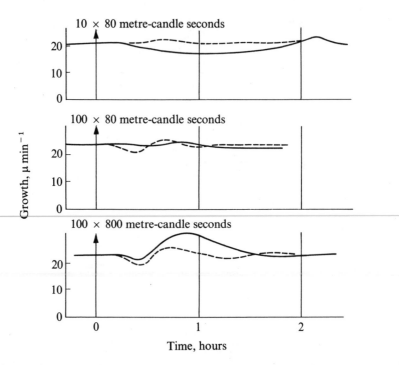

Figure 7.8 Light-growth reactions of *Avena* coleoptiles after three illuminations, according to van Dillewijn (1927). Solid line: after illumination of the apical 2 mm zone. Broken line: after illumination of a 2 mm basal zone.

curvatures in unilateral blue presumably occur because more light is concentrated in the back side of the sporangiophore than in the front, due to the so-called 'lens effect'. UV leads to negative curvatures because it does not penetrate to the back side. The important point here is that action spectrum data for phototropism (see p. 249) show a smooth, continuous transition from positive to negative response and the light-growth reaction action spectrum is also continuous, with peaks matching the phototropism spectrum except for the inversion in sign. This supports the notion of a single photoreceptor and a single basic mechanism for both phototropism and light-growth reactions.

The effect of red light on phototropic sensitivity

While attempting to discover the source of large variations in phototropic sensitivity of *Avena* coleoptiles, Curry (1957) found that exposure to red light prior to a blue stimulus greatly affected the magnitude of curvature obtained. The usual effect was that red light given prior to a blue test exposure substantially increased the amount of blue light needed to get a given, pre-determined curvature; i.e., red light caused a pronounced drop in the coleoptile's sensitivity to unilateral blue light, sensitivity being defined as the energy requirement for a specific response. In dose-response determinations this shows up as a shifting of the entire curve toward the right along the log-dosage axis (Fig. 7.9). This

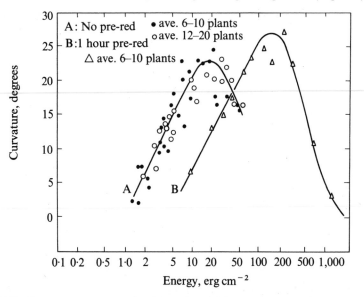

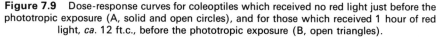

Figure 7.9   Dose-response curves for coleoptiles which received no red light just before the phototropic exposure (A, solid and open circles), and for those which received 1 hour of red light, *ca.* 12 ft.c., before the phototropic exposure (B, open triangles).

effect has been confirmed and studies of it greatly extended by Blaauw-Jansen (1959), Asomaning and Galston (1961), and Briggs (1963). Many additional complications have been uncovered, and there is not space to discuss them all here, but the central features seem to be the following:

(a) Red light given *prior* to phototropic stimulation shifts the first-positive range (system I, and probably system II also) toward the right; i.e., red light decreases system I and II sensitivity. It shifts the second positive range (system III) toward the left; i.e., it increases system-III sensitivity. (System-II negative responses may disappear as a consequence of overlap between I and III).

(b) Red light *after* phototropic stimulation has little effect on the position of the dose-response curve. It may lead to larger maximal curvatures.

(c) The precise schedule (duration and intensity) of red light exposure, especially prior to blue light stimulation, must be known in order to predict phototropic sensitivity. Red light may affect the subsequent situation if it is introduced at any time between germination and the phototropic sensitivity test; it has most pronounced effects when given an hour or two before the test.

All of these points are important in the interpretation of earlier experiments dealing with action spectra, dose-response curves, sensitivity location, and other sensitivity comparisons. Most earlier work was done using a red working-light (as were standard *Avena* curvature tests), and this was used in an undefined way in so far as intensity, duration, and schedule are concerned. For this reason alone it becomes necessary to repeat much of the earlier work in a critical way.

The mechanisms behind the red light effects are largely unknown. There are several possibilities, among which are the following:

(a) Red light may change the pigment composition, active or masking; it certainly initiates protochlorophyll conversion in coleoptiles and primary leaves. Asomaning and Galston (1961) report red-induced increases in both flavin and carotenoid content.

(b) Red light changes the distribution of growth in the coleoptile (Curry *et al.*, 1956; Mer, 1959).

(c) Red light affects auxin levels (Briggs, 1963).

(d) Red light may affect sensitivity of cells to auxin.

(e) Red light may have a variety of (additional) effects mediated through the phytochrome system.

A red-induced change in pigment concentration is an attractive hypothesis in view of the established effects of red illumination prior to the blue stimulus as opposed to the lack of effect of post-stimulus illumination. Other effects are undoubtedly involved; red light received in one part of the plant may affect the phototropic behaviour of a completely different part; red light affects geotropism (Wilkins, 1965), which certainly implies that the consequences of red illumination for phototropism go far beyond changes in blue-absorbing pigments. For an intriguing analysis of the interplay between red and blue light in systems I, II, and III the reader is referred to Briggs (1964). The red light effect, although introducing almost intolerable analytical difficulties, provides us with a powerful tool for testing hypotheses about phototropism.

Some recent observations on the problem

In our laboratory attention has recently been focused on a re-examination of problems in three main aspects of phototropism: sensitivity distribution, pigment distribution, and redistribution of growth during phototropic response. These studies are still in progress, but the results so far have an important bearing on interpretations; some of the more significant findings will therefore be discussed below before discussing a current working model for the mechanism of phototropism.

Sensitivity distribution

In checking Lange's observations (p. 251), but using continuous low intensity blue illumination to avoid some of the dose-response ambiguities and the subjective evaluation of 'threshold', we have confirmed that sub-apical regions of the *Avena* coleoptile are phototropically sensitive. Thus, when the apical 5-mm region is covered by an opaque cap, continuous unilateral illumination of the rest of the coleoptile gives small, but unmistakable, positive curvatures. The double shadowgraph technique, which records initial and final (post 100 minutes) position, clearly reveals such curvatures. Their small magnitude undoubtedly accounts for the fact that they were mostly overlooked by Darwin and subsequent workers. In any event, the simple fact here is that sub-apical regions *are* phototropically sensitive, and therefore *must* contain photoreceptor material.

When we tried to find out what the response would be if we kept a wedge of light just grazing the extreme apex, we observed very large curvatures developing in 90 minutes, which seemed to confirm the very high sensitivity of the apical surface. But we also observed large curvatures developing when the wedge of light was kept a millimetre or more above the extreme apex, even though the observer could not visually detect any light striking the coleoptile. Apparently, even with the best collimated light beam we could arrange, the small amounts of light scattered or diffracted from the slit (or the slit image) were sufficient to elicit response when the blue light operated continuously. This raises serious questions about the significance of results when attempts are made to illuminate very narrow regions of the coleoptile, especially if the slit or collar arrangement is more than 1 mm away from the tissue.

We next devised a simple system for illuminating one zone of the coleoptile from one side, and then reflecting the wedge of light back on to the same or a different zone of the coleoptile from the opposite side. The resulting curvature, if any, would presumably indicate which zone was most sensitive. This method has the advantage that any fluctuations in the stimulus intensity will occur simultaneously on both sides of the plant; furthermore, a single plant, having a single history, is used for the assay, so that complications due to red light history and other factors are minimized. Since the coleoptile is growing during the light exposure, its position relative to the light was adjusted downwards at 10 minute intervals to keep the zonal configuration nearly constant. Alternatively, the coleoptile tip was inserted through a folded black paper mask. The apex supported the paper, so that its position remained essentially unchanged during growth. Now, light from the lamp side could be directed solely at the tip and then reflected back from the other side onto a lower region of the coleoptile, or vice versa. Light scattering complications are minimized or balanced. This is, in essence, a modification of Darwin's tip painting experiments.

The main conclusions from these experiments are as follows:

(a) The 1 mm apex is more sensitive than any 1 mm zone lower down, but not vastly more so. (Frequently, a balance was observed.)

(b) Sensitivity falls off gradually from apex toward base.

(c) Continuous illumination of a 3 or 4 mm subapical zone from the right can over-ride identical illumination of a 1 mm apical zone from the left; i.e., the coleoptile bends toward the right.

We conclude from this that the photoreceptor is present throughout the length of the coleoptile, but it is especially concentrated in the apical few millimetres.

Pigment distribution

Examination of simple 'squash' preparations of coleoptiles reveals the presence of two yellow bands surrounding the vascular strands and running up almost to the apex. In corn coleoptiles the yellow bands often seem to arch right over the apex. In totally etiolated coleoptiles the bands are a very pale yellow; in coleoptiles grown under continuous dim red light, the bands are more conspicuous and they are pale green. Under high power the colour is found to be restricted to clumps of plastids or proplastids in several layers of cells around the vascular bundles. No coloured plastids are apparent in the extreme apex, but they begin to appear in cells 50 μ from the apex, and in the region from 100 to 500 μ many of the cells near the bundles seem to be more than half full. The number of yellow plastids per unit length decreases rapidly below the 2 mm apex, but even in cells near the base of the coleoptile, a few plastids are found scattered about individually. The parallel between plastid distribution and sensitivity distribution seems striking to us.

Using a microspectrophotometer, *in vivo* absorption spectra of these yellow plastids have been obtained. Such spectra agree qualitatively with those obtained using the etiolated primary leaf (Fig. 7.10). The pigments chiefly responsible for these spectra are carotenoids and protochlorophyll.

Redistribution of growth

In view of the new information about the effects of red light on phototropic sensitivity and growth redistribution, we decided to re-examine, under carefully controlled conditions, the following direct question: 'Do phototropic curvatures in *Avena* result principally from inhibition of growth on the illuminated (front) side, promotion of growth on the shaded side, or a combination of both?'

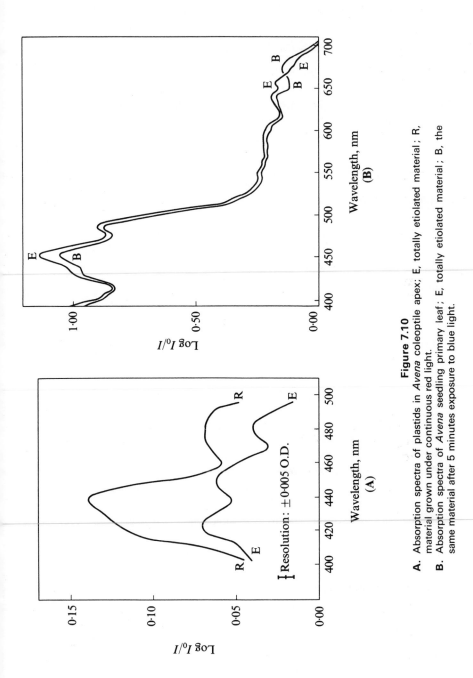

Figure 7.10

A. Absorption spectra of plastids in *Avena* coleoptile apex; E, totally etiolated material; R, material grown under continuous red light.

B. Absorption spectra of *Avena* seedling primary leaf; E, totally etiolated material; B, the same material after 5 minutes exposure to blue light.

For the present we have restricted our attention to curvatures induced by blue light dosages in the middle of the first-positive range, and to curvatures induced by continuous blue light (total energy in second-positive range).

The procedure may be outlined as follows: sets of six coleoptiles are tagged near the base and at the tip with small markers. The plants are grown under continuous red light (ca. 500 erg cm^{-2} s^{-1}) and all operations proceed under this light. At time zero the plants are shadowgraphed on a photographic plate, using an intense point source through a yellow filter. Ninety minutes later (t_{90}) a second shadowgraph is superimposed on the first. This plate establishes the 'initial' growth rates. Another shadowgraph is made on a new plate, and then the unilateral blue light is turned on (for 60 seconds, or continuously). After an additional 30 minutes (t_{120}), a second shadowgraph is again superimposed. Finally, a third double-shadowgraph is made for the interval from 30 to 120 minutes after stimulus onset (t_{120} to t_{210}). The choice of these intervals is governed by the fact that most of the curvature occurs in the last interval. The point source for the shadowgraphs produces interference fringes on the negatives, sharply defining the markers. The double shadowgraphs are projected on a wall, lengths of the front and back flanks of the coleoptiles (between base and tip markers) are measured, and growth rates in each time interval are calculated. The pooled results from six sets of such experiments are shown in Fig. 7.11A and B.

Our data suggest the following conclusions:

(a) The growth rate changes are surprisingly large. (Surprising because they are larger than the reported auxin differentials for first- and second-positive curvatures and because light-growth reactions reported in the literature are smaller in amplitude. They are not surprising from a geometrical point of view; in fact, it can be shown that they are just about right to account for the curvatures observed.)

(b) Growth inhibition of the front side is important and perhaps primary. It is common for both short and continuous stimuli.

(c) Growth promotion of the shaded side is very pronounced under continuous stimulation.

(d) After a short stimulus, total growth is less than it would have been in the absence of the stimulus; with a continuous stimulus, total growth is nearly 'compensated'.

A working hypothesis

Based on our current findings and on the earlier work, we have adopted a working hypothesis for the mechanism of phototropism. It is, essentially, an extension of the Cholodny–Went theory. We have attempted to keep the scheme as simple as we can, because of a growing conviction that most of the complexities can be accounted for in terms of a single, basic mechanism rather than a

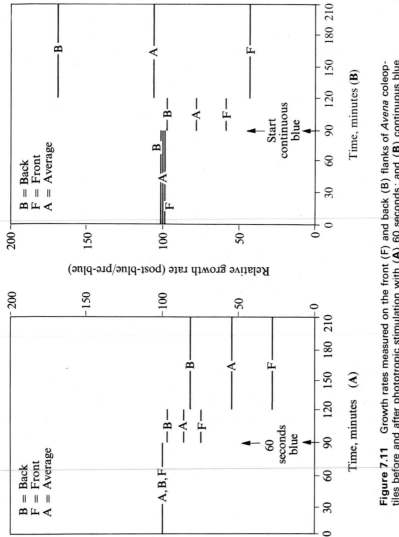

Figure 7.11 Growth rates measured on the front (F) and back (B) flanks of *Avena* coleoptiles before and after phototropic stimulation with (**A**) 60 seconds; and (**B**) continuous blue light, intensity 1,200 erg cm^{-2} s^{-1}.

multiplicity of different mechanisms. Much of the scheme is speculative, but it at least provides a unifying framework for further attacks on the problem.

We assume that there is a single photoreceptor for phototropism, specifically, a yellow pigment (probably carotenoid) located in plastid or proplastid-like organelles. Light absorption by the photoreceptor initiates a sequence of events (largely unknown) which changes the permeability properties of the cells containing the photoreceptor. These changes in turn affect the basipetal transport of auxin. An asymmetrical light stimulus causes asymmetrical changes in auxin transport, and hence curvature.

All of the elements of this hypothesis have been suggested earlier by various workers, and much of the evidence in its favour has already been discussed in this chapter. Some additional lines of evidence may be mentioned here. Blue light certainly can affect plastid morphology (Muhlethaler and Frey-Wyssling, 1959; Hongladarom and Honda, 1966); it also affects plastid position (Zurzycki, 1962). The latter point led Thimann and Curry (1961) to suggest that light-induced migration of plastids to the lateral walls might account for lateral transport of auxin. We have not yet observed such a migration of the yellow plastids *in vivo*, although there are some indications of positional changes during stimulation. Blue light-induced changes in protoplasmic viscosity, in protoplasmic streaming and in electrical polarity are known (Virgin, 1954; Bottelier, 1934; Schrank, 1957). Very recently Naqvi and Gordon (1966) have found that gravity modifies the basipetal auxin transport capacity in *Zea* coleoptile segments (but see chapter 4); they suggest that this is the primary geoeffect on auxin translocation, and that lateral movement of the hormone 'may be largely a consequence of that alteration'. This closely parallels our hypothesis for phototropism.

Naqvi and Gordon (1967) have also found that transport capacity is modified by light. Bilateral illumination significantly reduces the amount of auxin transported through *Zea* coleoptile segments. With labelled IAA they found that light caused a 'piling-up' of counts in the apical 10 mm, accompanied by a substantial decrease of counts in the basal parts of the seedling. Thornton and Thimann (1967), following the export rate of IAA-^{14}C from 4 mm coleoptile sections at five-minute intervals, found a transient depression of export after blue (but not red) illumination; the timing of this depression closely matches the timing of the light-growth reaction elicited by the same illumination. These results reinforce the idea that light acts primarily by modifying the transport system.

Based on the hypothesis the simplest model is constructed as follows. We assume that hormone (A) is produced in the apex by conversion of a precursor (X) at a constant (steady) rate, q. This hormone is in a pool at the head of the transport system. The flow rate out of the pool is assumed to be proportional to the amount of hormone in the pool and the transport capacity below it. In the absence of light, the hormone is symmetrically distributed to all sides of the coleoptile. Light decreases the transport capacity (C) below the pool. If the light is asymmetrical, the transport capacity on the front side, C_f, is decreased more than the capacity on the back side, C_b, and the hormone is distributed asymmetrically. Schematically, this is:

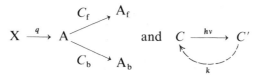

The mathematical development of this scheme becomes very elaborate, especially when additional assumptions are made about the nature of the capacity changes; space limitations do not permit its inclusion here. Castle (1966a) has recently published an elegant analysis of a very similar model for the light responses of *Phycomyces*, and the interested reader is referred to that paper for treatment of the differential equations.

Fortunately, the most interesting consequences of the model are intuitively apparent, especially under steady-state conditions of continuous illumination. In the symmetrical case, the model predicts a symmetrical decrease in transport capacity, causing a decrease in the rate of flow of hormone from the pool. Since hormone enters the pool at a constant rate, its concentration in the pool will then increase. But this, in turn, will tend to increase the rate of outflow in spite of the decreased transport capacity. In the steady-state condition, the outflow rate will be precisely the same as the initial rate. As a result of this sequence, growth rate will show a transient decrease, and then a return to the initial rate, the typical light growth reaction.

In the asymmetrical case, the same sequence of events occurs, but the final steady state is different. The growth rate on each side goes through a transient decrease and then comes back toward the initial rate. However, transport capacity in the front half remains slightly lower than that in the back half (because of the intensity gradient) and since outflow is from a common pool, outflow (and growth) in the front half will come to a steady-state value *below* the initial rate, while that in the back half will come to a rate *above* the initial rate. This predicts, therefore, that after an initial inhibitory phase, growth of the back half will be promoted to an extent matching inhibition of the front half, so that total or midline growth proceeds at the initial rate. This fits the observed situation strikingly (Fig. 7.11).

The non-steady-state formulations are not so obvious and will only be touched on here. The same arguments apply but further assumptions are required, particularly about the time course of capacity changes during and after the light stimulus. If we assume that there is a thermal back reaction with rate constant k, tending to restore transport capacity, we can account for the greater than initial growth rate which follows the transient depression after a symmetrical light pulse. The assumption seems reasonable on several counts, chief among which is the observation that photosensitivity is restored within half an hour after a pulse. Some kind of back reaction is a necessary assumption, in any case, to account for steady state levels of C greater than zero under continuous illumination. Under these conditions if the back reaction is slow compared to the light-driven forward reaction, the model predicts that curvature will be essentially independent of light intensity; i.e., ungraded. This agrees in an interesting way with observations from Darwin's time on. Furthermore, if we assume an exponential decay of C with light energy, we can generate equations such as those previously mentioned on p. 255. Thus, the model can account, at least qualitatively, for first- and second-positive curvatures. If one invokes a second light reaction following the first, it even becomes possible to generate negative curvatures from the model. At this point, however, the suspicion arises that the model may be more complex than the system it attempts to depict.

A simpler view takes into consideration the spatial distribution of the three major components of the system: photoreceptor, auxin production-distribution, and the effector cells which respond to the auxin. On this view the suggestion is made that first-positive responses are a consequence of high photoreceptor concentration in the apical half millimetre, just below the auxin production centre. It is precisely here in the solid apex that small, light-induced changes in transport symmetry would be expected to have large, sustained effects on the growing regions below. These effects would appear slowly and migrate basipetally which is just the case for first-positive or tip responses. The apex is easily light-saturated, destroying most of the asymmetry and leading to the descending arm of the dose-response curve in the first-positive range.

Second-positive responses, on the other hand, can be accounted for in terms of photoreception, transport asymmetry, and growth asymmetry along the entire length of the coleoptile. Effects should appear quite rapidly along the whole length, because receptor and effector regions are superimposed; (cf. base response). Small quantities of visible light would not yield obvious curvatures because the photoreceptor is less concentrated, any small asymmetry is relatively far 'downstream' from the auxin pool, and, perhaps of greatest importance, there is no easy path from one side of the hollow cylinder to the other. Prolonged or continuous stimulation, however, would establish the small asymmetry for a long period of time, and curvature would follow. The system would not saturate, except at extremely high intensities, because the primary leaf provides a large light gradient. This view, therefore, can account for many of the features of first- and second-positive curvatures, and tip and base responses, under a single scheme. We are currently studying the intriguing possibility that negative curvatures may be due to growth acceleration on the front side caused by a 'piling-up' of hormone in the apical 5 mm.

Conclusion

The discussion in this chapter has centred on the idea that phototropism can be explained in terms of a single, fundamental mechanism. The Cholodny–Went theory still remains at the heart of the matter. At the present stage, further progress with this problem depends on identification of the photoreceptor and discovery of the connections between it and the auxin transport system. In the author's view, a first and necessary step is the confirmation or elimination of the hypothesis that the photoreceptor is a carotenoid pigment in plastids.

The author's recent work discussed in this chapter was supported in part by grants G14266 and GB2795 from the National Science Foundation, U.S.A.

Bibliography

Selected further reading

Briggs, W. R., 1963. The phototropic responses of higher plants. In *Annual Review of Plant Physiology*, **14**, 311–352.
Briggs, W. R., 1964. Phototropism in higher plants. From *Photophysiology*, Vol. I, ed. A. C. Giese, Academic Press, New York.
Castle, E. S., 1966b. Light responses of *Phycomyces*. *Science*, **154**, 1416–1420.
Darwin, C., 1880. *The Power of Movement in Plants* (assisted by F. Darwin). John Murray, London.
Thimann, K. V. and G. M. Curry, 1960. Phototropism and phototaxis. In *Comparative Biochemistry*, Vol. **1**, eds. M. Florkin and H. S. Mason. Academic Press, New York, pp. 243–309.
Thimann, K. V. and G. M. Curry, 1961. Phototropism. From *Light and Life*, eds. W. D. McElroy and B. Glass. Johns Hopkins Press, Baltimore.

Papers cited in text

Asomaning, E. J. A. and A. W. Galston, 1961. Comparative study of phototropic response and pigment content in oat and barley coleoptiles. *Plant Physiol.* **36**, 453–464.
Blaauw, A. H., 1914. Licht and Wachstum. I. *Z. f. Bot.* **6**, 641–703.
Blaauw, A. H., 1915. Licht and Wachstum. II. *Z. f. Bot.* **7**, 465–532.
Blaauw-Jansen, G., 1959. The influence of red and far-red light on growth and phototropism of the *Avena* seedling. *Acta Botan. Neerl.* **8**, 1–39.
Bottelier, H. P., 1934. Über den Einfluss ausserer Faktoren auf die Protoplasmastromung in der *Avena*-Koleoptile. *Rec. trav. bot. neerl.* **31**, 474–582.
Briggs, W. R., 1963. Red light, auxin relationships, and the phototropic responses of corn and oat coleoptiles. *Am. J. Bot.* **50**, 196–207.
Briggs, W. R., 1963. Mediation of phototropic responses of corn coleoptiles by lateral transport of auxin. *Plant Physiol.* **38**, 237–247.
Bünning, E., 1937a. Phototropismus and Carotinoide. I. Phototropische Wirksamkeit von Strahlen verschiedener Wellenlänge und Strahlungsabsorption im Pigment bei *Pilobolus*. *Planta*, **26**, 719–736.

Bünning, E., 1937b. Phototropismus und Carotinoide. II. Das Carotin der Reizauf-nahmezonen von *Pilobolus, Phycomyces,* und *Avena. Planta,* **27,** 148–158.

Bünning, E., 1937c. Phototropismus und Carotinoide. III. Weitere Untersuchungen an Pilzen und höheren Pflanzen. *Planta,* **27,** 583–610.

Buy, H. G. du and E. L. Nuernbergk, 1934. Phototropismus und Wachstum der Pflanzen. Zweite Teil. *Ergebnisse der Biologie,* **10,** 207–322.

Castle, E. S., 1966a. A kinetic model for adaptation and the light responses of *Phy-comyces. J. Gen. Physiol.* **49,** 925–935.

Curry, G. M., 1957. *Studies on the spectral sensitivity of phototropism.* Ph.D. thesis, Harvard University.

Curry, G. M. and H. E. Gruen, 1957. Negative phototropism of *Phycomyces* in the ultra-violet. *Nature, Lond.* **179,** 1028–1029.

Curry, G. M. and H. E. Gruen, 1959. Action spectra for the positive and negative phototropism of *Phycomyces* sporangiophores. *Proc. Nat. Acad. Sci.* (U.S.A.) **45,** 797–804.

Curry, G. M. and H. E. Gruen, 1961. Dose-response relationships at different wave lengths in phototropism of *Avena.* In *Progress in Photobiology,* eds. B. C. Christensen and B. Buchmann. Elsevier Publishing Company, Amsterdam.

Curry, G. M., K. V. Thimann, and P. M. Ray, 1956. The base curvature response of *Avena* seedlings to the ultraviolet. *Physiol. Plant.* **9,** 429–440.

Delbrück, M. and W. Shropshire, 1960. Action and transmission spectra of *Phycomyces. Plant Physiol.* **35,** 194–204.

Dillewijn, C. van, 1927. Die Lichtwachstumsreaktionen von *Avena. Rec. trav. bot. neerl.* **24,** 307–581.

Galston, A. W. and R. S. Baker, 1949. Studies on the physiology of light action. II. The photodynamic action of riboflavin. *Am. J. Bot.* **36,** 773–780.

Hongladarom, T. and S. I. Honda, 1966. Reversible swelling and contraction of isolated spinach chloroplasts. *Plant Physiol.* **41,** 1686–1694.

Johnston, E. S., 1934. Phototropic sensitivity in relation to wave length. *Smithsonian Misc. Coll.* **92** (11), 1–17.

Lange, S., 1927. Die Verteilung der Lichtempfindlichkeit in der Spitze der Haferkoleop-tile. *Jahrb. f. wiss. Bot.* **67,** 1–51.

Mer, C. L., 1959. A Study of the growth and photoperceptivity of etiolated oat seedlings. *J. Exp. Bot.* **10,** 220.

Mills, K. S. and A. R. Schrank, 1954. Electric and curvature responses of the *Avena* coleoptile to unilateral ultraviolet irradiation. *J. Cell. Comp. Physiol.* **43,** 39.

Muhlethaler, K. and A. Frey-Wyssling, 1959. Entwicklung und Struktur der Proplasti-den. *J. Biophys. Biochem. Cytol.* **6,** 507–513.

Naqvi, S. M. and S. A. Gordon, 1966. Auxin transport in *Zea mays* L. coleoptiles. I. Influence of gravity on the transport of indoleacetic acid-2-^{14}C. *Plant Physiol.* **41,** 1113–1118.

Naqvi, S. M. and S. A. Gordon, 1967. Auxin transport in *Zea mays* coleoptiles. II. Influence of light on the transport of indoleacetic acid-2-^{14}C. *Plant Physiol.* **42,** 138–143.

Pickard, B. G. and K. V. Thimann, 1964. Transport and distribution of auxin during tropistic response. II. The lateral migration of auxin in phototropism of coleoptiles. *Plant Physiol.* **39,** 341–350.

Shropshire, W. and R. B. Withrow, 1958. Action spectrum of phototropic tip-curvature of *Avena. Plant Physiol.* **33,** 360–365.

Sorokin, H. P. and K. V. Thimann, 1960. Plastids of the *Avena* coleoptile. *Nature, Lond.* **181,** 1038–1039.

Thornton, R. M. and K. V. Thimann, 1967. Transient effects of light on auxin transport in the *Avena* coleoptile. *Plant Physiol.* **42,** 247–257.

Virgin, H. I., 1954. Further studies of the action spectrum for light-induced changes in the protoplasmic viscosity of *Helodea densa*. *Physiol. Plant.* **7**, 343–353.

Went, R. W., 1925. Over het verschil in gevoeligheid van top en basis der coleoptielen van *Avena* voor licht. *Proc. Koninkl. Ned. Akad. Wetenschap.* Amsterdam. **34** (8), 1–7.

Wilkins, M. B., 1965. Red light and the geotropic response of the *Avena* coleoptile. *Plant Physiol.* **40**, 24–34.

Zimmerman, B. K., 1962. *An analysis of phototropic curvature in oat coleoptiles.* Ph.D. thesis, Stanford University.

Zimmerman, B. K. and W. R. Briggs, 1963a. Phototropic dosage-response curves for oat coleoptiles. *Plant Physiol.* **38**, 248–253.

Zimmerman, B. K. and W. R. Briggs, 1963b. A kinetic model for phototropic responses of oat coleoptiles. *Plant Physiol.* **38**, 253–261.

Zurzycki, J., 1962. *Acta Soc. Botan. Polon.* **31**, 489–538.

8. Nastic responses

Nigel G. Ball

Introduction

As described in chapters 6 and 7, growing regions of plants which have been subjected to unilateral stimulation often show *tropic* responses in which the direction of curvature bears a definite relationship to the direction from which the stimulus is received. But appropriate stimulation can also induce so-called *nastic* responses in which the direction of bending is determined by the internal organization of the plant and is unaffected by the direction of the stimulus. Nastic movements may result even from diffuse stimuli such as changes in temperature or in the general intensity of illumination. They may also be induced by the action of endogenous factors which are probably chemical in nature and often operate with a definite periodicity (see chapter 18).

Nastic responses are of two kinds: *growth movements*, resulting from a difference between the rates of extension growth on opposite sides of an organ, and *variation movements*, which are due to changes in the turgor of certain cells. Growth curvatures are very often permanent although they can be eliminated, and sometimes subsequently restored, in organs which are still growing. Variation movements, on the other hand, usually occur after extension growth has ceased. They may be repeated again and again.

Growth movements

Epinasty and hyponasty

When the adaxial or morphological upper side of an organ grows more rapidly than the abaxial or morphological lower side, the resulting curvature is termed *epinastic*; when the opposite holds good, the movement is said to be *hyponastic*. Epinasty and, to a lesser extent, hyponasty play a part in the development of certain plant organs.

Lateral branches

The oblique orientation often assumed by lateral shoots has long been regarded as the result of an equilibrium between epinasty and negative geotropism. When the action of the latter is nullified by rotating the plant on a horizontal klinostat, the epinastic curvature is considerably increased. Leike and von Guttenberg (1962) studied the epinastic curvature of lateral branches of decapitated plants of *Coleus blumei* Benth. rotated on a horizontal klinostat. After 24 hours the mean curvature of intact lateral branches was 171·5°. With lateral shoots from which the terminal bud had been removed the mean curvature was 131·3°, while that of laterals which had been both decapitated and defoliated only reached 67·8°. That the decline in curvature was due to a shortage of auxin was indicated by other experiments in which indoleacetic acid (IAA) was supplied to the shoots. A preliminary application of 0·1 per cent IAA in lanolin to the apical cut surface of the decapitated and defoliated lateral shoots increased the curvature on the klinostat to 156·2° after 24 hours; when the IAA paste was applied not only to the apex of the laterals, but also to the cut petioles, the curvature was 172·0°. The exogenous IAA therefore sufficed for the full development of epinastic curvature and replaced the auxin which, in intact lateral shoots, would have been provided by the younger leaves.

Leike and von Guttenberg (1962) failed to eliminate or reverse the epinastic curvature which occurred on the klinostat when they changed the position at which the IAA was applied. This was so even when they placed it on the lower side of a decapitated and defoliated branch just below its tip. They therefore concluded that although auxin was necessary for the curvature an asymmetry in the distribution of auxin was not the cause of the inequality of growth on the two sides of the branch.

However, evidence that a suitable asymmetry in the distribution of auxin is indeed produced was provided shortly afterwards by Lyon (1963a). Radioactive IAA, labelled with ^{14}C in the 2-position of the molecule, was applied as a 1 per cent paste in lanolin to decapitated leafless branches of 15 plants of *Coleus blumei* Benth. in place of the terminal bud. After rotating the plants on a horizontal klinostat for 42–45 hours in a darkened room the amounts of radioactivity in extracts from the upper and lower halves of the curved portions of the branches were determined. In three separate experiments with vigorous branches the ratios of the radioactivity of the upper to that of the lower half were approximately 66:34, 61:39 and 67:33. Bio-assays of chloroform extracts of the upper and lower halves of the epinastically curved portions of leafy branches after 12 to 20 hours horizontal rotation afforded rather similar values. Calculated for a 20 g sample, the degrees of *Avena* curvature for the upper half were 10·9 ± 1·1 and for the lower one 5·6 ± 1·1, a ratio of 66:34.

In agreement with the results of Leike and von Guttenberg (1962), Lyon (1963a) found that when IAA was applied to the lower side of the branch, near to the disbudded tip, marked epinastic curvature developed on the horizontal klinostat (90·0 ± 6·8° after 24 hours). When a patch of the antiauxin triiodo-benzoic acid (TIBA), in the form of a 2 per cent emulsion in lanolin, was applied to the upper side of the branch at a point directly opposite the patch of IAA, the epinastic curvature was reduced to 33·4 ± 13·7°. A greater reduction, to

15·8 ± 4·8°, occurred when the TIBA was applied to the upper side about 2·5 cm from the tip, the IAA being placed on the lower side nearer the tip. As Lyon points out, this indicates that some of the auxin moves across the branch in a slanting direction, probably as a result of a combination of the known basipetal transport with some unknown mechanism for movement towards the upper side.

Although the epinasty of lateral shoots appears to operate independently of gravity, Leike and von Guttenberg (1962) have shown that the physiological dorsiventrality which controls it may be sensitive to gravity since it undergoes a slow reversal when the plant is inverted. They kept plants with auxin-deficient lateral shoots in the inverse position for 0–4 days and then applied IAA in lanolin to the tips of these shoots. The plants were then rotated on a horizontal klinostat and the curvatures measured after 24 hours rotation. The results are shown in Table 8.1. Three days in the inverse position sufficed to eliminate the epinastic tendency. After four days, the laterals curved hyponastically on the klinostat; that is, they showed an epinastic curvature in the sense of a reversed polarization.

Table 8.1 Epinastic curvature after various periods of induction in the inverse position.

Induction period (days)	Mean curvature after 24 h on klinostat (degrees)
0	+150
1	+27·6
2	+ 8·9
3	0
4	−24·8

Curvature in accordance with the original dorsiventrality (+); in the sense of a reversed polarization (−). Each value is derived from 8–16 lateral shoots. After Leike and von Guttenberg (1962).

When lateral shoots are detached from the parent plant they usually lose their epinastic properties and become orthotropic. But exceptions are known in which the epinasty is retained and continues to oppose the negative geotropic response. For example, according to Rawitscher (1923), cuttings derived from lateral shoots of *Tradescantia viridis* and *Zebrina pendula* tend to place themselves at angles of about 20° and 10° respectively from the vertical.

Leaves

The preferential diversion of auxin to the morphological upper surface, which Lyon (1963a) had shown to be the cause of epinasty in lateral shoots, was also demonstrated by him (1963b) to be responsible for the epinasty of petioles. Intact leaves of *Coleus blumei* Benth. produced an epinastic curvature of $39.6 \pm 1.6°$ after 24 hours on the horizontal klinostat, but debladed petioles showed no significant bending. But when these were provided with a terminal cap of 1 per cent IAA in lanolin they gave a curvature of $34.7 \pm 3.3°$ after 24 hours on the klinostat. When the IAA in lanolin was applied as a patch on the upper side of the petiole near its tip the resultant curvature was $58.4 \pm 5.8°$; when the patch was on the lower side it was $32.4 \pm 3.4°$. Therefore an epinastic curvature occurred regardless of the position near the distal end of the petiole at which the IAA had entered.

By the use of radioactive IAA, Lyon (1963b) showed conclusively that the epinastic curvature of the petioles was due to lateral transport of auxin. A lanolin paste of 1 per cent IAA-2-^{14}C was applied to the lower epidermis of leaf blades of *Coleus* and other genera. After rotation on the klinostat for about 40 hours, significantly more radiocarbon was extracted from the non-vascular tissues of the upper side of the petioles than from those of the lower side. With petioles of *Coleus blumei* the ^{14}C ratio of the upper to the lower side was approximately 69:31; with tomato petioles it was 58:42.

It is still not known how the basipetally moving auxin is diverted to the morphological upper side of the petiole. As Lyon (1963b) points out, the bilateral symmetry of the petiole contrasts with the radial symmetry of the internal tissues of the branch, so it cannot be assumed that the mechanism for upward lateral transport is necessarily the same in both leaf and branch. It has long been known that exposure of certain plants to low concentrations of ethylene may cause a marked increase in the epinastic curvature of the petioles. Whether this can be attributed to an effect of the gas on the polar transport of auxin is uncertain. Morgan and Gausman (1966) found that exposure to ethylene inhibited the polar transport of labelled IAA in stems and petioles of *Gossypium hirsutum* L., but Abeles (1966), using a lower concentration and a shorter exposure, found no observable difference in the polar transport of IAA in a variety of plants.

Although the *nyctinastic* movements ('sleep movements') of petioles are commonly the result of turgor changes, such rhythmical movements can also be caused by changes in the relative growth rates of the upper and lower sides. For example, it was shown by Yin (1941) that the nyctinastic movements of the younger leaves of *Carica papaya* are due to changes of this kind in a growing zone near the distal end of the petiole. He assayed the concentrations of auxin in the lamina and petiole by means of the *Avena* test and found that during the day auxin accumulates in the leaves, especially in the basal lobes of the lamina, so that by late afternoon these lobes possess more auxin than the apical ones.

Since the basal lobes are in vascular connection with the upper side of the petiole, more auxin is transported to that side than to the lower one which is supplied by the apical lobes. Consequently, during the evening, the upper half of the petiole elongates more than the lower one so that the leaf inclines downwards to the night position. Owing perhaps, as Yin (1941) suggests, to depletion of auxin or some other growth factor, the rapid growth on the upper side soon ceases. Since the growth of the lower side continues, the petiole gradually bends upwards, and by early morning the day position has again been assumed.

Flowers

In most flowers the divergence of the corolla during anthesis is due to a general, and sometimes rapid, expansion of pre-formed cells, with epinasty playing a prominent part especially towards the base of the petals. Although evidence has been obtained from comparatively few species, it seems likely that when the opening of ephemeral flowers tends to occur at a definite time of day the timing is controlled by an internal rhythm. This is shown, for example, by the night-flowering cactus, *Cereus grandiflorus*, studied by Schmucker (1928). He found that in July the opening of mature buds commenced about 4 p.m., the flowers being fully open by 9 p.m. Closure began in darkness at 3 a.m. the following morning and was complete at 4 a.m. Even when the buds had been kept in complete darkness since 8 a.m., anthesis occurred the same evening at the normal time. When the plants were kept in a dark chamber during the day and illuminated artificially by night, reversal of the rhythm commenced on the third day of inverse illumination, and by the fifth day the original rhythm had completely disappeared. After three days operation, the inverse rhythm was so strongly fixed that it persisted for at least two days in continuous light.

The opening of ephemeral flower buds of species of *Oenothera*, which takes place in the late evening, is likewise controlled by an internal rhythm. Stomps (1930), who worked with a hybrid between *O. hookeri* and *O. lamarckiana* var. *nanella*, found that plants which had been covered by a portable dark chamber retained for 6–8 days in darkness the power of opening their flower buds at about the normal time.

In some ephemeral short-day flowers which normally open in the early morning complete anthesis may not occur unless the mature buds have been subjected during the preceding night to a sufficiently long period of darkness. An example of this is afforded by the Ivy, *Hedera helix* L., which produces its flowers about October when the nights have become longer, the opening of the mature flower-buds occurring during the first hours after daybreak. Sigmond (1929) found that when the ripe flower-buds were illuminated during the night by a 100-candlepower electric lamp at a distance of 0·5 m they failed to open. For anthesis to occur, a period of darkness of at least $5\frac{1}{2}$ hours was necessary and with this minimum period only one bud out of nine opened next morning.

When the dark period was extended to $8\frac{1}{2}$–12 hours, the great majority of the ripe buds opened fully.

Another example is found in *Turnera ulmifolia* L. var. *elegans* Urb., a tropical perennial herbaceous plant, which was studied by Ball (1933) in Ceylon. Under normal conditions flower buds are produced throughout the year. As soon as a bud is mature the petals begin to emerge from the calyx in the evening and then elongate rapidly during the night. About one hour after dawn, the corolla, which is still rolled up, commences to unroll and the petals become fully expanded $1-1\frac{1}{2}$ hours later. Large amounts of starch are present in both mesophyll and epidermis of the petals during the day preceding that on which the buds are due to open. This starch becomes hydrolysed before and during the opening of the flowers. Closure of the flowers commences about noon and is followed rapidly by the death of the corolla, but the initial closing is apparently due to active hyponasty, since flowers exposed to chloroform vapour die in an expanded condition.

If the flowering shoots are illuminated during the night, or throughout the second half of the night, the corolla emerges from the calyx as usual but remains tightly rolled up, presumably owing to suppression of epinastic expansion. It is still in this condition when the flowers become withered. Unlike the flowers which open normally, those in which expansion of the corolla is suppressed by nocturnal illumination retain a large amount of starch in their petals. A considerable amount of this starch is still present in the mesophyll even after the death of the petals. Although direct evidence is lacking, there may well be a causal relation between the inhibition of starch hydrolysis in the corolla and the absence of the epinastic expansion of the petals.

The effect on subsequent anthesis of illuminating *Turnera* plants during the night with light of different wavelengths was later investigated by Ball (1936). Light between 650 and 700 nm was found to be as effective as white light in preventing the flowers from opening. Beyond 700 nm there was a considerable diminution in efficiency. Light between 400 and 600 nm was much less effective than red light and caused only a slight reduction in the expansion of the corolla.

The perianth segments of certain flowers, notably those of species of *Tulipa* and *Crocus*, are extremely sensitive to changes of temperature and can respond repeatedly by opening or closing movements. These have been shown by careful measurements (see Pfeffer, 1906) to be epinastic or hyponastic growth reactions resulting in a permanent increase in length. Opening movements have been detected by Andrews (1929) after very slight increases in temperature: 1°C in *Tulipa* and 0·2°C in *Crocus*. Closing movements require a decrease in temperature of at least 1–3°C. Since temperature is the controlling factor the movements may be denoted as *thermonastic*.

Working with *Tulipa gesneriana*, Bünning (1929) found that as regards the magnitude of the opening movement caused by a rise in temperature of 2–3°C, it was immaterial whether the rise was effected in 5 s or 20 min. In contrast, he found that a fall in temperature of about 10°C occurring in 30 s produced a closing movement with an amplitude nearly twice as great as that resulting from a fall of 20°C spread over half an hour.

According to Wood (1953), the mesophyll cells near the outer surface of the perianth segments of *Tulipa* have a temperature optimum for growth about 10°C lower than that of the cells near the inner surface. This difference also holds good for strips of tissue composed of mesophyll and epidermis taken from the inner and outer sides. When the temperature was raised from 7° to 17°C, Wood found that the extension growth of the inner strip showed a marked temporary increase, while that of the outer strip showed little change (Fig. 8.1A). On the other hand, lowering the temperature from 20° to 10°C caused the outer strip to grow at a greatly increased rate for about an hour while the inner strip practically ceased to grow (Fig. 8.1B). Similar, though smaller, differences were found in strips of epidermis removed from the inner and outer surfaces of the perianth. The differences in the response to temperature changes shown by the two sides of the perianth serves to explain the thermonastic reactions of tulip and crocus flowers, although the underlying cause of the difference is still obscure. The problem has recently been reviewed by Crombie (1962) who points out that it is generally agreed that there is no significant change in the osmotic pressure of the cell contents during movement. She favours the possibility that

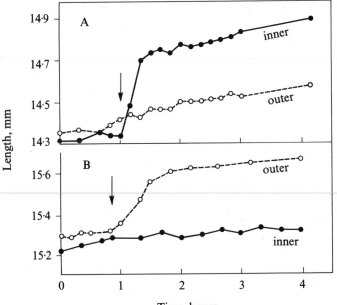

Figure 8.1   Effects of changes in temperature on the extension growth of strips of inner and outer cells of perianth members of *Tulipa* flowers. (A) Temperature raised from 7° to 17°C at the time indicated by the arrow. (B) Temperature similarly lowered from 20° to 10°C. (After Wood, 1953, by permission of the Clarendon Press, Oxford.)

11

changes in temperature may control water uptake by affecting the extensibility of the cell walls.

Variation movements

Introduction

Variation movements result from changes in the turgor of certain specialized cells with consequent alteration in their size. Normally, there is no permanent stretching of the cell walls so that complete restoration of the original condition is feasible. Movements of this type occur, for instance, in the guard cells of stomata and lead to the opening or closing of the pore. These movements are dealt with in chapter 9 and will not be discussed here.

Nastic movements of whole leaves or of separate leaflets are commonly due to turgor changes in the cells of pulvini situated at the base or apex of petioles or sub-petioles. In each pulvinus the conducting and strengthening tissues are aggregated to form a central strand leaving a wide zone of cortical parenchyma which can undergo localized changes in volume as a result of alteration in the turgor of the component cells. Expansion and contraction occur mainly in the longitudinal direction, and curvature of the pulvinus results when there is a difference between the reactions on opposite sides. Sometimes the changes in length are facilitated by the infolding of the surface of the pulvinus in the transverse direction so as to provide a kind of bellows structure.

Nyctinasty

Nyctinastic movements, operated by pulvini, frequently occur in compound leaves, especially those of members of the Leguminosae. In the evening the leaflets tend to fold together and the whole leaf may droop downwards. The following morning the leaf rises and the leaflets become expanded. Although the movements are primarily a response to changes in illumination, an internal rhythm is often established which continues to initiate the movements at the usual times, even in continuous darkness (see chapter 18). This aspect has been reviewed by Bünning (1959).

There is still some uncertainty regarding the cause of the turgor changes involved in nyctinastic movements. De Groot (1938) measured the decrease in length of strips of tissue taken from the upper and lower halves of the pulvinus at the base of leaves of *Phaseolus multiflorus* when they were transferred from water to a hypertonic solution of glucose. The results, in agreement with those of earlier investigators, indicated that when the leaf was in the day position the osmotic value of the cells in the upper half of the pulvinus was minimal and that of the lower half was maximal. In the night position the ratio of the osmotic values in the two halves was reversed.

The variations in osmotic value were ascribed by de Groot (1938) to changes in the sugar content of the cells, the amount of sugar being decreased by respiration and diffusion, and increased by the hydrolysis of starch. But this can hardly be an adequate explanation since Mosebach (1944) found that of a total osmotic pressure of 8–9 atm in the pulvini of *Phaseolus multiflorus* only about 2 per cent could be attributed to dissolved sugars. In passing from the day to the night position the osmotic value of the sugars in the upper half of the pulvinus rose from 0·21 to 0·25 atm, while that of the lower side decreased from 0·30 to 0·29 atm but the corresponding total changes in osmotic pressure at limiting plasmolysis were a rise from 8·7 to 12·2 atm and a fall from 14·1 to 9·4 atm. Since he found that 5·5–6 atm of the total osmotic value was due to electrolytes, Mosebach (1944) considers that the periodic variation in osmotic value must depend mainly on these, but the mechanism which causes the change is still obscure.

Explanations of nyctinasty based on the action of auxin on pulvini have been suggested by von Guttenberg and Kröpelin (1947), and by Brauner and Arslan (1951). When indoleacetic acid (IAA) in lanolin is applied unilaterally to the pulvinus of the primary leaves of *Phaseolus multiflorus* it induces negative curvature. Brauner and Arslan (1951) applied a paste of 0·1 per cent IAA in lanolin to the dorsal surface of the pulvinus. After 24 hours there was a downward curvature of 45·9° relative to controls treated with water and lanolin only. Similar applications to the ventral surface produced a net upward curvature of 82·1° due to the auxin. When the IAA was applied symmetrically to both dorsal and ventral surfaces of the pulvinus the net effect of the auxin was an upward curvature of 41·0° after 24 hours, the response to IAA in lanolin being a rise of 24·9 ± 3·5° and to water in lanolin a fall of 16·1°. The extent to which the auxin reaction was due to an increase in turgor was estimated by immersing the pulvini for 2 min in water at 70°C. The heating released the turgor and reduced the curvature by 24·1 ± 2·4°; an amount equivalent to 96·8 per cent of the original auxin curvature.

Von Guttenberg and Kröpelin (1947) applied IAA in lanolin to both faces of the midrib of the leaf of *Phaseolus multiflorus* and found that it caused the lamina to rise. They therefore attributed the diurnal upward movement to an auxin which developed in the lamina in daylight and passed back through the midrib to the pulvinus. They suggested that either because the lower side of the pulvinus is more sensitive to auxin than the upper, or because the auxin moves downwards owing to gravity, the lower side of the pulvinus expands more than the upper and the leaf rises. The downward movement was thought by them to be due to the expansion of the previously compressed cells on the upper side of the pulvinus as soon as the supply of auxin ceased or became very small.

Brauner and Arslan (1951), on the other hand, attribute both the upward and downward movements to the action of auxin. They confirmed that when 0·1 per cent IAA in lanolin is applied to the midrib of the leaf of *Phaseolus multiflorus* the pulvinus curves upwards and raises the lamina. But, in addition, they observed that when the auxin is applied to the lateral veins at the base of the leaf the lamina falls. At the end of 24 hours the upward curvature amounted to $10.6 \pm 1.9°$ and the downward to $27.5 \pm 4.9°$. By introducing a dyestuff into the veins of the leaf they showed that the vascular strands in the midrib are continuous with those of the ventral region of the pulvinus, while the strands in the dorsal and lateral regions of the pulvinus are connected with the lateral veins. They therefore postulate that the nyctinastic movements of the leaf are due to a rhythmical periodicity in auxin production linked with a diurnal variation in the distribution of active auxin in the distal and basal areas of the lamina. More recently, Williams and Raghavan (1966) have shown that low concentrations of IAA and of gibberellic acid promote the opening of the leaflets of *Mimosa pudica*, while higher concentrations prevent their closure at night.

Working with leaf pinnae of the tropical tree, *Samanea saman* (Jacq.) Merrill, from which either the upper or the lower half of the pulvinus had been removed, Palmer and Asprey (1958) found that even in continuous darkness a diurnal rhythm of movement occurred in each half pulvinus, although that of the upper one was very slight. They conclude that the diurnal movements result from one half of the pulvinar cortex undergoing expansion while the other simultaneously, but independently, is becoming flaccid.

A new aspect of the mechanism of nyctinastic variation movements was revealed by Fondeville, Borthwick, and Hendricks (1966) who discovered that the coming together of the pinnules of *M. pudica* in response to darkness may be dependent on the presence of phytochrome in the far-red absorbing form. They found that if the plants were irradiated in the far-red region of the spectrum beyond 700 nm the leaflets remained in the expanded condition for many hours in darkness, but if the far-red radiation was followed by red radiation the pinnules of the darkened plants closed within 30 min. In a later paper, Fondeville *et al.* (1967) reported that in normal light periods the leaflets remained open irrespective of the form of phytochrome which was present, and also that those which had closed in darkness regularly reopened in light.

By tests on a number of plants with pinnate leaves, Hillman and Koukkari (1967) proved that participation of phytochrome in the nyctinastic response was not confined to leaves such as those of *M. pudica* which possess a special sensitivity. Using pairs of pinnules excised from leaves of *Albizzia julibrissin* they showed that closure of the pinnules on darkening was rapid after red illumination and slow following exposure to far-red. Under good conditions the difference between the two treatments was obvious within 10 min. The control exerted by phytochrome was most marked when the experiment was carried out early in the daily 12-hour light period. Towards the end of the light period the effect was absent, or nearly so. Similar results were obtained by Jaffe and Galston (1967). Both they and Hillman and Koukkari (1967) suggest that another controlling factor, possibly an endogenous circadian rhythm, becomes active later in the day. Jaffe and Galston (1967) showed that the photoinduced opening of the pinnules after closure in darkness is induced by blue light alone, indicating that phytochrome is not concerned in this response. They also found that the closure response to pre-treatment with red light is accompanied by an increased efflux of electrolytes from the cut pinna base. They consider that this observation, coupled with the fact that rapid nyctinastic movement is not affected by actinomycin D which inhibits RNA synthesis in cells, suggests that the movement is not mediated through effects of RNA metabolism but through changes in membrane permeability. A rather similar conclusion was reached by Hillman and Koukkari (1967).

Seismonasty

In so-called 'sensitive plants', of which probably the best-known is *M. pudica* L., the leaf pulvini not only perform ordinary nyctinastic movements but also respond by rapidly becoming curved when the leaves are shaken or subjected to a mechanical disturbance. Such responses are usually denoted as *seismonastic*, although similar reactions can be induced by heat, chemical, or electrical stimuli. Two distinct and interesting problems are presented by these movements: the nature of the mechanism involved in the rapid pulvinar response, and the means by which a stimulus applied to one part of the plant can induce reactions in pulvini some distance away.

Pulvinar mechanisms

The study of rapid pulvinar responses has been confined almost entirely to the leaf of *M. pudica* L. The leaf is compound with two or four pinnae, each of which bears numerous pairs of pinnules. Primary and secondary pulvini are situated at the base of the petioles and sub-petioles, and a tertiary pulvinus occurs at the base of each of the pinnules. When the leaf is stimulated the various pulvini become curved in such a way that the main petiole droops, the pinnae come closer together, and the pinnules turn upwards so that the members of each pair are brought into contact.

The primary pulvinus of *M. pudica*, like most other pulvini, possesses a central vascular strand surrounded by a thick layer of parenchyma. Towards the centre, the parenchymatous cells are separated by well-developed intercellular spaces. Further out, these spaces diminish in size and disappear altogether just below the epidermis. In the unstimulated pulvinus the parenchyma is highly turgid but longitudinal extension is resisted by the central strand.

When the primary pulvinus is stimulated, the lower side, in which the cell walls are thinner than those of the upper one, responds rapidly by a partial loss of turgor. This upsets the balance between the two sides and enables the highly compressed cells on the upper side to expand in the longitudinal direction. Consequently, the whole pulvinus, including the central strand, becomes curved and the petiole falls. If the upper half of a pulvinus is excised, the petiole rises steeply but the remaining half is still able to give a slight response when stimulated. Asprey and Palmer (1955) have shown that this curvature no longer occurs if the leaf is positioned so that any movement would have to be in the horizontal plane. The downward movement which takes place in the vertical plane when the upper side of the pulvinus is absent must therefore be due to the weight of the leaf. This factor becomes effective as soon as the lower side has responded to stimulation by becoming less turgid.

Although a leaf of *M. pudica* after seismonic stimulation shows a general resemblance to one in the 'sleep' position, certain observations suggest that the mechanisms of the nyctinastic and seismonastic responses are somewhat different. It has long been known that the main pulvinus is much less flexible at night than it is after responding to a shock stimulus. Other differences have been illustrated in photographs by Weintraub (1952). These show that the primary petioles after 5 hours in the dark formed almost the same angle with the stem as they did in the unstimulated daytime position. On the other hand, the four pinnae of each leaf had moved closer together in darkness than they had done when the plant was given a shock stimulus during the day.

The chief difficulty in providing an explanation of the seismonastic movements of plants such as *M. pudica* is to account satisfactorily for the rapidity of the response. Under favourable conditions the reaction may commence about 0·1 second after stimulation and the whole movement be completed within a few seconds. Various explanations have been suggested. (References to some of the earlier literature are given, e.g., by Weintraub, 1952.) It has long been known that contraction of the stimulated cells is accompanied by ejection of liquid into the intercellular spaces. The displacement of air from these spaces can sometimes be detected by a change in colour of the main pulvinus and, if a leaf which has been detached by cutting through the base of the pulvinus is stimulated again after it has recovered, the curvature is usually accompanied by a slight exudation of liquid from the cut surface of the pulvinus. An exudation of water from the cells, with consequent loss of turgor, could occur if the osmotic pressure of the cell sap were to be decreased by a diminution in the number of molecules dissolved in it. Theoretically, this might result from adsorption or chemical combination, but there is no evidence to support such hypotheses and the rapidity of the response has generally been regarded as making them untenable, at any rate in relation to substances freely dispersed in the vacuoles.

From time to time it has been surmised that the ejection of liquid from the cells in the lower half of the main pulvinus might be due to a sudden increase in the permeability of the protoplasts to solutes. This would allow some of the osmotically active substances to escape from the cells and enable contraction of the stretched cell walls to force out some of the water of the cell sap.

The validity of this hypothesis was tested about fifty years ago by Blackman and Paine (1918). They used the primary pulvinus of *M. pudica*, separated from the stem but still attached to the petiole. Part of the external layer was excised from both flanks of the pulvinus which was then suspended by the petiole in a very small vessel of water. This also contained two electrodes so that changes in the electrical conductivity of the liquid could be measured. After the pulvinus had recovered, it was stimulated by momentarily withdrawing it from the water. Exosmosis of electrolytes, which before stimulation had been taking place at a slow steady rate, increased by 2–3 times on stimulation, but the slight increase in electrical conductivity was much too small to suggest that the loss of turgor by the lower side of the pulvinus was caused by a sudden increase in the permeability of the protoplasts to electrolytes. The authors indeed suggested that it might have been due merely to the contraction of the pulvinus squeezing out some material from the intercellular spaces or cell walls. Although no evidence was produced, they assumed that the lack of a significant increase in permeability also held good for non-electrolytes. Blackman and Paine (1918) also found that the isolated pulvinus could respond to stimuli repeated at 15-minute intervals over a period of 2 hours. As pointed out by them, it is difficult to see how the contractions, if they were due to loss of osmotic substances, could have continued to be produced in the isolated pulvinus, since these substances would have had to be repeatedly replaced at a fairly rapid rate.

Detailed comparison of the microscopic structure of sensitive pulvinar tissues before and after stimulation could presumably throw some light on the mechanism of contraction. Buvat (1946) compared the structure of the motor cells of *M. pudica* in leaves which had been lightly anaesthetized by chloroform vapour with that of others which had responded to stimulation. The motor cells from anaesthetized leaves contained spherical, ellipsoidal or cylindrical chondriosomes (mitochondria) with a diameter of 0·25 to 0·33 μ. In contrast, those from motor cells which had been subjected to mechanical stimulation had diameters between 0·6 and 1·5 μ. This difference was observed in all three types of pulvinus. Buvat, who considered that the swelling of the chondriosomes indicated that they absorbed water at the time the pulvini were stimulated, concluded that stimulation results in the appearance of free water in the cytoplasm. But unless this water was eliminated from the cells, there seems to be no reason why there should have been any significant change in their turgidity.

Weintraub (1952) found that when a portion of a pinna of *M. pudica* was floated on water in bright light, the pinnules soon opened out and were sensitive to shock stimuli. But if they were left floating on the water for two days or longer they completely lost their sensitivity and became 'senile'. When freehand sections of fresh material, both senile and sensitive, were studied, it was seen that the cytoplasm of the motor cells of the senile tertiary pulvini invariably contained numbers of globoid inclusions each about 1 μ in diameter. In sections of active pulvini which had reacted to the shock of sectioning the inclusions were much less numerous and in occasional cells they even appeared to be absent. Weintraub (1952) considered that these inclusions were vacuoles, since they were invariably globular and gave no reactions to iodine. He also found that neutral red and methylene blue accumulated in the inclusions in the same way as they did in the central vacuoles of the cells.

Weintraub (1952) later developed techniques which enabled microscopic observations of the sensitive cells to be made during the process of contraction. Microtome sections, preferably 130 μ in thickness, of a piece of a pinna in a sensitive condition were floated on water. Under a bright light 30–70 per cent of the sections re-opened, and these opened sections were sensitive to shock stimuli. If one was transferred to a drop of water and prodded with a needle in the region of the tertiary pulvinus, it closed promptly (Fig. 8.2). Recovery took place after about 15 min, and closure and recovery could be repeated a number of times. Transferring sections to a 1–2 per cent solution of sodium chloride also caused stimulation without affording any indication of plasmolysis.

An improvement in the technique was to float a section on water in a shallow well on a microscope slide and to cover it with a cover-slip. The water was then gradually replaced by a 2 per cent salt solution. This induced a slow response.

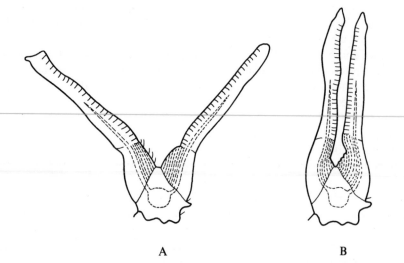

A   B

Figure 8.2   Longitudinal section of tertiary pulvini and pinnule blades of *Mimosa pudica*. (A) In the 'open' condition. (B) After closing in response to a needle prick. Stippled portion marks the sensitive upper half of each pulvinus. (After Weintraub, 1952.)

The first stage, through an angle of about 45°, lasted 10–15 min. During the process of contraction, observation through an oil-immersion objective showed that not only did the small vacuoles already present disappear but also that others formed and disappeared in large numbers during the initial phase of the closure. Towards the end of the movement most of the vacuoles already present disappeared and few, if any, new ones formed. But at the end of the contraction there were still some vacuoles left which had not got rid of their contents (Fig. 8.3).

Weintraub (1952) also observed that during the movement the central vacuole of the contractile cells diminished in size. He concluded that the initial loss of

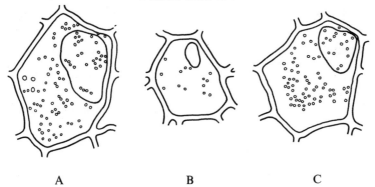

A   B   C

Figure 8.3   Camera lucida drawing of a cell from the sensitive upper half of a tertiary pulvinus of *Mimosa pudica*. × 700. (A) The cell in water. (B) In dilute sodium chloride solution. (C) Replaced in water. The small circles represent small vacuoles. The outline of the central vacuole appears in the right-hand corner of each cell. (From Weintraub, 1952.)

turgidity of the motor cells was due to the active contraction of the numerous small vaculoles already present in the cytoplasm. Further decrease in turgor then resulted from the formation and subsequent contraction of additional vacuoles which derived their contents either from the cytoplasm itself or indirectly from the central vacuole.

The minute vacuoles observed by Weintraub were about the same size as the chondriosomes described by Buvat (1946). It is possible that there has been some misinterpretation of the identity of these extremely small organelles, but an important difference in the two accounts is that while the swelling of the chondriosomes was considered by Buvat to be merely an indication of the liberation of water from the cytoplasm, the contractile vacuoles described by Weintraub (1952) were thought by him to play an active part in removing liquid from the turgid cells.

Dutt (1957), in a short account which gives few details, states that in the turgid condition each parenchymatous cell of the motor tissue of the main pulvinus of *M. pudica* contained a large tannin vacuole and smaller vacuoles in the cytoplasm. On stimulation, the smaller vacuoles were observed to disappear and the tannin vacuole to contract in volume. At the same time a liquid containing potassium salts was found in the intercellular spaces. The presence of potassium in the intercellular spaces after stimulation has also been reported by Toriyama (1955).

While contractile vacuoles as small as those described by Weintraub (1952) might play some part in lowering the turgidity of the motor cells, it hardly seems possible that they would be capable of removing a relatively large volume of liquid from the cells within a few seconds. A rough estimate of the size of the cell drawn by Weintraub (Fig. 8.3A) indicates that its volume in water would have been between 23,000 and 45,000 μ^3, and that a decrease in volume of about

9,000–18,000 μ^3 occurred after stimulation by 2 per cent sodium chloride (Fig. 8.3B). To transport the ejected liquid would have required about 17,000–34,000 vacuoles each 1 μ in diameter. The surface area of the turgid cell is estimated as 5,000–10,000 μ^2, but only part of this surface would abut on intercellular spaces. The area of the surface of the cell available for the rapid discharge of the required number of contractile vacuoles would therefore appear to be inadequate. The effectiveness of vacuoles which formed and disappeared within the body of the cytoplasm seems doubtful.

A suggestion that is perhaps more credible is that the appearance and disappearance of minute vacuoles in the cytoplasm are merely the result of fluctuations in its state of hydration, and that the operating mechanism of the turgor changes is to be found in some form of protoplasmic contraction. As pointed out by Weintraub (1952) this is an old idea first advocated by Gardiner in 1883 and 1888 and later contested by Pfeffer. It is true that the occurrence in *Mimosa* of cytoplasmic contraction has not been established, but it still seems worthy of consideration (see Umrath, 1959b). The rapidity of the seismonic response, and its partial or complete suppression by ether, unsuitable temperature, darkness, and senescence (Wallace, 1931a and b) point to an active participation of the cytoplasm, although the nature of the process is still a problem. Umrath (1959b) suggests that diminution in the volume of the cell contents could be caused by a decrease in the water-binding capacity of the cell colloids. This could lead to an ejection of liquid and probably to a contraction of the colloidal residue. The hypothesis is attractive but sufficient evidence to justify its adoption is not yet available.

Transmission of stimulation

It must be a very long time since it was first noticed that when part of a plant of *M. pudica*, or a related species, is subjected to cutting or burning, the effect of the stimulus may be transmitted to other parts where it causes leaf movement, but the earliest recorded attempt to elucidate the process of conduction seems to have been that of Dutrochet (1824). He removed a ring of cortex from the stem and found that the stimulus caused by burning a leaf was able to pass through the decorticated region and therefore concluded that the passage of the stimulus was connected with movement of liquid in the wood. Later workers, e.g., both Sachs and Pfeffer, impressed by the fact that when a slight cut is made in the stem a drop of liquid emerges, assumed that this came from the vessels and therefore inferred that transmission of the stimulus was effected by turgor changes in the xylem.

The most important paper on the subject published in the nineteenth century was that of Haberlandt (1890) who showed that the drops which emerge from the stem when it is cut do not come from the xylem but from special cells, the so-called tube-cells, which are situated in the phloem. These tube-cells resemble large sieve-tubes, but have a single large pit in the transverse walls. They also possess a nucleus.

Since he found that the stimulus could pass through a zone of the petiole which had been killed by heat, Haberlandt concluded that transport of the stimulation did not take place through the protoplasmic connections between the cells but was effected by means of a disturbance in the hydrostatic equilibrium in the tube-cells. Such a disturbance was believed to result when a cut was made into these cells, or when a part of the plant was exposed to a high enough temperature for the water in them to be turned into steam. Haberlandt's theory was fairly generally accepted for some twenty-five years. But there was some criticism. For example, Linsbauer (1914), while conceding the possibility that stimuli could be transmitted by means of hydrostatic changes in the tube-cells of the phloem, concluded that since they could also be transmitted through stems in which all tissues external to the wood had been removed in one or more internodes, an alternative mode of transport must be by sap movement through the wood.

A method by which such a movement of sap could be utilized was demonstrated by Ricca (1916). Working mainly with another species of *Mimosa*, *M. spegazzinii*, he showed that the effect of a stimulus could pass through a long region of stem which had been killed by heat, and also through a discontinuity produced by cutting through the stem and joining the two portions by a small water-filled tube; a fundamental experiment which was later confirmed by Seidel (1923), Snow (1924), and Ball (1926). According to Ricca (1916) some chemical substance, liberated from cells in the region where the stimulus is applied, can be carried in the transpiration stream through the xylem and even across a narrow water-filled discontinuity. When this substance, which may be denoted as a hormone, reaches the leaves it induces contraction of the motor cells of the pulvini. Evidence of the actual existence of this hormone was obtained by pounding up tissues of *Mimosa* sp. in a little water and applying some of the extract to the cut end of a shoot of the same species. A little later, the leaves responded in succession as the hormone reached them.

In a series of papers between 1927 and 1938, Umrath and co-workers have given an account of the distribution and efficacy of such substances in plants belonging to different families, and of attempts to determine the chemical nature of the hormone of the Mimosoideae. (For a full survey of this work, with references, see Umrath, 1959b.) Extracts from *M. pudica*, *M. spegazzinii*, and *Neptunia plena* were found to be equally effective at the same dilution on each of these three plants. Extracts of other Mimosoideae, some of which were not sensitive, were also effective when tested on *M. pudica* and *N. plena*. On the other hand, the stimulating substance occurring in *M. pudica* and *N. plena* appeared to be different from that of *Biophytum sensitivum* (Oxalidaceae). Extracts from each of these plants had a much greater effect on its own, or the related species, than on the unrelated one. In general, extracts from members of the same subfamily as the test plant were equally effective, while extracts from sub-families different from that of the test plant were fully effective, partly effective or ineffective. Umrath (1959b) concludes that extracts which induced responses in unrelated test plants contained unspecific stimulating substances, since these were very often effective not only on members of one but of many different sub-families. Owing to the difficulties of purification, the chemical nature of the stimulating substance of the Mimosoideae was not fully determined, but it was found to be thermostable and non-volatile with a molecular weight which was probably between 300 and 450. It was thought to be a monobasic oxyacid with one atom of nitrogen in the molecule. The best preparation was effective on *M. pudica* at a dilution of $1:5 \times 10^8$.

Ricca's (1916) demonstration that a stimulating hormone can be carried in the transpiration stream in *M. pudica* and *M. spegazzinii* has been fully confirmed and must be accepted, but that the stimulation can also be transmitted under conditions which make such a mode of transport totally inadequate is equally certain. Snow (1924) observed that rapid basipetal conduction of excitation occurred in leaves of *M. pudica* which were in nearly saturated air or had even been submerged under water for three hours. Later (1925), he showed that when the tip of a pinna of *M. spegazzinii* was cut under water containing a stain, the rate at which the excitation moved basipetally was 2·4–10 times greater than that at which the stain was transported through the xylem. He also found that the excitation travelled much more rapidly in leaves which had been submerged under water for some hours than in others which had remained in air and were still attached to the plant.

Ball (1927) found that in addition to the method of conduction described by Ricca (1916) a much more rapid one can occur in stems of *M. pudica* when the plants are very turgid. Such conduction can be observed under natural conditions in growing plants during wet weather, or in cut shoots submerged under water in the laboratory. It is readily started by electrical stimulation, cutting, burning, or even rubbing the stem with the back of a knife. The excitation can pass rapidly along the stem in either direction but it only affects the primary pulvini and does not appear to enter the pinnae. On the other hand, if a pinna is stimulated the excitation only passes back as far as the main pulvinus and does not enter the stem. The velocity of 'rapid conduction' in the stem is about 1–2 m min^{-1} whereas the highest velocity for movement of a hormone in the transpiration stream recorded by Snow (1924) for *M. pudica* under conditions designed to provide a high tension in the water columns of a cut shoot was about 0·52 m min^{-1}. Usually the velocity was much less. Conditions which promote one form of transport do not favour the other. There is no gradual transition between the two.

Rapid conduction in the stem was found by Ball (1927) to be unaffected by removal of all tissues outside the xylem in the internodes, but it failed to pass through a zone of the stem where the tissues on the inner side of the wood had been removed, leaving the xylem, phloem, and cortex intact. It did not pass through a discontinuity, or through a zone killed by steam, except when living cells in the same or an adjacent internode were stimulated by cutting the stem with a pair of scissors or by squeezing it with a forceps. In such cases it is likely that sap containing the stimulating substance released from the injured cells was forced through the xylem in the killed region, or across the liquid-filled discontinuity, and induced rapid conduction in the other half of the turgid shoot.

The logical deduction from these experiments is that rapid conduction of excitation in the stem of *M. pudica* takes place in living cells situated within the cylinder of wood, including the narrow elongated cells on the inner side of the protoxylem. The fact that this tissue is reduced to a very few cells, or even disappears altogether, when the petiolar bundles come together in the centre of the main pulvinus probably explains why the passage of the excitation, coming from either direction by this form of conduction, is blocked in the main pulvinus, although the motor tissue of the pulvinus itself responds.

Before discussing the mechanism of rapid conduction in *M. pudica* mention may be made of a third form of conduction which appears to be different from the other two. Snow (1924) found that when a cut was made with a razor in an internode just deep enough to reach the phloem, the well-known drop of liquid emerged and the next leaf above the cut fell almost immediately although the leaflets did not close. This very rapid transport of excitation he designated as 'high-speed conduction'.

Some further experiments on this form of conduction were carried out by Ball (1927). Occasionally, in addition to the leaf just above the cut, the next leaf on the same side reacted about 1 second later, but a single stimulus was never observed to affect more than two leaves. High-speed conduction did not occur in a decorticated zone. It took place equally well in shoots which were transpiring rapidly and in those which were submerged under water. In the latter, however, both 'high-speed' and 'rapid' conduction could be initiated by the same stimulus, the 'high-speed' affecting only one or two leaves, while the 'rapid' had a longer range but a smaller velocity. The limited range, the comparatively high velocity, and the exudation of liquid from a cut which penetrates the phloem, strongly suggest that high-speed conduction of excitation is effected by a loss of turgor propagated in the tube-cells in a direction away from the cut, in contrast to the direction of flow of the contained liquid which would be towards the cut. This hypothesis, suggested by Houwink (1935), is similar to that of Haberlandt (1890) but is more restricted in its application.

It has long been recognized that the passage of excitation through plant tissues may be associated with changes in electrical potential. Some of the earlier work was unconvincing and the contention of Bose (1926) that conduction of stimuli in *M. pudica* was due to the operation of a nervous mechanism was unacceptable to many, including the writer (Ball, 1927). However, unequivocal evidence has accumulated that rapid conduction of excitation in *M. pudica* and *M. spegazzinii* is dependent on the propagation of an action potential, as happens during the passage of nervous impulses in animals. Owing to limitation of space only a brief mention of the salient facts can be given here, but comprehensive surveys, with full references, have been provided by Umrath (1937, 1959a), whose studies in this field over a long period have afforded much valuable information.

If one electrode is inserted into the stem or petiole of a sensitive plant such as *M. pudica*, or even laid on the surface so as to make a moist contact, and a second is connected to the lower part of the stem or inserted into the surrounding earth, a change of potential can be detected when the excitation passes the first electrode. The change consists of a pulse of negativity, the action potential, amounting to 20–100 mV. It rises rapidly to a maximum and then declines. Houwink (1935), who used ice as a stimulant, found that when a petiole was stimulated, the main pulvinus responded a little later but the excitation did not pass beyond it. A second stimulus applied to the stem produced an action potential which was detected when it reached the electrode further up the stem (Fig. 8.4).

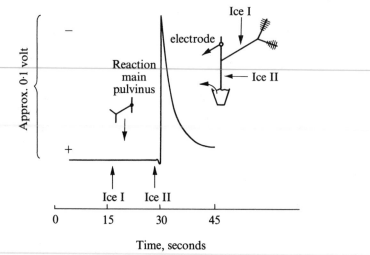

Figure 8.4   The effects of applying ice first to a petiole and then to the stem of *Mimosa pudica*. The approximate times of stimulation and pulvinar response are indicated by arrows. For further details see text. (After Houwink, 1935.)

Umrath (1928) found that in *M. pudica* and *M. spegazzinii* an action potential could be induced in a region of the stem from which all the tissues outside the xylem had been removed, but not in a region where half of the stem had been cut away and all the tissues on the inner side of the xylem removed by scraping. This supports the view that 'rapid' conduction in the stem takes place through tissues on the inner side of the xylem and not through the phloem. More recently, Sibaoka (1962) has investigated the location of the excitable cells in the petiole of *M. pudica*. By inserting microelectrodes into the cells of various tissues he found that the elongated parenchyma cells in the phloem and protoxylem regions had a resting potential of about -160 mV, while that of all the other cells was only about -50 mV. When basipetal conduction of excitation was started by stimulating the petiole electrically at its apex, a membrane action potential was always induced in the cells which had a larger resting potential. This suggests that it is these cells which probably provide the pathway for the conduction of excitation.

Although under certain conditions the conduction of stimulation in *Mimosa* is due to the transport of a hormone in the transpiration stream, and not to the propagation of an action potential, the passage of the hormone can give rise to a very irregular variation of potential which may last for several minutes. This effect was studied by Houwink (1935), who referred to it as 'the variation' (Fig. 8.5). Both he and Sibaoka (1953) consider that this is indeed the action

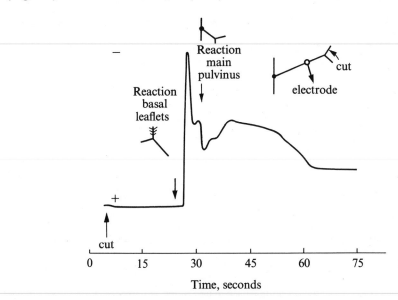

Figure 8.5 Cutting through a pinna of *Mimosa pudica* is followed by the reaction of the secondary and primary pulvini. An action potential is recorded when the excitation reaches the electrode and is followed by a variation of potential caused by the hormone coming from the injured cells. Ordinate scale as in Fig. 8.4. (After Houwink, 1935.)

potential of cells which have been stimulated by the hormone as it passes through the tissues. Umrath (1959a), who agrees with these views, makes the additional suggestion that the liberation of further supplies of hormone from the stimulated cells along the path of conduction might facilitate the process of conduction, but whether or not this occurs is still unknown.

Bibliography

Selected further reading

Bose, J. C., 1926. *The Nervous Mechanism of Plants*. Longmans, Green & Co., London.

Bünning, E., 1959. Tagesperiodische Bewegungen. In *Handbuch der Pflanzenphysiologie —Encyclopedia of Plant Physiology*, ed. W. Ruhland, Vol. 17(1). Springer, Berlin-Göttingen-Heidelberg, pp. 579–656.

Crombie, W. M. L., 1962. Thermonasty. In *Handbuch der Pflanzenphysiologie—Encyclopedia of Plant Physiology*, ed. W. Ruhland, Vol. 17(2). Springer, Berlin-Göttingen-Heidelberg, pp. 15–28.

Pfeffer, W., 1906. *Physiology of Plants*, Vol. III. English translation by A. J. Ewart, Clarendon Press, Oxford.

Umrath, K., 1937. Der Erregungsvorgang bei höheren Pflanzen. *Ergebn. Biol.* **14**, 1–142.

Umrath, K., 1959a. Der Erregungsvorgang. In *Handbuch der Pflanzenphysiologie— Encyclopedia of Plant Physiology*, ed. W. Ruhland, Vol. 17(1). Springer, Berlin-Göttingen-Heidelberg, pp. 24–110.

Umrath, K., 1959b. Mögliche Mechanismen von Krümmungsbewegungen. In *Handbuch der Pflanzenphysiologie—Encyclopedia of Plant Physiology*, ed. W. Ruhland, Vol. 17(1), Springer, Berlin-Göttingen-Heidelberg, pp. 111–118.

Papers cited in text

Abeles, F. B., 1966. Effect of ethylene on auxin transport. *Plant Physiol.* **41**, 946–948.

Andrews, F. M., 1929. The effect of temperature on flowers. *Plant Physiol.* **4**, 281–284.

Asprey, G. F. and J. H. Palmer, 1955. A new interpretation of the mechanism of pulvinar movement. *Nature, Lond.* **175**, 1122.

Ball, N. G., 1926. Transmission of stimuli in plants. *Nature, Lond.* **118**, 589–590.

Ball, N. G., 1927. Rapid conduction of stimuli in *Mimosa pudica*. *New Phytol.* **26**, 148–170.

Ball, N. G., 1933. A physiological investigation of the ephemeral flowers of *Turnera ulmifolia* L. var. *elegans* Urb. *New Phytol.* **32**, 13–36.

Ball, N. G., 1936. The effect of nocturnal illumination by different regions of the spectrum on the subsequent opening of flower-buds. *New Phytol.* **35**, 101–116.

Blackman, V. H. and S. G. Paine, 1918. Studies in the permeability of the pulvinus of *Mimosa pudica*. *Ann. Bot.* **32**, 69–85.

Brauner, L. and N. Arslan, 1951. Experiments on the auxin reactions of the pulvinus of *Phaseolus multiflorus*. *Rev. Fac. Sci. Univ. Istanbul*, **16/B**, 257–300.

Bünning, E., 1929. Über die thermonastischen und thigmonastischen Blütenbewegungen. *Planta* (Berl.) **8**, 698–716.

Buvat, R. M., 1946. Sur les chondriosomes des organes moteurs des feuilles de *Mimosa pudica*. *Compt. rend. Acad. Sci. Paris*, **223**, 1017–1019.

Dutrochet, R., 1824. *Recherches anatomiques et physiologiques sur la structure intime des animaux et des végétaux*. Paris.

Dutt, A. K., 1957. Vacuoles and movement in the pulvinus of *Mimosa pudica*.—Vacuoles of the pulvinus and the mechanism of movement. *Nature, Lond.* **179**, 254.

Fondeville, J. C., H. A. Borthwick, and S. B. Hendricks, 1966. Leaflet movement of *Mimosa pudica* L. indicative of phytochrome action. *Planta* (Berl.) **69**, 357–364.

Fondeville, J. C., M. J. Schneider, H. A. Borthwick, and S. B. Hendricks, 1967. Photocontrol of *Mimosa pudica* leaf movement. *Planta* (Berl.) **75**, 228–238.

Groot, G. J. de, 1938. On the mechanism of periodic movements of variation. *Rec. Trav. bot. néerl.* **35**, 758–833.

Guttenberg, H. v. and L. Kröpelin, 1947. Über den Einfluss des Heteroauxins auf das Laminargelenk von *Phaseolus coccineus*. *Planta* (Berl.) **35**, 257–280.

Haberlandt, G., 1890. *Das Reizleitende Gewebesystem der Sinnpflanze*. W. Engelmann, Leipzig.

Hillman, W. S. and W. L. Koukkari, 1967. Phytochrome effects in the nyctinastic leaf movements of *Albizzia julibrissin* and some other legumes. *Plant Physiol.* **42**, 1413–1419.

Houwink, A. L., 1935. The conduction of excitation in *Mimosa pudica*. *Rec. Trav. bot. néerl.* **32**, 51–91.

Jaffe, M. J. and A. W. Galston, 1967. Phytochrome control of rapid nyctinastic movements and membrane permeability in *Albizzia julibrissin*. *Planta* (Berl.) **77**, 135–141.

Leike, H. and H. v. Guttenberg, 1962. Die Rolle des auxins bei der epinastischen Krümmung plagiotroper, Seitensprosse von *Coleus blumei* Benth. *Planta* (Berl.) **58**, 453–470.

Linsbauer, K., 1914. Zur Kenntnis der Reizleitungsbahnen bei *Mimosa pudica*. *Ber deut. botan. Ges.* **32**, 609–621.

Lyon, C. J., 1963a. Auxin factor in branch epinasty. *Plant Physiol.* **38**, 145–152.

Lyon, C. J., 1963b. Auxin factor in leaf epinasty. *Plant Physiol.* **38**, 567–574.

Morgan, P. W. and H. W. Gausman, 1966. Effects of ethylene on auxin transport. *Plant Physiol.* **41**, 45–52.

Mosebach, G., 1944. Untersuchungen über die tagesperiodische Bewegung der Blattgelenke von *Phaseolus*. *Jahrb. wiss. Bot.* **89**, 20–88.

Palmer, J. H. and G. F. Asprey, 1958. Studies in the nyctinastic movement of the leaf pinnae of *Samanea saman* (Jacq.) Merrill. II. The behaviour of upper and lower half-pulvini. *Planta* (Berl.) **51**, 770–785.

Rawitscher, F., 1923. Epinastie und Geotropismus. *Z. Bot.* **15**, 65–100.

Ricca, U., 1916. Soluzione d'un problema di fisiologia—La propagazione di stimolo nella *Mimosa*. *Nuovo Giorn. Bot. Ital.* N.S. **23**, 51–170.

Schmucker, T., 1928. Die Bedingungen des nachtlichen Blühens von *Cereus grandiflorus*. *Planta* (Berl.) **5**, 549–559.

Seidel, K., 1923. Versuche über die Reizleitung bei *Mimosa pudica*. *Beitr. allg. Bot.* **2**, 557–575.

Sibaoka, T., 1953. Some aspects on the slow conduction of stimuli in the leaf of *Mimosa pudica*. *Sci. Rep. Tôhoku Univ. 4th ser. (Biol.)*, **20**, 72–88.

Sibaoka, T., 1962. Excitable cells in *Mimosa*. *Science*, **137**, 226.

Sigmond, H., 1929. Über das Aufblühen von *Hedera helix* L. und die Beeinflussung dieses Vorganges durch das Licht. *Beih. z. botan. Centralbl.* **46**, 68–92.

Snow, R., 1924. Conduction of excitation in stem and leaf of *Mimosa pudica*. *Proc. Roy. Soc. Lond.* **B96**, 349–374.

Snow, R., 1925. Conduction of excitation in the leaf of *Mimosa spegazzinii*. *Proc. Roy. Soc. Lond.* **B98**, 188–201.

Stomps, T. J., 1930. Die Entfaltung der Oenotheraknospe. *Ber. deut. bot. Ges.* **48**, 432–436.

Toriyama, H., 1955. Observational and experimental studies of sensitive plants VI. The migration of potassium in the primary pulvinus. *Cytologia*, **20**, 367–377.

Umrath, K., 1928. Über die Erregungsleitung bei sensitiven Pflanzen, mit Bemerkungen zur Theorie der Erregungsleitung und der elektrischen Erregbarkeit im allgemeinen. *Planta* (Berl.) **5**, 274–324.

Wallace, R. H., 1931a. The effect of animal anaesthetics and certain other compounds upon seismonic sensitivity. *Am. J. Bot.* **18**, 215–235.

Wallace, R. H., 1931b. The effect of temperature, humidity and certain other factors upon seismonic sensitivity. *Am. J. Bot.* **18**, 288–307.

Weintraub, M., 1952. Leaf movements in *Mimosa pudica* L. *New Phytol.* **50**, 357–382.

Williams, C. N. and V. Raghavan, 1966, Effects of light and growth substances on the diurnal movements of the leaflets of *Mimosa pudica*. *J. Exp. Bot.* **17**, 742–749.

Wood, W. M. L., 1953. Thermonasty in tulip and crocus flowers. *J. Exp. Bot.* **4**, 65–77.

Yin, H. C., 1941. Studies on the nyctinastic movement of the leaves of *Carica papaya*. *Am. J. Bot.* **28**, 250–261.

9. The movements of stomata

O. V. S. Heath
and T. A. Mansfield

Structure and mechanism

Stomata are found on most types of epidermal surface of flowering plants and gymnosperms but are especially numerous on the leaves. In most Angiosperm trees and shrubs the leaves have stomata restricted to the lower surface and are then termed *hypostomatous*; in a few water plants with floating leaves stomata are restricted to the upper surface (*epistomatous*), or as in most herbaceous plants they may be found on both leaf surfaces (*amphistomatous*).

The precise structure of the stomatal apparatus can vary considerably from one species to another but changes in pore width can usually if not always be ascribed to changes in the turgor difference between guard cells and subsidiary cells; a volume increase of the guard cells, or a volume decrease of the subsidiary cells, brings about an enlargement of the stomatal pore. The current views of the mechanics of guard cell movement are based on the theorizing of Schwendener (1881, 1889) some of whose conclusions have been summarized by Heath (1959a). The importance of guard cell turgor in determining the stomatal pore width was demonstrated experimentally by von Mohl (1856), who showed that stomatal closure could be caused by floating pieces of leaf on sugar solutions, and by Heath (1938) who found that puncturing single guard cells with a micro-needle caused closure of the pore on that side, so that the pore width was halved; if one subsidiary cell was punctured the stoma opened more widely on that side showing that the pore width depended on the turgor difference between guard cell and subsidiary cell. This effect can be observed on pieces of epidermis stripped from an onion leaf and mounted in tap water with 1/10,000 neutral red, which accumulates in the vacuoles of living cells only. The inner surface may be lightly scraped to remove adhering mesophyll. Where stomata are adjacent to living subsidiary cells they close tightly but they open widely where both subsidiary cells have been killed; if only one has been killed the stoma opens on that side. Changes in turgor of the subsidiary cells and hence in the pressure they exert upon the guard cells explain the so-called 'passive' stomatal opening that occurs with a sudden water deficit and also the 'passive' closure when water is supplied to a leaf under water strain. Sudden water loss from the mesophyll reduces the turgor of the ordinary epidermal and subsidiary cells before the guard cells are affected. The stomata then open widely for a short time (a matter of minutes) before the guard cells also lose turgor and closure occurs. Conversely, a renewed water supply reaches the epidermal and subsidiary cells before the guard cells which are thus pressed tightly together. These effects are easy to observe experimentally by depriving a detached leaf of its water supply or by watering a plant growing in rather dry soil.

Guard cells apparently always contain chloroplasts; these are generally smaller and contain less chlorophyll than those of mesophyll cells but they have been shown by autoradiography to fix carbon dioxide (Shaw and Maclachlan, 1954b) and they contain both chlorophylls *a* and *b* (Yemm and Willis, 1954). Guard cells often appear to be isolated from the other epidermal cells, for although there are numerous ectodesmata in the ventral walls adjacent to the pore (Fig. 9.1), plasmodesmata have been reported to be absent from the dorsal walls between guard cells and neighbouring cells (Sheffield, 1936). However, a few such plasmodesmata have been recorded for *Aponogeton distachyus* by Mouravieff (1959), and some are figured by Franke (1962). In senescing leaves the guard cell chloroplasts remain green after those in the mesophyll have lost most of their chlorophyll (Zucker, 1963) and the guard cells remain alive longer than the other epidermal cells (Hagen, 1918); this again suggests that they are more

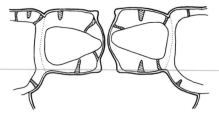

Figure 9.1   The distribution of ectodesmata in guard cells of *Plantago major*. Note the lack of plasmodesmata between guard cells and neighbouring cells. (After Franke, 1961.)

or less isolated. There are also some striking differences in metabolism between guard cells and ordinary epidermal or mesophyll cells. In the chloroplasts of mesophyll cells starch accumulates through the day as photosynthesis goes on and is removed at night but in those of guard cells it usually tends to disappear during the morning, to accumulate in the late afternoon and to remain through the night. These diurnal rhythms (see chapter 18) of guard cell starch content may be connected with the diurnal rhythms of movement often shown by stomata even under constant conditions. The guard cells of onion stomata apparently never form starch (Parkin, 1899; Heath, 1952) but there exists the possibility of a rhythm in hydrolysis and condensation of soluble polysaccharides.

Methods of observation

We shall mention briefly only some of the more important techniques; a more detailed account of the various methods that have been used is available elsewhere (Heath, 1959a).

The most obvious method is direct microscopic observation of stomata on a living leaf or portion of a leaf. Mouravieff (1954) uses so-called 'stomates isolés', which are stomata near the edges of surface sections where they are surrounded by dead epidermal cells cut through by the razor. Stålfelt (1964) has recently strongly advocated the use of leaf discs floated on water, claiming that this eliminates the risk of artifacts due to the water strain that can develop in intact leaves. He measures the stomatal apertures directly with a microscope with paraffin immersion, but another method is to make a 'negative' impression of the epidermis with silicone-rubber and obtain a 'positive' by painting the negative with cellulose acetate; this can be removed as a film and examined under the microscope (Zelitch, 1961). Although leaf discs floated on water are unlikely to be subject to water strain it seems likely that the high leaf turgor may cause partial 'passive' stomatal closure, as observed for some species by von Mohl (1856); this effect was described by Stålfelt (1955) as 'sub-optimal water deficit' and has been shown by Glinka and Meidner (1968) to occur in leaf discs of tobacco.

The more important indirect methods of observation are (a) transpiration measurements and (b) leaf porometers.

(a) The measurement of transpirational water loss can, under carefully controlled ambient conditions, provide an indication of the degree of stomatal opening (Virgin, 1956; Brun, 1962). If, however, the internal supply of water becomes limited the pressure of aqueous vapour surrounding the mesophyll cells may be below saturation (Meidner, 1965a) and this will lead to inaccuracies in the measurement. Consequently, transpiration methods should not be used in experiments involving different water deficits in the leaf. Meidner and Spanner (1959) introduced an ingenious device known as the 'differential transpiration porometer' in which two small jets of air, at the same temperature but at slightly different controlled humidities (such as 75 per cent and 85 per cent R.H.) are directed side by side onto two small areas on the lower leaf surface. This results in a difference in transpiration and hence in temperature; the latter can be measured using fine thermocouples. The temperature difference is a function of stomatal diffusive conductance (see below).

Van Bavel (1965) developed another type of transpiration porometer which is in principle similar to the classical cobalt chloride paper method (Livingston and Shreve, 1916; Henderson, 1936); the latter, though called a method for measuring transpiration, in fact gives a measure of stomatal diffusive conductance by substituting a standard 'sink' for the prevailing evaporating conditions (Milthorpe, 1955). In the van Bavel instrument a vessel containing a lithium chloride-impregnated resistor is dried out with an air stream and then sealed onto the leaf. Water vapour diffuses from the leaf to the hygroscopic surface of the resistor and the time for the electrical resistance to fall between two predetermined values is found. The instrument is calibrated in absolute terms by taking similar readings over a wet surface with tubes of various lengths interposed between the surface and the resistor; the leaf resistance for diffusion of water vapour can then be specified. Temperature corrections are necessary for diffusion rate and for the resistance meter sensitivity but the method has such obvious advantages for field work that there is little doubt that it will be extensively adopted and improved.

(b) 'Porometer' was originally the name for a device in which a small cup was attached to the leaf surface and air 'sucked' through the leaf under a known pressure difference (Darwin and Pertz, 1911). The (measured) rate of flow depended upon the resistance to air flow of the leaf tissue as a whole, but was to a large extent determined by the apertures of the stomata. There have since been many modifications of this simple device. The three most widely used today are the 'resistance porometer' of Gregory and Pearse (1934) and the 'Wheatstone bridge porometer' of Heath and Russell (1951), both of which permit the comparison of the resistance of the leaf to air flow with that of a standard capillary resistance, and the Alvim (1965) field porometer, in which the principle is essentially the same as in the Darwin and Pertz porometer. Figure 9.2 shows a simple and inexpensive field porometer (Meidner, 1965b) which also operates on this principle. In these types of porometer air is forced through the leaf by suction or pressure and the measurement represents the resistance or conductance for mass flow; this is a function, though sometimes a complex one (Heath, 1941), of stomatal aperture. In nature, however, gases diffuse in and out of the leaf under differences of partial pressure, and when the relations between stomatal aperture and assimilation or transpiration are being considered, it is the resistance to *diffusive* flow that is of interest. Several instruments, known as diffusion porometers, have been devised to measure this. The Spanner (1953) diffusion porometer makes use of a phenomenon known as the Dufour effect, that when two gases at the same temperature interdiffuse, temperature differences arise where they mix. Spanner directed a small jet of hydrogen against the upper surface of the leaf (which must be amphistomatous), and measured the temperature change close to the opposite surface as the hydrogen diffused through and met the air. Transpiration porometers (discussed above) can also be considered to be diffusion porometers since they depend on the diffusion of water vapour out of the leaf.

One aspect of porometer design which must be stressed is the importance of not permanently enclosing the portion of leaf within the cup. Stomatal behaviour is greatly affected by carbon dioxide concentration (see p. 309) and in light a portion of leaf enclosed shows abnormally wide stomatal opening due to the photosynthetic depletion of the carbon dioxide (Heath and Williams, 1948). The sensitivity to carbon dioxide concentration is so great that precautions must also be taken not to breathe near the plant while stomatal movements are being investigated (Heath and Mansfield, 1962; Mansfield, 1965a). An automatic recording porometer is therefore advantageous as it can be set up and the experiment left to run; it also allows of long-period experiments lasting for 24 hours and more. Heath and Mansfield (1962) have described a recording version of the 'resistance porometer' in which the portion of leaf is only enclosed for the brief period necessary to take a record. Four automatically detachable cups operate one after another on different leaves at $7\frac{1}{2}$ minute intervals so that a record for each is obtained once in every half hour. The use of four leaves has proved unexpectedly advantageous and given great encouragement for future

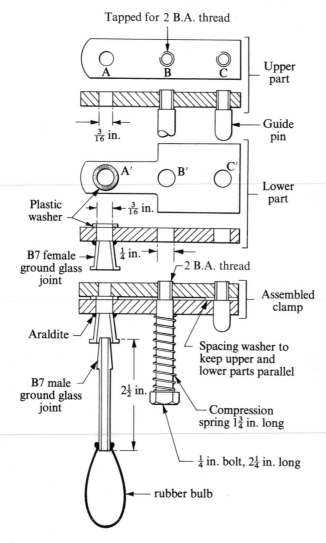

Figure 9.2 Constructional details of a simple portable porometer. The leaf is clamped between the upper part and the washer on the lower part, under the pressure of the spring. The rubber bulb is compressed and fitted, and air is sucked through the leaf, the time taken for the bulb to inflate being proportional to the resistance of the leaf to air flow. (After Meidner, 1965b.)

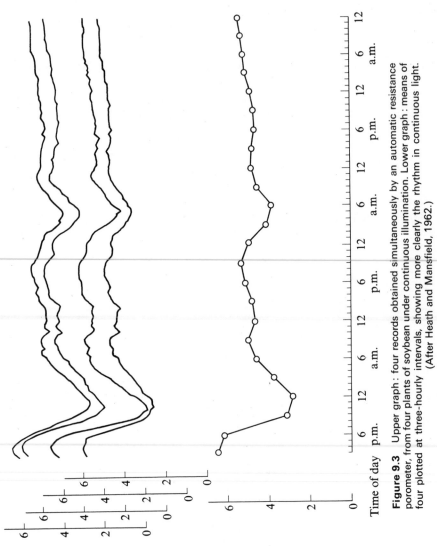

Figure 9.3 Upper graph: four records obtained simultaneously by an automatic resistance porometer, from four plants of soybean under continuous illumination. Lower graph: means of four plotted at three-hourly intervals, showing more clearly the rhythm in continuous light. (After Heath and Mansfield, 1962.)

investigation. Figure 9.3 shows records for four soybean plants under constant illumination. These plants had been grown side by side in the glasshouse (under fluctuating conditions) and therefore all had the same previous history. It is seen that the stomatal behaviour under supposedly constant conditions was complex but was almost exactly the same for all four plants. This shows that the behaviour of stomata is not capricious, as is sometimes supposed, but is an extremely sensitive response to previous and present external conditions and to endogenous rhythms (see chapter 18). The practical lesson for experimentation is that it is much better to compare treatments carried out on different leaves (of similar previous history) at the same time than on the same leaf at different times.

Aspects of the physiology of stomatal movements

Responses to carbon dioxide

Of recent years the most intensively studied aspect of stomatal physiology has been the response to changes in carbon dioxide concentration. As long ago as 1916 the work of Linsbauer demonstrated that removal of carbon dioxide from the atmosphere surrounding leaves caused stomata to open more widely. However, in spite of Linsbauer's contribution and several other early investigations (which gave conflicting results), the importance of the role played by carbon dioxide in determining stomatal movements has been recognized only comparatively recently. There is now an abundance of evidence that stomata respond markedly to changes in carbon dioxide concentration at physiological levels, that is, concentrations of the same order of magnitude as the 0·03 per cent in normal air (e.g., Freudenberger, 1940; Heath, 1950; Heath and Russell, 1954b). Experiments in which air of known carbon dioxide content was forced through the leaf showed that wheat stomata closed appreciably in light when the concentration was increased between 0·01 and 0·084 per cent (Heath and Russell, 1954b). Figure 9.4 is an isometric projection which shows, as if in three dimensions, stomatal aperture as affected by carbon dioxide concentration at three light intensities, and in darkness. It will be noticed that the stomata became capable of opening more widely in a given concentration as the light intensity increased, which would be expected to happen as the photosynthetic rate in the guard cells increased and the carbon dioxide was consumed more rapidly. A most important point to notice from Fig. 9.4 is that even in darkness the removal of carbon dioxide maintained stomatal opening. Later, Mansfield (1965d) found with *Xanthium pennsylvanicum* that carbon-dioxide-free air could cause stomatal opening in darkness comparable with that in bright light and ordinary air. This observation that light is not essential in the opening due to removal of carbon dioxide is important because it enables us to rule out the participation of those biochemical mechanisms for which illumination is necessary, at least for most of the effect.

Although the removal of carbon dioxide from the atmosphere surrounding a leaf can lead to stomatal opening, this acts by reducing the concentration at the internal surfaces of the guard cells, in the substomatal cavity. This was shown by the fact that in darkness carbon-dioxide-free air prevented closure of stomata which were still partly open, but would not re-open those which had already

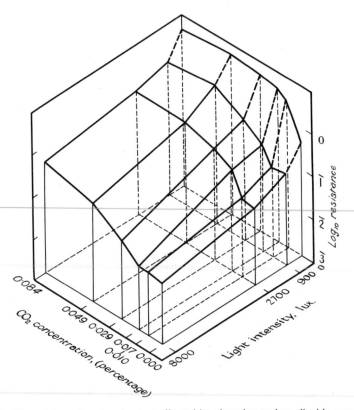

Figure 9.4   Stomatal aperture in wheat as affected by changing carbon dioxide concentration and light intensity. Note that an increasing value of the dependent variate ($\log_{10}$ resistance) indicates stomatal closure. (After Heath and Russell, 1954b.)

closed completely in ordinary air (Heath, 1950; Scarth and Shaw, 1951). Presumably only the inner walls of the guard cells are permeable to carbon dioxide.

The above information enables us to appreciate the important contribution towards stomatal opening in light made by removal of carbon dioxide. In light of sufficient intensity, photosynthesis in the leaf mesophyll, and also within the guard cells themselves (Heath and Russell, 1954b; Shaw and Maclachlan, 1954b) lowers the carbon dioxide concentration in the latter and leads to stomatal opening; in reduced light intensity or in darkness an accumulation of carbon dioxide due to respiration leads to stomatal closure. This simple statement of the contribution made by photosynthesis and respiration covers a most important aspect of the physiology of stomata, but it only partially defines the way in which light and darkness determine stomatal movements, as will be seen from subsequent sections.

Responses of stomata to other environmental factors, such as temperature and water strain, can also be explained in terms of their sensitivity to carbon dioxide. Investigators of the course of stomatal behaviour in certain plants (such as *Coffea arabica*) growing under natural conditions in warm climates often found that the stomata were open in the morning and late afternoon, but closed partially for a time in the middle of the day. It is now thought that this closure is due to the temperature increase around mid-day which in some plants causes the carbon dioxide concentration in the intercellular spaces of the leaves to rise very steeply (Heath and Orchard, 1957). In both *Coffea* and onion, Heath and Orchard found that the minimum intercellular space carbon dioxide concentration was only just below that in the ambient air at 35°C, whereas at 25°C it was less than 0·01 per cent. Meidner and Heath (1959) found that at 35°C high-temperature closure in onion could be prevented if the internal cavity of the hollow leaf was ventilated with air free from carbon dioxide. The rate of stomatal opening in light of 3000 lux was considerably greater at 35°C than at 26°C (Q_{10} = 2·2) when the leaf cavity of onion was ventilated in this way. Thus the opening process is actually favoured by high temperature but in some plants this fact is obscured by the overriding influence of the carbon dioxide concentration inside the leaf.

Stomata usually close if the plant suffers water strain. Heath and Meidner (1961) found that the minimum intercellular space carbon dioxide concentration in wheat leaves increased during water strain, but this increase was quite small, being only from about 0·007 to 0·012 per cent at 9000 lux. Note from Fig. 9.4 that this increase would not account for any appreciable stomatal closure in a high light intensity. There is evidence, however, that the *sensitivity* of the stomata to carbon dioxide increases greatly when leaves are under water strain (Heath and Mansfield, 1962). It seems likely that such enhanced sensitivity to carbon dioxide represents part of the plant's protective mechanism against excessive water loss during dry conditions. Even in the absence of water strain, the normal sensitivity to carbon dioxide might protect the plant against excessive water loss in wind, owing to the increase in the amount of carbon dioxide brought close to the leaf surface (Heath, 1950).

The mechanism by which changes in carbon dioxide concentration affect stomata is not known. The only information of any positive value is the observation that a reduction in carbon dioxide concentration may lead to hydrolysis of starch in the guard cells. This was suggested by the work of Williams (1952) and Williams and Barratt (1954) and confirmed by Mouravieff (1956) who observed stomatal opening and a reduction in guard cell starch content in carbon-dioxide-free air both in light and darkness.

Stomatal opening independent of carbon dioxide concentration

Light-induced opening. Wilson (1948) found stomatal opening in light in etiolated leaves of sweet potato which were apparently chlorophyll-free both in the mesophyll and guard cells, and which were therefore unlikely to be capable of photosynthesis, so that the carbon dioxide concentration in the leaf would not be lowered in light. The carbon dioxide response curves at different light intensities obtained by Heath and Milthorpe (1950) and Heath and Russell (1954b) provided further evidence that opening in light could take place independently of the carbon dioxide concentration in the leaf. Opening occurred as the light intensity increased both over carbon-dioxide-saturated portions of the response curves (between 0·049 and 0·084 per cent) and also in the range zero to 0·010 per cent where carbon dioxide concentration was without detectable effect. This will be evident from a careful consideration of Fig. 9.4 (a more detailed summary of this evidence will be found in Heath, 1959b, p. 448). Mouravieff (1956) applied photosynthetic inhibitors to his 'isolated' stomata, and found that starch hydrolysis and rise of osmotic pressure leading to stomatal opening were not prevented. Later, Mouravieff (1958) investigated the effectiveness of red, green, and blue light in producing stomatal opening, and found blue to be by far the most efficient. Some of his results are presented in Table 9.1.

Table 9.1   Light intensities in different wavelengths for producing comparable stomatal apertures in *Veronica beccabunga* (after Mouravieff, 1958).

Wavelength (nm)	Intensity producing half-open stomata $(J\ m^{-2}\ s^{-1})$	Intensity producing full opening $(J\ m^{-2}\ s^{-1})$
464 (blue)	0·2	0·7
526 (green)	0·6	10·5
660 (red)	0·6	11·9

The greater effectiveness of blue light than red would not be expected if photosynthetic consumption of carbon dioxide were the only factor involved. Recently Heath, Mansfield, and Meidner (1965) have shown for *Xanthium pennsylvanicum* that leaves placed in a closed system and illuminated in red or blue light show considerably greater opening in blue even when the carbon dioxide concentration within and surrounding the leaf is the same as in red. This has established beyond doubt the existence of a carbon-dioxide-independent opening effect of blue light. Mouravieff (1958) observed that blue light at an intensity well below the compensation point led to a decrease in guard cell starch, an increase of 2·5 atmospheres in the osmotic potential of the guard cells and stomatal opening, whereas with red light at the compensation point the stomatal starch content increased. He postulated that blue light caused starch hydrolysis or prevented condensation of sugar to starch, whereas red light had no such action or did the reverse. This seems to be a possible explanation of the way the carbon-dioxide-independent opening effect of blue light operates. Thus blue light could have a similar action to carbon dioxide removal in causing starch hydrolysis (see previous section).

Temperature-induced opening.　We have seen above (page 311) that in some plants increased temperature can lead to stomatal closure by producing a rise in the carbon dioxide concentration within the leaf. The effect of temperature which is independent of carbon dioxide concentration is, however, to produce stomatal opening. It was found many years ago by Francis Darwin (1898) that increased temperature could of itself produce stomatal opening in complete darkness, and a recent investigation has completely confirmed this (Mansfield, 1965b). Such opening must occur in spite of increased respiratory production of carbon dioxide, and consequently the opening effect of high temperature overrides the closing response to carbon dioxide. The relative importance of carbon-dioxide-induced closure and temperature-induced opening differs from species to species; in the onion leaf, which contains much chlorophyll-free tissue, carbon-dioxide-induced closure was found to predominate so that considerable closure occurred between 25° and 35°C even in light of 9000 lux (Meidner and Heath, 1959), whereas in *Xanthium pennsylvanicum* an increase from 27° to 35°C in darkness produced wide opening (Mansfield, 1965b). The high Q_{10} for rate of opening of onion stomata, when carbon-dioxide-induced closure was eliminated by flushing the leaf cavity with carbon-dioxide-free air (see p. 311 above), suggested that the rate was primarily controlled by a chemical dark reaction; Meidner and Heath suggested that as onion stomata do not contain starch an enzyme-controlled hydrolysis of a soluble polysaccharide might perhaps be involved in producing an increase in osmotic potential. In other species this might be brought about by a temperature-induced hydrolysis of starch. In light, this might operate in conjunction with other factors which appear to influence starch hydrolysis, such as carbon dioxide removal and blue light.

Transmission of stimuli

Several indications have been obtained that stimuli affecting the stomata can be transmitted within a leaf, or from one part of a plant to another. Thomas (1949) kept the lower half of a metre-long branch of *Cissus sicyoides* in the dark, yet observed stomatal opening on the lower leaves when the upper half was illuminated. It seems likely that this effect was due to changes of pressure in the vascular system rather than the transport of substances, since opening on illuminated and darkened portions occurred almost simultaneously. The capacity for transmission within a leaf seems to be very variable. Several workers (e.g., Scarth, 1932) observed stomatal opening in light on the white portions of variegated leaves, especially in the stomata near the green tissue. This might be explained in terms of a lowering of carbon dioxide concentration in the intercellular spaces due to photosynthesis in the green areas. Heath and Russell (1954a) found transmission along normal green leaves of wheat. They illuminated a control area of leaf with 2700 lux and varied the intensity in another area between 900, 2700, and 8000 lux. The distance between the two areas was 17 mm and this part of the leaf constituted a 'light barrier' which was kept at 8000 lux: the intercellular space system was continuously swept with carbon-dioxide-free air. The result of this experiment is shown in Fig. 9.5, from which

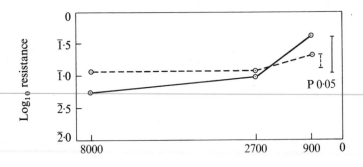

Light intensity, lux

Figure 9.5   The experiment of Heath and Russell (1954a) on wheat showing the transmission of a light effect from one part of a leaf to another. Continuous line: experimental area, light intensity varied. Broken line: control area, light intensity 270° lux throughout.

it will be seen that the change in light intensity led to a significant change in stomatal aperture in the control area. Since transmission due to removal of carbon dioxide was unlikely to have been involved, it seemed likely that an electrical or chemical stimulus, not travelling in the intercellular space system, was operative. Kuiper (1961) working with *Pelargonium* concluded that stomata could affect each other's light responses, but Zelitch (1961) using a less sensitive technique could not find any appreciable transmission across partly shaded leaf discs of tobacco. The stomata in the darkened portions remained closed, the line dividing open and closed stomata being within one millimetre of the line separating the illuminated and darkened regions. It seems likely, therefore, that there may be some variation from species to species.

The most likely means of transmission within a leaf is the translocation of substances, probably via the phloem; Williams (1948) found that heat shock stimuli causing stomatal closure were probably transmitted along the phloem in *Pelargonium*. Zelitch (1963) found inhibitors of stomatal opening in the epidermis of leaves kept in the dark, and these might be translocated from a darkened to an illuminated area to reduce the aperture in light. The observations of Heath and Russell might be explained in this way, especially as the only significant effect transmitted was that of the low light intensity (900 lux—see Fig. 9.5).

Endogenous rhythms and the associated light reaction

Like many other processes in the plant, the movements of stomata exhibit marked rhythmic behaviour (see chapter 18). Rhythmic movements occur both in light and in darkness, but those in light are the most readily followed. Figure 9.6 indicates how large is the amplitude of the rhythm in light in wheat and Fig. 9.3 shows a less regular rhythm in soybean. Until recent years stomatal

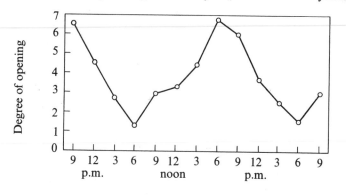

Time of day

Figure 9.6 Rhythm of stomatal aperture in wheat, in continuous light of 20,000 lux. (After Meidner and Mansfield, 1965.)

rhythms have been studied relatively little, and there is still a dearth of useful information about the behaviour in continuous light. There has been, for example, no systematic study which establishes beyond doubt that the rhythm is endogenous in nature (on the criteria listed in chapter 18). Since stomata are so sensitive to changes in the environment it is especially important to ensure that the movements under apparently constant conditions are not in fact being brought about by some fluctuating external factor, such as carbon dioxide concentration. In none of the studies of stomatal behaviour in continuous light has carbon dioxide concentration been under control.

The rhythm in darkness has been more thoroughly investigated and there is little doubt of its endogenous nature, for it exhibits many of the characteristics shown by the well-established endogenous rhythms in plants (see chapter 18).

Rhythmic stomatal opening does occur in darkness ('night opening'), but it is normally small in magnitude and difficult to observe (Stålfelt, 1963). The most sensitive indication of the progress of the rhythm has been obtained by switching on the light after different times in the dark and measuring either the rate of opening (Mansfield and Heath, 1961, 1963), or the delay before the onset of transpiration (Brun, 1962). The rhythm in *Xanthium pennsylvanicum* which was observed in this way is shown in Fig. 9.7. These and other studies have proved

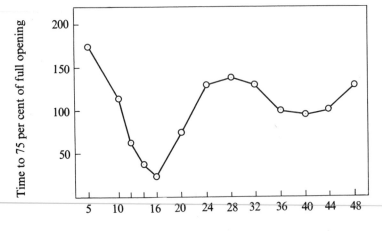

Figure 9.7   Rhythm in stomatal 'opening ability' in *Xanthium pennsylvanicum*. Ordinate is the time in minutes taken in reaching 75 per cent of full aperture in light of 15,000 lux. (After Mansfield and Heath, 1963.)

how important the endogenous rhythm in darkness is in determining the stomatal opening response to light. Some curves demonstrating the effect on the rate of opening in light are shown in Fig. 9.8.

The phase of the rhythm in darkness is determined by a low-intensity light reaction distinct from the high-intensity effects which result immediately in stomatal opening; this reaction is brought about by the red region of the spectrum, the most active wavelength being in the region of 700 nm (Mansfield, 1964). The existence of this low-light reaction complicates the way in which light determines stomatal behaviour, and it may be many years before the contribution made by different reactions towards the movements under natural conditions is understood. There are clearly several different pigments involved in determining the light responses of stomata: the opening due to carbon dioxide removal by photosynthesis depends on chlorophyll *a* and its accessory pigments such as chlorophyll *b*; the carbon-dioxide-independent opening reaction to blue light operates through an unknown pigment system—perhaps a carotenoid; the low-intensity reaction may operate through the phytochrome system, but the balance of evidence seems to be against this (Mansfield, 1965c).

Dynamics of stomatal movement

Many observations of stomatal behaviour have been concerned with the more or less steady apertures achieved under different conditions and there has been less emphasis on the study of rates of movement, partly because these are so much affected by the choice of a measure (Heath and Orchard, 1956). This is nonetheless an interesting topic, and one or two observations of importance must be mentioned.

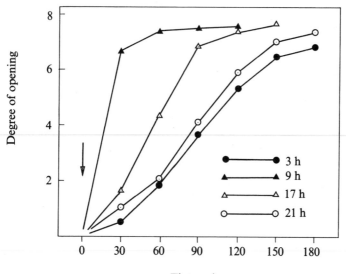

Figure 9.8   Effect of the length of the preceding night on the rate of opening of stomata of *Xanthium pennsylvanicum*. (After Mansfield and Heath, 1963.)

Both physiological and biochemical evidence has indicated that there are fundamental differences between the processes concerned in opening and closing of stomata. This was first noticed by Gregory and Pearse (1937) from the curves of opening and closing in *Pelargonium* in terms of leaf resistance: the curve of opening in light had a sigmoid form, whereas that of closure in darkness resembled a logarithmic decrement curve. Subsequent work has shown that there is an oxygen requirement for stomatal opening (Walker and Zelitch, 1963) but not for stomatal closure (Heath and Orchard, 1956; Walker and Zelitch, 1963). This could mean that opening depends upon an active, oxygen-requiring process, while closure does not (see also p. 323). The observation that stomata come to different steady state apertures at different light intensities (Fig. 9.4) shows that in light there must be two simultaneous processes, one tending to cause opening and the other closure; otherwise opening would proceed to full aperture at any light intensity though more slowly at lower intensity. The two processes might perhaps be an active uptake of solutes into the guard cells against a concentration gradient and a passive 'leak' out, or assimilation and respiration in the guard cells controlling the internal carbon dioxide concentration, as suggested by Heath (1959a). Walker and Zelitch also studied the effects of the inhibitor sodium azide on stomatal movements. They made the important observation that lower concentrations (10^{-4} M) caused closure of stomata which were open in light, whereas higher concentrations (about 10^{-3} M) did not. They suggested that this was best explained if one considered that opening and closing occurred by different reactions and not by reversal of the same reaction. Azide at a concentration of 10^{-4} M was thought to affect the opening more than the closing reaction, so that closure occurred; at higher concentrations both reactions were inhibited and so the stomata remained open. Walker and Zelitch considered from this evidence that a balance between the opening and closing reactions determined the aperture in light.

There is some evidence that the reaction concerned in the production of opening differs from that involved in maintaining it. Walker and Zelitch (1963) found in tobacco that there was an oxygen requirement for opening in light, but that lack of oxygen did not cause closure of stomata which were already open. This result is not, however, entirely conclusive because lack of oxygen in light reduces the carbon dioxide concentration within the leaf (Tregunna, Krotkov, and Nelson, 1964) and this might be sufficient to *maintain* stomatal opening, in conditions in which some energy could be derived from anaerobic photophosphorylation.

Hypotheses of the mechanisms involved in the stomatal response to light and other factors

Photosynthetic production in the guard cells

The earliest suggestion of the mechanism by which guard cells gain turgor in light was made by von Mohl (1856), who thought that the chloroplasts they contained manufactured osmotically active substances by photosynthesis, which increased turgor and produced stomatal opening. This simple suggestion is not now accepted, for if it were correct, increasing the carbon dioxide concentration round the leaf in bright light should lead to increased opening, whereas in fact it causes partial closure. The hypothesis that photosynthetic production in the guard cells plays an important part in stomatal opening in light has, however, recently been revived. It is known that the nature of the products of photosynthesis is greatly affected by the carbon dioxide concentration, and at low concentrations glycollic acid ($CH_2OH.COOH$) is an important product (Tolbert, 1958). Zelitch (1963) has suggested that production of glycollic acid in the guard cells is an important factor in stomatal opening. He suggests that the glycollate gives rise to carbohydrate, thus raising the osmotic pressure and also that it could participate in the production of ATP, which might provide the energy required for the opening process:

(a) $$NADP^+ + ADP + P_i + H_2O \xrightarrow{\text{light}} NADPH + H^+ + ATP + \tfrac{1}{2}O_2$$
(non-cyclic photophosphorylation)

(b) $$\text{glyoxylate} + NADPH + H^+ \longrightarrow \text{glycollate} + NADP^+$$
(glyoxylate reductase)

(c) $$\text{glycollate} + \tfrac{1}{2}O_2 \longrightarrow \text{glyoxylate} + H_2O_2$$
(glycollate oxidase)

(d) $$H_2O_2 \longrightarrow H_2O + \tfrac{1}{2}O_2$$
(catalase)

Under conditions of limited carbon dioxide supply less NADPH will be consumed in carbon dioxide reduction and it will tend to accumulate, slowing reaction (a). The glycollate-glyoxylate shuttle (reactions b and c) would achieve the reoxidation of NADPH, and maintain the rate of (a) and the production of ATP. The overall balance of (a)—(d) is simply:

$$ADP + P \xrightarrow{\text{light}} ATP,$$

and the sequence of four reactions is a suggested mechanism for cyclic photophosphorylation (Butt and Peel, 1963). Zelitch believes that ATP produced by this mechanism might participate in the active pumping of water into the guard cells (see page 323 below).

This hypothesis is now a centre of controversy in the field of stomatal physiology. Its main attraction is that it takes account of opening in response to reduced carbon dioxide concentrations, yet does not eliminate photosynthetic production as a factor. However, it cannot account for the fact that stomata open widely in the dark in response to carbon dioxide removal (see p. 309), nor can the stomatal opening that occurs in carbon-dioxide-free air in light, when there can be no net carbon dioxide assimilation, be explained in terms of photosynthetic formation of carbohydrate via glycollate or by any other route. Whether Zelitch can reconcile his hypothesis with these important shortcomings will be seen in due course. If the postulated glycollate-glyoxylate shuttle is shown to result in the production of more ATP in carbon-dioxide-free air, the importance of ATP for stomatal opening will also need to be demonstrated (see p. 323 below).

Zelitch's hypothesis would be supported by the views of Stålfelt (1964), who recognizes that carbon dioxide removal of itself leads to stomatal opening, but also believes that under conditions of complete carbon dioxide starvation stomatal opening is suppressed. His view implies that there should be an optimum carbon dioxide concentration for stomatal opening below and above which closure occurs. This is not, however, borne out by the experimental observations of other workers (see p. 309).

The starch ⇌ sugar hypothesis

An hypothesis which has figured prominently in the literature on stomata attempts to explain opening in light as being due to the action of an enzyme causing hydrolysis of the starch in the guard cell plastids to sugar, the resulting increase in osmotic pressure leading to stomatal opening. This has often been presented in elementary textbooks as an established fact, but it is most unlikely that it represents the main mechanism behind light-operated stomatal movements, although it probably plays a part.

Lloyd (1908) developed the hypothesis from a suggestion originally made by Kohl (1895). He supported it with observations of stomatal starch content and aperture on strips of leaf epidermis taken throughout the day from plants under natural conditions—starch content was usually low during the day when the stomata were open, and high at night when they were closed. Similar observations of diurnal changes were presented by Loftfield (1921). This is not, of itself, very good evidence in favour of the hypothesis because the changes in starch content and the stomatal movements could both be due to diurnal changes in the environment, or to endogenous rhythms, without being causally related. Some other experiments of Lloyd, in which the plants were illuminated and darkened at other than the normal times of day, did not provide a convincing negative correlation between aperture and starch content (see Heath, 1949).

The hypothesis was developed further by Sayre (1926) as a result of his detailed investigation of the stomatal physiology of *Rumex patientia*, also carried out on diurnal changes. He found, with the indicator bromophenol blue, that the pH in the guard cells increased during the day when the stomata were open, and decreased at night. This observation was supported by experiments on pieces of detached epidermis in buffer solutions, in which stomatal movements occurred in response to pH changes. He suggested that opening was set in motion by light increasing the pH within the guard cells, which might be due to removal of carbon dioxide by photosynthesis or to light-induced changes in organic acid content. The increase in pH was thought to favour the conversion of starch to sugar. This hypothesis was taken up by several later workers, and further evidence in support of it was claimed, especially with regard to the changes in pH occurring in light and darkness (Scarth, 1932; Small and Maxwell, 1939; Small, Clarke, and Crosbie-Baird, 1942). Hanes (1940) found an enzyme in plant tissues which catalysed the reaction:

$$\text{starch} + \text{inorganic phosphate} \xrightarrow{\text{phosphorylase}} \text{glucose-1-phosphate}$$

and it became accepted that this was operative in stomatal movements (Small, Clark, and Crosbie-Baird, 1942; Alvim, 1949), the now much-modified hypothesis being supported by Yin and Tung's (1948) finding that phosphorylase was present in guard cells.

One important fact was overlooked, namely that the reaction catalysed by phosphorylase cannot of itself lead to any change in osmotic pressure, since the total number of soluble molecules remains the same. It is therefore necessary to suppose either that inorganic phosphate is transported from a source outside the guard cell (in which case the starch → sugar change would not, of itself, be responsible for raising the osmotic pressure) or that further reactions lead to hydrolysis of glucose-1-phosphate. Steward (see Heath, 1959a, p. 216) has pointed out that energy would then be needed for the re-formation of starch for stomatal closure, which should not therefore occur in darkness in the absence of oxygen. Anaerobic conditions do not, however, prevent stomatal closure in darkness (Heath and Orchard, 1956; Walker and Zelitch, 1963). The hypothesis had already received a setback when it was shown that the stomata of onion, which do not contain any starch at all, show normal responses to light and darkness (Heath, 1952).

Heath (1949) found for *Pelargonium zonale* that if effects of diurnal rhythms were eliminated by comparing strips of epidermis taken at the same time of day from plants in light and dark respectively, there were no consistent differences in guard cell starch content between stomata relatively open in light or relatively closed in darkness. Williams (1952), using an ingenious quantitative method for estimating stomatal starch (Williams and Spencer, 1950) found an increase of guard cell starch in the afternoon under constant illumination, also in *Pelargonium*.

The above discussion shows that the evidence in favour of starch ⇌ sugar interconversion playing a major part in the stomatal light response is at best equivocal. It is a weakness that practically all the evidence relates to changes of guard cell starch, while it is the sugar that is the osmotically active component. Photosynthesis in the guard cells might result in sugar or starch varying independently of the other and translocation between guard cells and subsidiary cells might also falsify the picture obtained from starch alone. The hypothesis must not be dismissed entirely, however, for we have seen above that there is evidence that carbon-dioxide-free air (p. 311) and blue light (see p. 312), both of which promote stomatal opening, can lead to a reduction in guard cell starch content; the blue light effect was, moreover, accompanied by an increase in osmotic potential. It has also been found that water strain, which even in light causes stomatal closure and an increase in the sensitivity to carbon dioxide (p. 311), causes rapid and considerable increases in stomatal starch content (Iljin, 1914). The pronounced diurnal rhythm in guard cell starch content, independent of illumination (Heath, 1949; Williams, 1952) could be a factor involved in the endogenous rhythms of stomatal behaviour which proceed in constant conditions (p. 315).

In conclusion, therefore, we can say that there is some evidence that starch hydrolysis and re-formation contribute to the stomatal responses to light but they are probably more important in controlling movements due to water strain and in operating endogenous rhythms. The magnitude of their role in the light response is difficult to determine, but it is clear that other internal factors are also involved and that light-induced stomatal movements can occur independently of starch $\rightleftharpoons$ sugar changes.

The permeability hypothesis

Linsbauer (1916) suggested that stomatal movements might be explained by changes in guard cell permeability in light and darkness. Kissilew (1925) made some measurements of the rates of entry of the dye indigo-carmine into guard cells and found that it entered more rapidly into closed stomata in the dark than into open ones in the light. This suggested a greater permeability in darkness, and consequently it was thought that closure resulted from a loss of osmotically active molecules. This was not necessarily an alternative to the starch $\rightleftharpoons$ sugar hypothesis; Kissilew suggested that carbohydrate changes would help to stabilize the changes in turgor brought about by the permeability mechanism. The case for the hypothesis was later weakened by the completely conflicting evidence that guard cell permeability increased in light (Linsbauer, 1927).

Recently Zelitch (1964) has shown that substances which cause an increase in the permeability of cell membranes can lead to stomatal closure in light, and thus the mechanism originally suggested was probably quite feasible. There is, however, the problem of how osmotically active molecules lost from the guard cells in darkness could be regained in the light—a process of active uptake of the molecules might have to participate.

The turgor change—other possible mechanisms

A stomatal opening movement is achieved when the guard cells increase their turgor sufficiently relative to that of the subsidiary cells. If osmotically active molecules are involved in producing this turgor difference, they must either be transported against a gradient from the subsidiary cells into the guard cells, or they must be produced or released within the guard cells from a source already there. Under normal circumstances the only substance of any significance likely to gain access to the guard cells other than via the subsidiary cells is carbon dioxide (only when leaves are wet can water and dissolved substances enter via the outer walls). From the accounts on pp. 318 and 319, it will be clear that experimental evidence rules out the possibility that substances manufactured from carbon dioxide within the guard cells play a major part in opening in light, although it would seem that they probably make some contribution, and consequently one is compelled to think in terms of the other possibilities.

We discussed above the suggestion that starch $\rightarrow$ sugar conversion provides the motive force for opening. There are other possible ways in which the turgor increase might be produced from sources within the guard cells. Scarth (1929) suggested that opening in light was due to a rise of guard cell pH which caused swelling of amphoteric colloids in the vacuole (they could equally well have been in the cytoplasm) and that 'night opening' was due to the fall of guard cell pH taking them past the isoelectric point on the acid side so that they swelled again. Such imbibition by the colloids would lower the water potential in the guard cells and cause inflow of water from the subsidiary cells. However, the latter part of the hypothesis was disproved by Mansfield and Heath's (1963) observation that if stomata in a state of 'night opening' were illuminated they did not at first close, with shrinkage of the colloids to their isoelectric point, and then reopen but opened rapidly and continuously. Scarth's ingenious hypothesis unfortunately lacked direct evidence to support it and the disproof of the part concerned with night opening casts doubt on the whole.

Although the guard cells seem to be somewhat isolated from the other epidermal cells in having few or no plasmodesmatal connections, there is no doubt that movement of substances between guard cells and epidermis can take place. Sugars are transported along the epidermis to the guard cells (Pallas, 1964) and substances sprayed on to leaves have been shown to reach the epidermal cells after being taken in by the guard cells (Sargent and Blackman, 1962; Franke, 1964). If substances were transported against a concentration or electrochemical gradient into the guard cells, energy would be required to effect the process which would then be termed 'active'. There is some evidence that stomatal opening may be an energy-requiring process. Heath and Orchard (1956) found that lack of oxygen slowed down opening and accelerated closure in wheat, and Walker and Zelitch (1963) have shown for tobacco that there is an oxygen requirement for opening in light. The latter authors carried out experiments with sodium azide, an inhibitor of respiration, but the results are rather difficult to interpret. There was little doubt that it inhibited the opening process, but it could also prevent closure, depending on the concentration (see also p. 318). Stomatal closure occurs in the absence of oxygen (Heath and Orchard, 1956; Walker and Zelitch, 1963) and consequently it is most unlikely that this is an active process (Williams, 1954). Zelitch (1963) has postulated that ATP manufactured in light by the system outlined on p. 319 would maintain a 'pumping' of water into the guard cells, but this must be regarded as highly speculative in view of the evidence available (Meidner and Mansfield, 1965). Allaway and Mansfield (1967) found that stomata on leaves in which photophosphorylation was strongly inhibited could open widely in response to a low carbon dioxide concentration, which suggests that ATP formed directly from photophosphorylation is not essential in stomatal opening. On balance, the evidence seems to suggest the participation of an active transport mechanism using ATP derived from oxidative phosphorylation. It has long been known that ions of group I metals stimulate opening of stomata on epidermal strips, and recently, evidence has come to light indicating that when epidermal strips are immersed in a solution containing potassium ions, these accumulate in the guard cells as the stomata open (Fujino, 1967; Fischer, 1968). Thus, it seems a real possibility that ionic movements are responsible for increasing the osmotic potential of guard cells in the intact leaf, although so far we have no direct evidence of this. Even if such evidence is forthcoming, a big physiological problem will remain: how do the environmental factors that stimulate stomatal opening, for example low carbon dioxide concentrations and blue light, stimulate active transport into the guard cells?

The mechanism of action of carbon dioxide—dark fixation?

The most promising key to the stomatal mechanism appears to be the response to carbon dioxide. Stomata close markedly in response to increases in carbon dioxide concentration at physiological levels, and another process in the plant which shows great sensitivity to the same range of concentrations is dark fixation (Bonner and Bonner, 1948). Dark fixation of carbon dioxide is a process which occurs very widely in plant tissues, being especially important in succulent plants of the *Crassulaceae* where the amounts of acid formed in darkness can be considerable. There are several reactions by which fixation of carbon dioxide into organic acids can take place. The following, catalysed respectively by malic enzyme, isocitric enzyme, and phosphoenolpyruvic carboxylase could be involved:

(a)   pyruvate + CO_2 + $NADPH_2$ $\rightleftharpoons$ malate + NADP

(b)   α-oxoglutarate + CO_2 + $NADPH_2$ $\rightleftharpoons$ isocitrate + NADP

(c)   phosphoenolpyruvate + CO_2 + H_2O $\rightleftharpoons$ oxaloacetate + inorganic phosphate

The equilibrium position of reaction (a) favours decarboxylation and the incorporation of carbon dioxide requires high concentrations of both pyruvate and carbon dioxide. A similar difficulty applies to reaction (b). Reaction (c) is more promising since there is strong evidence that carbon dioxide is fixed by this means in members of the Crassulaceae, malate being subsequently formed from oxaloacetate. Walker (1966) has pointed out that by this means 18 molecules of carbon dioxide could be incorporated into 12 molecules of malate, but that the energy required for this would be 42 ATP equivalents. Thus energy would be consumed if carbon dioxide incorporation into malate took place in guard cells commencing with reaction (c). If in darkness this energy had to be provided by oxidative phosphorylation a dependence of stomatal closure on oxygen might be expected, but there is evidence that stomatal closure can proceed in the absence of oxygen (see p. 323). It appears at first sight therefore that observed stomatal responses to carbon dioxide are not readily accommodated by known carboxylation reactions, but the question undoubtedly merits further consideration. The difficulty over the oxygen requirement is not insuperable, for if maintenance of stomatal opening were to depend upon a process requiring energy that process might be stopped by either dark fixation or anaerobic conditions—closure then occurring as a result of a passive loss of guard cell turgor.

Shaw and Maclachlan (1954a) showed by means of autoradiography that dark fixation does occur in guard cells, and they, like Heath (1950), considered that the carbon dioxide effect might operate through this process.

One most important point must be stressed. 'Dark' fixation is probably a misnomer since there is evidence that fixation by the same mechanism can occur in light. Ranson and Thomas (1960) state that carbon dioxide is preferentially used for photosynthesis, and it is only the shortage of carbon dioxide at carboxylation centres which stops or retards the 'dark' process. Guard cells are not rich in chlorophyll and with photosynthesis operating less efficiently, 'dark' fixation may remain relatively active in guard cells in the light, and consequently there would be no difficulty in explaining guard cell responses to carbon dioxide in light in terms of dark fixation.

One interesting corollary to the hypothesis is that increases in the amount of dark fixation would lead to the production of organic acids in the guard cells, and there have been several observations of reduced pH accompanying stomatal closure in the dark (see p. 320).

The review by Meidner and Mansfield (1965) should be consulted for a slightly more detailed account of the 'dark fixation hypothesis'. So far it has received comparatively little attention in the literature, most of the discussion relating to it being largely speculative. This therefore represents one of the lines of investigation which may prove profitable in future researches into stomatal physiology.

Multi-system control

In this section postulated mechanisms have been discussed primarily with reference to light, but some of them would be operated also by changes in the other important external factors: carbon dioxide concentration and temperature. Further, cell turgor is not only an important component in the mechanics of stomatal movement (p. 303) but also an 'internal factor' operating other parts of the mechanism (by changing the rate of net photosynthesis and so altering internal carbon dioxide concentrations—p. 311); it is itself controlled by the two external factors of water potential in the soil and vapour pressure deficit in the air and also by the resistance to water movement within the plant. Of the factors: light, internal carbon dioxide concentration, leaf temperature, and water-strain, light intensity is the only one that is unaffected by changes in the others and internal carbon dioxide concentration is especially dependent upon changes in the other three. Many of these effects of one factor on another occur because of changes in stomatal aperture and in turn cause further changes in aperture. This may be expected to result in 'negative feedback' in many cases and so prevent sudden very wide stomatal opening or complete closure. For example, an opening movement in light will reduce the carbon dioxide gradient across the stomata (increasing the carbon dioxide concentration in the substomatal cavities), increase transpiration and lower leaf temperature. The first two will tend to cause closure and the last opening or closure according to the type of leaf and probably the light intensity. A fifth factor which should be mentioned is time, that most independent of independent variates; this affects the course of endogenous rhythms and also the ageing of leaves which results in changes of photosynthetic activity and stomatal responsiveness.

Some of these different factors apparently act on the same stomatal mechanism, either directly or indirectly. For instance there is good evidence that light causes stomatal opening by reducing the carbon dioxide concentration in the guard cells and that very low carbon dioxide concentration causes disappearance of guard cell starch. Blue light, which is very efficient in causing opening independently of carbon dioxide also causes disappearance of starch whereas red (which is more efficient in photosynthesis) does not (Mouravieff, 1958). If it is assumed that disappearance of starch is accompanied by an increase of sugar it would seem that a starch $\rightleftharpoons$ sugar mechanism must account for at least some of the light effect—unless indeed there were no change in osmotic pressure because phosphorylase was involved (p. 321) but Mouravieff (1958) has found concomitant increases in osmotic pressure (p. 312). However, by suitable choice of times of day it is possible to observe smaller opening in darkness than in light accompanied by lower guard cell starch contents (Heath, 1949); in onion leaves light responses occur in the complete absence of guard cell starch (Heath, 1952). The starch $\rightleftharpoons$ sugar mechanism cannot therefore be the only one involved in opening responses to light. The low-intensity light effect which controls endogenous rhythms and is most effective at a wavelength of 700 nm must surely involve yet another mechanism. There seems little doubt that there are several mechanisms which not only vary in importance in different species, but are operated to different extents by various factors, both internal and external. An attempt to show some of these postulated effects diagrammatically is presented in Fig. 9.9. Some of these effects are extremely hypothetical and others based on good evidence; it seems obvious, however, that the stomatal opening achieved under any set of circumstances will be the resultant of many different effects and that the only hope of disentangling the responses to the various factors sufficiently to yield information on the different mechanisms is to carry out factorial experiments under very carefully controlled conditions.

Stomatal behaviour under natural conditions

We saw earlier (pp. 309–318), and have just discussed, how stomatal movements are elicited by a variety of environmental factors. The way in which the course of stomatal behaviour is determined in a plant growing under natural conditions must necessarily pose a complex problem, because of the many possible interactions between different factors. There has so far been very little specific investigation of these interactions.

As an example of the complexity of the situation we can point to stomatal opening in the morning under natural conditions. We know that blue light and

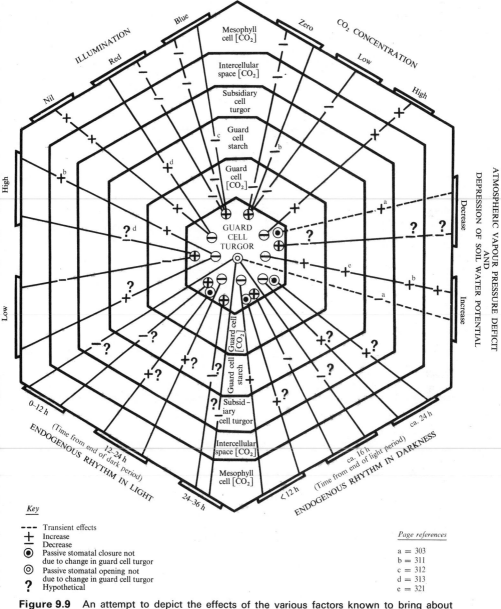

Figure 9.9 An attempt to depict the effects of the various factors known to bring about stomatal movements.

Key

--- Transient effects
+ Increase
− Decrease
⊙ Passive stomatal closure not due to change in guard cell turgor
◎ Passive stomatal opening not due to change in guard cell turgor
? Hypothetical

Page references

a = 303
b = 311
c = 312
d = 313
e = 321

carbon dioxide removal by photosynthesis make independent contributions to stomatal opening in light. In addition, stomatal opening occurs (even in darkness) under control of the endogenous rhythm, and wide temperature-induced opening in darkness is also possible. All these will contribute to the actual opening movement taking place in the morning, but their effects are most unlikely to be simply additive and the movement will be the result of a complex interaction between them.

These and other complexities have frequently been overlooked. The simple assumption that 'stomata open in light and close in darkness' has unfortunately not been confined to elementary textbooks; even authors of advanced texts, review and research papers have failed to recognize that stomata can open in darkness and close in light under both endogenous and environmental control.

It is beyond the scope of this chapter to discuss in detail the question of stomatal behaviour under natural conditions and there is, in any case, little useful information in the literature on which to base any discussion. We hope the evidence we have presented will lead the student to realize that the problems of stomatal physiology are of great complexity, and that simple all-embracing statements cannot be made concerning behaviour under natural conditions, nor are the problems of the mechanisms behind the movements likely to be open to simple solution.

Bibliography

Selected further reading

Heath, O. V. S., 1959a. The water relations of stomatal cells and the mechanisms of stomatal movement. In *Plant Physiology, a treatise*, ed. F. C. Steward, Vol. II. Academic Press, New York, pp. 193–250 and 727–730.

Heath, O. V. S., 1959b. Light and carbon dioxide in stomatal movements. In *Handbuch der Pflanzenphysiologie*, ed. W. Ruhland, Vol. 17/1. Springer, Berlin, pp. 415–64.

Meidner, H. and T. A. Mansfield, 1965. Stomatal responses to illumination. *Biol. Rev.* **40**, 483–509.

Meidner, H. and T. A. Mansfield, 1968. *Physiology of Stomata*. McGraw-Hill, London.

Ketellapper, H. J., 1963. Stomatal physiology. *Ann. Rev. Plant. Physiol.* **14**, 249–270.

Papers cited in text

Allaway, W. G. and T. A. Mansfield, 1967. Stomatal responses to changes in CO_2 concentration in leaves treated with 3-(4-chlorophenyl)-1, 1-dimethylurea. *New Phytol.* **66**, 57–63.

Alvim, P. de T., 1949. Studies on the mechanism of stomatal behaviour. *Am. J. Bot.* **36**, 781–91.

Alvim, P. de T., 1965. A new type of porometer for measuring stomatal opening and its use in irrigation studies. In *Symp. on 'Methodology of Eco-Physiology'*. U.N.E.S.C.O., Montpelier, France, 1962, pp. 325–329.

van Bavel, C. H. M., F. S. Nakayama and W. L. Ehrler, 1965. Measuring transpiration resistance of leaves. *Plant Physiol.* **40**, 535–540.

Bonner, W. and J. Bonner, 1948. The role of carbon dioxide in acid formation by succulent plants. *Am. J. Bot.* **35**, 113–117.

Brun, W. A., 1962. Rhythmic stomatal opening responses in banana leaves. *Physiol. Plant.* **15**, 623–630.

Butt, V. S. and M. Peel, 1963. The participation of glycollate oxidase in glucose uptake by illuminated *Chlorella* suspensions. *Biochem. J.* **88**, 31P.

Darwin, F., 1898. Observations on stomata. *Phil. Trans. Roy. Soc.* (Lond.) **B190**, 531–621.

Darwin, F. and D. F. M. Pertz., 1911. On a new method of estimating the aperture of stomata. *Phil. Trans. Roy. Soc.* (Lond.) **B84**, 136–154.

Fischer, R. A., 1968. Stomatal opening: role of potassium uptake by guard cells. *Science* **160**, 784.

Franke, W., 1961. Ectodesmata and foliar absorption. *Am. J. Bot.* **48**, 683–691.

Franke, W., 1962. Dienen Ektodesmen als Transportbahnen bei Stoffaufnahme und Stoffabgabe der Blätter? *Umschau*, **62**, 501–504.

Franke, W., 1964. Role of guard cells in foliar absorption. *Nature, Lond.* **202**, 1236–1237.

Freudenberger, H., 1940. Die Reaktion der Schliesszellen auf Kohlensäure und Sauerstoffentzug. *Protoplasma*, **35**, 15–54.

Fujino, M., 1967. Role of adenosinetriphosphate and adenosinetriphosphatase in stomatal movement. *Sci. Bull. Fac. Educ., Nagasaki Univ.* **18**, 1–47.

Glinka, Z. and Meidner, H., 1968. The measurement of stomatal responses to stimuli in leaves and leaf discs. *J. Exp. Bot.* **19**, 152–166.

Gregory, F. G. and H. L. Pearse, 1934. The resistance porometer and its application to the study of stomatal movement. *Proc. Roy. Soc.* (Lond.) **B114**, 477–493.

Gregory, F. G. and H. L. Pearse, 1937. The effect on the behaviour of stomata of alternating periods of light and darkness of short duration. *Ann. Bot.* N.S. **1**, 3–10.

Hagen, F., 1918. Zur Physiologie des Spaltöffnungsapparates. *Beitr. allgem. Bot.* **1**, 260–291.

Hanes, C. S., 1940. Enzymatic synthesis of starch from glucose-1-phosphate. *Nature, Lond.* **145**, 348.

Heath, O. V. S., 1938. An experimental investigation of the mechanism of stomatal movement, with some preliminary observations upon the response of the guard cells to 'shock'. *New Phytol.* **37**, 385–395.

Heath, O. V. S., 1941. Experimental studies of the relation between carbon assimilation and stomatal movement, II. The use of the resistance porometer in estimating stomatal aperture and diffusive resistance, Part I. *Ann. Bot.* N.S. **5**, 455–500.

Heath, O. V. S., 1949. Studies in stomatal behaviour. II. The role of starch in the light response of stomata, Part I. *New Phytol.* **48**, 186–211.

Heath, O. V. S., 1950. Studies in stomatal behaviour. V. The role of carbon dioxide in the light response of stomata, Part I. *J. Exp. Bot.* **1**, 29–62.

Heath, O. V. S., 1952. Studies in stomatal behaviour. II. The role of starch in the light response of stomata, Part 2. *New Phytol.* **51**, 30–47.

Heath, O. V. S., 1959a. The water relations of stomatal cells and the mechanisms of stomatal movement. In *Plant physiology, a treatise*, ed. F. C. Steward, Vol. II. Academic Press, New York, pp. 193–250.

Heath, O. V. S., 1959b. Light and carbon dioxide in stomatal movements. *Handbuch der Pflanzenphysiologie*, ed. W. Ruhland, Vol. 17/1. Springer, Berlin, pp. 415–464.

Heath, O. V. S. and T. A. Mansfield, 1962. A recording porometer with detachable cups operating on four separate leaves. *Proc. Roy. Soc.* (Lond.) **B156**, 1–13.

Heath, O. V. S., T. A. Mansfield, and H. Meidner, 1965. Light-induced stomatal opening and the postulated role of glycollic acid. *Nature, Lond.* **207**, 960–962.

Heath, O. V. S. and H. Meidner, 1961. The influence of water strain on the minimum intercellular space carbon dioxide concentration and stomatal movements in wheat leaves. *J. Exp. Bot.* **12**, 226–242.

Heath, O. V. S. and F. L. Milthorpe, 1950. Studies in stomatal behaviour. V. The role of carbon dioxide in the light response of stomata, Part 2. *J. Exp. Bot.* **1**, 227–243.

Heath, O. V. S. and B. Orchard, 1956. Studies in stomatal behaviour. VII. Effects of anaerobic conditions upon stomatal movements—a test of Williams's hypothesis of the stomatal mechanism. *J. Exp. Bot.* **7**, 313–325.

Heath, O. V. S. and B. Orchard, 1957. Midday closure of stomata. *Nature, Lond.* **180**, 180–181.

Heath, O. V. S. and J. Russell, 1951. The Wheatstone bridge porometer. *J. Exp. Bot.* **2**, 111–116.

Heath, O. V. S. and J. Russell, 1954a. Studies in stomatal behaviour. VI. An investigation of the light responses of wheat stomata with the attempted elimination of control by the mesophyll. Part I. Effects of light independent of carbon dioxide and their transmission from one part of the leaf to another. *J. Exp. Bot.* **5**, 1–15.

Heath, O. V. S. and J. Russell, 1954b. Studies in stomatal behaviour. VI. An investigation of the light responses of wheat stomata with the attempted elimination of control by the mesophyll. Part 2. Interactions with external carbon dioxide, and general discussion. *J. Exp. Bot.* **5**, 269–292.

Heath, O. V. S. and W. T. Williams, 1948. Studies in stomatal action. Adequacy of the porometer in the investigation of stomatal aperture. *Nature, Lond.* **161**, 178–179.

Henderson, F. Y., 1936. The preparation of 'three-colour' strips for transpiration measurements. *Ann. Bot.* **50**, 321–324.

Iljin, W., 1914. Die Regulierung der Spaltöffnungen im Zusammenhang mit Veränderung des osmotischen Druckes. *Beih. bot. Zbl.* **32**, 15–35.

Kissilew, N., 1925. Veränderung der Durchlässigkeit des Protoplasma der Schliesszellen im Zusammenhange mit stomatären Bewegungen. *Bieh. bot. Zbl.* **41**, 287–308.

Kohl, F. G., 1895. Über Assimilationsenergie und Spaltöffnungsmechanik. *Botan. Centr.* **64**, 109–110.

Kuiper, P. J. C., 1961. The effects of factors on the transpiration of leaves, with special reference to the stomatal light response. Meded. Landbouhoogesch. *Wageningen*, **61**, 1–49.

Linsbauer, K., 1916. Beiträge zur Kenntnis der Spaltöffnungsbewegungen. *Flora*, N.F. **9**, 100–143.

Linsbauer, K., 1927. Weitere Beobachtungen an Spaltöffnungen. *Planta* **3**, 527–561.

Lloyd, F. E., 1908. The Physiology of stomata. *Carnegie Inst. Wash. Publ.* **82**, 1–142.

Livingston, B. E. and Shreve, E. R., 1961. Improvements in the method for determining the transpiring power of plant surfaces by hygrometric paper. *Plant World*, **19**, 287–309.

Loftfield, J. V. G., 1921. The behaviour of stomata. *Carnegie Inst. Wash. Publ.* **314**, 1–104.

Mansfield, T. A., 1964. A stomatal light reaction sensitive to wave-lengths in the region of 700 mμ. *Nature, Lond.* **201**, 470–472.

Mansfield, T. A., 1965a. Responses of stomata to short duration increases in carbon dioxide concentration. *Physiol. Plant.* **18**, 79–84.

Mansfield, T. A., 1965b. Studies in stomatal behaviour. XII. Opening in high temperature in darkness. *J. Exp. Bot.* **16**, 721–731.

Mansfield, T. A., 1965c. The low-intensity light reaction of stomata—effects of red light on rhythmic stomatal behaviour in *Xanthium pennsylvanicum*. *Proc. Roy. Soc.* (Lond.) **B162**, 567–574.

Mansfield, T. A., 1965d. Glycollic acid metabolism and the movements of stomata. *Nature, Lond.* **205**, 617–618.

Mansfield, T. A. and O. V. S. Heath, 1961. Photoperiodic effects on stomatal behaviour in *Xanthium pennsylvanicum*. *Nature, Lond.* **191**, 974–975.

Mansfield, T. A. and O. V. S. Heath, 1963. Studies in stomatal behaviour. IX. Photoperiodic effects on rhythmic phenomena in *Xanthium pennsylvanicum. J. Exp. Bot.* **14**, 334–352.

Meidner, H., 1965a. Stomatal control of transpirational water loss. *Symp. Soc. Exp. Biol.* **19**, 185–203.

Meidner, H., 1965b. A simple porometer for measuring the resistance to air flow offered by the stomata of green leaves. *School Science Review*, **47**, 149–151.

Meidner, H. and O. V. S. Heath, 1959. Stomatal responses to temperature and carbon dioxide concentration in *Allium cepa* L. and their relevance to mid-day closure. *J. Exp. Bot.* **10**, 206–219.

Meidner, H. and T. A. Mansfield, 1965. Stomatal responses to illumination. *Biol. Rev.* **40**, 483–509.

Meidner, H. and D. C. Spanner, 1959. The differential transpiration porometer. *J. Exp. Bot.* **10**, 190–205.

Milthorpe, F. L., 1955. The significance of the measurement made by the cobalt chloride paper method. *J. Exp. Bot.* **6**, 15–19.

Mohl, H. von. 1856. Welche Ursachen bewirken die Erweiterung und Verengung der Spaltöffnungen? *Bot. Z.* **14**, 697–704, 713–720.

Mouravieff, I., 1954. Actions des solutions tampons et de leur constituents sur les cellules stomatiques d'*Aponogeton distachyus* L. *Ann. de l'Univ. de Lyon.* Sect. C. **8**, 87–115.

Mouravieff, I., 1956. Expériences avec les stomates maintenus sur les milieux complexes. *Botaniste*, **40**, 195–212.

Mouravieff, I., 1958. Action de la lumière sur la cellule végétale. Production de mouvement d'ouverture stomatique par la lumière des diverses régions du spectre. *Bull. Soc. Bot. Fr.* **105**, 467–475.

Mouravieff, I., 1959. Observations sur la plasmolyse des cellules annexes de l'appareil stomatique. *Bull. Soc. Bot. Fr.* **106**, 301–305.

Pallas, J. E., 1964. Guard cell starch retention and accumulation in the dark. *Bot. Gaz.* **125**, 102–107.

Parkin, J., 1899. Contributions to our knowledge of the formation, storage and depletion of carbohydrates in monocotyledons. *Phil. Trans. Roy. Soc.* (Lond.) **B191**, 35–79.

Ranson, S. L. and M. Thomas., 1960. Crassulacean acid metabolism. *Ann. Rev. Plant Physiol.* **11**, 81–110.

Sargent, J. A. and G. E. Blackman, 1962. Studies on foliar penetration. I. Factors controlling the entry of 2,4-dichlorophenoxyacetic acid. *J. Exp. Bot.* **13**, 348–368.

Sayre, J. D., 1926. Physiology of stomata of *Rumex patientia. Ohio J. Sci.* **26**, 233–266.

Scarth, G. W., 1929. The influence of H ion concentration on the turgor and movement of plant cells with special reference to stomatal behaviour. *Proc. Int. Congr. Plant. Sci. Ithaca,* **1926**, 1151–1162.

Scarth, G. W., 1932. Mechanism of the action of light and other factors on stomatal movement. *Plant Physiol.* **7**, 481–504.

Scarth, G. W. and M. Shaw, 1951. Stomatal movement and photosynthesis in *Pelargonium.* I. Effects of light and carbon dioxide. *Plant Physiol.* **26**, 207–225.

Schwendener, S., 1881. Über bau und Mechanik der Spaltöffnungen. *Monatsber. königlichen Akad. Wiss. Berlin.* **46**, 833–867.

Schwendener, S., 1889. Die Spaltöffnungen der Gramineen and Cyperaceen. *S.B. preuss, Akad. Wiss. Berlin,* **6**, 1–15.

Shaw, M. and G. A. Maclachlan, 1954a. Chlorophyll content and carbon dioxide uptake of stomatal cells. *Nature, Lond.* **173**, 29–30.

Shaw, M. and G. A. Maclachlan, 1954b. The physiology of stomata. I. Carbon dioxide fixation in guard cells. *Can. J. Bot.* **32**, 784–794.

Sheffield, F. L. M., 1936. The role of plasmodesm in the translocation of virus. *Ann. Appl. Biol.* **23**, 506–508.

Small, J. and K. M. Maxwell, 1939. pH phenomena in relation to stomatal opening. I. Coffea arabica and some other species. *Protoplasma*, **32**, 272–288.

Small, J., M. I. Clarke, and J. Crosbie-Baird, 1942. pH phenomena in relation to stomatal opening. II–V. *Proc. Roy. Soc. Edin.* **B61**, 233–266.

Spanner, D. C., 1953. On a new method for measuring the stomatal aperture of leaves. *J. Exp. Bot.* **4**, 283–295.

Stålfelt, M. G., 1955. The stomata as a hydrophotic regulator of the water deficit of the plant. *Physiol. Plant.* **8**, 572–593.

Stålfelt, M. G., 1963. Diurnal dark reactions in the stomatal movements. *Physiol. Plant.* **16**, 756–766.

Stålfelt, M. G., 1964. Reactions participating in the photoactive opening of stomata. *Physiol. Plant.* **17**, 828–838.

Thomas, J. B., 1949. Experiments on the water household of tropical plants. V. Water balance and stomatal movements in *Cissus sicyoides* L. *Ann. Bot. Gardens, Buitenzorg*, **5**, 167–176.

Tolbert, N. E., 1958. Secretion of glycollic acid by chloroplasts. *Brookhaven Symposia in Biology*, **11**, 271.

Tregunna, E. B., G. Krotkov, and C. D. Nelson, 1964. Further evidence on the effects of light on respiration during photosynthesis. *Can. J. Bot.* **42**, 989–997.

Virgin, H. I., 1956. Light-induced stomatal movements in wheat leaves recorded as transpiration. Experiments with the corona-hygrometer. *Physiol. Plant.* **9**, 280–303.

Walker, D. A., 1966. Carboxylation in plants. *Endeavour*, **25**, 21–26.

Walker, D. A. and I. Zelitch, 1963. Some effects of metabolic inhibitors, temperature and anaerobic conditions on stomatal movement. *Plant Physiol.* **38**, 390–396.

Williams, W. T., 1948. Studies in stomatal behaviour. I. Stomatal movement induced by heat-shock stimuli, and the transmission of such stimuli across the leaves of *Pelargonium zonale*. *Ann. Bot.* N.S. **12**, 35–51.

Williams, W. T., 1952. Studies in stomatal behaviour. II. The role of starch in the light response of stomata. Part 3. *J. Exp. Bot.* **3**, 110–127.

Williams, W. T., 1954. A new theory of the mechanism of stomatal movement. *J. Exp. Bot.* **5**, 343–352.

Williams, W. T. and F. A. Barratt, 1954. The effect of external factors on stomatal starch. *Physiol. Plant.* **7**, 298–311.

Williams, W. T. and G. S. Spencer, 1950. Quantitative estimation of stomatal starch. *Nature, Lond.* **166**, 34.

Wilson, C. C., 1948. The effect of some environmental factors on the movements of guard cells. *Plant Physiol.* **23**, 5–37.

Yemm, E. W. and A. J. Willis, 1954. Chlorophyll and photosynthesis in stomatal guard cells. *Nature, Lond.* **173**, 726.

Yin, H. C. and Y. T. Tung, 1948. Phosphorylase in guard cells. *Science*, **108**, 87–88.

Zelitch, I., 1961. Biochemical control of stomatal opening in leaves. *Proc. Nat. Acad. Sci. Wash.* **47**, 1423–1433.

Zelitch, I., 1963. The control and mechanism of stomatal movement. In Stomata and water relations in plants. *Conn. Agric. Expt. Sta. Bull.* **664**, 18–42.

Zelitch, I., 1964. Reduction of transpiration of leaves through stomatal closure induced by alkenylsuccinic acids. *Science*, **143**, 692–693.

Zucker, M., 1963. Experimental morphology of stomata. *Conn. Agric. Expt. Sta. Bull.* **664**, 1–17.

10. Photosynthesis

Bessel Kok

Introduction

Photosynthesis is the light-driven evolution of O_2 from water and the storage of the resulting reducing power in the numerous carbon components which constitute living matter. Chlorophyll a and accessory pigments, which absorb about half of the solar radiation ($\lambda < 700$ nm), sensitize two primary energy conversions in two different photosystems. In each system many 'harvesting' pigment molecules feed their excitations into a photochemical conversion centre. In these, the energy of the photons is used to move electrons against the thermogradient: in photosystem I from an agent with medium reducing power to one with strong reducing power, thermodynamically capable to reduce CO_2; in photosystem II from water with release of O_2, to an agent with mild oxidizing power. The two photoproducts of intermediate potential react together. In addition, some of the light energy is retained in a high energy state which can yield ATP.* This sequence of events is illustrated in the scheme of Fig. 10.1, used as a framework of discussion in this chapter. The four primary products generated in the two photoacts are processed in consecutive enzymatic steps. Various chloroplast intermediates known to participate in these steps are also shown in the scheme of Fig. 10.1, arranged according to their respective midpoint potentials. The metabolic end products of the energy conversion system: NADPH and ATP are utilized in the reduction of CO_2. The latter process and many other facets of photosynthesis are given only scant attention in this chapter, which is largely restricted to biochemical and physiological aspects of the conversion of light energy.

Primary events †

Absorption

To interact with matter, light must be absorbed. To this end the photosynthetic apparatus is equipped with densely packed layers of light-gathering pigments (Smith and French, 1963). Predominant are the chlorophylls, characterized by

---

* Abbreviations: ADP, adenosine diphosphate; ATP, adenoside triphosphate; CMU, 3-(4-chloro phenyl)–1,1–dimethylurea; DADH$_2$, 2,3,5,6-tetramethyl-p-phenylenediamine; DCMU, 3-(3,4-dichlorophenyl)-1,1-dimethylurea; DPIP, 2,6-dichlorophenolindophenol; EDTA, ethylenediaminetetraacetic acid; GAP, glyceraldehyde-3-phosphate; NADP, nicotinamide adenine dinucleotide phosphate; NADPH, reduced form of NADP; PGA, phosphoglyceric acid; PMS, phenzaine methosulphate; RuDP, ribulose diphosphate.

† See Clayton (1965) Duysens (1964).

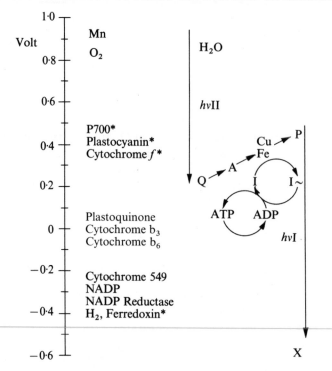

Figure 10.1   Two series-connected photoacts (arrows) drive reducing power from water to weak reductant Q in system II and from intermediate P to strong reductant X in system I. The scale on the left shows potential changes involved and the midpoint potentials of various inter-mediates. Electron transfer from Q → P is mediated by iron-containing cytochrome(s) (Fe), plastoquinone (A) and copper-containing plastocyanin (Cu). Intermediates labelled with an asterisk occur in a concentration equal to that of the photochemical trapping centres, i.e., a concentration of about 1 mole per 400 moles of chlorophyll. Following reduction of X, electron transfer from X → NADP is mediated by ferredoxin and NADP reductase. High energy inter-mediate I∼, generated by the light-driven electron transport, yields ATP from ADP.

two strong absorption bands: one in the red, the other in the blue, which represent the first and second excited singlet state respectively. Weaker bands occur between these two major peaks and in the near UV (see Fig. 10.2). The exact wavelength location of the absorption bands varies not only with the chemical structure of the molecule (chlorophyll *a*, *b*, *c*, etc.) but also with its environment and density of packing. In various organic solvents the red absorption maximum of chlorophyll *a* ranges between 660 and 670 nm; *in vivo* it occurs in several functionally different binding states so that the red band is broad and complex (Fig. 10.3).

The absorption power of a pigment at a given wavelength is characterized by its extinction coefficient, which has the dimension of a 'capture area'. The molar extinction of the main absorption bands of chlorophyll *a* and *b* is about 10^5 l

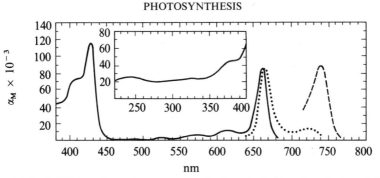

Figure 10.2 Solid line: absorption spectrum of chlorophyll *a* in ether (Holt and Jacobs, 1954). Dashed line: red absorption band of chlorophyll *a* in microcrystalline state, note the shift of 80 nm towards longer wavelength. Dotted line: fluorescence emission spectrum of the ether solution.

mole-cm^{-1}. Similarly high extinction coefficients characterize other photosynthetic pigments such as the carotenoids, a variety of which occurs in all photosynthetic organisms and the bilin pigments which occur as accessory pigments in blue-green and in red algae.

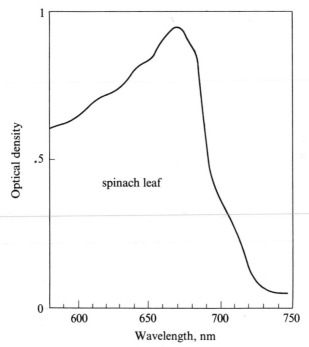

Figure 10.3 Absorption spectrum of a whole spinach leaf, cooled to $-196°$C. Note maxima due to chlorophyll *b* at 650 nm, two major binding states of chlorophyll *a* ($\sim$670 and 680 nm respectively) and a minor, long wave chlorophyll *a* fraction ($\sim$705 nm).

When a pigment molecule absorbs a photon, it goes into an excited state, because one of its electrons at ground state is moved to a different orbital. The difference in energy of the two orbital levels is exactly equal to the energy of the absorbed photon. According to Planck's formula, $E = hv$ (in which $h = 6.55 \times 10^{-27}$ erg-s and v is the frequency of the light) the energy of a photon is inversely proportional to the wavelength. In photochemical reactions single quanta interact with single molecules and it is convenient to know the energy represented by a mole (Avogadro's number) of quanta, or an Einstein. An Einstein of blue (~ 430 nm) light represents about 60 kcal, one of red (~ 700 nm) light only 40 kcal. In photosynthesis under optimal conditions slightly more than two Einsteins are used to move an equivalent of reducing power from water to CO_2, i.e., the quantum requirement is $\geqslant 2\ hv/\text{eq}$ or the quantum efficiency $\leqslant 0.5$ eq/hv (or about 0.1 oxygen molecules $(hv)^{-1}$).

Light absorbed in the blue (Soret) band of chlorophyll yields an extremely unstable excited state (the second singlet). In less than 10^{-12} s this state transforms into the first excited singlet, the one corresponding to the red absorption band—which has a much longer lifetime ($\sim 10^{-9}$ s). About one-third of the energy of blue light is then degraded into heat within the chlorophyll molecule. Similarly, with a corresponding loss of energy, all light absorbed in the other, weaker bands of chlorophyll ends up in the lowest singlet state.

From this state the excitation energy can leave the molecule in one of several possible ways. (a) *Radiationless transition* to the ground state. In this case the energy is wasted as heat, i.e., vibration of the molecule and its surroundings. (b) *Fluorescence*, and (c) by *transfer* to a neighbouring molecule.

Fluorescence

In fluorescence the excited electron returns to its ground state with emission of a light quantum of nearly the same, but in general, a slightly lower energy (longer wavelength) than that of the excited state. Since the transitions from other excitation levels to the lowest excited singlet occur extremely rapidly ($\sim 10^{-12}$ s), the chance is negligible that fluorescence emission occurs from any but the much longer-lived red absorption band (10^{-8} to 10^{-9} s). Therefore, regardless of what wavelength is used to irradiate a solution of chlorophyll or a chloroplast suspension, the fluorescence is always red ($\lambda > 680$ nm, see Fig. 10.2). For this same reason one can assume that all photochemical events are initiated by the first singlet excited state; all light used in photosynthesis is 'red' (680–700 nm).

Transfer

The third way in which an excitation can leave a molecule is transfer of the excited state, to a neighbouring pigment molecule. Transfer of excitation energy can occur between like as well as unlike molecules—as long as there is close proximity and the receiver absorbs at the same or a slightly longer wavelength than the emitter which originally absorbed the energy. There must be a good overlap between the fluorescence spectrum of the donor and the absorption spectrum of the receiver. *In vivo*, or in a dense solution of the two pigments, one can irradiate with a wavelength which is mainly absorbed by chlorophyll *b* (480 or 650 nm *in vivo*) and observe the fluorescence emission of chlorophyll *a* exclusively. The *a* component has a slightly lower excitation level (its red absorption band is at ~ 675 nm) and drains the energy from the *b* component. Transfer in the opposite direction, to a shorter wavelength, is very unlikely since it would require an input of energy to raise the frequency of the quantum. The dense pigment layers in the chloroplast allow efficient (80–100 per cent) energy transfer so that within its lifetime an excited state can migrate through several hundred chlorophylls (Duysens, 1952, 1964). This co-operation of many pigments comprises the so-called 'photosynthetic unit'.

Trapping

Distributed through the pigment bed a few ($\leqslant 0.3$ per cent) special chlorophyll molecules occur: so-called trapping centres. When visited by a photon these attain an excited state which has lower energy than that of the surrounding donor molecules so that the quantum is trapped. The trapping centres are the endpoint of quantum flow and the loci where photochemistry occurs. In effect, the harvesting pigment which collects energy into a trap acts like a lens which increases the intensity 100–300 fold (the number of quanta arriving per unit time) at its focal point, the conversion centre. We thus arrive at the most important possibility: that a quantum initiates a photochemical reaction. Photosynthesis is driven by two different light reactions (denoted I and II) so there must be two types of conversion centres. Possibly these two types of centres are sensitized to some extent by the same harvesting pigment molecules; we will presently assume a complete separation between the two photoacts, each having its own units of pigments and traps.

System I

Photo-oxidant and photoreductant

The scheme of Fig. 10.4 illustrates the primary events in system I as presently conceived. The trap itself is a kind of chlorophyll *a* molecule which, together with its reaction partners, is bound in the chloroplast matrix in a special fashion so that its red absorption band is at a relatively long wavelength; 700 nm (Kok, 1956). 'P700' probably receives most of its energy from a small group (~ 10) of surrounding, long-wave chlorophyll *a* molecules absorbing at 695–700 nm (Butler, 1962). These 'C700' molecules are photochemically inactive but mediate

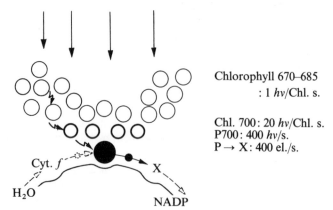

Figure 10.4 Two-step concentration of intensity (10–20 times in each step) in a system I pigment unit. Quanta absorbed by bulk pigment (light circles) flow towards minor pigment C700 (heavy circles) and from these to trap P700 (dot). Upon excitation P700 yields an electron to X, and receives one from ferrocytochrome f (or cupro plastocyanin).

the quantum flow from the shorter-wave absorbing bulk pigments to the trap. Quanta which arrive in this long-wave pigment group cannot escape (either as fluorescence or by returning to the bulk chlorophyll, cf. p. 343). They either find an open P700 trap, or are degraded into heat. Therefore, if the intensity absorbed in system I pigment equals $\alpha_1 I$, the photochemical rate will be

$$R_1 = \alpha_1 I(P^-/P_{total}) \qquad (1)$$

Where α_1 is the fraction of incident intensity I which is absorbed by pigment absorbed in photosystem I.

If a photon arrives at P700 while in its photochemically active (reduced, absorbing) form, the excitation results in a photoreaction: the transfer of an electron from the chlorophyll to an as yet unidentified reaction partner denoted X in Figs. 10.1 and 10.4. The essence of this primary photoact is the stability of the new state in which P is oxidized and X is reduced: an activation barrier prevents the immediate return of the electron to the original, thermodynamically much more likely, state.

In its oxidized state P700 is colourless, both the red and the blue absorption bands have disappeared—it cannot receive another quantum unless it is reduced in a subsequent reaction. This colour change upon oxidation or reduction allows in situ spectroscopic observations of its behaviour. Photobleaching has been observed even at $1°K$—which is indicative of a purely photochemical event.

P700 is a mild oxidant which behaves as a single electron transfer agent. The midpoint potential of the couple P^-/P is $+ 0.43$ V.

X^- is a very strong reductant; the couple X/X^- has a normal potential of $\geqslant -0.55$ V, significantly below that of the hydrogen electrode. Thus, in the photoact, an electron has been driven against the thermogradient and gained nearly 1 V of chemical potential. Since one electron volt corresponds to 23 kcal per mole, we calculate that the 700 nm light, which represents 40 kcal per Einstein (1·8 V), has been converted into chemical potential with an efficiency of about 50 per cent. Both primary products undergo subsequent reactions in the dark, which involve energy losses; thus, regardless of what type of products are ultimately formed, the overall efficiency will be always less than this 50 per cent.

Secondary reductants.   *In vivo* system I reduces carbon dioxide, with the aid of NADPH. The reduction of NADP is mediated by two chloroplast enzymes, ferredoxin and NADP reductase (San Pietro and Black, 1965). Ferredoxin, presumably the natural oxidant of X^- in green plants, is a non-haem iron protein, present in a concentration of $\sim 1/500$ [Chl.]. Its midpoint potential is close to that of the hydrogen electrode $(-0·43$ V) (Arnon, 1963). The NADP reductase is a flavoprotein which also exhibits diaphorase and transhydrogenase activity (Avron and Jagendorf, 1956). Chloroplasts isolated from whole cells can still reduce NADP, provided ferredoxin, which leaves the particles during preparation, is added back to the suspension.

In addition to NADP, isolated chloroplasts can photoreduce a host of components, such as dyes, quinones, and ferricyanide, and including some very low potential agents such as viologen dyes $(E_m < -0·6$ V), without needing ferredoxin.

Reduced low potential dyes are auto-oxidizable, their reaction with molecular oxygen yielding hydrogen peroxide. Equations (2) to (5) illustrate the sequence of events:

$$H_2O \xrightarrow{h\nu} 2[H] + \tfrac{1}{2}O_2 \tag{2}$$

$$2[H] + 2 \text{ Viol.} \longrightarrow 2 \text{ Viol. H} \tag{3}$$

$$O_2 + 2 \text{ Viol. H} \longrightarrow 2 \text{ Viol.} + H_2O_2 \tag{4}$$

$$H_2O_2 \xrightarrow{\text{catalase}} H_2O + \tfrac{1}{2}O_2 \tag{5}$$

$$H_2O_2 + C_2H_5OH \xrightarrow{\text{catalase}} CH_3CHO + 2H_2O \tag{6}$$

If no catalase were present, or if the endogenous catalase activity of the chloroplasts were inhibited by cyanide, the net result of eqs. (2)–(4) would be the uptake of $\tfrac{1}{2}O_2$. If catalase is present in sufficient concentration so that reaction (5) can keep pace with the production of H_2O_2, no net O_2 exchange is observed at all in the light. The addition of ethanol (eq. 6) results in its peroxidatic conversion into acetaldehyde, no O_2 is evolved from H_2O_2 so that again a net uptake of $\tfrac{1}{2}O_2$ results.

This reaction system in which O_2 is the electron acceptor of photosynthetic reducing power is often named the Mehler reaction, (Mehler, 1951); its analysis requires the use of isotope techniques (^{18}O). The simultaneous O_2 uptake and release occurs also in illuminated chloroplasts in the absence of an added auto-oxidizable substrate, be it at a slower rate. This is obviously due to the oxidation of X^- directly or via an endogenous mediator. The same is observed in whole cells under conditions in which the CO_2 reduction cycle fails to keep pace with photochemical reactions, such as in low $[CO_2]$ or during the first moments of illumination after a dark period (p. 360).

Secondary oxidants.   In the chloroplast lamellae two enzymes occur in the same concentration as P700 and closely associated with it: cytochrome f (Davenport and Hill, 1952) and the copper protein plastocyanin (Katoh, 1960). These two components have midpoint potentials (0.37 and 0.39 V respectively) slightly below that of P700 and reduce the latter component after it has been photo-oxidized in the photoact. Light induced oxidation of cytochrome f has been observed at temperatures considerably below freezing, which indicates a purely electronic event, in a complex of rigidly fixed reaction partners (Chance and Nishimura, 1960; Müller and Witt, 1961). This also agrees with the observations that the electron transfer from cytochrome f to P700 can be extremely rapid (<1 ms). The relatively small absorbance change of plastocyanin upon oxidation or reduction makes its direct spectroscopic observation more difficult. In strong light P_{700}, cytochrome f and probably also plastocyanin accumulate in the oxidized state. Although the precise mode of interaction between the three components in the oxidized side of system I is still being debated, present evidence seems to favour the flow sequence Cyt. $\rightarrow$ PC $\rightarrow$ P_{700} (Gorman and Levine, 1966).

System II

Photosystem II evolves oxygen from water and donates electrons to system I. Much of our present knowledge of the early events in system II rests on observations of fluorescence. Unlike the situation in system I, where incoming quanta are converted either into chemical potential (trap open) or into heat (trap closed), a variable fraction, some 3–6 per cent of the light absorbed in system II units escapes as fluorescence. If, within its lifetime, an excitation on its journey through the pigment bed fails to meet an open trap, its chance to be re-emitted as fluorescence is enhanced. With all traps 'open', the rate of photochemistry is high and the fluorescence yield is low; therefore we denote the agent which acts as a trap Q, for Quencher (Duysens, 1964). It proved that fluorescence is low when Q is in its oxidized state, apparently the state required for electron acceptance in the photochemical act (see Fig. 10.6).

One thus would expect the rate of system II to be proportional to the intensity absorbed by its pigment ($\alpha_2 I$) and the fraction of Q which is the oxidized state:

$$R_2 = \alpha_2 I \, Q^+/Q_{\text{total}} \qquad (7)$$

Whereas the corresponding prediction proved valid for system I (eq. 1, p. 340), in system II at low values of Q^+ the observed rates were higher than expected. This may be attributed to differences in photon-trapping mechanisms: In system I, 'C700' molecules prevent exit and a subsequent search for 'open' P700; system II contains no such limitation on siteseeking (Joliot and Joliot, 1964).

O_2 evolution

While excitation of the reaction centre results in the photo-reduction of Q, an unknown electron donor, Z, becomes oxidized. In an as yet unknown fashion, strong oxidant Z^+ is able to liberate oxygen. Photosystem II was shown to be absolutely dependent upon the presence of one and possibly two manganese catalysts occurring in a concentration of 6–8 Mn equivalents per trapping centre (Cheniae and Martin, 1968). Because of the high oxidizing potential of some valency states of this transition metal, we tend to suspect the participation of these states in the O_2 evolving process.

A rapid polarographic procedure revealed that the evolution of oxygen is very fast, the delay between photochemical excitation of system II and the liberation of O_2 being less than a millisecond, (Joliot, 1966).

Allen and Franck (1955) noticed that a brief light flash given after a few minutes of darkness yields no oxygen; however, a second flash given shortly thereafter, does. Joliot (1966) showed that in the dark the O_2 evolution capacity of system II is inactivated (half time ~ 3 s at 25°C in *Chlorella* and much slower in isolated chloroplasts). Upon illumination the first one or two photons produce the reduced form of Q without, however, evolving O_2. Apparently, O_2 evolution occurs by the concerted action of more than one relatively unstable Z^+ equivalents per photochemical centre. Especially at higher temperatures, the rate of the deactivation process becomes appreciable and as a result in very weak light (~ 100 erg cm^{-2} s^{-1}), the rate of quantum flux cannot balance the rate of deactivation and photosynthesis cannot proceed efficiently. Figure 10.5 illustrates how, after a prolonged dark time, light is initially used for activation instead of O_2 evolution, which results in a gradual onset of the rate.

The photoreductants

Primary photoreductant Q—as yet unidentified—occurs in the chloroplast in a concentration of $\sim 1/500$ [Chl.], like the traps of system I. Q appears intimately associated with its immediate reaction partner (electron acceptor) which we denoted A in scheme 1. The concentration of A is (1/30–1/50 [Chl.]), 10–20 fold greater than that of Q; the reaction $Q + A \rightleftharpoons Q^+ + A^-$ is rapid, electron transfer occurs in less than 1 ms if all Q is reduced and all A oxidized. The kinetic behaviour of Q and A suggests a low equilibrium constant of the reaction system, i.e. that the two partners have a similar midpoint potential; redox

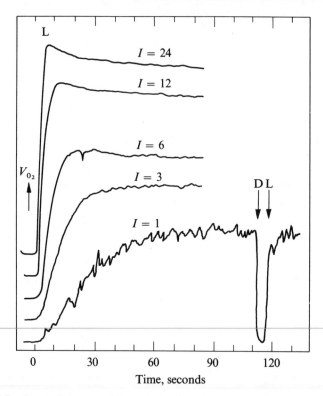

Figure 10.5   Onset of the rate of O_2 evolution in various light intensities after a four-minute dark period (*Chlorella* cells, 30°C). The ordinate V_{O_2} is rate/intensity (relative efficiency). In the weakest light (I = 1) the rate is initially zero and slowly accelerates to the maximal value: activation of system II. In a subsequent brief dark period (10 s) again some deactivation takes place. In higher intensities activation is proportionally faster. Detection of this small O_2 activation loss requires a sensitive technique for measuring O_2 evolution rate (Joliot, 1966).

titration of the fluorescence yield indicated a value $E_m > +200\,\text{mV}$ (Kok *et al.*, 1966). Intermediate A, is probably plastoquinone A: Chloroplasts contain a large amount (~ 30 per per cent) of lipid and lipid-soluble materials (Zill and Cheniae, 1962; Benson, 1964). While most of these seem to serve a structural role, one or more of the several quinones (~ 10 per cent of the total fraction) are involved in electron transport (Ogren *et al.*, 1965). Extraction of chloroplasts with hexane results in a loss of O_2 evolution (Lynch and French, 1957), as well as photophosphorylation. Both activities can be restored by recondensation of the extract, the active component of which proved to be plastoquinone A (Bishop, 1958). This quinone whose midpoint potential *in vitro* is $\simeq 115\,\text{mV}$ (Carrier, 1967) appears to be responsible for the light-induced absorption changes at $\sim 260\,\text{nm}$. Long-wave light causes its oxidation while short-wave light results in its reduction (Amesz, 1964; Witt *et al.*, 1966).

After a period of darkness or far red light, in which pool A has become oxidized, photosystem II can develop, in a 'gush', a stoichiometric amount of oxygen. This gush can be clearly observed in whole algae, where after a dark period the onset of CO_2 reduction is delayed, or in isolated chloroplasts without added oxidant, so that reoxidation of A by system I is slow. During this depletion of oxidized A, quencher Q also becomes reduced, so that the fluorescence rises from an initial minimum to a high value (see Fig. 10.6). In whole cells a subsequent decrease of the fluorescence and increase of O_2 rate occurs, due to the onset of CO_2 reduction which allows system I to reoxidize A and Q. A similar transient has been observed spectroscopially at the 260 nm plastoquinone band (Schmidt-Mende and Witt, 1968).

Besides cytochrome f (α band 555 nm, E_m + 0·370 V) the chloroplast contains two b cytochromes (b_6 and b_3) (α bands 563, 559 nm, E_m − 0·06 V, +0·06 V respectively) (Hill and co-workers, 1953). The concentration of each is about twice that of cytochrome f.

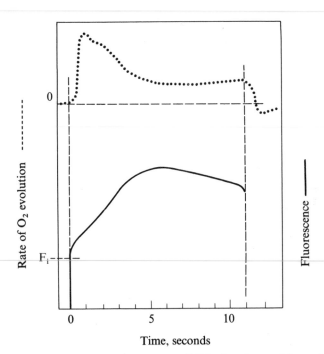

Time, seconds

Figure 10.6   Time course of fluorescence yield (full line) and rate of O_2 evolution (dotted line) observed with *Chlorella* cells, during the first seconds of illumination after two minutes darkness. Oxygen evolution is initially zero, then accelerates to a maximum (activation) and decays again when pool A becomes depleted. The fluorescence is initially low (all Q in the quenching form), the first rise reflects activation, the second rise a simultaneous reduction of A and Q (Delosme *et al.*, 1959).

Several workers found evidence that one of the *b* cytochromes operates close to the reducing side of system II. As yet, however, no clear picture has emerged about the role of the *b* cytochromes (reviewed, e.g., by Hind and Olson, 1968). The early events in system II are far from being resolved, for the moment we may consider pool A as the reduced product of system II; from which electrons flow to system I.

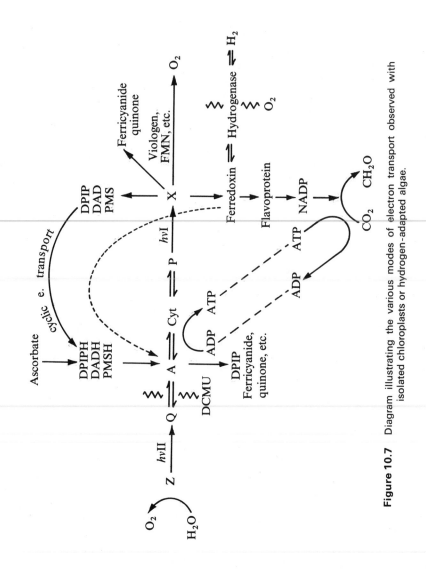

Figure 10.7 Diagram illustrating the various modes of electron transport observed with isolated chloroplasts or hydrogen-adapted algae.

Isolation and co-operation of the two systems

Isolation of system I

System II can be inhibited by numerous poisons, some of the most potent ones being the herbicides DCMU, CMU and s-triazines. Also ageing, brief heating, or detergent will destroy system II preferentially and, to feed electrons to system I, other electron donors instead of water must be used (Fig. 10.7). The most commonly used artificial donor is ascorbic acid, in combination with, e.g., indophenol dye. The dye kept reduced by excess ascorbate, can react more rapidly with the electron carriers of the chloroplast than ascorbic acid itself (Vernon and Zaugg, 1960). In detergent-treated chloroplasts ferrocytochrome c can act as electron donor—in this case either plastocyanin or cytochrome f must be added to carry electrons to $P700^+$. Upon illumination, low-potential oxidants such as NADP or viologen are now reduced at the expense of ascorbate, etc. In the first case, reduced NADP accumulates, in the second case, the reoxidation of photoreduced viologen leads to an uptake of O_2 (eqs. 3 and 4). Under appropriate conditions and in strong light very high rates of electron transport ($\geqslant 5,000$ eq/Chl. hour) can be sustained in these so-called 'open' systems; i.e., systems which drive a net reduction of one and oxidation of another compound.

As also illustrated in Fig. 10.7, partially reduced DPIP as well as some other agents can mediate a so-called 'cyclic' operation of system I. Their oxidized form is reduced by X^- and in turn their reduced form reduces one of the mediators which donate electrons to $P700^+$. In fresh chloroplasts, both the 'open' and 'cyclic' operation of system I are still accompanied by the generation of ATP and ADP (see later).

With system II inactivated, it is possible to maintain the rigid anaerobic conditions, that are required for the operation of the enzyme hydrogenase. This enzyme allows molecular hydrogen to react with ferredoxin and thus indirectly with other electron carriers; in its presence illuminated chloroplasts can evolve H_2 gas in the light (Arnon, 1963).

Gaffron and Rubin (1942) observed that under anaerobic conditions whole green algae develop hydrogenase activity. Such 'hydrogen adapted' algae reveal a number of interesting reactions. One of these is a DCMU insensitive reduction of CO_2 by molecular hydrogen in the light. As shown in Fig. 10.7 one can conceive that in this process H_2 reduces NADP, while in addition, (system I) light is needed to provide the ATP required for CO_2 assimilation. If this interpretation is correct, whole algae must possess a mechanism in which ATP is formed in a cyclic operation of system I. Several other observations indicate that *in vivo* ATP can be generated in a cycling of system I. This process presumably underlies a partial suppression of dark respiration by far-red light (see Fig. 10.14), as well as a stimulation of the ATP consuming uptake of glucose

13

and other organic substrates by algae (Kandler, 1950). It is unknown whether and to which degree such a cycling of system I still occurs under conditions that also system II is operating.

Isolation of system II

No treatment is known to inhibit specifically system I, which, although particle bound, is quite stable *in vitro*. One of the many artificial mutants of photosynthetic organisms lacks C700 and P700 (cf. Figs. 10.3 and 10.4) and shows no system I activity (Bishop, 1964). While producing O_2, it can reduce only weak oxidants such as ferricyanide or DPIP, which have a higher midpoint potential than A or Q ($E_m > 0.2$ V) (Kok and Datko, 1965). Other mutants lack one or more of the components which connect the two photoacts (Levine, 1966); these fail to carry out a complete electron transport, but can show either system I or II activity. Sonic treatment is reported to remove plastocyanin from isolated chloroplasts and thus break the chain at this point (Katoh and Takamiya, 1964). A significant degree of physical separation of the two photosystems has been achieved by fractionating particles obtained from chloroplasts which were mildly treated with detergent (Boardman and Anderson, 1964). While the two photosystems can be considered as essentially separate entities, as we shall see later, their interaction in the complete process shows the features of an integrated, structure bound apparatus.

Pigment composition

During a steady rate of photosynthesis, the rate of electron flow through the two series-connected photosystems is equal. Weak, rate-limiting intensities will therefore be used with optimal efficiency if the photosystems absorb an equal number of photons per unit time and convert these with equal efficiency. If in one of the two systems the rate of absorption is lower than in the other, the overall rate cannot exceed that of the limiting system.

In vivo the concentration of the two conversion centres is probably nearly equal (~ 1 per 500 [Chl]$_{tot}$) and so is the size of the units, ~ 200 molecules of harvesting pigments feeding into each centre. The pigment composition of the two types of units differs, however. In practically all classes of plants, system II contains a larger percentage of the so-called 'accessory pigments': chlorophyll *b* in green plants, fucoxanthol in brown algae, phycobilins in red and blue-green algae. Generally less than half of the total chlorophyll *a* is connected to system II; it collects quanta from the accessory pigments and feeds into the reaction centres.

Due to this difference in pigmentation, the distribution of absorbed light between the two photoacts (α_1 and α_2) varies with wavelength so that the action spectra of the two individual photosystems are different (see Fig. 10.8) and the efficiency of the overall process varies with wavelength.

The lower curve in Fig. 10.9 shows the quantum yield of photosynthesis as a function of wavelength measured by Emerson and Lewis (1943) in green algae. A significant dip in the blue, indicates that carotenoids do not transfer their energy to chlorophyll with more than ~ 60 per cent efficiency; a much smaller dip at 650 nm is due to the excess absorption of chlorophyll b in system II. With the exception of the small 650 dip, in the red region the yield is quite con-

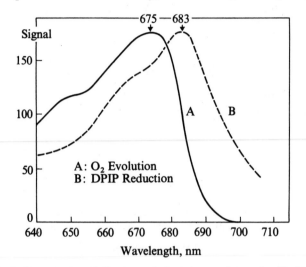

Figure 10.8 Action spectra of the two photosystems, observed with isolated spinach chloroplasts; the respective maxima are normalized. The system II spectrum is the modulated rate of O_2 evolution induced by a very weak modulated light of varying wavelength which was added to an unmodulated background light of 720 nm. Note the shoulder at 650 nm, due to participation of chlorophyll b and the maximum activity at 675 nm. The system I spectrum is the modulated rate of viologen reduction induced by a weak modulated light, of varying wavelength superimposed on a background of continuous 650 nm light. Note the lack of a 650 nm band, the maximum at 683 nm and the extended activity in long wavelengths. (Joliot, *et al.*, 1968.)

stant until 690 nm and then drops severely at longer wavelengths where the relatively long-wave absorption of the harvesting chlorophyll a and minor pigment C700 (Fig. 10.3) of system I becomes predominant. The quantum yields of curve A (Fig. 10.9, top) were obtained with isolated chloroplasts, measuring NADP reduction with concomitant O_2 evolution and show the similar wavelength dependence of complete photosynthesis. Curve B shows the yields of NADP reduction, this time observed under conditions where photosystem II was inhibited, and ascorbate rather than water functioned as electron donor. Now the quantum yield rises at long wavelength, indicating that a greater percentage of the light is absorbed by system I (for $\lambda > 700$ nm, $\alpha_1 > \alpha_2$). Curve

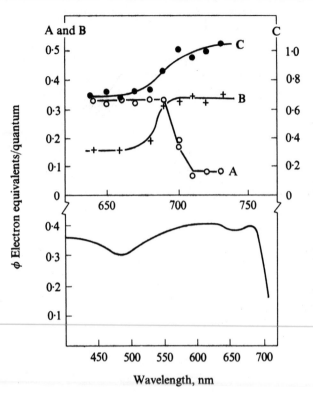

Figure 10.9 Quantum yields as a function of wavelength. Lower figure: photosynthetic O_2 evolution by *Chlorella* cells (Emerson and Lewis, 1942). Top figure: curve A: NADP reduction with concomitant O_2 evolution; curve B: NADP reduction with concomitant ascorbate oxidation (in the presence of DCMU). Both curves were measured with freshly prepared spinach chloroplasts (left-hand ordinate scale). Curve C (right-hand ordinate scale): photo-oxidation of ferrocytochrome *c* with concomitant viologen reduction, measured with small particles prepared from digitonin treated chloroplasts. (Schwartz, 1967.)

C again shows system I activity, now measured with small particles separated from digitonin-treated chloroplasts, which are enriched in system I. The quantum yield approaches 1 eq/hv in far red and drops but little at shorter wavelengths —indicating the absence of interfering (system II) pigment.

Chromatic effects

The fact that in some wavelengths system I absorbs predominantly and in others system II predominates, underlies a large number of phenomena, the study of which has greatly aided in resolving photosynthetic electron transport (cf. Duysens, 1964; Witt *et al.*, 1965). 'Long-wave light' (which in blue-green and red algae comprises also the 680 maximum of chlorophyll *a*, here quite predominantly associated with system I), generates strong reducing power. It oxidizes all electron carriers between the photoacts: P700, cytochrome f, plastocyanin, plastoquinone (pool A), Q and lowers the fluorescence yield of system II. 'Short-wave light' (excess system II), besides evolving O_2, brings about the reduction of all these intermediates and enhances the fluorescence yield by bringing Q in its reduced, non-quenching, state. An example of such a push-and-pull effect of different wavelengths upon electron carriers is shown in Fig. 10.10.

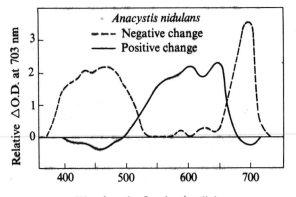

Figure 10.10 Effect of wavelength upon the redox state of P700 in whole cells of the blue-green alga *Anacystis*. Negative change (photo-oxidation) was measured without positive change (photo-reduction) with a background light of 700 nm. (Kok and Gott, 1960.)

Another consequence of the unequal pigment distribution is the so-called 'enhancement' between wavelengths in reaction systems which require the co-operation of the two photoacts (Emerson, 1958). Long-wave light is used inefficiently, because it is mostly absorbed by the pigments of system I. On the other hand, some short wavelengths (such as 650 nm, the red band of chlorophyll b in green plants) are used suboptimally because they yield a shortage of system I and excess system II excitations. If two such beams are given together, these excess excitations in each light can co-operate so that the resulting rate will exceed the sum of the individual rates.

Finally, we should mention the so-called 'chromatic transients', initial rate fluctuations observed upon switching from a long-wave to a short-wave beam and vice versa (Blinks, 1957). These rest upon the fact that in each wavelength, an equilibrium state of P, A, and Q, etc. is attained, determined by the ratio of absorptions in the two photosystems. This ratio is rather independent of intensity as long as it is not rate-saturating. The large concentration of A prevents an immediate readjustment so that the rate in the second wavelength initially is affected by the preceding one.

'Spill over' vs. 'separate package', equilibrium constant

Although the action spectra of the individual systems differ significantly, the quantum yield of the overall process varies relatively little with wavelength $<$ 690 nm. For instance, although system I contains practically no chlorophyll b, the curves in Fig. 10.9 show only a slight drop of efficiency at 650 nm. To explain this, the possibility has been considered that excess quanta absorbed in system II might 'spill over' into the pigments units of system I (Myers, 1963). Actually, the observed quantum yield spectra are also compatible with the assumption of non-interfering pigment packages, provided one assigns a low equilibration constant to the interconnecting reaction chain. The explanation for this is as follows: In steady-state photosynthesis, or in steady-state chloroplast reactions which involve both photoacts, the rates of the two series connected photosystems must be equal: $R_1 = R_2$ or $\alpha_1 P^- = \alpha_2 Q^+$ (eqs. 1, p. 370, and 7, p. 342). At wavelengths where the two systems absorb equally ($\alpha_1 = \alpha_2$), the efficiency in weak light will be optimal. However, this optimal value will be equal to the theoretical maximum (2 $h\nu$/eq.) only if all traps are open ($P^- = P_{total}$ and $Q^+ = Q_{total}$) (Ely and Myers, 1967). The degree of opening of the traps will be higher, the higher the equilibrium constant K of the dark reaction in which the photoconverted states are restored by each other:

$$P + Q \underset{k_2}{\overset{k_1}{\rightleftharpoons}} P^- + Q^+ \qquad k_1/k_2 = K \qquad (8)$$

If K were as low as one, only half of the traps would be open in wavelengths where $\alpha_1 = \alpha_2$ and the optimal efficiency would be only half of the theoretical value.

One can compute that low K values imply, besides some loss of efficiency, a relatively small effect of the ratio α_1/α_2 and thus of wavelength, upon efficiency. Joliot et $al.$ (1967) obtained good fit with the observed spectra—and other kinetic aspects of reaction sequence eq. (8) on the basis of separate pigment packages and a value $K \approx 10$. Strong, rate saturating light, regardless of its wavelength fully oxidizes P and fully reduces Q. In this case the dark reactions (8) cannot keep pace with the quantum flux and limit the rate of the overall process (cf. later).

Photophosphorylation

ATP formation

The photoreactions produce chemical energy not only in the form of reducing power, but in addition and simultaneously, in the form of high-energy phosphate bonds. In isolated chloroplasts this light-driven generation of ATP (Arnon et $al.$, 1954) can only be observed when electron transport takes place. Under appropriate conditions one finds a rate-independent stoichiometric relation between electron transport and phosphorylation. In O_2-evolving systems a ratio of 1 ATP per two equivalents (2 ATP/O_2) has been observed by most workers (e.g., Arnon, 1959) although some report higher ratios, (Winget et al., 1965).

In analogy to mitochondrial systems, one tends to conclude from this ratio P/O = P/2 eq = 1, that the transport chain contains only one 'coupling site', in which one ATP is generated by the successive passage of two electrons.

In freshly prepared chloroplasts, photophosphorylation can also accompany cyclic and noncyclic electron transport in which only photosystem I is operative (see Fig. 10.7). The cyclic electron transport mediated by PMS and the open transport, in which viologen serves as electron acceptor and $DADH_2$ (Trebst, 1963) as a donor, support, in strong light, very high rates of ATP formation, well exceeding those observable in O_2 evolving systems (1,000–2,000 ATP Chl^{-1} h^{-1} compared to <400 ATP Chl^{-1} h^{-1}) (Avron, 1960; Izawa et al., 1966). However, the P/2e ratios which can be observed in open systems using NADP or viologen (i.e., O_2, see Fig. 10.7) as electron acceptors do not exceed 0·3–0·5 and are even lower with some donors. This is one of the reasons why the scheme of Fig. 10.7 hedges about the exact point of entry of the various electron donors in the electron transport chain. To a variable degree, they appear to 'bypass' the phosphorylation step, which itself is not localized with certainty.

Proton and ion movement

Hind and Jagendorf (1963) observed that upon illumination of a chloroplast suspension, containing a mediator of electron transport, the pH of the medium rises, indicating an uptake of protons by the particles. A counterflow of cations (Mg^{2+}, K^+) offsets, at least partially, the resulting charge (Dilley and Vernon, 1965). The pH gradient attains a steady-state value in the light and returns in subsequent darkness; slowly at pH 6·5, fast at high pH 8. This dark decay is accelerated by uncouplers. At low pH as much as 1 proton per chlorophyll molecule can accumulate within the chloroplasts vesicles. The size of this pool considerably exceeds that of known chloroplast electron carriers and thus it takes a few seconds to build up in the light. The mechanism of this proton pump is as yet not clarified.

The light-induced transport of protons and other ions and the association and dissociation of weak acids in the vesicles are accompanied by movement of water in and out of the chloroplasts. Thus, as illustrated in Fig. 10.11, depending on the ionic composition of the medium, illumination can induce swelling and shrinkage of the particles, resulting in drastic changes of their volume and light-scattering properties (Packer et al., 1966).

These ion transport phenomena reflect changes of energy. If immediately after illuminating chloroplasts in the presence of an electron transport mediator, ADP and inorganic phosphate are added, ATP is generated in the dark; the amount thus formed can approach 1 ATP per 10 chl. Apparently, while carrying out electron transport in the light, chloroplasts accumulate a long-lived high-energy intermediate capable of generating ATP in darkness. The kinetics of formation and decay of this high-energy intermediate parallel the accumulation and loss of protons in the vesicles, suggesting that the proton gradient is closely related or identical to the driving force of phosphorylation. A mechanism of this type was proposed earlier by Mitchell and Moyle (1962). Accordingly, Jagendorf and Uribe (1966) could generate ATP in total darkness—by first equilibrating a chloroplast suspension at pH $\approx$ 4 and subsequently neutralizing the medium to pH $\approx$ 8. The ATP formed after neutralization could amount to as much as $\frac{1}{3}$ [Chl.].

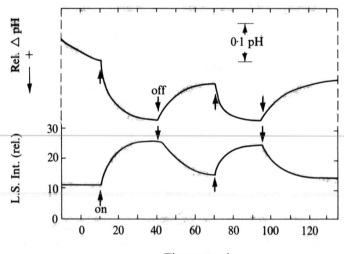

Figure 10.11   Uptake of H$^+$ during illumination and excretion in subsequent darkness (top trace) is accompanied by shrinking and swelling of the chloroplasts—observed as changes of their light scattering properties (bottom trace). The reaction mixture contained 20 ml 0·1 M KCl, 0·13 mM TMQH$_2$, and 20 µg chlorophyll equivalent chloroplasts per ml. The pH at the instant the first light cycle was initiated was 6·2 (Dilley and Vernon, 1965).

If a concentration difference of protons and possibly other ions on the inside and outside of the chloroplast membrane were the driving force of phosphorylation, one would predict a requirement for a finite minimum gradient, thermodynamically sufficient to drive the process: Accordingly, Shen and Shen (1962) reported a lack of ATP formation in weak intensities, Sakurai et al. (1965) a time lag at the onset of illumination, while Schwartz (1968) found the rate to be zero when the steady state proton gradient was below a critical level, and proportional to the gradient at values beyond this level. Electron transport appears to drive the proton gradient via another, as yet unidentified intermediate, having a pool size of $\sim\frac{1}{20}$ [Chl.], one proton being translocated per electron transfer and two being consumed from the gradient per ATP formed (Schwartz and Ebert, 1968).

Rate limitation, coupling site

Oxygen evolution by chloroplasts is affected by variations in phosphorylation. If ADP, inorganic phosphate, and magnesium, needed for ATP formation, are omitted from the reaction medium, O_2 production in strong light can be several fold lower than in the presence of these reagents. This interaction, 'coupling', can be relieved by suboptimal preparative procedures; and especially by the addition of so-called 'uncoupling' agents which permit maximum rates of electron transport without concomitant phosphorylation. Figure 10.12 shows an actual demonstration of these interactions.

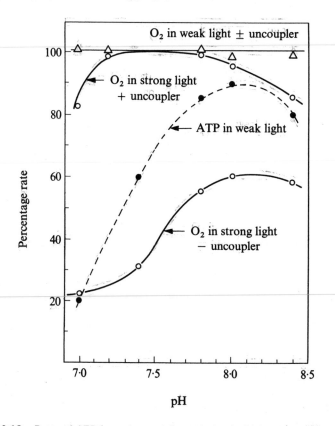

Figure 10.12 Rates of ATP formation and O_2 evolution by isolated spinach chloroplasts in weak light and in a strong, rate-saturating intensity, as a function of pH in presence or absence of 10^{-3} M ammonia. (Schwartz, 1965.)

13*

The points connected by a dashed line show that phosphorylation accompanying O_2 evolution proceeds optimally at pH $\geqslant 8$: at pH 7 the rate is $\sim \frac{1}{5}$ maximal. Still, the rate of O_2 evolution in weak light is unaffected by pH, i.e., by the occurrence or absence of ATP formation. Also, uncoupling by ammonia has no effect. The rate in strong light, however, is limited by the phosphorylation step as can be seen by the dependence of rate upon pH; relieved in the presence of an uncoupler.

A great variety of substances act as uncouplers and their mode of action can be quite different (Izawa *et al.*, 1966). The uncoupling effect of EDTA is brought about by the release from the chloroplasts of a high molecular weight coupling factor, which upon readdition, can restore activity (Avron, 1963; McCarty and Racker, 1966).

Specific uncouplers of photophosphorylation are ammonium ion, and various amines (Krogmann and Jagendorf, 1959). These agents prevent, besides the formation of ATP also the build up of the proton gradient—apparently enhancing the permeability of the membrane. Other uncouplers such as phlorizin prevent exclusively the final conversion from ADP → ATP (Izawa *et al.*, 1966).

As shown in eqs. 9, we can ascribe the rate effects to the participation of an intermediate (I) in the electron transfer step which provides the energy for phosphorylation:

$$A^- + C^+ + I \overset{k_1}{\underset{k_2}{\rightleftharpoons}} A + C + I \sim \qquad k_1/k_2 = K \tag{9}$$

$$I \sim + [H] \rightleftharpoons I + [H] \sim, \qquad [H] \sim + ADP \rightleftharpoons [H] + ATP \tag{9A}$$

$$I \sim \rightarrow I, \qquad [H] \sim \rightarrow [H], \qquad ATP \rightarrow ADP \tag{9B}$$

Part of the energy available in the reduction of C^+ by A^- is retained in state $I \sim$ (eq. 9) which drives proton translocation ($[H] \sim$) and ATP formation (eq. 9A).

Reaction (9) cannot proceed unless I is regenerated, either in a useful way (9A) or via a back reaction such as might be accelerated by uncoupling agents (9B). Because of a spontaneous slow decay of the gradient, $I \sim$ does not accumulate significantly in weak light so that in this case uncouplers have no effect.

We might now consider the question which reaction step in the electron transport chain yields the energy for phosphorylation. Since one ATP is generated per two electron transfers through the coupling site, an energy requirement of ~ 10 kcal mole^{-1} ATP implies that the midpoint potentials of the two intermediates which comprise the site must differ by $\simeq 200$ mV—if the equilibrium constants of all steps in eq. 9 and 9A are low—and more if the sequence were irreversible. In 1960 Hill and Bendall proposed that, in analogy with the mitochondrial process ATP formation was coupled to the electron transfer from cytochrome b_6, reduced by system II, to cytochrome f, oxidized by system I. In eq. 9 we located the site between plastoquinone (pool A) and cytochrome f (C): The observation that in strong light Q and A accumulate in the reduced form and P_{700} and cytochrome f in the oxidized state, implies a rate limitation in the transfer between A and C.

Figure 10.13 shows the rate of reduction of P_{700}^+ by photoreduced Q and A, immediately following exposures of different duration to bright light. The transfer rate increases with the amount of A^- made during the flash, becoming maximal at 0·1 sec when all A is reduced; this shows participation of A in the limiting step. In the presence of an uncoupling agent this rate is high and unaffected by further increase of illumination time. In the absence of an un-coupler of phosphorylating reagents the rate is restrained, except after very brief flashes, which reduce only a fraction of A. With increase of illumination time, retardation develops in a biphasic fashion reflecting the consecutive filling of a small pool $(I \sim)$ and a much larger one $([H] \sim)$ which requires a few seconds of illumination.

As we will discuss later (p. 366), the production of 1 ATP/2 eq. or 2ATP/O_2, which is observed in isolated chloroplasts and compatible with a single coupling

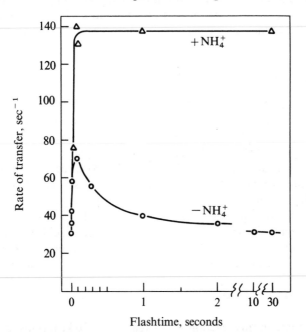

Flashtime, seconds

Figure 10.13 Spectroscopic observations of the rate with which P_{700} is reduced after being oxidized by an intensity saturated light flash of variable length. The flashes were given after a long dark period in which Q and A were oxidized and P was reduced. Following a very brief flash which converts the primary traps only, reduction of P^+ by Q^- (via A and cytochrome f) occurs with a halftime of ~ 22 ms (k = 32 sec^{-1}) unaffected by ammonium ion. Longer flashes reduce an increasing fraction of the A pool and the subsequent reduction of P^+ becomes proportionally faster in the uncoupled system. In the coupled system the rate slows down again with increase of illumination time. Isolated spinach chloroplasts pH 7·7, electron acceptor 10^{-4}M methyl viologen (Kok *et al.*, 1968).

site in the reaction chain, seems insufficient to fill the requirements of photosynthesis in the whole cell. This discrepancy underlies speculations concerning the generation of additional ATP, for instance (Hoch and Owens, 1963) in a cyclic operation of system I which might still comprise the same coupling site as discussed above.

Photosynthesis in the whole cell

In the foregoing we discussed the most unique aspect of photosynthesis, the conversion of light into semi-stable metabolic energy carriers such as $NADPH^+$ and ATP. In the cell most of this energy is further stabilized and stored in stable carbon compounds—available on demand for further syntheses or for the generation of energy carriers in respiration. The following discussion concerns some aspects of CO_2 assimilation and the interactions between photosynthesis and other cellular energy conversions.

Fixation and reduction of CO_2

A detailed description of the path of carbon in photosynthesis, the main features of which were elucidated several years ago can be found in Bassham and Calvin (1957). The essential reactions in which CO_2 is reduced to the level of sugar with the aid of ATP and NADP, generated by the light reactions are the following: a 5-carbon sugar, ribulose 5-phosphate (RuMP), reacts with ATP to yield ribulose 1,5 diphosphate (RuDP). This facilitates the subsequent carboxylation reaction, which yields two molecules of 3-phosphoglyceric acid (PGA), the primary CO_2 fixation product. Two additional ATP's are used to convert the two PGA molecules into phosphoryl-3-phosphoglyceric acid, which allows their subsequent reduction by NADPH to glyceraldehyde-3-phosphate (GAP). The net result of this primary, energy-storing series of reactions thus is:

$$CO_2 + RuMP + 3ATP + 2NADP \rightarrow 2GAP + NADP^+ + 3ADP + 2H_3PO_4$$

The energy stored in this reaction is 118 kcal, while 20 kcal is lost in driving the synthesis, so that the efficiency of the carbon reduction process is $\approx$ 85 per cent. Part of the triose phosphate enters a regenerative cycle to regenerate the CO_2 acceptor. We thus can summarize the process as follows*:

*Abbreviations: 3-phosphoglyceric acid (PGA); triose phosphates (C_3); hexose mono- or diphosphate (C_6P_1, C_6P_2); C_4—erythrose phosphate pentose phosphates (C_5); sedoheptulose phosphate (C_7).

Carboxylation: $3C_5 + 3ATP + 3CO_2 \longrightarrow 6PGA + 3ADP$
Reduction: $6PGA + 6ATP + 12H \longrightarrow 6C_3 + 6ADP + 6P_i$

Net: $3C_5 + 9ATP + 12H + 3CO_2 \longrightarrow 6C_3 + 9ADP + 6P_i$

$4C_3 \longrightarrow 2C_6P_2$

$2C_6P_2 \longrightarrow 2C_6P_1 + 2P_i$

$C_6P_1 + C_3 \longrightarrow C_5 + C_4$

Regeneration: $C_6P_1 + C_4 \longrightarrow C_7 + C_3$

$C_7 + C_3 \longrightarrow 2C_5$

Net: $5C_3 \longrightarrow 3C_5 + 2P_i$

Overall: $3CO_2 + 9ATP + 12H \longrightarrow C_3 + 9ADP + 8P_i$

This carbon reduction cycle is not restricted to photosynthetic organisms, but occurs generally in autotrophic organisms. It leads to a great variety of products among which amino acids and proteins can predominate in dividing algae and young leaves; in older leaves of most plants polysaccharides are the major product. Among the questions which still might have to be clarified are the detailed mechanisms of the primary reduction and the low affinity for CO_2, at least *in vitro* of the ribulose diphosphate carboxylase (Racker, 1957). Recently, techniques have been developed for the isolation of chloroplasts that can photoreduce CO_2 with equivalent O_2-evolution at rates approaching those of the normal leaf (Jensen and Bassham, 1966; Baldry *et al.*, 1966). These may provide further answers concerning the mechanism of the process, as well as its regulation and compartmentalization.

Glycolate metabolism, alternate carboxylations

While the Calvin–Benson cycle is well established as the major path of CO_2 reduction, we should also mention some additional, physiologically important aspects of CO_2 metabolism.

In high light and in low, rate-limiting CO_2 concentrations, kinetic experiments have shown that glycolic acid can be a predominant early product, accounting for more than 50 per cent of all carbon fixed (Tolbert and Zill, 1956; Warburg and Krippahl, 1960; Whittingham *et al.*, 1962).

Most workers have assumed that glycolate arises from one or more intermediates of the Calvin–Benson cycle. If this assumption were valid, a kinetic study would reveal that the initial specific activity of glycolate was less than that of the carboxyl-carbon of PGA, the initial product of carboxylation. This has been established for algal cells carrying out photosynthesis under 'high' CO_2-tensions (~ 3–5 per cent) but not, however, under low $[CO_2]$. Zelitch (1964) found with tobacco leaves, that the specific activity of glycolate exceeded that of the carboxyl-carbon of PGA and, moreover, equalled the specific activity of the $C^{14}O_2$ administered. Since both carbons of glycolate were of equal specific activity, it was postulated that a 'back to back' carboxylation occurs. From these results Zelitch suggested that this carboxylation was primary to the carboxylation of ribulose di-phosphate under the conditions of low $[CO_2]$.

It appears entirely possible that under some conditions carboxylations other than that of ribulose diphosphate may be manifested. The phosphoenolpyruvate carboxylase reaction:

$$\text{COOH—COPO(OH)}_2\text{=CH}_2 + CO_2 + H_2O \xrightarrow{Mg^{2+}} \text{COOH—CO—CH}_2\text{COOH} + H_3PO_4$$
Phosphoenolpyruvic acid   Oxaloacetic acid

has a high affinity for CO_2. This reaction which is predominant in Crassulacean plants, was observed in spinach leaves by Bandurski and Greiner (1953). It might be significant that in sugar-cane leaves, which operate more effectively in low $[CO_2]$ than most other leaves, Kortschak et al. (1965) found oxaloacetic acid to be produced before phosphoglyceric acid.

In order to be useful, such carboxylations, for example, the proposed carboxylation to glycolate, should operate with good efficiency and the product be metabolized rapidly into compounds which upon oxidation yield metabolic energy.

Most leaves and many algae contain glycolate oxidase, a flavin enzyme, which oxidizes glycolate to glyoxylate, resulting in peroxide formation. Glyoxylate is rapidly assimilated into glycine and serine or is further oxidized completely to CO_2. The total oxidation of glycolate appears to be a useless combustion since it is not coupled to an energy-conserving reaction (e.g., ATP formation). At the present time, therefore, the significance of glycolate to autotrophic CO_2-fixation and its relation with the Calvin–Benson cycle is not clear.

In the presence of an inhibitor of the oxidase, glycolic acid accumulates. The observation that such an inhibitor also prevents O_2 uptake in 'photorespiration', discussed in the next section, as well as opening of the stomata (chapter 9), suggests a correlation of these processes with glycolate metabolism, and thus low $[CO_2]$ (Zelitch, 1965).

Rate regulation

Among the many regulating phenomena in the photosynthetic apparatus are concentration effects of substrate (reducing power) and product (assimilates).

McAllister (1937) observed that the onset of CO_2 consumption upon illumination is delayed ($\frac{1}{2}$–2 min). An induction loss (of 1–5 CO_2 Chl^{-1}) occurred not only upon the transition from dark to light, but also upon a sudden increase of intensity. Such a delay after a few minutes darkness is absent in O_2 evolution; apparently the rate of the carbon cycle is regulated by that of the energy conversion system. The pools of the Calvin cycle are rapidly emptied in darkness and require ATP, generated through electron flow, to reattain their steady state levels in the light. Conversely one can visualize a regulation of electron transport by ATP consuming processes (the third reaction in eq. 9B on p. 356).

The accumulation of assimilates in leaves appears to restrict the rate of photosynthesis. Most probably this effect is the result of feedback control on one or more of the steps involved in regeneration of the CO_2 acceptor in much the same way as feedback control is observed in glycolysis and other metabolic pathways. For example, 11 days after a newly formed ear was removed from a maize plant, the net photosynthesis rate in this plant was decreased fourfold and two days after removing the fruits from a tomato plant its photosynthetic rate was only 12 per cent of the initial value. In both instances the sugar concentration in the leaves had doubled (Moss, 1962). It is uncertain to what extent this same inhibition contributes to the so-called 'midday depression' of photosynthesis (Stålfelt, 1960). The mechanism of the control exerted by accumulated products is unknown. Also completely unclarified are other regulatory mechanisms, such as the one which, in algae, underlies a sometimes pronounced rhythmic rate fluctuation induced by alternating light-dark cycles, which tends to continue in subsequent continuous light (see chapter 18; Tamiya, 1966).

O_2 uptake and CO_2 release in the light

In dark, leaves and algae adjust slowly (halftime $\sim$ 8 hours) to a low 'endogenous' level of O_2 uptake in which at 25°C some 0·1 per cent of their dry weight is lost per hour (Audus, 1947; French et al., 1934; Kok, 1952). A period of photosynthesis can increase this rate by an order of magnitude and, since rise and decay kinetics are relatively slow, the value of 'dark respiration' is rather undefined. If respiration is measured after prolonged periods of photosynthesis in various intensities, the rates of the two processes tend to be correlated, the latter exceeding the first process ten- to twenty-fold.

Observation of the rate of respiration *during* photosynthesis requires the use of isotope techniques, since the two processes follow the same overall reaction in opposite directions (Brown, 1950). In the experiment shown in Fig. 10.14, ^{18}O was used in the gas phase to monitor respiratory uptake. In weak illumination the O_2 uptake is slower than in darkness; at high intensities it increases and can exceed the dark rate 3–4 times with readjustments occurring within $\sim$ 1 minute.

The partial suppression of O_2 uptake in weak light proved to be sensitized by long wavelengths. It was ascribed to a cyclic operation of system I generating ATP, which presumably limited the ADP available for respiratory phosphorylation and thus suppressed the latter process. On the other hand, short-wave light, most effective in photosynthetic O_2 evolution, sensitized the increased oxygen uptake in strong light.

Using labelled CO_2 and O_2, Ozbun et al. (1964) observed, with soybean leaves with ample supply of CO_2, that light intensities > 300 ft.c., about doubled respiratory O_2 uptake and halved respiratory CO_2 evolution. This decrease of CO_2 release in the light also seen by Brown and Weis (1959) in algae, indicates an inhibition of glycolysis or Krebs cycle activity.

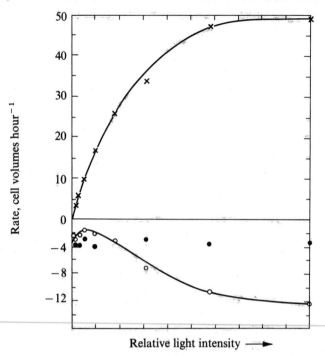

Figure 10.14 Production (crosses) and uptake (open circles) of O_2 by *Anacystis* cells in various light intensities observed using ^{18}O and mass spectrometry. Filled circles indicate rate of dark O_2 uptake before each illumination. Note the two different scales on the ordinate. (Hoch, *et al.*, 1963.)

After CO_2 depletion, the soybean leaf continued to evolve oxygen at a rate > 30 per cent of the previous rate, but now it showed an equal rate of O_2 uptake, exceeding the dark uptake as much as fifteenfold. In all likelihood this high uptake rate represented a direct reoxidation of light-generated reducing power (eqs. 3–6, p. 341). The smaller extra O_2 uptake in the light seen in Fig. 10.14, and also by Ozbun *et al.*, in the presence of CO_2, was interpreted by Hoch *et al.* (1963) as a non-cyclic electron transport (i.e., involving O_2 release and uptake) coupled to the formation of extra ATP; the latter presumably being used to assist CO_2 assimilation. However, no such energy coupling can be assumed if also under these conditions glycolate oxidation (discussed below) were responsible for the extra O_2 uptake in the light.

CO_2 compensation point, photorespiration

If a leaf is placed in a closed chamber and illuminated, the CO_2 concentration in the chamber will reach an equilibrium value $[CO_2]_C$ which is called the CO_2-compensation point. For most leaves $[CO_2]_C$ ranges between 50 and 100 p.p.m. (Moss, 1962). Under these conditions one would expect that, except in very weak light, the rate of photosynthesis P is limited by $[CO_2]$ and equal to the rate of CO_2 production in respiration R, so that $P = k_p[CO_2]_C$. According to this picture, $[CO_2]_C$ will be lower, the lower R and the greater k_p, the rate constant for the light-driven carboxylation.

Forrester *et al.* (1966a) found with soybean leaves a linear increase of $[CO_2]_C$ (5–180 p.p.m.) with increased oxygen concentration (1–100 per cent). In the same leaves, illuminated with a moderate intensity at two rate-limiting CO_2 concentrations, the net rate of CO_2 absorption declined strongly with increasing

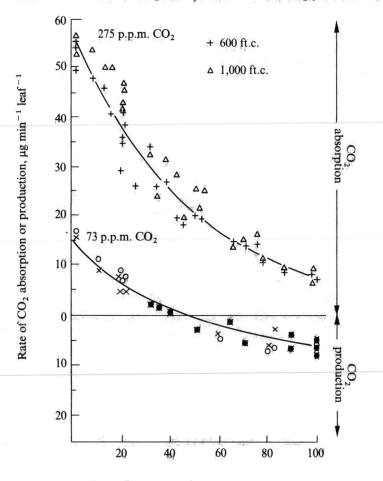

Figure 10.15   Effect of O_2 upon the rate of apparent photosynthesis in detached soybean leaves, measured as CO_2 uptake or release in two intensities and at two CO_2 concentrations. (Forrester *et al.*, 1966.)

$[O_2]$ (see Fig. 10.15). Especially in high oxygen a brief CO_2 release ('burst') was observed upon darkening, indicative of enhanced respiration in light.

The authors concluded from analysis of these data that there is a dual action of oxygen: (a) a direct effect upon photosynthesis, i.e., a threefold decrease of k_p in 100 per cent oxygen, probably due to photo-oxidation and (b) an enhancement of respiration in the light: 'photorespiration'. They assumed that in the light photorespiration replaced dark respiration, its rate being about equal to that of the latter process at 10–20 per cent O_2 and exceeding it about threefold in 100 per cent O_2. The oxidase with low affinity for oxygen, which mediates photorespiration is probably glycolate oxidase.

Zelitch (1966) observed that the oxidation of glycolate can severely depress the net rate of photosynthesis at high temperatures: with tobacco leaves illuminated in air at 0·1 per cent CO_2, a rise in the ambient temperature from 25 to 35°, resulted in a slight decrease rather than an increase of the net rate of CO_2 uptake (from 6·5 to 5·5 mg dm^{-2} h^{-1}); the rate of glycolate oxidation, however, increased fourfold. An inhibitor of the glycolate oxidase did not affect CO_2 uptake in the light at 25°, but tripled it (to 15 mg dm^{-2} h^{-1}) at 35°. This implied that without inhibitor at the high temperature $\geqslant$ 60 per cent of the generated photosynthetic reducing power was reoxidized via glycolate.

Maize leaves, on the other hand, while able effectively to convert glycolate into other carbon compounds, did not evolve CO_2 from it at either temperature and apparently lack the oxidase. This is in agreement with the fact that in leaves of corn and sugar cane, the CO_2 compensation point is negligible ($<$ 10 p.p.m.), even in 100 per cent oxygen. Instead of a CO_2 burst upon darkening, corn leaves show a delayed onset of respiratory CO_2 uptake, indicating a suppression of dark respiration in light. The only effect of high O_2 concentration, also seen in soybean leaves, is a partially reversible suppression of photosynthesis, due to photo-oxidation (Forrester et al., 1966b).

The regulation and functional significance of glycolate mediated photorespiration are not clear; it seems at least partly the cause (or the consequence) of differences in response to CO_2 concentration among species. These are illustrated in Table 10.1, which shows that the photosynthetic rate of tobacco approaches that of maize as CO_2 concentration is increased.

Table 10.1 Effect of CO_2 concentration upon net CO_2 uptake (gCO_2 m^{-2} h^{-1}) by tobacco and maize leaves. Light: 2·4 cal cm^{-2} min^{-1}; Temp.: 25–31° (from Hesketh, 1963).

$[CO_2]$ p.p.m.	150	300	600	1,000
Maize	2·7	4·7	7·4	9·2
Tobacco	1·0	2·4	4·8	6·9
Ratio: Maize/Tobacco	2·7	2·0	1·5	1·3

A deeper understanding of the matters discussed in the above sections is highly desirable since, as we shall see below, effective use of atmospheric CO_2 is a crucial aspect of photosynthetic production in the field.

Efficiency of light utilization

Photosynthesis provides all the energy for the organic world. Domesticated in agriculture and related ventures, it provides food for mankind, which is likely to be in short supply for many years to come. In domesticated plant culture most environmental factors can often be controlled to a considerable degree. However, the low atmospheric concentration of CO_2 and the variable intensity of sunlight escape such control and a main limitation of natural growth yields is the inability of plants to use effectively high (solar) intensities. This section discusses some facets of photosynthetic energy conversion in natural light fields.

Efficiency in weak light

A great number of investigations have been devoted to the question of the best possible yield of energy conversion in photosynthesis (reviewed in Emerson, 1958; Kok, 1960). Throughout this chapter we have assumed a minimum requirement of 10 absorbed quanta, for the evolution of one O_2 molecule ($> 2\ hv$/eq, see Fig. 10.9); the value obtained by most workers. The number of quanta used per evolved O_2 is a direct measure of the caloric efficiency, since regardless of the type of carbon product formed, about 110 kcal of chemical energy are fixed per mole O_2 liberated. The only exception to this rule is encountered in the photoreduction of NO^-_3, in which only 50 kcal are fixed per mole O_2 evolved. In algae, where about half of the immediate product of photosynthesis is protein, one-third of the O_2 evolved can be due to NO^-_3 reduction (assimilatory quotient $CO_2/O_2 \sim 0.7$–0.8). The use of ammonia or urea instead of NO^-_3 as a nitrogen source ($CO_2/O_2 \sim 0.9$–1.0) does not change the quantum requirement (hv/O_2) and thus leads to a higher caloric efficiency, nitrate reduction being a rather wasteful process (van Oorschot, 1955).

A requirement of 10 hv/O_2 corresponds to a caloric efficiency of $110/400 = 27$ per cent in red light (40 kcal/Einstein), and to an efficiency of 22 per cent for $\lambda = 575$ nm, the weighted average of the shortwave solar radiation ($\lambda < 700$ nm) which can be absorbed and used by plants. About 45 per cent of the total solar emission is in this usable wavelength region, $\geqslant 80$ per cent of which is absorbed by a single leaf of average thickness. For practical purposes one can therefore assume that a leaf or a vegetation absorbs half of the total incident sunlight and all 'visible' radiation of $\lambda < 700$ nm. The ascending slope of the light curves in Fig. 10.16 (left) indeed illustrates this conversion by a leaf of $\leqslant 20$ per cent of *incident* visible radiation. Similar yields have been observed for the growth of green algae, the material actually used for most quantum yield determinations. One should realize therefore, that the quantum efficiency of 0.1 O_2/hv pertains to the net result of numerous metabolic processes which lead to a diversity of products. In these conversions the cell must spend a considerable amount of energy which does not show up in the O_2 production and the caloric energy balance. One estimates a total requirement of ~ 6 ATP per CO_2 fixed—3 for its reduction to the carbohydrate level and 3 for further conversions and growth (Bauchop and Elsden, 1960). The close correspondence between the 'theoretical' requirement according to the scheme of Fig. 10.1 and the observed requirements leaves an uncomfortably small margin, not only for imperfections of the system, but also for the needed ATP (p. 357).

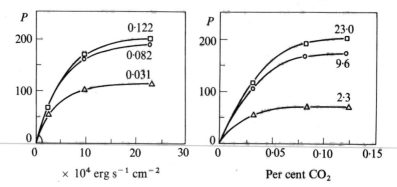

Figure 10.16 Photosynthesis of a turnip leaf as a function of intensity (left) and CO_2 concentration (right). Rate unit: μl CO_2 cm^{-2} h^{-1}. (Gaasstra, 1959.)

Unlike photosynthesis *per se*, the net energy balance of the plant drops to zero at the intensity compensation point, where respiration is balanced, and it is negative in still weaker light. As discussed in the foregoing, respiratory O_2 uptake varies with the illumination, the compensation point tending to decrease in weaker light, so that the lower intensity limit of energy fixation is difficult to define. Moreover, this limit must vary dramatically in the plant kingdom; while exceeding 100 ft.c. in some heliophilic species, it must be negligible in those growing in deep shade where illumination at noon may remain below 5 ft.c.

Photo-oxidation and photoinhibition

In biological systems most ingredients given in too high a concentration can act in inhibitory fashion. This is especially true for light, which produces potent high-energy states and intermediates. In carotenoid deficient mutants of plants, and also of bacteria devoid of O_2 evolving capacity, the photosynthetic apparatus is destroyed (photo-oxidized) even in weak light (Griffith *et al.*, 1955). The main function of carotenoid pigments, ubiquitous in all photosynthetic cells, might well be the harmless conversion into heat of excess photons; however, a specific mechanism for such an 'overflow system' has not been visualized.

'Photo-oxidation' is a direct photo-induced O_2 uptake, strongly dependent upon oxygen concentration which results in the oxidation of cellular intermediates and the photosynthetic pigment system. Its interference appears more pronounced after another effect of strong light, 'photoinhibition', described below, has taken its course (Kandler and Sironval, 1959).

In the experiment of Fig. 10.17, algae, grown in dark or moderate illumination, were exposed to various intensities. While in high light all rates were initially identical, those in the stronger lights declined following an exponential

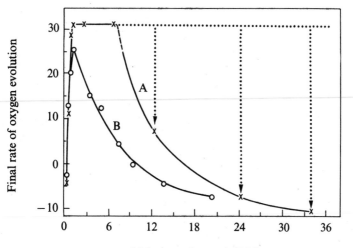

Light intensity, × 1,000 ft.c.

Figure 10.17   Steady-state rate of O_2 evolution as a function of intensity, observed with *Chlorella* cells grown in the presence of glucose in light (curve A) or dark (curve B). Plotted are the final rates attained in each intensity. (Myers and Burr, 1940).

time course until they reached a final constant value. With increasing intensity the decline became faster and the final rate lower, until photosynthesis was inactivated. The data imply that this photoinhibition is counteracted by a dark 'repair' process: the rate remains high as long as restoration can take care of the damage, it decays to zero when inhibition is overriding; intermediate final rates represent a balance between destruction and restoration. Photoinhibition decreases the photochemical efficiency (a protective measure?); especially in light-adjusted cells, it can lower the quantum yield to a considerable degree while the saturation rate remains unaffected. Accordingly, after an exposure to strong light, the rate in weak light is often temporarily decreased.

Photoinhibition comprises several photoprocesses, of which the rate is proportional to intensity and unaffected by temperature and oxygen concentration.

Ultraviolet light (λ_{max} 250 nm), possibly absorbed by quinone is deleterious to most photosynthetic tissues. It acts with a high quantum yield, destroying O_2 evolution and ATP formation but not system I. Wavelengths absorbed by the two photosystems, act with a low quantum yield and inactivate the respective trapping centres (Jones and Kok, 1966). As exemplified by curve B in Fig. 10.17, observed with algae grown in darkness, the sensitivity to photoinhibition changes with pretreatment; it also varies greatly among plant species.

CO_2 limitation, yields in natural light fields

The data of Fig. 10.16 and Table 10.1 illustrate that in all plants the low CO_2 content of the atmosphere (300 p.p.m.) limits the rate at high intensity. In the more effective species, CO_2 limitation interferes at relatively higher intensities. Most reports agree that an increase of CO_2 beyond 0·1 per cent yields no further increase of the rate in turbulent air (Hoover et al., 1933; see Rabinowitch, 1951).

The regulation of transpiration in land plants by stomata (see chapter 9) poses a variable barrier in the flow of CO_2 to the chloroplast. Since diffusion through many tiny openings in an impervious membrane can be as high as 70 per cent of the rate observed without any such barrier, the fully opened stomata in mesophytic leaves offer a relatively small resistance (Brown and Escome, 1900). However, in fully closed condition, the flow of CO_2 is restricted to diffusion through the cuticle, which can severely limit the rate of photosynthesis.

The flux of CO_2 through stomata, $F(gCO_2 \text{ mm}^{-2} \text{ s}^{-1})$ can be expressed in terms of Fick's law of diffusion:

$$F = \frac{D_{CO_2}([CO_2]_{air} - [CO_2]_{Chl.})}{L + S + M} \tag{a}$$

$$\frac{1}{F} = k_1 + k_2 S \tag{b}$$

in which D_{CO_2} is the coefficient of diffusion of CO_2 in air (mm^2 s^{-1}), $[CO_2]_{air}$ and $[CO_2]_{Chl.}$ (g mm^{-3}) are the CO_2 concentrations in the outside air and at the site of fixation in the chloroplast respectively. The denominator is the sum of the apparent path lengths of diffusion (mm): L from the air to the leaf, S through the stomata (S_{max}, the path through the cuticle, assumed ∞), while M is the diffusion path through the leaf, the air-water interfaces, and the aqueous mesophyll cells to the fixation sites. Zelitch and Waggoner (1962) controlled the opening of the guard cells of tobacco leaves with an inhibitor, which did not affect the rate of CO_2 fixation. The relation between F and S then followed eq. (b) if S was calculated from the measured stomatal width according to Penman and Schofield's (1951) formula. From these experiments, performed in a turbulent atmosphere, the ratios $L:S:M$ were estimated as $1·4:1:1$ for a stomatal width of 10 microns

($S = 5$ mm) and $1\cdot4:4:1$ for a width of 1 micron ($S = 2$ mm). In still air, a CO_2 gradient builds up at the surface of the leaf and the atmospheric resistance L can become predominant.

Since $[CO_2]_{air}$ and L (eq. (a)) are fixed quantities, differences of $[CO_2]_{Chl.}$, the CO_2 concentration at the fixation site, and of mesophyll and stomatal resistance contribute to the variation among plants in their capacity to use CO_2 effectively.

If the intensity of full sunlight ($0\cdot6$ cal cm.$^{-2}$ min $^{-1}$, $\lambda400$–700 nm) were used with the maximal efficiency of 20 per cent, one would observe a rate of ~ 25 g CO_2 per hour and per square metre irradiated area. Instead, Fig. 10.16A shows that in the turnip leaf, which is representative of many crop plants, the rate levels off at one tenth of this value (~ 2 g CO_2 m^{-2} h^{-1}) in 0·03 per cent CO_2 and $\sim \frac{1}{5}$ of it in saturating $[CO_2]$. The best rates, observed with the more effective leaves of corn, sugar cane (Fig. 10.18), and sunflower are ~ 4 g CO_2

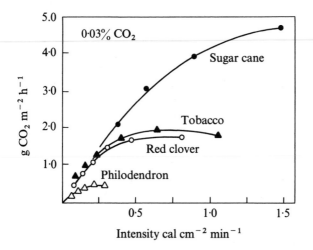

Figure 10.18 Rates of photosynthetic CO_2 uptake as a function of intensity observed with leaves of various plants. Intensity is total incident radiation, approximately half of which is absorbed by the leaf pigments. (Hesketh and Moss, 1963.)

m^{-2} h^{-1} in atmospheric $[CO_2]$, a value which can again be doubled by increasing the carbon dioxide concentration to a rate of > 300 O_2 Chl^{-1} h^{-1}, corresponding to a conversion yield of ~ 6–7 per cent of absorbed radiation, $\sim \frac{1}{3}$ of the maximum efficiency (Wilstatter and Stoll, 1918; Hesketh and Moss, 1963).

For cultures of algae, in which most environmental factors, including CO_2 supply can be controlled similar optimal performance values are reported: rates up to 400 O_2 Chl^{-1} h^{-1} and conversion efficiencies of full sunlight intensity ~ 7 per cent (Wassink, 1959).

The efficiency with which in the field the daily energy total can be converted by a vegetation is only indirectly related to the above figures. One tends to suspect an improvement, due to the diurnal variation of light and temperature, to be partly offset by the relatively greater effect of respiratory loss upon the overall yield.

Fixation of 20 per cent of the approximately 3,000 kcal photosynthetically useful radiation, which are available on a bright summer day per m^2 horizontal surface, would yield 600 kcal chemical energy per m^2, corresponding to $\sim$ 150 g of carbohydrate (4 kcal g^{-1}) or 100 g algae, which represent $\sim$ 5·5 kcal per gram dry weight.

As illustrated in Fig. 10.19, under favourable temperature conditions, optimal

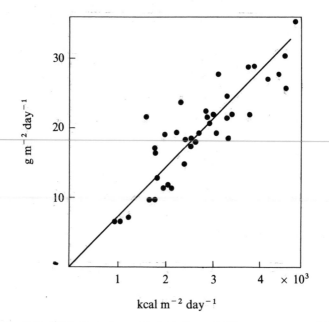

Figure 10.19   Daily harvests (g dry weight) of *Scenedesmus* cells in a dense outside culture plotted vs. daily insolation. Slope indicates 8 per cent conversion efficiency, assuming that half of the incident light (λ < 700 nm) was absorbed by the algae. (Kok and V. Ooorschot, 1954.)

daily harvests of algae (up to 30 g m^{-2}) were roughly proportional to the total absorbed energy and represented $\sim$ 8 per cent of it. Maximum daily yields observed with crop plants are reported to approach similar values ($\sim$ 6 per cent) (Talling, 1961). A discussion of the generally much inferior efficiency with which the energy available during the year or a growing season is converted into actual crops is beyond the scope of this chapter. We should, however, recall the several peculiar features (see carboxylation, photorespiration, maximum rate) of corn and sugar cane, which might underlie their relatively high production yields.

Efficiency in strong light

We conclude this chapter with some additional remarks concerning the utilization of strong light. In the foregoing sections, especially those dealing with photorespiration and photoinhibition we saw that the fate of quanta and the response of photosynthesis in rate saturating intensities is complex and actually poorly understood. It thus should be clear that kinetic analyses of 'light saturation' in terms of one or more rate-limiting dark steps are simplifications, significant only for short term observations; consequently rate vs. intensity curves as shown in Figs. 10.16 and 10.18 escape rigid mathematical formulation.

Emerson and Arnold's (1932) original data showed that brief (10^{-5} s) strong flashes could generate at most 1 O_2 per ~ 2000 chl. (or fix 1 quantum per 200 chl., the size of a unit). Half this amount was produced in each of a series of flashes, if the interposed dark time was ~ 0.01 s, presumably the half time of the rate-limiting dark reaction. Though later observations indicated additional (co)limiting steps, and somewhat higher flash yields (Kok and Cheniae, 1966) we assume for the following argument a single time constant of ~ 10 ms (30°C) which implies that each trap can at most convert ~ 100 excitations per second.

On the other extreme, very weak intensities cannot be used efficiently either. The spontaneous deactivation of system II ($t_{\frac{1}{2}} < 3$ s at 30°C, p. 343) demands a minimum flux in each trap of 1 hv per ~ 3 s. These two limits yield a non-linear (S shaped) rate vs. intensity curve in which the efficiency exceeds 50 per cent of maximum over a 'dynamic range' of about 300. This dynamic range is largely independent of temperature since the two limiting steps have similar temperature coefficients.

A single chlorophyll molecule, exposed to full sunlight (0.6 cal. cm^{-2} min^{-1}, $\lambda < 700$ nm, $\sim 8,000$ ft.c.) absorbs some 10 quanta per second so that photosynthesis would proceed quite ineffectively if the traps were to collect their own excitations, unassisted by harvesting pigment.

If each trap were fed by 100 chlorophylls, a light flux of 1 hv/chl s. (~ 0.1 of full sunlight) would just saturate a rate of 3,600 hv/chl hour or 360 O_2/chl hour.

In full sunlight, however, $\sim 1,000$ hv s^{-1} would arrive in a centre, some 10 times more than can be processed, so that the conversion yield would be only $0.1 \times 20 = 2$ per cent, resulting in a rate of only 2.5 g rather than the 25 g CO_2 m^{-2} h^{-1} which would correspond to maximal conversion efficiency. On the other extreme, the minimum flux is 0.003 hv/chl s, or < 0.1 per cent of full sunlight, well below the intensity compensation point so that respiration loss rather than deactivation restricts the net assimilation in weak light.

A smaller unit would favour the efficiency in bright light and reduce photoinhibition, conversely a larger one would allow the use of lower intensities.

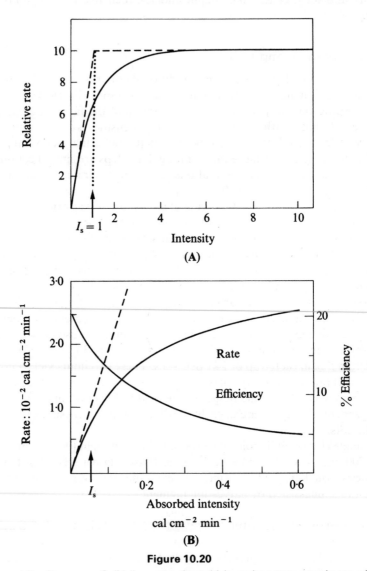

Figure 10.20

A. Rate vs. intensity curves. Solid line: rate observed in a short-term experiment with an optically thin suspension of algae. The ascending asymptote, broken line, represents the efficiency in weak light, the horizontal one the light-saturated rate. At the intersection $I_s(=1)$ is the 'saturation intensity'.

B. Dependence of conversion rate and conversion efficiency upon intensity, predicted for a dense suspension of algae whose light response is shown in Fig. A. Assumptions: $I_s = 0.05$ cal cm^{-2} min^{-1}, maximal efficiency 20 per cent, fractional absorption 95 per cent. (From Van Oorschot, 1955)

The virtually complete absorption of incident radiation by a leaf, or a dense culture of algae, results in a strong internal intensity gradient. Directly exposed chloroplasts may be light saturated, others receiving less light, use it more effectively; while those deep in the tissue may be in virtual darkness. As a result such a system approaches light saturation much more gradually than an optically thin layer of algae or chloroplasts. Figure 10.20B illustrates the predicted intensity dependence of rate and efficiency of a dense population of chloroplasts or algae, of which the individual partners show the rate-intensity relation of Fig. 10.20A. In Fig. 10.20A the unit of intensity is I_s, the abcissa of the intersection of the two asymptotes of the rate curve. In Fig. 10.20B we assumed, as in our above example, I_s to be 0·05 cal cm^{-2} min^{-1}, $\sim$ 0·1 of full sunlight. It appears that in the dense system full sunlight is used $\sim$ 3 times more efficiently than in the optically thin one ($\sim$ 6 per cent vs. 2 per cent).

Actually, in a leaf the situation may be much more favourable than predicted on the basis of a homogeneous suspension. Special devices such as intensity-regulated phototactic movements of chloroplasts (Fig. 10.21), light-pipe effects

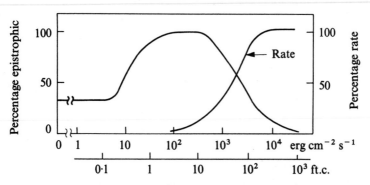

Figure 10.21 Correlation between orientation of the chloroplasts and the rate of photosynthesis in *Lemna*. In darkness distribution is random; in weak light maximum absorbing power is established when in the epistrophe position the irradiated front and back walls are covered. Strong, rate saturating light induces movement to the sidewalls. (Zurzycki, 1962.)

in the palisade cells and differentiation among chloroplasts in different cell layers and leaves might approach optimal scattering and usage of the light.

A similar 'intensity dilution' is again encountered if we now consider a stand of plants, rather than an individual leaf, illuminated with a perpendicular light beam as discussed so far. The total leaf area in such a stand or a tree, can exceed the projected area of the covered soil up to eightfold. Oblique angles of incidence, scattering, and shading effects lower the average light flux per chloroplast so that the intensity required to saturate the vegetation is increased. The resulting lower rate per unit leaf area effectively ameliorates CO_2 concentration limitation.

Increased complexity of leaf structure and area, however, carries the penalty of an enhanced energy balance point, due to greater respiratory loss; which limits the dynamic range of net growth at the lower end of the intensity scale. The compromise found in different plants to cope with the large diurnal and local variation of natural insolation, varies greatly with species and habitat as is illustrated in the light curves in Fig. 10.18.

Dr I. Zelitch (Connecticut Agricultural Station, New Haven) contributed considerably to several parts of this chapter. The author gratefully acknowledges his help and that of Dr G. Cheniae who critically read the manuscript.

Bibliography

Selected further reading

Bassham, J. A. and M. Calvin, 1957. *The Path of Carbon in Photosynthesis*, Prentice Hall, Englewood Cliffs, New Jersey.

Clayton, R. K., 1965. *Molecular Physics in photosynthesis*, Blaisdell Co., New York.

Duysens, L. N. M., 1964. Photosynthesis. *Progress in Biophysics*, Vol. 14, Pergamon Press, pp. 1–104.

Rabinowtich, E. I., *Photosynthesis*, Vol. I (1945); Vol. II, p. 1 (1951); Vol. III, p. 2 (1956); Wiley, Interscience, New York.

Photosynthetic Mechanisms of Green Plants. Publication 1145. National Academy of Science—National Research Council, Washington, D.C. (1963).

Energy Conversion by the Photosynthetic Apparatus. Brookhaven Symposia in Biology Number 19. U.S. Atomic Energy Comm. (1966).

San Pietro, A. and F. A. Greer (eds.), 1967. *Harvesting the Sun: Photosynthesis in Plant Life*. Academic Press, New York.

Papers cited in text

Allen, F. L. and J. Franck, 1955. Photosynthetic Evolution of Oxygen by Flashes of Light. *Arch. Biochem. Biophys.* **58**, 124–143.

Amesz, J., 1964. Spectrophotometric Evidence for the Participation of a Quinone in Photosynthesis of Intact Blue-Green Algae. *Biochim. Biophys. Acta.* **79**, 257–265.

Arnon, D. I., M. B. Allen, and F. R. Whatley, 1954. Photosynthesis by Isolated Chloroplasts. *Nature, Lond.* **174**, 394–396.

Arnon, D. I., 1963. Localization in subcellular particles of the energy conversion process of photosynthesis. In *La Photosynthese*. Centre. Natl. Rech. Scient. Paris, pp. 509–541.

Arnon, D. I., 1963. Photosynthetic electron transport and phosphorylation in chloroplasts. *Photosynthesis Mechanisms in Green Plants*, Publ. 1145 NAS-NRC, pp. 195–212.

Audus, L. J., 1947. The effects of illumination on the respiration of shoots of the cherry laurel. *Ann. of Bot.* **11**, 165–201.

Avron, M. and A. T. Jagendorf, 1956. A TPNH Diaphorase from chloroplasts. *Arch. Biochem. Biophys.* **65**, 475–490.

Avron, M., 1960. Photophosphorylation by swiss-chard chloroplasts. *Biochim. Biophys. Acta.* **40**, 257–272.

Avron, M., 1963. A coupling factor in photophosphorylation. *Biochim. Biophys. Acta.* **77**, 699–702.

Baldry, C. W., C. Bucke, and D. A. Walker, 1966. Temperature and photosynthesis. I. Some effects of temperature on carbon dioxide fixation by isolated chloroplasts. *Biochim. Biophys. Acta.* **126**, 207–213.

Bandurski, R. S. and C. M. Greiner, 1953. The enzymatic synthesis of oxalacetate from phosphoryl-enolpyruvate and carbon dioxide. *J. Biol. Chem.* **204**, 781–786.

Bassham, J. A., 1964. Kinetic studies of the photosynthetic carbon reduction cycle. *Ann. Rev. Plant Physiol.* **15**, 101–120.

Bauchop, T. and S. R. Elsden. 1960. The growth of micro-organisms in relation to their energy supply. *J. Gen. Microbiol.*, **23**, 457–469.

Benson, A. A., 1964. Plant Membrane Lipids. *Ann. Rev. Plant Physiol.* **15**, 1–16.

Bishop, N. I., 1958. Vitamin K, an essential factor for the photochemical activity of isolated chloroplasts. *Proc. Nat. Acad. Sci. U.S.* **44**, 501–504.

Bishop, N. I., 1964. Mutations of unicellular green algae and their application to studies on the mechanism of photosynthesis. *Rec. Chem. Prog.* **25**, 181–195.

Blinks, L. R., 1957. Chromatic transients in photosynthesis of red algae. In *Research in Photosynthesis.* Interscience Publ., Inc., New York, ed. H. Gaffron. pp. 444–449.

Boardman, N. K. and J. M. Anderson, 1964. Isolation from spinach chloroplasts of particles containing different proportions of chlorophyll *a* and chlorophyll *b* and their possible role in the light reactions of photosynthesis. *Nature, Lond.* **203**, 166–167.

Brown, H. T. and F. Escombe, 1900. Static diffusion of gases and liquids in relation to the assimilation of carbon and translocation in plants. *Phil. Trans. Roy. Soc.* Lond. **B193**, 223–291.

Brown, A. H. and D. Weis, 1959. Relation between respiration and photosynthesis in the green alga, ankistrodesmus braunii. *Plant Physiol.* **34**, 224–234.

Butler, W. L., 1962. Effects of red and far-red light on the fluorescence yield of chlorophyll *in vivo*. *Biochim. Biophys. Acta.* **64**, 309–317.

Carrier, J. M., 1967. Oxidation-reduction potential of plastoquinones. *Biochemistry of chloroplasts*, Vol. II. Academic Press, New York, pp. 551–557.

Chance, B. and M. Nishimura, 1960. On the mechanism of chlorophyll-cytochrome interaction: The temperature insensitivity of light-induced cytochrome oxidation in chromatium. *Proc. Nat. Acad. Sci. U.S.* **46**, 19–24.

Cheniae, G. M. and I. F. Martin, 1966. *Studies on the function of manganese in photosynthesis.* Brookhaven Symp. in Biology No. 19, pp. 406–417.

Cheniae, G. M. and I. F. Martin. 1968. Site of Manganese Function in photosynthesis. *Biochim. Biophys. Acta*, **153**, 819–837.

Davenport, H. E. and R. Hill, 1952. The preparation and some properties of cytochrome *f. Proc. Roy. Soc.* **139**, 327–346.

Delosme, R., P. Joliot and J. Lavorel, 1959. Sur la complimentarite de la fluorescence et de l'emission d'oxygene pendant la periode d'induction de la photosynthese. *C.R. Acad. Sci.* **249**, 1409–1411.

Dilley, R. A., and L. P. Vernon, 1965. Ion and water transport process related to the light-dependent shrinkage of spinach chloroplasts. *Arch. Biochem. Biophys.* **111**, 365–375.

Duysens, L. N. M., 1952. *Transfer of excitation energy in photosynthesis.* Ph.D. Thesis, Utrecht, Netherlands.

Eley, J. H. and J. Myers, 1967. Enhancement of photosynthesis by alternated light beams and a kinetic model. *Plant Physiol.* **42**, 598–607.

Emerson, R. and W. Arnold, 1932. A separation of the reactions in photosynthesis by means of intermittent light. *J. Gen. Physiol.* **15**, 391–420.

Emerson, R., and C. H. Lewis, 1943. The dependence of the quantum yield of Chlorella photosynthesis on wave length of light. *Am. J. Bot.* **30**, 165–178.

Emerson, R., 1958. The quantum yield of photosynthesis. *Ann. Rev. Plant Physiol.* **9**, 1–24.

Forrester, M. L., G. Krotkov, and C. D. Nelson, 1966. Effect of oxygen in photosynthesis, photorespiration and respiration in detached leaves. I. Soybean. *Plant Physiol.* **41**, 422–427.

Forrester, M. L., G. Krotkov, and C. D. Nelson, 1966. Effect of oxygen on photosynthesis, photorespiration and respiration in detached leaves. II. Corn and other monocotyledons. *Plant Physiol.* **41**, 428–531.

French, C. S., H. I. Kohn, and P. S. Tang, 1934. Temperature characteristics for the metabolism of Chlorella. II. The rate of respiration of cultures of Chlorella pyrenoidosa as a function of time and of temperature. *J. Gen. Physiol.* **18**, 193–207.

Gaastra, P., 1959. Photosynthesis of crop plants as influenced by light, carbon dioxide, temperature, and stomatal diffusion resistance. *Med. Landbouwhogeschool te Wageningen, Nederland*, **59**, 1–68.

Gaffron, H. and J. Rubin, 1942. Fermentative and photochemical production of hydrogen in algae. *J. Gen. Physiol.* **26**, 219–241.

Good, N., S. Izawa, and G. Hind, 1966. Uncoupling and energy transfer inhibition in photophosphorylation. In *Current Topics in Bioenergetics*, Vol. I., ed. D. R. Sanadi. Academic Press, pp. 75–112.

Gorman, D. S. and R. P. Levine. 1966. Photosynthetic electron transport chain of chlamydomonas reinhardi. V. Purification and properties of cytochrome 553 and ferredoxin. *Plant Physiol.*, **41**, 1643–1647.

Griffiths, M., W. R. Sistrom, G. Cohen-Bazire, and R. Y. Stanier, 1955. Function of carotenoids in photosynthesis. *Nature, Lond.* **176**, 1211–1214.

Hesketh, J. D. 1963. Limitations to Photosynthesis Responsible for Differences Among Species, *Crop Science* **3**, 493–496.

Hesketh, J. D. and D. N. Moss, 1963. Variation in the response of photosynthesis to light. *Crop Science*, **3**, 107–110.

Hill, R. and E. F. Hartree, 1953. Hematin compounds in plants. *Ann. Rev. Plant. Physiol.* **4**, 115–150.

Hill, R. and F. Bendall, 1960. Function of the two cytochrome components in chloroplasts. A working hypothesis. *Nature, Lond.* **186**, 136–137.

Hind, G. and A. T. Jagendorf. 1963. Separation of light and dark stages in photophosphorylation. *Proc. of the Natl. Acad. of Sci.*, **49**, 715–722.

Hind, G. and J. M. Olson. 1968. Electron transport pathways in photosynthesis. *Annual Reviews of Plant Physiology*, Vol. 19, p. 249–282.

Hoch, G., O. v. H. Owens and B. Kok, 1963. Photosynthesis and respiration. *Arch. Biochem. Biophys.* **101**, 171–180.

Hoch, G. and O. v. H. Owens. 1963. Photoreactions and respiration. In: Photosynthesis Mechanisms in Green Plants, Publ. 1145, Natl. Acad. Sci.—Natl. Res. Council, p. 409–420.

Holt, A. S. and E. E. Jacobs, 1954. Spectroscopy of Plant Pigments. I. Ethyl chlorophyllides A and B and their pheophorbides. *Am. J. Bot.* **41**, 710–722.

Hoover, W. H., E. S. Johnston, and F. S. Brackett, 1933. Carbon dioxide assimilation in a higher plant. *Smithsonian Inst. Publ. Misc. Coll.* **87**, No. 16, pp. 1–21.

Izawa, S., T. N. Connolly, G. D. Winget, and N. E. Good, 1966. Inhibition and uncoupling of photophosphorylation in chloroplasts. In *Energy Conversion by the Photosynthetic Apparatus*, Brookhaven Symposia in Biology No. 19, pp. 169–187.

Izawa, S., C. D. Winget and N. E. Good. 1966. Phlorizin, a specific inhibitor of photophosphorylation and phosphorylation-coupled electron transport in chloroplasts. *Biochem. Biophys. Res. Commun.*, **22**, 223–226.

Jagendorf, A. T. and E. Uribe, 1966: Photophosphorylation and the chemiosmotic hypothesis. In *Energy Conversion by the Photosynthetic Apparatus*. Brookhaven Symposia in Biology, No. 19, pp. 215–245.

Jensen, R. G. and J. A. Bassham, 1966. Photosynthesis by isolated chloroplasts. *Proc. Nat. Acad. Sci. U.S.* **56**, 1095–1101.

Joliot, A. and P. Joliot, 1964. Etude cinetique de la reaction photochimique liberant l'oxygene au cours de la photosynthese. *C.R. Acad. Sc. Paris.* **258**, 4622–4625.

Joliot, P., 1966. Oxygen evolution in algae illuminated by modulated light. In *Energy Conversion by the Photosynthetic Apparatus*. Brookhaven Symposia in Biology, No. 19, pp. 418–433.

Joliot, P., A. Joliot and B. Kok, 1968. Analysis of the interactions Between the Two Photosystems in Isolated Chloroplasts, *Biochim. Biophys. Acta*, **153**, 635–652.

Jones, L. W. and B. Kok, 1966. Photoinhibition of chloroplast reactions. II. Multiple Effects. *Plant Physiol.* **41**, 1044–1049.

Kandler, O., 1950. Über die Beziehungen zwischen Phosphathaushalt und Photosynthese. I. Phosphatspiegelschwankungen bei Chlorella Pyrenoidosa als folge des licht-dunkel-wechsels. *Z. Naturf.* **5b**, 423–437.

Kandler, O. and C. Sironval, 1959. Photoxidation processes in normal green Chlorella cells. II. Effects on metabolism. *Biochim. Biophys. Acta.* **33**, 207–215.

Katoh, S. and A. Takamiya, 1965. Restoration of NADP photoreducing activity of sonicated chloroplasts by plastocyanin. *Biochim. Biophys. Acta.* **99**, 156–160.

Kok, B. and J. L. P. van Oorschot, 1954. Improved yields in algal mass cultures. *Acta. Bot. Neerl.* **3**, 533–546.

Kok, B., 1956. On the reversible absorption change at 705 mμ in photosynthetic organisms. *Biochim. Biophys. Acta.* **22**, 399–401.

Kok, B., 1960. Efficiency of photosynthesis. *Encyclopedia of Plant Physiology*. Springer-Verlag, Berlin, pp. 566-633.

Kok, B. and W. Gott, 1960. Activation Spectra of 700 mμ Absorption Change in Photosynthesis, *Plant Physiol.*, **35**, 802–808.

Kok, B. and G. M. Cheniae, 1966. Kinetics and intermediates of the oxygen evolution step in photosynthesis. In *Current Topics in Bio-energetics*, ed. D. Sanadi. Academic Press, pp. 1–47.

Kok, B., S. Malkin, O. Owens, and B. Forbush, 1966. Observations on the reducing side of the O_2 evolving photoacts. In *Energy Conversion by the Photosynthetic Apparatus*. Brookhaven Symposia in Biology, No. 19, pp. 446–459.

Kok, B., P. Joliot and M. P. McGloin. 1968. Electron transfer between the photoacts. *Proc. Int. Cong. Photosynthesis*, Freudenstadt, Germany.

Kortschak, H. P., C. E. Hartt, and G. O. Burr, 1965. Carbon dioxide fixation in sugarcane leaves. *Plant Physiol.* **40**, 209–213.

Krogmann, D. W. and A. T. Jagendorf, 1959. Comparison of ferricyanide and 2,3′,6-Trichlorophenol indophenol as Hill reaction oxidants. *Plant Physiol.* **34**, 277–282.

Levine, R. P., D. S. Gorman, M. Avron, and W. L. Butler, 1966. Light-induced absorbance changes in wild-type and mutant strains of *Chlamydomonas reinhardi*. In *Energy Conversion by the Photosynthetic Apparatus*. Brookhaven Symposia in Biology No. 19, pp. 143–148.

Lynch, V. H. and C. S. French, 1957. β-carotene, an active component of chloroplasts. *Arch. Biochem. Biophys.* **70**, 382–391.

Mehler, A. H., 1951. Studies on reactions of illuminated chloroplasts. II. Stimulation and inhibition of the reaction with molecular oxygen. *Arch. Biochem. Biophys.* **34**, 339–351.

Malkin, S. and B. Kok, 1966. Fluorescence induction studies in isolated chloroplasts. I. Number of components involved in the reaction and quantum yields. *Biochim. Biophys. Acta.* **126**, 413–432.

McAllister, E. D., 1937. Time course of photosynthesis for a higher plant. *Smithsonian Inst. Pub. Misc. Coll.* **95**, 1–19.

McCarty, R. E. and R. Racker, 1966. Effect of a coupling factor and its antiserum on photophosphorylation and hydrogen ion transport. In *Energy Conversion by the Photosynthetic Apparatus.* Brookhaven Symposia in Biology No. 19, pp. 202–214.

Mitchell, P., 1966. *Chemiosmotic coupling in oxidative and photosynthetic phosphorylation.* Glynn Research Labs., Bodmin, Cornwall, England Research Report No. 66/1.

Moss, D. N., 1962. The limiting carbon dioxide concentration for photosynthesis. *Nature, Lond.* **193**, 587.

Müller, A. and H. T. Witt, 1961. Trapped primary product of photosynthesis in green plants. *Nature, Lond.* **189**, 944–945.

Myers, J. and G. O. Burr, 1940. Studies on photosynthesis some effects of light of high intensity on Chlorella. *J. Gen. Physiol.* **24**, 45–67.

Myers, J., 1963. Enhancement. In *Photosynthetic Mechanisms in Green Plants.* Publ. 1145 NAS-NRC, Washington, D.C., pp. 301–317.

Ogren, W. L., J. J. Lightbody, and D. W. Krogmann, 1965. Functional lipids in the photosynthesis of higher plants. In *Biochemical Dimensions of Photosynthesis*, ed. D. W. Krogmann and W. H. Powers. Wayne State Univ. Press, Detroit, pp. 84–94.

Ozbun, J. L., R. J. Volk and W. A. Jackson, 1964. Effects of light and darkness on gaseous exchange of bean leaves. *Plant Physiol.* **39**, 523–527.

Packer, L., D. W. Deamer, and A. R. Crofts, 1966. Conformational changes in chloroplasts. In *Energy Conversion by the Photosynthetic Apparatus.* Brookhaven Symposia in Biology No. 19, pp. 281–302.

Penman, H. L. and R. K. Schofield, 1951. Some physical aspects of assimilation and transpiration. In *Symp. Soc. Exp. Biol.* Vol. V. Academic Press, pp. 115–129.

Racker, E., 1957. The reductive pentose phosphate cycle. I. Phosphoribulokinase and ribulose diphosphate carboxylase. *Arch. Biochem. Biophys.* **69**, 300–310.

Sakurai, H., M. Nishimura and A. Takamiya. 1965. Studies on photophosphorylation. I. Two-step excitation kinetics of photophosphorylation. *Plant and Cell Physiol.*, **6**, 309.

Schmidt-Mende and H. T. Witt. 1968. Zur plastochinonoxydation bei der photosynthese. *Z. fur Naturf.*, **23b**, 228–235.

Schwartz, M. 1967. Wavelength-Dependent Quantum Yield of ATP Synthesis and NADP Reduction in Normal and Dichlorodimethylphenylurea-Poisoned Chloroplast. *Biochim. Biophys. Acta*, **131**, 559–570.

Schwartz, M., 1965. Light effects on oxygen evolution and phosphorylation in spinach chloroplasts. *Biochim. Biophys. Acta*, **102**, 361–372.

Schwartz, M. 1968. Light induced proton gradient as the link between electron transport and phosphorylation in spinach, *Nature*, **219**, 915–919.

Schwartz, M. and G. A. Ebert. 1968. Control mechanism for electron transport and phosphorylation in photosynthesis. Submitted to *Science*.

Shen, Y. K. and G. M. Shen. 1962. Studies on photophosphorylation. II. The 'Light Intensity Effect' and intermediate steps of photophosphorylation. *Sci. Sinica*, **11**, 1097–1106.

Smith, J. H. C. and C. S. French, 1963. The major and accessory pigments in photosynthesis. *Ann. Rev. Plant Physiol.* **14**, 181–224.

Stalfelt, M. G., 1960. Die Abhängigkeit von zeitlichen Faktoren. In *Encyclopedia of Plant Physiology*, Vol. 2, ed. A. Pirson. Springer-Verlag, Berlin, pp. 226–254.

Talling, J. F., 1961. Photosynthesis under natural conditions. *Ann. Rev. Plant Physiol.* **12**, 133–154.

Tamiya, H., 1966. Synchronous cultures of algae. *Ann. Rev. Plant Physiol.* **17**, 1–26.

Tolbert, N. E. and L. P. Zill, 1956. Excretion of glycolic acid by algae during photosynthesis. *J. Biol. Chem.* **222**, 895–906.

Trebst, A., H. Eck, and S. Wagner, 1963. Effects of quinones and oxygen in the electron transport system of chloroplasts. In *Photosynthetic Mechanisms of Green Plants.* Publ. 1145, NAC-NRS, pp. 174–194.

Van Oorschot, J. L. P., 1955. Conversion of light energy in algal culture. *Mededelingen van de landbouwhogeschool te Wageningen (Nederland)* **55**(5), 225–276.

Vernon, L. P. and W. S. Zaugg, 1960. Photoreductions by fresh and aged chloroplasts: Requirement for ascorbate and 2,6–dichlorophenolindophenol with aged chloroplasts. *J. Biol. Chem.* **235**, 2728–2733.

Warburg, V. O., and G. Krippahl, 1960. Glykolsäurebildung in Chlorella. *Z. Naturf.* **15b**, 197–199.

Wassink, E. C., 1959. Efficiency of light energy conversion in plant growth. *Plant Physiol.* **34**, 356–361.

Whittingham, C. P., M. Birmingham, R. G. Hiller, and G. G. Pritchard, 1962. The use of isonicatinyl hydrazide as a metabolic inhibitor in photosynthesis. In *La Photosynthese.* Centre Natl. Rech. Scient. Paris, pp. 571–588.

Wilstatter, R., and A. Stoll, 1918. *Untersuchungen Über die Assimilation der Kohlensäure.* Springer, Berlin.

Winget, G. D., S. Izawa, and N. E. Good, 1965. The stoichiometry of photophosphorylation. *Biochem. Biophys. Res. Comm.* **21**, 438–443.

Witt, H. T., B. Rumberg, P. Schmidt-Mende, U. Siggel, B. Skerra, J. Vater, and J. Weikard, 1965. Über die Analyse der Photosynthese mit Blitzlicht. *Angew. Chem.* **19**, 821–876.

Witt, H. T., G. Döring, B. Rumberg, P. Schmidt-Mende, U. Siggel, and H. H. Stiehl, 1966. Electron transport in photosynthesis. In *Energy Conversion by the Photosynthetic Apparatus.* Brookhaven Symposia in Biology No. 19, pp. 161–168.

Zelitch, I., 1964. Environmental and biochemical control of stomatal movement in leaves. *Biol. Rev.* **40**, 463–482.

Zelitch, I., 1965. The relation of glycolic acid synthesis to the primary photosynthetic carboxylation reaction in leaves. *J. Biol. Chem.* **240**, 1869–1876.

Zelitch, I., 1966. Increased rate of net photosynthetic carbon dioxide uptake caused by the inhibition of glycolate oxidase. *Plant Physiol.* **41**, 1623–1631.

Zill, L. P. and G. M. Cheniae, 1962. Lipid metabolism. *Ann. Rev. Plant Physiol.* **13**, 225–264.

Zurzycki, J., 1962. The action spectrum for the light dependent movements of chloroplasts in *Lemma triscelca* L. *Acta Soc. Bot. Poloniae*, **XXXI**, 489–538.

11. Translocation of nutrients

Martin Zimmermann

Introduction (types of transport in plants)

The requirement for translocation is a result of specialization of plants into a higher organizational level. Land plants, especially tall ones, have their two supply centres separated: the leaves, which take up carbon dioxide from the atmosphere and carry out photosynthesis, are remote from the roots, which take up water and mineral nutrients from the soil. The leaves have to supply carbohydrates necessary for metabolism and growth of roots. Roots in turn, have to supply the leaves with water and mineral nutrients. This bidirectional distribution of food in vascular land plants is achieved by means of a grand circulation system. The present chapter discusses one aspect of this, long-distance transport in the living tissues.

The need for translocation in a land plant can best be appreciated by comparison with a marine alga. The morphological organization of some seaweeds is fairly complex, yet their environment is much more uniform than that of a land plant. All parts of a seaweed are ordinarily submerged in sea water, the whole environment is therefore a nutritional solution. Water and mineral nutrients can be taken in by any part of the plant. Under most conditions all plant parts are more or less evenly illuminated and can carry out photosynthesis. Exceptions to this rule are of interest. Very large kelps such as *Macrocystis* have developed transport tissue very similar to that of higher plants and translocate carbohydrates from well illuminated fronds at the surface to poorly illuminated parts at greater depths and also to rapidly growing apical regions (Parker, 1965). It is probable that transport of assimilates is more common in seaweeds than we believe, although it is only pronounced in the larger algae. We shall return to this during the course of our discussion.

Apart from diffusion as the simplest method of transport in vascular plants, there are several additional methods. There is passive movement along the transpiration stream through the *apoplast* which includes dead spaces outside the living parts of the cells. This consists largely of the xylem together with the walls of all cells. The mechanism of this water movement is discussed in chapter 12 and need not be considered further here, except to note that substances which are carried along by the transpiration stream can be absorbed as they pass along the channels and so taken into living cells (cf. Bollard, 1960). This apoplast movement (xylem transport) often involves very long distances.

There are a number of transport phenomena which may be summarized by using the term 'secretion'. They involve the *symplast* (i.e., the living parts of plant cells). A characteristic feature of secretion is that movement can take place against a concentration gradient, a feat which is achieved at the expense of meta-bolic energy. Some of these secretion phenomena are described in other chapters (e.g., chapters 4 and 13). Most plant cells can accumulate nutrients from very dilute solutions. This occurs in the phloem where carbohydrates are accumulated in the sieve tubes in concentrations exceeding that of the surrounding photo-synthesizing tissue where they are produced. This secretion into or 'loading' of sieve tubes (Barrier and Loomis, 1957), will be discussed later. Secretion of sugars from the phloem to the outside of the plant body via special secretory tissues as in floral and extrafloral nectaries (cf. Schnepf, 1965) may be regarded as an example of secretion.

Protoplasmic streaming is another and more striking type of transport in the plant (cf. Kamiya, 1960). In combination with diffusion, streaming is a factor in the distribution of nutrients. Whether it accounts for transport over long distances in the sieve tubes is another question which we shall later deal with more carefully.

The central problem to be discussed in this chapter is the long-distance trans-port in the phloem of higher plants. Phloem transport, typically, takes place between the following centres:

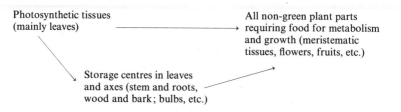

Other important characteristics of phloem transport are:

(a) Large amounts of substances move over long distances (in contrast to secretion where transport distances are very short).
(b) Special organization of the conducting tissues.

There is possibly no other aspect of plant physiology which compares with translocation in the way in which the problem is tied to structure as well as organization of the plant as a whole. Space does not allow more than an occa-sional reference to phloem structure, but an understanding of anatomy is an absolute prerequisite for discussion of translocation phenomena. Any reader without this understanding is therefore urged to acquire an adequate working knowledge of phloem anatomy from a good anatomy text (e.g., Esau, 1965).

Channel of long-distance transport in the symplast

As early as 200–300 years ago it was found that water ascends in the wood of trees, whereas assimilated foodstuffs move down from leaves through the bark. These early findings were the result of surgical experiments such as the removal of a ring of bark from the stem, a treatment called 'girdling' (Malpighi, 1675; Hales, 1727; Duhamel du Monceau, 1768; citations in Crafts, 1961 and Münch, 1930). These basic experiments were later repeated many times under a variety of conditions and with different plants. The results were always essentially the same. When the bark of a woody plant is interrupted by a girdle, leaves above the girdle do not wilt showing that water transport continues. On the other hand growth continues above the girdle but is greatly reduced below. This in turn means that the downward transport of assimilates from leaves is stopped, i.e., takes place through the bark.

Further understanding was achieved during the last century largely through the work of the German forest botanist Theodor Hartig. He discovered sieve tubes (Hartig, 1837) and sieve-tube exudation (Hartig, 1860). Hartig was unfortunately overshadowed by the famous plant physiologist Sachs who somewhat dogmatically (but we know now wrongly) declared diffusion to be the main principle involved in nutrient distribution in plants.

Schneider-Orelli's (1909) paper marks another considerable step forward but is unfortunately very poorly known. Schneider-Orelli, a Swiss entomologist, studied the mining moth *Lyonetia clercella* L. This insect lays its eggs in the tissue of apple leaves. The developing larva eats its way through the leaf tissue until it finally escapes as a fully developed insect via a hole eaten through the leaf surface. Schneider-Orelli's observations clearly show two different types of transport: cell-to-cell transport towards the conducting bundles, and phloem transport out of the leaves via vascular bundles. Wherever the mining larva had eaten its way through phloem, carbohydrate transport out of the leaf was interrupted. Schneider-Orelli's illustrated paper should be studied by every serious student of translocation.

Mason and Maskell began translocation work with cotton in the 'twenties in Trinidad. They found diurnal fluctuations of sugar concentrations in leaves; these fluctuations appeared with some delay also in the bark but never in the wood. When the bark was divided into three concentric layers it was found that the fluctuations were located in the innermost layer. Sucrose was mostly involved in the fluctuations. Wax paper between bark and wood did not interfere with transport. These experiments represent quantitative evidence of sucrose transport via the innermost layer of the bark (Mason and Maskell, 1928).

Further progress came with the work of Schumacher (1933) with *Pelargonium*. The petiole of this plant contains a large central and a number of smaller peripheral vascular bundles. Schumacher, in painstaking surgical work, removed the peripheral parts of the petiole so that the leaf blade was connected to the stem only by means of the central bundle. Translocation still took place. He then eliminated either the xylem or the phloem. Translocation took place only when the phloem was present. As an indicator of translocation, Schumacher used nitrogen transport during induced senescence (nitrogenous substances are exported from yellowing leaves when plants are put in the dark). Much of Schumacher's further work dealt with the microscopic observation of movement of fluorescent dyes. Early attempts with eosin failed but the reason for this

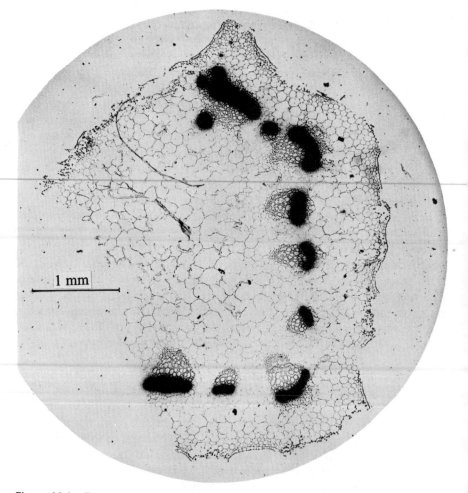

1 mm

Figure 11.1   Tissue autoradiograph of the water-soluble, ^{14}C-labelled compounds in sugar beet petiole, 30 minutes after photosynthetic assimilation of $^{14}CO_2$ by the blade. The blackened phloem marks the path of translocation. (From Mortimer, 1965, reproduced by permission of the National Research Council of Canada.)

failure is notable: eosin caused closure of sieve plates with callose. No translocation occurred after this closure (Schumacher, 1933). This experiment is still the only direct evidence that the sieve tubes are the conducting channels, although there is a great deal of additional but indirect evidence. Schumacher's further work (1933, 1937, etc.) discovered dyes which are transported in the sieve tubes (the most widely used is potassium fluorescein), but it ended with a controversy which was apparently never satisfactorily resolved. Schumacher reported fluorescein movement in the cytoplasm, Bauer maintained that it moved in the vacuole and was secondarily absorbed by the cytoplasm (Bauer, 1952; Schumacher and Hülsbruch, 1955).

The introduction of autoradiography brought a confirmation of all this older work. If translocation sugars are labelled with radioactive tracers, as for example if a leaf photosynthesizes $C^{14}O_2$, autoradiograms of petiole or stem sections through which transport has occurred show the radioactivity to be concentrated in the phloem (Fig. 11.1) (cf., e.g., Mortimer, 1965; Neeracher, 1965, Nelson et al., 1959; and others). Attempts to refine the technique so as to show transport within sieve tubes themselves have not yet been successful. The extreme sensitivity of sieve tubes towards injury and the solubility of their contents in water represent difficulties which have not yet been fully overcome.

Much of the evidence which shows that sieve tubes are the translocation channels is indirect. Some cell types can be discounted because they are not always present in translocating tissues. Fibres, for example, can be excluded because there are many plants which have no phloem fibres. Companion cells are often discontinuous from one end of a sieve element to the other, furthermore there are occasional sieve elements which have no companion cells (Cheadle and Esau, 1958). The choice is thus narrowed to parenchyma cells and sieve tubes. Only the latter are highly specialized and it is only rational to conclude that they are the translocating cells. *Macrocystis* provides telling evidence. It has evolved a translocation capacity as extensive as that of higher plants; it also has evolved cell files which are very similar to sieve tubes. Further evidence discussed in the next section confirms the conclusions of Hartig and Schumacher.

Nature of translocated substances

Methods

Analysis of extracts. Identifiable translocated material can be obtained by applying radioactively labelled substances to a leaf and subsequently analysing various plant parts for radioactive compounds which had been exported from the treated leaf. This method is very useful if reasonable care is taken in interpretation of results. The weakness of this method is the difficulty of distinguishing moving from stationary material. One would expect that isotopes allow a short enough experimental time so that a moving (i.e., radioactive) substance can be distinguished from the stationary (non-radioactive) material. Unfortunately, the situation is not quite so simple, especially in a young, growing plant where translocated substances are rapidly exchanged with, and incorporated into, substances in the surrounding tissues. Vernon and Aronoff (1952) found for example that in a bean seedling part of the moving ^{14}C-labelled sugar was incorporated into cell-wall cellulose along its transport path before the detectable front of radioactivity had even reached the roots. It is therefore never quite certain what proportion of the tracer is actually moving and what is stationary.

Analysis of sieve-tube exudate.   Sieve-tube exudate can be obtained by making an incision into the inner bark of a woody plant, or from the cut stump of the stylet bundle of an aphid which had been feeding on the plant. Both these methods can be used from the time after the break of winter dormancy when callose seals on sieve plates are dissolved, until the leaves have begun to grow. No more exudate can then be obtained until the leaves are fully mature and exporting sugars. There seems to be a time of little downward transport in the stem between these two periods (Shiroya *et al.*, 1966; Zimmermann, 1964). Not all species are equally suitable for tapping of the phloem, because slime-plugging may prevent copious exudation. Furthermore, in species with resin ducts or latex canals contamination may be a serious problem. There is thus a limitation to certain 'suitable' species. Experimental work which involves the taking of a large number of samples from bark incisions during the course of an experiment is further restricted to relatively large individual trees, because each time an incision is made some of the phloem tissue is interrupted.

Many aphids obtain food by inserting their mouth parts into the conducting sieve tubes of plants. Some years ago the entomologist Mittler developed a method which has subsequently acquired appreciable importance in translocation studies (Mittler, 1957, 1958). The following procedure is involved. A few individual aphids are allowed to feed on a branch; if desirable they can be caged at a predetermined point. Honeydew production indicates when they are feeding, that is, when their stylets are inserted in the phloem. They are then anaesthetized and one of them is severed from its stylet bundle. If the operation is successful the crystal-clear exudate emerges from the stump of the cut stylet bundle and has to be caught immediately in a pipette with a micro manipulator. Exudation continues for several days under favourable circumstances, pipettes are renewed periodically and the samples are analysed by paper chromatography. The aphid-stylet method has the advantage that the sieve-tube content is uncontaminated by traces of non-mobile substances washed from the cut surface of the bark. On the other hand, it may not be exactly identical with the translocated material, because the local turgor drop in the punctured sieve-tube may cause not only a dilution of the exudate but also a certain amount of local mobilization of reserve material. This point will be discussed in a later section of this chapter (pp. 396–397).

Evidence to show that 'sieve-tube exudate' is the translocated material. The sieve-tube exudation itself provides perhaps the strongest evidence indicating that it is the translocated solution. Rates up to and exceeding 5 mm^3 per hour are obtained. Stylet tips are inserted in a single sieve element (Zimmermann, 1961) so we know that each element has to be refilled 3–10 times per second. Weatherley *et al.* (1959) have shown that this movement is axial (i.e., in the direction of normal transport). Furthermore, all circumstantial evidence leads to the same conclusion. The same substances, found with tracer experiments to be moving, are also found in the exudate. Nitrogenous substances appear in the exudate during natural as well as induced senescence. The available velocity figures and exudate concentrations give precisely the right mass transfer values (see p. 403). Furthermore, exudation takes place only during those times when translocation goes on, and when the incision (or aphid stylet bundle) reaches the innermost (conducting) sieve tubes. Aphids feeding below a leafy crown release radioactive honeydew soon after $^{14}CO_2$ is applied to leaves (e.g., Canny, 1962a).

Results

Carbohydrates. Analysis of sieve-tube exudates as well as the results of tracer experiments indicate that the bulk of translocated material in the majority of higher plants is *sucrose*. In experiments in which the natural translocation products were labelled with ^{14}C, small amounts of labelled hexoses glucose and fructose were often also found in extracts. The fact that the ratio labelled hexoses/labelled sucrose decreases with distance from the supply leaf indicates that the hexoses are secondary break-down products (Swanson and El-Shishiny, 1958). In double grafts (Jerusalem artichoke/sunflower/Jerusalem artichoke, or sunflower/Jerusalem artichoke/sunflower) radioactive label, fed to one of the upper leaves, was found to move down through the plant and across the grafts in the form of sucrose. The ratio labelled hexoses/labelled sucrose also decreased along the way in these experiments, and there was no abrupt change from one plant species to the other in spite of the fact that the sugar metabolism of the two species is quite different (Kursanov *et al.*, 1958).

Other sugars, translocated as well as sucrose, are the raffinose family of oligosaccharides, *raffinose*, *stachyose*, and *verbascose* (Fig. 11.2). They consist of sucrose with one, two, or three molecules of galactose attached and are therefore non-reducing. In very many plants small amounts of raffinose and sometimes stachyose can be found as translocation sugars besides sucrose, but there are certain plant families in which the larger oligosaccharides are most abundant. These sugars were identified by Zimmermann (1957a) in sieve-tube exudates of trees, and later found with tracer translocation experiments by various other authors like Pristupa (1959) in *Cucurbita* and Trip *et al.* (1965) in *Fraxinus* and *Syringa*. A survey made by the writer of exudates from some 250 tropical tree species showed these higher sugars to be prevalent in a number of

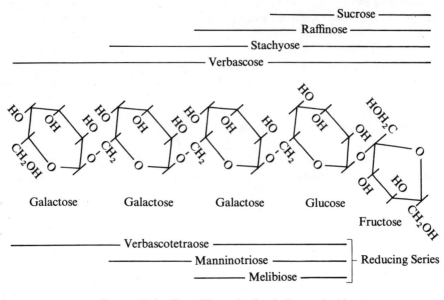

Figure 11.2   The raffinose family of oligosaccharides.

plant families (Bignoniaceae, Celastraceae, Combretaceae, Myrtaceae (partly), Oleaceae, Verbenaceae, etc.).

Sieve-tube exudate analyses of *Fraxinus americana* L. have indicated the presence of relatively large amounts of the sugar alcohol D-*mannitol* (Zimmermann, 1957b). This was also confirmed with tracer experiments (Trip *et al.*, 1965). The survey mentioned above revealed further species with D-mannitol in the exudate, but taxonomically the distribution seems to be inconsistent. In the Combretaceae, for example, there are large amounts of mannitol in a few species but none in others. This even occurs within the single genus *Terminalia*. The physiological as well as the taxonomic meaning of this distribution remains to be elucidated. An interesting and probably significant finding was recently reported by Parker (1966). The giant kelp *Macrocystis* translocates D-mannitol but no sugars. This is at present the only example known where mannitol is translocated in the absence of sugars. It is perhaps not so very surprising because seaweeds are known to produce mannitol in large quantities (e.g., Bidwell *et al.*, 1958).

Sorbitol is another transport sugar-alcohol. It is found as such in some members of the Rosaceae. About half the dry weight of exudate from several cherry species consists of sorbitol (the other half mostly sucrose) (Zimmermann, 1961). Tracer translocation experiments with apple trees showed very similar results (Webb and Burley, 1962).

In summary we can say that the bulk of photosynthetic products are translocated in the form of non-reducing oligosaccharides of the raffinose family (including and often exclusively in the form of sucrose). In some species there is in addition D-mannitol or sorbitol; in the brown alga *Macrocystis* D-mannitol is the main translocate. The fact that all these substances are non-reducing may have some unknown physiological meaning. As far as the sugars are concerned, transport of the oligosaccharide bond energy may be an advantage.

Other substances. *Serine* and *malic acid* have been reported to move in considerable quantities in certain seedling stages of the soybean (Nelson *et al.*, 1961). Otherwise quantities of other moving substances are relatively small. The fact that almost any experimentally applied substance is translocated in the phloem makes a mere listing meaningless at this point. The matter will be taken up in various later sections. It is sufficient to say here that some of these other substances may be translocated predominantly at certain times of plant development (e.g., export of nitrogenous and phosphorus compounds from senescent leaves and flowers).

Proposed mechanisms of translocation

Earlier botanists were—and today we still are—very much impressed by the amount of material translocated throughout plants over long distances without any apparent special device such as a pumping heart (see p. 403). It is therefore not surprising that translocation problems occupy much space in botanical literature. Mere diffusion, effective as it may be within very short distances as within a single cell, becomes inadequate as the distances increase even to a few centimetres as can easily be seen from Fick's diffusion equation:

$$\frac{\mathrm{d}m}{\mathrm{d}t} = \text{const.} \times TS \times \frac{\Delta_{\text{conc.}}}{\text{distance}}$$

$\mathrm{d}m/\mathrm{d}t$ = mass transferred per unit time

TS = tranverse-sectional area through which diffusion takes place

$\Delta_{\text{conc.}}$ = concentration difference between the two points

Mason and Maskell (1928) found with cotton plants that the diffusion constant would have to be 40,000 times as great as that of a 2 per cent solution, to explain the observed transport rate of sucrose. Even in cotton plants these authors dealt with relatively short distances when compared with distances of 100 metres and more in the tallest trees and longest tropical vines! Much thought and speculation has gone into this problem and numerous hypotheses have been put forward over the past 100 years in attempts to explain phloem transport. The more important of these theories will be described and discussed briefly in this section. A more detailed discussion will follow in the next section (beginning p. 394) which reviews experimental translocation research.

Protoplasmic streaming hypothesis

Protoplasmic streaming is one of the oldest methods suggested to account for translocation; it was originally proposed by DeVries in 1885. It found an advocate in Curtis (1935) and others and has been re-emphasized during recent years (Thaine, 1962; Canny, 1962). Protoplasmic streaming is thought to promote, activate, or cause translocation of solutes in plants. The concepts described in the literature of how this could be accomplished differ widely. It has been suggested that the streaming protoplast stirs or mixes the vacuolar sap, and that solutes then diffuse through the sieve pores to the next element. The most recent concept envisages streaming of protoplasm from element to element through the sieve pores (Thaine, 1962). The concentrations of dissolved solutes in the vacuole are thought to equilibrate with those of the strand by diffusion. At points of higher concentration of vacuolar solutes net diffusion would be into the strands, at places of lower concentration net diffusion would be out of the strand. The overall effect would be translocation of solutes down a gradient (Canny, 1962). It should be noted that transport in this model has to go always along a concentration gradient from higher to lower concentration. With this model bidirectional transport in a single sieve tube is not only feasible but would be very common. A radioactive tracer, applied somewhere in the middle of a transport path, would have to move both up and down regardless of net movement.

Thaine (1962) describes his observation of transcellular streaming in sieve tubes. His motion picture, which the writer has had the opportunity to see twice, is, however, not too convincing. The main difficulty is the thickness of the photographed sections which make it difficult to recognize the focal plane clearly (see also Esau *et al.*, 1963; and the discussion following Thaine and Preston, 1964). Thaine's observations have received a certain amount of support from electronmicroscopic studies such as those of Evert and Murmanis (1965). Physiological support is based primarily upon Canny's experimental and theoretical work. Some of this will be discussed in the following sections.

Interfacial flow hypothesis

Van den Honert (1932) found a very rapid spreading of oleate in an ether–water interface. This gave rise to the idea that transported molecules might move along interfaces (such as between the vacuole and the protoplast). This hypothesis has found occasional supporters, but so far, there is no evidence in support

of such a mechanism in plants. It is difficult to see how a wide variety of substances, such as inorganic ions, sugars, and whole virus particles (Kluge, 1967,
see also the literature cited there) could move equally well through such a
system. We also know that membranes are not static, but constantly changing.

Mass flow hypothesis

Many earlier botanists thought of phloem transport as an actual flow of solution.
Such a theory was outlined by Münch in 1930 and has been re-emphasized many
times by other authors, though in a somewhat modified form. The principle is
quite simple. All living cells are surrounded by the water of the apoplast. The
whole symplast acts like an osmotic cell. In places where photosynthesis or
mobilization of reserve material takes place, solute molecules are introduced
into the vacuoles and water must follow for osmotic reasons (this point is called
the 'source'). In areas where solutes are eliminated from the vacuole through
the incorporation of sugar molecules into polysaccharides (so-called 'sinks'),
solute concentration is lowered and water is lost. The result is a turgor gradient
from source to sink. If the whole vacuole system is differentially permeable towards the apoplast but permeable within (from one cell to the next), there must
be a mass flow of solution along the turgor gradient from source to sink. Figure
11.3 illustrates the principle diagrammatically. It should be noted that a growing
area where turgor is lowered physically can become a sink without lowering of
concentration.

  It soon became apparent that the mass-flow hypothesis had to be restricted
to the sieve-tube system. The only cells whose vacuoles are in permeable contact
are the sieve-tube elements. Transport into and out of the sieve tubes is a metabolic step. This conclusion was unavoidable when it was learned that secretion

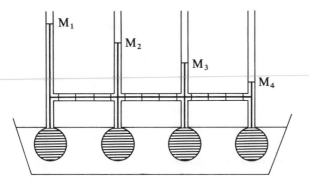

Figure 11.3 Model of Münch's mass-flow mechanism. Four osmotic cells containing decreasing solute concentrations from left to right (osmotic pressures in each indicated by mano-
meters M_1 to M_4). A flow of solution takes place in such a system from left to right, water
enters the cells on the left and is lost from cells on the right. (From Münch, 1930.)

into sieve tubes often takes place against a concentration gradient (Curtis and Asai, 1939; Röckl, 1949). It must be emphasized that the whole translocation process is highly dependent upon the plant's metabolism. Even though the axial flow of solutes may be a passive one, it is driven by the active transport into and out of the sieve tubes. Furthermore, solutes have to be kept in sieve tubes by metabolic energy throughout the whole of the translocation path.

Electro-osmotic theory

Liquid flow through small pores, based upon electro-osmotic effects, can be very much greater per unit solute concentration than those based upon osmosis. Phloem transport theories based upon such phenomena have been proposed independently by Fensom (1957) and by Spanner (1958). The driving force for solute movement is thought to be an electric potential across the sieve plates. Such a mechanism could work if the sieve plates permit the passage of highly hydrated ions and if the sieve-plate potential is (metabolically) maintained. Spanner suggested a circulation of K^+ at the sieve plates as a possibility for potential maintenance. The pressure gradient along the path of transport would be a 'saw-tooth' gradient; the liquid would be 'pumped' across each sieve plate. The electro-osmotic theory is very attractive; the main difficulty is the fact that it is—so far at least—not much more than a speculation.

Other suggestions

Many workers in the field of translocation do not support any of the above theories. In fact, it sometimes appears as though there are as many concepts as there are workers. There are a few points of agreement, but many of disagreement. Metabolism is recognized as the driving force of translocation, but there is no agreement of how metabolism does the trick. Some authors think that metabolism somehow propels the translocated molecules directly, others, for example Choi and Aronoff (1966) merely suggest that 'the leaves provide a direct force requiring structural continuity, or a translocation carrier'.

Discussion of translocation theories in the light of experimental evidence

Is transport a flow of solution?

One of the most fundamental and immediate questions is whether or not translocation involves a flow of a solution. Does solvent water flow together with the translocated molecules? Once this question can be answered satisfactorily the choice of transport mechanisms is narrowed considerably. Considered superficially the question appears to be a very simple one. If one could experimentally have THO and ^{14}C-labelled photosynthates move out of a leaf simultaneously at the same velocity, the question would seem to be answered. Biddulph and Cory (1957) performed such an experiment successfully, but their results could not be confirmed by Gage and Aronoff (1960). The difficulty with this experiment is twofold. First, in supplying (labelled) water one has to compete with the phloem's natural source of water, the xylem, a tissue immediately adjacent to the phloem. Second, even if water has been successfully introduced into the sieve tubes, it is subject to lateral loss—much more so than solutes—because of diffusional exchange with surrounding tissue.

The most direct evidence in support of a solution flow is seen in sieve-tube exudation and its response to changes of the osmotic equilibrium. These aspects will be discussed in a later section (p. 404). The exudation from an aphid stylet bundle originates in a single sieve element (Zimmermann, 1961) and is maintained by axial transport (Weatherley et al., 1959). The rate of exudation is, as mentioned before, 5 mm^3 per hour and more from *Tilia* branches, and this corresponds to 3–10 times the volume of a single sieve element per second. It is difficult to conceive of any mechanism other than flow along the sieve tube which could keep up this rate of exudation.

Münch (1930) reported that when he pried strips of bark off the trees in such a way that the upper end was still attached, and enclosed these strips in a container to prevent evaporation, new wood was formed on the inside of the flaps and at the same time water dripped from the flap (cf. Fig. 25 in Münch, 1930). Münch considered this to be solvent water lost from sieve tubes after solute removal. Xylem is excluded as a source of this water because transpiration prevents xylem pressures from increasing above +1 atm (this happens only in certain species and only before the leaves emerge). Münch estimated that this solvent water liberated from sieve tubes contributes about 5 per cent to the transpiration stream; in other words, about 5 per cent of the water ascending in the xylem descends again in the phloem. These experiments have been repeated by Brown (1964) with *Populus* with similar results. It is difficult to explain them in any other way.

Metabolism and translocation

In contrast to xylem transport living cells are essential for transport in the phloem. If a stem section is killed, for example with superheated steam, xylem transport continues but phloem transport stops. This treatment, called 'steam girdling' is often used to distinguish between xylem and phloem transport.

Willenbrink (1957) isolated the central vascular bundle of *Pelargonium* petioles and found with Schumacher's method (see p. 386) phloem transport to be sensitive to metabolic poisons. Interestingly enough, anaerobic conditions did not affect transport, a finding other authors had reported before. On the basis of my own work with trees I would like to speculate that the phloem might obtain oxygen dissolved in the moving xylem water. If this is the case, experiments on local application of a N_2 atmosphere to the axis would check phloem transport only if the transpiration stream were stopped.

Some emphasis has been placed by various authors upon enzymes found in sieve tubes by histochemical methods or analysis of exudates. This is of no consequence in the translocation mechanism; even vacuoles of parenchyma cells (corn rootlets) have an appreciable complement of enzymes as Matile (1966) has so elegantly shown. Even though sieve elements may have a reduced metabolism which might depend to a certain extent on companion cells, they cannot merely be regarded as inert tubes. Sensitivity of phloem transport to locally applied poisons at the source, the sink, or anywhere between the two points is no evidence against the mass-flow concept as often stated. An osmotic system as described on p. 393 can only function if metabolism maintains differential permeability all along the translocation path.

The effect of temperature upon phloem transport is interesting. Experiments are not easy to carry out. If the whole plant is subjected to various temperatures the results are meaningless because direct effects cannot be distinguished from indirect ones, as on the rate of photosynthesis. Even if a temperature jacket is applied locally between source and sink, for example around a petiole or short length of stem, the transpiration stream moves the applied heat into the leaf blade. If transpiration is suppressed by placing the plant in a humid atmosphere, the transpiration stream may be maintained by root pressure as in herbaceous plants. If one considers these facts, it becomes evident that some of the older literature reports on temperature effects are not very reliable.

Three recent papers may be cited as examples of temperature effects. Webb and Gorham (1965) subjected the primary leaf node of *Cucurbita* to temperatures between 0° and 55°C. Neither assimilation nor transpiration was affected between 0 and 45°C. Sugar export from the leaf, however, was at a maximum at about 25°C and decreased with both lowered and raised temperatures to become negligible at 0° and 55°C. The inhibitory effect at 55° was irreversible, but that of 0° was reversible. When the nodal temperature was raised from 0° to 25° slow export from the leaf began immediately, but recovery to normal rate required about an hour. Temperature dependence of translocation did not appear to be correlated with that of protoplasmic streaming. Geiger (1966) found a (reversible) decline of translocation of photosynthate when the sink region was cooled. This emphasizes that uptake into storage and growing areas is a metabolic process. Perhaps the most illuminating of the three papers is that of Ford and Peel (1966), who studied the contributory length of sieve tubes. The 'contributory length' is the minimum axial length in an isolated stem segment necessary to maintain maximum flow rate from an aphid stylet bundle. Previously, Weatherley *et al.* (1959) had determined this to be about 160 mm in willow. This (somewhat variable) distance must be great enough to allow sufficient mobilization of storage carbohydrates along the translocation path to permit maximum supply to the sink (the aphid stylet bundle). Ford and Peel (1966) found that this distance is increased when the stem segment between sink and supply end is chilled. This is taken to mean that entry of carbohydrates into sieve tubes is much more dependent upon metabolism than longitudinal movement.

Direction and polarity of movement

The question of polarity.   Although phloem transport is directional in the sense that movement is always from source to sink, it is not polar in respect to phloem tissue (this is in contrast to auxin movement; cf chapter 4). An example of a very clear-cut experiment which shows this is given in Fig. 11.4. Assimilates

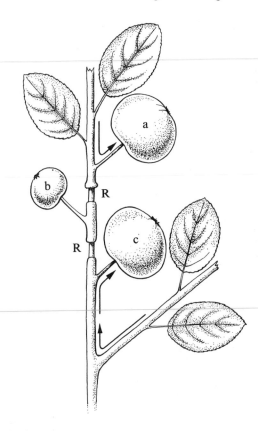

Figure 11.4   The growth of apples on a branch girdled at R indicates that phloem transport can take place in either direction. The apple b, cut off from assimilate supplies by the two girdles, does not grow. Apples a and c grow equally well; a is supplied by basipetal, c by acropetal movement in the phloem of the branch. (From Münch, 1930.)

can be transported into a growing apple in either direction (Münch, 1930). If the locations of source and sink are changed, the direction of transport in the phloem is reversed. This happens every time a leaf matures: as long as the leaf is very young and vigorously growing, it imports assimilates via the phloem. When photosynthetic production exceeds consumption by growth and respiration, export begins, i.e., the direction of transport is reversed. Old leaves never import photosynthates via phloem even when darkened, but they can be forced to do so: application of kinetin can create a local sink in an old leaf (Mothes, 1961; Müller and Leopold, 1966).

Hartt and Kortschak (1964) reported basipetal polarity of sugar movement in leaf blades of the sugar cane. This may perhaps be explained as an indirect effect. The equilibrium of sugars entering and leaving the sieve tubes may differ in older (distal) and younger (basal) parts of (monocotyledonous!) leaves. If older parts are stronger sources than younger parts, polar transport could result in detached leaves.

Lateral movement.   It is difficult to decide to what extent phloem transport is strictly vascular, or, to put it into other terms, how easily solutes are transferred from one channel to a neighbouring one. Injury to the vascular system of most plants is followed by renewal and re-location of vascular tissue around the wound, an important survival factor for plants. Experiments on lateral transport must be kept short in order to rule out formation of new conducting tissue. For example, labelled sugars fed to a single sunflower leaf accumulated in a single radial row of seeds. However, if one or more leaves were removed from one side of the plant, seeds were supplied from other leaves (Prokofyev *et al.*, 1957). It is uncertain whether in this case a reorientation of vascular tissue was involved.

Weatherley *et al.* (1959) found remarkable axial movement towards an aphid stylet bundle. Zimmermann (1960) defoliated one half of a Y-shaped tree and found that photosynthetic products from the foliated side spread tangentially less than 1° towards the phloem belonging to the defoliated side. These two reports seem to concern axial transport, i.e., the degree of tangential cross connections of sieve tubes. Peel (1964, 1966) studied lateral transport under conditions in which source and sink were located on different sides of the stem. Movement of sucrose from one stem side to the other did occur, probably involving considerable lateral transfer from sieve tube to sieve tube.

Palms and other monocotyledons have a very extensive system of cross connections between neighbouring vascular bundles in the form of vascular bridges between leaf traces and axially running bundles (Zimmermann and Tomlinson, 1965). This interconnection may be vital in most monocotyledons because old stems are unable to produce re-located vascular tissue in response to wounding (Tomlinson, pers, comm.).

Bidirectional movement

The question of bidirectional movement is a fundamental one from the point of view of translocation mechanism. Movement of substances in two different directions within a single sieve tube at any one time is not possible with liquid flow. On the other hand, Canny's hypothesis requires that any applied substance which moves in the phloem *must* be distributed in opposite directions from the point of entry, regardless of net movement. Demonstration of bidirectional movement has been claimed many times in the past, but the majority of the reported experiments were of a duration which would have allowed substances ample time to be recirculated via phloem and xylem. Even in recently reported movement of photosynthates into old leaves (e.g., Neeracher, 1965) xylem transport (and secondary uptake into phloem) is not excluded. Eschrich (1967) recently reported bidirectional transport in a single sieve tube. [14]C was applied above and fluorescent dye below feeding aphids, and both tracers were found in honeydew. This is transport towards an aphid from two sides and confirms what Weatherley et al. (1959) have shown. It has nothing to do with the controversial question of bidirectional movement in a single channel.

Bidirectional movement does occur, however, in seedlings and rapidly growing shoots. This is illustrated in Fig. 11.5. Leaves near a shoot apex almost invariably export assimilates both up and down the stem (Fig. 11.5a). As the shoot grows, new leaves mature which also export in both directions (Fig. 11.5b). In the segment between these two sets of leaves phloem transport is bidirectional.

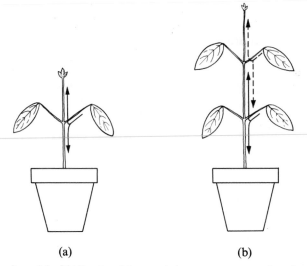

(a)   (b)

Figure 11.5 A model of bidirectional transport. Leaves near a growing shoot apex export assimilates in two directions from the point of petiole attachment (a). In the internode between two such sets of leaves (b), transport is bidirectional.

The question now arises as to whether opposing movement is in the same or in different channels. From an anatomical point of view we would assume that the development of new phloem in the stem is continuously associated with the development of leaves. Translocation in the two directions would then automatically go through different layers of phloem. A very careful study of this situation was made by Biddulph and Cory (1960). In most cases bidirectional transport was in different bundles, in a few cases it was in the same bundle. The authors suggested that two different transport mechanisms might be involved, causing bidirectional movement in younger, unidirectional movement in older phloem layers. However, one could equally well assume movement in opposite directions in the two respective layers. Be this as it may, Biddulph and Cory (1965) fully realize the importance of recognizing the nature of vascular connections (the plumbing system of the plant), a fact many other authors tend to ignore.

Bidirectional transport of an applied substance is essential for the protoplasmic streaming hypothesis but evidence in support is negative so far. Schumacher (1933) found that applied potassium fluorescein moved basipetally in the petiole. Ziegler (1956) made similar observations with applied monosaccharides. These results are decidedly in conflict with the concept of protoplasmic streaming. The question is very important and similar experiments should therefore be repeated with applied radioactive tracers.

Translocation into fruits

In his book (which everybody cites but nobody seems to read) Münch (1930) discusses translocation into fruits. He suggests that during growth certain dry fruits might receive more water via phloem than they can transpire or retain. In such cases, he suggests, water might return to the plant via xylem. This statement triggered a certain amount of experimental work. Doepp (1939) found that *Cucurbita* fruits did not properly develop when the xylem of the fruit stalk was removed. From this he concluded that there was no return of xylem water. Ziegler (1963), on the other hand, did obtain evidence for xylem water movement from the fruit back into the plant. The basis of such conflicts is the fact that the dry weight of most fruits is quite comparable to that of sieve-tube exudate, this suggests that xylem water does not have to move exclusively one way or the other. It is conceivable that movement via xylem is at some times into, and at other times, out of the fruit.

Translocation of applied substances

Many substances, when applied to the phloem, are taken into the sieve tubes and transported in the direction of general sugar movement. This is of some practical importance as, for example, in the application of mineral nutrients to leaves (Bucovac and Wittwer, 1957) and of herbicides in weed control (Crafts and Yamaguchi, 1964). Mitchell and Brown (1946), Hay and Thimann (1956) and others have made the interesting observation that when a transportable substance such as a herbicide or mineral is applied to a leaf, it is exported with translocated sugars via the phloem. If the leaf is placed in the dark long enough for its food reserve to be exhausted, export does not take place. If the leaf is placed in the light or dipped into a sugar (but not mannitol or urea) solution, export takes place. Metabolic energy is necessary to take the substance into the phloem, and a flow of sugar out of the leaf also seems necessary to 'sweep' the substance along.

Quantitative aspects of translocation

Translocation velocities. Reported velocities usually range from 10–150 centimetres per hour, higher values (up to approximately 3 m h^{-1}) are also occasionally reported as well as lower ones (10–20 mm h^{-1}). It is very likely that velocities in the phloem are graded across the conducting channels so that there are simultaneously high and low velocities. Therefore the meaning of any report has to be very carefully assessed. For example, the velocity distribution of liquid flow through a capillary is paraboloid. Velocities range from zero at the capillary walls to a maximum in the centre of the tube (cf., e.g. Reiner, 1960). This means that peak velocities of flow are twice as great as average velocities. This example is not meant to imply that phloem transport is directly comparable with liquid flow in capillaries, it merely emphasizes the complexity of graded velocities. Reported phloem transport velocities are often maximum values, or, expressed more precisely, minimal peak velocities, obtained from the timing of the arrival, at a given point, of a radioactive front.

A comparison of velocities of simultaneously moving substances would be very useful, because any mass flow concept requires that all substances move together in a single solution. Attempts to do such experiments have been made (e.g., Swanson and Whitney, 1953) but interpretation is far from simple (Horwitz, 1958). The velocities measured are apparent velocities; the more of a substance is removed from the sieve tubes along the way, the smaller is its apparent velocity. If removal rates of two substances are different their apparent velocities differ also, even though their actual velocities of movement are the same. Similarly, the rate of entry into the sieve tubes of an applied substance affects the apparent velocity. Obviously these difficulties are particularly serious in seedlings where rapid growth of the axis makes strong demands on translocates.

Translocation velocities conflict with the protoplasmic streaming hypothesis. The highest streaming velocities in cells of higher plants are of the order of 50 mm h^{-1} (Kamiya, 1960). Average translocation velocities are about 10 times greater. Protoplasmic streaming seems to be insufficient to move solutes at the reported rates.

Nelson *et al.* (1959) reported very rapid translocation (ca. 50 m h^{-1}) of ^{14}C from leaf to roots in soybean seedlings. Only part of this transported isotope passed steam girdles. In ungirdled plants some of the radioactive label accumulated in living tissue of the root. From this the authors concluded that there are two fast transport mechanisms, one in the xylem, the other in living tissue, besides the relatively slow (1 m h^{-1}) phloem transport. An alternative and simpler explanation of these experiments can be given if it is assumed that fast transport took place only in the xylem, whereas accumulation in living tissues of the root was secondary. The plants were steam girdled the day before the experiments so that roots were starved and metabolic energy for uptake into living tissues was not available. The authors have good evidence that movement to roots was not in the form of carbon dioxide. The matter merits reinvestigation.

Mass transfer. Numerous attempts have been made in the past to assess the translocation process quantitatively by measuring the dry-weight increase of fruits or tubers and by estimating the transverse-sectional area of tissue through which translocation to these areas has to take place (e.g., Clements, 1940; Crafts, 1933; Dixon and Ball, 1922). The usual purpose of such measurements was to calculate how fast a solution of a certain concentration would have to flow in order to account for the amount of transferred dry material. Canny (1960) collected and reviewed many of these results and made the interesting discovery that the translocation capacity is a rather well defined quantity. The reported measurements included (a) mass transfer, i.e., the amount of dry weight transferred per unit time (with a correction for respired material), and (b) the transverse-sectional area of phloem tissue or sieve tubes. Canny found that it is very useful for comparative purposes to express the results of different authors as 'specific mass transfer':

$$\text{Specific mass transfer (SMT)} = \frac{\text{Mass transfer}}{\text{Transverse-sectional area}}$$

If we consider the possibility of translocation being a flow of a solution, SMT would have to be equal to the average flow velocity multiplied by the concentration of the moving solution:

$$\text{SMT } [\text{g h}^{-1} \text{ cm}^{-2}] = \text{conc. } [\text{g cm}^{-3}] \times \text{average velocity } [\text{cm h}^{-1}]$$

This simply equates mass transfer to a liquid cylinder moving at a certain speed. The reader is reminded that liquid flow never takes place in such a way (see p. 401). Even in ideal capillaries flow is paraboloid. This means that peak velocities would have to be twice the average velocity of the above equation.* To our knowledge only xylem vessels of vines are sufficiently long and uniform to permit ideal capillary flow. Sieve-tube transport, even if considered mass flow,

*The volume of a cylinder is $r^2\pi h$, that of a paraboloid $r^2\pi h/2$.

would hardly be of a simple paraboloid nature. However, it is emphasized that maximum velocities in sieve tubes are very likely considerably higher than the average velocity in the above equation.

Collecting all available data, Canny (1960) found that during times of peak translocation SMT values of various plants are quite uniform, namely ca. 2–5, based upon the transverse area of the whole phloem (SMT_{ph}). Based upon sieve tubes only (SMT_{st}), the figures would have to be multiplied by a factor of 5 because sieve tubes occupy about 20 per cent of the transverse phloem area. For petioles, values seem to be consistently lower, namely $SMT_{ph} = 0.56$–0.7. Münch (1930) reported mass transfer studies of a number of tree species. His results are based upon the transverse area of sieve tubes. They range from 9–14 for dicotyledonous trees. This compares well with Canny's survey ($SMT_{st} = 10$–25). An interesting point is that Münch's figures for pine ($SMT_{sieve\ cells} = 2.22$–$2.45$) resemble figures for dicotyledonous petioles, i.e., is considerably lower than for dicotyledonous stems.

Trees offer the unique opportunity to study the right-hand side of the specific mass transfer equation. This enables us to see if the concept of mass flow is reasonable in terms of specific mass transfer. Indeed, Münch's sieve-tube exudate concentrations which he reported for the same individual trees used for SMT calculations, yield required average velocities of 40–70 cm h^{-1} for the dicotyledons and 18–20 cm h^{-1} for pine. These are indeed very reasonable values in view of our knowledge of measured velocities in trees. In an ideal experiment all factors of the equation should be measured simultaneously in a plant. This has never been done, but would certainly be a very worthwhile project.

Mass transfer measurements can be assessed in terms of Fick's diffusion equation (see p. 391). In a diffusion process, SMT has to be equal to the diffusion coefficient times the solute concentration gradient along any given path. Making calculations with some of the known quantities (SMT per phloem or per sieve tube, diffusion coefficient of sugar in water, solute gradients in sieve tubes (p. 405), etc.) is very illuminating indeed. By doing a few exercises himself, the reader will understand much better what has been said in the introductory paragraph to the translocation mechanisms (p. 391).

Mathematical models of translocation. The reliability of mathematical calculations has led to more elaborate treatment of the translocation problem in the setting up of mathematical models of translocating mechanisms (e.g., Horwitz, 1958). Instructive as these are, it is dangerous to rely too heavily on them because there are many variables about which we have no information. If too many assumptions are made, such models become mere 'statements of faith'. As more is learned about translocation it might be possible to base these models more and more upon facts and they might become more reliable. Up to the present, however, they have not decisively influenced our way of thinking.

Osmotic considerations. Gradients

Osmotic phenomena in sieve tubes. Exudation indicates good longitudinal permeability within the sieve tubes. At the same time there is a remarkable lateral differential permeability, that is, sieve tubes do not leak. The sieve-tube system retains its turgescence for days after the supply of photosynthetic products to the phloem is interrupted by defoliation (Zimmermann, 1958). When sieve-tube turgor is released by means of an incision, the concentration drops gradually to a lower level, indicating entry of water from the surrounding tissue. This is exactly what one would expect from an osmotic system (Zimmermann, 1960). Another indication is the dependence of the concentration of the sieve-tube exudate upon the relative pressure in the apoplast (xylem). When the tension in the xylem goes up, the concentration in the sieve tubes rises, water is 'pulled' out of the sieve tubes (Weatherley et al., 1959; Peel and Weatherley, 1962), when pressure is applied to the xylem, water appears to be 'pushed' into sieve tubes (Peel, 1965).

Exudation from aphid stylet bundles on isolated sections of branches indicates two things: (a) sieve tubes are sealed off near the cut ends of the branches (see p. 408), and (b) the lowered sieve-tube turgor caused by the drain results in a mobilization of reserve sugars. There seems to be a simple relationship: lowered turgor causes net water entry and thus dilution of the internal concentration; lowered concentration triggers (enzymic) net entry of sugars into sieve tubes.

A discussion of osmotic phenomena must contain comments about translocation in wilting plants. It is well known that phloem transport can go on in plants which are beginning to wilt, and this is often taken as evidence against mass flow. It is equally well known, however, that secretion into sieve tubes is often against a concentration gradient. It is therefore clear that the sieve-tube turgor does not have to drop to zero even if the leaf parenchyma is in a state of wilting. This assumption is confirmed by continued exudation from aphid stylet bundles in wilting plants (Weatherley et al., 1959).

Radioactivity gradients along the translocation path. When radioactively labelled material is exported from a leaf, the distribution pattern in extracts from axial tissue (including isotope from inside and outside the translocation channel!) often decreases logarithmically from the point of origin (e.g., Biddulph and Cory, 1957; Swanson and Whitney, 1953). In other words, if the logarithm of radioactivity is plotted against distance, individual points are more or less on a straight line. Such a drastic decrease of isotope along the translocation path is an expression of sieve-tube 'unloading'; it is typical for seedlings in which utilization is pronounced along the axis. Sugar-beet petioles do not seem to consume much of the sugar that moves through them, the gradient of activity of extracts is more nearly a linear one (Mortimer, 1965). Canny (1962a) who studied the translocation profile in *Salix* found that only the advancing front of radioactivity shows a logarithmic slope, behind this frontal slope, radioactivity remains more or less constant.

Nelson *et al.*, (1959) reported an interesting observation concerning the radioactivity front. Beyond the point where the front reaches the level of background radiation (further away from the supply leaf) there is another front of irregular activity, causing the profile to be discontinuous. This discontinuity is not shown in results plotted by other workers; it may have something to do with the entry of isotope into the sieve tubes, and should be investigated further before any significance can be attached to it.

Concentration gradients in the sieve tubes. An osmotically driven mass-flow requires a turgor gradient in the direction of flow. If flow is towards a meristem where turgor is lowered by growing cell walls, the turgor gradient does not have to be reflected by a molar concentration gradient. However, where the translocation path is quite uniform as in a tree stem for example, a decrease of molar concentration in the direction of transport would be anticipated. Such gradients are invariably found in trees; their magnitude is of the order of 0.01 mol m^{-1}. Moreover, when the sugar supply to the phloem is eliminated by defoliation, some individual gradients remain positive, others become negative and the gradient of total molarity disappears (Zimmermann, 1958). This is in agreement with the mass-flow hypothesis but in disagreement with the protoplasmic streaming theory which requires that all individual gradients disappear.

Of particular interest is Canny's (1962) comparison of the radioactivity profile along a stem with a time study of arrival of radioactivity at an aphid stylet bundle. The two curves correspond very well, indicating translocation in the sieve tubes. The velocities so obtained were rather low (10–22 mm h^{-1}) possibly because the phloem was not functioning at full capacity in the plants used. It would be interesting to do these experiments under conditions of known peak mass transfer. Advancing wave profiles are evidence in support of Canny's protoplasmic streaming theory (Canny and Phillips, 1963).

Structural aspects

Once more it is emphasized that space for this chapter does not permit a full description of phloem anatomy. The following discussion assumes that the reader is familiar with the general anatomy of phloem, and has studied one of the recent summaries of submicroscopic structures as well (e.g., Esau, 1966; Evert *et al.*, 1966; Kollmann, 1964). This section merely discusses those structural aspects which are of immediate interest from a functional point of view.

Sieve-element protoplast. The crucial question about translocation is whether it involves a flow of solution or movement of solute molecules independent of solvent water. Naturally, this question hinges on structural features. Developmental studies of the sieve-element cell wall of *Cucurbita* have shown that the pores of the sieve plates are broken through, i.e., skeleton material is eliminated to make room for connecting strands (Frey-Wyssling and Müller, 1957). The sieve element protoplast undergoes a series of characteristic modifications of which the two most conspicuous are disappearance of the nucleus and tonoplast (Esau, 1966). This series of events has been described in various ways. Many years ago the term 'pre-mortal' was coined for the mature state of sieve elements. This term has caused a great deal of misunderstanding. It is undoubtedly an unfortunate one because sieve tubes of tall monocotyledons like palms remain functional for 50 and more years. The reduction of metabolic activity in sieve elements is merely a sign of very high specialization expressed in various ways like lack of nucleus, reduced elaboration of mitochondria, scarcity of ribosomes, etc. That this specialization is correlated with conduction is shown by giant kelps which, although 'lower' plants and therefore generally considered non-vascular, have evolved very similar files of translocation cells.

Plasmolysis of mature sieve elements is possible but considered very difficult (Currier et al., 1955). This difficulty of plasmolysing sieve elements is often explained as the result of loss of differential permeability during maturation of sieve elements. In saying this, one has to be more specific. It is certain that lateral differential permeability is not lost, because the sieve tubes remain highly turgescent during their functional life. The permeability from one sieve element to the next via sieve plates is a longitudinal one. Lateral loss of water in a plasmolysis experiment at any one point can therefore be corrected by a longitudinal flow of cell sap. This is the most reasonable explanation of why it is difficult to plasmolyse sieve elements.

Connecting strands have been examined closely by electron microscopists during recent years. But the sieve tubes are very sensitive to injuries. When a sieve tube is cut, cell sap rushes towards the point of injury and fine structures are disturbed. Chemical fixation often causes partial closure of the pores by callose. Early electron micrographs were very crude illustrations of artifacts; it is only during recent years that methods of preparation have become more refined, but interpretation of sieve pore content still remains uncertain. Images depend upon the type of fixation. Pores of lateral sieve areas resemble plasmodesmata somewhat, except that they are larger. Pores of sieve plates may contain cytoplasmic components or membrane structures, sometimes interpreted as tubules and possibly deriving from the endoplasmic reticulum (cf., e.g., Evert et al., 1966; Kollmann, 1964). The function of connecting strands is of critical importance. They have been interpreted as transcellular strands or as structures providing the basis for an interfacial flow of solutes; they have been regarded as the site of a sieve-plate potential, or merely as a place of increased resistance to flow.

Resistance to flow in sieve tubes.   This question is a complex one. Numerous authors have made calculations of 'pressures necessary if pores were open', etc.

Almost invariably the Poiseuille equation has been used, or rather misused, for such calculations. The problem must be investigated step by step as follows. Let us first imagine sieve tubes without any sieve plates, and consider flow through them as through a capillary. Poiseuille's equation is experimentally supported for glass capillaries of a diameter of 0·01–0·3 mm (Reiner, 1960). At 20°C, a pressure gradient of about 0·4 atm m^{-1} would push a 20 per cent sucrose solution through a capillary of 10 microns diameter at a peak velocity of 0·5 m h^{-1}. If we increase the capillary diameter to 30 microns, 0·1 atm m^{-1} would be sufficient for a peak velocity of 1 m h^{-1}. If the capillary is 40 microns in diameter, 0·05 atm m^{-1} would push it at 1 m h^{-1} peak velocity. This simple situation is readily comprehended. Mentally introducing the sieve plates gives us a complex structure which cannot be handled theoretically with the Poiseuille equation, because a length equivalent to about 50 capillary diameters is needed for the flow paraboloid to develop. Sieve pores are very much shorter than this required length. Fortunately, there is a model for flow in such a complex system: coniferous wood. An appreciation of the fine structure of tracheids as described for example by Frey-Wyssling and Bosshard (1953) and Liese and Bauch (1964) will be useful to the reader in understanding the following. Tracheids are dead, spindle-shaped cells, water-filled when functional. Neighbouring tracheids are in contact via bordered pits which are essentially membranes with many tiny pores a fraction of a micron in diameter. The tracheid skeleton is therefore quite comparable with the structure of sieve cells or sieve tubes with 'open pores': flow would have to go through each cell but pass from cell to cell via many tiny pores. Experimental flow studies have been made with coniferous wood and show that the pressure necessary to push water through tracheids is about three times as great as the pressure needed for pushing water through 'endless' capillaries with the same diameter as the tracheids (Münch, 1943). If sieve areas are similarly considered, theoretical values given at the beginning of this paragraph would have to be multiplied by a factor of 3 as sieve plates are introduced. Such pressure requirements are still reasonable.

The theoretical situation is further complicated, however, because some of the sieve pores show structural material in electron micrographs. The nature of this material is still debated but it is perfectly clear that calculations under these conditions involve wild speculations. Nevertheless, in my opinion mass flow is not excluded if the following considerations are made. The 'structures' seen in the sieve pores in electron micrographs may be quite 'fluid' and part of the moving stream. This would explain why, in certain preparations, some pores are 'filled', others not (Esau, 1966). Stylet exudate viewed under the electron microscope might illuminate this point. Indeed, if these structures are moving, perhaps some sort of unification of the transcellular protoplasmic streaming and the mass-flow hypothesis could be found. If the structures are stationary, they could provide the site for an electro-osmotic potential which drives the solution across the sieve plates. This is at least a possibility, although it is difficult to see under such conditions, how flow towards an aphid stylet bundle could be instantaneously reversed.

At this stage of our knowledge we simply have to admit that the rheology of the sieve plate area is entirely unknown. Liquids can be in contact with solids via wall layers acting as 'lubricants'. Furthermore there are 'slippage' phenomena of liquid flow (Reiner, 1960). All these possibilities are still unexplored. In conclusion let us remember exudation from aphid stylet bundles: solution *does* flow!

The phloem's protection against injury.   One of the striking features of sieve tubes is their extreme sensitivity to injury. This is not surprising in view of their turgescence and high longitudinal permeability. When sieve tubes are severed their content suffers a shock-like displacement towards the point of injury. This results in a plugging of sieve plates near the wound with so-called 'slime'. During sieve-element development the cytoplasm undergoes modification (Esau, 1966) and slime is considered to be derived during this cytoplasmic change. Species differ in the effectiveness of this plugging and in some plants it is incomplete so that it is possible to collect exudate from an incision. Aphids have 'learned' to tap the sieve tubes 'gently'. The resistance to flow in the stylet bundle is sufficient to maintain the turgor in the sieve tube at a level which does not trigger slime plugging.

A second mechanism of sieve-plate closure is the formation of callose in the sieve pores. Callose is a $\beta,1$-3, glucose polysaccharide (Kessler, 1958) which reacts with resorcin blue and other stains and can thus be made visible in microscopic preparations of phloem tissue. Callose formation can be very rapid and follows slime plugging as a secondary wound reaction. A comprehensive review on the physiology of callose formation has been published by Eschrich (1965). Some years ago, callose cylinders were considered a normal feature of functioning sieve tubes but more recently it has been found that they rapidly appear as fixation artifacts. They are formed as a wound response prior to death of tissue before and during the process of tissue fixation.

An important function of callose is the closure of sieve pores during the dormant period such as winter in temperate climates. Callose is re-dissolved with the break of dormancy and the sieve tubes, thus re-activated, may function again before new ones are formed (Esau, 1948; Hill, 1962; Kollmann and Schumacher, 1962). It is apparent that complete closure of sieve plates is correlated with cessation of transport. At least this is what aphid stylet experiments indicate (Hill, 1962).

A peculiar finding, reported recently by Eschrich *et al.* (1965) was that restriction of sieve pores seemed to accelerate transport. The meaning of this is unknown and the matter is certainly worth further investigation. In terms of mass flow, increased velocity due to narrowed channels does not seem unreasonable.

Closure of sieve plates upon injury is obviously of vital importance to the plant. Without this mechanism wounding would cause serious drains upon the plant's reserves (see p. 412). Botanists have often wondered why sieve plates have not disappeared during the course of evolution as have vessel-element end walls in many species. If sieve plates represented merely resistance to flow one could argue that they would have disappeared during evolution. But one could argue with equal justification that the plates' protective function had a greater survival value than disadvantage due to their resistance.

Longevity of sieve tubes.   Longevity of sieve tubes is quite limited in dicotyledons. Even in woody plants which may reach an age of hundreds or even thousands of years, vascular tissues (xylem and phloem) are regularly renewed. Renewal may be continuous or periodic, as in the tropics, or annual as in temperate climates. Therefore phloem seldom functions more than one year as a conducting tissue. This can be seen in transverse sections of many different kinds of bark where older sieve tubes are collapsed (Huber, 1939). It has been suggested repeatedly that it may be primarily, or even exclusively, the young sieve tubes which function in conduction (e.g., Kollmann, 1965). This suggestion, even if it has some experimental support, cannot be generalized, because the plant kingdom contains a great number of arborescent monocotyledons (the palm family alone contains about 2,500 species) in which vascular tissue in the stem is not renewed. The sieve tubes in the lower part of the stem have to function throughout the whole lifetime of these plants which in many species may be over 50 years (Parthasarathy and Tomlinson, 1967).

Conclusions

Both rates of exudation from cut aphid stylets and specific mass transfer calculations, are strong evidence in favour of the mass-flow concept. Interfacial spreading of solute molecules is an interesting suggestion, but so far there is no experimental support for such a mechanism. Movement of widely differing molecules and particles, as well as much of the evidence indicating mass flow, speaks directly against interfacial spreading. The protoplasmic streaming theory cannot be dismissed entirely because streaming does occur in many cells and short-distance transport in such tissues is almost certainly accelerated by protoplasmic streaming. The question which we have discussed in this chapter is the long-distance transport through sieve tubes and here several reasons make it very unlikely that protoplasmic streaming is a factor. First, streaming in mature sieve tubes does not seem to have been convincingly reported; second, velocities of protoplasmic streaming as measured in various cells are at least 10 times too slow for rapid long-distance transport; third, with protoplasmic streaming mass transfer of large amounts is impossible over distances of metres; fourth, fluorescent dyes should be moved bidirectionally by protoplasmic streaming from the point of application, but they are not.

We are then left with the possibility of mass flow, driven by osmosis, or possibly by electro-osmosis, or with an entirely unknown mechanism. The crucial difficulty with the mass-flow concept is the significance of the structures in the sieve pores seen on electron micrographs. Are they artifacts? Are they passively or actively moving, or are they stationary structures? Do they possess some rheological significance? We do not know. Nor do we have any idea of how resistance to flow could be measured or calculated. Calculations made in the past are highly speculative and rheologically unfounded. Critics of the mass-flow hypothesis have often failed to appreciate the special features of the concept: metabolic energy is required all along the way to maintain differential permeability and to exchange solute molecules with the surrounding tissue. Therefore it can be concluded that the mass-flow concept is still by far the best single explanation of long-distance transport in the phloem, although this statement must not be made dogmatically but with caution and readiness to revise it if future evidence requires this. It is also conceivable that more than one mechanism is responsible for phloem transport. This suggestion has been made by a number of workers during the past years.

Storage and circulation of nutrients

Storage and mobilization

Plants have peak periods of production and of consumption both in regard to short-term as well as long-term periods. Photosynthetic capacity generally exceeds the phloem's capacity to export assimilates from leaves. Transport goes on during the day, at the same time excess photosynthate is stored in leaf parenchyma, mostly in the form of starch. As the day ends, and with it photosynthesis, the accumulated starch in the leaf is mobilized again and transported away throughout the night (e.g., Schneider-Orelli, 1909).

Woody plants which grow periodically accumulate reserve materials in the parenchyma of wood and bark during times of photosynthetic activity. They then go through a period of dormancy such as winter in temperate climates, a dry season or merely an internally imposed dormant period in the tropics. As growth is resumed, the stored reserves allow the plant to produce new shoots and leaves quickly. Storage in tubers of perennial plants, or in seeds, serves similar purposes. Storage is also important if periods of especially high demands arise. Many tree species flower at the end of the dormant period before the leaves emerge. In doing this they draw heavily upon their reserve materials. Other plants conclude their life cycle by flowering. They store reserve material during earlier vegetative growth, which may be several decades in an *Agave* or in palms like *Corypha* or *Metroxylon*, and mobilize it all during the brief flowering time.

Recovery of nutrients from organs prior to abscission

The change in colour of many leaves in autumn is to some extent the sign of breakdown of certain substances such as chlorophyll. During senescence of leaves and flowers some substances, especially those containing nitrogen and phosphorus, are broken down into soluble forms and translocated back into the plant before the organ abscisses. At such times the sieve-tube exudate contains numerous amino acids, amides, and other substances which are not present in detectable amounts during other times of the year (Mittler, 1958; Hill, 1962; Ziegler, 1956; see also the discussion in Zimmermann, 1964). In this recovery of nitrogenous substances from flowers prior to their abscission, some plants are 'wasteful', others are 'thrifty' (Schumacher, 1931).

Circulation of nutrients within the plant

Long-distance transport takes place simultaneously through xylem and phloem. Xylem transport is primarily upwards from roots to leaves, phloem transport is primarily downwards, from leaves to roots. But there are circumstances when translocation in the two systems is unidirectional, for example into developing flowers and growing shoot tips. The two systems are largely (though not entirely) independent of each other.* Rates and direction of movement in both are regulated by existing demands. This explains the plants' remarkable ability to distribute and re-distribute nutrients efficiently (cf. Biddulph *et al.*, 1958; Bachofen and Wanner, 1962; etc.).

The mobility of substances in xylem as well as phloem is quite variable. Certain minerals are poorly mobile in the xylem because they are absorbed along the way. Therefore, even if taken up they may in effect be unavailable to the plant. Mobility of iron, for example, can be greatly improved by 'chelating' it, i.e., binding it to an organic molecule so that it is not precipitated along its way through the xylem (cf. Bollard, 1960). Mobility in the phloem is similarly variable. We know for example that calcium moves very poorly in the phloem. It is rapidly removed from the sieve tubes and deposited in certain cells in the form of oxalate crystals. Thus, not all substances are equally easily circulated in the plant.

Circulation in plants has been investigated quite thoroughly by Kursanov and his co-workers in connection with root metabolism. Nitrogen ascends in the plant primarily in the form of amino acids and amides. The carbon skeletons for these substances are sent to roots via phloem. This phloem transport is directly regulated by the demand of the roots. Sugar transport towards the roots is depressed reversibly when the roots are subjected to mineral deficiency, in

*The dependence of the two primarily involves water. As transpiration increases and the tension in the xylem goes up, water is withdrawn from the phloem (e.g., Peel and Weatherley, 1962). This can ultimately lead to a considerable reduction of phloem transport (e.g., Roberts, 1964).

other words the demand for carbon skeletons in the roots stimulates translocation towards these points (Kursanov, 1961, 1963, see also the papers cited therein).

Regulation of phloem transport

Phloem transport is regulated, as pointed out repeatedly before, by supply at the source and demand at the sink. In some examples, many of them discussed on the previous pages, the situation is quite clear. The mechanism described on p. 404 with the following sequence: turgor drop—water entry—dilution—enzymic sugar entry, is a direct demand type regulation from the sink end. This mechanism appears to work in the intact plant wherever turgor is lowered by rapid growth or concentration is lowered by polymerization of sugars into storage products. Another type of regulation is the control of the entry and removal balance. Experimentally this has been achieved by the application of kinetin (see chapter 3) (shift of balance towards increased removal from sieve tubes). During recent years some evidence has emerged indicating that there might be hormone-like messengers from the leaves controlling loading and unloading of sieve tubes. This effect could, of course, be direct or indirect. Mechanisms are still obscure and the different experimenters may have been dealing with a wide variety of effects (Eschrich, 1966; DeStigter, 1961; Zimmermann, 1958). Regulation of phloem transport is one of the least investigated, but at the same time most fascinating, aspects of this field.

Bibliography

Selected further reading

Crafts, A. S., 1961. *Translocation in plants*. Holt, Rinehart and Winston, New York.
Esau, K., 1961. *Plants, viruses, and insects*. Harvard University Press, Cambridge, Massachusetts.
Kursanov, A. L., 1963. Metabolism and transport of organic substances in the phloem. In *Adv. in Bot. Research*, Vol. 1. Academic Press, London.
Münch, E., 1930. *Die Stoffbewegungen in der Pflanze*. Fischer, Jena.
Nelson, C. D., 1962. The translocation of organic compounds in plants. *Can. J. Bot.* **40**, 757–770.
Swanson, C. A., 1959. Translocation of organic solutes. In *Plant Physiology*, ed. F. C. Steward, Vol. II. Academic Press, New York, pp. 481–551.

Papers cited in text

Bachofen, R. and H. Wanner, 1962. Transport und Verteilung von markierten Assimilaten VI. *Ber. Schweiz. Bot. Ges.* **72**, 272–279.
Barrier, G. E. and W. E. Loomis, 1957. Absorption and translocation of 2,4-dichlorophenoxyacetic acid and P^{32} by leaves. *Plant Physiol.* **32**, 225–231.
Bauer, L., 1952. Zur Mechanik der Stoffwanderung in den Siebröhren. *Naturwiss.* **39**, 207.

Biddulph, O., S. Biddulph, R. Cory, and H. Koontz, 1958. Circulation patterns for phosphorus, sulfur and calcium in the bean plant. *Plant Physiol.* **33**, 293–300.

Biddulph, O. and R. Cory, 1957. An analysis of translocation in the phloem of the bean plant using THO, P^{32}, and C^{14}. *Plant Physiol.* **32**, 608–619.

Biddulph, O., and R. Cory, 1960. Demonstration of two translocation mechanisms in studies of bidirectional movement. *Plant Physiol.* **35**, 689–695.

Biddulph, O. and R. Cory, 1965. Translocation of C^{14} metabolites in the phloem of the bean plant. *Plant Physiol.* **40**, 119–129.

Bidwell, R. G. S., J. S. Craigie, and G. Krotkov, 1958. Photosynthesis and metabolism in marine algae III. *Can. J. Bot.* **36**, 581–590.

Bollard, E. G., 1960. Transport in the xylem. *Ann. Rev. Plant Physiol.* **11**, 141–166.

Brown, C. L., 1964. The influence of external pressure on the differentiation of cells and tissues cultured *in vitro*. In *The formation of wood in forest trees*, ed. M. H. Zimmermann. Academic Press, New York.

Bukovac, M. J. and S. H. Wittwer, 1957. Absorption and mobility of foliar applied nutrients. *Plant Physiol.* **32**, 428–435.

Canny, M. J., 1960. The rate of translocation. *Biol. Rev.* **35**, 507–532.

Canny, M. J., 1962a. The translocation profile: sucrose and carbon dioxide. *Ann. Bot.* **26**, 181–196.

Canny, M. J., 1962b. The mechanism of translocation. *Ann. Bot.* **26**, 603–617.

Canny, M. J. and O. M. Phillips, 1963. Quantitative aspects of a theory of translocation. *Ann. Bot.* **27**, 379–402.

Cheadle, V. I. and K. Esau, 1958. Secondary phloem of Calycanthaceae. *Univ. of Calif. Publ. in Botany*, **29**, 397–510.

Choi, I. C. and S. Aronoff, 1966. Apetiolar photosynthate translocation. *Plant Physiol.* **41**, 1130–1134.

Clements, H. F., 1940. Movement of organic solutes in the sausage tree, *Kigelia africana*. *Plant Physiol.* **15**, 689–700.

Crafts, A. S., 1933. Sieve-tube structure and translocation in the potato. *Plant Physiol.* **8**, 81–104.

Crafts, A. S., 1961. *Translocation in plants*. Holt, Rinehart & Winston, New York.

Crafts, A. S. and S. Yamaguchi, 1964. *The autoradiography of plant materials*. Div. of Agricult. Sciences, Univ. of Calif. Manual 35.

Currier, H. B., K. Esau, and V. I. Cheadle, 1955. Plasmolytic studies of phloem. *Am. J. Bot.* **42**, 68–81.

Curtis, O. F., 1935. *The translocation of solutes in plants*. McGraw-Hill Book Co., New York.

Curtis, O. F. and G. N. Asai, 1939. Evidence relative to the supposed permeability of sieve tube cytoplasm. *Am. J. Bot.* **26**, 16s–17s.

De Stigter, H. C. M., 1961. Translocation of C^{14}-photosynthates in the graft muskmelon/ *Cucurbita ficifolia*. *Acta Botanica Neerl.* **10**, 466–473.

De Vries, H., 1885. Ueber die Bedeutung der Circulation und der Rotation des Protoplasmas für den Stofftransport in der Pflanze. *Bot. Z.* **43**, 1–6, 18–26.

Dixon, H. H. and N. G. Ball, 1922. Transport of organic substances in plants. *Nature, Lond.* **109**, 236.

Doepp, W., 1939. Beiträge zur Frage der Stoffwanderung in den Siebröhren. *Jahrb. wiss. Bot.* **87**, 679–705.

Esau, K., 1948. Phloem structure in the grapevine, and its seasonal changes. *Hilgardia*, **18**, 217–296.

Esau, K., 1965. *Plant anatomy*, 2nd edn. John Wiley & Sons, New York.

Esau, K., 1966. Exploration of the food conducting system in plants. *Am. Scientist*, **54**, 141–157.

Esau, K., E. M. Engleman, and T. Bisalputra, 1963. What are transcellular strands? *Planta*, **59**, 617–623.

Eschrich, W., 1965. Physiologie der Siebröhrencallose. *Planta*, **65**, 280–300.

Eschrich, W., 1966. Translokation ¹⁴C-markierter Assimilate im Licht und im Dunkeln bei *Vicia faba*. *Planta*, **70**, 99–124.

Eschrich, W., 1967. Bidirektionelle Translokation in Siebröhren. *Planta*, **73**, 37–49.

Eschrich, W., H. B. Currier, S. Yamaguchi, and R. B. McNairn, 1965. Der Einfluss verstärkter Callosebildung auf den Stofftransport in Siebröhren. *Planta*, **65**, 49–64.

Evert, R. F. and L. Murmanis, 1965. Ultrastructure of the secondary phloem of *Tilia americana*. *Am. J. Bot.* **52**, 95–106.

Evert, R. F., L. Murmanis and I. B. Sachs, 1966. Another view of the ultrastructure of *Cucurbita* phloem. *Ann. Bot.* **30**, 563–585.

Fensom, D. S., 1957. An electrokinetic theory of transport. *Can. J. Bot.* **35**, 573–582.

Ford, J. and A. J. Peel, 1966. The contributory length of sieve tubes in isolated segments of willow, and the effect on it of low temperatures. *J. Exp. Bot.* **17**, 522–533.

Frey-Wyssling, A. and H. H. Bosshard, 1953. Ueber den Feinbau der Schliesshäute in Hoftüpfeln. *Holz als Roh- und Werkstoff*, **11**, 417–420.

Frey-Wyssling, A. and H. R. Müller, 1957. Submicroscopic differentiation of plasmodesmata and sieve plates in *Cucurbita*. *J. Ultrastruct. Res.* **1**, 38–48.

Gage, R. S. and S. Aronoff, 1960. Translocation III. Experiments with carbon 14, chlorine 36, and hydrogen 3. *Plant Physiol.* **35**, 53–64.

Geiger, D. R., 1966. Effect of sink region cooling on translocation of photosynthate. *Plant Physiol.* **41**, 1667–1672.

Hartig, Th., 1837. Vergleichende Untersuchungen über die Organisation des Stammes der einheimischen Waldbäume. *Jahresber. Fortschr. Forstwiss. u. forstl. Naturk.* **1**, 125–168.

Hartig, Th., 1860. Beiträge zur physiologischen Forstbotanik. *Allg. Forst- und Jagdztg.* **36**, 257–261.

Hartt, C. E. and H. P. Kortschak, 1964. Sugar gradients and translocation of sucrose in detached blades of sugarcane. *Plant Physiol.* **39**, 460–474.

Hay, J. R. and K. V. Thimann, 1956. The fate of 2,4-dichlorophenoxyacetic acid in bean seedlings. II. Translocation. *Plant Physiol.* **31**, 446–451.

Hill, G. P., 1962. Exudation from aphid stylets during the period from dormancy to bud break in *Tilia americana* L. *J. Exp. Bot.* **13**, 144–151.

Horwitz, L., 1958. Some simplified mathematical treatments of translocation in plants. *Plant Physiol.* **33**, 81–93.

Huber, B., 1939. Das Siebröhrensystem unserer Bäume und seine jahreszeitlichen Veränderungen. *Jahrb. wiss. Bot.* **88**, 176–242.

Kamiya, N., 1960. Physics and chemistry of protoplasmic streaming. *Ann. Rev. Plant Physiol.* **11**, 323–340.

Kessler, G., 1958. Zur Charakterisierung der Siebröhrenkallose. *Ber. Schweiz. Bot. Ges.* **68**, 5–43.

Kluge, M., 1967. Viruspartikel im Siebröhrensaft von *Cucumis sativus* L. nach Infektion durch das Cucumisvirus 2A. *Planta* (in press).

Kollmann, R., 1964. On the fine structure of the sieve element protoplast. *Phytomorphology*, **14**, 247–264.

Kollmann, R., 1965. Zur Lokalisierung der funktionstüchtigen Siebzellen im sekundären Phloem von *Metasequoia glyptostroboides*. *Planta*, **65**, 173–179.

Kollmann, R. and W. Schumacher, 1962. Ueber die Feinstruktur des Phloems von *Metasequoia glyptostroboides* und seine jahreszeitlichen Veränderungen III. Die Reaktivierung der Phloemzellen im Frühjahr. *Planta*, **59**, 195–221.

Kursanov, A. L., 1961. The transport of organic substances in plants. *Endeavour*, **20**, 19–25.

Kursanov, A. L., 1963. Metabolism and the transport of organic substances in the phloem. *Adv. in Bot. Res.* **1**, 209–278.

Kursanov, A. L., M. Kh. Chailakhian, O. A. Pavlinova, M. V. Turkina, and M. I. Brovchenko, 1958. Translocation of sugars in grafted plants. *Soviet Plant Physiol.* **5**, 3–15 (translation).

Liese, W. and J. Bauch, 1964. Ueber die Wegsamkeit der Hoftüpfel von Coniferen. *Naturwiss.* **51**, 516.

Mason, T. G. and E. J. Maskell, 1928. Studies on the transport of carbohydrates in the cotton plant I and II. *Ann. Bot.* **42**, 189–253, 571–636.

Matile, Ph., 1966. Enzyme der Vakuolen aus Wurzelzellen von Maiskeimlingen. *Z. Naturforsch.* **21b**, 871–878.

Mitchell, J. W. and J. W. Brown, 1946. Movement of 2,4-dichlorophenoxyacetic acid stimulus and its relation to the translocation of organic food materials in plants. *Bot. Gaz.* **107**, 393–407.

Mittler, T. E., 1957. Studies on the feeding and nutrition of *Tuberolachnus salignus* (Gmelin) (Homoptera, Aphididae). I. The uptake of phloem sap. *J. Exp. Biol.* **34**, 334–341.

Mittler, T. E., 1958. Studies on the feeding and nutrition of *Tuberolachnus salignus* (Gmelin) (Homoptera, Aphididae). II. The nitrogen and sugar composition of ingested phloem sap and excreted honeydew. *J. Exp. Biol.* **35**, 74–84.

Mortimer, D. C., 1965. Translocation of the products of photosynthesis in sugar beet petioles. *Can. J. Bot.* **43**, 269–280.

Mothes, K., 1961. Aktiver Transport als regulatives Prinzip für gerichtete Stoffverteilung in höheren Pflanzen. In *Biochemie des aktiven Transports*. 12. Colloquium Ges. Physiol. Chem. Springer Verlag, Berlin, Göttingen, Heidelberg.

Müller, K. and A. C. Leopold, 1966. The mechanism of kinetin-induced transport in corn leaves. *Planta*, **68**, 186–205.

Münch, E., 1930. *Die Stoffbewegungen in der Pflanze*. Fischer, Jena.

Münch, E., 1943. Durchlässigkeit der Siebröhren für Druckströmungen. *Flora*, **136**, 223–262.

Neeracher, H., 1965. Transportuntersuchungen an *Zea mays* mit Hilfe von THO und Microautoradiographie. *Ber. Schweiz. Bot. Ges.* **75**, 303–345.

Nelson, C. D., H. Clauss, D. C. Mortimer, and P. R. Gorham, 1961. Selective translocation of products of photosynthesis in soybean. *Plant Physiol.* **36**, 581–588.

Nelson, C. D., H. J. Perkins, and P. R. Gorham, 1959. Evidence for different kinds of concurrent translocation of photosynthetically assimilated C^{14} in the soybean. *Can. J. Bot.* **37**, 1181–1189.

Parker, B. C., 1965. Translocation in the giant kelp *Macrocystis*. I. Rates, direction, quantity of C^{14}-labelled products and fluorescein. *J. Phycol.* **1**, 41–46.

Parker, B. C., 1966. Translocation in *Macrocystis*. III. Composition of sieve tube exudate and identification of the major C^{14}-labelled products. *J. Phycol.* **2**, 38–41.

Parthasarathy, M. V. and P. B. Tomlinson, 1967. Anatomical features of metaphloem in *Sabal* and *Cocos* and two other palms. *Am. J. Bot.* **54**, 1143–1151.

Peel, A. J., 1964. Tangential movement of ^{14}C-labelled assimilates in stems of willow. *J. Exp. Bot.* **15**, 104–113.

Peel, A. J., 1965. The effect of changes in the diffusion potential of xylem water on sieve-tube exudation from isolated stem segments. *J. Exp. Bot.* **16**, 249–260.

Peel, A. J., 1966. The sugars concerned in the tangential movement of ^{14}C-labelled assimilates in willow. *J. Exp. Bot.* **17**, 156–164.

Peel, A. J. and P. E. Weatherley, 1962. Studies in sieve-tube exudation through aphid mouth parts: the effect of light and girdling. *Ann. Bot.* **26**, 633–646.

Pristupa, N. A., 1959. On the transportable form of carbohydrates in pumpkin plants. *Soviet Plant Physiol.* **6**, 26–32 (translation).

Prokofyev, A. A., L. P. Zhdanova, and A. M. Sobolev, 1957. Certain regularities in the flow of substances from leaves into reproductive organs. *Soviet Plant Physiol.* **4**, 402–408 (translation).

Reiner, M., 1960. *Deformation, strain and flow*. H. K. Lewis & Co., London.

Roberts, B. R., 1964. Effects of water stress on the translocation of photosynthetically assimilated carbon-14 in yellow poplar. In *The formation of wood in forest trees*, ed. M. H. Zimmermann. Academic Press, New York.

Röckl, B., 1949. Nachweis eines Konzentrationshubs zwischen Palisadenzellen und Siebröhren. *Planta*, **36**, 530–550.

Schneider-Orelli, O., 1909. Die Miniergänge von *Lyonetia clercella* L. und die Stoffwanderung in Apfelblättern. *Zbl. f. Bact.* **24**, II, 158–181.

Schnepf, E., 1965. Physiologie und Morphologie sekretorischer Pflanzenzellen. In *Sekretion und Exkretion. 2. wiss. Konf. der Ges. Deutsch. Naturforscher und Aerzte.* Springer Verlag, Berlin, Heidelberg, New York.

Schumacher, W., 1931. Ueber Eiweissumsetzungen in Blütenblättern. *Jahrb. wiss. Bot.* **75**, 581–608.

Schumacher, W., 1933. Untersuchungen über die Wanderung des Fluoreszeins in den Siebröhren. *Jahrb. wiss. Bot.* **77**, 685–732.

Schumacher, W., 1937. Weitere Untersuchungen über die Wanderung von Farbstoffen in den Siebröhren. *Jahrb. wiss. Bot.* **85**, 422–449.

Schumacher, W. and M. Hülsbruch, 1955. Zur Frage der Bewegung fluoreszierender Farbstoffe im Pflanzenkörper. *Planta*, **45**, 118–124.

Shiroya, T., G. R. Lister, V. Slankis, G. Krotkov, and C. D. Nelson, 1966. Seasonal changes in respiration, photosynthesis and translocation of ^{14}C labelled products of photosynthesis in young *Pinus strobus* L. plants. *Ann. Bot.* **30**, 81–91.

Spanner, D. C., 1958. The translocation of sugar in sieve tubes. *J. Exp. Bot.* **9**, 332–342.

Swanson, C. A. and E. D. H. El-Shishiny, 1958. Translocation of sugars in the Concord grape. *Plant Physiol.* **33**, 33–37.

Swanson, C. A. and J. B. Whitney, Jr., 1953. Studies on the translocation of foliar-applied P^{32} and other radioisotopes in bean plants. *Am. J. Bot.* **40**, 816–823.

Thaine, R., 1962. A translocation hypothesis based on the structure of plant cytoplasm. *J. Exp. Bot.* **13**, 152–160.

Thaine, R. and R. D. Preston, 1964. The role of transcellular streaming in phloem transport. In *The formation of wood in forest trees*, ed. M. H. Zimmermann. Academic Press, New York.

Trip, P., C. D. Nelson, and G. Krotkov, 1965. Selective and preferential translocation of C^{14}-labeled sugars in white ash and lilac. *Plant Physiol.* **40**, 740–747.

Van den Honert, T. H., 1932. On the mechanism of transport of organic materials in plants. *K. Akad. Wet. Amsterdam Proc.* **35**, 1104–1112.

Vernon, L. P. and S. Aronoff, 1952. Metabolism of soybean leaves. IV. Translocation from soybean leaves. *Arch. Biochem. Biophys.* **36**, 383–398.

Weatherley, P. E., A. J. Peel, and G. P. Hill, 1959. The physiology of the sieve tube. *J. Exp. Bot.* **10**, 1–16.

Webb, J. A. and P. R. Gorham, 1965. The effect of node temperature on assimilation and translocation of C^{14} in the squash. *Can. J. Bot.* **43**, 1009–1020.

Webb, K. L. and J. W. A. Burley, 1962. Sorbitol translocation in apple. *Science*, **137**, 766.

Willenbrink, J., 1957. Ueber die Hemmung des Stofftransports in den Siebröhren durch lokale Inaktivierung verschiedener Atmungsenzyme. *Planta*, **48**, 269–342.

Ziegler, H., 1956. Untersuchungen über die Leitung und Sekretion der Assimilate. *Planta*, **47**, 447–500.

Ziegler, H., 1963. Verwendung von ^{45}Calcium zur Analyse der Stoffversorgung wachsender Früchte. *Planta*, **60**, 41–45.

Zimmermann, M. H., 1957a. Translocation of organic substances in trees I. *Plant Physiol.* **32**, 288–291.

Zimmermann, M. H., 1957b. Translocation of organic substances in trees II. *Plant Physiol.* **32**, 399–404.

Zimmermann, M. H., 1958. Translocation of organic substances in trees III. *Plant Physiol.* **33**, 213–217.

Zimmermann, M. H., 1960. Longitudinal and tangential movement within the sieve-tube system of white ash (*Fraxinus americana* L.). In Festschrift Prof. Frey-Wyssling. *Beih. Zeitschr. des Schweiz. Forstv.* **30**, 289–300.

Zimmermann, M. H., 1961. Movement of organic substances in trees. *Science*, **133**, 73–79.

Zimmermann, M. H., 1964. The relation of transport to growth in dicotyledonous trees. In *The formation of wood in forest trees*, ed. M. H. Zimmermann. Academic Press, New York.

Zimmermann, M. H. and P. B. Tomlinson, 1965. Anatomy of the palm *Rhapis excelsa* I. *J. Arnold Arb.* **46**, 160–178.

12. The water relations of plants

J. Dainty

Introduction

In the space available this is necessarily a limited and, in many ways, an over-simplified account of the water relations of plants. I have concentrated on giving, again in a simplified though I hope correct form, an account of the fundamental theory of water movement. This inevitably involves an elementary feeling for thermodynamics and I make no apology for bringing this in; students of plant physiology can no longer 'get by' with a negligible knowledge of physical chemistry. An excellent recent text on Plant–Water Relationships by R. O. Slatyer (1967) has recently been published. The reader is referred to this to fill in the details and overcome the deficiencies of this present brief account.

General theory

Like anything else, water moves in response to a force. It is the purpose of this section to consider the nature of this force and the quantitative relations between the movement of water and the force. Since the force is 'thermodynamic' in nature a few remarks about certain aspects of thermodynamics must be made.

Free energy and chemical potential

The free energy, G, of a system, which might be a whole plant with its immediate environment or even a part of an individual plant cell, is the capacity of that system for doing *useful* work. It is not the total energy of the system, for the second law of thermodynamics limits the amount of useful work that can be obtained from the heat energy of the system. If the system undergoes a transition from one state to another such that the free energy decreases, in symbols ΔG is negative, then useful work can be done during this transition. Such a transition is therefore a natural or spontaneous one, for it does not require any supply of work from the surroundings; it is exactly analogous to a weight falling from a height to the ground, from a state of high potential energy to one of lower potential energy. On the other hand if during the transition the free energy increased, i.e., ΔG were positive, then this could only be achieved by work being done on the system by its surroundings; the analogy here would be that the weight would have to be lifted from the ground by some external agency. If there is no change in free energy during the transition, $\Delta G = 0$, then the system is in equilibrium; this would be analogous, for example, to the weight moving horizontally along a smooth surface. Thus the changes in free energy of a system indicate whether the system is spontaneously approaching equilibrium, is at equilibrium, or is being forced away from equilibrium. And the magnitudes of the changes give the maximum useful work that can be got out of the system (ΔG negative) or the minimum work that has to be put into the system (ΔG positive).

A biological system is made up of many components: water, ions, nonelectrolyte solutes, macromolecules, and so on. Each component contributes something to the free energy. Roughly speaking the amount contributed per mole of any component is called the chemical potential of that component. The chemical potential of a component, symbol μ, is also called the partial molar free energy of the component and is accurately defined as the partial differential coefficient $\partial G/\partial n$; this means the increase in G, caused by a small increase dn in the number of moles of the component, divided by dn, it being understood that there is no change in the number of moles of any other component nor in the temperature and pressure of the system. It will be seen in the next section that the chemical potential is the parameter which determines the movement of a component.

The driving force on water

Consider now a system consisting of two parts; this might be two adjacent regions of soil, or the soil adjacent to the root and the root itself, or a plant cell bathed by an external solution, and so on. If dn moles of water are removed from one part (A), the free energy of that part is decreased by μ_w^A dn, where μ_w^A is the chemical potential of water in part A; this follows from the definition of chemical potential given in the previous paragraph. If these dn moles of water are added to the other part (B) of the system then the free energy of this part is increased by μ_w^B dn, where μ_w^B is the chemical potential of water in part B. Thus considering the system as a whole there will be a change in free energy equal to $(\mu_w^B - \mu_w^A)$ dn. If μ_w^A is greater than μ_w^B, this change in free energy will be a decrease and thus water will naturally flow from A to B. If μ_w^A is less than μ_w^B the change will be an increase and work will have to be done to move water from A to B; the work necessary would be equal to $(\mu_w^B - \mu_w^A)$ per mole of water moved. If μ_w^A were equal to μ_w^B then there would be no change in free energy and the water in one part of the system would be in equilibrium with the water in the other part of the system. Equality of the chemical potentials of a component between various parts of a system is a necessary and sufficient condition for that component to be in equilibrium between the various parts of the system.

Now suppose that the parts A and B of the system are separated by a very small distance, dx. Suppose $\mu_w{}^A$ is greater than $\mu_w{}^B$ so that the movement of water is from A to B. A force will be driving this movement of water; call it F per mole of water. The work done by this force in moving a mole of water over a distance dx is equal to the product of the force times the distance moved, i.e., $F\,dx$. This work must be equal to the decrease in free energy which is $(\mu_w{}^A - \mu_w{}^B)$. Since the parts are separated by an infinitesimal distance, the difference in the chemical potentials must also be infinitesimal and will be called $-d\mu_w$. (The minus sign is inserted because of a mathematical convention.) Thus $F\,dx = -d\mu_w$, hence $F = -d\mu_w/dx$; i.e., the driving force on water (or indeed any component) in a system is the negative gradient of the chemical potential. We have perforce frequently, indeed usually, to deal with systems in which the gradient of the chemical potential cannot be determined at every point in the system. A tacit averaging of the chemical potential gradient is then made and the driving force is then taken to be the actual chemical potential difference between the two parts of a system, that is $\mu_w{}^A - \mu_w{}^B$ or more compactly $\Delta\mu_w$.

It should be stated at this point that the above conclusion is strictly only true if water is the only moving component in the system. When other components move they can exert frictional drags on the water which constitute additional forces on the water. Taking into account these frictional drags is the province of the theory of irreversible thermodynamics, which I feel is beyond the scope of this account. Where necessary, modifications of formulae due to this frictional drag will be indicated. The application of the theory of irreversible thermodynamics to water movement in plants has been discussed by Dainty (1963). An excellent account of thermodynamics in general, suitable for biologists, has been written by Spanner (1964).

Factors affecting the chemical potential of water

The chemical potential of water, μ_w, is the contribution made by water, per mole, to the free energy of the system. As always when discussing energy we are really concerned with the energy difference between one system and another. When dealing with water it is convenient to take as a standard system pure water at a pressure of one atmosphere and at the same temperature as the system we are interested in. The chemical potential of water in a given system, for example soil or plant tissue, will differ from that of pure free water because the water in the system will be acted upon by a number of forces which do not exist in the standard pure water system. If any solute molecules are present in the system, they exert forces on neighbouring water molecules; if the hydrostatic pressure is, for example, greater than one atmosphere then the water molecules are pushed slightly closer together than they are in the standard pure water system and hence the repulsive forces between the water molecules will be slightly greater; if the system contains an extensive array of solid–liquid or gas–liquid interfaces as in soil, or cell walls, or protoplasm, then the water molecules in or near these interfaces will be subject to special force fields which do not exist in the standard pure water system. We would therefore expect the chemical potential of water in a system to depend on the hydrostatic pressure, the presence of solutes, and the presence of interfaces. This dependence is expressed in the following formula, the derivation of which by thermodynamic arguments has been clearly discussed by Slatyer (1967):

$$\Psi = (\mu_w - \mu_w{}^*)/\bar{V}_w = P - \pi - \tau \tag{1}$$

In this formula, for convenience, the difference in the chemical potential of water between the system, μ_w, and the standard pure water system, $\mu_w{}^*$, has been divided by the effectively constant partial molar volume of water, $\bar{V}_w$, which is equal to 18 cm^3 mole^{-1}. This quantity Ψ (Psi) is called the water potential and its units are basically energy per unit volume which are equivalent to force per unit area, i.e., pressure. The water potential can thus be expressed in erg cm^{-3}, joule cm^{-3}, dyne cm^{-2}, atmospheres or bars (1 bar $= 10^6$ dyne cm$^{-2} = 0.987$ atm); I shall express it in atmospheres, this being the most widely used unit at present. P, on the right-hand side of eq. (1), is the pressure in excess of one atmosphere exerted on the system; its contribution to μ_w is positive because the effect of pressure is to press the molecules closer together and hence increase the free energy of the system. The term π (Pi) arises from the presence of dissolved solutes in the system. It is in fact the conventional osmotic pressure and for dilute, ideal, solutions is given by RTC_s, where C_s is the solute concentration in the water of the system. The presence of dissolved solutes lowers the chemical potential of water, hence the 'osmotic' contribution to Ψ is minus π. τ (tau) is called the matric potential and is a measure of the interactions of water molecules in solid- and gas-liquid interfaces; since this contribution arises from extra attractive forces it is negative, i.e., minus τ.

This discussion may so far have been somewhat unfamiliar to many botanists. But an easy correspondence with the older terminology of water relations in plant physiology can be made. For instance in the vacuole of a mature plant cell there are usually few macromolecules and so the matric potential, τ, is negligible. Then in the vacuole the water potential is given by

$$\Psi = P - \pi \tag{2}$$

i.e., the hydrostatic, or turgor pressure minus the osmotic pressure of the vacuolar solution. The older plant physiology textbooks call $\pi - P$, i.e., $OP - TP$, the diffusion pressure deficit, DPD, or suction potential. Thus, at least for the mature vacuolated cell, plant physiologists were using a thermodynamic potential for the driving force. It is now thought unnecessary to give it a very special name, which no other kind of scientist uses, and it is becoming more common to use water potential, Ψ, which is a shorthand way of saying the difference in water chemical potential between the system under consideration and pure water at atmospheric pressure and the same temperature (divided by $\bar{V}_\text{w}$ to get it into more familiar and useful units).

Water flow between systems separated by a semipermeable membrane

It was shown earlier that the driving force on water is the (negative) gradient of the chemical potential. Usually only the difference, $\Delta\mu_\text{w}$, between the chemical potentials of two systems is known; in this case, by means of a tacit integration, the driving force can be taken as this difference of the chemical potentials. Since water potential, Ψ, is defined as in eq. (1), the driving force can also clearly be taken as the difference, $\Delta\Psi$, in water potentials of the two systems. It should be stressed, however, that $\Delta\Psi$ is the sole driving force on water only if water is the only moving component between the two systems, i.e., if the two systems are separated by a membrane permeable to water only—a semipermeable membrane.

The flow of water between two such systems is proportional to the driving force. We write:

$$J_\text{v} = L_\text{p}\,\Delta\Psi \qquad (3)$$

where J_v is the volume flux of water in $\text{cm}^3\ \text{cm}^{-2}\ \text{s}^{-1}$, i.e., it is the number of cm^3 flowing across 1 cm^2 of membrane per second. L_p is the so-called hydraulic conductivity of the membrane in $\text{cm}\ \text{s}^{-1}\ \text{atm}^{-1}$. Substituting for $\Delta\Psi$ in terms of its three components (eq. (1)):

$$J_\text{v} = L_\text{p}(\Delta P - \Delta\pi - \Delta\tau) \qquad (4)$$

Water flow between systems in general

If two systems are separated by a membrane which is permeable to solutes as well as to water, then the water potential difference is not the sole driving force on the water; the moving solutes can exert a frictional drag on the water and these drags constitute extra forces. The way to take account of this has been worked out from the theory of irreversible thermodynamics (see Dainty, 1963). The result of the theory is to modify the osmotic pressure term in eq. (4), for permeating solutes cannot express their full osmotic pressure. The flow becomes:

$$J_\text{v} = L_\text{p}(\Delta P - \Delta\pi_\text{imp} - \sigma\,\Delta\pi_\text{p} - \Delta\tau) \qquad (5)$$

In this equation the solutes have been divided into two classes: permeating and non-permeating. So far as the non-permeating solutes are concerned the membrane is semipermeable and they exert their full osmotic pressure, $\Delta\pi_\text{imp}$. The permeating solutes do not exert their full osmotic pressure; $\Delta\pi_\text{p}$ is multiplied by a new parameter, σ (sigma), which is a quantitative expression of this fact.

σ is called the reflection coefficient of the membrane for the particular solutes. It is always less than 1, becoming zero for a membrane which does not distinguish between solutes and water. It can even become negative if the membrane is sufficiently more permeable to the solutes than to water. Although a theoretical formula for the reflection coefficient can be worked out (Dainty and Ginzburg, 1963), it is best to treat it as a coefficient which has to be determined by experiment.

It is worth running through, at this stage, various situations in the soil-plant-air complex to see how eq. (5) applies.

If we are dealing with a vacuolated plant cell and considering the water flow between an external solution and the vacuole, then it is customary (in our present state of ignorance) to treat the protoplast, with its two membranes, as a single membrane. The usual solutes in the external solution and the vacuole are such that they permeate the membrane very slowly, much more slowly than water does. Thus the membrane can normally be considered effectively semipermeable, i.e., $\sigma = 1$ for all solutes. Also the matric potential components in both external solution and vacuole are negligible. We thus get the familiar equation for water flow:

$$J_v = L_p(\Delta P - \Delta \pi) \qquad (6)$$

This gives the water volume flux, in $cm^3 cm^{-2} s^{-1}$, out of the cell in terms of the hydrostatic and osmotic pressure differences between the inside and outside of the cell.

As the water (solution) flows along the xylem vessels of a plant there is no effective membrane separating various parts of the xylem vessels in the longitudinal direction. Thus for all solutes there is no discrimination between solutes and water as they flow along the xylem; this means that for all solutes $\sigma = 0$. The matric potential component is also negligible and therefore the equation for flow along the xylem vessels is simply

$$J_v = L_p \Delta P \qquad (7)$$

i.e., the flow is solely a hydrostatic pressure-driven flow, as expected for flow along a pipe.

In soil there are no membranes separating one part of the soil from another. Thus again we would expect no discrimination between solutes and water, hence $\sigma = 0$ for all solutes in the soil water. In soil hydrostatic pressure differences play a minor role in water flow, but matric potential differences are very important and constitute the major driving force. Thus the equation of flow in soils will be:

$$J_v = L_p \Delta \tau \qquad (8)$$

Obviously these examples can be multiplied. Equation (5) is the basic equation of flow and in each case of flow between two systems, it is necessary to consider the value of σ which is appropriate.

One contributory force has so far been neglected: gravity. It is sometimes important in the movement of water in soils and of course must be considered in the movement of water up tall trees. Some people include a gravitational term in the chemical potential, others treat it as an extra field of force. In either case it contributes a term Mgh to the chemical potential and therefore ρgh to the water potential. (M is the molecular weight of water, ρ the density of water, g the acceleration due to gravity, and h the height of the water above some arbitrary zero level.) Taking gravity into account eq. (5) will become

$$J_v = L_p(\rho g\,\Delta h + \Delta P - \Delta\pi_{imp} - \sigma\,\Delta\pi_p - \Delta\tau) \qquad (9)$$

Here Δh is the difference in heights between the two systems, between which water flow is taking place.

Water relations of single plant cells

The vacuolated plant cell is rather a complex object; even from the simplest possible point of view it consists of a cell wall—a concentrated weak acid, inhomogeneous ion-exchange resin under considerable tension—surrounding the protoplast which, in turn, encloses the vacuole. The protoplast is considered to be bounded by two membranes, the plasmalemma and the tonoplast, and it may swell or shrink when gaining or losing water (see Briggs, 1957). Because little is known about the water permeabilities of the individual protoplast membranes or about the swelling and shrinking of the protoplast, it is customary to treat the whole protoplast as a single membrane surrounding the vacuole and itself being enclosed by an elastic box—the cell wall. It is recognized that this treatment is grossly oversimplified, but until more is known about the properties of the protoplast and its individual membranes, it is the best that can be done.

It has already been shown (eq. (6)) that, because the matric potentials are negligible in the vacuole and an external solution, and because the normal solutes are effectively non-permeating, the equation for water flow from the vacuole to the external solution is:

$$J_v = L_p(\Delta P - \Delta\pi) \qquad (6)$$

In this equation ΔP and $\Delta\pi$ are the differences betwen the internal and external hydrostatic and osmotic pressures. The term $(\Delta P - \Delta\pi)$ is of course the water potential difference, $\Delta\Psi$, between the vacuole and the external solution.

The cells of algae and other plants living under water are probably in equilibrium, i.e., they are neither gaining nor losing water except in so far as there will be a small influx of water associated with growth. Thus, practically,

$$J_v = 0 \quad \text{and} \quad \Delta\Psi = \Delta P - \Delta\pi = 0 \qquad (10)$$

However, for the cells of the aerial parts of higher plants the condition (10) rarely applies. This is partly because there is a daily cycle of gain and loss of water by the plant and, perhaps, only during the night would J_v and $\Delta\Psi$ be approximately equal to zero. Another factor obscuring the application of eq. (10) to the cells of higher plants is that the internal hydrostatic pressure is taken to be the turgor pressure and the external hydrostatic pressure is assumed to be one atmosphere. But the external hydrostatic pressure on a cell *in situ* is partly due to the pressure exerted by the surrounding cells and this can be considerably greater than one atmosphere. When a cell or assembly of cells (piece of tissue) is excised from the plant, the pressure of the surrounding cells is largely released; therefore ΔP, and therefore $\Delta\Psi$, for a cell in an excised tissue is different from what it was in the complete plant. So when $\Delta\Psi$ is measured for an excised tissue, as it frequently is, it may well be different from zero even though it was zero in the plant.

For these two reasons, one real and one apparent, it is usually found that the water potential difference for the cell of a higher plant is different from zero and is, of course, given by

$$\Delta\Psi = \Delta P - \Delta\pi \qquad (11)$$

It is useful to have a visual image of this relationship for a single cell as a function of the volume of the cell. If we assume that the external solution is standard pure water at one atmosphere pressure then, by definition, the water potential difference, $\Delta\Psi$, becomes the water potential, Ψ, in the vacuole and eq. (11) becomes

$$\Psi = P - \pi \qquad (12)$$

all quantities referring to the vacuole. The relation between turgor pressure, P, and cell volume, V, can be quite complicated (see Crafts, Currier, and Stocking, 1949). I will assume, just for the sake of illustration, that the relationship is a linear one, that is

$$P = k(V - V_0)/V_0 \qquad (13)$$

where V_0 is the cell volume when the turgor pressure is zero, i.e., the actual hydrostatic pressure in the vacuole is one atmosphere. The quantity represented by k is a kind of elastic modulus of the cell wall; the bigger it is the less the cell expands due to its own turgor pressure. The relation between osmotic pressure and cell volume is well known; the product of osmotic pressure and volume is constant to a first approximation. Therefore:

$$\pi = \pi_0 V_0/V \qquad (14)$$

where π_0 is the osmotic pressure at zero turgor pressure. All the quantities, Ψ, P, and π can be plotted in one diagram (Fig. 12.1) which summarizes the static water relations of a single plant cell. The curve representing Ψ in the diagram shows the water potential in the cell vacuole or the water potential differ-

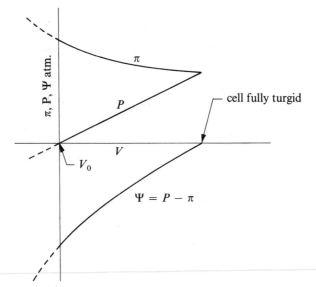

Figure 12.1 Diagram of relationship between π, P, Ψ and cell volume V. The scales are exaggerated; they are drawn as if the change in volume between incipient plasmolysis, V_0, and full turgor were about 30 per cent, whereas 5 per cent would be a more usual figure. The relationship between pressure and volume has been assumed linear. The dotted lines on the left of the $V = V_0$ axis, indicate the relationships when the protoplast adheres to the cell wall.

ence between the vacuole and pure water. Ψ varies from a maximum (negative) value when there is no turgor pressure to zero when the cell is fully turgid, i.e., when the cell water is in equilibrium with pure water at one atmosphere pressure as the external solution.

If the cell is surrounded by a solution of non-permeating solutes of osmotic pressure π_e and is in water flux equilibrium, then from eq. (11)

$$\Delta\Psi = 0 = P - \pi + \pi_e \qquad (15)$$

If π_e is equal to π_0, the osmotic pressure of the cell at zero turgor pressure then, of course, $P = 0$ and there is no hydrostatic pressure difference between the inside and outside of the cell. If the external osmotic pressure were somewhat increased *and* the protoplast adhered to the inner surface of the cell wall, then there would be little change in cell volume and therefore little change in internal osmotic pressure, π. Thus π_e would exceed π and P would therefore become negative, i.e., the internal hydrostatic pressure would become less than the external. There is some evidence that this can occur but it needs a considerable force of adhesion between the protoplast and the cell wall. Usually the necessary adhesive force is absent, the negative hydrostatic pressure cannot therefore be sustained and the protoplast separates from the cell wall. The cell is said to have undergone *plasmolysis*. A plasmolysed protoplast, i.e., one not connected with the cell wall, probably cannot sustain any pressure difference across its surface, hence eq. (15) becomes

$$0 = \pi - \pi_e \tag{16}$$

which means that the plasmolysed protoplast adjusts its volume and hence its osmotic pressure to equal the external osmotic pressure. When $\pi_e = \pi_0$, i.e., the cell is just at zero turgor pressure, it is said to be at the point of incipient plasmolysis.

Methods of measuring the parameters π, P, and Ψ

The most difficult of these parameters to measure is undoubtedly the turgor pressure, P, and few direct measurements have been attempted. Bennet-Clark and Bexon (1940) estimated the turgor pressure of cells in leaf tissue by squeezing the leaf tissue in a hydraulic press and measuring the hydrostatic pressure necessary to cause sap exudation. It is not at all clear what is happening when tissue is crushed in a hydraulic press and some arbitrary decisions have to be made about the volume of sap expressed by the pressure. It seems unlikely that this method is quantitatively adequate. Tazawa (1957) has measured the turgor pressure in the large cylindrical internodal cells of *Nitella* by measuring the force, applied to a rod resting across the cell, necessary to produce a certain deformation of the cell. Unfortunately the relationship between force, deformation, and turgor pressure cannot be adequately worked out on a theoretical basis. Thus the method has to be calibrated by using external solutions of known osmotic pressure which produce assumed turgor pressures according to eq. (15). Tazawa's method is therefore not, strictly speaking, an independent one. Recently Green and Stanton (1967) have achieved a truly independent method of measuring turgor pressure in the large *Nitella* cells. In effect they have inserted a pressure recording manometer directly into the cells. The manometer consists of a piece of capillary tube, 40 μ in diameter, sealed at one end and with the other end drawn down to a tip diameter of perhaps 5 μ (see Fig. 12.2). When placed in water a small amount of water enters the narrow open end, due to surface tension, and thus a volume of air is enclosed in the manometer. The tip is then inserted into a *Nitella* cell and the enclosed volume of air is compressed by the hydrostatic pressure inside the cell. A simple calculation using Boyle's Law gives the turgor pressure. More such direct measurements of turgor pressure are needed and it is hoped that the work of Green and Stanton will stimulate others to do likewise.

The most frequently measured parameter has been the vacuolar osmotic pressure, π. Two main procedures have been used, one involving the extraction of the vacuolar sap and the other an estimation of the point of incipient plasmolysis with intact cells and tissues. Both have their uncertainties.

Vacuolar sap can be extracted, by micropipette or other methods, in a relatively uncontaminated state only from giant algal cells such as occur in the

Characeae, in *Valonia*, *Halicystis*, *Hydrodictyon*, and other species. From the cells of higher plants it can only be obtained by crushing tissue in a hydraulic press, frequently after freezing and thawing it or applying other procedures to disrupt the cell membranes. Thus it must always be contaminated with fluid from the intercellular spaces, the cell walls, and the protoplasm. What the degree of contamination is is not accurately known, but a reasonable sample of sap is probably obtained from tissues in which the cells have vacuoles containing highly concentrated sap (see Bennet-Clark, 1959). After the sap has been obtained it is a simple matter to determine the osmotic pressure of the sample by measuring either the vapour pressure or the depression of the freezing point. Commercial instruments are available to do this on a very small scale.

The other major procedure, the plasmolytic method, involves the microscopic observation of large numbers of cells in a tissue. A slice of tissue is exposed to solutions of various osmotic pressures near the internal osmotic pressure. The fraction of cells plasmolysed is plotted as a function of the external osmotic pressure. From this graph the external osmotic pressure which would plasmolyse half the cells is estimated and this is taken as the average internal osmotic pressure, π_0, at the point of incipient plasmolysis. If the relation between cell volume and turgor pressure is known, π_0 can be corrected to the volume it would have at full turgor (or to any other cell volume). There are at least two possible sources of error in this method. The solutes in the external solution must be non-permeating and this is not always easy to ensure; mannitol is the usual solute used. The other possible source of error is the neglect of any adhesive forces between the protoplast and the cell wall; their presence would result in an over-estimation of the vacuolar osmotic pressure.

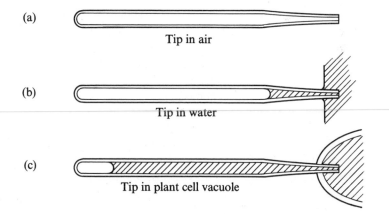

(a)

Tip in air

(b)

Tip in water

(c)

Tip in plant cell vacuole

Figure 12.2 Three stages in the use of the manometer of Green and Stanton (1967) to measure the turgor pressure of a plant cell. The hydrostatic pressure inside the cell is given by the ratio of the volume of air in (b) to the volume of air in (c), times atmospheric pressure. (Adapted from Green and Stanton (1967), by permission of the authors and *Science*. Copyright 1967 by the American Association for the Advancement of Science.)

A variant of the plasmolytic method is the so-called plasmometric method (see Stadelmann, 1966). In this procedure the cells of a tissue are exposed to a solution of such a high known osmotic pressure that they are all plasmolysed. The volumes of the plasmolysed protoplasts are then measured. If the volumes of the fully turgid protoplasts are known then a simple inverse proportionality gives the osmotic pressure in the vacuole of the fully turgid cell. Although this method avoids the possible error from the adhesion of the protoplast to the cell wall, the volume measurements involved are very difficult to make accurately. Also the method loses accuracy if the protoplasm occupies a significant fraction of the cell volume.

Perhaps most attention has been paid to the problem of measuring the water potential of plant tissue. The basis of all the methods (Slatyer, 1967) is that if the tissue is in water flux equilibrium with another aqueous phase, either vapour or liquid, then no net gain or loss of water by the tissue can take place. Then the water potential of the tissue is equal to the water potential in the other phase, which is measurable by straightforward methods.

The simplest method, in principle, is that of confining the tissue in a small enclosed space and directly measuring the water vapour pressure built up in the space. The space should be as small as possible so that little water is lost in building up the equilibrium vapour pressure and so that equilibrium is achieved as rapidly as possible. The vapour pressure is usually measured, essentially, by a pair of tiny thermocouples acting as wet and dry bulbs—the so-called thermocouple psychrometer.

The other commonly used method is to expose the tissue, or several samples of the tissue, to either the vapour in equilibrium with various solutions of known water potentials or to the solutions themselves, providing that the solutes in the solutions are non-permeating. Then the sample which has not changed in length, volume, weight, etc., has a water potential equal to that of the vapour or solution to which it was exposed. In one variant of this method samples of the tissues are exposed to a range of solutions of various water potentials and, instead of the tissue itself being tested for 'no change', the refractive indices of the solutions are measured, the 'correct' solution being that in which no change occurs; i.e., there has been no loss or gain of water by it.

Slatyer (1967) discusses well and relatively briefly these methods of measuring water potential. The main sources of error would appear to arise from penetration of the cells of the tissue by the supposed non-permeating solutes and, in vapour pressure methods, slowness in reaching equilibrium.

Water movement in plant cells and tissues

This section will be concerned with the fundamentals of the 'dynamic' water relations of plant cells and tissues. In the space available only a brief outline can be given. More detailed, critical, accounts are given by Dainty (1963), Briggs (1967), and Slatyer (1967).

The hydraulic conductivity, L_p, of plant cell membranes

When a plant cell is bathed by an external solution and the conditions are such that a difference of hydrostatic pressure, ΔP, and a difference of osmotic pressure, $\Delta \pi$, due to non-permeating solutes, exists between the vacuole and the external solution, then the flux of water from the vacuole to the external solution is given by

$$J_V = L_p \Delta \Psi = L_p(\Delta P - \Delta \pi) \tag{6}$$

In this equation J_V is in $cm^3 cm^{-2} s^{-1}$, ΔP and $\Delta \pi$ are in atmospheres and L_p, the hydraulic conductivity, is in $cm \ s^{-1} \ atm^{-1}$. Note that, as defined, L_p is the hydraulic conductivity of the total barrier—cell wall, plasmalemma, protoplasm, tonoplast, all in series. It is usually thought that the cell wall and the protoplasm have a very high conductivity to water and that L_p therefore is a measure of the combined (in series) conductivities of the plasmalemma and tonoplast. However, suberized cell walls may have a very low conductivity and some authors (e.g., Dick, 1966) think that the conductivity to water of the protoplasm may be quite low. These considerations should be kept in mind during the subsequent discussion.

All the methods of determining L_p are of course based on eq. (6). In the most widely used method (see Stadelmann, 1963) the cells in a tissue are plasmolysed. If the osmotic pressure, due to non-permeating solutes, of the external plasmolysing solution is π_e^1 and if time is allowed to reach water flux equilibrium then, since $\Delta P = 0$ for a plasmolysed protoplast and $J_v = 0$, the vacuolar osmotic pressure, π^1 say, must be equal to the external osmotic pressure, π_e^1. The plasmolysed tissue is then rapidly transferred to another external solution of osmotic pressure π_e^2, of such a magnitude that the cells would still be plasmolysed when they come to equilibrium with the new external solution. After this rapid transfer water enters or leaves the cell, depending on whether π_e^2 is less than or greater than π_e^1, and the plasmolysed protoplast swells or shrinks at a rate dependent upon the magnitude of $(\pi_e^2 - \pi_e^1)$ and upon the value of L_p. An equation can be derived from eq. (6) which quantitatively describes the course of swelling or shrinking (Stadelmann, 1966) and this enables L_p to be determined from the experimental observations. This method has been applied, particularly by the Vienna school of Höfler, to a wide range of plant tissues. The values obtained are in the range of 0.5×10^{-7} to $2.0 \times 10^{-6} \ cm \ s^{-1} \ atm^{-1}$.

There are several objections to this method. Perhaps the most serious one is that the membranes, and particularly the plasmalemma, of a plasmolysed protoplast may be quite different from those of the normal, more or less turgid, cell. Another objection is that the volume, and therefore the rate of change of volume, of a plasmolysed protoplast is difficult to measure with any high degree of accuracy; further the protoplasm occupies a much larger fraction of the plasmolysed protoplast than it does of the turgid cell and this leads to errors in estimating the volume of the vacuole. A further source of error, which will be discussed in detail later, is common to all methods. This is that no matter how rapid the transfer from one solution to another, the new solutes have to diffuse up to the surface of the cell being observed. This may take a long time and indeed be the rate-determining step in the swelling and shrinking of the protoplast; in this latter case no value of L_p can be calculated from the observations!

It is clearly important to try and measure L_p for normal turgid cells. In general this is more difficult because the volume changes occurring consequent on changing the external osmotic pressure, *while still keeping the cell turgid*, are very small—2 to 5 per cent. They are small because the elastic modulus of the cell wall is very high and the cell wall therefore only expands a little in response to changes in internal hydrostatic pressure. These volume changes are too small to be measurable for individual cells of plant tissue, but they can be measured for certain giant algal cells. Kelly, Kohn, and Dainty (1963) have used this technique on the giant internodal cells of *Nitella translucens*. They observed the rate of swelling and shrinking of the cell volume when a single cell was transferred from one external solution to another, neither of whose osmotic pressures were great enough to plasmolyse the cell. They were able to correct for the time lag associated with the transfer, due to the finite rate of diffusion of the new solutes up to the cell surface. Their result, which involved measuring the elastic modulus of the cell wall, agreed with the result of a different method of measuring L_p: by transcellular osmosis, to be described later.

Although this method cannot be applied to the typically small individual plant cells of, say, higher plant tissue, it can be applied, in principle, to the aggregate of cells in a tissue. If the rate of loss or gain of water (measured by weighing the tissue at intervals of time) consequent on changing the external solution bathing the tissue is observed, this can in principle give the average value of L_p for the cells of the tissue. It is of course necessary to ensure that the external solution concentrations are such that the cells of the tissue are never plasmolysed. However, on transferring from one solution to another, the new solute molecules have to diffuse a long way through the intercellular spaces to reach the surfaces of the cells and, in fact, this slow diffusion is the rate-controlling factor in the gain or loss of weight (Kohn and Dainty, 1966). Thus it appears particularly difficult to measure L_p for small turgid cells and new ideas for making this measurement are very much needed.

There is one method which is apparently free from serious objection but can only be applied to cells such as the long, cylindrical, giant internodal cells of the Characeae. This is the method of transcellular osmosis, first used in a quantitative manner by Kamiya and Tazawa (1956). In this method the cell is placed in a double chamber (Fig. 12.3), usually with half the cell in one chamber and

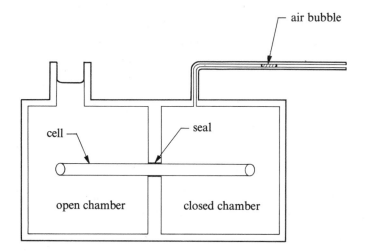

Figure 12.3 Diagrammatic representation of the method of transcellular osmosis. A capillary, with an air bubble in it, is connected to the closed chamber on the right. The chamber on the left is open, to ensure quick changes of solution in this chamber. There is a water-tight seal where the cell passes through the dividing wall.

the other half in the other chamber. The cell passes through a water-tight seal from one chamber to the other. One chamber is closed and has a capillary tube attached to it so that any volume changes in that chamber can be accurately measured. The experiment is carried out as follows. Initially both chambers, closed and open, contain tap water or some artificial dilute culture solution, the closed chamber being completely filled to some level in the capillary tube. After temperature stabilization has been reached (the apparatus is immersed in a thermostatted water bath), the solution in the open chamber is rapidly replaced by a solution of sucrose, assumed a non-permeating solute. (The concentration of the sucrose solution is such that the cell is not plasmolysed.) The whole cell then acts as a semipermeable membrane, effectively, between the almost pure water in the closed chamber and the sucrose solution in the open chamber. Water flows from the closed chamber into one end of the cell, along the cell vacuole and out of the other end of the cell into the sucrose solution. This rate of flow is very easily measured from the rate of movement of the meniscus in the capillary tube attached to the closed chamber. From the *initial* rate of flow, J, the areas of the cell in the two chambers, A_1 and A_2, and the osmotic pressure difference, $\Delta\pi$, between the liquids in the two chambers, L_p can be very simply calculated from the following formula (Kamiya and Tazawa, 1956):

$$J = L_p \frac{A_1 A_2}{A_1 + A_2} \Delta\pi \tag{17}$$

The rate of flow falls off with time because the flow of water from one end of the cell to the other polarizes the solutes in the vacuole and this reduces the effective driving force on the water. The whole chain of events in this experiment is rather more complicated than described above and is discussed more fully by Dainty and Ginzburg (1964b). This method seems to be free of serious errors and has given values for L_p for various Characean internodal cells of 1 to 3×10^{-5} cm s^{-1}atm^{-1}. Note that these are nearly two orders of magnitude greater than the values obtained on the plasmolysed protoplasts of higher plant cells. All large algal cells do not show this high permeability to water. Gutknecht (1968) has recently measured L_p for the giant cell of *Valonia ventricosa* by an unexceptionable method and has obtained the value of 1.8×10^{-7} cm s^{-1}atm^{-1}.

The diffusional permeability, P_d, of plant cell membranes to water

The hydraulic conductivity, L_p, is an expression for the permeability of cell membranes to water. Another way of measuring the water permeability would be to take a cell which is in water flux equilibrium with its external solution, add labelled water (e.g., tritiated water) to the external solution and observe the rate at which the labelled water entered the cell. The flux of labelled water into the cell would be given by a typical permeability, Fick's Law, type of equation:

$$J_w = P_d \, \Delta C_w^* \tag{18}$$

where J_w is the flux of labelled water in mole cm^{-2} s^{-1}, ΔC_w^* is the difference in concentrations of labelled water between the outside and inside of the cell and P_d in cm s^{-1}, would be the permeability coefficient for water in the cell membrane. The suffix d stands for *diffusional*, it being assumed that, since the experiment is carried out under conditions of water flux equilibrium, the only mechanism of movement of labelled water across the membrane must be by diffusion, in exchange for ordinary water molecules. It can be shown (Dainty, 1963) that if the mechanism of movement of water due to a difference of water potential, $\Delta\psi$, is also diffusional then L_p and P_d are related by the following expression:

$$L_p RT / \bar{V}_w = P_d \tag{19}$$

$\bar{V}_w$ is the partial molar volume of water, R and T have their usual meanings.

It would seem an easy matter to measure P_d for plant cell membranes by a labelled water experiment, using eq. (18). However, the experiments face one very serious obstacle: the presence of unstirred layers of solution outside and, to some extent, inside the cells. (The unstirred layers inside the cells are only important for large cells.) These layers can be very thick, from 30 to 300 μ, and the situation is much exacerbated for cells in a tissue because of the long diffusion paths through the intercellular spaces. These layers in effect add an extra resistance to the diffusion of labelled water between the cell and an external solution. In fact this extra resistance is usually so much greater than the resistance (inverse of permeability) of the cell membranes that it dominates the exchange of labelled water between the cell and its surroundings. Thus the unstirred layer is the rate-controlling factor in this exchange and no measurement of P_d can be made. Until recently all attempts to measure P_d for plant cells have failed for this reason. However, taking advantage of the low water permeability (high resistance) of the cell membranes of *Valonia ventricosa*, Gutknecht (1968) has been able to correct for the unstirred layer effect and has obtained a value for P_d of $2\cdot4 \times 10^{-4}$ cm s^{-1} for this cell. L_p for this cell is $1\cdot78 \times 10^{-7}$ cm s^{-1}atm^{-1} and Gutknecht was able to verify that eq. (19) applied to *Valonia ventricosa*.

If water traverses cell membranes by way of narrow pores, as has frequently been suggested, then it might be expected that the equality shown in eq. (19) would not hold. For a net flux of water, driven by a water potential difference, would have the nature of a bulk flow through the pores. It would be a pressure-driven, hydrodynamic flow which would depend, approximately, on the fourth power of the radius of the pores as in Poiseuille's Law. On the other hand in a labelled water experiment to determine P_d there is no net flux of water; the only process going on is diffusion and this would be expected to depend on the area of the pores, i.e., the square of the radius. It can be shown that, if water-filled pores do exist in cell membranes as the main pathways of water transport, then

$$L_p RT/\bar{V}_w > P_d \tag{20}$$

Many experiments with animal cells have indeed shown that this inequality is almost invariably true. However, in most cases no correction has been made for the effect of unstirred layers on the value of P_d; this means that the investigators have underestimated P_d. Had they made the correction the inequality could well have been replaced by an equality, which would imply that there were no pores in the membrane, i.e., the diffusion mechanism operated in both water potential driven flow and in labelled water exchange. Only for erythrocytes has the above inequality been reasonably well proved and hence the existence of pores. As stated in the previous paragraph, no good measurements of P_d have been made on plant cells until recently. Gutknecht's (1968) results—the only good ones—for *Valonia* show that the equality between P_d and $L_p RT/\bar{V}_w$ holds in this species, implying therefore that there are no aqueous pores in the cell membranes of this cell.

The reflection coefficient of plant cell membranes

It was mentioned earlier that if the cell membranes are permeable to some of the vacuolar or external solution solutes then frictional interaction can occur between the flows of water and these permeating solutes and the equation of flow of water becomes:

$$J_v = L_p(\Delta P - \Delta \pi_{imp} - \sigma \Delta \pi_p - \Delta \tau) \tag{5}$$

σ, the reflection coefficient, which multiplies the osmotic pressure due to the permeating solutes, is an important parameter of the membrane. So far little attention has been paid to measuring it for plant cells.

Dainty and Ginzburg (1964a) modified the method of transcellular osmosis so as to measure σ for various solutes for the membranes of *Nitella translucens*. For details of the theory and method the original paper should be consulted but, in principle, the technique is quite easy. The transcellular flows in an apparatus such as that illustrated in Fig. 12.3 are measured first with a non-permeating solute like sucrose and then with a permeating solute. The ratio of the latter flow to the former gives the reflection coefficient for the permeating solute. The values obtained for solutes such as methanol and ethanol were of the order of 0·5. Recently Gutknecht (1968) has measured σ for methanol for the cell membranes of *Valonia ventricosa*. The value obtained was −1·3; this negative value is consistent with the higher permeability of these cell membranes to methanol than to water.

An interesting application of the reflection coefficient concept is to the collapse of turgor in the pulvinar cells of the sensitive plant, *Mimosa pudica* (see chapter 8). When the pulvinus is turgid the cells will be in water flux equilibrium. When the *Mimosa* leaves are touched, some stimulus is passed down the petiole to the pulvinus. A rapid loss of water occurs from the cells in the lower half of the pulvinus; the cells lose their turgidity and the leaf and its petiole collapse in a dramatic fashion. What is happening can be understood from the following equation:

$$J_v = L_p(\Delta P - \sigma \Delta \pi) \tag{21}$$

In the normal turgid state of the pulvinar cells, there will be water flux equilibrium ($J_v = 0$) and the vacuolar solutes will be non-permeating ($\sigma = 1$). Thus $\Delta P = \Delta \pi$. The stimulus to cause collapse must suddenly make some or all of the vacuolar solutes permeating and therefore σ less than 1. This immediately causes a driving force $(\Delta P - \sigma \Delta \pi) = \Delta \pi (1 - \sigma)$, driving water out of the cell and causing the loss of turgidity which leads to leaf collapse. Only a change in reflection coefficient could result in such an event.

Active water movements

So far I have discussed the movement of water in plant cells and tissues as if it were completely *passive*, i.e., the movement was solely in response to the thermodynamic force of chemical potential gradient or difference. It is well known, however, that the movement of many solutes—ions, sugars, amino-acids—is, as we say, at least partly *active*. This means that the transport of these solutes is frequently against their chemical potential gradients. Such active transport needs energy which is supplied by metabolism. If active transport of solutes occurs, why not active transport of water? In general there are two reasons why active transport of water is unlikely; these are that water is a bulk constituent of cells, in far greater concentration than any other component, and that the permeability of cell membranes to water is orders of magnitude greater than the permeabilities to other normal components. These two facts imply that a very great amount of metabolic energy would be required to move water against its chemical potential gradient and it is unlikely that cells have any such energy to spare for such a task. Active transport of water may possibly occur in situations where water is scarce and water permeabilities are low. It might be looked for therefore in xerophytic species.

Water relations of whole plants

In the soil-plant-air system water flows through the soil to the plant root surface, across the root to the xylem vessels, up the xylem vessels to the leaf, across the leaf to the evaporating surfaces of the leaf and finally via the vapour phase to the turbulent air. Throughout this pathway various resistances to movement are encountered and there must, of course, be forces driving the water across these resistances. Clearly, in the steady state, the amount of water flowing across these resistances must be the same. By a reasonable analogy with electric current flowing through a number of resistances in series, van den Honert (1948) pointed out that the flow could be expressed, through any part of the system, as a driving force divided by a resistance. Since in the steady state the flow is the same through all the resistances, van den Honert wrote:

$$\text{flow} = \frac{\Delta\Psi_{1,2}}{r_{1,2}} = \frac{\Delta\Psi_{2,3}}{r_{2,3}} = \frac{\Delta\Psi_{3,4}}{r_{3,4}} = \text{etc.} \tag{22}$$

In this equation $\Delta\Psi_{1,2}$ represents the water potential difference at step $1 \to 2$, which might be in the soil for example, and similarly for the other $\Delta\Psi$'s; $r_{1,2}$ is the resistance of the step $1 \to 2$ and similarly for the other r's.

This eq. (22) has been subjected to a great deal of criticism, some of it rather ill-informed. For instance the equation does *not* imply that the resistances are constants independent of the driving forces; they are more or less constant in some parts of the system and in others vary very greatly with the driving forces. I believe this feeling that the resistances ought to be constants is responsible for the frequently-repeated remark that the vapour phase pathway cannot be included in van den Honert's equation because the 'real driving force' is the vapour pressure difference and not the chemical potential difference. This exclusion is quite unnecessary; to be thermodynamically consistent, the appropriate chemical potential differences must be used throughout the equation.

One warning should, however, be made about the use of van den Honert's expression. Because the steps do not all involve the passage of water through semi-permeable membranes, solute reflection coefficients should be included in the $\Delta\Psi$ terms, i.e., $\Delta\Psi$ is of the form:

$$\Delta\Psi = \Delta P - \Delta\pi_{imp} - \sigma\Delta\pi_p - \Delta\tau \tag{23}$$

where the symbols have been explained before.

The van den Honert expression, if used with care, is a very valuable way of looking at whole plant water relations. For instance it says that the maximum resistance to flow will occur where there is the maximum water potential difference; this is in the leaf to air vapour phase. I will now briefly discuss each part of the soil–plant–air system; a most excellent and much more detailed account is given by Slatyer (1967).

Water movement through the soil to the plant root surface

Soil is a complex mixture of solid, liquid, and gaseous phases whose relative proportions vary greatly from one soil to another and, within the same soil, with the dryness of the soil. The water in the soil is in close contact with the solids, either intimately part of the solid—swelling the soil colloids—or in narrow pores. It is also in close contact with the air interfaces. Thus the matric potential contribution to the water potential would be expected to be very high. In fact it is the dominating contribution, for the concentration of dissolved solutes in the soil water is usually only in the millimolar range. When water moves from one part of the soil to another it does not of course pass through any semi-permeable membranes. Therefore the solute reflection coefficients are zero and the osmotic pressure part of the water potential makes no contribution to the driving forces on water; this is true even in highly saline soils which have an appreciable quantity of dissolved solutes. The driving force on water in the soil is therefore the difference of the matric potential, not the difference of the total water potential, between one part of a soil and another An equation can thus be written for the rate of movement of water from, say, a place a few centimetres away from the root up to the root surface, as follows:

$$\text{flow, } Q, = \frac{\Delta\Psi}{r_{soil}} = \frac{\Delta\tau}{r_{soil}} \tag{24}$$

Since water moves *down* a water potential gradient, the water potential at the surface of the root must be lower than the water potential in the soil as a whole. If the soil is wet, so that its water (strictly matric) potential is close to zero, then the resistance of the soil to water movement is also low and the water potential at the root surface need probably be not less than -1 atmosphere in order to maintain an adequate flow of water to the root (Gardner, 1960). As the soil dries out and its water potential decreases, the soil resistance rapidly increases. This is mainly because the larger pores, the channels of least resistance, become empty leaving only the narrow pores of high resistance as channels of water movement. As the soil resistance increases, the water potential difference required to supply an adequate amount of water to the roots increases quite markedly. For instance when the soil water potential is -15 atm., i.e., the soil has become somewhat dry, the water potential at the root might be as low as -30 atm. This is of course quite a low water potential for the root to extract water from; it corresponds to the osmotic pressure of a $1 \cdot 3$ molar sucrose solution!

This brief discussion illustrates, I hope, the main point about water movement in the soil, that is the strong dependence of the soil resistance on the water potential in the soil. This can lead to very low water potentials at the root surface, which the root might not be able to work against.

Water movement across the root

The water absorbing zone of the root appears to extend from just behind the meristematic tip for several centimetres up to where cutinization and suberization of the epidermis begins to be important. Root hairs are usually present in this absorbing zone and they may effectively increase the surface area for water absorption, perhaps by a factor of 3 or 4. A transverse section of a root in the water absorbing zone is shown in Fig. 12.4. The cells are differentiated into

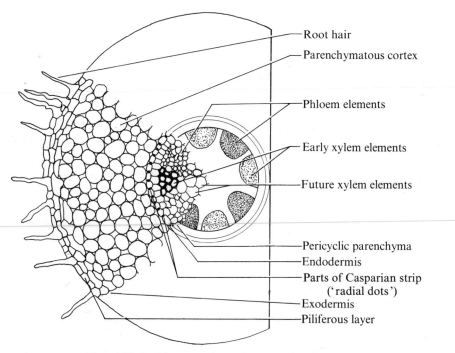

Figure 12.4   Diagrammatic transverse section of root.
(Adapted, by permission of publishers, from Fig. 10 in 'The Plant in Relation to Water' by R. O. Knight, published by Heinemann Educational Books Ltd. London (1965).)

several layers and the water has to pass across these layers, either through the cells themselves or through the intercellular spaces and cell walls of the layers, to reach the xylem vessels near the centre of the root. The outer layer, the epidermis, may be slightly cutinized but is probably permeable to water; some of the epidermal cells develop root hairs, which may, by increasing surface area, increase the effective water permeability. Just within the epidermis, the cortex starts. This usually constitutes the bulk of the root volume and is composed of 5 to 15 layers of parenchyma cells, which are quite loosely packed and have large intercellular spaces. The cortex is sharply bounded from the inner root tissues, the stele, by a single layer of closely packed cells—the endodermis. A characteristic feature of the endodermis is that part of the cell walls are suberized in such a way as to form a barrier to the passage of water from the cortex to the stele *through the endodermal cell walls*. However, the barrier is probably not complete, there being numbers of unsuberized endodermal cells—so-called passage cells—which would allow some cell wall passage of water to the stele. The suberized endodermal cell wall barrier is called the Casparian strip. Inside the endodermis the stele consists of one or two layers of thin-walled cells, the pericycle, and inside that the vascular tissue, xylem and phloem, surrounded by thin-walled parenchyma cells.

Water moves through the root tissues from just outside the root surface to the xylem vessels. The driving force is the modified water potential difference:

$$\Delta\Psi = \Delta P - \Delta\pi_{imp} - \sigma\,\Delta\pi_p - \Delta\tau \qquad (23)$$

Slatyer (1966) has suggested that the membrane, or membranes, encountered during this water movement may be such that $\sigma < 1$ for some of the solutes. This may be so, but it needs further verification, and I will assume in this discussion that the normal solutes involved are sufficiently non-permeating to ensure that $\sigma = 1$. With this assumption the above equation can be rewritten

$$\Delta\Psi = -P^i - (\pi^o - \pi^i) - \tau^o \qquad (25)$$

where the superscripts i and o refer to the inside (xylem) and outside (soil) solutions respectively. It has been reasonably assumed that the external pressure is zero (i.e., 1 atm.) and that in the xylem fluid the contribution of any matric potential to the water potential is negligible. Equation (25) can also be perhaps more clearly put:

$$\Delta\Psi = (-\pi^o - \tau^o) - (P^i - \pi^i) \qquad (26)$$

where the first term in the brackets is the water potential just outside the root surface and the second term is the water potential inside the xylem vessels.

The above water potential difference will drive water from just outside the root surface into the xylem vessels against the resistance of the tissues in the pathway. The main question involved is the nature of this resistance and specifically where the chief barrier to water movement is anatomically located. Two places for this barrier have been suggested in the past: the outer surface of the epidermal cells and the endodermis, with its Casparian strip. The preponderance of opinion is in favour of the endodermis as the critical barrier (see Slatyer, 1967), because the presence of the Casparian strip would seem to force the water at this stage to go through the protoplasts of the endodermal cells. The only real evidence for this view, however, is this anatomical argument; there seems to be no good physiological evidence.

Another view has recently been put forward by House and Findlay (1966a and b) which suggests that the barrier is actually at the surface of the xylem vessels themselves. They worked with de-topped four-day-old maize roots and studied exudation rates, and solute concentrations in the exudate, from the top of the excised roots. The roots were immersed in aqueous solutions. In this system, τ^o and P^i are zero and therefore the water potential difference and the water flow into the xylem are given by

$$\Delta\Psi = -\pi^o + \pi^i \tag{27}$$

$$J_v = L_p \, \Delta\Psi = L_p(-\pi^o + \pi^i) \tag{28}$$

When the external solution is a dilute salt solution, e.g., KCl, salt is 'pumped' into the xylem vessels and π^i becomes greater than π^o, the difference between which provides the driving force on water as illustrated in eqs. (27) and (28). House and Findlay (1966a) were able to show that eq. (28) was obeyed, with a constant value of L_p, over a range of water potential differences. Their value for L_p, calculated on the basis of the *external* surface area of the root, was 0.6×10^{-6} cm s^{-1}atm^{-1} in agreement with other measurements (see Slatyer, 1967).

The important part of their experiments was, however, to observe the transient behaviour of the exudation rate when the external solution concentrations were rapidly changed by the addition of a non-permeating solute. The type of response obtained is illustrated in Fig. 12.5. In this response the delay in the exudation rate reaching its minimum value (part A of the curve in Fig. 12.5) represents the time required for the new solution to diffuse up to the barrier to water movement. This therefore gives some indication of the location of the barrier. Unfortunately the indication is not sufficiently accurate to say whether the location is at the endodermis or further inside the stele, but it does prove that the barrier is not at the epidermis. A more important piece of evidence about the location is given by the next part of the transient curve, part B on Fig. 12.5. What is happening here is that the flow of water into the compartment bounded by the barrier has been severely cut down by adding the non-permeating solute to the external solution. Thus the solute concentration inside the compartment rises, for the salt continues to be pumped into the xylem. That is π^i rises and so increases the driving force on the water, which increases the flow into the xylem until a new steady state is reached. The important point is that the rate of rise of

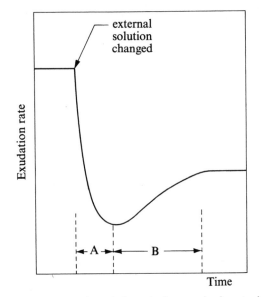

Figure 12.5   Transient response of exudation rate from excised root when the external osmotic pressure, due to a nonpermeating solute, is suddenly increased. In phase A the new solute concentration is diffusing towards the osmotic barrier. In phase B a readjustment of the osmotic flow is taking place consequent on changes in concentration within the osmotic barrier. (After House and Findlay, 1966b.)

solute concentration inside the compartment is determined by the volume of the compartment and thus provides a measure of this volume. (The theory of this can be looked up in the paper by House and Findlay (1966b).) It turns out that the volume of the compartment bounded by the water barrier corresponds to about *half* the volume of the xylem elements in the root and is roughly equal to the volume of those xylem vessels which contain living protoplasts. Thus this evidence indicates that the water barrier is at the surface of the xylem vessels themselves. A similar earlier experiment leading to the same conclusion had been performed by Arisz, Helder, and van Nie (1951) on tomato plants, but this experiment had regrettably not attracted the attention it deserved.

There are many other interesting and important aspects of water movement across roots which deserve discussion. Given that the main barrier to water movement is, say, at the xylem vessels themselves, does the water travel to this barrier through the cell walls and intercellular spaces or through the cells of the epidermis, cortex, etc., themselves? There seems to be no clear-cut answer to this, though I favour the idea that the principal mode of travel is through the cell walls and the intercellular spaces. Another feature of the water movement across the root is that the rate is markedly affected by metabolic inhibitors or low temperature. Although some investigators have interpreted this as evidence for active water uptake, it is much more likely to be a secondary effect, the primary one being an effect on the active salt uptake into the xylem and hence on the water potential difference. These and other aspects of water movement across the root are well discussed by Slatyer (1967) to whom the reader is referred.

In the whole plant, as distinguished from the excised root, the water potential in the xylem vessels is partly due to the (negative) hydrostatic pressure caused by transpiration and this should enhance the water flow. In the absence of transpiration, water flow still occurs (and can lead to such phenomena as guttation) because the pumping of salts into the xylem leads to a value of π^i greater than $(\pi^o + \tau^o)$. π^i can be greater by one or two atmospheres and this results in the appearance of so-called root pressure.

Water flow in the xylem vessels

The path of water from the roots of a higher plant to the leaves is mainly via the xylem vessels of angiosperms and the tracheids of gymnosperms. It would seem necessary that there should be continuous liquid columns in these pathways; any break in a column would lead to cessation of flow. In many plants the pressure in these columns must be less than atmospheric pressure, i.e., negative in our convention, and this must certainly be true in tall trees for in these the length of the columns is much greater than could be supported by one atmosphere pressure.

To take an extreme case, in a tree 100 metres high the pressure difference just to maintain a liquid column of this length stationary is 10 atm., i.e., if the pressure is 1 atmosphere at the bottom it must be -9 atm. at the top. When the water moves, as it does when the tree transpires, the pressure differential between top and bottom may rise to as much as 30 atm.; thus the negative pressure in the xylem vessels at the top of a tall tree could be about -30 atm. How can a column of water stand such tensions and negative pressures? In one sense there is no difficulty; the theoretical cohesive strength of water is many hundreds of atmospheres and special experiments with water in sealed capillary tubes have verified that it can withstand the kinds of tensions likely to be met with in plants. So providing that the pull of up to 30 atm. can be produced, the cohesive strength of water will ensure that the liquid columns in the xylem vessels move up the tree (see Dixon, 1914).

The pull is produced by evaporation of water from the surfaces of parenchyma cells in the leaf and a liquid column will extend continuously from these surfaces to the root. From the thermodynamic point of view evaporation from leaf cell walls lowers the water potential, which is largely a matric potential, in these cell walls. This lowers the water potential, largely a pressure potential, at the top of the xylem columns and this produces the pressure (or suction if preferred) pull up the xylem. One can ask a different kind of question; if there is a negative pressure at the top of a column of liquid of 30 atm., it must be produced by a curved air–water interface; what is the value of the radius of curvature? This is simply given by a well-known law of surface tension in physics which states that the excess pressure, P, over a curved surface, the radius of curvature, R, and the surface tension, T, are related by:

$$P = 2T/R \qquad (29)$$

If $P = 30$ atm or 3×10^7 dyne cm^{-2} and $T = 75$ dyne cm^{-1}, then $R = 250$ Å. There is no difficulty about finding equivalent capillary tube radii in plant cell walls where the cellulose microfibrils are separated by distances of this order of magnitude.

Thus far there seems to be no difficulty about the physical mechanisms involved in the movement of water up the xylem, even in very tall trees. However, when a liquid is under tension it is in a rather unstable condition. Any gases in solution in the xylem water will tend to come out of solution and form a bubble, thus breaking the continuity of the liquid column. One might expect any disturbance, due to strong winds for instance, to produce this cavitation. An extra-strong transpiration pull might have the same effect. Milburn (private communication) has detected, using a sensitive microphone, cavitation to occur quite freely even in herbaceous plants and small trees such as willow, (the snapping of the columns due to cavitation can actually be heard). Since cavitation would appear to occur so readily, how is it possible for sufficient liquid columns to be maintained to supply water to the leaves of a plant? In the first place the snapping of one column is almost certainly limited to that particular xylem vessel. The anatomy of the vascular elements is such that the bubble of gas formed cannot spread to other xylem vessels. Secondly in plants not more than, say, 20 metres in height the vessels can probably be refilled with liquid during the night by root pressure. Finally in all plants, and this must be the major effect in tall trees, new filled xylem vessels are being differentiated by the secondary meristems and these new vessels probably provide an adequate pathway for the water (Preston, 1958). There are still a lot of unsolved problems in this field; even the present brief survey discloses this.

Movement of water through the leaf

When the vascular bundle enters the leaf, it splits up into many branches in a pattern which is characteristic of the species. As they split up, the xylem bundles get progressively smaller until the terminal bundles are single xylem vessels. The leaf is so thoroughly 'invaded' by the xylem system that at no place in the leaf are the terminal bundles separated from the intercellular air spaces by more than one or two cells of the parenchyma tissue. Thus the pathway from the water inside the xylem to the parenchyma cell (spongy and palisade mesophyll) walls, where evaporation into the intercellular air spaces takes place, is very short and probably of low resistance. Only a very small water potential difference will be necessary to drive water through this low resistance and therefore the water potential in the evaporating cell walls will be almost equal to (slightly less than) the water potential in the adjacent xylem vessels. As explained in the previous section, the water potential in the cell walls will be almost all matric potential, whereas the water potential in the xylem vessels will be almost all pressure potential. It seems probable (Weatherly, 1963) that the water travels from the xylem vessels to the intercellular spaces via the cell walls of the intervening cells, rather than through the protoplasts.

Movement of water as vapour from the leaf to the external air

Water moves as vapour from the evaporating surfaces of the leaf to the bulk air outside. There are essentially two evaporating surfaces: the wet cell walls of the mesophyll cells from which evaporation takes place into the intercellular air spaces of the leaf and from there via the stomata (see chapter 9) to the bulk air, and the cuticle of the leaf itself. Water vapour from the cuticle or from the stomata has to diffuse through the stationary boundary layer of air to reach the turbulent bulk air. Thus water vapour at the external surface of the leaf has to traverse, whatever its source, an external resistance—the boundary layer— and this we will consider first.

It has become customary in discussing the movement of water vapour to express the driving force over a given pathway as the difference in water vapour concentrations over the pathway. The flow of water vapour through unit area is then written:

$$E = \Delta C/r \tag{30}$$

where E is the flow in $g\ cm^{-2}\ s^{-1}$, ΔC is the concentration difference in $g\ cm^{-3}$ and r is the resistance to flow; its units will be $s\ cm^{-1}$. ΔC can be expressed in terms of Δe, the vapour pressure difference, by the following formula (Slatyer, 1967):

$$\Delta C = \frac{273}{PT} \rho_v\ \Delta e \tag{31}$$

where P is atmospheric pressure, T is absolute temperature and ρ_v is the density of water vapour at standard temperature and pressure. Vapour pressure, e, is related to water potential by the expression

$$\Psi = \frac{RT}{\bar{V}_w} \ln (e/e^0) \tag{32}$$

The reason why it has become customary to use ΔC, or Δe, as the driving force rather than $\Delta \Psi$ is that a simple kinetic picture of gaseous diffusion naturally leads to a Fick's Law type of equation and hence naturally to ΔC as a useful expression for the driving force. Also the use of ΔC results in the resistance, r, of eq. (30) being more or less constant independent of driving force. These considerations do not invalidate the use of $\Delta\Psi$ as the vapour driving force; in fact when using van den Honert's expression (22), $\Delta\Psi$ must be used. However the resistance associated with the use of $\Delta\Psi$, will vary with $\Delta\Psi$. I will, in this section, use the customary formulation (30) since this is more or less universal in the literature.

If the stationary boundary layer adjacent to the leaf is d cm thick then the Fick's Law equation referred to above would give for the flux of water down a concentration gradient $\Delta C/d$:

$$E = D_w(\Delta C/d) \tag{33}$$

where D_w is the diffusion coefficient of water vapour in normal air; this is approximately 0.24 cm^2 s^{-1}. Comparing eqs. (33) and (30), it can be seen that the resistance of the boundary layer, r_a, is given by

$$r_a = d/D_w \tag{34}$$

This boundary layer of air is a well-known hydrodynamic phenomenon. It is equivalent to what has previously been called in this article the unstirred layer at solid–liquid interfaces. It arises because near the solid surface the viscous forces in the fluid predominate and force the fluid to flow tangentially to the surface. Because the fluid flow is tangential, it is only by diffusion that materials can move perpendicular to the surface (see Bircumshaw and Riddiford, 1952, and Dainty, 1963). The equivalent thickness, d, of the air boundary layer at the surface of a leaf depends on the leaf's linear dimensions and the wind speed. It varies, for instance, from about 0.24 mm for a grass leaf 1 cm wide at a wind speed of 10 m s^{-1} to about 7 mm for a cotton leaf 10 cm wide at a wind speed of 10 cm s^{-1}. The corresponding resistances, r_a, would be 0.1 s cm^{-1} for the grass leaf in a high wind to 3.0 s cm^{-1} for the cotton leaf in a very low speed wind. It will be seen that these external resistances are very important, often controlling, in determining the rate of water loss from a leaf. Some classical experiments on water loss, e.g., Brown and Escombe, 1900, were quite wrongly interpreted because of the failure to recognize this air resistance.

The air resistance is in series with resistance in the two other pathways, cuticular and stomatal, which themselves are in parallel. The cuticle is a term covering the suberized cell walls of the epidermis plus the waxy layers which occur superficially on this surface. Because of the lipid nature of the cuticular barrier its resistance, r_c, to water flow is quite great. It seems to vary from about 20 s cm^{-1} for shade plants to more than 200 s cm^{-1} for xerophytes (Slatyer, 1967). If all the water in the leaf had to be lost through the cuticle, then this would be the dominant resistance. However, when the stomata are open the cuticle resistance is 'short-circuited' by the lower stomatal resistance r_s.

The main pathway of water vapour movement is by evaporation from the mesophyll cell walls, diffusion through the intercellular air spaces and diffusion through the stomatal pores. There are three resistances in series in this pathway; thus what is usually referred to as the stomatal resistance, r_s, is given by

$$r_s = r_w + r_i + r_p \qquad (35)$$

where r_w, r_i, r_p are the resistances of the mesophyll cell walls, the intercellular spaces, and the stomatal pores. The cell wall resistance probably need not be taken into consideration. Occasionally it has been suggested that under conditions of severe water stress the water menisci in the cell walls will withdraw into the depths of the cell wall and thus vapour will have to diffuse in a long tortuous path until it reaches the outside of the cell walls. Most authors consider this unlikely to produce a significant extra resistance, so we will take r_w to be negligibly small. The intercellular resistance, r_i, is also a somewhat unknown factor, but it may turn out to be a significant proportion of the total resistance, r_s. The only direct measurement, by a diffusion porometer by Jarvis, Rose, and Begg (1967), gave a value of 3 s cm^{-1} for cotton leaves, which was 75 per cent of the total resistance, r_s, with open stomata. If bulk flow of air occurs in the intercellular spaces, which seems rather unlikely to me, then r_i would be greatly reduced.

The magnitude of the stomatal pore resistance, r_p, depends of course on the degree of opening of the stomata. It can vary from as low as perhaps 0·5 s cm^{-1} when fully open to infinity when the stomata are closed. The mechanism and control of stomatal pore opening and closing is discussed in detail in chapter 9. The response is a change of turgor of the guard cells associated, in some way not yet known, with the level of carbon dioxide concentration in the intercellular spaces. What seems to be clear is that the control of the degree of opening of the stomata is practically the only active control the plant has on the rate of loss of water. Even this control is only partial, since the stomatal resistance is in series with the air resistance, r_a. A well-known experiment by Bange (1953) illustrates this point. Bange measured the transpiration of *Zebrina* leaves in still air, when $r_a = 2\cdot0$ s cm^{-1}, and in moving air, when $r_a = 0\cdot1$ s cm^{-1}. It will be seen from Fig. 12.6 that whereas the stomata exercised good control over water loss in the moving air, only when the stomata were practically closed did they control the loss in still air. The quantitative reason for this was that fully open stomata had a resistance, r_s, of about 1·0 s cm^{-1}. This is about 10 times the air resistance in moving air and therefore at all degrees of opening the stomata were the rate-limiting factors. But in still air, r_s for the fully open stomata was only half the air resistance; the stomata therefore had to close considerably before the air resistance became negligible.

The essence of the problems of the vapour pathway from the inside of the leaf to the turbulent air can be summarized in the electrical analogue diagram of Fig. 12.7. Here the air resistance, r_a, is shown in series with r_c and r_s which themselves are in parallel. With care the other resistances in the soil–plant–air system can be added in according to van den Honert's ideas. The final step in

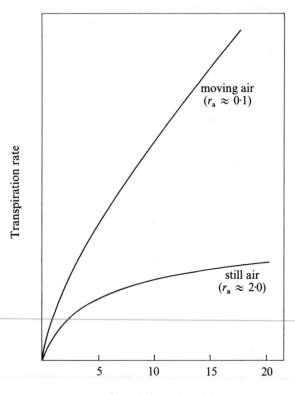

Figure 12.6 Diagrammatic representation, after Bange (1953), of the transpiration rate of *Zebrina* leaves as a function of the degree of opening of the stoma. In the upper curve the experiment was performed in moving air, when the boundary layer resistance was negligible. In the lower curve the experiment was performed in still air, when the boundary layer resistance was high and rate-controlling for large stomatal apertures. (Adapted from Figs. 13 and 14 of Bange (1953) by permission of the publishers, Royal Botanical Society of the Netherlands.)

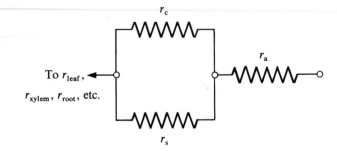

Figure 12.7 Electrical analogue diagram of vapour pathway in plant leaves. r_a, r_c, and r_s are the resistance of the boundary layer, the cuticle and the intercellular air spaces plus stomatal pores, respectively.

the pathway, the vapour phase, has much the greatest resistance and therefore is the rate-controlling step in the water relations of plants in the steady state. This can be seen from the following considerations. From soil to mesophyll cells in the leaf the drop in water potential is rarely greater than 50 atm. But the drop from mesophyll to external atmosphere can easily be 1,000 atm.; this corresponds to an atmospheric relative humidity of about 50 per cent. Where the biggest driving force is, there is the biggest resistance.

Bibliography

Selected further reading

Bennet-Clark, T. A., 1959. Water relations of cells. In *Plant Physiology—A Treatise*, ed. F. C. Steward, Vol. 2. Academic Press, New York and London.
Briggs, G. E., 1967. *Movement of Water in Plants*. Blackwell, Oxford.
Dainty, J., 1963. Water relations of plant cells. *Adv. Bot. Res.* **1**, 279–326.
Dick, D. A. T., 1966. *Cell Water*. Butterworth, London and Washington, D.C.
Slatyer, R. O., 1967. *Plant-Water Relationships*. Academic Press, New York and London.
Spanner, D. C., 1964. *Introduction to Thermodynamics*. Academic Press, New York and London.

Papers cited in text

Arisz, W. H., R. J. Helder, and R. van Nie, 1951. Analysis of the exudation process in tomato plants. *J. exp. Bot.* **2**, 257–297.
Bange, G. G. I., 1953. On the quantitative explanation of stomatal transpiration. *Acta bot. neerl.* **2**, 255–297.
Bennet-Clark, T. A., 1959. Water relations of Cells. In *Plant Physiology—A Treatise*, ed. F. C. Steward, Vol. 2. Academic Press, New York and London.
Bennett-Clark, T. A. and D. Bexon, 1940. Water relations of plant cells. *New Phytol.* **39**, 337–361.
Bircumshaw, L. L. and A. C. Riddiford, 1952. Transport control in heterogeneous reactions. *Quart. Rev. chem. Soc.* **6**, 157–185.
Briggs, G. E., 1957. Osmotic pressure of vacuolar sap by the plasmometric method. *New Phytol.* **56**, 258–260.
Briggs, G. E., 1967. *Movement of Water in Plants*. Blackwell, Oxford.
Brown, H. T. and F. Escombe, 1900. Static diffusion of gases and liquids in relation to the assimilation of carbon and translocation in plants. *Phil. Trans. Roy. Soc.* **B. 193**, 223–291.
Crafts, A. S., H. B. Currier, and C. R. Stocking, 1949. *Water in the Physiology of Plants*. Chronica Botanica Company, Waltham, Mass.
Dainty, J., 1963. Water relations of plant cells. *Adv. Bot. Res.* **1**, 279–326.
Dainty, J. and B. Z. Ginzburg, 1963. Irreversible thermodynamics and frictional models of membrane processes, with particular reference to the cell membrane. *J. Theoret. Biol.* **5**, 256–265.
Dainty, J. and B. Z. Ginzburg, 1964a. The reflection coefficient of plant cell membranes for certain solutes. *Biochim. Biophys. Acta.* **79**, 129–137.

Dainty, J. and B. Z. Ginzburg, 1964b. The measurement of hydraulic conductivity (osmotic permeability to water) of internodal characean cells by means of transcellular osmosis. *Biochem. Biophys. Acta.* **79**, 102–111.

Dick, D. A. T., 1966. *Cell Water.* Butterworth, London and Washington, D.C.

Dixon, H. H., 1914. *Transpiration and the Ascent of Sap in Plants.* MacMillan, London.

Gardner, W. R., 1960. Dynamic aspects of water availability to plants. *Soil. Sci.* **89**, 63–73.

Green, P. B. and F. W. Stanton, 1967. Turgor pressure: direct manometric measurement in single cells of *Nitella*. *Science,* **155**, 1675–1676.

Gutknecht, J., 1968. Permeability of *Valonia* to water and solutes: apparent absence of aqueous membrane pores. *Biochim. Biophys. Acta.* **163**, 20–29.

Heath, O. V. S., 1959. The water relations of stomatal cells and the mechanisms of stomatal movement. In *Plant Physiology—A Treatise,* ed. F. C. Steward, Vol. 2. Academic Press, New York and London.

van den Honert, T. H., 1948. Water transport in plants as a catenary process. *Disc. Faraday Soc.* **3**, 146–153.

House, C. R. and N. Findlay, 1966a. Water transport in isolated maize roots. *J. exp. Bot.* **17**, 344–354.

House, C. R. and N. Findlay, 1966b. Analysis of transient changes in fluid exudation from isolated maize roots. *J. exp. Bot.* **17**, 627–640.

Jarvis, P. G., C. W. Rose, and J. E. Begg, 1967. An experimental and theoretical comparison of viscous and diffusive resistances to gas flow through amphistomatous leaves. *Agric. Meteorol.* **4**, 103–117.

Kamiya, N. and M. Tazawa, 1956. Studies of water permeability of a single plant cell by means of transcellular osmosis. *Protoplasma,* **46**, 394–422.

Kelly, R. B., P. G. Kohn, and J. Dainty, 1963. Water relations of *Nitella translucens.* *Trans. Proc. Bot. Soc. Edin.* **34**, 373–391.

Knight, R. O., 1965. *The Plant in Relation to Water.* Heinemann, London.

Kohn, P. G. and J. Dainty, 1966. The measurement of permeability to water in disks of storage tissue. *J. exp. Bot.* **17**, 809–821.

Preston, R. D., 1958. The ascent of sap and the movement of soluble carbohydrates in stems of higher plants. In *The Structure and Properties of Porous Materials,* eds. D. H. Everett and F. S. Stone. Butterworth, London.

Slatyer, R. O., 1966. An underlying cause of measurement discrepancies in determinations of osmotic characteristics in plant cells and tissues. *Protoplasma,* **62**, 34–43.

Slatyer, R. O., 1967. *Plant-Water Relationships.* Academic Press, New York and London.

Spanner, D. C., 1964. *Introduction to Thermodynamics.* Academic Press, New York and London.

Stadelmann, E., 1963. Vergleich und Umrechnung von Permeabilitäts-Konstanten für Wasser. *Protoplasma,* **57**, 660–678.

Stadelmann, E., 1966. Evaluation of turgidity, plasmolysis and deplasmolysis of plant cells. In *Methods in Cell Physiology,* ed. D. M. Prescott, Vol. II. Academic Press, New York and London.

Tazawa, M., 1957. Neue Methode zur Messung des osmotischen Wertes einer Zelle. *Protoplasma,* **48**, 342–359.

Weatherley, P. E., 1963. The pathway of water movement across the root cortex and leaf mesophyll of transpiring plants. In *The Water Relations of Plants,* eds. A. J. Rutter and F. H. Whitehead. Blackwell, London.

13. The ionic relations of plants

J. Dainty

Introduction

In a manner similar to that of the previous chapter on the water relations of plants, this account will be devoted largely to the fundamentals of ionic movement across plant cell membranes, including some discussion of the movement in tissues. Again the theory is based on thermodynamics, the essential basis of which was discussed in the last chapter. Recent accounts of ionic relations, which should be read for further developments and details, will be found in Dainty (1962), Brouwer (1965), Briggs, Hope, and Robertson (1961), Sutcliffe (1962), and Jennings (1963).

General theory

The driving force on ions

By precisely the same kinds of arguments as were used for water in the last chapter, it can be shown that the driving force on an ion is the gradient of the chemical potential of that ion. The chemical potential of an ion, because it carries an electric charge, has an additional term in it over and above that for water or a non-electrolyte. If the system is at an electric potential, ψ, compared with some standard system, then there is an extra free energy term due to the electrical energy (per mole of ion), $zF\psi$. Thus the ion chemical potential, which is usually called the electrochemical potential and symbolized by $\bar{\mu}$, is given by

$$\bar{\mu} = \bar{\mu}^* + P\bar{V} + RT \ln C + \tau + zF\psi \qquad (1)$$

In this expression $\bar{\mu}^*$ is the chemical potential of the ion in its standard state, i.e., in a system in which the atmospheric pressure is unity, the concentration is unity, there is no matric potential contribution—i.e., no interfaces—and in which the electric potential is zero. P is the pressure in excess of atmospheric, $\bar{V}$ is the partial molar volume of the ion, C is the molar concentration, τ is a contribution from ions which are in or near solid– or gas–liquid interfaces, z is the (algebraic) valency of the ion and F is the Faraday, i.e., the number of coulombs of charge carried by one gram-equivalent of the ion. Equation (1) assumes that the solution is ideal. This is usually far from the truth, but the equation can be modified by the inclusion of an activity coefficient in the $RT \ln C$ term, that is making it $RT \ln \gamma C$. In order to keep discussions of principle clear, I will retain the assumption of ideality in this account.

The full eq. (1) for the chemical potential of an ion need rarely be used. In practice both the $P\bar{V}$ and τ terms can be ignored; the $P\bar{V}$ term is negligible except at pressures of hundreds of atmospheres and the τ term can be included, if necessary, in the activity coefficient multiplying the concentration term. So for discussion purposes I will take the electrochemical potential of an ion j to be given by:

$$\bar{\mu}_j = \bar{\mu}_j{}^* + RT \ln C_j + z_j F \psi \tag{2}$$

If R is expressed in J mol^{-1} deg^{-1} and ψ in volts, $\bar{\mu}_j$ will be in J mol^{-1}.

The driving force on an ion is the (negative) electrochemical potential gradient, $-\mathrm{d}\bar{\mu}_j/\mathrm{d}x$, given by differentiation of eq. (1) as:

$$-\frac{\mathrm{d}\bar{\mu}_j}{\mathrm{d}x} = -\frac{RT}{C_j}\frac{\mathrm{d}C_j}{\mathrm{d}x} - z_j F \frac{\mathrm{d}\psi}{\mathrm{d}x} \tag{3}$$

Note that the driving force consists of two terms, one depending on the concentration gradient and one on the electric potential gradient. It is extremely important to realize that both these terms are involved. Too often in the past has it been assumed that only the concentration gradient term exists and this assumption has led many authors into making false conclusions about ionic movements. It might also be stressed here that at this stage I am referring only to passive ion movements, i.e., I am assuming that the ions are moving as ions (not bound to any other molecules) and that they are moving independently of each other and of other ions and molecules; this assumption implies that the only force acting on the ion concerned is its electrochemical potential gradient.

The Nernst equation

Before considering the passive fluxes of ions, it is important to look at the situation of an ion in physico-chemical, thermodynamic, equilibrium on the two sides of a membrane. In strict analogy with what was shown in the chapter on water, an ion which can permeate a membrane will be in equilibrium between the two sides of a membrane when its electrochemical potential is the same on the two sides of the membrane. Remembering that chemical potential is free energy per mole of ions, this means that no change in free energy occurs on transferring these ions from one side of the membrane to the other; no work is done, the ions are at the same free energy level. In mathematical terms, labelling the ions on one side (the 'outside') with the superscript o and the ions on the other side (the 'inside') with the superscript i:

$$\bar{\mu}_j{}^i = \bar{\mu}_j{}^o \tag{4}$$

that is,

$$\bar{\mu}_j^* + RT \ln C_j^i + z_j F \psi^i = \bar{\mu}_j^* + RT \ln C_j^o + z_j F \psi^o \tag{5}$$

The standard electrochemical potentials, $\bar{\mu}_j^*$, are the same on both sides because both solutions are referred to the same standard state. Cancelling the $\bar{\mu}_j^*$ and collecting the potential terms on the left-hand side and the concentration terms on the right-hand side leads to the equation:

$$\psi^i - \psi^o = \frac{RT}{z_j F} \ln \left(\frac{C_j^o}{C_j^i} \right) \tag{6}$$

Putting $(\psi^i - \psi^o) = E$, inserting the actual numerical values for R and F and assuming T is about 20°C, eq. (5) becomes

$$E = \frac{58}{z_j} \log \left(\frac{C_j^o}{C_j^i} \right) \text{ millivolts (mV)} \tag{7}$$

Equation (6) (or (7)) is an extremely important expression for the electric potential in terms of the ion concentrations when the ions are in *passive flux equilibrium* across a membrane. The equation is called the Nernst equation, the potential is frequently referred to as the Nernst potential and I shall often symbolize it by E_j^N.

It is worth the reader's while to insert some figures into this equation to get some feeling for it. For instance an ion may be concentrated 100 fold inside the membrane compared with outside, $C_j^i = 100 \ C_j^o$. If it is a cation the Nernst equation predicts that the potential should be -116 mV at 20°C if the ion is in passive flux equilibrium. Thus if it is experimentally found that indeed the inside is 116 mV more negative than the outside, then the ion is in passive flux equilibrium. Therefore no work is done on that ion despite the fact that it is 100 times more concentrated inside than outside. This situation can also be looked at in the following way; the outwardly directed 'concentration gradient force' (see eq. (3)) is exactly balanced by the inwardly directed electric potential gradient force. It is very important to realize that ion accumulation does not necessarily imply work done on that ion.

Passive ion fluxes

The development of equations for the passive ion fluxes from the basic driving force of eq. (3) can be a somewhat long and involved process. I have inadequate space to develop these equations here and the details are not sufficiently important, at this level, to justify using this space. Briggs, Hope, and Robertson (1961) and Dainty (1962) discuss the most common way of deriving the flux equations from the basic driving forces and the reader is referred to these accounts which were written for plant physiologists. In summary, after making the rather arbitrary assumptions that the electric potential is a linear function of the distance inside the membrane (constant electric field assumption) and that the electric potential steps at the two boundaries of the membrane with the bathing aqueous solutions can be ignored or are equal and opposite, the flux of an ion j is worked out to be given by the following expression:

$$J_j = -P_j \frac{z_j FE/RT}{1 - \exp(z_j FE/RT)} \left[C_j^{\,o} - C_j^{\,i} \exp(z_j FE/RT) \right] \tag{8}$$

In this expression P_j is the permeability coefficient of the ion in the membrane, $C_j^{\,o}$ is the concentration in the outside solution and $C_j^{\,i}$ is the ion concentration in the inside solution. J_j, the ion flux, is in mol cm^{-2} s^{-1} and is in the direction from outside to inside. The other symbols have been explained before. This looks a complicated expression for flux, but in practice it is rarely used as it stands. Instead it should be noted that the equation can be written

$$\text{net flux } (J) = \text{influx } (J_{in}) - \text{efflux } (J_{out}) \tag{9}$$

where

$$J_{in} = -P_j \frac{z_j FE/RT}{1 - \exp(z_j FE/RT)} C_j^{\,o}. \tag{10}$$

and

$$J_{out} = -P_j \frac{z_j FE/RT}{1 - \exp(z_j FE/RT)} C_j^{\,i} \exp(z_j FE/RT) \tag{11}$$

By dividing eq. (10) by eq. (11), the ratio of the two fluxes for an ion is given by:

$$\frac{J_{in}}{J_{out}} = \frac{C_j^{\,o}}{C_j^{\,i} \exp(z_j FE/RT)} \tag{12}$$

By taking logarithms of the two sides of eq. (12) and by comparing the result with the expression (2) for electrochemical potential, eq. (12) can be written:

$$RT \frac{J_{in}}{J_{out}} = \bar{\mu}_j^{\,o} - \bar{\mu}_j^{\,i} \tag{13}$$

where $\bar{\mu}_j^{\,o}$ and $\bar{\mu}_j^{\,i}$ are the electrochemical potentials of the ion j in the outside and inside solutions bathing the membrane, respectively.

This equation, either (12) or (13), was derived independently by Ussing (1949) and Teorell (1949). It is usually referred to as the Ussing–Teorell equation and is the basis of a most important test for the independent, passive, movements of ions across a membrane. Note that when $J = 0$ or $J_{in} = J_{out}$, eq. (8), (12), and (13) all reduce to the Nernst equation (6) for ions in passive flux equilibrium, so that the Ussing–Teorell equation can be looked upon as a generalization of the Nernst equation. Although the actual point-to-point driving force is the gradient of the electrochemical potential (eq. (3)), the integrated total driving force from one side of a membrane to the other can be looked upon as the electrochemical potential difference, $\Delta\bar{\mu}$. Thus the eq. (13) form of the Ussing–Teorell equation gives a relationship between the flux ratio and the driving force that will be found extremely useful.

The nature of the membrane potential

The electric potential difference across a membrane separating two aqueous salt solutions is a so-called diffusion potential. It arises because the permeating ions tend to carry electric charge across the membrane at different rates. They cannot in fact do this because the solutions on the two sides must remain in electrical neutrality. What happens is that an electric potential arises which slows down the faster moving ions and speeds up the slower moving ions until they are moving at the same rate and therefore no net electric charge is carried across the membrane, i.e., there is no electric current across the membrane. Suppose, for example, that a membrane separates a dilute solution of sodium chloride from a concentrated one. And suppose that, as in water, the mobility or diffusion coefficient of chloride in the membrane is greater than that of sodium. Then as NaCl diffuses through the membrane from the more concentrated solution to the more dilute, the Cl^- tends to get ahead of the Na^+. This separation of charge will immediately lead to an electric field, i.e., an electric potential gradient, which will slow down the faster moving Cl^- and speed up the slower moving Na^+. The electric field and speeds of the two ions will be thereby adjusted until Na^+ and Cl^- are travelling with the same velocity and their entry into the more dilute side will therefore not lead to any departure from electrical neutrality.

Exactly the same principles as the above apply to the more complicated situation of many ions moving across living cell membranes. It should be clear that to find an expression for the membrane potential, one has simply to say that there must be no net electric current through the membrane, that is,

$$\sum z_j J_j = 0 \qquad (14)$$

Applying this to flux equations of the type of eq. (8) and assuming that the principal moving ions are Na^+, K^+, Cl^- one gets, after solving the resulting equation for E:

$$E = \frac{RT}{F} \ln \left(\frac{P_K[K^o] + P_{Na}[Na^o] + P_{Cl}[Cl^i]}{P_K[K^i] + P_{Na}[Na^i] + P_{Cl}[Cl^o]} \right) \qquad (15)$$

This is the well-known constant-field equation for the membrane potential in terms of the external, $[K^o]$ etc., and internal, $[K^i]$ etc., ion concentrations and the ion permeability coefficients, P_K etc.

According to eq. (15) the membrane potential is determined by the passively moving ions, in this case K^+, Na^+, and Cl^-. Anticipating future discussion it may be asked whether active transport of ions makes any contribution to the potential. It certainly makes an *indirect* contribution, for the internal ion concentrations are usually largely determined by active transport processes, i.e., by the ion pumps. However, a more interesting question is: do the ion pumps contribute *directly* and immediately, i.e., should there be a term in the constant field equation for the ion pumping. The answer to this depends on the nature of the pump. If the pump is neutral, that is, assuming the carrier hypothesis of a pump, if the ion-carrier complex is uncharged in all its journeys across the membrane, then since there is no moving charge there can be no direct contribution from the pump to the membrane potential. In some systems it appears that the Na-K pump is such a neutral pump. If, however, the ion-carrier complex is charged on one or both of its journeys, then this moving charge will contribute, directly, to the membrane potential. Such a pump is called *electrogenic* and many examples of electrogenic pumps are now being found.

Experimental methods for single plant cells

The most definitive and clear-cut work on the ionic relations of plant cells has been done on giant coenocytic algal cells: on *Valonia*, *Halicystis*, *Hydrodictyon*, and the internodal cells of various members of the Characeae. Although these cells are very large and therefore easier to study than typical higher plant cells, they still have a much more complicated anatomy than animal cells and our knowledge of their ionic relations is still relatively fragmentary. Instead of a single plasma membrane dividing the protoplasm from the external solution there are, in a plant cell, two plasma membranes separating the protoplasm from the external solution on the one hand and from the vacuole on the other. The outside membrane is called the plasmalemma and the inside one the tonoplast. In addition there is the plant cell wall to which the protoplast is closely pressed, and which is quite a concentrated weak-acid ion-exchange resin. Its presence can give rise to serious experimental difficulties in the study of the ionic relations of the protoplast.

Electric potential measurements

The potential difference between the inside and outside of a cell is experimentally determined by inserting the tip of a microelectrode into the cell and measuring the voltage between that and a similar electrode in the external solution. A microelectrode consists of a glass tube a few millimetres in diameter whose tip is drawn down to a diameter of the order of a micron. It is usually filled with 3 N KCl. It is in fact a micro salt bridge. To the wide end of the tube a Ag/AgCl or calomel electrode is attached. The external electrode is similar except that the tip is not usually drawn down to such a small diameter. The arrangement described above is illustrated in Fig. 13.1. 3 N KCl is used in the salt bridge to minimize the diffusion potential at the electrode tip. The potential difference is measured with a high-impedance voltmeter, preferably with an input impedance of at least 10^{10} ohms, because the microelectrode sometimes has an impedance of as much as 10^8 ohms.

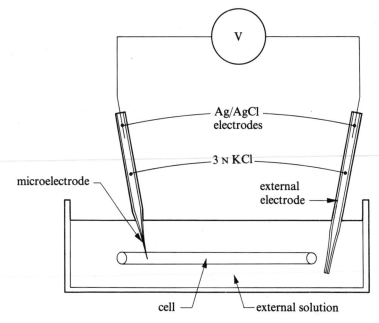

Figure 13.1 Arrangement for measuring electric potential difference between inside of a plant cell and outside. Plant cell is drawn as if it were a Characean internodal cell. V is a high impedance voltmeter. Calomel electrodes are also often used instead of Ag/AgCl electrodes to connect the salt bridges to the voltmeter.

There are many difficulties associated with inserting microelectrodes into even large plant cells. The plant cell wall is fairly tough and the microelectrode tips have to have relatively thick walls in order to penetrate easily. Microelectrodes often have tip potentials, that is conditions at the tip are such as to modify the mobility of either K^+ or Cl^- and thus produce a significant and unwanted diffusion potential between the salt bridge and the solution into which the tip is dipping. Insertion of a microelectrode tip through a plant cell wall often greatly enhances the tip potential, for the electrode tip frequently becomes blocked with ion-exchange material, probably pectins. Finally there is some difficulty in knowing where the tip of the microelectrode actually is: whether, in a plant cell, it is in the vacuole or in the cytoplasm for instance.

The tip is so narrow that it is difficult to see and, even if the tip seems to be in the centre of the vacuole it is difficult to see whether the tonoplast has formed a sheath over it. However, most investigators have given this and other difficulties full consideration and their results are probably quite reliable (see Spanswick and Williams, 1964).

Concentration measurements

These form one 'half' of the necessary measurements to get electrochemical potentials, the other half being the above-described electric potential measurements. In large algal cells, measurement of the concentrations of the principal ions in the vacuole presents no difficulty: flame photometry for the alkali metal cations, absorption spectrophotometry for the divalent cations, electrometric titration for chloride, and so on. There is plenty of solution to work with even from a single cell. The concentrations in the cytoplasm present greater experimental difficulties in their determination. Members of the Characeae, e.g., *Nitella translucens* and *Chara australis* to mention the two most studied species, have two types of cytoplasm in their internodal cells. About half the cytoplasm, next to the plasmalemma and cell wall, is in a gel state; the chloroplasts are in this stationary layer. The inner layer of cytoplasm is fluid and streams in a long helical path around the cell. By gentle centrifugation at a few g this flowing layer can be made to settle at one end of the cell, from where it can be collected and the ion concentrations measured. The gel layer can be obtained by cutting off the ends of the cell, inserting a syringe into one end and gently blowing out with liquid paraffin the vacuolar contents and the flowing cytoplasm. The gel layer left can then be analysed. MacRobbie (1964) and Spanswick and Williams (1964) describe these methods in detail. Raven (1967a) and Gutknecht (1966) describe methods of doing these same analyses on *Hydrodictyon africanum* and *Valonia ventricosa* respectively.

In these giant algal cells, at least, reliable potential and concentration, and therefore electrochemical potential, measurements have been made in the chief phases of the cell. The results of these basic measurements will be considered later.

Ion flux measurements

The other basic measurements to make are of the fluxes, both influxes and effluxes, of the principal ions across both membranes of the plant cell. This can be very difficult even with giant algal cells.

The principles of the method used are simplest to see when it is applied to measuring the fluxes into and out of an animal cell, which can be treated as a single compartment (see Fig. 13.2). Let the volume of the cell be V cm^3 and the area A cm^2, the internal and external concentrations of the ion studied be C_i

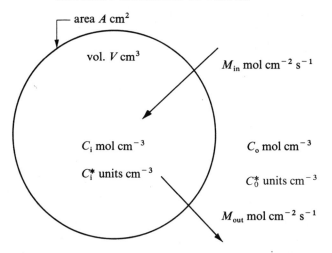

Figure 13.2 The parameters involved in flux measurements into and out of a single compartment. The concentrations of radioactive ions, C_i^* and C_o^*, can be measured in any convenient units providing they are the same for both concentrations.

and C_o, the corresponding concentrations of an isotope of the ion be C_i^* and C_o^*, both expressed per cm^3. Let the influx and the efflux of the ion be M_{in} and M_{out} both in mol cm^{-2} s^{-1}. Then the rate of increase of radioactivity inside the cell is given by:

$$V\frac{dC_i^*}{dt} = A\left[M_{in}\frac{C_o^*}{C_o} - M_{out}\frac{C_i^*}{C_i}\right] \tag{16}$$

In this equation the term, for example, $AM_{in}C_o^*/C_o$ is explained as follows. The total rate of entry of ion is the flux, M_{in} in mol cm^{-2} s^{-1}, times the area, A cm^2. A fraction C_o^*/C_o of the entering ions is radioactive. Therefore the rate of entry of radioactive ions is $AM_{in}C_o^*/C_o$. Similarly the rate of exit of radioactive ions is $AM_{out}C_i^*/C_i$. The net rate of entry is the rate of increase of total inside radioactivity, $d(VC_i^*)/dt$.

An influx experiment is performed by measuring the *initial* rate of increase of internal radioactivity, when an unlabelled cell is placed in a radioactive solution, i.e., C_i^* is negligible throughout the experiment. Then from eq. (16), M_{in} is given by

$$M_{in} = \frac{V(dC_i^*/dt)}{A(C_o^*/C_o)} \tag{17}$$

An efflux experiment is performed as follows. The cell is allowed to take up the radioactive ion from its normal bathing solution until the internal specific activity, $S_i = C_i^*/C_i$, has reached some convenient value. The cell is then transferred to the same bathing solution, but non-radioactive and kept that way, i.e., $C_o^* = 0$, by frequent solution changes, and the internal radioactivity is measured as a function of time. In this case, in eq. (16) $C_o^* = 0$ all the time and the equation becomes a simple differential equation whose solution can be written in the form:

$$\log (VC_i^*) = \log (VC_i^*)_{t=0} - 0.4343 \frac{AM_{out}}{VC_i} \cdot t \tag{18}$$

i.e., the logarithm of the quantity of radioactivity in the cell is a linear function of time. From the slope M_{out} can be determined.

The situation, even in a single plant cell, is much more complicated. There are three compartments—cell wall, cytoplasm, and vacuole. Even if the non-membrane-bounded cell wall compartment is suitably corrected for, not always an easy matter, there are two membrane-bounded compartments in series—a small cytoplasmic compartment and a large vacuolar one. MacRobbie (1964) has carried out a detailed theoretical analysis of this situation and her paper should be referred to for complete details. Some of her conclusions will be described here. If the ion fluxes across the plasmalemma are much greater than those across the tonoplast, in other words if the plasmalemma is much more permeable than the tonoplast, then during an influx experiment the cytoplasm fills up with radioactivity first and after it is full the vacuole starts to fill up. Thus the separate influxes into the cytoplasm and the vacuole can be measured. If the whole cell is loaded with radioactivity and then placed in unlabelled solution for an efflux experiment, three phases of loss of radioactivity can be seen, as in Fig. 13.3. For each phase—cell wall, cytoplasm, vacuole—loses its

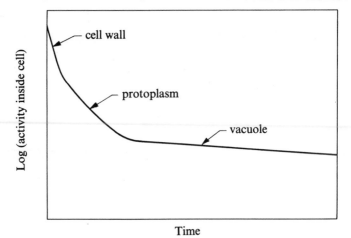

Figure 13.3   Type of three-phase efflux curve obtained for plant cell when the plasmalemma is much more permeable than the tonoplast. From the slopes of the protoplasm and vacuole phases the effluxes across the plasmalemma and the tonoplast can be calculated. From the intercepts of the straight lines on the ordinate the amounts of ion in each phase can be calculated.

radioactivity in turn before the preceding phase starts to lose any. Thus the slopes of the efflux lines will give the effluxes, M_{out}, according to eq. (18).

The situation is very different when the tonoplast is much more permeable than the plasmalemma. In this case as fast as radioactivity enters the cytoplasm, it moves into the vacuole. Thus only one rate of entry can be seen and this essentially gives M_{in} for the plasmalemma. Similarly an efflux curve is rate-controlled by the plasmalemma and only gives M_{out} for the plasmalemma. To get the fluxes across the tonoplast some very difficult measurements of the small differences in the specific activities in the cytoplasm and vacuole have to be carried out. MacRobbie (1964) did this successfully for the fluxes of potassium and chloride in *Nitella translucens*. Her results will be given later.

Ionic relations of giant algal cells

The electrochemical potentials

The first items of basic information required, so as to understand the ionic relations of a cell, are the electrochemical potential differences for the ions between the various compartments of the cell. One of the earliest studies from this point of view was by MacRobbie and Dainty (1958) on the internodal cells of *Nitellopsis obtusa*. Since that time more detailed studies have been made on two other Characeae, *Nitella translucens* and *Chara australis*, as well as somewhat less detailed studies on other giant algal cells. In order to confine my discussion to principles, I will concentrate on the work done on *N. translucens*.

Spanswick and Williams (1964) have made careful measurements of the electric potentials and the ion concentrations of Na, K, and Cl in the vacuole and flowing cytoplasm of *N. translucens*. Their results are shown in Fig. 13.4. MacRobbie (1962, 1964) has made similar, though not quite so complete analyses, and gets similar results except that she finds Na to be somewhat more concentrated (54 mM) in the flowing cytoplasm. She has also analysed the ion concentrations in the gel, chloroplast, layer of the cytoplasm and finds higher concentrations of all three ions than in the flowing cytoplasm; this is probably due to higher concentrations in the chloroplasts themselves.

From the data of Fig. 13.4 the electrochemical potential differences between cytoplasm and external solution and between cytoplasm and vacuole can be calculated. It has become customary to express electrochemical potential differences in volts; the reasoning behind this is as follows: From eq. (2) the electrochemical potential difference, $\Delta\bar{\mu}_j$, between a phase i and a phase o is given by

$$\Delta\bar{\mu}_j = \bar{\mu}_j^i - \bar{\mu}_j^o = RT \ln C_j^i + z_j F\psi^i - RT \ln C_j^o - z_j F\psi^o$$

$$= z_j F(\psi^i - \psi^o) - RT \ln (C_j^o/C_j^i)$$

(19)

	C_o mM	Plasmalemma	E_{co}^N mV	$\dfrac{\Delta\bar{\mu}_{co}}{zF}$ mV	C_{cyt} mM	$\dfrac{\Delta\bar{\mu}_{cv}}{zF}$ mV	E_{cv}^N mV	Tonoplast	C_{vac} mM	
Na	1·0		−66	−72	14	−57	+39		65	Na
K	0·1		−178	+40	119	−6	−12		75	K
Cl	1·3		+99	−237	65	+5	−23		160	Cl

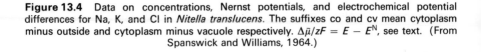

138 mV

Electric potential profile (E)

18 mV

Figure 13.4 Data on concentrations, Nernst potentials, and electrochemical potential differences for Na, K, and Cl in *Nitella translucens*. The suffixes co and cv mean cytoplasm minus outside and cytoplasm minus vacuole respectively. $\Delta\bar{\mu}/zF = E - E^N$, see text. (From Spanswick and Williams, 1964.)

Now $(\psi^i - \psi^o)$ is the observed electric potential difference, E, between phases i and o and $(RT/z_j F)/\ln(C_j^0/C_j^1)$ is equal to the Nernst potential, E_j^N, for the ion concerned, i.e., it is what the potential would be if the ion j were in passive flux equilibrium. Thus eq. (19) can be written

$$\Delta\bar{\mu}_j = z_j F(E - E_j^N) \qquad (20)$$

and we speak of the electrochemical potential difference as being so many volts $(E - E_j^N)$, meaning that it strictly is so many volts times $z_j F$. It is a very useful convention, not unique to this field of study for it is also the convention of redox potentials. In Fig. 13.4 the Nernst potentials for the ions and the electrochemical potential differences are given, both in millivolts.

To take one example, the Nernst potential for Na between cytoplasm and external solution is -66 mV; this means that the electric potential difference required to give passive flux equilibrium is -66 mV. But the observed electric potential difference between cytoplasm and outside is -138 mV. Thus $\Delta\bar{\mu}_{Na} = F(-138 + 66) = F(-72)$ or, as we say, -72 mV. Sodium is thus at a lower electrochemical potential in the cytoplasm than it is in the external solution despite the fact that it is 14 times more concentrated in the cytoplasm.

From Fig. 13.4 it can be seen that potassium is at a higher electrochemical potential inside the cytoplasm than it is outside, and so is chloride—at a considerably higher value in fact—for it must be remembered that the number of millivolts has to be multiplied by $z_j F$ to get the actual $\Delta \bar{\mu}$; z_j for chloride is -1. At the tonoplast the electrochemical potential differences for K and Cl are approximately zero, i.e., the observed electric potential difference across the tonoplast is approximately equal to the Nernst potentials for these ions. However $\Delta \bar{\mu}$ for Na at the tonoplast is quite large, -56 mV, and the electrochemical potential for sodium is lower in the cytoplasm than it is in the vacuole.

The only other cell for which the electrochemical potential differences are known in comparable detail is *Hydrodictyon africanum*, studied by Raven (1967a). In this cell the situation at the plasmalemma is similar to that in *N. translucens*; Na is at a lower electrochemical potential (by about 20 mV) in the cytoplasm than in the outside solution, K and Cl are at a higher electrochemical potential (by about 50 mV and 150 mV respectively) in the cytoplasm than outside. At the tonoplast while Na and K appear to be at the same electrochemical potential in the cytoplasm as in the vacuole, Cl is at a higher electrochemical potential (by about 40 mV) in the cytoplasm than it is in the vacuole. Thus the pattern of a lower electrochemical potential for Na in the cytoplasm than in the external solution and higher electrochemical potentials for K and Cl in the cytoplasm than outside is a common one between *Nitella* and *Hydrodictyon*. Gutknecht (1966) finds a similar pattern too in *Valonia ventricosa*, though the electrochemical potential difference for Cl is not so pronounced.

In other giant algal cells not so much information is available. In *Nitellopsis obtusa* (MacRobbie and Dainty, 1958) the electrochemical potential differences between the vacuole and the external solution are known: for Na, -105 mV; for K, $+10$ mV; for Cl, -165 mV. It is not known how these differences are distributed across the plasmalemma and tonoplast, but it seems likely that the same pattern of a low $\bar{\mu}$ for Na and higher $\bar{\mu}$'s for K and Cl in the cytoplasm as compared with the external solution will also be found to be present in this species. In *Chara australis* (Hope and Walker, 1960) and *Halicystis ovalis* (Blount and Levedahl, 1960) qualitatively somewhat similar electrochemical potential differences to those in *Nitellopsis* occur and it is surmised that the situation at the plasmalemma is similar to what it is in *Nitella translucens* and *Hydrodictyon africanum*.

These giant algal cells probably all, then, have similar electrochemical potential difference patterns across the plasmalemma: Na at a lower electrochemical potential in the cytoplasm than outside; K and Cl at higher electrochemical potentials inside than outside. At the tonoplast no definite statement can be made, except that usually the electrochemical potential differences are smaller than those across the plasmalemma, i.e., the ions are closer to passive flux equilibrium across the tonoplast. This is not so, however, in *Valonia* (Gutknecht, 1966). Much more work is needed to establish these patterns better in a large range of cells, particularly with respect to the tonoplast.

Ion fluxes

It has been established in the previous paragraphs that at the plasmalemma of *Nitella translucens* and *Hydrodictyon africanum* the electrochemical potential differences are such that the physicochemical driving forces are: on Na, inwards; on K, outwards; and on Cl, outwards. Thus if the ions move passively, i.e., under the action of these forces only, the influx of Na should be greater than the efflux and the effluxes of K and Cl should be greater than their influxes —all fluxes being across the plasmalemma. Using the Ussing–Teorell equation, the actual ratio of the passive fluxes can be calculated, for the actual electrochemical potential differences are known. For instance from the data of Spanswick and Williams (1964) for *N. translucens* given in Fig. 13.4, J_{in}/J_{out} for the three ions should be: for Na, 12/1; for K, 1/5; for Cl, 1/10,000. Similarly from the data of Raven (1967a) for *H. africanum*, J_{in}/J_{out} should be: for Na, 2/1; for K, 1/10; for Cl, 1/1,000.

The measured plasmalemma fluxes do not at all support these predictions. MacRobbie (1962, 1964) has done a great deal of careful and difficult work on the fluxes in *N. translucens*. She finds that all the fluxes (of Na, K, and Cl) across the plasmalemma are of the order of magnitude of 10^{-12} mol cm^{-2} s^{-1}, and that for each ion *the influx is approximately equal to the efflux*. Raven (1967a) obtained very similar results to these for the plasmalemma fluxes in *H. africanum*.

In other giant algal cells where only the electrochemical potential differences between vacuole and external solution are known, it can be predicted that the influx of Na into the vacuole from the external solution should be greater than the efflux. In fact, as in *Nitellopsis obtusa*, the fluxes are approximately equal. For K, the electrochemical potential differences between vacuole and external solution in *Chara australis*, *N. obtusa*, and *Halicystis ovalis* are near zero and therefore the fluxes should be about equal, in and out, as is experimentally found. In *Valonia ventricosa* a much greater efflux of K from the vacuole than influx is predicted; equal fluxes are found (Gutknecht, 1966). Except for *V. ventricosa*, the electrochemical potential differences for Cl between vacuole and external solution are such that there should be a much greater efflux than influx; in fact again, experimentally, the fluxes are approximately equal. In *V. ventricosa* equality of fluxes is predicted and found.

Thus the general picture is that the ion fluxes do not obey the rules for passive behaviour, with occasional exceptions for K and Cl. We can explore this situation a little more deeply by taking the case of Cl in *Nitella translucens*. If Cl moves passively across the plasmalemma, the efflux should be about 10,000 times greater than the influx. We can put this another way; the barrier to inward movement of Cl is so high (236 mV, or 5,450 cal mol^{-1} high) that it is very rare for an ion meeting the barrier to have enough energy to get over it. Yet in practice a large number do get over it (10,000 times as many as expected). Where do they get the large amount of energy required (5,450 cal mol^{-1}) to surmount this barrier? There is only one source of energy: the metabolic energy of the cell. Inward movement of chloride across the plasmalemma of *N. translucens* must be somehow linked with metabolic energy; so we say that chloride is actively transported inwards across the plasmalemma, or that there is an inwardly directed chloride pump at the plasmalemma. Similarly there is an inwardly directed chloride pump at the plasmalemma of *H. africanum*.

Exactly the same argument applies in the other cases. Na is actively transported outwards across the plasmalemma of both *N. translucens* and *H. africanum*; there is a sodium extrusion pump at the plasmalemma of these species. K is actively transported inwards across the plasmalemma of these cells. In *Chara australis*, *Nitellopsis obtusa*, and *Halicystis ovalis* there are outwardly directed sodium pumps and inwardly directed chloride pumps somewhere between the vacuole and external solution. (We only know the overall situation, not the exact situation at each membrane.) It is fairly clear that the sodium and chloride pumps are at the plasmalemma of *C. australis*. In the original papers the sodium pumps in *N. obtusa* and *H. ovalis* were placed, probably correctly, at the plasmalemma but the chloride pumps were tentatively placed at the tonoplast. This latter assignment is probably incorrect and the situation ought to be reinvestigated. Potassium pumps are not strictly needed in these three species; they probably exist, however, but are 'short-circuited' by the passive fluxes.

From two basic pieces of experimental information—the electrochemical potential differences and the ion fluxes—we have argued the existence of active transport in specific directions and even of a definite amount. The next section is concerned with testing these predictions and asking some questions about the supply of metabolic energy to the pumps.

Metabolic energy and ion fluxes

It has been argued in the previous section that, in the well-investigated giant algal cells *Nitella translucens* and *Hydrodictyon africanum*, there occur sodium extrusion pumps and inwardly directed potassium and chloride pumps, all at the plasmalemma. If these pumps exist they need a supply of energy; if the energy supply is shut off the pumps should stop. This section is concerned with testing this prediction and discussing the nature of the energy supply.

In a green plant cell there are two potential metabolic energy sources: photosynthesis and respiration. The photosynthetic source can easily be cut off: by taking away all light sources and having the plant in darkness. This procedure has a profound effect on some of the ion fluxes. In *N. translucens* (MacRobbie,

1962) the inward plasmalemma flux of K is reduced from a mean value of 0.9×10^{-12} mol cm^{-2} s^{-1} in the light to 0.3×10^{-12} mol cm^{-2} s^{-1} in the dark; the K efflux at the plasmalemma, on the other hand, is the same in both light and dark. This is qualitatively what would be expected from the previous arguments; the plasmalemma K influx should be at least partly active and the efflux should be wholly passive. The Na efflux at the plasmalemma, which is 0.55×10^{-12} mol cm^{-2} s^{-1} in the light, is reduced to 0.10×10^{-12} mol cm^{-2} s^{-1} in the dark; the Na influx is the same in the light and the dark. This again is what one would expect if a Na extrusion pump operates and the Na influx, down the electrochemical potential gradient, is passive. The Cl influx at the plasmalemma, 0.85×10^{-12} mol cm^{-2} s^{-1} in the light for the particular batch of cells used in the MacRobbie (1962) paper, is reduced to 0.05×10^{-12} mol cm^{-2} s^{-1} in the dark. Hope, Simpson, and Walker (1966) found that the Cl efflux actually goes up in the dark by a factor of 2 to 4; this observation does not of course contradict the idea that the Cl efflux is passive. Thus for chloride the observations of the effect of light and dark on the influx strongly support the contention that the influx is almost wholly active.

Very similar results to these have been obtained by Raven (1967a) on *H. africanum*. Transferring from light to dark scarcely affects the Na influx, K efflux, and Cl efflux all of which were predicted to be passive, down the corresponding electrochemical potential gradients. But the Na efflux is reduced to about 45 per cent of its light value, the K influx to about 13 per cent of its light value and the Cl influx to about 35 per cent of its light value. Thus the predicted active fluxes are all affected by cutting off the supply of energy from photosynthesis. In addition to this kind of experiment, Raven also examined the temperature dependence of the various ion fluxes across the plasmalemma of *H. africanum*. He found that over the range of temperatures used, 2°C to 17°C, the predicted passive fluxes—Na influx, K and Cl effluxes—were little dependent on temperature, while the predicted active or partly active fluxes—Na efflux, K and Cl influxes—were strongly dependent on temperature. As might be expected, because it has to surmount the largest electrochemical potential barrier, the highest Q_{10} was found for chloride influx.

Thus, for these thoroughly investigated cells, the predictions from the thermodynamic data have been verified. We can now speak with some confidence of Na, K, and Cl pumps and discuss other aspects of their properties.

One of the cardiac glycosides, ouabain, has been found to be a specific inhibitor of sodium and potassium pumps in animal cell membranes. It is thought to inhibit, not by interfering with the supply of metabolic energy to the pumps, but by binding on the presumed macromolecule carrier of the pump and interfering with the binding of, perhaps competing with, the cations. MacRobbie (1962) found that 5×10^{-5} M ouabain reduced the K influx in the light to near its dark value. She found no effect of ouabain on the Cl pump. Raven (1967a) has studied the effect of ouabain on the ion pumps in *Hydrodictyon*. He had to use rather high concentrations of ouabain, the maximum inhibitory effect being at 0.5×10^{-3} M concentration in the bathing fluid, but this does not seem to affect the validity of his results. He found that ouabain had no significant effect on any fluxes except K influx and Na efflux. It had no effect also on photosynthesis as measured by the fixation of $^{14}CO_2$. Ouabain had an inhibitory effect on the K influx in the light; it reduced it to about 35 per cent of control value. It also reduced the Na efflux in the light to about 30 per cent of its control value.

These experiments suggest that the sodium and potassium pumps in these plant cell membranes are similar to those in animal cell membranes. If so it would be expected that the outward, pump-driven, movement of sodium is linked to the inward, pump-driven, movement of potassium. It has been suggested in fact that the same carrier molecule is involved; on its outward journey it takes Na and, after a change in specificity, on its inward journey it brings K into the cell. (This picture cannot be quite correct because the linkage is not always one Na for one K.) If there is some such linkage then the efflux of Na should depend on the external K concentration; indeed if there is no external K the carrier should not be able to make its return journey to pick up more Na. Raven (1967a) has found with *Hydrodictyon* that the active sodium efflux does depend on the external K concentration in the predicted way. The relationship between Na efflux and $[K^\circ]$ is an enzyme kinetics type of curve with a 'Michaelis constant' for the potassium of about 0·1 mM. It is interesting that the active K influx depends on $[K^\circ]$ in exactly the same way, with about the same 'Michaelis constant'. These are excellent supporting data to the concept of a Na–K linked pump at the plasmalemma of a type similar to that generally found in animal cells.

One of the most interesting questions concerning the ion pumps is their source of supply of energy. There have been two major views; Lundegårdh (1960) and Robertson (1960) consider that at least some of the ion pumps are closely linked with electron transfer processes; many others, e.g. Laties (1957) and Sutcliffe (1959) think that the ion pumps are energized by ATP. In the Lundegårdh–Robertson picture, as the electrons move down, say, a cytochrome chain anions can be envisaged moving in the opposite direction; the energy to move the ions is directly provided by the redox potential differences between adjacent members of the cytochrome chain. The active ion fluxes of Na, K, and Cl in giant algal cells provide an opportunity for looking at this question of energy supply in a situation where practically all the energy comes from photosynthetic processes (see chapter 10).

Active ion transport in green plants in the light is not affected by the complete removal of CO_2 from the air. Thus the totality of photosynthetic reactions, including carbon fixation, is not required for energizing the ion pumps (van Lookeren Campagne, 1957, Raven, 1967b). Since carbon fixation is not involved, the photosynthetic reactions from the oxidation of water to the reduction of NADP are presumably the relevant ones for providing energy for ion transport (see chapter 10). A fairly widely accepted scheme (Duysens, 1964; Clayton, 1965) for these reactions is shown in Fig. 13.5, in outline. Fig. 13.5 is a rough redox potential energy diagram; it can be looked upon as an electron flow diagram from water to NADP. (Reduced NADP plus ATP carry out the process of carbon fixation not shown in this diagram.) An electron travels from water—thereby liberating oxygen—first to an electron acceptor, Z, which is probably a form of chlorophyll a. A light quantum, hv_2, is absorbed at this stage and this essentially transfers the electron to a higher redox potential stage (more reducing level), Q. From Q the electron travels through a series of compounds, including one or more cytochromes, to another chlorophyll a system, called here P. Another light photon, hv_1, is absorbed here lifting the electron to a still higher redox potential. The highly reducing compound here is very closely linked with ferredoxin, which I have shown in Fig. 13.5 as the actual electron acceptor. From ferredoxin the electron finally travels 'downhill', eventually to reduce NADP.

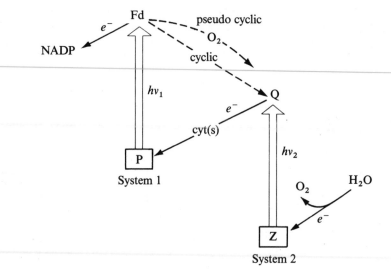

Figure 13.5 Outline scheme of electron pathway in photosynthesis, from water to NADP. It is thought that ATP is made through some kind of coupling with electron transfer from Q to P. Fd = ferredoxin. (Adapted from Fig. 1 of MacRobbie (1965) by permission of the author and the publishers of *Biochim. et Biophys. Acta.*)

The above is the through pathway for an electron from water to NADP. It can be seen to comprise two photochemical reactions, referred to as systems 1 and 2. System 1, from P to ferredoxin, is similar in many respects to the total photosynthetic system of bacteria. In general it absorbs light up to a maximum wavelength of about 730 nm. System 2 is unique to green plants and is closely associated with oxygen production. It needs higher energy quanta and will only absorb light up to a maximum wavelength of about 705 nm. Under certain conditions an electron having reached ferredoxin and thereby reduced it, will, instead of going on to NADP, return to P via the cytochrome system between Q and P. Under certain other conditions it will go to oxygen perhaps via the same cytochrome system. The first of these is referred to as the cyclic path and the second, to oxygen, as the pseudocyclic path. It is thought that ATP is formed during the electron transfer along the cytochromes between Q and P in a manner analogous to oxidative phosphorylation in mitochondria. Thus ATP can be formed in any of the three pathways: straight through from H_2O to NADP, in the cyclic pathway, and in the pseudocyclic pathway. In the latter two cases we speak of cyclic and pseudocyclic phosphorylation. Thus chloroplasts have machinery for reducing NADP and making ATP, through a series of electron transfer reactions.

If red light of wavelengths between 705 nm and 730 nm is used to illuminate a green plant, then it can excite system 1 only. Full photosynthesis (effectively the reduction of NADP) cannot be sustained and there is no oxygen production. But ATP can still be made by cyclic phosphorylation. MacRobbie (1965) and Raven (1967b) found that under these conditions, in *Nitella translucens* and *Hydrodictyon africanum*, the active Cl influx was inhibited, but the active K and Na fluxes were unaffected. Here then is an immediate suggestion that the cation pumps are ATP driven whereas the chloride pump does not need ATP as an energy supply, but is somehow dependent on system 2. Many other experiments were carried out to check this hypothesis. If the plant cells are exposed to white light (systems 1 and 2 operative) in the presence of the uncoupling agent carbonyl cyanide *m*-chlorophenyl hydrazone (CCCP), which suppresses the formation of ATP, then the active Cl influx is unaffected while the active cation fluxes are totally inhibited. Imidazole, another uncoupler of ATP formation from electron transfer, has the same effect. Dichlorophenyl dimethyl urea (DCMU), which inhibits system 2, has the effect of inhibiting the active Cl influx while not affecting the cation fluxes, at least not at the low concentration (10^{-7}M) at which it inhibits the Cl influx.

All these, and other, experiments on *N. translucens* and *H. africanum*, support the idea that the Na-K pump is energized by ATP produced by photophosphorylation. (This fits in with the generally accepted notion that animal cell Na-K pumps, which seem to be closely associated with an ATPase, are powered by ATP, in this case mostly produced by oxidative phosphorylation or in glycolysis.) But the Cl pump is different; it is not energized by ATP and is closely associated with system 2 of photosynthesis. Perhaps it is indeed a Lundegårdh type of anion pump, closely linked with electron transfer.

Conclusions from work on giant algal cells

I have devoted so much space to discussing the work on giant algal cells, because this work is primarily concerned with the fundamentals of ion transport in plant cells. Because of the size of the cells, the experimental difficulties are minimized, though they are still formidable. These cells have sodium extrusion pumps and inwardly directed potassium and chloride pumps at the plasmalemma, which in general seems to be the less permeable of the two membranes. There may also be pumps at the tonoplast for some of the ions, but the evidence for these is not so good as for the plasmalemma pumps. The cation pumps would seem to be powered by ATP, and they seem in general to be rather similar to the animal cell cation pumps. The chloride pump on the other hand, which is a particularly powerful pump, is not dependent on ATP but seems to be closely linked with electron transfer.

With these general principles in mind, we can now turn our attention to the problem of ion transport in higher plant tissue. Though we need not expect the above facts to be absolutely universal, we should expect the principles evolved in the giant algal cell studies to be applicable to higher plants.

Ionic relations of higher plant cells and tissues

Sutcliffe (1962) has fully discussed the kinds of experimental materials and methods used in higher plant studies. The two most common tissues studied have been disks of storage tissue, carrot, beet, artichoke, potato, etc., and excised plant roots, barley, maize, wheat, bean, etc. There are several difficulties associated with working with these multicellular systems. The cells are inevitably a heterogeneous collection; this is especially true of excised roots and it is at least partly so for storage tissue although storage tissue, if properly chosen, can be reasonably homogeneous. Part of a tissue consists of an extracellular space, chiefly composed of the intercellular spaces and the cell walls. Solutes have to diffuse through this extracellular space to reach the protoplasm and vacuoles of the individual cells. This could be a rate-limiting factor for highly permeable solutes, though fortunately ions do not as a whole come into this class. Also the ions in the extracellular space (often called the free space) have to be corrected for before conclusions about ions in the protoplasm and vacuoles can be drawn. However, the chief difficulty of studying the ionic relations of higher plant tissue is the smallness of the cells and, particularly, the tiny amount of protoplasm in each cell. This means, for instance, that it is quite difficult to make one of the essential basic measurements: the electric potential difference between the inside and outside of the cell. A few such measurements have been made recently by Etherton and Higinbotham (1960), Poole (1966), Macklon and MacDonald (1966). In most cases the microelectrode tip can only be inserted into the cell with not much idea of exactly where it is, because of the small volume occupied by the protoplasm. However, Etherton and Higinbotham (1960) took advantage of the special disposition of the protoplasm in root hair cells to measure the electric potential difference across the tonoplast in these cells. It is also extremely difficult to measure the requisite ion concentrations. The total ion concentrations in grams per fresh weight of tissue are of course easy to measure by standard techniques. But the partitioning of these concentrations between the various compartments is very difficult. The extracellular space concentrations can be estimated by various techniques, chiefly by the fact that extracellular ions ought to exchange more rapidly with ions in the external solution, particularly at low temperatures or in the presence of metabolic inhibitors. The concentrations in the protoplasm are particularly difficult to obtain and estimates have so far been made only from isotopic uptake and exchange studies.

Work with higher plant tissue therefore faces greater difficulties than work with giant algal cells and our knowledge is, in general, in a more primitive state.

Free space in plant tissue

Briggs (1957), see also Briggs, Hope, and Robertson (1961), coined the expressions free space, apparent free space (AFS), water free space (WFS), and Donnan free space (DFS) in an attempt to characterize quantitatively those parts of a tissue, the solutes in which were in free diffusive connection with an external solution bathing the tissue. The free space is precisely that volume of the tissue which is in free diffusive connection with the external solution. An estimate of the free space, by whatever method or solute used, is called the apparent free space (AFS). Briggs (1957) defined the AFS as follows. 'If U is the estimated free space uptake of a solute by a tissue originally devoid of this solute from a solution of concentration a_0, then AFS is U/a_0.'

The estimate of free space, or the AFS, clearly depends very much on the nature of the test solute. If labelled water were used then the AFS would be practically the whole of the tissue! If a small non-electrolyte molecule, such as mannitol, to which the cell membranes were scarcely permeable were used then, providing the uptake times were not too long, the free space would essentially be the volume of water external to the cell membranes, i.e., that in the intercellular spaces, cell walls, any cut or dead cells, and so on. If a labelled monovalent cation, say ^{24}Na, were used as the test solute then it would enter not only the general water outside the cell membranes, but it would exchange with some of the cations, mostly Ca, neutralizing the fixed negative charge of the uronic acids in the cell walls. Thus the uptake would be relatively higher than that of a non-electrolyte like mannitol. However, if a labelled divalent cation, like ^{45}Ca, were used as the test solute, this would exchange much more effectively with the divalent ions neutralizing the cell wall negative charge and hence there would be a higher uptake and a higher AFS. If this experiment were properly carried out the radioactive calcium could be made to exchange completely with the cations sitting opposite the negative charges of the cell wall; this would give

17

a very large AFS—the maximum possible for a solute which does not permeate the plant cell membranes themselves. Thus AFS can be looked upon as being made up of two parts: one comprises most of the water outside the living cell membranes, water which is not under the electrical influence of the fixed uronic acid negative charges, in fact water in which the two ions of, say, a uni-univalent salt are in equal concentrations; this is called the water free space (WFS). The other comprises that volume which is under the electrical influence of the fixed negative charges, in which therefore the cations of a mobile salt are at a greater concentration than the anions; this is called the Donnan free space (DFS), because in this volume Donnan equilibrium conditions approximately apply. A properly conducted labelled divalent cation uptake experiment would give the sum of both the WFS and the DFS, as explained above. The uptake of an anion, on the other hand, would be only into the WFS, for it would be excluded from the region (DFS) influenced by the uronic acid negative charges. The AFS measured by mannitol, or some similar solute, would also approximately give the WFS, although the mannitol is not excluded from the DFS and therefore the mannitol free space would be a little higher than the water free space.

At one time, see Briggs (1957), it was thought that the free space of a plant tissue included the protoplasm as well as the intercellular spaces and cell walls; that is, it was thought that the plasmalemma was a negligible barrier to the free diffusion of many solutes, including salts. One of the reasons for this belief was that early values of the WFS were of the order of 20 to 30 per cent of the tissue and this was too high to be accommodated in the intercellular spaces and cell walls alone. These early measurements overestimated the WFS, because inadequate account was taken of the film of solution left on the outside of a tissue when it was taken out of the uptake solution. It is now believed (Brouwer, 1965) that the free space does not include any protoplasm and that free space is simply another term for extracellular space, i.e., space outside the living cell membranes. It is perhaps not worth using the term free space, in fact. However the introduction of the concept by Briggs (1957) and the controversies over its location did stimulate many experiments to characterize its properties and, as a result, we know a good deal more about the extracellular space of plant tissues than perhaps we would otherwise have done.

Ion uptake by storage tissue

Many plant physiologists interested in salt uptake have worked with disks of storage tissue. I refer to Robertson and his co-workers on carrot tissue (see Briggs, Hope, and Robertson, 1961) and to Sutcliffe and co-workers on beet tissue (see Sutcliffe, 1962) though there have been many others as perusal of the volumes of the *Annual Review of Plant Physiology* will show. The reasons for choosing storage tissue were doubtless those of convenience and greater homogeneity of cell type than in other plant parts. After cutting the disks, usually about 10 to 20 mm in diameter and 1 mm thick, they are washed in aerated and frequently changed distilled water for a day or two before use. During this period they develop a capacity for ion uptake, perhaps due to a metabolic change (Laties, 1959), which they do not possess when freshly cut. They are then put into dilute salt solutions and the ion uptake followed either by chemical or by tracer means.

The initial uptake, in the first hour or so, is largely into the free space of the tissue and I will assume that this has been corrected for in the subsequent description and discussion. If the uptake of the ions of a salt like KCl is followed over several days, it shows a gradually decreasing tempo as illustrated in Fig. 13.6 and finally reaches a steady value. In beet disks, taking up K and Cl from

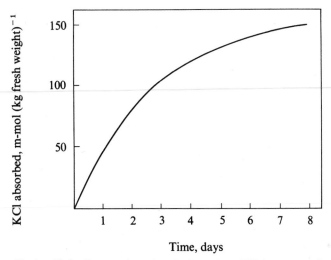

Figure 13.6 Tempo of uptake of salt, such as KCl, by storage tissue.

external solutions of 1 to 5 mM, the final concentrations of both ions are about 150 to 200 mM. It has been verified that these ions are largely in free solution, presumably in the cell vacuoles, by osmotic pressure measurements (by the method of incipient plasmolysis). It might be commented here that since *both* ions of the salt have been accumulated to much higher concentrations (of free ions) inside the cells than outside, then at least one of the ions must have been actively transported. Measurements using radioactive isotopes of the ions concerned show that the cells are not in general in flux equilibrium, the influx is always greater than the efflux and both decline as the vacuoles fill with salt. The reason for this decline is not known. The electric potentials between vacuole (probably) and external solution have been measured in certain situations for beet by Poole (1966) and for potato by Macklon and MacDonald (1966). The potentials are generally negative by about -80 mV to -140 mV, which suggests that at some stage in the accumulation process the anion must certainly be actively transported. The net fluxes, during the early stages of accumulation are of the order of 10^{-12} mol cm^{-2} s^{-1}.

When the initial rates of uptake by the living cells are plotted against external ion concentrations, the relationship is not linear. Usually the relationship resembles a Michaelis–Menten enzyme-kinetic type as shown in Fig. 13.7. Occa-

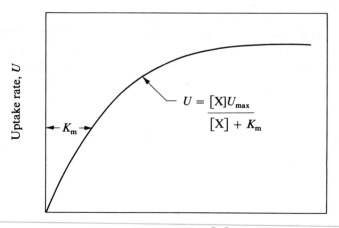

$$U = \frac{[X]U_{max}}{[X] + K_m}$$

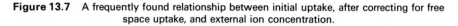

Figure 13.7   A frequently found relationship between initial uptake, after correcting for free space uptake, and external ion concentration.

sionally the relationship is such that the uptake increases faster than proportional to the external concentration (Macklon and MacDonald, 1966). A Michaelis–Menten type relationship is frequently cited as evidence for active transport (see next section) but in fact passive entry of ions because of its dependence on electric potential difference and on a labile permeability factor can fit almost any kind of relationship between net flux and external concentration.

It is found that when a disk, previously in distilled water, is placed in a salt solution the respiration of the disk increases to a new level. This additional respiration, over and above the ground respiration, has been called salt respiration. It appears from this observation that the act of accumulation of the ions of a salt needs energy and the tissue respires faster to provide the extra energy. As might be expected anything that cuts down or stops respiration also decreases the rate of ion accumulation; for instance the addition of low concentrations of cyanide or azide cuts down both respiration and ion accumulation. Robertson (see Briggs, Hope and Robertson, 1961), following Lundegårdh, considers that ion accumulation is closely coupled with electron transfer during respiration. If this is so then for every electron moved down the cytochrome chain towards oxygen one anion should move in the opposite direction into the cell—or at least not more than one anion if some inefficiency in the mechanism is allowed. Since 4 electrons are needed to reduce oxygen to the level of water, the ratio of equivalents of ion accumulated to moles of oxygen consumed in salt respiration should not be greater than 4 to 1. Robertson and his coworkers in a careful study of this relation in carrot storage tissue did indeed find that this ratio never exceeded 4, though it frequently closely approached it. One disturbing observation was that the addition of dinitrophenol (DNP), which is supposed to uncouple phosphorylation from respiration and thereby suppress the production of ATP without affecting electron transfer or even enhancing it, reduced the ion accumulation while not affecting or even increasing the respiration. This observation of course suggests that the ion pumps in storage tissue cells are energized by ATP rather than linked to electron transfer as in the Robertson-Lundegårdh hypothesis. However, Robertson (1960) maintains that the action of DNP is not that of a simple uncoupler, but that it also interferes with the separation of charge which is necessary to produce the ion accumulation on his picture.

To summarize at this stage, there is little doubt that in general active transport of one or both ions of a simple salt like KCl is necessary for the accumulation of this salt in storage tissue cells. What little evidence there is suggests that the anion is more likely to be the primarily pumped ion. Where the pumps are is not known, though Robertson thinks that the primary pumping action must be into the mitochondria where oxidation is mainly carried out; the mitochondria would then discharge into the vacuoles by some unknown mechanism. Whether the pumps are closely connected with ATP or electron transfer is still debated; it may be from analogy with the giant algal cell work that the anion pump is linked to electron transfer and any cation pumps are energized by ATP. Sutcliffe (1962), on the basis of inhibition of salt uptake by the antibiotic chloramphenicol, thinks that salt accumulation is closely associated with protein synthesis. However, chloramphenicol inhibits processes other than protein synthesis and Sutcliffe's hypothesis remains unproven.

A few remarks should be made about the relationship between organic acid production and salt accumulation. In higher plant cells the usual anion in the cell vacuoles is organic acid anion, in most cells malate. (Chloride is only essential to higher plant growth in trace amounts.) From a salt like KCl, the anion and cation are usually taken up in equal amounts and there is no increase or decrease in the organic acid content of the vacuoles. However, from a salt like K_2SO_4, the cation is predominantly taken up and to preserve electrical neutrality organic acid is synthesized in the cells, presumably in the cytoplasm, and organic acid anion put into the vacuole; whether or not it is pumped across the tonoplast is not known. Similarly from a salt like $CaCl_2$, the anion is predominantly taken up and there is a decrease in the organic acid anion content of the vacuoles. Though the above facts are well known, there is no clear picture of

what is happening. How is the production or destruction of organic acid triggered off? What is happening at the tonoplast with respect to the transport of
organic acid anion? We do not know the answers to these or any other detailed
questions.

Ion uptake by excised roots and the carrier hypothesis

Excised roots are favourite experimental materials for salt uptake studies despite their manifest disadvantages due to the very heterogeneous cell population.
Of course one very important reason for studying them is that the root is the
chief vehicle of entry of salts into higher plants. Despite the difficulties of working with such a mixed population of small cells, some very important ideas have
come from such studies chiefly from the work of Epstein and his colleagues
(see Epstein, 1966).

In their initial studies on excised barley roots with salts at low external concentration, the uptake of both cations and anions as a function of external
concentration followed a Michaelis–Menten enzyme-kinetics type of curve such
as is illustrated in Fig. 13.7. This suggested a carrier model of uptake which can
be represented as follows:

$$X + C \underset{}{\overset{(A)}{\rightleftharpoons}} XC \xrightarrow{(B)} C + X$$

X represents the ion; on the left it is outside the membrane and on the right it is
inside the membrane. C represents the carrier molecule, possibly a protein,
which can be thought of as being confined to the membrane; on the left it is at
the outside surface of the membrane and on the right it is at the inside surface
of the membrane. Reaction (A) is an equilibrium between ion and carrier at the
outside surface of the membrane. Reaction (B) represents the irreversible movement of the ion-carrier complex to the inside surface of the membrane and its
release there of the ion to the inside solution. This formulation of ion uptake is
exactly the same as that for enzyme-substrate interaction, C being the enzyme
and X the substrate. It can easily be shown that the rate of entry of ion is given
by an equation identical in form with that for the rate of an enzyme-catalyzed
reaction, viz.

$$U = \frac{[X]\, U_{max}}{[X] + K_m} \tag{21}$$

U is the rate of ion uptake at external ion concentration $[X]$. U_{max} is the maximum rate of uptake. K_m—the 'Michaelis constant'—is the equilibrium constant
for the reaction (A). Fig. 13.7 shows graphically how uptake is related to external concentration.

It is rather surprising that such a mixed cell population should obey such a
law as (21), with a single value of K_m. Note also that the model assumes that the

cell membranes are impermeable to free ions, i.e., there is no passive ion diffusion even of the cations. Some passive diffusion might certainly have been expected. However, Epstein and his co-workers have accumulated other evidence in favour of the model. For instance in experiments with mixed solutions of potassium and rubidium chlorides, the relative uptakes of K and Rb suggest that they compete for the carrier site in exactly the same way that substrates compete for the same enzyme site. In other words potassium inhibits the uptake of rubidium and vice versa in a competitive manner.

In recent experiments with a wider range of salt concentrations, Epstein (1966) has found evidence for multiple carriers, which have different Michaelis constants and different specificities. For instance in excised barley roots potassium is taken up at low concentrations by a single carrier system with a high affinity for K. The Michaelis constant of this carrier is 0·02 mM; Rb competitively inhibits K uptake while Na does not. Above 0·2 mM K concentration however, the ion seems to be taken up by other carriers which have a low affinity for K. These carriers which operate at a higher concentration are not so specific to K; sodium is here a competitive inhibitor, whereas it does not inhibit at all the low K, high-affinity uptake mechanism. Epstein (1966) has suggested that the low-concentration, high-affinity carriers are at the plasmalemma. As the concentration increases passive diffusion across the plasmalemma increases and then the low-affinity carriers, *at the tonoplast*, come into operation. This is an attractive idea, but it needs much more experimentation to verify it.

Ion uptake into the xylem vessels of plant roots

Perhaps the most important problems in the ionic relations of higher plants are those associated with the entry of ions into the xylem vessels of the roots. The anatomy of the root was briefly described and a diagram given (Fig. 12.4) in the previous chapter. In discussing the transport of ions across roots, one anatomical feature has been stressed by many investigators though it is particularly associated with Arisz (1953). This is the fact that the protoplasm of all the cells forms a continuum, for each cell is connected to its neighbours by protoplasmic threads—the plasmadesmata. This protoplasmic continuum, which stretches from the epidermis to the xylem vessels, has been called by Arisz the *symplasm*. Great importance has been attached to the symplasm because, in the ion-absorbing region of the root, the Casparian strip of the endodermis seems as if it ought to be a barrier to the movement of ions from the cortex to the stele through the free space. The symplasm goes right through this barrier.

Probably the most widely accepted theory of ion movement to the xylem is the one involving the symplasm (Arisz, 1953, see also Brouwer, 1965). In this theory the external solution is in free communication with the intercellular spaces and cell walls of the epidermis and cortex, but not beyond these tissues. Cations of course will be at a higher concentration in the cell walls because of the fixed uronic acid negative changes there, though this probably does not enhance their rate of absorption. From the free space of the cortex, ions will be taken up across the plasmalemmas of the cortical cells into the symplasm, probably by an at least partly active transport process. The ions will thus then be in the symplasm and from there they can move either into the vacuoles of the cortical and other cells or proceed inwards to the xylem vessels. Entry into the cell vacuoles is looked upon as a loss to the transport system, though there is evidence that ions once in the vacuoles can get back into the transport system and hence into the xylem vessels. Ion transport across the tonoplast and into vacuoles may also be active. Those ions which are not lost to the cell vacuoles move inwards and by some unknown mechanism are passed to the xylem vessels and hence, via the transpiration stream to the aerial parts of the plant. This final step of passing from the symplasm to the lumens of the xylem vessels may be similar to passage across a tonoplast and may also be an active transport process.

This theory is based quite strongly on anatomical considerations. It seems perfectly reasonable except for a vagueness about what actually happens at the xylem end of the symplasm. It may be, as House and Findlay (1966) think (see preceding chapter), that the operative xylem vessels for ion and water uptake are those that contain living protoplasts. In which case the symplasm will extend right up to the tonoplasts of these living xylem cells and the final step will be a secretion across the tonoplast of these xylem cells.

An alternative theory, which does not exclude the symplasm as an important pathway, is that the free space is not seriously interrupted by the endodermal Casparian strip and extends right into the cell walls of the xylem vessels. In this case ions can move in the free space up to the plasmalemma of the living xylem vessels and be transported across this membrane and across the tonoplast into the lumen of the xylem vessels. We can think then of two parallel ion transport pathways to the xylem: one via the symplasm and the other via the free space and of course there are side connections between these pathways. The symplasm path may be the more important one at low external ion concentrations and the free space pathway more important at high external ion concentrations.

That active transport is involved at some stage in this transport seems undoubted; metabolic inhibitors cut down the transport and respiratory substrates enhance it. Again the anion of the salt seems more likely to be pumped than the cation. Bowling and Spanswick (1964) have measured the electrochemical potential difference for various ions between the bleeding sap of a detopped *Ricinus* plant and the external solution. They find, in general, that the anions are at a much higher electrochemical potential in the bleeding sap than in the external solution, whereas the cations are at approximately the same electrochemical potential in the two solutions. This suggests strongly, even proves, that the anions are being actively transported.

There has been much debate, particularly between Hylmö and Brouwer (see Brouwer, 1965), on the relation between water and ion uptake into the xylem vessels. Hylmö (1953) found, at high transpiration rates, a linear relation between ion uptake into the shoot of pea plants and the rate of transpiration. He suggested that the stream of water passing through the roots, largely through the free space, into the xylem dragged ions with it in a passive manner. Since the calculated ratio of ions to water in the xylem transpiration stream was less, by a factor of about 4, than the ratio of ions to water in the external solution, he concluded that there was some sieving effect which filtered out about three-quarters of the ions, Brouwer (1965) contends that the undoubted effect of transpiration rate on ion uptake to the shoot is quite different. He suggests that the effect of a high transpiration rate is to sweep away ions out of the small volume of the xylem vessels in the root, thus lowering their concentration at the place of initial accumulation. This would enable the ion active transport mechanism to work faster, for one would expect its rate of working to depend on the concentration of ions in the root xylem vessels. It must be said that most of the evidence is in favour of Brouwer's suggestions, though it does not exclude a small passive component of ion movement moving according to Hymlö's ideas.

Conclusions

It will have been noticed that there is a large gap between the kinds of detailed knowledge of ion pumps at individual membranes that is now accumulating for giant algal cells, and the rather less basic knowledge we have of ion transport processes in higher plants. It must be the aim of future investigation to fill in this gap as far as possible and to interpret processes in higher plants on the same kind of basis that is being developed in the giant algae.

Bibliography

Selected further reading

Briggs, G. E., A. B. Hope, and R. N. Robertson, 1961. *Electrolytes and Plant Cells.* Blackwell, Oxford.

Brouwer, R., 1965. Ion absorption and transport in plants. *Ann. Rev. Plant Physiol.* **16**, 241–266.

Dainty, J., 1962. Ion transport and electrical potentials in plant cells. *Ann. Rev. Plant Physiol.* **13**, 379–402.

Jennings, D. H., 1963. *The Absorption of Solutes by Plant Cells.* Oliver and Boyd, Edinburgh.

Sutcliffe, J. F., 1962. *Mineral Salts Absorption in Plants.* Pergamon, New York, Oxford, London, Paris.

Papers cited in text

Arisz, W. H., 1953. Significance of the symplasm theory for transport across the root. *Protoplasma*, **66**, 5–62.

Blount, R. W. and B. H. Levedahl, 1960. Active sodium and chloride transport in the single celled marine alga *Halicystis ovalis*. *Acta Physiol. Scand.* **49**, 1–9.

Bowling, D. J. F. and R. M. Spanswick, 1964. Active transport of ions across the root of *Ricinus communis*. *J. exp. Bot.* **15**, 422–427.

Briggs, G. E., 1957. Some aspects of free space in plant tissue. *New Phytol.* **56**, 305–324.

Briggs, G. E., A. B. Hope, and R. N. Robertson, 1961. *Electrolytes and Plant Cells*. Blackwell, Oxford.

Brouwer, R., 1965. Ion absorption and transport in plants. *Ann. Rev. Plant Physiol.* **16**, 241–266.

Clayton, R. K., 1965. *Molecular Physics in Photosynthesis*. Blaisdell Publishing Company, New York.

Dainty, J., 1962. Ion transport and electrical potentials in plant cells. *Ann. Rev. Plant Physiol.* **13**, 379–402.

Duysens, L. N. M., 1964. Photosynthesis. *Progr. Biophys. Mol. Biol.* **14**, 1–104.

Etherton, B. and N. Higinbotham, 1960. Transmembrane potential measurements of cells of higher plants as related to salt uptake. *Science*, **131**, 409–410.

Epstein, E., 1966. Dual pattern of ion absorption by plant cells and by plants. *Nature, Lond.* **212**, 1324–1327.

Gutknecht, J., 1966. Sodium, potassium, and chloride transport and membrane potentials in *Valonia ventricosa*. *Biol. Bull.* **130**, 331–344.

Hope, A. B., A. Simpson and N. A. Walker, 1966. The efflux of chloride from cells of *Nitella* and *Chara*. *Australian J. Biol. Sci.* **19**, 355–362.

Hope, A. B. and N. A. Walker, 1960. Ionic relations of cells of *Chara australis*. III. Vacuolar fluxes of sodium. *Australian J. Biol. Sci.* **13**, 277–291.

House, C. R. and N. Findlay, 1966. Analysis of transient changes in fluid exudation from isolated maize roots. *J. exp. Bot.* **17**, 627–640.

Hylmö, B., 1953. Transpiration and ion absorption. *Physiologia Plantarum*, **6**, 333–405.

Jennings, D. H., 1963. *The Absorption of Solutes by Plant Cells*. Oliver and Boyd, Edinburgh.

Laties, G. G., 1959. Active transport of salt into plant tissue. *Ann. Rev. Plant Physiol.* **10**, 87–112.

van Lookeren Campagne, R. N., 1957. Light-dependent chloride absorption in *Vallisneria* leaves. *Acta bot. neerl.* **6**, 543–582.

Lundegårdh, H., 1960. Salts and respiration. *Nature, Lond.* **185**, 70–74.

MacRobbie, E. A. C., 1962. Ionic relations of *Nitella translucens*. *J. gen. Physiol.* **45**, 861–878.

MacRobbie, E. A. C., 1964. Factors affecting the fluxes of potassium and chloride ions in *Nitella translucens*. *J. gen. Physiol.* **47**, 859–877.

MacRobbie, E. A. C., 1965. The nature of the coupling between light energy and active ion transport in *Nitella translucens*. *Biochim. Biophys. Acta*. **94**, 64–73.

MacRobbie, E. A. C. and J. Dainty, 1958. Ion transport in *Nitellopsis obtusa*. *J. gen. Physiol.* **42**, 335–353.

Macklon, A. E. S. and I. R. MacDonald, 1966. The role of transmembrane electrical potentials in determining the absorption isotherm for chloride in potato. *J. exp. Bot.* **17**, 703–717.

Poole, R. J., 1966. The influence of the intracellular potential on potassium uptake by beetroot tissue. *J. gen. Physiol.* **49**, 551–563.

Raven, J. A., 1967a. Ion transport in *Hydrodictyon africanum. J. gen. Physiol.* **50,** 1607–1625.

Raven, J. A., 1967b. Light stimulation of active transport in *Hydrodictyon africanum. J. gen. Physiol.* **50,** 1627–1640.

Robertson, R. N., 1960. Ion transport and respiration. *Biol. Rev.* **35,** 231–264.

Spanswick, R. M. and E. J. Williams, 1964. Electrical potentials and Na, K, and Cl concentrations in the vacuole and cytoplasm of *Nitella translucens. J. exp. Bot.* **15,** 193–200.

Sutcliffe, J. F., 1959. Salt uptake in plants. *Biol. Rev.* **34,** 159–220.

Sutcliffe, J. F., 1962. *Mineral Salts Absorption in Plants.* Pergamon, New York, Oxford, London, Paris.

Teorell, T., 1949. Membrane electrophoresis in relation to bio-electric polarisation effects. *Arch. Sci. Physiol.* **3,** 205–219.

Ussing, H. H., 1949. The distinction by means of tracers between active transport and diffusion. The transfer of iodide across the isolated frog skin. *Acta Physiol. Scand.* **19,** 43–56.

14. Phytochrome

H. W. Siegelman

Introduction

The discovery of phytochrome

The discovery of the photoreversible control of plant growth and development and the subsequent isolation of the photoreceptor pigment, phytochrome, constitutes one of the most fascinating chapters in plant physiology. Primarily, the concerted efforts of a group of investigators at the Plant Industry Station, U.S. Department of Agriculture, Beltsville, Maryland, working over approximately a twenty-five year period have provided most of the current information. This work was led by two experimentalists, H. A. Borthwick and S. B. Hendricks. Borthwick is a plant morphologist by training and a keen student of plant behaviour. Hendricks is a physical chemist and a perceptive analyst of experimental findings. Successful pursuit of the overall problem required collaboration with plant physiologists, electronic engineers, biophysicists, and biochemists. Few pursuits in plant physiology have successfully utilized such a diversity of disciplines.

A fundamental advance in plant photobiology was the discovery that growth responses elicited by red light are reversible by far-red light. It was previously known that the germination of lettuce seed, which requires light, is promoted by red light and suppressed by far-red light (Flint and McAlister, 1935, 1937). Detailed action spectra for light-sensitive lettuce seed germination showed maximum promotion near 660 nm and maximum suppression near 730 nm (Borthwick *et al.*, 1952). The important idea derived from the action spectra studies was that the seeds are potentiated for germination by red light and the potentiation is reversed by far-red light. If this were true, seed repeatedly exposed to red and far-red light would show a response dependent on the spectral region of the final light exposure. Simple experiments confirmed the idea of a photoreversible photoresponse. The action spectra and energy requirements for seed germination and for control of flowering in cocklebur were found to be similar (Fig. 14.1).

The simplest interpretation of the photoresponse was that a single photoreceptor pigment could exist in two photoconvertible forms:

$$P_r \underset{\text{far-red light}}{\overset{\text{red light}}{\rightleftharpoons}} P_{fr}.$$

phytochrome : photoreversible proteaceous plant pigment functioning as enzyme to initiate germination growth roots stem leaves flowers

490   THE PHYSIOLOGY OF PLANT GROWTH AND DEVELOPMENT

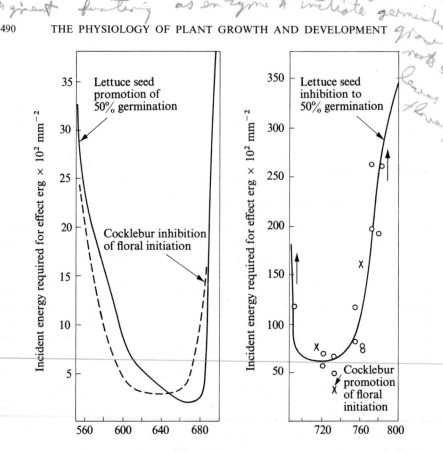

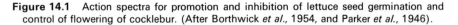

Figure 14.1 Action spectra for promotion and inhibition of lettuce seed germination and control of flowering of cocklebur. (After Borthwick *et al.*, 1954, and Parker *et al.*, 1946).

The pigment P_r was found to have an action maximum near 660 nm, and P_{fr}, near 730 nm. Absorption of light by either form converted it to the other form. Implicit from the action spectra studies is that a pigment must be acting as a photoreceptor and that its absorption maxima must be close to the action maxima. It could thus be predicted that the pigment would be blue in colour, and approximate absorption wavelength maxima could be specified. The low energy requirement suggested that the photoaction brings about a change in molecular configuration rather than a transfer of energy to another system.

The detection of phytochrome

A number of years passed between the description of the properties of phytochrome deduced from the action spectra studies and the physical measurement of the pigment. The existence of the yet unseen pigment was regarded with considerable scepticism.

A principal problem in measuring phytochrome *in vivo* was the apparent low concentration of the pigment in plant tissues. The molar extinction coefficients of P_r and P_{fr} were determined and found to be quite high (Hendricks *et al.*, 1956). Nonetheless, a pigment with the required properties was not apparent even in albino seedlings which showed typical phytochrome-controlled growth responses (Borthwick *et al.*, 1951). Since the catalytic or regulating function of phytochrome is still unknown, an assay was conceivable based on the spectral properties defined by the action spectra. Direct physical measurement of phytochrome required the development of spectrophotometric equipment which could measure small absorbance changes in optically dense samples. The overriding absorbance of chlorophyll has obscured the measurement of phytochrome in green tissues. The proper selection of plant material used for measurement is paramount since phytochrome has not been detected in many phytochrome-containing tissues. The plant response is frequently a more sensitive assay for phytochrome than current methods of physical measurement.

The first tissue in which phytochrome was physically detected was the cotyledons of dark-grown turnip seedlings grown in the presence of chloramphenicol (Butler *et al.*, 1959). This tissue was selected because of its low protochlorophyll content and its capacity to show responses to light. The cotyledons were placed in a cuvette and irradiated with red light. The radiation transformed the P_r to P_{fr} and protochlorophyll to chlorophyll. The absorbance of the tissue was measured from 570 to 850 nm. The tissue was then irradiated with far-red light and the absorbance again measured from 570 to 850 nm. The presence of the pigment was clearly evident from the absorbance changes. Far-red radiation caused an absorbance decrease in the far-red region of the spectrum with maximal change at 735 nm and an absorbance increase in the red region with maximal change at 655 nm. Red radiation had the opposite effect.

With the assurance that the specific characteristics of phytochrome derived from action spectra were confirmed by photoreversible absorbance changes at the proper wavelengths, a rapid method was devised for detection. A dual-wavelength difference photometer (Fig. 14.2) was devised by Birth (1960) which measured directly the absorbance difference (ΔA) between 660 and 730 nm, $\Delta A = A_{660} - A_{730}$. The relative amount of phytochrome in a sample was calculated from the difference between the ΔA readings after irradiation of the sample with actinic sources of red and far-red light: $\Delta(\Delta A) = \Delta A_{fr\ irrad} - \Delta A_{r\ irrad}$. The assay based on absorbance changes at 660 and 730 nm measures the sum of the photoreversible P_r and P_{fr} absorbances. P_r and P_{fr} can be determined separately by measuring the ΔA between 660 and 800 nm and between 730 and 800 nm, respectively. In some cases the 730 vs 800 nm measurement may be the most reliable assay.

The availability of a simple, sensitive, rapid, and specific physical assay permitted the examination of a large number of plant materials which might serve as useful sources for extraction and purification of phytochrome. A survey (Butler *et al.*, 1965) indicated that phytochrome could be detected in several dark-grown seedlings and the cauliflower head. Hillman (1964) found a high content of phytochrome in several non-green plant tissues. The distribution of phytochrome in several dark-grown seedlings was examined *in vivo* (Briggs and Siegelman, 1965). The concentration of phytochrome in monocotyledon seedlings was highest at the apex of the coleoptile and at the junction of the mesocotyl and node. In legume seedlings the maximum concentration of phytochrome occurred in the growing tip and hypocotyl arch. Dark-grown seedlings,

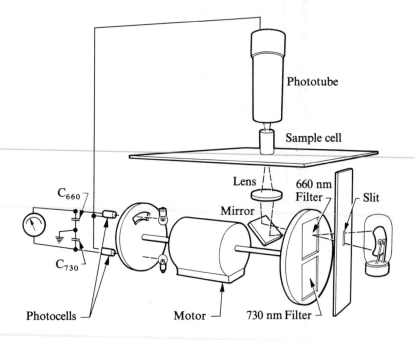

Figure 14.2   Diagram of the dual-wavelength difference photometer used for phytochrome assay. (From Birth, 1960.)

which can be grown reproducibly in relatively large amounts, are an enriched source of the pigment. They have a higher content of phytochrome than light-exposed seedlings and a minimum of chlorophyll contamination. Phytochrome has now been detected in all parts of higher plants including roots, hypocotyls, cotyledons, coleoptiles, stems, petioles, leaf blades, vegetative buds, floral receptacles, inflorescences, and developing fruits, and in both primary and secondary growth.

Isolation and purification

The isolation and purification of phytochrome was dependent on the photo-metric assay and the augmented content of phytochrome in dark-grown seed-lings. Methods for extracting proteins were used in the isolation. Consideration of some properties of phytochrome suggested that it was a protein with a chromophoric (pigment) group. The very low radiant energy needed to bring about a physiological response suggested a need for amplification analogous to the functioning of an enzyme. Substances present in trace amounts are usually physiologically active by virtue of binding to a protein. Initial attempts to isolate phytochrome as a protein were successful, but subsequent purification required considerable time. Attempts to isolate the phytochrome chromophore directly from plant tissue by methods used in pigment chemistry have not yet been successful.

The purification of phytochrome has been satisfactorily achieved to date only from dark-grown monocotyledonous seedlings. Dark-grown five-day-old seedlings of *Avena sativa* (oats) provided a useful source of easily extractable phytochrome (Siegelman and Firer, 1964). The amount of phytochrome ex-tracted is largely dependent on the final pH after the initial extraction. At pH values above 7·3, the supernatant solution after filtration and centrifugation contained the maximal extractable pigment. At pH 6·2 or below, the phyto-chrome was only in the sediment. The intercellular location of phytochrome has not been systematically pursued.

Satisfactory methods for the purification of many plant proteins have only recently become available. There are several problems peculiar to the purifi-cation of plant proteins. The growing plant cell is highly vacuolated and has a low content of protein. Protein precipitants, acidic substances, and phenolase substrates are frequently found in considerable concentration in plant vacuoles. These substances must be rapidly and effectively counteracted or the protein will be denatured or rendered insoluble. Some of these can be inhibited by the use of reduced sulphydryl compounds such as 2-mercaptoethanol. Sufficient amounts of the reduced sulphydryl compound should be used to ensure that a detectable amount remains after the initial extraction. Polyvinylpyrrolidone may prove useful in the isolation of some plant proteins (Loomis and Battaile, 1966). Low temperature during protein isolation reduces protein denaturation and possible proteolysis.

The low proportion of proteins extracted from plants often requires con-centration before purification methods can be applied. Concentration by ultra-filtration is simple and convenient.

In the purification procedure for phytochrome from oats at least two proper-ties of proteins are utilized. Ultrafiltration and gel filtration on Sephadex and agar columns separate proteins on the basis of molecular size. Calcium phosphate and ion-exchange separation methods are based on the charge of the protein. Alternating these kinds of separation processes proved to be a useful procedure. The purification scheme now used is outlined in Fig. 14.3 (Siegelman and Hendricks, 1965). A similar procedure has been effective with dark-grown corn and barley seedlings.

An improved preparation of phytochrome from dark-grown oat seedlings has recently been described (Mumford and Jenner, 1966). Phytochrome has also

Grind 6·0 kg dark-grown oats with 6 litres
of buffer and 300 g of cellulose; filter
Centrifuge (16,000g, 25 min)
↓
Ultrafiltrate
↓
Centrifuge (37,000g, 20 min)
↓
Gel filtrate (7 per cent agar)
↓
Calcium phosphate chromatography
↓
Ultrafiltrate
↓
Gel filtrate (7 per cent agar)
↓
Calcium phosphate chromatography
↓
0·5 satd $(NH_4)_2SO_4$ precipitation of active fractions
↓
Centrifuge (23,000g, 10 min)
↓
Redissolve precipitate in 20 ml buffer
containing 0·5 M sucrose
↓
Centrifuge (23,000g, 15 min)
↓
Store at −15°C

Figure 14.3 Procedure for the isolation and purification of phytochrome from dark-grown oat seedlings. (From Siegelman and Hendricks, 1965.)

been purified from dark-grown rye seedlings (Correll *et al.*, 1966) and isolated from the following green plants: spinach (Lane *et al.*, 1963) and *Mesotaenium* (a green alga) and *Sphaerocarpus* (a liverwort) (Taylor and Bonner, 1967).

Visible detection of phytochrome can be achieved in about 2 hours from dark-grown oat seedlings provided all the required materials are prepared in advance (Miller *et al.*, 1965) An extract of 100 g of seedlings is chromatographed on a column of dicalcium phosphate. The phytochrome can be seen on development of the chromatogram as a narrow blue-green band whose blue colour increases with far-red radiation and decreases with red radiation.

In vitro properties

Action spectra

The action spectra for the overall photochemical transformations of P_r and P_{fr} were determined by irradiating purified phytochrome solutions with various wavelengths of monochromatic light (Butler *et al.*, 1964a). The degree of conversion was measured with the dual-wavelength difference photometer. The phototransformations of P_r and P_{fr} were known to be first order (Butler, 1961). Therefore, the products of the molar extinction coefficients and the quantum yields at any wavelength could be determined from the first-order rate constants. The value of the first-order rate constant was determined experimentally from the slope of a semilogarithmic plot of per cent conversion versus time of irradiation. The absorption spectra of phytochrome (Fig. 14.4) indicated that far-red

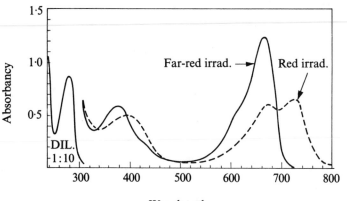

Figure 14.4 Absorption spectra of oat phytochrome following irradiation with red and far-red light. The absorbance at wavelengths shorter than 300 nm was on a solution diluted ten-fold. (From Butler *et al.*, 1964a.)

irradiation drives the phytochrome strongly toward P_r, while the red source established an intermediate state since there was a considerable spectral overlap. At the photostationary state in red light, 81 per cent P_{fr} and 19 per cent P_r were found (Butler *et al.*, 1964a).

The action spectra for the photochemical conversion of P_r to P_{fr} are shown in Fig. 14.5 as the product of the molar extinction coefficient, α, and the quantum efficiency, ϕ, versus wavelength, λ. The extinction coefficient, ε, of P_r at 665 nm is at least 1.6×10^4 litre mol^{-1} cm^{-1} or greater if the quantum efficiency is less than one.

The action spectra have bands in the blue and near ultraviolet regions consistent with the absorption bands. The overlap between these bands of P_r and

P_{fr} is such that light in the blue near ultraviolet regions will not drive phytochrome very far to either extreme. Action spectra have been determined for the *in vivo* phototransformation of phytochrome in maize coleoptile tips (Pratt and Briggs, 1966). The results obtained were similar to the *in vitro* action spectra.

Absorption spectra

The absorption spectra of a purified solution of phytochrome are shown in Fig. 14.4. P_r and P_{fr} have principal absorption bands near 665 and 725 nm, respectively, and subsidiary absorption bands near 370 and 400 nm. The absorption band at 280 nm is due to the aromatic residues of the protein and does not change on red or far-red irradiation. The ratios of the absorbances of

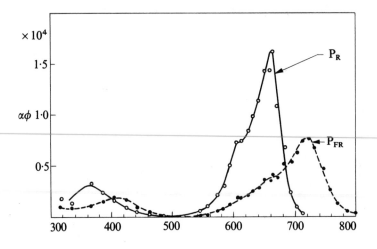

Wavelength, nm

Figure 14.5   Action spectra of photochemical transformations of P_r and P_{fr}. The extinction coefficient, α, is litre mol^{-1} cm^{-1} and the quantum yield, ϕ, is in mol Einstein^{-1}. (From Butler *et al.*, 1964a.)

the various bands after far-red irradiation to that of the 280 nm band provide an index of the purity of the solution. The ratio of the absorbance of the 665 nm band to that of the 280 nm band in Fig. 14.4 is 1 : 7. A purified oat phytochrome solution with a 665 nm : 280 nm absorbance ratio of 1 : 1 was recently described (Mumford and Jenner, 1966). This preparation was reported to be homogeneous, with molecular extinction coefficients of 7.63×10^4 and 4.61×10^4 litre mol^{-1} cm^{-1} for the 664 and 724 nm bands, respectively.

The colour of a solution of purified phytochrome was observed to change from blue-green after irradiation with far-red light, to light green after irradiation with red light.

Fluorescence of the P_r form has been measured with an excitation maximum near 670 nm and an emission maximum near 690 nm. The fluorescence is enhanced at liquid nitrogen temperatures. No fluorescence has been measured for P_{fr} (Hendricks et al., 1962).

Flash spectroscopy

The phototransformation of phytochrome is usually written in its most simple form as

$$P_r \underset{725 \text{ nm}}{\overset{665 \text{ nm}}{\rightleftharpoons}} P_{fr}.$$

Detailed examination of the phototransformation by rapid methods of irradiation and measurement (flash spectrophotometry) showed that the apparently simple photoreaction above proceeds through several intermediate forms (Linschitz et al., 1966).

The procedure used for the flash spectrophotometric experiments is simple in principle. Phytochrome is transformed to P_r or P_{fr} and then irradiated, by flash lamps, with filtered light of high intensity and short duration, about 5×10^{-6} second. Following the flash irradiation, the absorption of the sample at many selected wavelengths is monitored by a spectrophotometric apparatus, with an oscilloscope used to follow the absorbance changes at various sweep speeds. Within the time resolution of the instrument, red-flash irradiation of P_r causes an immediate bleaching at 664 nm but no corresponding absorption increase at 724 nm. The absorption at 724 nm begins to appear about 1×10^{-3} second after the flash. Following the initial bleaching at 664 nm, there is a further bleaching with a lifetime of about 2×10^{-2} second which is closely matched by an increased absorbance at 724 nm. There follows a third bleaching with a lifetime of $2 \cdot 5 \times 10^{-1}$ second and an additional absorbance increase at 724 nm. After bleaching at 664 nm is completed, the increase in absorbance at 724 nm continues with a lifetime of about 4 seconds. Flash spectroscopy revealed the presence of four intermediates in the $P_r \rightarrow P_{fr}$ conversion.

In the photoconversion of P_{fr} to P_r, a far-red flash results in an immediate bleaching at 724 nm and increased absorbance at 664 and 638 nm. Following the initial changes, two further absorbance increases are found with lifetimes of 1×10^{-3} and 5×10^{-3} second. These absorbance changes show the presence of two intermediates in the conversion. Analysis of the flash spectroscopy experiments indicates that the following reaction sequence occurs:

$$P_r \xrightarrow{h\nu} P_r^* \longrightarrow P_{r1} \underset{\searrow}{\overset{\nearrow}{\underset{P_{r4}}{\overset{P_{r2}}{\longrightarrow}}}} P_{r3} \longrightarrow P_{fr}$$

The first step, from P_r to a P_r^*, the metastable excited state, is the only photo-chemical step in the transformation. The time resolution of the flash apparatus used did not permit the observation of P_r^*. The remainder of the steps are dark reactions.

The sequence of reactions in the conversion of P_{fr} to P_r is

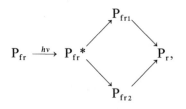

and the P_{fr}^* state was not observed. Briggs and Fork (1965) have measured the longer-lived intermediates following irradiation by wavelengths absorbed by both P_r and P_{fr}.

The molecular changes involved in these reactions are as yet unknown. It appears that absorption of light by the chromophore brings about a change in conformation of the protein, which passes through several intermediates to the final form. The resemblance of the phototransformation of P_r to P_{fr} to the rhodopsin phototransformation is striking, although the chromophores are of completely different types.

Protein nature

Many properties of phytochrome are indicative of its protein nature. It is de-natured by heat, strong acid or alkali, urea, and proteolytic enzymes. Phyto-chrome was purified by methods of protein chemistry, with the biuret test used to assay the protein concentration. After considerable purification, the 280 nm absorption correlated with biuret determinations. Amino acid analysis of the purified protein provides conclusive evidence of the protein nature (Mumford and Jenner, 1966).

The weight average molecular weight of phytochrome from equilibrium centrifugation was found to be either 90,000 or 150,000 with an $s_{20,w}$ value of 4·5 (Siegelman and Firer, 1964). A recent estimate of the molecular weight obtained by gel filtration gives a value of about 60,000 (Mumford and Jenner, 1966). The number of chromophores per molecule of phytochrome can be estimated. The molecular extinction coefficient of the P_r form of homogeneous phytochrome is reported as $7·63 \times 10^4 \ 1 \ mol^{-1} \ cm^{-1}$ (Mumford and Jenner, 1966). The molecular extinction coefficient of biliverdin is $2·3 \times 10^4$ (Gray et $al.$, 1961). Therefore the number of chromophores per molecule is probably 3 and need not exceed 4. This is in accord with an enzymatic rather than an energy transferring function.

Denaturation

That the photoreversible properties of phytochrome are lost on denaturation is indicative of a dependence on interaction of the chromophore with the native protein. Denaturation with trichloroacetic acid gave an insoluble product which retained a blue colour with a wavelength maximum near 630 nm.

The susceptibility of phytochrome to denaturation is dependent on the form of the pigment. The P_{fr} form is more susceptible to urea, proteolytic enzymes, and sulphydryl reagents (Butler et al., 1964b) than is P_r. The denaturation studies suggested that the phototransformation might also be accompanied by a protein conformational change. The change in protein structure may be related to the photoactivation of the enzymatic properties of phytochrome as was early predicted.

Chromophore

Similarities between the in vivo phytochrome action spectra and the absorption spectra of the algal chromoproteins C-phycocyanin and allophycocyanin provided an early clue to the nature of the phytochrome chromophore. The chromophores of the algal biliproteins are known to be bile-type pigments (Lemberg and Legge, 1949). Purification of phytochrome permitted a direct comparison of the absorption spectra of phytochrome and the algal bile pigments. The close correspondence in the absorption spectra of the P_r form of phytochrome and of C-phycocyanin and allophycocyanin is clearly evident in Fig. 14.6.

The availability of purified phytochrome provided a source of the chromophore free of other pigments having absorbance in the red region of the spectrum.

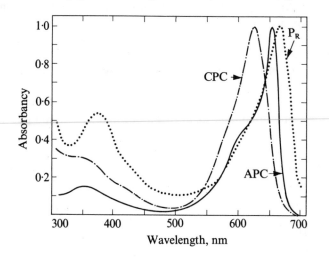

Figure 14.6 Absorption spectra of the P_r form of phytochrome, allophycocyanin (APC), and phycocyanin (CPC). (From Siegelman et al., 1966.)

Successful cleavage of the chromophore (Siegelman *et al.*, 1966) depended on two seemingly unrelated pieces of information. (a) It was observed that trichloroacetic acid denatured phytochrome, but the denatured protein remained blue in colour. The use of alkali or acid to cleave the trichloroacetic acid denatured protein (Lemberg and Legge, 1949; ÓhEocha, 1965), which is successful with the algal biliproteins, was not effective with phytochrome. (b) Fujita and Hattori (1962) found that certain blue-green algal cells refluxed in 1 per cent ascorbic acid in methanol yielded a blue pigment. They recognized the product as a bilitriene type of bile pigment, but, because of the poor yield and the mild conditions for cleavage, they suggested that the pigment obtained was not the chromophore of phycocyanin but a precursor of it. Purified C-phycocyanin served as a model substance in devising chromophore cleavage procedures and as a source of bile-type pigments for comparative studies. Experiments revealed that trichloroacetic acid denatured C-phycocyanin and phytochrome would both release their chromophores on refluxing in methanol alone although yields were only of the order of 10 per cent. The observation that the chromophores were released from the two denatured proteins by the same procedure suggested that the chromophores were similarly bound to their respective proteins.

Comparisons of the properties of the chromophores of phytochrome and phycocyanin could now be made with authentic bile pigments. The absorption spectra of the phycocyanin and phytochrome chromophores were similar but not identical to those of the bilitrienes biliverdin and glaucobilin, but were markedly different from those of the bilidienes, mesobiliviolin and mesobilirhodin. The dimethyl esters of the phytochrome and phycocyanin chromophores were separable on thin-layer chromatograms and were distinctly different from authentic bile pigments. The zinc complexes of the phytochrome and phycocyanin chromophores and of biliverdin and glaucobilin were not visibly fluorescent on ultraviolet irradiation.

The chromophore of phytochrome is a bile pigment of the bilitriene class closely similar but not identical to the chromophore of C-phycocyanin. It is a diacid but must differ in some yet unknown details. The complete chemical structures of the easily obtainable algal chromoprotein chromophores must be determined before the much less readily available phytochrome chromophore structure can be elucidated. The surmise from the action spectra studies that phytochrome was probably a bile pigment is a credit to the perception of the original investigators.

In vivo properties

The spectrophotometric assay allows examination of phytochrome properties *in vivo* in dark-grown seedlings and several tissues in which chlorophyll is low or absent. The *de novo* synthesis of phytochrome in dark-grown maize seedlings increases linearly with time for about 8 days. The phytochrome content of dark-grown soybean cotyledons increases rapidly during the first 5 days of imbibition, then levels off, and finally decreases during the next 5 days (Butler *et al.*, 1965). An apparent synthesis of phytochrome in pea seedlings has been found following a specific light régime (Clarkson and Hillman, 1967).

Phytochrome in dark-grown seedlings was found to be present entirely as P_r (Butler *et al.*, 1963; Delint and Spruit, 1963). When the seedlings were irradiated with red light, P_{fr} was found to be unstable *in vivo* at room temperature (Butler *et al.*, 1963). During several hours of darkness following a red irradiation, there was apparently a complete dark destruction of P_{fr} and the total amount of phytochrome decreased to about 20 per cent of the initial value. The phytochrome remaining at the conclusion of the dark period was all in the P_r form. Since red light can convert no more than 80 per cent of P_r to P_{fr}, the 20 per cent remaining was the P_r which had not been initially converted to P_{fr}. When a portion of the P_{fr} had undergone dark destruction, irradiation with red light could convert a portion of the remaining P_r to P_{fr} (Butler *et al.*, 1964a; Delint and Spruit, 1963).

Dark reversion of P_{fr} to P_r was first demonstrated with cauliflower head tissue (Butler *et al.*, 1963). The total amount of phytochrome did not undergo dark destruction on phototransformation of P_r to P_{fr}, and Hillman (1964) reported similar results with parsnip root and artichoke receptacle. Closer examination of the parsnip (Koukkari and Hillman, 1967) showed that dark destruction of P_{fr} was slow and had to compete with a rapid initial reversion. In dark-grown oat and maize seedlings, dark reversion has not been demonstrated. However, in pea epicotyls and hypocotyls of radish, bean, and soybean, dark destruction of P_{fr} was slower than in corn or oat coleoptiles, and apparently reversion occurs (Hopkins and Hillman, 1965). The mechanisms of phytochrome destruction and reversion reactions are unknown.

Attempts to relate photometric measurements of phytochrome in dark-grown seedlings to physiological responses have given conflicting results which have recently been reviewed in detail (Hillman, 1967). The difficulties may arise in part from the greater (by an order of magnitude or more) phytochrome content of dark-grown seedlings than of light-grown plants. There is no assurance that the greater production of phytochrome in these tissues is directly coupled to growth.

Direct photometric measurements of phytochrome in green plants would be particularly useful in exploring the physiological activity of the pigment. Such measurements are difficult by current methods because of the low phytochrome and high chlorophyll concentrations but new spectrophotometric developments may make them feasible.

Physiological responses

The higher plant is a captive of its environment and must be provided with sensors to appraise variables of its milieu. Garner and Allard (1920) clearly perceived that light was one of the most important variables. Physiological studies have shown that many aspects of the growth and development of higher plants are under control of phytochrome (see chapter 15). Flowering (photoperiodism), one of the most obvious manifestations of phytochrome action, is discussed in chapter 16. The economic importance of flowering has resulted in an inordinate emphasis on this aspect of phytochrome regulation. Furthermore, flowering studies appear to provide only indirect access to an understanding of phytochrome action because of the involvements with growth, differentiation, periodic phenomena, and the long times required for completion. The relationship of phytochrome to other aspects of growth, development, and biochemistry of the plant will be discussed here.

The photocontrol of etiolation in beans was found to be dependent on phytochrome by Downs (1955). Etiolation is not a simple phenomenon but consists of an assemblage of several responses including control of hypocotyl length, leaf expansion, and plumular hook opening. Red light induced a reduction in hypocotyl length—a growth inhibition—and expansion of the leaves—a growth promotion. The results can be clearly measured on the fourth day after the beginning of the light treatment, and photoperiodic phenomena are not involved.

Downs *et al.* (1967) showed that the internode length of several plants was strongly influenced by phytochrome action. These results are similar to those for reversal of etiolation and are indicative that etiolation-type phenomena can occur in light-grown plants. The location of the photoreceptor for the regulation of internode length was in the internode itself.

Several examples of biochemical reactions under phytochrome control are known. The coloration of the tomato cuticle is clearly a biochemical synthesis under phytochrome control (Piringer and Heinze, 1954) resulting from the deposition of a yet unknown pigment into the cuticle of the fruit. Flavonoid synthesis in many plants requires light and is under phytochrome control (Siegelman, 1964). The synthesis is enhanced if the irradiation period is followed by a period in which phytochrome is present in the P_{fr} form. The formation of anthocyanin in apple skin is an instructive case of phytochrome control (Downs *et al.*, 1965). Apple skin forms anthocyanin when exposed to light and accumulates aldehyde and alcohol when held in darkness (Siegelman and Hendricks, 1958). Apparently the substrate for anthocyanin synthesis in light, if not utilized, forms aldehyde and alcohol. It is suggestive that pyruvate or a close derivative is the probable substrate for either the anthocyanin or aldehyde and alcohol formation.

The development and differentiation of plastids of etiolated leaves of higher plants is light dependent. The synthesis of NADP triose phosphate dehydro-genase (Marcus, 1960) and chloroplast protein (Mego and Jagendorf, 1961) were shown to be under phytochrome control. Margulies (1965) confirmed these observations and further showed that prolonged far-red irradiation brought about the same response as brief exposures to red light. The lag phase for chlorophyll accumulation in dark-grown plants is eliminated by a prior illumi-nation with red light, and the effect is reversed by far-red light (Virgin, 1964).

The phytochrome interaction with several tropistic responses is now becoming clarified. The geotropic response of dark-grown oat seedlings was immediately stimulated after exposure to red light and was reversed by far-red light (Wilkins, 1965). A one-second exposure to red light was sufficient to bring about almost the maximal response. The closing movements of mimosa leaflets on changing from light to darkness is under phytochrome control (Fondeville et al., 1966, see chapter 8). The response is evident in 5 minutes and completed in 30 minutes. The transport of the stimulus for leaf unrolling of wheat seedlings, completed in one minute, was found to be a phytochrome response (Wagne, 1965).

Haupt (1965) has recently reviewed the light-induced chloroplast movements in algae. In *Mougeotia* and *Mesotaenium*, the low intensity movement of the single chloroplast is a phytochrome reaction (Haupt, 1959). From polarized microbeam studies, the phytochrome was found to be oriented in a screw similar in shape to the spiral chloroplast in the related *Spirogyra*. The movement of the plastid is evident in less than 10 minutes and is completed in 30 minutes.

Mechanism of action

The wide variety in types of phytochrome responses of plants do not immediately provide useful clues to the mechanism of action. Two different mechanisms of action have been suggested. Hendricks and Borthwick (1965) speculate that control over a diversity of expressions can be achieved by regulation of a basic metabolic reaction or of cell permeability. Mohr (1966) proposes control by gene repression or activation (see chapter 15). In neither case is it clear how phytochrome interacts with the control system proposed.

Conclusions

It has been shown that the growth and development of all parts of the plant are under phytochrome control. The methods by which phytochrome regulates many diverse and seemingly unrelated processes are not understood at present. Further studies are needed to elucidate the mechanism of phytochrome regula-tion from the organismal to the molecular level.

The work described in this chapter was carried out at Brookhaven National Laboratory under the auspices of the U.S. Atomic Energy Commission.

504 THE PHYSIOLOGY OF PLANT GROWTH AND DEVELOPMENT

Bibliography

Selected further reading

Butler, W. L., S. B. Hendricks, and H. W. Siegelman, 1965. Purification and properties of phytochrome. In *Chemistry and Biochemistry of Plant Pigments*, ed. T. W. Goodwin. Academic Press, London.

Siegelman, H. W. and W. L. Butler, 1965. Properties of Phytochrome. *Ann. Rev. Plant Physiol.* 16, 383–392.

Hendricks, S. B. and H. A. Borthwick, 1965. The physiological functions of phytochrome. In *Chemistry and Biochemistry of Plant Pigments*, ed. T. W. Goodwin. Academic Press, London.

Hendricks, S. B. and H. W. Siegelman, 1967. Phytochrome and photoperiodism in plants. In *Comparative Biochemistry*, Vol. 25, ed. M. Florkin and E. H. Stotz. Elsevier, Amsterdam.

Hillman, W. S., 1967. The physiology of phytochrome. *Ann. Rev. Plant Physiol.* 18, in press.

Siegelman, H. W. and S. B. Hendricks, 1964. Phytochrome and its control of plant growth and development. *Advan. Enzymol.* 26, 1–33.

Siegelman, H. W. and S. B. Hendricks, 1965. Purification and properties of phytochrome: a chromoprotein regulating plant growth. *Fed. Proc.* 24, 863–867.

Papers cited in text

Birth, G. S., 1960. Agricultural applications of the dual-monochromator spectrophotometer. *Agr. Eng.* 41, 432–435.

Borthwick, H. A., S. B. Hendricks, and M. W. Parker, 1951. Action spectrum for inhibition of stem growth in dark-grown seedlings of albino and non-albino barley (*Hordeum vulgare*). *Bot. Gaz.* 113, 95–105.

Borthwick, H. A., S. B. Hendricks, M. W. Parker, E. H. Toole, and V. K. Toole, 1952. A reversible photoreaction controlling seed germination. *Proc. Nat. Acad. Sci. U.S.* 38, 662–666.

Borthwick, H. A., S. B. Hendricks, M. W. Toole and V. K. Toole. 1954. Action of light on lettuce seed germination. *Botan. Gaz.* 115, 205–225.

Briggs, W. R. and D. C. Fork, 1965. Studies on phytochrome transformations *in vitro*. *Carnegie Inst. Yearbook*, 64, 406–412.

Briggs, W. R. and H. W. Siegelman, 1965. Distribution of phytochrome in etiolated seedlings. *Plant Physiol.* 40, 934–941.

Butler, W. L., 1961. Some photochemical properties of phytochrome. In *Proc. 3rd Intern. Congr. Photobiol.*, eds. B. C. Christensen and B. Buchmanns. Elsevier, Amsterdam.

Butler, W. L., S. B. Hendricks, and H. W. Siegelman, 1964a. Action spectra of phytochrome *in vivo*. *Photochem. Photobiol.* 3, 521–528.

Butler, W. L., H. C. Lane, and H. W. Siegelman, 1963. Nonphotochemical transformations of phytochrome. *Plant Physiol.* 38, 514–519.

Butler, W. L., K. H. Norris, H. W. Siegelman, and S. B. Hendricks, 1959. Detection, assay, and preliminary purification of the pigment controlling photoresponsive development of plants. *Proc. Nat. Acad. Sci. U.S.* 45, 1703–1708.

Butler, W. L., H. W. Siegelman, and C. O. Miller, 1964b. Denaturation of phytochrome. *Biochemistry*, 3, 851–857.

Clarkson, D. T. and W. S. Hillman, 1967. Apparent phytochrome synthesis in *Pisum* tissue. *Nature, Lond.*, in press.

Correll, D. L., E. Steeres, J. L. Edwards, J. R. Suriano, and W. Shropshire, 1966. Phytochrome: isolation and partial characterization. *Fed. Proc.* **25**, 736.

DeLint, P. J. A. L. and C. J. P. Spruit, 1963. Phytochrome destruction following illumination of mesocotyls of *Zea mays* L. *Mededel. Landbouwhogeschool, Waageningen*, **63**, 1–7.

Downs, R. J., 1955. Photoreversibility of leaf and hypocotyl elongation of dark-grown red kidney bean seedlings. *Plant Physiol.* **30**, 468–473.

Downs, R. J., S. B. Hendricks, and H. A. Borthwick, 1957. Photoreversible control of elongation of Pinto beans and other plants under normal conditions of growth. *Bot. Gaz.* **118**, 199–208.

Downs, R. J., H. W. Siegelman, W. L. Butler, and S. B. Hendricks, 1965. Photoreceptive pigments for anthocyanin synthesis in apple skin. *Nature, Lond.* **205**, 909–910.

Flint, L. H. and E. D. McAlister, 1935. Wavelengths of radiation in the visible spectrum inhibiting the germination of light-sensitive lettuce seed. *Smithsonian Inst. Misc. Coll.* **94**, 1–11.

Flint, L. M. and E. D. McAlister, 1937. Wavelength of radiation in the visible spectrum promoting the germination of light-sensitive lettuce seed. *Smithsonian Inst. Misc. Coll.* **96**, 1–80.

Fondeville, J. C., H. A. Borthwick, and S. B. Hendricks, 1966. Leaflet movement of *Mimosa pudica* L. indicative of phytochrome action. *Planta*, **69**, 357–364.

Fujita, Y. and A. Hattori, 1962. Preliminary note on a new phycobilin pigment isolated from blue-green algae. *J. Biochem.* **51**, 89–91.

Garner, W. W. and H. A. Allard, 1920. Effect of the relative length of day and night and other factors of the environment on growth and reproduction in plants. *J. Agr. Res.* **18**, 553–606.

Gray, C. H., A. Lichtorowicz-Kulczycka, D. C. Nicholson, and Z. Petryka, 1961. The chemistry of the bile pigments, Pt. II. The preparation and spectral properties of biliverdin. *J. Chem. Soc.* **1961**, 2264–2268.

Haupt, W., 1959. Die Chloroplastendrehung bei *Mougeotia*. I. Über den quantativen und qualitativen Lichtbedarf bei Schwachlichtbewegung. *Planta*, **53**, 484–502.

Haupt, W., 1965. Perception of environmental stimuli orienting growth and movement in lower plants. *Ann. Rev. Plant Physiol.* **16**, 267–290.

Hendricks, S. B., H. A. Borthwick, and R. J. Downs, 1956. Pigment conversion in the formative responses of plants to radiation. *Proc. Nat. Acad. Sci. U.S.* **42**, 19–26.

Hendricks, S. B., W. L. Butler, and H. W. Siegelman, 1962. A reversible photoreaction regulating plant growth. *J. Phys. Chem.* **66**, 2550–2555.

Hillman, W. S., 1964. Phytochrome levels detectable by *in vivo* spectrophotometry in plant parts grown or stored in light. *Am. J. Bot.* **51**, 1102–1107.

Hopkins, W. G. and W. S. Hillman, 1965. Phytochrome changes in tissue of dark-grown seedlings representing various photoperiodic classes. *Am. J. Bot.* **52**, 427–432.

Koukkari, W. L. and W. S. Hillman, 1967. Effects of temperature and aeration on phytochrome transformations in *Pastinaca sativa* L. root tissue. *Am. J. Bot.*, **54**, 1118–1122.

Lane, H. C., H. W. Siegelman, W. L. Butler, and E. M. Firer, 1963. Detection of phytochrome in green plants. *Plant Physiol.* **38**, 414–416.

Lemberg, R. and J. W. Legge, 1949. *Hematin compounds and bile pigments*. Interscience, New York.

Linschitz, H., V. Kasche, W. L. Butler, and H. W. Siegelman, 1966. The kinetics of the phytochrome conversion. *J. Biol. Chem.* **241**, 3395–3403.

Loomis, W. P. and J. Battaile, 1966. Plant phenolic compounds and the isolation of plant enzymes. *Phytochemistry*, **5**, 423–438.

Marcus, A., 1960. Photocontrol of formation of red kidney bean leaf triosephospho-pyridine nucleotide linked triosephosphate dehydrogenase. *Plant Physiol.* **35**, 126–128.

Margulies, M. M., 1965. Relationship between red light mediated glyceraldehyde-3-phosphate dehydrogenase formation and light dependent development of photosynthesis. *Plant Physiol.* **40**, 57–61.

Mego, J. L. and A. T. Jagendorf, 1961. Effects of light on growth of Black Valentine bean plastids. *Biochim. Biophys. Acta,* **53**, 237–254.

Miller, C. O., R. J. Downs, and H. W. Siegelman, 1965. A rapid procedure for the visible detection of phytochrome. *Bio Science*, **15**, 596–597.

Mohr, H., 1966. Differential gene activation as a mode of action of phytochrome 730. *Photochem. Photobiol.* **5**, 469–483.

Mumford, F. E. and E. L. Jenner, 1966. Purification and characterization of phytochrome from oat seedlings. *Biochemistry* **5**, 3657–3662.

ÓhEocha, C., 1965. Phycobilins. In *Chemistry and Biochemistry of Plant Pigments*, ed. T. W. Goodwin. Academic Press, London.

Parker, M. W., S. B. Hendricks, H. A. Borthwick and N. J. Scully. 1946. Action spectrum for the photoperiodic control of floral initiation of short-day plants. *Botan. Gaz.* **108**, 1–26.

Piringer, A. A. and P. H. Heize, 1954. Effect of light on the formation of a pigment in the tomato fruit cuticle. *Plant Physiol.* **29**, 467–472.

Pratt, L. H. and W. R. Briggs, 1966. Photochemical and non-photochemical reactions of phytochrome *in vivo*. *Plant Physiol.* **41**, 467–474.

Siegelman, H. W., 1964. Physiological studies on phenolic biosynthesis. In *Biochemistry of Phenolic Compounds*, ed. J. B. Harborne. Academic Press, London.

Siegelman, H. W. and E. M. Firer, 1964. Purification of phytochrome from oat seedlings. *Biochemistry*, **3**, 418–423.

Siegelman, H. W. and S. B. Hendricks, 1958. Photocontrol of alcohol, aldehyde, and anthocyanin production in apple skin. *Plant Physiol.* **33**, 409–413.

Siegelman, H. W., B. C. Turner, and S. B. Hendricks, 1966. The chromophore of phytochrome. *Plant Physiol.* **41**, 1289–1292.

Taylor, A. O. and B. A. Bonner, 1967. Isolation of phytochrome from the algae *Mesotaenium* and liverwort *Sphaerocarpus*. *Plant Physiol.* **42**, in press.

Virgin, H. I., 1964. Some effects of light on chloroplasts and plant protoplasm. In *Photophysiology*, Vol. I, ed. A. C. Giese. Academic Press, New York.

Wagne, C., 1965. The distribution of the light effect from irradiated to non-irradiated parts of grass leaves. *Physiol. Plant.* **18**, 1001–1006.

Wilkins, M. B., 1965. Red light and the geotropic response of the *Avena* coleoptile. *Plant Physiol.* **40**, 24–34.

15. Photomorpho-genesis

H. Mohr

Introduction

Definitions

All living systems on this planet depend on a narrow band in the electromagnetic spectrum which we call 'light'. By the term 'light' we normally mean that range of the electromagnetic spectrum which affects the human eye. Physically this range can be located between about 390 and 760 nm. When we deal with plants we will also use the term 'light' in the sense of visible radiation but we usually include the near ultraviolet down to about 320 nm and the very near infrared up to about 800 nm.

Quanta of this spectral range have a relatively low energy of about 90 down to 36 kcal Einstein^{-1}, depending on the particular wavelength. This energy is sufficient to alter the outer electronic energy levels of atoms and molecules but not sufficient to induce ionization. Living systems have adapted themselves to this sort of radiation in course of evolution. Quanta of this size can be absorbed in a plant or in an animal only by a very few types of molecules which are characterized by extended π-electron systems as, e.g., the chlorophylls or the carotenoids. Most molecules which occur in the cell—water, proteins, nucleic acids, lipids, carbohydrates—cannot absorb quanta in the spectral range of light with the result of an electronic excitation.

By the term 'photomorphogenesis' we designate the control which may be exerted by light over growth, development, and differentiation of a plant, independently of photosynthesis. In order to conceive the full importance of this phenomenon we have to remember that the specific development of a living system depends on the genetical information of the particular system and on the environment. The most important factor of the environment is light, at least in all higher plants. It is important, however, to realize that light does not carry any *specific* information. Light can only be regarded as an 'elective' factor which influences the manner in which those genes which are present in the particular organism are used.

Phenomena

Figures 15.1 and 15.2 illustrate some characteristic phenomena of photomorphogenesis. The two mustard seedlings in Fig. 15.1 have virtually the same genes and both were grown on the same agar medium which contained an

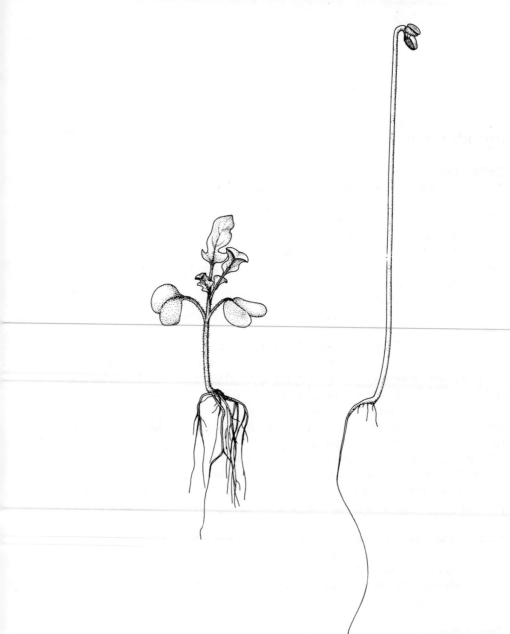

Figure 15.1   These two mustard seedlings (*Sinapis alba* L.) have the same chronological age and are genetically virtually identical. Right, a dark-grown seedling; left, a light-grown seedling.

abundance of organic molecules. 'Light' must be responsible for their different appearance. The dark-grown seedling on the right shows the characteristics of 'etiolation', whereas the light-grown seedling has developed in a way which we designate by the term 'normal'. It is not difficult to demonstrate that the influence of light is not due to photosynthesis. We have rather to conclude that in a seedling some photochemical reaction must occur independently of photosynthesis in order to make possible what we are accustomed to call normal growth and development. This statement is probably true for all higher plants. A second example is shown in Fig. 15.2. The sprouts of a potato tuber will etiolate in darkness whereas in the light the normal potato plant will develop. A characteristic of the etiolation is that the internodes grow rapidly while the leaves remain rudimentary.

The biological significance of etiolation is obvious. As long as a plant has to grow in darkness it uses the limited supply of storage material predominantly for axis growth. In this way the probability is highest that the tip of the plant will reach the light before the storage material is exhausted. The physiological problem is to understand how light can force a young plant to develop in a 'normal' manner instead of becoming etiolated. Apparently two things are involved. First, a photochemical reaction, and second, a change in the manner in which the genes of the particular organism are used. We recognize that the causal analysis of photomorphogenesis may contribute to our knowledge of development and lead to a better understanding of a basic problem of biology: control of differentiation and development in multicellular systems.

Seedlings of dicotyledonous plants (e.g., *Sinapis alba* L., *Lactuca sativa* L.) turned out to be excellent objects for the causal analysis of photomorphogenesis. The first stages of vegetative growth after seed germination are especially useful for the following reasons. First, the seedling (e.g., of *Sinapis alba* L.) consists during this phase only of three parts, cotyledons, hypocotyl, and radicle. The plumule is hardly developed. Second, the seedling contains so much storage material—mainly fat and protein—in the cotyledons that it is completely independent of photosynthesis or of the external supply of organic molecules for a number of days after germination. Third, the seedling can be grown on a medium which supplies only water. An external supply of ions is not required. The seedling can therefore be regarded as a closed system for nitrogen or phosphate. This is an advantage if we want to study light-induced changes of protein or nucleic acid synthesis.

The problem

The problem is to understand the causalities of photomorphogenesis in a modern, preferentially 'molecular' terminology. In order to approach this problem we ask the following questions: what type of pigment absorbs the effective light, what kinds of photochemical reactions occur, and what are the causal

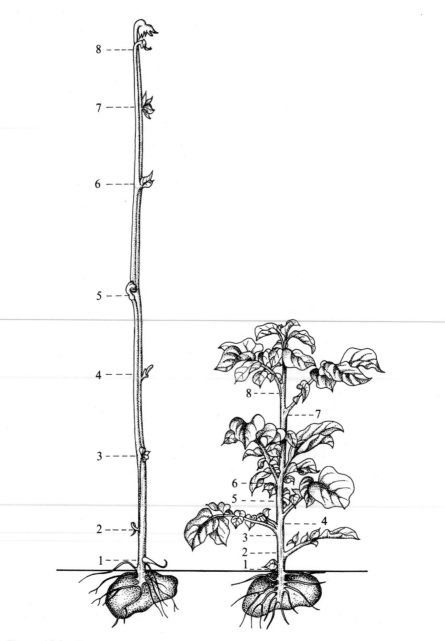

Figure 15.2 These two potato plants (*Solanum tuberosum* L.) are genetically identical. Left, a dark-grown plant; right, the corresponding light-grown plant. (After Pfeffer, 1904. p. 98.)

relationships between the photochemical reactions and the final photoresponses which we can plainly see and measure, e.g., the light-mediated growth of the cotyledons, or the light-mediated synthesis of anthocyanin, or the light-mediated inhibition of hypocotyl lengthening. And the final question: how can the integration of the photoresponses of the different cells, tissues, and organs be understood. This integration is essential: photomorphogenesis is a complex but harmonic process, precisely regulated in space and time.

Phytochrome-mediated photomorphogenesis

Some properties of phytochrome

What type of pigment absorbs the effective light? The answer is, phytochrome (at least above 550 nm). The history of phytochrome has been described in a previous chapter. We mention only a few results of phytochrome research in order to create some basis for the further question we shall discuss in more detail: how does phytochrome act?

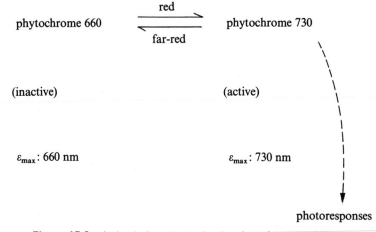

phytochrome 660 $\xrightleftharpoons[\text{far-red}]{\text{red}}$ phytochrome 730

(inactive)   (active)

ε_{max}: 660 nm ε_{max}: 730 nm

photoresponses

Figure 15.3 A simple formulation for the phytochrome system.

The phytochrome system is widely distributed in the plant kingdom somewhat as are chlorophyll a and b, namely among the potentially green plants, and it is as basically important for photomorphogenesis as chlorophyll is for photosynthesis. Phytochrome (Fig. 15.3) is a bluish chromoprotein having two interconvertible forms: phytochrome 660, or P_{660}, with an absorption maximum in the red around 660 nm and phytochrome 730, or P_{730}, with an absorption maximum in the far-red around 730 nm. P_{730} is the physiologically active form of the phytochrome system. The function of P_{730} must be supposed to be to initiate basic changes in the metabolism and in the energetics of the cells and tissues leading finally to the photoresponses, e.g., growth of a leaf or formation of compounds like anthocyanins. In a dark-grown seedling only P_{660} will be present. The formation of P_{730} involves a photochemical step.

The absorption spectra of P_{660} and P_{730} overlap considerably. This overlap is the reason why under conditions of saturating irradiations photostationary states are characteristic for the status of the phytochrome system in the cell. It has been shown (Pratt and Briggs, 1966) that at the most about 80 per cent of the total phytochrome can be present as P_{730}, for instance, if we irradiate the living system with pure red light around 660 nm. If we irradiate with pure far-red around 730 nm only about 1 per cent of the total phytochrome will be present as P_{730}. The photostationary state of the phytochrome system is rapidly established at least in the red and far-red where the extinction coefficients of the phytochromes and the quantum efficiencies of the photochemical conversions are high. A few minutes of irradiation with medium quantum flux density are satisfactory for establishing the photostationary state. In brief, only 'short time irradiations' are required to establish photostationary states in the phytochrome system *in vivo* as well as in the photochemically active extract. While P_{660} is stable in darkness, P_{730}—the physiologically active species of phytochrome—is not. It can disappear in two ways: it is readily destroyed by an irreversible reaction or it reverts slowly to P_{660} in a thermal reaction. In a seedling the irreversible destruction of P_{730} seems to be the dominant process.

'High energy responses' and phytochrome

For some years it seemed necessary to postulate even in the wavelength range above 550 nm the existence of a separate photoreceptor besides phytochrome in order to account for the phenomena of photomorphogenesis (e.g., Mohr, 1964). We briefly repeat some of the basic observations which have led to the concept of a separate 'high energy reaction' besides phytochrome action. We choose the light-dependent enlargement of the cotyledons of the mustard seedling as an example of a graded photoresponse which can easily and accurately be measured. This response can furthermore be regarded as a reliable gauge of the photomorphogenic light effect on a seedling as a whole. The results of the studies on action spectra* are confusing in so far as one gets very different types of action spectra with irradiation periods of different durations (Fig. 15.4). The dashed curve indicates the action spectrum for light-dependent cotyledon expansion in the case of 'short-time irradiation', i.e., at a maximum, 5 min or

---

* Action spectra usually specify the efficiency of the different parts of the spectrum with respect to a particular photoresponse. On the basis of experimental dose-response-curves one can calculate the number of quanta (or Einsteins) per cm^2 $(N\lambda)$ which are required at each wavelength to produce the same physiological response. The reciprocal values as a function of wavelength are called the 'action spectrum' of the particular response measured. An action spectrum represents the extinction or absorption spectrum of the photoreceptor involved if several prerequisites which can be checked are fulfilled. Coincidence of the action spectra of different photoresponses means that they are mediated by the same photoreceptor.

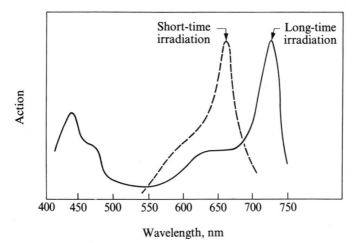

Figure 15.4   Characteristics of action spectra of photomorphogenesis in the cases of short-time irradiation and long-time irradiation; based on data for the photoresponse 'enlargement of the cotyledons' of the mustard seedling (*Sinapis alba* L.). (After Wagner and Mohr, 1966a.)

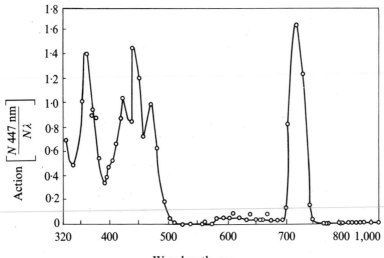

Figure 15.5   An action spectrum for control of hypocotyl lengthening by light in lettuce seedlings (*Lactuca sativa* L., cv. Grand Rapids). Wavelengths above 760 nm are completely ineffective. The action spectrum was determined between 54 and 72 hours after sowing. During this period lengthening of the hypocotyl is almost exclusively due to cell lengthening at least in the cortex (Häcker, Hartmann, and Mohr, 1964). Individual points were determined with interference line filters (DIL, IL). The effectiveness at the reference wavelength (447 nm) was measured with a bandpass filter (DAL) and by definition is equal to 1·0. This reference point is not shown. (After Hartmann, 1967a.)

thereabouts of irradiation. This action spectrum reflects the effectiveness of the different wavelengths in transforming P_{660} into P_{730}. The solid curve represents the type of action spectrum we observe in the case of 'long-time irradiation', i.e., 6, 8 or more hours of continuous irradiation. In this case the extent of the response is much greater than in the case of 'short-time irradiation', and the action spectrum is extremely different. But, nevertheless, there are now good arguments that the active photoreceptor is in both cases exclusively phytochrome at least above 550 nm.

The experimental evidence according to which the far-red peak of action is due to the formation and maintenance of P_{730}, has been elaborated primarily by Hartmann (1966, 1967a, b, c). He has been using photoresponses of the lettuce seedling (*Lactuca sativa* L., cv. Grand Rapids, tip burn resistant strain). Figure 15.5 shows an action spectrum for control of hypocotyl lengthening in lettuce under long-time irradiation. If we exclude at the moment the blue part of the spectrum from our consideration we recognize that quanta between 500 and 700 nm are hardly effective, far-red above 760 nm is ineffective whereas in the near far-red we observe a symmetric band of action close to 716 nm where the photoequilibrium of the phytochrome system contains about 3 per cent P_{730} ($[P_{730}]/[P] \approx 0.03$), ($[P_{730}] + [P_{660}] = [P]$). The problem is whether or not the far-red peak of action can be understood on the basis of phytochrome. Figure 15.6 illustrates one successful type of experimental approach. Lettuce seedlings are irradiated simultaneously with 768 nm, a wavelength which is ineffective if applied separately, and with virtually ineffective

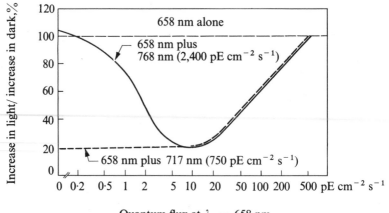

Figure 15.6 Hypocotyl lengthening in lettuce seedlings (*Lactuca sativa* L., cv. Grand Rapids) under continuous simultaneous irradiation with different quantum flux densities in the red (658 nm) at a constant high quantum flux density background of different far-red bands (717 nm, 768 nm). Ordinate: ratio between 'increase in light' and 'increase in dark', measured between 54 and 72 hours after sowing. (After Hartmann, 1967b.)

red light—658 nm—at different quantum flux densities. We recognize that there is an intensity ratio of the wavelengths 768 and 658 nm which leads to an optimum 'high energy effect'. On the other hand the typical 'high energy effect' of the wavelength 717 nm can be nullified by a simultaneous irradiation with 658 nm at a suitable quantum flux density. This sort of result indicates that the strong photomorphogenic effect of prolonged far-red might be due to a favourable ratio between the photostationary concentrations of P_{660} and P_{730}. The other view, namely, that a separate photoreceptor is responsible for the far-red band of action under long-time irradiation, is difficult to reconcile with these experimental results. Similar experiments to that illustrated in Fig. 15.6 have been performed using other wavelengths in the far-red range (e.g., 706, 745, 801 nm). In all cases the action of the high intensity far-red band can be modified (as shown in principle in Fig. 15.6) by a simultaneous irradiation with red light (658 nm) which is virtually ineffective if applied separately. Taking into account the most reliable data on the photoequilibrium in phytochrome (Butler, Hendricks, and Siegelman, 1964; Pratt and Briggs, 1966), Hartmann (1966) has calculated that a maximum 'high energy response' with respect to hypocotyl lengthening will always occur when the ratio $[P_{730}]/[P]$ is about 0·03. In the case of lettuce seed germination the most favourable photoequilibrium for the 'high energy response' is about 0·10. Correspondingly the peak of the far-red band of action is close to 707 nm (Hartmann, 1966, 1967a). If the ratio $[P_{730}]/[P]$ is above 0·3 or below 0·002 the action of light on hypocotyl lengthening is below 10 per cent deviation from the dark control.

The 'high energy responses' show a characteristic intensity dependence (Hartmann, 1966; Wagner and Mohr, 1966a) whereas the photoequilibrium of the phytochrome system at a particular wavelength does not depend on intensity. Hartmann (1967b) has provided experimental evidence that the photoequilibrium as well as the total absorption of quanta in the phytochrome system are essential for the degree of the 'high energy response'. This latter finding explains the intensity dependence. At the moment all available data are consistent with the hypothesis that the typical 'high energy response', inhibition of hypocotyl lengthening (Fig. 15.5), is a consequence exclusively of phytochrome action, at least above 550 nm. The further problem is to identify the 'effector molecule' of the high energy response. Possible 'candidates' are P_{730} and the shortlived intermediates between P_{660} and P_{730} (Linschitz, Kasche, Butler, and Siegelman, 1966; Linschitz and Kasche, 1967; Spruit, 1966). If P_{730} is the effector one can predict that the far-red band of action (around 716 nm; Fig. 15.5) can be nullified not only by a simultaneous irradiation with red light (Fig. 15.6) but also through a simultaneous irradiation with a high intensity far-red band which is only absorbed by P_{730} and therefore depresses the steady state concentration of P_{730} close to zero. One of the pertinent experiments is illustrated in Fig. 15.7. We find that a quantum flux density of 175 pE cm^{-2} s^{-1} at 716 nm depresses hypocotyl lengthening by about 60 per cent. On the other hand the wavelength

759 nm is virtually ineffective up to 28 nE cm^{-2} s^{-1}. We further see that simultaneous irradiation with 175 pE cm^{-2} s^{-1} at 716 nm and, e.g., 300 pE cm^{-2} s^{-1} at 759 nm depresses hypocotyl lengthening by 80 per cent. This fact indicates that the photoequilibrium which is established by 716 nm is somewhat higher than the optimum value. Maximum action is expected to be close to 720 nm. The peak of the extremely refined action spectrum (Hartmann; still unpublished) is indeed located at 720 nm. The decisive point, however, is the following: The simultaneous irradiation with 27 nE cm^{-2} s^{-1} at 759 nm virtually nullifies the action of 716 nm. A calculation of the phytochrome photoequilibrium under this condition leads to the result $[P_{730}]/[P]$ about 0·0005. The conclusion seems

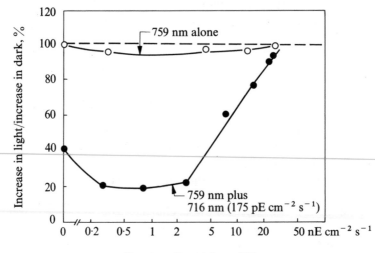

Quantum flux at $\lambda_m = 759$ nm

Figure 15.7   Hypocotyl lengthening in lettuce seedlings (*Lactuca sativa* L., cv. Grand Rapids) under continuous simultaneous irradiation with different quantum flux densities at 759 nm at a constant quantum flux density background of the wavelength 716 nm. Ordinate: ratio between 'increase in light' and 'increase in dark', measured between 54 and 72 hours after sowing. (After Hartmann, 1967b.)

to be justified that phytochrome is the photoreceptor of the 'high energy response' in the far-red range with P_{730} as the specific effector molecule. 'Without P_{730} there is no far-red high energy response' (Hartmann, 1967b, c). If we accept Hartmann's conclusion (there seems to be no alternative at the moment) we are left with the following problem: Why is it that in the lettuce seedling, under conditions of long time irradiation, a photoequilibrium of $[P_{730}]/[P] = 0·03$ is much more effective than a photoequilibrium of $[P_{730}]/[P] = 0·8$? Hartmann (1966) has advanced two different hypotheses to explain the paradoxical situation: First, the peculiarities of prolonged irradiation could be connected with the fact that P_{730} decreases in seedlings through irreversible destruction. Hartmann has calculated the decrease of the pool size of P_{730} with time for a number of photoequilibria. It turns out, of course, that the pool size of P_{730} approaches zero faster the higher the photoequilibrium $[P_{730}]/[P]$. In the range in which between 1 and 5 per cent of the total P is present as P_{730} the decrease of the pool size of P_{730} is only small. This is the sort of photostationary state (= photoequilibrium) which is maintained by the most effective wavelength in the near far-red range.

As an alternative hypothesis Hartmann has discussed a mechanism of 'competitive self-inhibition' of the P_{730} molecules. This mechanism does not require the involvement of P_{730} destruction.

At the moment it is still an open problem which mechanism will explain the 'high energy responses' exerted by P_{730}. Nevertheless recent checks of Hartmann's first hypothesis (involvement of phytochrome destruction) have already led to important results. To take one example it has been reported (Pratt and Briggs, 1966) that in the coleoptiles of dark-grown corn seedlings the destruction of P_{730} follows zero order kinetics until about 10 per cent of the P_{730} initially present after a 5 min exposure to red light ($[P_{730}]/[P] = 0.8$) is left, at which point the curve becomes non-linear. This result confirms the observation by Butler, Lane, and Siegelman (1963) that only about 10 per cent of the phytochrome in a dark-grown corn seedling needs to be present as P_{730} to saturate the destruction reaction. These data seemed to exclude Hartmann's first hypothesis. It turned out, however, that in tissues of dicotyledonous seedlings the situation is quite different. The destruction of P_{730} follows essentially first-order kinetics (Klein, Edwards, and Shropshire, 1967; Fig. 15.8). Furthermore the phytochrome concentration in dark-grown seedlings of *Pisum sativum* L., *Phaseolus aureus* Roxb. and *Sinapis alba* L. was found to remain constant under continuous far-red irradiation for periods of many hours (Clarkson and Hillman, 1967; Fig. 15.8). Similar far-red treatment of corn seedlings reduced the phytochrome concentration by more than 60 per cent.

With respect to dicotyledonous seedlings (e.g., *Sinapis alba* L.) the conclusion seems to be justified that under continuous far-red light ($[P_{730}]/[P]$ in the range of 0.02–0.05) a low but virtually stationary concentration of the effective P_{730} can be maintained over a considerable period of time (steady state conditions with respect to $[P_{730}]$).

Looking at Fig. 15.4 we find, however, that the situation in the case of the mustard seedling is obviously less clear-cut than in the case of the lettuce seedling. The mustard seedling is different from the lettuce seedling inasmuch as red light has a pronounced effectiveness, even in the case of prolonged irradiation, and long wavelength far-red (e.g., 765 nm) still exerts a measurable effect. However, in experiments with simultaneous irradiation with two wavelengths one gets in principle the same results as with lettuce (Wagner, 1967). The effectiveness of the far-red band 765 nm on anthocyanin synthesis can, for example, be increased by a simultaneous irradiation with red—658 nm—to about the extent one might expect. The optimum photoequilibrium for anthocyanin synthesis has not yet been determined with the same accuracy as with hypocotyl lengthening in lettuce; the location of the peak of the far-red band of action (Fig. 15.4) indicates, however, that the optimum ratio is low. Why is prolonged red fairly effective in the case of the mustard seedling (Fig. 15.4) and apparently in many other seedlings as well? Successive irradiations with far-red and red in connection with kinetic studies of several photoresponses of the mustard seedling, among them anthocyanin synthesis, have led to the conclusion (Wagner and Mohr, 1966a) that in the mustard seedling there are two types of P_{730} (Butler and Lane, 1965). The larger part is unstable towards irreversible destruction (Fig. 15.8). A small part, however, not detectable with the 'Ratiospect'

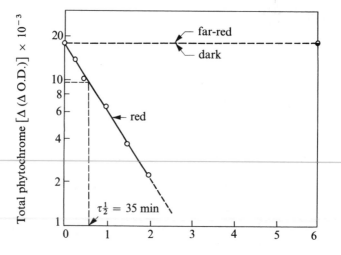

Time after onset of irradiation, hours

Figure 15.8   Changes of total phytochrome under continuous red or far-red light. Material: hypocotyl tissue of *Sinapis alba* L. Under continuous far-red no loss of phytochrome can be detected within the experimental period. Under continuous red light the phytochrome content decreases with a half-life of about 35 min. Four hours after the onset of red light the phytochrome level is below 0.001 Δ (Δ O.D.) and can no longer be detected by the instrument used ('Ratiospect'; noise level $\pm$ 0.001 Δ (Δ O.D.). (Unpublished data; obtained by P. Schopfer in W. S. Hillman's laboratory at the Brookhaven National Laboratory.)

(see chapter 14) seems to be resistant to destruction. Left in darkness it reverts to P_{660}. The stability of the small fraction might be related to the difference in the protein moiety or to a specific localization in the cell (Siegelman and Butler, 1965). If we assume that both types of P_{730} can exert the same physiological function we are able to understand, at least in principle, the effects of long-time irradiation on the basis of phytochrome (Wagner and Mohr, 1966b). A similar conclusion has been reached by Vince and Grill (1966) with respect to anthocyanin synthesis in turnip. In their opinion anthocyanin synthesis in both red and far-red light depends on the absorption by phytochrome.

To summarize: Using only assumptions which seem to be justified it is possible to interpret the 'high energy responses' of the mustard seedling (*Sinapis alba* L.) on the basis of phytochrome, at least as far as wavelengths above 550 nm are concerned. Consequently we shall adhere in the following pages to the conception that photomorphogenesis in the wavelength range above 550 nm is exclusively due to the formation of P_{730}. We shall deal with the blue and near ultraviolet in a later section.

The ecological outlook

The conclusion that an optimum action of phytochrome requires a favourable, relatively low ratio $[P_{730}]/[P]$ is supported by the results of yield experiments where different types of fluorescent lamps were compared. The greatest yield of tomato plants (*Lycopersicum esculentum* Mill., cv. 'Valiant') in the vegetative stage were produced by a lamp combining a small amount of blue light with a large amount of red light. The yield could be enhanced by the addition of some emission in the photosynthetically ineffective far-red region, above 700 nm (Thomas and Dunn, 1967a). It is interesting to note that natural light to which the plant is adapted is characterized by similar amounts of red and far-red light.

The effect of the spectral composition of the fluorescent light on growth and reproduction of mature plants was evaluated using bean plants (*Phaseolus vulgaris* L., cv. N.H.No.HRI Bush Horticultural Bean; a dwarf variety). In general the results with bean agree with those for tomato in that best growth was obtained with a lamp high in red light emission, a moderate amount in the far-red, and very little in the blue part of the spectrum (Thomas and Dunn, 1967b).

Phytochrome content and the extent of the physiological response

Before we can proceed to the problem of how P_{730} acts at the molecular level we have to discuss briefly our knowledge with regard to the relation between P_{730} content and the extent of the physiological response. This is a controversial matter. Freshly excised roots of pea seedlings (*Pisum sativum* L., cv. 'Alaska') contain spectrophotometrically detectable amounts of phytochrome but failed to respond in a measurable way to red or far-red light. After 2 weeks of sterile culture in the dark, however, red light strongly suppressed lateral root initiation whereas phytochrome was no longer detectable (Furuya and Hillman, 1964; Furuya and Torrey, 1964). There are several other cases of discrepancy between the physiological response of plants or plant parts to light and the known properties of phytochrome in these plant materials, e.g., the red-induced inhibition of growth of stem section from etiolated peas, can be reversed by far-red long after all measurable P_{730} has disappeared (Hillman, 1965). Likewise the red-induced opening of hypocotyl hooks of *Phaseolus vulgaris* L. can be reversed with far-red light in the absence of detectable P_{730} (Edwards and Klein, 1964). Hillman (1965) has proposed a segregation of active from bulk phytochrome to account for these observations of significant far-red reversibility of the red-induced response long after P_{730} is photometrically undetectable. The most extreme example of the discrepancies between measurable phytochrome and physiological effectiveness has been described by Briggs and Chon (1966). They compared the influence of red light in altering the phototropic sensitivity of corn coleoptiles (*Zea mays* L., cv. Burpee Barbecue Hybrid) with the spectrophotometric status of the phytochrome they contain. Among other things, they found that dosages of red light sufficient to saturate the physiological system are two orders of magnitude too small to induce measurable phytochrome transformations. The data call for the following hypothesis: The phytochrome which mediates the physiological response must be functionally separate from the bulk which one can measure and must be present in a concentration too small to detect with currently available instrumentation. This active phytochrome must be sequestered in some manner that facilitates its photochemical transformations. It is attractive to assume this active phytochrome to be packaged in a stable structure in the cell. With the help of linearly polarized red and far-red light it has been shown in physiological experiments with several types of cells (e.g., the green alga *Mougeotia* (Haupt, 1960, 1965); fern sporelings of *Dryopteris filix-mas* (L.) Schott, (Etzold, 1965; Fig. 15.22) that molecules of phytochrome can be oriented in a dichroic structure in the outer cytoplasm, possibly along the plasmalemma. In another system, however, which contains phytochrome (Taylor and Bonner, 1967) and responds to polarized light in the blue and near ultraviolet range of the spectrum, no equivalent dichroic orientation of the phytochrome molecules could be found (Experiments with filamentous sporelings of the liverwort *Sphaerocarpus donnellii* Aust. (Steiner, 1967; Figs. 15.21, 15.22)). We must further take into account that within the same cell phytochrome molecules may be located in different 'compartments'.

Some data, however, look simpler. Loercher (1966), for example, reported that in oat seedlings (*Avena sativa* L., cv. Victory) the extent of mesocotyl inhibition, observed 5 days after a brief irradiation, is directly proportional to the logarithm of P_{730} concentration in the mesocotyl tissue at the time of light exposure. This fact—a semilogarithmic relation between the amount of P_{730} and the relative growth response—means that minute changes at low P_{730} concentrations result in comparatively large changes of the growth response. Loercher's data do not exclude the possibility that only a fraction of the total P_{730} is physiologically effective. However, it must be postulated then that a definite constant ratio exists between this active fraction and the total measurable quantity of P_{730}.

There are two major difficulties in fully evaluating most of these data. First, in many papers data on kinetics (degree of a graded response versus time after onset of light) are not available. Loercher's approach is characteristic: brief irradiation of the plant material; after the light treatment the seedlings were incubated in the dark for another 5 days. After that time the seedlings were excised and the length of the mesocotyl was measured. The effect of light was calculated as the difference between length of dark-grown and light-treated mesocotyls. Secondly, in most papers no attempt is made to follow the response under conditions where the P_{730} concentration is virtually the same over an extended period of time. It seems that the experimental approach has to be improved as follows: The kinetics of graded photoresponses must be followed under steady state conditions of P_{730}. In the following chapter an attempt to fulfil this requirement will be described.

Kinetics under continuous far-red light: a justification

One of the conclusions drawn from Fig. 15.8 was that under continuous far-red a low but virtually stationary concentration of the active P_{730} can be maintained over a considerable period of time. Consequently we are dealing with steady-state conditions of $[P_{730}]$ if we investigate the process of photomorphogenesis under continuous far-red. Furthermore, we avoid any interaction of photosynthesis since under far-red light virtually no chlorophyll will be formed. The plastids are completely inactive as far as photosynthesis is concerned. These points of view convinced us (Mohr, 1966a) that it might be best to study the causalities of phytochrome-mediated morphogenesis under continuous far-red light. We remember (Figs 15.4, 15.5) that this part of the spectrum exerts the strongest effect on morphogenesis in the case of 'long-time irradiation'.

The far-red radiation applied must be strictly standardized. The ratio $[P_{730}]/[P]$ is determined by the spectral distribution of the far-red light. The physiological effectiveness of the $[P_{730}]$ depends on the quantum flux density of the far-red applied. A quantitative comparison between experiments can only be made if the far-red sources are identical. The standard far-red source which has been used in our laboratory has been described on several occasions (e.g., Mohr, 1966a).

Phytochrome and the diversity of photoresponses: the problem of differentiation

If we desire to understand the phenomenon of photomorphogenesis we have to ask the following question: What are the causal relationships between P_{730} and the final photoresponses, e.g., light-mediated growth of the cotyledons of a seedling, or light-mediated synthesis of anthocyanin, or light-mediated inhibition of hypocotyl lengthening? To put the question more precisely, let us again consider the two mustard seedlings in Fig. 15.1. All photoresponses of the light-grown seedling are consequences of the formation of P_{730} in the cells of the seedling. Some of the many photoresponses are enumerated in Table 15.1. We have good reasons for stating that P_{730} seems to be the same in all the cells of the seedling in which it occurs. But we observe that the different organs and tissues of the seedling respond differently to the formation of P_{730}. P_{730} exerts a 'multiple action'. Which response can take place apparently depends on the 'specific status of differentiation' of the cells and tissues. This point of view is emphasized in Fig. 15.9 which shows segments of cross sections through the

Table 15.1   Photoresponses of the mustard seedling, *Sinapis alba* L. (supplemented after Mohr, H., 1966b).

Inhibition of hypocotyl lengthening
Inhibition of translocation from the cotyledons
Enlargement of cotyledons
Unfolding of the lamina of the cotyledons
Hair formation along the hypocotyl
Opening of the hypocotylar ('plumular') hook
Formation of leaf primordia
Differentiation of primary leaves
Increase of negative geotropic reactivity of the hypocotyl
Formation of tracheary elements
Differentiation of stomata in the epidermis of the cotyledons
Formation of plastids in the mesophyll of the cotyledons
Changes in the rate of cell respiration
Synthesis of anthocyanin
Increase in the rate of ascorbic acid synthesis
Increase in the rate of chlorophyll accumulation
Increase of RNA synthesis in the cotyledons
Increase of protein synthesis in the cotyledons
Changes in the rate of degradation of storage fat
Changes in the rate of degradation of storage protein

hypocotyl of mustard seedlings. Under the influence of P_{730} certain cells of the epidermis have formed long hairs and all the cells of the subepidermal layer— but no other cells— have formed anthocyanin. It is obvious from these simple drawings that P_{730} functions only as a 'trigger'; the specificity of the response— e.g., hair formation or anthocyanin synthesis—must depend on the 'specific status of differentiation' of the cells and tissues at the moment when P_{730} is formed in the seedling. This specific status of differentiation may be called 'primary status of differentiation'. By this term we have been designating in connection with photomorphogenesis the pattern of differentiation of a plant before the first formation of P_{730} within this plant. This primary pattern of differentiation is not static; rather it is dynamic, that means, it will change with time. The causalities of 'primary differentiation' are virtually unknown in plant as well as in animal embryology. Phytochrome research can only help us to

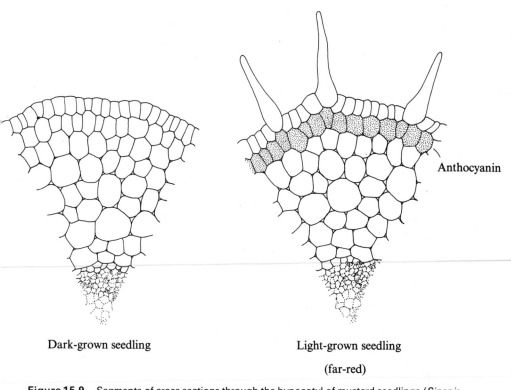

Anthocyanin

Dark-grown seedling   Light-grown seedling

(far-red)

Figure 15.9   Segments of cross sections through the hypocotyl of mustard seedlings (*Sinapis alba* L.). Left, from a dark-grown seedling; right, from a light-grown seedling. (After Wagner and Mohr, 1966c.)

understand certain aspects of 'secondary differentiation', namely those processes which are set in motion by P_{730}. We are dealing with an analogous situation in the case of hormone research in both plants and animals—and in human beings.

Positive, negative, and complex photoresponses

To approach the problem of phytochrome action further it proved to be useful to divide the many photoresponses of the mustard seedling into three distinct categories (Mohr, 1966a): positive, negative, and complex photoresponses. 'Positive' photoresponses are those which are characterized by an initiation or an increase of biosynthetic or growth processes. An example would be phytochrome-mediated synthesis of anthocyanin (Fig. 15.10). The long lag-phase after the first onset of light—about 3 hours—is characteristic of this type of photoresponse. (Remember that continuous far-red is supposed to maintain in

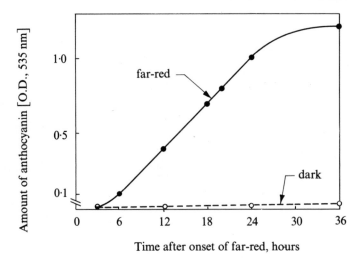

Figure 15.10   Time course of anthocyanin accumulation in mustard seedlings (*Sinapis alba* L.) under continuous far-red. The lag-phase is about 3 hours. Onset of far-red: 36 hours after sowing. (After Mohr, 1966b.)

the seedling a low but stationary concentration of P_{730} over a considerable length of time. The photostationary state is virtually established within minutes after the onset of far-red.) 'Negative' photoresponses are those which are characterized by an inhibition of growth processes or other physiological processes like translocation. In hibition of hypocotyl lengthening (Fig. 15.11) is a typical response of this sort. A characteristic feature of a negative photoresponse is the short lag-phase. In the present case we see that within an hour after the onset of far-red irradiation a new steady-state rate of lengthening is established. The corresponding lag-phase is probably less than 30 min. 'Complex' photoresponses are those which are characterized during the first part of the kinetics by an inhibition and later by a promotion of the response compared with the corresponding dark controls. The control by phytochrome of the rate of O_2 uptake of the cotyledons is an example of this sort of a photoresponse (Fig. 15.12) Complex photoresponses can possibly be explained—as we have concluded in the case of O_2 uptake and fat degradation (Hock, Kühnert, and Mohr, 1965)—by an interaction of positive and negative photoresponses. As a matter of fact we are left with positive and negative photoresponses which are different in all respects except that both types are mediated by P_{730}.

Phytochrome and differential gene activity: a unifying hypothesis

We now again raise the question: How does P_{730} act? We have good arguments and some experimental evidence (e.g., Mohr, 1966a, b, Schopfer, 1967a, b)

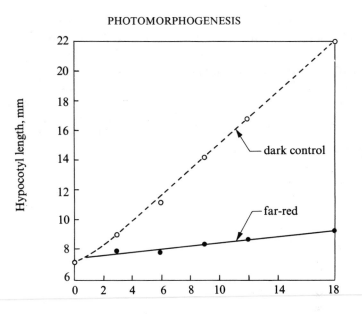

Time after onset of far-red, hours

Figure 15.11 Inhibition of hypocotyl lengthening of the mustard seedling (*Sinapis alba* L.) as a typical example of a 'negative' photoresponse. The lag-phase after the onset of far-red is short. Onset of far-red: 36 hours after sowing. (After Mohr, 1966b.)

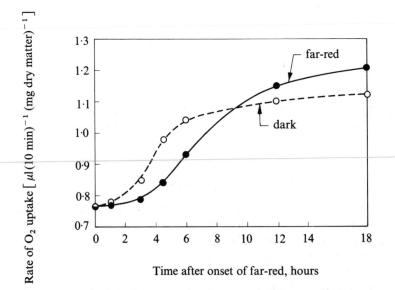

Time after onset of far-red, hours

Figure 15.12 Control by phytochrome of the rate of O_2 uptake of the cotyledons of the mustard seedling (*Sinapis alba* L.) as a typical example of a 'complex' photoresponse. Onset of far-red: 36 hours after sowing. (After Mohr, 1966b.)

that the hypothesis illustrated in Fig. 15.13 describes in principle the action of P_{730} on morphogenesis in a developing plant. It is now generally understood that all living cells of a particular plant—except perhaps the sieve tube elements —contain the total complement of genetic DNA that is characteristic for the individual; in other words, all the genes are present in all cells but only a fraction of them are active at a given time in a given cell. Genes are thus turned 'off' and 'on'. Such differential gene activity results in differences among cells that have the same set of genes. In connection with photomorphogenesis we have to

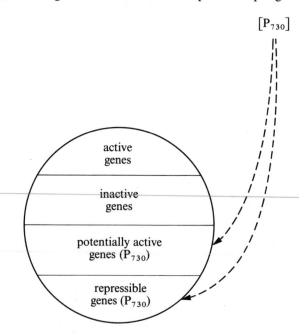

Figure 15.13   This scheme illustrates the hypothesis of differential gene activation and differential gene repression through P_{730}. (The hypothesis is explained in some detail in the text). (After Schopfer, 1967b.)

refine this general scheme in the following way: The total genes of each particular cell of a dark-grown seedling which is able to respond to P_{730} must be divided into at least four functional types: active, inactive, potentially active, and repressible genes. Active genes are those which function the same way in an etiolated plant as they do in the light-grown plant; inactive genes are active neither in the dark-grown seedling nor in the seedling exposed to light (e.g., flowering genes). Potentially active genes with an index P_{730} are those which are ready to function and whose activity can be started or increased in some way by P_{730}. The activation of potentially active genes leads to 'positive' photoresponses. Repressible genes with an index P_{730} are those which can be repressed by P_{730}. The repression of repressible genes leads to 'negative' photoresponses. Which genes are active, inactive, potentially active or repressible in a particular cell at a particular moment is determined by those regulating factors which determine the dynamic pattern of primary differentiation in a multicellular seedling. The nature of these regulating factors of primary differentiation is virtually unknown in both plant and animal embryology.

This sort of reasoning is supported by the evidence available at the moment. Figure 15.14 can serve as a general justification of the hypothesis. We recognize that, for example, an epidermal cell can respond to P_{730} with the formation of a hair, a 'positive' photoresponse, as well as with the reduction of the rate of

Hypocotyl
(longitudinal section)

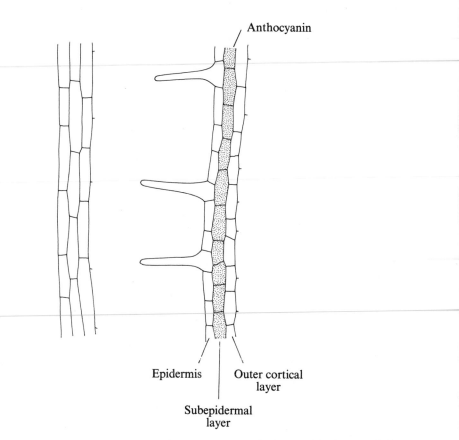

Anthocyanin

Epidermis | Outer cortical
layer

Subepidermal
layer

Figure 15.14 These drawings represent the three outer cellular layers of the hypocotyl of a mustard seedling (*Sinapis alba* L.). Left, from a dark-grown seedling; right, from a light-grown seedling. (Adapted from Häcker, unpublished observations.)

lengthening, a 'negative' photoresponse. A subepidermal cell can respond with the formation of anthocyanin as well as with the reduction of lengthening. If we include the common 'credo' of molecular biology, gene-mRNA-protein, in our reasoning the hypothesis illustrated in Fig. 15.13 can be subjected to experimental verification at the molecular level. A great number of pertinent experiments have been performed (e.g., Mohr, 1966a; Schopfer, 1967a, b), which support the hypothesis.

Control of RNA and protein synthesis by phytochrome

If the function of P_{730} is connected with a de-repression of certain genes one would expect that RNA and protein synthesis would be activated if P_{730} has been formed in a system—like the mustard seedling—which shows predominantly 'positive' photoresponses. Table 15.2 shows that both the RNA and the protein content of a whole mustard seedling increase under the influence of P_{730}. In these 'pioneer' experiments repeated 'short-time irradiations' were used.

Table 15.2   Action of phytochrome on RNA and protein synthesis of whole mustard seedlings, *Sinapis alba* L. (after Weidner, Jakobs, and Mohr, 1965).

Programme of irradiation	
'Red'	36 h dark + 6 × (5 min red + 355 min dark)
'Far-red'	36 h dark + 6 × (5 min far-red + 355 min dark)
'Reversion'	36 h dark + 6 × (5 min red + 5 min far-red + 350 min dark)
'Control'	72 h dark

Percentage increase of RNA and protein contents above control

	'Red'	'Far-red'	'Reversion'
RNA	19·5	10·0	7·3
Protein	10·1	3·4	4·4

Under these conditions it will be remembered that, if phytochrome is involved, an induction with red can be reversed by far-red down to about the level which is induced by far-red itself. Short-time irradiation can give us only the information that phytochrome is involved. In order to understand the changes in RNA and protein more in detail one has to run kinetic experiments under steady state conditions of phytochrome in a system which responds 'positively' to phytochrome, such as cotyledons. The results can be summarized as follows: Histochemical studies have shown that P_{730}—maintained in the plant by continuous far-red irradiation—enhances the degradation of storage protein in the cotyledons; at the same time, however, P_{730} will stimulate in the cotyledons a strong *de novo* synthesis of soluble and structural protein. The lag-phase seems to be less than 3 hours (Häcker, 1967, Jakobs and Mohr, 1966). Part at least of the structural protein can be identified as 'plastids'. These proteinaceous bodies are hardly distinguishable from chloroplasts under the light microscope except that they do not contain chlorophyll. The electron microscope reveals that the plastids formed under continuous far-red are homologous to chloroplasts. Only the final arrangements of the thylakoids as grana seem to require the formation of the photosynthetic pigments (Häcker, 1967). The strong phytochrome-mediated increase of RNA content in the cotyledons after a lag-phase of about 6 hours (Weidner and Mohr, 1967a) comprises probably all fractions of the RNA. ^{14}C-uridine was applied to the seedlings for a number of hours (e.g., 3 or 6 hours) either in the dark or under continuous far-red radiation. The RNA was extracted and fractionated on MAK* columns. All seven fractions of RNA which can be identified are more or less homogeneously labelled with ^{14}C in the dark as well as in far-red light. However, the degree of labelling, that is, the specific activity based on UV absorption, is higher in seedlings irradiated with far-red (Dittes, unpublished results).

Obviously P_{730} causes an increase of RNA synthesis even during that period after the onset of far-red light—about 6 hours—where there is no net increase of RNA content. This conclusion is confirmed by experiments on total RNA which show that P_{730} increases the specific activity of the total RNA in the mustard cotyledons to a remarkable extent (31 per cent) within 3 hours (Weidner and Mohr, 1967a). Within the period of experimentation which has been used in connection with RNA and protein synthesis (usually 24 hours), phytochrome has hardly any influence on the DNA content of the mustard cotyledons and hypocotyls (Weidner, 1967). It seems that control by P_{730} of replication of DNA does not have to be taken into account in trying to understand the 'molecular biology' of photomorphogenesis in the mustard seedling.

Phytochrome-mediated anthocyanin synthesis a 'positive' photoresponse in mustard seedlings (*Sinapis alba* L.)

A dark-grown etiolated mustard seedling will form hardly any anthocyanin whereas under continuous far-red, for example, the seedling will synthesize a large amount of the red pigment (Fig. 15.10). Conventional induction-reversion-experiments have been done previously. They demonstrate that the response is phytochrome controlled (Mohr, 1957). We find (Fig. 15.10) that for at least 20 hours after the initial lag-phase the accumulation of anthocyanin is linear with time. One may conclude that over this period of time a virtually stationary concentration of P_{730} will lead to a constant rate of anthocyanin synthesis. The final termination of anthocyanin accumulation is independent of the level which has been reached. It is a function of time or, in other words, it is a matter of the

* MAK = Methylated Albumin coated on Kieselgur.

changing pattern of differentiation which we cannot discuss yet in molecular terms. We may describe this progressive differentiation as 'primary differentiation' as well, because P_{730} has, as far as we know at the moment, no influence on the pattern and time course (Wagner and Mohr, 1966c).

If the hypothesis that phytochrome-induced anthocyanin synthesis is due to a differential gene activation is justified, the effect of phytochrome on anthocyanin synthesis should be modified in a predictable manner by specific inhibitors, e.g., by Actinomycin D, a potent inhibitor of 'transcription' (i.e., DNA-dependent RNA synthesis), or by Puromycin, an antibiotic which specifically inhibits protein synthesis. This is indeed the case (Lange and Mohr, 1965, Mohr and Senf, 1966, Schopfer, 1967c). Only a few data can be mentioned (Fig. 15.15). When Actinomycin D is added before the onset of far-red light no

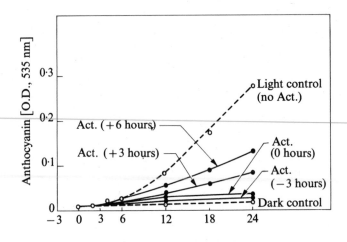

Time after onset of far-red, hours

Figure 15.15 Synthesis of anthocyanin in the mustard seedling (*Sinapis alba* L.) under the influence of continuous far-red and Actinomycin D. The concentration of the antibiotic in the solution around the seedling is not critical. One gets in principle similar curves with 200 μg ml^{-1} and 10 μg ml^{-1}. Act (-3 hours) means that Actinomycin D was applied 3 hours before the onset of far-red. (After Lange and Mohr, 1965.)

anthocyanin synthesis will occur. If, however, the antibiotic is added during or after the lag-phase, e.g., 6 hours after the onset of light, anthocyanin synthesis will proceed in essentially the same way as in the controls but with reduced intensity. These facts show that the total inhibition is not due to a non-specific effect of the Actinomycin D. Obviously the seedlings can synthesize anthocyanin in the presence of the antibiotic if this synthesis is already under way at the moment when the substance is added to the system. For a number of further reasons we had to conclude (Lange and Mohr, 1965; Mohr, Schlickewei, and Lange, 1965; Mohr and Bienger, 1967) that Actinomycin D at a suitable dose will block the formation of new kinds of mRNA but still permit the continued synthesis of mRNA already in production at the time the antibiotic is added. In the terminology of Fig. 15.13: Actinomycin D at relatively low concentrations can easily block potentially active genes but it can hardly influence those genes which are actively engaged in transcription.

In any case the facts (Mohr and Bienger, 1967) can hardly be interpreted without the assumption that the sensitivity of certain genes towards Actinomycin D is irreversibly altered under the influence of P_{730}.

No destructive influences of a general nature have ever been observed in these experiments with Actinomycin D. The seedlings look 'healthy' except that the 'positive' photoresponses are more or less depressed, not only anthocyanin synthesis but also such characteristic positive photoresponses as cotyledon enlargement (Mohr, Schlickewei, and Lange, 1965), opening of the plumular hook or unfolding of the lamina of the cotyledons (Schopfer, 1967a). Other biosynthetic processes, like carotenoid synthesis, which take place readily in darkness are not inhibited by the Actinomycin D treatment.

In the case of Puromycin the effect on phytochrome-mediated anthocyanin synthesis is independent of the time of application (Lange and Mohr, 1965). If Puromycin is applied at a concentration of 100 μg ml^{-1} in the solution around the seedlings after 12 hours of far-red irradiation, it is observed that anthocyanin synthesis continues at an unaltered rate for several hours, but it virtually stops after about 5 hours even in the case where far-red irradiation is continued (Mohr and Senf, 1966). This fact means that at least one of the enzymes which are involved in anthocyanin synthesis decays below a critical level during this period.

Initial lag-phase and secondary lag-phase in phytochrome-dependent anthocyanin synthesis. The initial (or primary) lag-phase after the onset of far-red is always in the order of 3 hours (Fig. 15.10). If, however, a seedling which was pre-irradiated (e.g., with 12 hours of far-red) is kept in darkness for an extended period and is then re-irradiated with far-red no appreciable lag-phase for the action of the second irradiation can be detected even when anthocyanin synthesis had already ceased during the preceding dark period (Fig. 15.16). Since the action of the second irradiation as measured by anthocyanin synthesis can be completely inhibited by low doses of Puromycin and Cycloheximide (a potent inhibitor of protein synthesis) we conclude that the re-appearance of P_{730} leads to *de novo* synthesis of enzyme protein which is required for a resumption of anthocyanin synthesis. Application of Actinomycin D (10 μg ml^{-1}) only partially inhibits the action of the second irradiation as measured by anthocyanin synthesis. This result was to be expected. The pertinent data were discussed in the preceding chapter.

The tentative interpretation of Fig. 15.16 can be summarized as follows: P_{730} exerts two functions during the initial lag-phase. First, it eliminates a 'barrier' around the potentially active genes in so far as it makes these genes accessible for the activating action of P_{730}. Second, P_{730} starts the activity of potentially active genes. To maintain gene activity the continuous presence of P_{730} is required. On the other hand, if a gene has once been 'opened' to the action of P_{730} it remains easily accessible for the activating action of P_{730} even in the case when P_{730} has disappeared and gene activity ceased for an extended period of time.

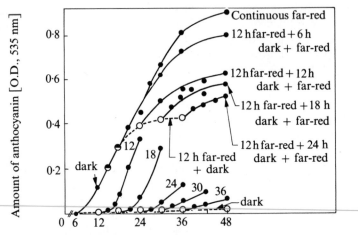

Time after initial onset of far-red, hours

Figure 15.16   Initial lag-phases (lower part) and secondary lag-phases (upper part) of far-red mediated anthocyanin synthesis. To determine the secondary lag-phases the mustard seedlings were irradiated for 12 hours with far-red, placed in darkness and re-irradiated with far-red after dark intervals of different duration as indicated in the figure. The numbers (lower part of the figure) indicate the time of onset of far-red light. (After Lange, Bienger, and Mohr, 1967.)

Phytochrome-mediated enzyme synthesis

The direct measurement of P_{730}-mediated enzyme synthesis would be a desirable verification of the hypothesis illustrated in Fig. 15.13. Pertinent experiments have been performed using phenylalanine ammonia-lyase (E C 4.3.1.5) which catalyses the formation of trans-cinnamic acid from phenylalanine. This reaction is probably the first step of the biosynthetic sequence which leads from phenylalanine—which must be regarded as the main storage compound of phenolic nature in the mustard cotyledons—to the phenyl-propane-moiety of the flavonoids, including the anthocyanidins. Phenylalanine ammonia-lyase activity can be induced by phytochrome in the mustard seedling (Fig. 15.17). The initial lag-phase after the onset of far-red is 1·5 hours, then the increase of activity is sharp and rapid, and the activity remains high for some time, decreasing later (Durst and Mohr, 1966). If we accept the 'credo' DNA-mRNA-enzyme, the sudden and linear increase of enzyme activity means that the mRNA involved has a rapid turnover since a steady-state concentration seems to be established within a short period of time. Phenylalanine ammonia-lyase is not the rate-limiting enzyme of phytochrome-mediated anthocyanin synthesis, for the following reason: the linear rate of anthocyanin synthesis is established within 6 hours after the onset of far-red (Fig. 15.10); phenylalanine ammonia-lyase activity, on the other hand, increases over at least 12 hours after the onset of far-red. Because the rate of anthocyanin synthesis is phytochrome-dependent

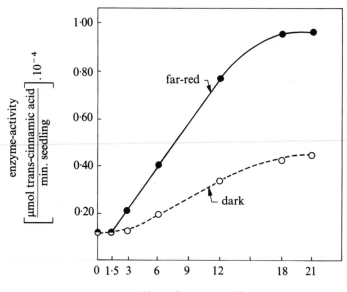

Time after onset of far-red, hours

Figure 15.17 The control by far-red light of the activity of phenylalanine ammonia-lyase (E C 4.3.1.5.) from mustard seedlings (*Sinapis alba* L.). (After Rissland and Mohr, 1967.)

during the linear phase, phenylalanine ammonia-lyase cannot be the rate-limiting enzyme under these conditions. The same phenomenon with respect to initial and secondary lag-phases which has been described in connection with phytochrome-mediated anthocyanin synthesis (Fig. 15.16) has been observed with phenylalanine ammonia-lyase as well (Fig. 15.18). The initial (or primary) lag-phase after the onset of far-red is always 1·5 hours. If, however, a seedling which has been pre-irradiated with 12 hours of far-red is kept in darkness for 6 hours and is then re-irradiated with far-red, no lag-phase for the action of the second irradiation can be measured. Since the action of the second irradiation, as measured by increase of enzyme activity, can be completely inhibited by relatively low doses of Puromycin and Cycloheximide we conclude (Rissland and Mohr, 1967) that the re-appearance of P_{730} leads to *de novo* synthesis of enzyme protein. Application of Actinomycin D (10 μg ml^{-1}) only partially inhibits the action of the second irradiation, as one would expect. The problem of how the initial lag-phase can be understood has been dealt with in more detail in the preceding section on anthocyanin synthesis.

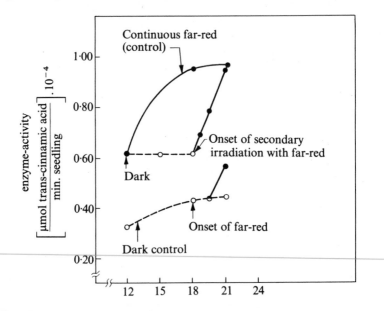

Time after initial onset of far-red, hours

Figure 15.18   Initial and secondary lag-phases of far-red mediated increase of phenylalanine ammonia-lyase activity in the mustard seedling (*Sinapis alba* L.). To determine the secondary lag-phase the seedlings were irradiated for 12 hours with far-red, placed in darkness for 6 hours and re-irradiated with far-red. The data of this figure cannot be compared except in principle with the preceding Fig. 15.17 since a different sample of seed was used. The response of the mustard seedling with respect to far-red induced phenylalanine ammonia-lyase activity depends to some extent on the age after harvest of the seed population. (Supplemented after Rissland and Mohr, 1967.)

Other enzymes which have been studied in the mustard seedling in connection with photomorphogenesis are not under the control of phytochrome, e.g., isocitritase (E C 4.1.3.1) (Karow and Mohr, 1967). P$_{730}$ has no influence on the activity of isocitritase in the mustard seedling, although the enzyme activity does show a strong increase and a following decline during the period of experimentation.

The behaviour of isocitritase as compared with that of phenylalanine ammonia-lyase supports our view that photomorphogenesis is not due to a general metabolic change in the cells; it rather seems to be due to very specific changes on the enzyme level. The bulk of enzyme activities in the cells is probably not affected by phytochrome.

Some difficulties

All findings we have been discussing so far are consistent with the hypothesis that in connection with positive photoresponses P_{730} acts through differential gene activation. There are positive photoresponses, however, which are more difficult to interpret. A characteristic photoresponse of the mustard cotyledons is the accumulation of ascorbic acid (Schopfer, 1966). The cotyledons synthesize some ascorbic acid in darkness, too, but the content of ascorbic acid is rapidly and enormously increased under the influence of P_{730}. The lag-phase after the onset of far-red is about 1·5 hours. What happens to the phytochrome-mediated increase in the rate of synthesis of ascorbic acid if we apply Actinomycin D at a concentration which virtually blocks phytochrome-mediated RNA increase in the cotyledons (Weidner and Mohr, 1967b) or phytochrome-mediated 'positive' photoresponses like anthocyanin synthesis, cotyledon enlargement, unfolding of the lamina, or opening of the plumular hook? Table 15.3 shows that

Table 15.3   Content of ascorbic acid (AS) in the cotyledons of mustard seedlings (*Sinapis alba* L.), 24 hours after the onset of far-red. Concentration of Actinomycin D (Act) in the solution around the seedlings: 10 μg ml^{-1} (adapted from Schopfer, 1967c).

Treatment	Concentration of AS n mol per 2 cotyledons
Dark	16·9
Dark + Act (−6 h)*	19·7
Far-red	35·2
Far-red + Act (−6 h)*	35·7

* (−6 hr) means that Actinomycin D was applied 6 hours before the onset of far-red.

Actinomycin D does not markedly influence the effect of P_{730} on ascorbic acid synthesis. Carotenoid synthesis in the mustard seedling responds to P_{730} in much the same way as ascorbic acid synthesis. The initial lag-phase, however, is 3 hours (Fig. 15.19). Actinomycin D (10 μg ml^{-1}) applied before the onset of exposure to far-red light, does not inhibit carotenoid synthesis, either in the dark or under the influence of far-red light. On the other hand carotenoid synthesis in the dark and under far-red light in the mustard seedling is very sensitive towards Cycloheximide, a potent inhibitor of protein synthesis.

An attractive interpretation of these results is that the genes which are involved in ascorbic acid (or carotenoid) synthesis are insensitive towards Actinomycin D. Ascorbic acid and carotenoids are already synthesized in darkness; all pertinent genes are thus active to some extent in darkness. We have discussed in connection with Fig. 15.15 that active genes are far less sensitive against Actinomycin D than genes which are not yet active. If we increase the concentration of Actinomycin D, synthesis of ascorbic acid and carotenoids will be depressed in light as well as in darkness (Schopfer, 1967c). The question remains of how P_{730} can regulate the *rate of activity* of one or several genes which are already active. Why is there a considerable initial lag-phase if the genes are already active? It is obvious that we need more factual information before a satisfactory interpretation can be advanced.

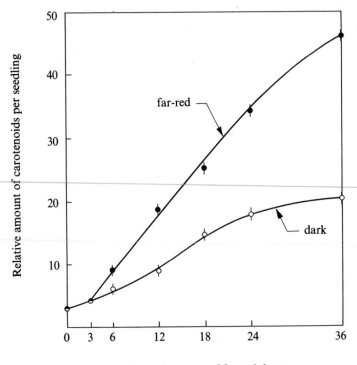

Figure 15.19   Time courses of carotenoid accumulation in the mustard seedling (*Sinapis alba* L.) in darkness and under the influence of far-red (i.e., under the control of P_{730}). After an initial lag-phase of 3 hours far-red increases the rate of carotenoid accumulation and it prolongs the period of synthesis. (After Schnarrenberger and Mohr, 1967b.)

Phytochrome-mediated germination of lettuce seeds is the classical example of a phytochrome-mediated photoresponse. It is sometimes believed that P_{730} may set in motion a mechanism which causes the formation or activation of hydrolytic enzymes. These weaken the seed coat to such an extent that the radicle can break through. This sort of argument fits nicely into the hypothesis that P_{730} activates potentially active genes. The data obtained with inhibitors, however, do not verify this idea (Schopfer, 1967a). Neither Actinomycin D nor Puromycin disturb the process of dark or of phytochrome-mediated germination. If we exclude the possibility that the antibiotics cannot enter the seed (Schopfer, 1967a) we have to conclude that apparently all enzymes which are needed for the germination process are already present in the dormant seed. We have no real idea at the moment of how P_{730} can make a lettuce seed germinate. Obviously we have to bear in mind the possibility that there are several sites or levels (e.g., transcription *and* translation) where P_{730} can influence the activity of a cell, just as there are several doors which can be opened by one and the same key. In any case, however, P_{730} can only be regarded as a 'trigger'. The specificity of the response of a particular cell is determined by the 'specific status of differentiation' at the moment when P_{730} appears. It is not determined by phytochrome.

This might be the best place to mention that the function of phytochrome in a *Mougeotia* cell (Haupt, 1965), in a filamentous fern sporeling (Etzold, 1965; Hartmann, Menzel, and Mohr, 1965) or in the pulvini of *Mimosa pudica* L. (Fondeville, Borthwick, and Hendricks, 1966; Fondeville, Schneider, Borthwick, and Hendricks, 1967) may not be connected with differential gene activity. In the case of the photonastic movement of the fully differentiated *Mimosa* pulvini it has been suggested that phytochrome action is on membrane permeability. These responses, however, are beyond the scope of the present article. At the moment we are trying to find a 'molecular' explanation for the problem of how P_{730} can control *differentiation* and *morphogenesis*. In this case we have to include differential gene activity in our considerations. It does not seem to be satisfactory to discuss the problem of how P_{730} may act in the cell only in connection with the 'primary (or initial) action of P_{730}'. By this term is understood the very first reaction in which P_{730} is involved after its formation. This reaction is completely unknown. Maybe the assumption is justified that this reaction is the same in all cells. (One can hold the position that phytochrome acts differently in different compartments.) In any case, if we desire to understand the causalities which govern the processes between P_{730} and the great number of final photoresponses in a seedling, we can hardly neglect the conception of differential gene activity. We must remember (Fig. 15.9) that the *specificity* of the phytochrome-mediated response does not depend on phytochrome but does depend on the 'specific status of differentiation' of the cells and tissues at the moment when P_{730} is formed in a seedling.

Phytochrome-mediated control of hypocotyl lengthening: a 'negative' photoresponse in mustard seedlings (*Sinapis alba* L.)

We choose phytochrome-mediated control of hypocotyl lengthening in the mustard seedling (Fig. 15.11) as a prototype of a negative photoresponse for a number of reasons. First, the response of the hypocotyl—namely, longitudinal growth—can be studied with high accuracy under well-established steady-state conditions of growth. The growth velocity is virtually constant for a satisfactory length of time in darkness as well as under the influence of a presumably stationary concentration of P_{730} (Fig. 15.11). It should be emphasized perhaps that steady-state conditions are not just an advantage but a prerequisite for a causal analysis of growth phenomena. It is obvious that any adequate kinetic analysis for interactions between physiological variables must be based on a steady-state process (Lockhart, 1965). Second, hypocotyl growth of the mustard seedling is—under our experimental conditions—almost exclusively due to cellular lengthening. There is virtually no increase in DNA content of the whole organ and hardly any more cell divisions occur, at least not in the cortex tissue which represents the larger part of the hypocotyl tissue (Fig. 15.9). This means, by the way, that many responses of the hypocotyl, like phytochrome-induced changes of RNA or protein synthesis, can be attributed without any gross error to the more or less homogeneous cortex tissue. Third, control of hypocotyl lengthening by phytochrome is independent of the absolute rate of growth and independent of whether the cotyledons are attached to the seedling or not (Table 15.4). Obviously phytochrome and the factors of growth which are

Table 15.4 The control of hypocotyl lengthening by far-red is independent of the absolute growth rate and of whether or not the cotyledons are attached to the axis system (seedlings of *Sinapis alba* L.) (after Schopfer, 1967b).

Hypocotyl lengthening	Dark	Far-red
Growth rate (mm h^{-1}):		
intact seedlings	0·79	0·14
seedlings without cotyledons	0·39	0·07
Inhibition by far-red (%):		
intact seedlings	0	82
seedlings without cotyledons	0	81
Inhibition due to the lack of cotyledons (%)	54	50

supplied by the cotyledons do not show any significant interaction in the mustard seedling. This fact gives us the opportunity to study the phytochrome-dependent responses of the axis in a system without cotyledons. In this way any translocation from the storage cotyledons to the axis is avoided and the way is open to correlate more strictly the growth response of the hypocotyl with chemical (or 'molecular') changes in this organ.

The hypothesis has been advanced (Schopfer, 1967b) that 'negative' photo-responses are the consequence of a differential gene repression which is exerted in some way or another by P_{730} (Fig. 15.13). In the special case of hypocotyl lengthening this hypothesis would mean that P_{730} very quickly represses a gene or genes, the function of which is essential for cell lengthening. We have to assume if we follow this model that mRNAs and growth rate-limiting enzymes which are coded by these repressible genes have a short lifetime. Remember that the lag-phase of the morphological response is probably less than 30 min (Fig. 15.11).

It seems that all experimental data which are available at the moment are in fact consistent with the hypothesis of differential gene repression through P_{730}. Since, however, direct evidence (enzyme repression by P_{730}) has still to be demonstrated in connection with control of hypocotyl lengthening the hypothesis is not very convincing. Only a few data are mentioned.

The total RNA content of the hypocotyl is strongly depressed by far-red (Weidner and Mohr, 1967a). It is obvious that total RNA content and longitudinal growth of the organ are not correlated. While the length of the hypocotyl increases under continuous far-red with a low but constant rate (Fig. 15.11) the RNA content of this organ will even decrease slightly below the original value under the influence of P_{730}. It seems that growth velocity of the hypocotyl is not limited by bulk RNA. It rather seems to be related to a small fraction of the total RNA, possibly mRNA.

RNA synthesis is much more sensitive to Actinomycin D in the hypocotyl than in the cotyledons. It seems that Actinomycin D at a moderate concentration ($10\mu g$ ml^{-1} in the solution around the seedling) can block net RNA synthesis completely in the hypocotyl while some RNA synthesis is still going on in the cotyledons (Weidner and Mohr, 1967b). One can argue that on the average the genes are more sensitive towards Actinomycin D in the hypocotyl than in the cotyledons.

To correlate more strictly the growth response of the hypocotyl with 'molecular' changes in this organ the axis system without cotyledons (Table 15.4) has been used (Mohr, Holderied, Link, and Roth, 1967). Even under these conditions the growth velocity of the hypocotyl is nearly constant in light (continuous far-red) and dark during the whole period of experimentation. Since the number of cells per hypocotyl is virtually constant, hypocotyl lengthening is almost exclusively due to cellular lengthening. If we follow the protein content of the hypocotyl we find that the total protein of the organ decreases steadily in spite of the fact that the organ grows at a constant velocity. There is no significant difference in protein contents between dark-grown and far-red grown hypocotyls although the growth velocities differ by a factor of 4. The situation is somewhat different with respect to total RNA. The RNA content eventually decreases in far-red as well as in dark-grown hypocotyls but the decrease is significantly faster in the far-red treated systems than in the dark controls. It has been concluded that only a very small part of the total RNA and the total protein of a cell can be considered as being related to the control of cellular growth. Changes in bulk RNA and bulk protein obviously do not necessarily reflect changes in the growth velocity or growth capacity of an organ or a cell.

K. Roth and W. Link have performed a great number of unpublished experiments, using ^{14}C-uridine and ^{14}C-leucine, to demonstrate a P_{730}-dependent decrease of the specific activity of total RNA and total protein within 30 min after the onset of far-red. P_{730}, however, does not exert any measurable influence on total uptake of radioactivity or on the linear incorporation of the label into RNA or protein. One explanation of the results is that the fraction of RNA and protein which is thought to be involved in P_{730}-mediated control of hypocotyl lengthening is so small that it cannot be detected this way.

Schopfer (1967b) has elaborated the action of far-red, Actinomycin D, and Puromycin on hypocotyl lengthening in the mustard seedling. Under steady-state conditions of hypocotyl lengthening far-red and Actinomycin D lower the growth velocity of the organ. The inhibitory action of Actinomycin D (on a percentage basis) is the same with and without far-red. This fact means that Actinomycin D and P_{730} act as two independent factors in a multiplicative system. Likewise Puromycin lowers the growth velocity of the hypocotyl. The inhibitory action of this substance (on a percentage basis) is the same with and without far-red. This fact means that Puromycin and P_{730} act as two independent factors in a multiplicative system. The interpretation of Schopfer's results is not unambiguous. In any case much more work is required to support the hypothesis (Fig. 15.13) that 'negative photoresponses' are due, in general, to a differential gene repression by P_{730}. Clearly, evidence must be presented eventually at the level of repressible enzymes.

Recent results obtained by H. Karow with the enzyme lipoxidase look very promising. The increase of lipoxidase activity in the mustard seedling is arrested immediately after the onset of far-red. Since the light effect can be shown by conventional induction-reversion experiments to be phytochrome-mediated, one may conclude that we are dealing with a specific repression of enzyme synthesis by P_{730}.

Phytochrome-mediated photomorphogenesis: a summary

We return to the question asked at the beginning of the article when we tried to find an approach to the problem of photomorphogenesis. The first question was: What type of pigment absorbs the effective radiation? The answer is phytochrome—at least above 550 nm. The second question was: What kinds of photochemical reactions occur? The answer is that P_{730}, the physiologically effective species of phytochrome, is formed. The third question can now be put in the following way: What are the causal relationships between P_{730} and the final photoresponses? The answer is that P_{730} exerts a 'multiple action'. Some of the 'positive' photoresponses can be understood in terms of differential gene activation through P_{730}. 'Negative' photoresponses may possibly be understood eventually in terms of differential gene repression. The final question we asked was the following: How can the integration of the different photoresponses of a seedling be understood? The answer here is that the status of 'primary differentiation' of a cell or a tissue determines the type of response which occurs when P_{730} appears in the plant. The extent of the response of a particular tissue or organ is determined, however, more or less by the interaction with other parts of the plant. The kinetics of 'complex' photoresponses in a seedling (Fig. 15.12) demonstrate very well the correlative integration of the different organs and tissues during photomorphogenesis.

Blue-light dependent photoresponses

Introduction

Both species of phytochrome, P_{660} as well as P_{730}, absorb in the blue and ultra-violet regions of the spectrum. The absorption coefficients and the relative quantum efficiencies of the photoconversions are much lower than in the red or far-red regions of the spectrum but, nevertheless, photoequilibria are established if the irradiation is continued for some time (Butler, Hendricks, and Siegelman, 1964; Pratt and Briggs, 1966). The ratio $[P_{730}]/[P]$, as measured *in vivo* (Pratt and Briggs, 1966), seems to be between 1 per cent and 35 per cent, depending on the particular wavelength. In any event P_{730} will be formed under the influence of blue and near ultraviolet light. There are, however, a considerable number of blue-light dependent photoresponses where an explanation on the basis of phytochrome seems to be excluded. We restrict our description to four clear-cut examples.

Light-dependent carotenoid synthesis in *Fusarium aquaeductuum* (Radlk. et Rabenh.) Lagerheim

Carotenoid synthesis in this system can be regarded as a prototype of a photo-response—including morphogenic responses—in fungi. The action spectrum (Fig. 15.20) shows that carotenoid synthesis is initiated only by light below 520 nm. The action spectrum has maxima at 375–380 nm and 450–455 nm, one shoulder at 430–440 nm and a further shoulder (or possibly a third maximum) between 470 and 480 nm. From this action spectrum carotenoids can be ruled out as possible photoreceptors. The spectrum rather resembles the absorption spectra of certain flavoproteins. Consequently the acting photoreceptor is thought to be a flavoprotein (Rau, 1967). The causalities between the light absorption and the carotenoid accumulation are only poorly understood so far. The assumption, however, seems to be justified that the light absorption causes the synthesis of enzymes which are required for carotenoid synthesis (Rau, 1967).

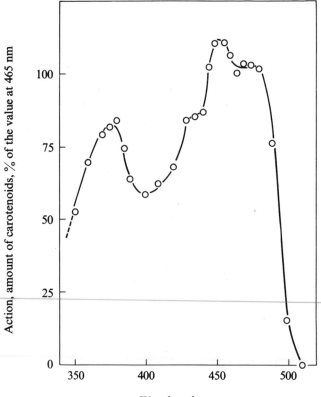

Wavelength, nm

Figure 15.20 An action spectrum of light-dependent carotenoid synthesis in the fungus *Fusarium aquaeductuum*. The amount of carotenoids which can be induced by $4 \cdot 2 \times 10^{-7}$ Einstein cm^{-2} at each wavelength is given as a function of wavelength. (After Rau, 1967.)

Polarotropism in filamentous gametophytes (= sporelings) of the liverwort *Sphaerocarpus donnellii* Aust.

Figure 15.21 illustrates the basic phenomena of polarotropism in this particular system. The sporelings will grow at an angle of 90° to the plane of vibration of the electrical vector of the linearly polarized light which is applied from above. If the plane of vibration is turned, the sporelings will rapidly and correspondingly change their direction of growth. The action spectrum of the polarotropic response (Fig. 15.22) shows that only short-wavelength light is effective. The details of the spectrum strongly suggest a flavoprotein as a photoreceptor. The photoreceptor molecules must be highly orientated in a dichroic structure presumably close to the surface of the cell.

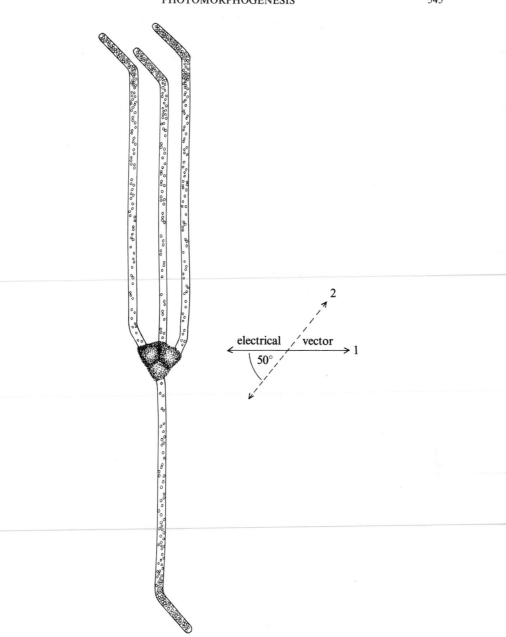

Figure 15.21 A basic phenomenon of 'polarotropism'. Linearly polarized light applied normal to the surface can strictly determine the direction of growth of a filamentous liverwort sporeling (*Sphaerocarpus donnellii* Aust.). The sporelings will grow normally to the plane of vibration of the electrical vector. (After Steiner, unpublished photographs.)

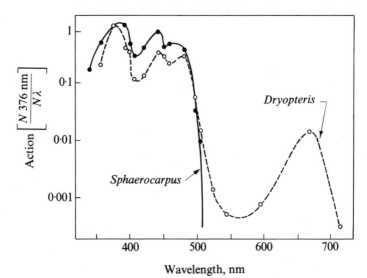

Figure 15.22   Action spectra for polarotropism in germlings of a fern (*Dryopteris filix-mas* (L.) Schott.) and a liverwort (*Sphaerocarpus donnellii* Aust.). The action spectrum for the protonema of *Dryopteris* was calculated from dose response curves at the constant irradiation time of 240 min for the response angle of 17·5°. The action spectrum for the germ tubes of *Sphaerocarpus* was calculated from dose response curves obtained at the constant irradiation time of 180 min for the response angle of 22·5°. Both angles represent 50 per cent maximum response. (After Steiner, 1967.)

The light-growth response in the sporangiophore of the fungus *Phycomyces* spec.

When sporangiophores are irradiated symmetrically about their long axis and at constant intensity they grow at a uniform rate. This rate is the same regardless of the absolute intensity. If, however, the intensity is changed the rate of elongation changes transiently and finally returns to the initial value. This phenomenon has been called 'light-growth response'.

Delbrück and Shropshire (1960) determined an action spectrum of the light-growth response using a null measuring system in which the spectral sensitivity was obtained by comparing the sensitivity at any wavelength to the sensitivity for a standard blue source. The symmetry of the irradiation was assured by rotating the specimen continuously at 2 rev min^{-1}. This speed is sufficiently fast to eliminate tropic effects and sufficiently slow to avoid mechanical problems. The action spectrum of the light-growth-response (Fig. 15.23) shows that the photoreceptor has absorption maxima around 280, 385, 455, and 485 nm. The conclusion seems to be justified that the photoreceptor involved might be a flavoprotein. However, despite many efforts the causalities at the 'molecular' level of the light-growth response are still unknown (Shropshire, 1963).

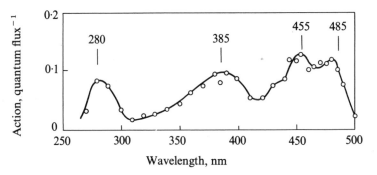

Figure 15.23   An action spectrum for light growth response in *Phycomyces* sporangiophore longitudinal growth. For each wavelength measured the ordinate gives the reciprocal quantum flux which matches the standard blue light (for further details see text). (After Delbrück and Shropshire, 1960.)

General conclusions

The action spectra mentioned (Figs. 15.20, 15.22, 15.23) point to a flavoprotein as the active photoreceptor. Other reliable action spectra of blue-light-dependent photoresponses, e.g., those of phototropic responses in fungi and in higher plants, can be interpreted the same way (but see chapter 7). Therefore it seems possible that the same photoreceptor might be acting in all cases in which development and movement are mediated by light of short wavelength (Rau, 1967). If indeed the physiologically effective short-wavelength light is absorbed in all systems by the same photoreceptor, the situation would be somewhat analogous to what has been discussed in connection with phytochrome: one and the same photoreceptor with a diversity of photoresponses.

In those responses where phytochrome is involved it is still difficult to evaluate the contribution, if any, of a blue-light-specific photoreceptor (e.g., Fig. 15.5). It is striking, however, that the short-wavelength part of the action spectrum in lettuce (Fig. 15.5) is very similar to the action spectra of responses where exclusively blue and ultraviolet light is effective (Figs. 15.20, 15.22, 15.23). In any event, if there is a strong effect of short-wavelength light and only a slight effect of long-wavelength light [e.g., anthocyanin synthesis in milo seedlings (Downs and Siegelman, 1963); synthesis of the enzyme cinnamic acid hydroxylase in hypocotyl segments of gherkin seedlings (Engelsma, 1966); control of flowering in mustard (Hanke, 1967)] it seems reasonable to postulate a simultaneous action of phytochrome and a flavoprotein. A typical response of this sort will be discussed now.

Photomorphogenesis in fern gametophytes

This section will be restricted to the young gametophytes (= sporelings) of the common male fern, *Dryopteris filix-mas* (L.) Schott. The results which have been obtained with these sporelings (Mohr, 1963, 1965) can be regarded as

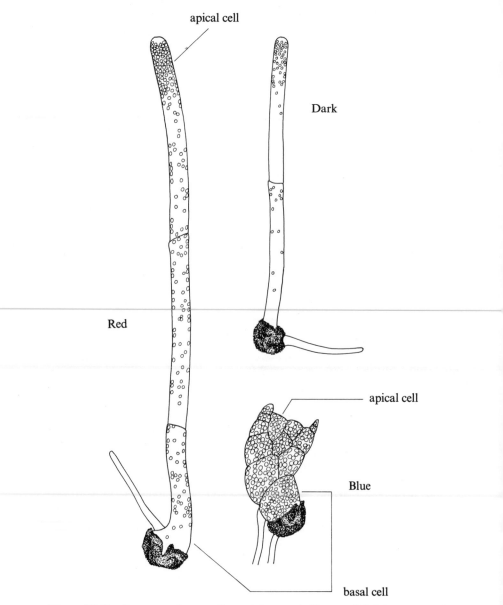

Figure 15.24   Representative sporelings of *Dryopteris filix-mas* (L.) Schott. grown in darkness, continuous standard red, or blue light. The sporelings were grown for 6 days after
germination on inorganic nutrient solution. The blue-grown and the red-grown sporelings have
about the same dry weight. (After Mohr, and Ohlenroth, 1962.)

being representative for the sporelings of many leptosporangiate ferns. The gonospores of *Dryopteris* germinate only in the light. This is a typical phytochrome response. The sporelings, however, can develop 'normally' from the very transient filamentous one-dimensional stage, called the 'protonema', to the two- or three-dimensional stage, called the 'prothallus' only if they receive enough short-wavelength visible light below 500 nm. Under long-wavelength visible light, e.g., red or far-red light of low or higher intensities, these sporelings may continue to grow for weeks as cellular filaments which are similar to the filaments of the dark control (Fig. 15.24). The problem is to understand in 'molecular' terms how blue light can cause the sporelings to switch over from filamentous to two- or three-dimensional growth. A good gauge of the photomorphogenic light effect on the sporelings has been the 'morphogenic index', i.e., the length of the protonema or prothallus divided by its maximal width. This index can be applied because the morphogenic effect of the blue light, i.e., the induction of two-dimensional growth, expresses itself in a very strong inhibition of the lengthening of the protonemal stage. A low morphogenic index is the result of a high morphogenic effectiveness of the light. The action spectrum (Fig. 15.25)—morphogenic index as a function of wavelength—shows that the

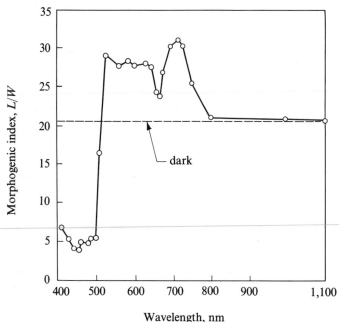

Figure 15.25 The morphogenic index (length/maximum width) of the protonema of the fern *Dryopteris filix-mas* (L.) Schott, as a function of wavelength. Measurements were made after 6 days culture in continuous monochromatic light at an intensity of 200 erg cm^{-2} s^{-1}. Under these conditions and within this period of time the sporelings throughout remained filamentous, even under blue light. (After Mohr, 1956.)

morphogenic effectiveness is very high below 500 nm. Additional structures appear in the red—a slight peak of action—and in the far-red, a dip of action. These latter structures can be attributed to phytochrome (Schnarrenberger and Mohr, 1967a). When the sporelings are irradiated simultaneously with red and far-red at suitable quantum flux densities the slight morphogenic effect of the red is cancelled by the simultaneous far-red. On the other hand the strong morphogenic effect of the blue light does not show any significant interaction with far-red. All data at present available indicate that the effect of blue light is due to a photoreceptor other than phytochrome. The photoreceptor involved is possibly a flavoprotein. This conclusion is supported by, among other things, the detailed action spectrum of polarotropism in this system (Fig. 15.22).

How can blue light—absorbed by a flavoprotein—initiate two-dimensional morphogenesis? The following hypothesis seems to be appropriate: The blue-light-dependent photoreaction eventually leads to an activation of those genes whose function is required for the 'normal', multi-dimensional development of the fern gametophytes. We thus assume that blue-light-specific morphogenesis requires the action of some enzymes which are not active, or not sufficiently active, in red-grown sporelings.

In an attempt to verify this hypothesis it was found that blue light strongly increases the protein contents of the sporelings.* The decrease of the morphogenic index was found to be inversely correlated with the increase of the relative protein content, etc. Detailed kinetics of the 'molecular' and the morphological responses mediated by blue light have shown, however, that the morphological response is much faster than the increase of the protein content. Morphological responses can be measured about 3 hours after the onset of blue light. Under the same conditions it takes about 9 hours before the protein content increases considerably above the level which is maintained in the red light (Kasemir and Mohr, 1965; Drumm and Mohr, 1967a). At least most of the additional protein which is accumulated under the influence of blue light cannot be regarded as being directly connected with morphogenesis. It is mainly located in the chloroplasts, probably to a large extent as structural protein. Bergfeld's (1963a) microscopic observations support the biochemical data. He found that the size of the chloroplasts in the sporelings is determined by the light quality. In blue light the chloroplasts are much larger than in red light. The process of blue-light-dependent plastid growth is reversible. The size of the chloroplasts adjusts at any given time to the value characteristic of the particular light quality.

If one offers $^{14}CO_2$ to red-grown or blue-grown sporelings one finds (Payer, 1967) that much less of the ^{14}C is present in free amino acids in the blue-grown than in the red-grown sporelings. On the other hand much more ^{14}C is incorpo-

---

*It should be emphasized, perhaps, that all these biochemical analyses—summarized by Mohr (1965)—were done under carefully controlled standard conditions of growth, i.e., the cultures grew logarithmically on inorganic liquid medium with nearly the same rate of photosynthetic dry matter accumulation in red and blue light. In general dry matter has been used, for good reasons, as a system of reference for the biochemical data (Drumm and Mohr, 1967b).

rated into protein in the blue as compared with the red light. Obviously the chloroplasts in the blue-grown sporelings synthesize protein more rapidly than chloroplasts in red-grown sporelings. If red-grown sporelings are placed under blue light the rate of protein synthesis—as measured by the rate of ^{14}C incorporation from $^{14}CO_2$ into protein—is rapidly increased. There is a significant increase within an hour, and after about 12 hours the rate of protein synthesis which is characteristic of the blue-grown controls is reached.

Obviously blue light directs many more primary products of photosynthesis into the channel of protein synthesis than does red light. Quantitative amino acid analysis of the protein of blue- and red-grown sporelings did not reveal any striking qualitative difference, however (v. Deimling and Mohr, 1967). The contents of all amino acids, obtained by protein hydrolysis, show about the same increase under the influence of blue light. The pools of the free amino acids are much smaller in the blue than in the red. This fact indicates that the rate of protein synthesis in the red light is not limited by the pool size of the free amino acids. Blue light seems to exert its promotive influence directly in connection with polypeptide synthesis.

To summarize: Blue light increases the rate of protein synthesis in the fern sporelings. As a result the relative protein content in blue-grown sporelings is higher than in red-grown sporelings. Most of the 'blue light dependent' protein is located in the plastids. According to Payer (1967) some blue-light dependent protein synthesis is going on in the cytoplasm as well. This latter conclusion is, however, based on indirect evidence.

The RNA content of sporelings of the same age is always higher in blue light than in red light (Drumm and Mohr, 1967a). The relative RNA content of red-grown sporelings increases rapidly after the onset of blue light. The conclusion that the rate of RNA synthesis is increased under these conditions is supported by the observation that there is a rapid blue-light-dependent increase of the specific activity of the RNA fraction after the application of ^{14}C-uridine. It is probable that part of the blue-light-dependent RNA is RNA of the plastids. The rapid increase of RNA synthesis after the onset of blue light seems to indicate, however, that the activity of the nuclear genes is involved as well. The blue-light-dependent increase of the RNA content occurs at least as fast as the blue-light-dependent morphological changes in the apical cell of the red-grown protonema (Fig. 15.24; left). In any case the blue-light-dependent increase of RNA occurs faster than the marked blue-light-dependent increase of the protein content. All available data (summarized by Drumm and Mohr, 1967a) are consistent with the hypothesis that in some way or another blue light initiates 'blue-light specific' RNA synthesis.

Cytochemical studies have shown that in the red-grown protonema there exists a steep gradient of RNA along the protonema. The highest concentration is close to the tip. On the other hand such a gradient cannot be found in blue-grown protonemas. Here the longitudinal distribution of RNA is more or less homogeneous (Menzel, unpublished data). An analogous situation exists with respect to nuclear and nucleolar volume (Bergfeld, 1967). The apical cells of the red-grown protonemas (Fig. 15.24) contain extremely large nuclei and nucleoli; in the fully differentiated basal cell, however, nucleus and nucleolus are small. In the case of blue-grown protonemas there are only slight differences between the apical and the basal cell as far as nuclear and nucleolar volumes are concerned. In the apical cell these structures are much smaller in the blue-grown sporeling than in the red-grown protonema; in the basal cell, however, nucleus and nucleolus are larger in the blue-grown sporeling. Since the nuclei in the basal cells (Fig. 15.24) no longer divide, it is of great interest whether or not they still respond to changes in the light conditions. This is indeed the case (Bergfeld, 1963b, 1967). The volume of the nuclei increases up to a certain level when the sporelings are placed under blue light instead of red light. The same is true for the nucleoli. These shifts in nuclear and nucleolar volume are reversible: if the sporelings are kept under red light instead of blue the volumes decrease to the level which is characteristic of the controls which have only received red light. One can argue that the change of nuclear and nucleolar volume reflects a change of nuclear activity. Since the characteristic changes of the nuclear and nucleolar volumes which follow a change in the light conditions are already detectable after 3 hours the conclusion is justified (Bergfeld, 1967) that the striking light-dependent changes in morphogenesis of the sporelings are preceded by fundamental changes in the synthetic activity of the nucleus and nucleolus.

Photomorphogenesis of fern sporelings may be summarized as follows. 'Normal' morphogenesis of young fern gametophytes (= sporelings) is blue-light dependent (Fig. 15.24). The data which are available at the moment [including those from inhibitor experiments (Mohr, 1965)] are consistent with the hypothesis that the differences in morphogenesis reflect differential gene activity. It has not yet been possible, however, to study the problem at the level of enzymes. There are indications that the photoreceptor whose absorption is responsible for the morphogenic effect of blue light might be a flavoprotein.

Bibliography

Selected further reading

Hendricks, S. B., 1964. Photochemical aspects of plant photoperiodicity. In *Photophysiology*, Vol. 1, ed. A. C. Giese. Academic Press, New York.
Hillman, W. S., 1967. The physiology of phytochrome. *Ann. Rev. Plant Physiol.* **18**, 301–324.
Mohr, H., 1964. The control of plant growth and development by light. *Biol. Rev.* **39**, 87–112.

Siegelman, H. W. and W. L. Butler, 1965. Properties of phytochrome. *Ann. Rev. Plant Physiol.* **16**, 383–392.

Vince, D., 1964. Photomorphogenesis in plant stems. *Biol. Rev.* **39**, 506–536.

Papers cited in text

Bergfeld, R., 1963a. Die Wirkung von hellroter und blauer Strahlung auf die Chloroplastenausbildung. *Z. Naturforsch.* **18b**, 328–331.

Bergfeld, R., 1963b. Die Beeinflussung der Zellkerne in den Vorkeimen von *Dryopteris filix-mas* durch rote und blaue Strahlung. *Z. Naturforsch.* **18b**, 558–562.

Bergfeld, R., 1967. Kern- und Nucleolusausbildung in den Gametophytenzellen von *Dryopteris filix-mas* (L.) Schott bei Umsteuerung der Morphogenese. *Z. Naturforsch.* **22b**, 972–976.

Briggs, W. R. and H. P. Chon, 1966. The physiological versus the spectrophotometric status of phytochrome in corn coleoptiles. *Plant Physiol.* **41**, 1159–1166.

Butler, W. L., S. B. Hendricks, and H. W. Siegelman, 1964. Action spectra of phytochrome in vitro. *Photochem. Photobiol.* **3**, 521–528.

Butler, W. L. and H. C. Lane, 1965. Dark transformations of phytochrome in vivo. II. *Plant Physiol.* **40**, 13–17.

Butler, W. L., H. C. Lane, and H. W. Siegelman, 1963. Non-photochemical transformation of phytochrome *in vivo*. *Plant Physiol.* **38**, 514–519.

Clarkson, D. T. and W. S. Hillman, 1967. Stability of phytochrome concentration in dicotyledonous tissues under continuous far-red light. *Planta* **75**, 286–290.

Delbrück, M. and W. Shropshire jr., 1960. Action and transmission spectra of *Phycomyces*. *Plant Physiol.* **35**, 194–204.

v. Deimling, A. und H. Mohr, 1967. Eine Analyse der durch Blaulicht bewirkten Steigerung der Proteinsynthese bei Farnvorkeimen auf der Ebene der Aminosäuren. *Planta*, **76**, 269–284.

Downs, R. J. and H. W. Siegelman, 1963. Photocontrol of anthocyanin synthesis in milo seedlings. *Plant Physiol.* **38**, 25–30.

Durst, F. and H. Mohr, 1966. Phytochrome mediated induction of enzyme synthesis in mustard seedlings (*Sinapis alba* L.) *Naturwiss.* **53**, 531–532.

Drumm, H. und H. Mohr, 1967a. Die Regulation der RNS-Synthese in Farngametophyten durch Licht. *Planta*, **72**, 232–246.

Drumm, H. und H. Mohr, 1967b. Die Regulation der DNS-Synthese in Farngametophyten durch Licht. *Planta*, **75**, 343–351.

Edwards, J. L. and W. H. Klein, 1964. Relationship of phytochrome concentration and physiological responses. *Plant Physiol.* **39**, Suppl. L.

Engelsma, G., 1966. The influence of light of different spectral regions on the synthesis of phenolic compounds in gherkin seedlings in relation to photomorphogenesis. III. Hydroxylation of cinnamic acid. *Acta Bot. Neerl.* **15**, 394–405.

Etzold, H., 1965. Der Polarotropismus und Phototropismus der Chloronemen von *Dryopteris filix-mas* (L.) Schott. *Planta*, **64**, 254–280.

Fondeville, J. C., H. A. Borthwick, and S. B. Hendricks, 1966. Leaflet movement of *Mimosa pudica* L. indicative of phytochrome action. *Planta*, **69**, 359–364.

Fondeville, J. C., M. J. Schneider, H. A. Borthwick, and S. B. Hendricks, 1967. Photocontrol of *Mimosa pudica* L. leaf movement. *Planta*, **75**, 228–238.

Furuya, M. and W. S. Hillman, 1964. Observations on spectrophotometrically assayable phytochrome in vivo in etiolated Pisum seedlings. *Planta*, **63**, 31–42.

Furuya, M. and J. G. Torrey, 1964. The reversible inhibition by red and far-red light of auxin-induced lateral root initiation in isolated pea roots. *Plant Physiol.* **39**, 987–991.

Häcker, M., 1967. Der Abbau von Speicherprotein und die Bildung von Plastiden in den Kotyledonen des Senfkeimlings (*Sinapis alba* L.) unter dem Einfluss des Phytochroms. *Planta*, **76**, 309–325.

Häcker, M., K. M. Hartmann, und H. Mohr, 1964. Zellteilung und Zellwachstums im Hypokotyl von *Lactuca sativa* L. unter dem Einfluss des Lichtes. *Planta*, **63**, 253–268.

Hanke, H. J., 1967. *Ein Wirkungsspektrum für die Blühinduktion der Langtagpflanze Sinapis alba L. (Cruciferae).* Ph.D. thesis, University of Freiburg i. Br.

Hartmann, K. M., 1966. A general hypothesis to interpret 'high energy phenomena' of photomorphogenesis on the basis of phytochrome. *Photochem. Photobiol.* **5**, 349–366.

Hartmann, K. M., 1967a. Ein Wirkungsspektrum der Photomorphogenese unter Hochenergiebedingungen und seine Interpretation auf der Basis des Phytochroms (Hypokotylwachstumshemmung bei *Lactuca sativa* L.). *Z. Naturforsch.* **22b**, 1172–1175.

Hartmann, K. M., 1967b. Photoreceptor problems in photomorphogenic responses under high-energy-conditions (UV-blue—far-red). *Book of Abstracts, European Photobiology Symposium, Hvar* (Jugoslavia), pp. 29–32.

Hartmann, K. M., 1967c. Phytochrome 730 (P_{fr}), the effector of the 'high energy photomorphogenic reaction' in the far-red region. *Naturwiss.* **54**, 544.

Hartmann, K. M., H. Menzel, und H. Mohr, 1965. Ein Beitrag zur Theorie der polarotropischen und phototropischen Krümmung. *Planta*, **64**, 363–375.

Haupt, W., 1960. Die Chloroplastendrehung bei *Mougeotia*. II. Die Induktion der Schwachlichtbewegung durch linear polarisiertes Licht. *Planta*, **55**, 465–479.

Haupt, W., 1965. Perception of environmental stimuli orienting growth and movement in lower plants. *Ann. Rev. Plant Physiol.* **16**, 267–290.

Hillman, W. S., 1965. Phytochrome conversion by brief illumination and the subsequent elongation of etiolated Pisum stem segments. *Physiol. Plant.* **18**, 346–358.

Hock, B., E. Kühnert, und H. Mohr, 1965. Die Regulation von Fettabbau und Atmung bei Senfkeimlingen durch Licht (*Sinapis alba* L.). *Planta*, **65**, 129–138.

Jakobs, M. und H. Mohr, 1966. Kinetische Studien zur phytochrominduzierten Proteinsynthese. *Planta*, **69**, 187–197.

Karow, H. und H. Mohr, 1967. Aktivitätsänderungen der Isocitritase (EC 4.1.3.1.) während der Photomorphogenese beim Senfkeimling. *Planta*, **72**, 170–186.

Kasemir, H. und H. Mohr, 1965. Die Regulation von Chlorophyll und Proteingehalt in Farnvorkeimen durch sichtbare Strahlung. *Planta*, **67**, 33–43.

Klein, W. H., J. L. Edwards, and W. Shropshire, 1967. Spectrophotometric measurements of phytochrome *in vivo* and their correlations with photomorphogenic responses of *Phaseolus. Plant Physiol.* **42**, 264–271.

Lange, H., I. Bienger, und H. Mohr, 1967. Eine neue Beweisführung für die Hypothese einer differentiellen Genaktivierung durch Phytochrom 730. *Planta*, **76**, 359–366.

Lange, H. und H. Mohr, 1965. Die Hemmung der phytochrominduzierten Anthocyansynthese durch Actinomycin D und Puromycin. *Planta*, **67**, 107–121.

Linschitz, H., V. Kasche, W. L. Butler, and H. W. Siegelman, 1966. The Kinetics of phytochrome conversion. *J. Biol. Chem.* **241**, 3395–3403.

Linschitz, H. and V. Kasche, 1967. Kinetics of phytochrome conversion: multiple pathways in the P_r to P_{fr} reaction, as studied by double-flash technique. *Proc. Nat. Acad. Sci., U.S.*, **58**, 1059–1064.

Lockhart, J. A., 1965. The analysis of interactions of physical and chemical factors on plant growth. *Ann. Rev. Plant Physiol.* **16**, 37–52.

Loercher, L., 1966. Phytochrome changes correlated to mesocotyl inhibition in etiolated *Avena* seedlings. *Plant Physiol.* **41**, 932–936.

Mohr, H., 1956. Die Abhängigkeit des Protonemawachstums und der Protenemapolarität bei Farnen vom Licht. *Planta*, **47**, 127–158.

Mohr, H., 1957. Der Einfluss monochromatischer Strahlung auf des Längenwachstum des Hypokotyls und auf die Anthocyanbildung bei Keimlingen von *Sinapis alba* L. (= *Brassica alba* Boiss.). *Planta* **49**, 389–405.

Mohr, H., 1963. The influence of visible radiation on the germination of archegoniate spores and the growth of the fern protonema. *J. Linn. Soc.* (Bot.) **58**, 287–296.

Mohr, H., 1964. The control of plant growth and development by light. *Biol. Rev.* **39**, 87–112.

Mohr, H., 1965. Die Steuerung der Entwicklung durch Licht am Beispiel der Farngametophyten. *Ber. dtsch. bot. Ges.* **78**, 54–68.

Mohr, H., 1966a. Untersuchungen zur phytochrominduzierten Photomorphogenese des Senfkeimlings (*Sinapis alba* L.). *Z. Pflanzenphysiol.* **54**, 63–83.

Mohr, H., 1966b. Differential gene activation as a mode of action of phytochrome 730. *Photochem. Photobiol.* **5**, 469–483.

Mohr, H. und I. Bienger, 1967. Experimente zur Wirkung von Actinomycin D auf die durch Phytochrom bewirkte Anthocyansynthese. *Planta*, **75**, 180–184.

Mohr, H., Ch. Holderied, W. Link, und K. Roth, 1967. Protein- und RNS-Gehalt des Hypokotyls beim stationären Wachstum im Dunkeln und unter dem Einfluss von Phytochrom (Keimlinge von *Sinapis alba* L.). *Planta*, **76**, 348–358.

Mohr, H. und K. Ohlenroth, 1962. Photosynthese und Photomorphogenese bei Farnvorkeimen von *Dryopteris filix-mas. Planta,* **57**, 656–664.

Mohr, H., I. Schlickewei, und H. Lange, 1965. Die Hemmung des phytochrom-induzierten Kotyledonenwachstums durch Actinomycin D. *Z. Naturforsch.* **20b**, 819–821.

Mohr, H. und R. Senf, 1966. Die Hemmung der phytochrominduzierten Anthocyansynthese durch Puromycin und 2-Thiouracil. *Planta*, **71**, 195–203.

Payer, H. D., 1967. *Der Einfluss der Lichtqualität auf die Aufnahme von* $^{14}CO_2$ *und auf die Verteilung des* ^{14}C *im Stoffwechsel (Untersuchungen an Farngametophyten, Dryopteris filix-mas (L.) Schott*. Ph.D. thesis, University of Freiburg i. Br.

Pfeffer, W., 1904. *Pflanzenphysiologie*. Wilhelm-Engelmann-Verlag, Leipzig, p. 98.

Pratt, L. H. and W. R. Briggs, 1966. Photochemical and nonphotochemical reactions of phytochrome *in vivo. Plant Physiol.* **41**, 467–474.

Rau, W., 1967. Untersuchungen über die lichtabhängige Carotinoidsynthese. I. Das Wirkungsspektrum von *Fusarium aquaeductuum. Planta*, **72**, 14–28.

Rissland, I. und H. Mohr, 1967. Phytochrom-induzierte Enzymbildung (Phenylalanindesaminase), ein schnell ablaufender Prozess. *Planta*, **77**, 239–249.

Schnarrenberger, C. und H. Mohr, 1967a. Die Wechselwirkung von Hellrot, Dunkelrot und Blaulicht bei der Photomorphogenese von Farngametophyten (*Dryopteris filix-mas* (L.) Schott). *Planta*, **75**, 114–124.

Scharrenberger, C. and H. Mohr, 1967b. Phytochrome-mediated synthesis of carotenoids in mustard seedlings (*Sinapis alba* L.). *Naturwiss.* **54**, 837.

Schopfer, P., 1966. Der Einfluss von Phytochrom auf die stationären Konzentrationen von Ascorbinsäure und Dehydroascorbinsäure beim Senfkeimling (*Sinapis alba* L.). *Planta*, **69**, 158–177.

Schopfer, P., 1967a. Die Hemmung der phytochrominduzierten Photomorphogenese ('positive' Photomorphosen) des Senfkeimlings (*Sinapis alba* L.) durch Actinomycin D und Puromycin. *Planta*, **72**, 297–305.

Schopfer, P., 1967b. Der Einfluss von Actinomycin D und Puromycin auf die phytochrominduzierte Wachstumshemmung des Hypokotyls beim Senfkeimling (*Sinapis alba* L.). *Planta*, **72**, 306–320.

Schopfer, P., 1967c. Weitere Untersuchungen zur phytochrominduzierten Akkumulation von Ascorbinsäure beim Senfkeimling, (*Sinapis alba* L.). *Planta*, **74**, 210–227.

Shropshire, W. jr., 1963. Photoresponses of the fungus, *Phycomyces. Physiol. Rev.* **43**, 39–67.

Siegelman, H. W. and W. L. Butler, 1965. Properties of phytochrome. *Ann. Rev. Plant Physiol.* **16**, 383–392.

Spruit, C. J. P., 1965. Absorption spectrum changes during dark decay of phytochrome 730 in plants. *Meded. Landbouwhogeschool Wageningen*, **65**–12, 1–6.

Spruit, C. J. P., 1966. Thermal reactions following illumination of phytochrome. *Meded. Landbouwhogeschool Wageningen*, **66**–15, 1–7.

Steiner, A. M., 1967. Action spectra for polarotropism in germlings of a fern and a liverwort. *Naturwiss.* **54**, 497–498.

Taylor, A. O. and B. Bonner, 1967. Isolation of phytochrome from the alga *Mesotaenium* and liverwort *Sphaerocarpus*. *Plant Physiol.* **42**, 762–766.

Thomas, A. S. and S. Dunn, 1967a. Plant growth with new fluorescent lamps. I. Fresh and dry weight yields of tomato seedlings. *Planta*, **72**, 198–207.

Thomas, A. S. and S. Dunn, 1967b. Plant growth with new fluorescent lamps. II. Growth and reproduction of mature bean plants and dwarf marigold plants. *Planta*, **72**, 208–212.

Vince, D. and R. Grill, 1966. The photoreceptors involved in anthocyanin synthesis. *Photochem. Photobiol.* **5**, 407–411.

Wagner, E., 1967. *Ein Beitrag zur phytochrominduzierten Photomorphogenese des Senfkeimlings (Sinapis alba L.).* Ph.D. thesis, University of Freiburg i. Br.

Wagner, E. and H. Mohr, 1966a. Kinetic studies to interpret 'high-energy phenomena' of photomorphogenesis on the basis of phytochrome. *Photochem. Photobiol.* **5**, 397–406.

Wagner, E. und H. Mohr, 1966b. Kinetische Studien zur Interpretation der Wirkung von Sukzedanbestrahlungen mit Hellrot und Dunkelrot bei der Photomorphogenese (Anthocyansynthese bei *Sinapis alba* L.). *Planta*, **70**, 34–41.

Wagner, E. und H. Mohr, 1966c. 'Primäre' und 'sekundäre' Differenzierung im Zusammenhang mit der Photomorphogenese von Keimpflanzen (*Sinapis alba* L.). *Planta*, **71**, 204–221.

Weidner, M., 1967. Der DNS-Gehalt von Kotyledonen und Hypokotyl des Senfkeimlings (*Sinapis alba* L.) bei der phytochromgesteuerten Photomorphogenese. *Planta*, **75**, 94–98.

Weidner, M., M. Jakobs, und H. Mohr, 1965. Über den Einfluss des Phytochroms auf den Gehalt an Ribonucleinsäure und Protein in Senfkeimlingen (*Sinapis alba* L.). *Z. Naturforsch.* **20b**, 689–693.

Weidner, M. und H. Mohr, 1967a. Zur Regulation der RNS-Synthese durch Phytochrom bei der Photomorphogenese des Senfkeimlings (*Sinapis alba* L.). *Planta*, **75**, 99–108.

Weidner, M. und H. Mohr, 1967b. Die Wirkung von Actinomycin D auf die phytochromabhängigen Veränderungen des RNS-Gehaltes im Senfkeimling (*Sinapis alba* L.). *Planta*, **75**, 109–113.

16. Photoperiodism and vernalization

W. S. Hillman

Introduction

The development of most plants is seasonal, often with the precision of a calendar. Although this was undoubtedly recognized even before the invention of agriculture, only in this century have we begun to understand how it comes about. The most obvious explanation for seasonal activities is that they are direct results of gross environmental conditions—heat, cold, drought, and the like—that permit or prevent further growth. This notion hardly accounts for the precision involved and is mistaken, except in an evolutionary sense. The yearly responses of plants are rarely directly coerced by the weather. Instead, plants in their native habitats actually appear to anticipate the great seasonal fluctuations. They do so by responding to other relatively subtle, but more reliable, characteristics of the changing environment. These characteristics are in effect taken as signals that set in motion complex patterns of development. Photoperiodism and vernalization are two major mechanisms underlying such responses.

First, a quick look at the words themselves. Photoperiodism combines the Greek roots for 'light' and 'a length of time' and is thus almost self-defining. Vernalization is slightly more obscure, being a translation of the Russian yarovizatzya, in which the root for 'spring' (Russian: yarov; Latin: ver) is combined with a suffix meaning 'to make', after the observation that cold treatments convert winter strains of cereals to the spring flowering habit. As initial definitions, then, the following will serve: Photoperiodism is a response to the length of the day, and vernalization is an effect on flowering brought about by a period of low temperature. It should not be hard to see how organisms armed with such devices might well appear to be watching the calendar. If one were designing mechanisms for the precise yearly scheduling of events, one could hardly do better. Such mechanisms probably evolved because some synchronization of the reproductive activities of populations has considerable survival value. In addition, it is probably advantageous for organisms to occupy specific temporal 'niches' in the changing pattern of the seasons in the same way as they occupy different physical locations in the environment.

Consider the factor of day length. Everywhere except precisely on the Equator, day length changes in a regular yearly cycle. The extremes, of course, depend on the latitude. At 45° N (London 51°, New York 41°) the maximum length on 21 or 22 June is about $15\frac{1}{2}$ hours and the minimum is 9 hours. At 30° N (Delhi, Shanghai) the range is about 14 to 10, and at 60° N (Stockholm, Leningrad) it is 19 to 6. The rate of change also varies, not only with the latitude, but with the month as well, being least at mid-summer and mid-winter and greatest in spring and autumn. Average rates of change for April and August are at 60°, 40 minutes per week; at 45°, 25 minutes per week; and at 30°, 12 minutes per week. Thus a process dependent upon day length need only discriminate to within these limits to time its response within a week. It is remarkable that the flowering of many fully tropical plants (for example, sugar cane) is often day-length dependent in spite of the fact that even at 15° latitude (Manila, Dakar, Guatemala City) the yearly extremes of day length are only 13 hours to 11 hours with the weekly rate of change in April roughly 5 minutes (see Withrow, 1959).

While a device sensitive to day length is thus excellent for the yearly scheduling of events, low temperature, the factor perceived in vernalization, can play an additional role. This is particularly obvious in plants of the so-called temperate regions in which the winters are essentially unfit for plant growth. Consider, for example, a plant in which the photoperiodic response is adjusted to start the train of events leading to flowering and fruiting in all day lengths longer than 13 hours. At a latitude of 45°, such a day length would be reached in early April, but will persist through most of the summer. If the flowering and fruiting occupies some months, plants which for some reason (for example, late germination) fail to attain the proper size early enough may initiate flowers too late in the season to complete the process. Suppose, however, that the plants only become photoperiodically sensitive after they have been exposed to at least six weeks of low temperature. Then, plants germinating in the summer will not initiate flowers at all until the following spring. Such a requirement for a long cold treatment, which we can call vernalization, here becomes, in a sense, a device to insure that the winter has passed and will not interfere with subsequent development. Thus, combined with photoperiodism, it can give a much greater precision to the flowering response. Though plants with photoperiodism alone might under some circumstances 'confuse' spring and autumn, the additional device of vernalization makes such confusion less likely.

Historical notes

Photoperiodism.   The discovery of photoperiodism is generally credited to W. W. Garner and H. A. Allard, plant physiologists in the U.S. Department of Agriculture, who characterized and named the phenomenon. While they by no means originated the suggestion that day length might affect plant growth (see Murneek, 1948) their two extensive papers (1920, 1922) provided the first systematic and critical evidence that it did so.

Garner and Allard were led into this work by a number of apparently un-related observations on flowering times. One of these is shown in Table 16.1,

Table 16.1 Growth and flowering of Biloxi soybeans near Washington, D.C. as a function of germination date (data from Garner and Allard, 1920).

Germination date	Date of first open blossoms	Maximum height (inches)	Days to blossoming
2 May	4 September	52	125
2 June	4 September	52	94
16 June	11 September	48	92
30 June	15 September	48	77
15 July	22 September	44	69
2 August	29 September	28	58
16 August	16 October	20	61

selected from a much larger set of data in the original paper. The Biloxi variety of soy beans (*Glycine max*) flowered in September or October whether it germinated in May, June, July, or August. A variation of 59 days in germination date during May and June caused a difference of only 11 days in the time to the first open flower. Both the number of days from germination to blossoming and the final height of the plants decreased as the season continued. The data in Table 16.1 thus exemplify in a cultivated plant the kind of seasonal timing discussed earlier and suggest that some factor, which we now know to be day length, becomes more favourable to flowering as the season progresses.

A second observation on flowering time, which has become equally 'classical', involved a mutant of tobacco (*Nicotiana tabacum*) that at first failed to flower at all, but simply grew to a very large size. This behaviour earned the varietal name 'Maryland Mammoth', but hardly contributed to the breeding programme in which it had appeared. However, when the season became too cool for field growth and the plants were thus propagated vegetatively in the greenhouse, even relatively small plants began to flower profusely during the winter time.

Garner and Allard were soon able to show that the explanation for both the Biloxi soy bean and the Maryland Mammoth flowering behaviour lay in the seasonal changes of day length. Both plants were short-day plants, which flowered only when the daily exposure to the light was reduced below a certain critical duration. From such beginnings and after many complex experiments, the conclusion was reached that 'the relative length of the day is a factor of the first importance in the growth and development of plants . . .' (Garner and Allard, 1920). Within a relatively short time this generalization was being confirmed, expanded, and analysed in laboratories and agricultural stations in all parts of the world. We should note here also that while photoperiodism was first discovered and has been most carefully studied in plants, it occurs in animals as well. Many examples are now known of birds, insects, and mammals in which, as in plants, photoperiodism is the major mechanism in the seasonal regulation of reproductive development and behaviour.

Vernalization.   Until the work of Garner and Allard, photoperiodism, when it was thought of at all, was at most a minor curiosity known only to a few professional botanists. In contrast to this, the fact that exposing plants or germinating seeds to low temperatures can modify subsequent flowering behaviour, has been known and used (at least occasionally) in horticultural and agricultural practices for more than a century and perhaps much longer (Whyte, 1948). Thus, no absolutely clear beginning can be cited for work on vernalization. However, if any single group of investigations can be regarded as central to the recent development of the field it is that of F. G. Gregory and O. N. Purvis of the Imperial College of Science and Technology, London, starting in the early 'thirties. Their analysis of seed vernalization and its relation to photoperiodism has in most respects not been improved upon and will be discussed later.

Photoperiodism: light, darkness, and time

Definitions; response types

Most definitions of photoperiodism have something misleading about them. For instance, viewing it as 'a response to the relative length of day and night' (Garner and Allard, 1920) suggests that a plant would respond to 14 hours of light, alternating with 10 hours of darkness in the same way as to 28 hours of light alternating with 20 hours of darkness, which is rarely if ever true. With such distinguished precedent perhaps ambiguity can hardly be avoided, but the following definition seems clear enough: Photoperiodism is a response to the timing of light and darkness. *Timing* is the central word here—total light energy and intensity above a relatively low level are secondary to the duration and succession of light and darkness.

Photoperiodism is easily confused with other phenomena if this definition is not carefully borne in mind. For example, the differences between plants grown with 8 or with 16 hours per day of artificial light are not necessarily due to photoperiodism. They may be, but the schedules in question also provide a twofold difference in total light energy so that there will be greater total photosynthesis under the longer photoperiod. A further test of the nature of the effects is to give both sets of plants in this example, 8 hours per day of bright light followed either by darkness for the short photoperiod, or by 8 hours more of dim light—1 to 10 ft c—for the longer photoperiod. If the differences are still observed, there is now more reason to regard them as photoperiodic, since timing and not total energy seems to be the main factor. In nature, of course, and under many experimental conditions, the effects of light timing and total light energy are frequently confounded, but it is always desirable to try making the distinction. In some plants, moreover, high light intensities may also be involved in the control of flowering, but such requirements seem to be secondary to the photoperiodic mechanism and can often be replaced by the products of photosynthesis. In a sense, the ideal organisms for the study of photoperiodism proper would be either non-photosynthetic or able to obtain their energy supply, at least temporarily, in some other manner. Unfortunately there are hardly any higher plants in the first category and few enough, such as the duckweeds, with characteristics amenable to the second.

Photoperiodic responses are generally classified with respect to flowering, but similar categories might be set up with respect to effects on other processes such as pigmentation, branching, tuberization, or dormancy. Most of these, however, have been studied far less than flowering. All known responses can be viewed as modifications or combinations of the three basic types: day-length-indifference, short-day response, and long-day response.

Day-length-indifference is perhaps the commonest response—that is to say, photoperiodism appears to have no effect on the process in question—but is obviously of no interest at the moment. In the typical short-day response, a plant flowers only, or flowers most rapidly, when illuminated less than a certain number of hours per day. The 'certain number of hours' depends on the plant in question and is referred to as its critical day length. Conversely, in the typical long-day plant, flowering occurs only, or occurs most rapidly, when the plant receives more than a certain number of hours of light per day—that is, when the length of illumination exceeds the critical day length of the plant in question. Certain plants show combined responses, flowering only as a result of a short-day response followed by a long-day response or vice versa.

The distinction between long-day and short-day responses is frequently misstated. The distinction is *not* that the critical day length of the one category is longer than the other. The distinction is *not* that the former flower with daily illumination longer than 12 hours and the latter with daily illumination shorter than 12 hours. In fact, the critical day length for one of the most widely studied short-day plants, the cocklebur, *Xanthium*, is about $15\frac{1}{2}$ hours, while the critical day length for a widely studied long-day plant, the henbane, *Hyoscyamus*, is about 11 hours. What, then, is the difference between long-day and short-day responses? It resides simply in the relationship between flowering and the critical day length for each plant. Flowering in long-day plants, whatever the absolute value of their critical day lengths, requires an illumination *longer than* their critical day length, while short-day plants require illumination *shorter than* their critical day length in order to flower. Perhaps if one renamed the response types longer-day or shorter-day, thus emphasizing the comparison of the flowering day length to the critical day length, there might be less confusion. But the terminology is in universal use and can hardly be changed now.

There is no basis on which to make even an intelligent guess concerning the relative frequency of various photoperiodic response types among higher plants.

As mentioned earlier, possibly the largest single group of plants is that in which the members are day length-indifferent with respect to flowering. One should not conclude from this that these plants are totally non-photoperiodic, however, since effects on vegetative growth are usually easy to demonstrate. Examples of such plants are *Lycopersicum* (tomato) and *Pisum* (peas).

Much more work has been done with short-day plants than with those of other response types, and there is little doubt that much of what we think we know about photoperiodism and flowering is biased in this regard. Examples of short-day plants include *Glycine max* variety Biloxi, the cocklebur *Xanthium pennsylvanicum*, and another common weed, *Chenopodium rubrum*. These three are *qualitative* or *obligate* short-day plants, unable to flower except under appropriate short-day treatment. Plants such as *Nicotiana tabacum*, variety Maryland Mammoth, various strains of morning glory—*Pharbitis* or *Ipomoea* —and the duckweed *Lemna perpusilla*, may be described as conditional short-day plants, since under certain conditions—low temperatures for the first two plants, and high levels of copper ion for the third—their short-day requirement is no longer absolute. There are also many *quantitative* short-day plants, in which flowering is hastened by the appropriate short-day conditions but occurs eventually without them. An example is the common sunflower, *Helianthus annuus*.

Obligate or qualitative long-day responses are found in certain strains of the black henbane, *Hyoscyamus niger*; in *Hordeum vulgare* (barley), in the common weed *Plantago lanceolata*, and in the grass *Lolium temulentum*. Spinach, *Spinacia oleracea*, is a common example of both conditional and quantitative long-day response. Long-short-day responses are found in species of *Bryophyllum*, while some varieties of *Triticum* (wheat) and *Secale* (rye) are short-long-day plants. These have been studied far less than the simple short-day or long-day plants.

Perception, induction, initiation, and development

Most of what follows will be clearer if certain important characteristics of the photoperiodic control of flowering are outlined now, even though they are not to be discussed until later in the chapter. In addition, a brief consideration of the kind of observations made in experiments on flowering is desirable.

Photoperiodic perception.   Photoperiodic conditions are usually perceived by the leaves. In many plants, exposure of a single leaf or even a small portion of a leaf to the appropriate photoperiod will bring about flowering. The experimental utility of being able to obtain effects by treatment of a single leaf per plant should be apparent. Such results also indicate some sort of communication between the leaves and the meristems that actually produce the flowers and there is a large body of work on hormonal and other explanations for such effects.

Photoperiodic induction. In many species the appropriate photoperiodic schedule need only be given for a few days in order for flowering to occur, even though the plants are subsequently maintained under photoperiods unfavourable for flowering. For some plants—*Xanthium*, *Lolium*, and *Pharbitis* among others—a single day of the proper treatment is sufficient to bring about some eventual flowering though more extensive treatment may elicit a more rapid and vigorous response. Such persistent after-effects of photoperiodic treatment are the results of *photoperiodic induction*. Like the question of photoperiodic perception, induction has of course been the subject of many investigations and is again obviously useful for experimental purposes since the actual treatments under investigation need often be continued no more than two or three days.

Floral initiation and subsequent development. Experiments on flowering are rarely conducted by waiting until the flowers are fully developed. For one thing, demands on time and space suggest that the sooner the results can be determined the better. Hence early phases of flower development, often including primordia observable only by dissection, are evaluated on both a quantitative—size—and qualitative—developmental stage—basis. A more theoretical reason has also been advanced for such procedures, namely that the process of the greatest interest is that of floral initiation—the basic change by which a meristematic apex becomes floral rather than vegetative—rather than subsequent stages of floral development.

Whether an arbitrary separation of initiation from development is truly tenable or not, it is worth while in focusing attention on early stages in the flowering process since these are less likely to be affected by conditions affecting growth in general. For example, the expansion of large flower parts must necessarily be strongly dependent on a good supply of carbohydrates and water, while the change in a meristem from the production of leaf initials to the production of flower primordia—all involving a few thousand cells—is not so obviously limited by gross nutrition.

It may require considerable care to separate general effects on growth from specific effects on flowering; this can be done by estimating the rates of both vegetative and floral development and then comparing the two estimates. Consider two groups of plants, one group now at a higher temperature than the other but all originally from the same conditions. The high-temperature group has buds averaging 12 mm in length, the other has 2 mm buds visible only on dissection. Should one conclude that high temperatures favour flowering? Yes, of course, if time is the only criterion. But one can also determine how many leaves the plants laid down before the flower buds began to appear. If the number of leaves before flowering is equal in both conditions, then neither condition can be regarded as having a specific effect on flowering. If, however, the plants at low temperature actually laid down fewer leaves before their buds than did those at high temperature, one could conclude that the low temperature favoured flower initiation relative to vegetative growth, though it slowed development in general.

Two experiments

Figures 16.1 and 16.2, showing the results of two photoperiodic experiments, should render the preceding discussion more concrete. Figure 16.1 illustrates the short-day response of a strain of *Pharbitis nil*, the Japanese morning glory. Plants raised in continuous light were given either one or three cycles of the indicated day length treatment and then returned to continuous light where they were maintained for several weeks until the effect of each treatment was determined. Obviously the precise value of the critical day length, as derived from such data, depends both on the definition used and on the number of treatment cycles employed. It might be defined as the shortest day length long enough to prevent all flowering, as the day length giving 50 per cent flowering, or as the longest day length short enough to permit 100 per cent flowering—the choice is arbitrary. Choosing the first definition, the one-cycle data give a

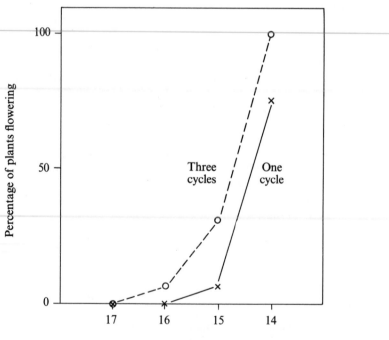

Figure 16.1   The effect of one or three days under the indicated day lengths on flowering of *Pharbitis nil* evaluated as a percentage of the plants flowering. Plants kept in continuous light before and after photoperiodic treatments. (Data from Imamura and Takimoto, 1955.)

value between 15 and 16 hours while the three-cycle data give 16 to 17 hours. This is not unusual; the critical day length of short-day plants generally appears longer the more experimental cycles are used, while that of long-day plants appears shorter. There are limits, however, if a plant is truly photoperiodic: photoperiodic induction of flowering in *Pharbitis* will not take place even with continual exposure to 18-hour day lengths.

Plant age is another important variable in photoperiodic experiments. Figure 16.2 shows the effect of age on the responsiveness of the long-day plant *Lolium*

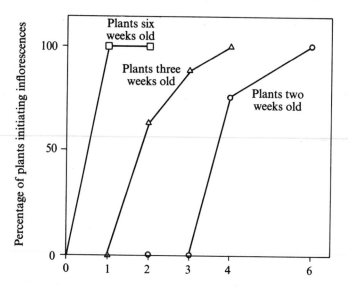

Figure 16.2 Response of 2-, 3-, and 6-week old *Lolium temulentum* plants to various numbers of long days (continuous light) evaluated as the percentage of plants initiating inflorescences. Plants kept in 8-hour photoperiods before and after long-day treatments. (Data from Evans, 1960.)

temulentum. For this experiment, plants of various ages were grown under 8-hour photoperiods, given one or more long-day (continuous light) cycles and returned to short days. The apices were eventually dissected and the percentage inflorescence initiation was scored. The older plants were fully responsive even to one long-day cycle, while two-week old plants were unresponsive even to three, requiring six for a maximum response. The reasons for the dependence of photoperiodic sensitivity on plant age, which in many cases is much greater than that shown, are probably manifold and vary from plant to plant. In *Lolium* it is probably due to the greater sensitivity of the leaves produced later rather than to the increase in the total leaf area (Evans, 1960).

Within the scope of this chapter, these two brief examples must suffice to indicate that a great many variables acting before, during, or after the experimental treatment, can modify photoperiodic responses. A more thorough analysis of photoperiodism, however, calls for closer examination of the prime factors of light, darkness, and time. Even when confined to these factors, our analysis will resemble nothing so much as a crossword puzzle to which half the clues are missing.

Light breaks and the dark period

On normal 24-hour cycles, changes in day length necessarily involve changes in night length. The question thus arises, is the day length (light period) or the night length (dark period) of primary importance in the photoperiodic response? The effects of brief interruptions of the dark periods with low energies of light were discovered relatively early in attempts to answer this question.

Night-interruptions can often prevent the effects that a given photoperiodic schedule would otherwise have. For many short-day plants, for example, schedules favourable to flowering because the light periods are short enough (dark periods long enough), will become completely unfavourable to flowering if the nights are interrupted near the middle by a few minutes of light. Conversely, for many long-day plants, a schedule unfavourable to flowering because the light period is too short (dark period too long) may become favourable to flowering if the dark periods are interrupted. Such results, coupled with the fact that similarly brief dark interruptions of the light period have no effect, led to the idea that short-day plants are actually 'long-night' plants—that is to say, that the controlling factor in photoperiodism was the length of the uninterrupted dark period. As we shall see, this view is no longer tenable. Nevertheless, interruptions of the dark period—so-called 'light-breaks'—have proved to be very useful tools in the experimental analysis of photoperiodism.

Phytochrome and the dark period.   Through the light-break phenomenon, phytochrome (chapter 14) has been implicated as the major, if not the only, pigment involved in photoperiodism. In both long-day and short-day plants, action spectra for light-break effects indicate much higher effectiveness in the red than in other wavelengths. In addition, these effects of red light are at least partially prevented by far-red given immediately afterwards, and repeated reversibility can be demonstrated in some systems in the classical phytochrome manner. An example is shown in Table 16.2A.

Such results suggest that in short-day plants P_{fr} inhibits flowering if it is produced during the dark period, while in long-day plants P_{fr} during the dark period must be regarded as promoting flowering. In short-day plants, at least, the action must be extremely rapid. Table 16.2B shows that in *Xanthium* interposition of even a 30-minute delay between the red and far-red exposures re-

Table 16.2 Red, far-red reversibility of light-break effect in *Xanthium* (data from Downs, 1956).

Treatment near middle of 12-hour dark period on three successive days*	Mean stage of floral development 10 days after plants were returned to long-day conditions†
A. D only (control)	6·0
R	0·0
R–F	5·6
R–F–R	0·0
R–F–R–F	4·2
B. D only	7·0
R	0·0
R–F	6·5
R–30 min D–F	1·8
R–60 min D–F	0·0
R–60 min D at 5°C–F	2·8

* D = darkness, R = 2 minutes red light, F = 2 or 3 minutes far-red light. Temperature 20°C except as noted. Red light source was cool white fluorescent light filtered through red cellophane. Unfiltered source was 1,000 ft.c. at plant level, filtered source delivered 10 kiloergs cm^{-2} s^{-1} at 650 nm. Far-red source was sunlight or incandescent light filtered through red and blue cellophane.

† Arbitrary index of floral primordium development scoring various stages from 0 (completely vegetative) to 7.

duces the reversal by more than 50 per cent, while one hour after the red treatment the entire process has 'escaped from photochemical control' since far-red causes no reversal at all. However, as shown also in Table 16.2B, the rapidity of this escape is reduced if the plants are placed at 5°C during the intervening period. Under these conditions some reversal is still possible after an hour, suggesting that the inhibitory reaction involving P_{fr} is slowed down at low temperatures.

One role of the dark period in photoperiodism then, according to these data, is the control of the level of P_{fr}. One can postulate that for a short-day plant, the dark period must be long enough so that through a substantial proportion of it there is little or no P_{fr}, while for a long-day plant the dark period must be kept short since high levels of P_{fr} seem to be required. While this analysis holds to some degree, it is not a sufficient explanation for all the effects of dark periods. Evidence in its favour can be found, for instance, in experiments showing that under the proper conditions the critical day length of *Xanthium* is about 2 hours longer (dark period for flowering is about 2 hours shorter) if the main light period is closed with a far-red rather than with a red treatment. However, under other conditions, quite the opposite results can be obtained, as we shall see later. In addition, flowering in a number of long-day plants is also strongly promoted by far-red given just before a critical or subcritical dark period, and this is not consistent with the simple notion that such plants require P_{fr} during the dark period. These and similar results suggest that the level of P_{fr} during the dark period, though important, is not the only factor in photoperiodism.

Timing of light breaks.   Another very active area of research is concerned with the time during a dark period at which a single light-break is most effective. One might expect the time of greatest effectiveness to be at the precise middle of the dark period, thus breaking that period up into two smaller ones with maximum efficiency, but this is only rarely true. No such simple rule applies to most of the plants studied, even with dark periods approximating normal terrestrial night lengths on 24-hour cycles.

The most interesting and complex results have been obtained in experiments using extremely long dark periods of 48–96 hours. Figure 16.3 shows a relatively recent example of such work, with seedlings of the short-day plant *Chenopodium rubrum*. These plants flower (even when very small) if they are given a single dark period and then returned to continuous light. In the experiments shown, all the plants received a 72-hour dark period, but for various groups this dark period was interrupted with a few minutes of red light at indicated times. The

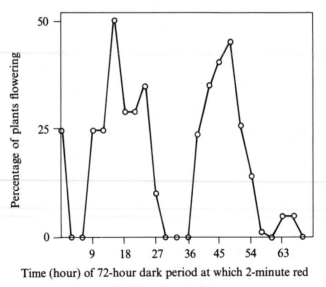

Figure 16.3   The effect of a red light interruption given at various times during a 72-hour dark period on the flowering of *Chenopodium rubrum*. Plants were kept in continuous incandescent light before and after the single dark period. (Modified from Cumming, Hendricks, and Borthwick, 1965.)

data show some of the variability and irregularity typical of work of this kind, but even allowing for this, obvious maxima and minima are evident. A complete understanding of such data is not yet possible. For example, if one thinks in terms of the length of the uninterrupted dark period, as the earlier workers did, one wonders why an interruption at hour 6, which still leaves a 66-hour dark period, inhibits flowering so effectively, while an interruption at hour 48, leaving a 48-hour period and a 24-hour dark period, actually enhances flowering. Results of this kind, as we shall see shortly, have been used as evidence that endogenous circadian rhythms constitute the basic timing mechanism in flowering. They serve here simply to indicate that no simple 'hour-glass' hypothesis of photoperiodic timing—for example, the notion that time measurement depends entirely on the reversion in darkness of the P_{fr} to the P_r form of phytochrome—is adequate.

Characteristics of the main light period

The definition of photoperiodism adopted earlier would suggest that the only significant characteristic of the light is whether it is on or off. While such a definition may be ideal, few plants live in the ideal world. At least two characteristics of the light period other than length (timing) can modify the photoperiodic response in many plants. These are intensity and quality (colour).

Intensity.   Many short-day plants will initiate flowers even in total darkness, so high-intensity light is not absolutely required for photoperiodic induction. Nevertheless, in the widely studied cocklebur, *Xanthium*, flowering is improved if a dark period long enough to be inductive is both preceded and followed by a period of high-intensity light. Under favourable circumstances both the 'first and second high-intensity light processes' can be partially replaced by carbohydrates fed to the leaves, which suggests that they are simply photosynthetic, supplying sources of energy for the inductive processes in the leaves and also perhaps influencing the movement of hormones in the stream of assimilates (see later). However, more recent work on *Xanthium* (Salisbury, 1965) indicates that another process, saturating at low intensity and probably depending on phytochrome, is also involved in these effects. Thus the high-intensity-light processes of the older literature may turn out to be mainly low-intensity light processes after all.

Among long-day plants the effect of supplementary light in prolonging the main photoperiod is often highly dependent on intensity. However, in such experiments the main photoperiod is often of natural or fluorescent light, while the supplementary light is incandescent, with a much higher proportion of far-red, so that these effects are probably more closely related to those in the following section.

20

Quality. It is not easy to produce relatively pure spectral bands of light at sufficiently high intensities over large enough areas to grow plants of any size. Nevertheless, several laboratories have found that the spectral quality of the main light period strongly affects photoperiodic response. A striking case is the discovery by Stolwijk and Zeevaart (1955) that the long-day plant *Hyoscyamus niger* flowers poorly, if at all, even in continuous red light (640–660 nm) or with day-lengthening supplements of red light. It appears to require either violet (370–430 nm), blue (410–500 nm), or far-red (700–1000 nm). Two sets of data are shown in Table 16.3. Like many long-day plants, *Hyoscyamus* is a 'rosette

Table 16.3   Effect of various light qualities on flowering and bolting of *Hyoscyamus niger* (data from Stolwijk and Zeevaart, 1955).

Repeating schedule, hours of indicated light*	Days of treatment	Result	
		Growing point dissection	Mean stem length (mm)
Continuous R	19	Veg	0
63 R–9 B	19	Floral	77
39 R–9 B	19	Floral	113
15 R–9 B	19	Floral	218
10 W–14 D	33	Veg	0
10 W–4 R–10 D	33	Veg	0
10 W–4 F–10 D	33	Floral	81
10 W–4 R–4 F–6 D	33	Floral	180
10 W–4 F–4 R–6 D	33	Floral	149

* R, red; B, blue; W, white fluorescent; F, far-red; D, darkness.

plant' in which the leaves are borne in a tight rosette at the ground level until flower induction has begun, at which time the stem elongates rapidly in the phenomenon known as 'bolting'. It is clear from the table that some blue or far-red—as little blue, for example as 9 hours in every three days—are required for both flowering and bolting. In a sense these results indicate that *Hyoscyamus* is photoperiodic only under violet, blue, or far-red and is essentially photoperiodically insensitive to red, orange, and green.

Results of an opposite kind have been obtained with a number of plants, notably the duckweed *Lemna perpusilla*. This plant shows a typical short-day response under red light but is essentially day length indifferent under either blue or far-red, flowering even under continuous illumination with these colours (Oda, 1962). However, under a long-day blue schedule that permits flowering, the addition of even a few minutes of red will prevent flowering—that is, will bring about a true long-day response. Thus blue (or far-red) is not merely inactive in this system, but interacts with red light (Hillman, 1965).

The only generalization that often holds for such work—and certainly holds for the two apparently opposite cases discussed above—is that if there are quality effects at all, the red region has one action and the blue and far-red regions another. At least two explanations for this have been proposed. One is that in addition to phytochrome, which is affected by brief exposures to both red and far-red, there is a second pigment, the so-called 'high energy reaction' pigment (see chapter 15) that responds to high energies of both blue and far-red, but not to red. Another possibility is that all these effects are mediated through phytochrome itself, the only difference being that red light brings about conversion of a very high proportion of the total phytochrome to P_{fr} while both blue and far-red maintain a much lower level of P_{fr}. Under neither explanation, of course, do we have as yet a general hypothesis that explains photoperiodism.

Photoperiodic timing and endogenous rhythms

The crucial aspects of photoperiodism lie neither in the dark period alone nor in the light period, but in the interactions of light and darkness. Since very few experiments as complex as those on such interactions are done without a hypothesis in mind, it may be helpful first to indicate some of the theoretical background.

As already noted, one hypothesis of photoperiodic timing is that the length of the dark period required for a given effect is related to the reversion of the P_{fr} form of phytochrome, made during the light period, to P_r. While some of the evidence against this as a sole timing mechanism has already been presented, it may well be an element of the overall system. Phytochrome reversion would provide a simple linear 'hour-glass' type of timer, but it is increasingly widely accepted that the overall timing system probably involves one or more oscillating timers depending on endogenous circadian rhythms (see chapter 18). This general idea has assumed various forms, none of them by any means proved. Perhaps the best known is the hypothesis of one of the earliest and most vigorous advocates of the rhythm mechanism in photoperiodism, Erwin Bünning. In its simplest form, Bünning's view is that the endogenous rhythm of the plant consists of—or is expressed as—a continual alternation of two phases, a photophile ('light-liking') phase and a scotophile ('dark-liking') phase, each lasting roughly twelve hours. Light given during the former promotes flowering, while light in the latter inhibits. The difference between short-day and long-day plants, in this view, resides in the different ways in which the daily onset of light and darkness regulates the succession of the two phases. Bünning has changed the details of this hypothesis many times over the years, but the concept of the various phases has remained (Bünning, 1963).

Various other hypotheses of photoperiodic timing based at least in part on endogenous rhythms have been proposed. An example of the complexities that can be envisaged is the suggestion by Takimoto and Hamner (1965), based on work with *Pharbitis nil*, that there are at least three kinds of timing mechanisms in the photoperiodic response. One is an 'hour-glass' type, linear in character and highly temperature sensitive. The two others are both rhythmic and relatively temperature insensitive, but one is set in motion by the 'light-on' signal provided by the start of a light period, while the other is similarly affected by the 'light-off' signal at the start of a dark period. Though it is impossible in this chapter to examine any specific mechanism in detail, some of the evidence in favour of the participation of endogenous circadian rhythms in photoperiodism can be briefly outlined.

A property of endogenous circadian rhythms relevant to photoperiodic timing is their temperature compensation (see chapter 18). From an evolutionary point of view, a photoperiodic mechanism that is not temperature compensated would be of little use. More to the point, in a number of plants various quantitative photoperiodic characteristics such as the critical day length, or the time during a dark period at which a light-break is maximally effective, are extremely similar even at widely differing temperatures. They never resemble simple chemical reactions, the rate of which would at least double with a 10°C rise in temperature. Since there are also plants in which a photoperiodic response is exhibited at certain temperatures but not at others, the temperature-independent character of photoperiodism should not be over-generalized. Nevertheless in certain cases it is striking, and recalls the similar characteristics of endogenous circadian rhythms.

The effects of light-breaks provide some of the best evidence in favour of the participation of endogenous rhythms in photoperiodic timing. The results in Fig. 16.3, for example, could be regarded as due to the interaction of the light-break with a succession of scotophile and photophile phases during the 72-hour dark period. The two maxima in flowering are about 30 hours apart, as are the three minima, which would suggest that the free-running circadian periodicity in this plant under the conditions used is about 30 hours. No general hypothesis of photoperiodic timing not based on rhythms has so far succeeded in providing a plausible explanation of results such as these.

In spite of such results in various plants, not all investigators are in agreement concerning the importance of endogenous circadian rhythms in photoperiodism. While most would probably agree that there is a close relationship between the two phenomena, the significant objection remains that explaining one unknown —photoperiodism—in terms of another—endogenous rhythms—hardly contributes to any real understanding of either. True enough, but if photoperiodism is indeed a special function of a more general 'biological clock', the fundamental explanation must come from understanding the latter. At any rate, most recent work on light-dark interactions is best understood with this background in mind.

Light-dark interactions

Light-period lengths and dark-period lengths. A striking example of the inter-action between light-and dark-period lengths was observed by Claes and Lang (1947) on the long-day plant *Hyoscyamus niger*. In light-dark cycles totalling 24 hours, *Hyoscyamus* flowers under any photoperiod longer than 11 hours and the maximum response requires 15–16 hours light in every 24. However, if the light-dark cycle totals 48 hours, flowering occurs with photoperiods longer than 9 hours—that is, even on cycles consisting of 9 hours light, 39 hours dark-ness—while the maximum response is given with 13 hours of light alternating with 35 hours of darkness. Thus the critical day length, 11 hours on the 24-hour cycle, is actually decreased to 9 hours in conjunction with longer dark periods. No simple hypothesis by which photoperiodism depends simply on the length of either the light or the dark period will explain such results.

Cycle lengths that are not simple multiples of 24 hours can be obtained by combining light and dark periods of various lengths. Studies of this kind on a number of plants indicate that flowering is best on cycles with total lengths closest to simple multiples of 24 hours and worse on cycles of intermediate length such as 36 hours. A very complex group of experiments of this type has been done with Biloxi soy beans. Blaney and Hamner (1957) combined photo-periods 8 hours long with dark periods ranging from 10 to 48 hours and obtained maximum flowering with cycles either 24 or 48 to 52 hours in total length. However, flowering did not occur on any cycles with dark periods less than 10 hours long, or with dark periods 26 hours long even when the latter were asso-ciated with light periods 22 hours long. Flowering also failed on cycles with light periods of 16 hours or longer, even with dark periods of 32 or 36 hours. Thus certain absolute effects of dark- or light-period length are not overridden by the use of cycles of 24- or 48-hour duration. Such experiments also indicate, of course, that the 'relative length of day and night' is not the major factor in photoperiodism, since Biloxi soy beans flower rapidly in 8 hours light–16 hours darkness, but not in either 16 light–32 darkness or 12 light–24 hours darkness:

Interruption of light periods. Dark interruptions of the light period are inactive if they are as short as the few minutes required for many light breaks, but longer interruptions can have interesting effects. Testing the hypothesis that the scoto-phile phase in long-day plants begins shortly after the daily onset of light, Bünning and Kemmler (1954) studied the effects of inserting a 2-hour dark inter-ruption 3 hours after the start of each $17\frac{1}{2}$ hour light period. Thus, their compari-son was between the schedules $17\frac{1}{2}$ hours light–$6\frac{1}{2}$ hours darkness and the schedule 3 hours light–2 hours darkness–$12\frac{1}{2}$ hours light–$6\frac{1}{2}$ hours darkness. There was no effect on either *Plantago* or *Spinacia* but the flowering of *Anethum* (Dill) was hastened by the dark interruption. Whether or not this partial con-firmation significantly strengthened the original hypothesis, the fact that the

Table 16.4 The effect of darkness during the light period on flowering in *Lemna perpusilla* (data from Hillman, 1963).

Light/dark schedule (hours) given for seven days	Relative flowering*
A. 1. $1\frac{1}{4}$L/$22\frac{3}{4}$D	1·00
2. 8L/16D	0·98
3. 9L/15D	0·96
4. 10L/14D	1·10
5. 11L/13D	1·09
6. 12L/12D	0·37
B. 1. $1\frac{1}{4}$L/$22\frac{3}{4}$D	1·00
2. 1L/$6\frac{3}{4}$D/$\frac{1}{4}$L/16D	0·85
3. 1L/$7\frac{3}{4}$D/$\frac{1}{4}$L/15D	0·25
4. 1L/$8\frac{3}{4}$D/$\frac{1}{4}$L/14D	0·02
5. 1L/$9\frac{3}{4}$D/$\frac{1}{4}$L/13D	0·00
6. 1L/$10\frac{3}{4}$D/$\frac{1}{4}$L/12D	0·00
C. 1. $1\frac{1}{4}$L/$22\frac{3}{4}$D	1·00
2. 1L/$8\frac{3}{4}$D/$\frac{1}{4}$L/14D	0·02
3. $\frac{1}{4}$L/2D/$\frac{1}{4}$L/2D/$\frac{1}{4}$L/$2\frac{1}{2}$D/$\frac{1}{4}$L/$1\frac{3}{4}$D/$\frac{1}{4}$L/14D	1·22

* As a fraction of the flowering in each experiment under the standard cycle $1\frac{1}{4}$L/$22\frac{3}{4}$D. This causes flowering of 25 to 60 per cent of the plants in each culture flask, depending on the experiment.

flowering of a long-day plant can be promoted by an appropriately timed dark interruption of the light period surely emphasizes the interacting roles of light and darkness.

A large effect of darkness during the light period is demonstrable with the short-day plant *Lemna perpusilla*, which can be grown with sucrose as a substrate and thus does not require photosynthesis for growth. As shown in Table 16.4A, flowering is essentially constant with dark periods of from $22\frac{3}{4}$ to 13 hours (light periods from $1\frac{3}{4}$ to 11 hours) and falls off with 12-hour alternations. In Table 16.4B, however, all schedules contain only $1\frac{1}{4}$ hours of light per day arranged so as to imitate the beginning and end of the various light periods in Table 16.4A. Such schedules leave the same uninterrupted long dark periods, but at the same time include long periods of darkness in the 'light periods'. Here, flowering even with the 15-hour dark periods is markedly reduced. There is no flowering at all with the 14- or 13-hour dark periods, though these allow maximum flowering when the light periods consist entirely of light. Thus, long dark interruptions in the light periods clearly inhibit the short-day promotion of flowering, if, say, treatment A4 is compared with treatment B4. On the other hand, comparing all the treatments in A with all those in B, one might rather conclude that interpolation of darkness in the light period in effect shifts the 'critical night length' from its value of slightly more than 12 hours with uninterrupted light periods, to more than 15 hours. Whatever the meaning of this or other interpretations, Table 16.4C makes it clear that the effect of the darkness can be abolished by a few brief light exposures, so that we are evidently dealing with a genuine effect of timing, not with the total amount of light. Later experiments on this effect (Hillman, 1964) indicate that it depends on an endogenous circadian rhythm; a possible mechanism has been proposed by Pittendrigh (1966).

An investigation in Bünning's laboratory indicates that interruptions of the main light period with varying light qualities, may also modify flowering. Using the short-day plant *Chenopodium amaranticolor*, Könitz (1958) reported that an extensive far-red exposure during the light period inhibited flowering on an otherwise inductive schedule. He interpreted this effect as due to an endogenous rhythm in phytochrome action such that flowering requires P_{fr} during the photophile phase but it is inhibited by P_{fr} during the scotophile. Though Könitz's results have been the subject of some controversy (Kofranek and Sachs, 1964) the conclusion that the effect of phytochrome conversion on flowering depends on concomitant events, such as the phase of an endogenous circadian rhythm, is consistent with an increasing amount of evidence including the (later) work on *Lemna* already discussed.

Effects of the light period on light-break action.   Recently, several investigations have indicated major interactions between the character of the light period and the action of light-breaks. This work began with some apparently anomalous responses of *Pharbitis nil*, in which it was reported that although mature plants responded to red and far-red during the dark period essentially as does *Xanthium*, seedlings showed no far-red reversibility. In addition, far-red exposures early in the dark period surprisingly inhibited, rather than promoted, flowering in this short-day plant (Nakayama, Borthwick, and Hendricks, 1966). A detailed study by Fredericq (1964) shows some of the factors involved in both phenomena. As shown in Fig. 16.4A, the length of the photoperiod has a powerful effect on far-red reversibility in *Pharbitis* seedlings. Reversibility fails with short, but not with longer photoperiods. With extremely short photoperiods, such as 2 hours, the inhibition by far-red early in the dark period is also easily demonstrated. This inhibitory action is less pronounced in conjunction with high-intensity light periods. The data in Fig. 16.4B, for example, show total inhibition by far-red at the start of a 22-hour dark period whether the light period is 800 or 1600 ft c, but under the latter conditions the inhibition is no longer obtainable by the sixth hour of darkness, while it is still evident in conjunction with the lower intensity photoperiod. In short, far-red reversibility of light-breaks is obtainable with long high-intensity light periods; it fails, and far-red inhibition at the start of the dark period appears, with short, low-intensity light periods.

These results have been interpreted as due to complex interactions of P_{fr} with its unknown substrate or substrates. They certainly confirm the suggestion that

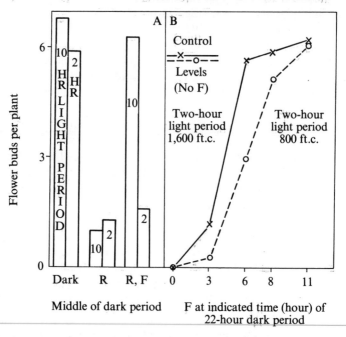

Figure 16.4   Effects of light-period length (A) and intensity (B) on far-red action in *Pharbitis nil*. Young plants were given three cycles of the indicated treatments, then evaluated after five more days in continuous light. All light was cool white fluorescent; intensity in (A) was 2,000 ft.c. Far-red and red in (A) were 30 seconds each; far-red in (B) 4 minutes. (Data from Fredericq, 1964.)

P_{fr}, though inhibitory to the flowering of short-day plants during the dark period, is required during the light period. In this view, cycles with very short dim photoperiods provide insufficient time and substrate for the P_{fr} 'light period' action and thus show inhibition by removal of P_{fr} (by far-red exposure) early in the dark period. However, this hypothesis alone still fails to explain the lack of reversibility observed later in the dark period—one can only postulate in general terms that the rate of the P_{fr} 'dark period' action may be related to the previous amount of light period action. Even using long, high intensity photoperiods, Fredericq (1964) found that the P_{fr} dark period action in *Pharbitis* is extremely rapid. While reversibility is demonstrable if 30 seconds of red are followed immediately by far-red, reversibility is lost if even 3 minutes of darkness are interposed between red and far-red. Thus the lack of reversibility in conjunction with short, dim light periods may well represent an even more rapid escape from photochemical control.

That the relation between characteristics of the photoperiod and the action of light-breaks is not peculiar to *Pharbitis* is indicated by the results of Borthwick and Downs (1964) with *Xanthium*. These show the same sort of relationship between light period length and intensity on the one hand, and the effect of far-red early in the dark period on the other hand. Reversibility in the middle of the dark period was not tested.

Lemna perpusilla exhibits quality-dependent interactions between the main light period and light-break action. In conjunction with a red main photoperiod, either red or blue light is effective as a light-break. In conjunction with a blue main photoperiod, however, only red will give a long-day (flower-inhibiting) effect; blue is inactive (Hillman, 1965). A further interaction is on reversibility. With a red main photoperiod, neither a blue nor a red light-break is far-red reversible, since far-red itself is inhibitory in the dark period irrespective of photoperiod length. However, with a blue main photoperiod, far-red in the dark period is no longer fully inhibitory and thus becomes capable of partially reversing the effect of a red light-break (Hillman, 1966).

Results of the kind discussed in this section must complete our outline of photoperiodism itself; they may well represent the closest approach so far to systems in which the basic mechanism of photoperiodism in plants will be elucidated. Although light responses other than those mediated by phytochrome may be involved, it is at least a reasonable hypothesis that there are only two basic controlling factors in photoperiodism. One would be the rate of the still unknown reaction in which P_{fr} participates, and which in turn would be affected by the amount of P_{fr}, the time during which it is presented and the level of substrate. The second would be the phase of the endogenous circadian rhythm according to which the product of the P_{fr} action might have differing effects. Alternatively, the rhythm might control the availability of the substrate in question. At least, the idea that photoperiodism is a simple function of either the light or the dark period must be discarded; a mechanism is required that explains interactions across time.

Vernalization

Cold treatments and flowering

The term vernalization is best used for the specific promotion of flower initiation by a previous cold treatment. Exposure to low temperatures can indeed affect many aspects of plant development, notably dormancy and seed germination (chapter 17), but at least for the purposes of definition we distinguish such effects from vernalization. Though the basic mechanisms of all actions of low temperature may well be related, there is still a difference, as indicated earlier, between specific promotions of flower initiation and general enhancements of growth.

In its narrowest sense, vernalization means the specific promotion of flowering in some ('winter') strains of cereals by cold treatments of the moistened or germinating seeds. Species with typical winter strains include *Secale* (rye) and

Triticum (wheat). Winter annual strains are also found in many non-cereals including *Pisum* (pea). Otherwise identical effects can be obtained in certain species only by cold treatment of already developed plants rather than seeds or very young seedlings, and the term vernalization is reasonably extended to cover these effects as well. Such plants include many typical biennials, including strains of *Daucus* (carrot), *Lunaria* (money-plant) and *Hyoscyamus*, as well as perennials such as *Chrysanthemum*. As with photoperiodism, the vernalization requirements of various species may be either quantitative—in that a plant will eventually flower even without being vernalized—or qualitative—that is, absolute. Though there is no necessary relationship between vernalization and a particular photoperiodic response-type, many vernalizable plants studied are plants in which vernalization confers responsiveness to long days, which then bring about flowering.

A great proportion of what is known about vernalization has been derived from work with the winter strain of Petkus rye (*Secale cereale* var Petkus) and the biennial strain of *Hyoscyamus niger*. Each of these appears to differ by a single gene from the corresponding strain that does not require cold. Petkus winter rye is a winter annual with a quantitative vernalization requirement, while the cold treatment of biennial *Hyoscyamus* is absolute. In both plants vernalization is required before the plants can respond to long days.

The situation in cereals has been analysed in a long series of papers by Gregory, Purvis, and their collaborators (see, e.g., Gregory and Purvis, 1938a, b). Petkus winter rye germinated at usual growing temperatures is essentially day length indifferent; it does not initiate flowers until it has laid down approximately 22 leaves no matter what the conditions. The spring strain, germinated under the same conditions, is a quantitative long-day plant: under short days it also initiates about 22 leaves before its inflorescence, but, under sufficiently long days, only 7 leaves precede inflorescence initiation. However, if moistened seeds of the winter strain are held at about $1°C$ for some time before planting, the seedlings produced respond more like the spring strain. The degree to which the response approaches that of the spring strain depends on the duration of the cold treatment. As little as 4 days has some effect, but a number of weeks of cold are required to cause flower initiation (under subsequent long days) after as few as 7–9 leaves have been initiated.

In most instances of vernalization the effect of a cold treatment depends in a fairly complex fashion on both its duration and the temperature employed. Figure 16.5 shows some data on the response of biennial *Hyoscyamus niger* obtained by Lang (1951), who exposed young plants to various cold treatments under short-day conditions and then transferred them to warm long days. A fully vernalized plant responds to long days exactly as does the annual—by flower initiation and bolting—so the intensity of the vernalization can be estimated by its reduction of the time required to detect visible bolting. Figure 16.5A indicates that even one or two weeks at 3 or $10°$ have some effect—since

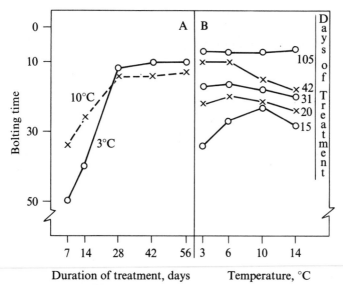

Figure 16.5 Effects of duration and temperature of cold treatments on subsequent response of biennial *Hyoscyamus niger* to long days. Bolting time—the number of days to the start of visible stem elongation—is related to floral initiation. The shorter the time, the greater the promotion of flowering. Photoperiod 8 hours during cold treatment, which was terminated by transfer of plants to 23°C under 16 hours photoperiod. A. Effects of 1 to 8 weeks at 3° or 10°. B. Effects of increasing durations of treatment at 3, 6, 10, or 14°. (Modified after Lang, 1951.)

untreated plants never flowered—but four or more weeks at these temperatures give a maximum effect (minimum bolting time). In an apparent paradox, the higher temperature seems more effective than the lower for exposures of one or two weeks. Figure 16.5B confirms and extends this apparent paradox, showing some relationships between the effects of exposure time and of temperature. For the 15 days exposure, 10° is optimal, but as exposure time is extended the optimum shifts to lower temperatures. Finally with 105 days of treatment, 3, 6, 10, or 14° are equally effective.

As indicated previously, the difference between winter annuals like rye, and biennials like *Hyoscyamus* is that the latter must reach a certain size before cold treatments are effective. The relationship between size and responsiveness to cold is shown very clearly in Table 16.5, based on a study of *Lunaria biennis* by Wellensiek (1958). Unlike the plants just discussed, *Lunaria* is photoperiodically indifferent even after vernalization. The cold requirement is absolute, however, and plants kept at high temperatures do not flower. Table 16.5A shows that a 12-week cold treatment is completely ineffective unless the plants are at least 7 weeks old at its start, while Table 16.5B gives some idea of the relationship between the age of a plant and the length of the cold treatment required to saturate its vernalization requirement. The possible complexities of vernalization are just hinted at by the observation in the same series of experiments that although cold treatment of the seeds alone will not lead to flowering, it does reduce the cold requirement of the young plants themselves. Thus, plants from vernalized seeds require slightly shorter cold treatments or can be vernalized at a slightly younger stage than plants from seeds germinated at higher temperatures.

Table 16.5 Effects of age on vernalizability in *Lunaria biennis* (data from Wellensiek, 1958).

A.	Age (wks) at start of cold treatment*	Percentage of plants flowering†	Mean number of days to budding
	17	100	15·2
	13	100	16·3
	9	100	20·0
	7	50	26·7
	5	0	—
	3	0	—

B.	Age (wks) at start of cold treatment*	Percentage of plants flowering with indicated number of weeks of cold				
		12	10	8	6	4
	12	100	100	96	61	0
	10	91	100	52	13	0
	8	94	71	26	0	0
	6	4	0	0	0	0
	4	0	0	0	0	0

* Cold treatments at 5°C with 12 hours of fluorescent light per day. Treatment in A was 12 weeks long. Days to budding calculated from end of cold treatment.

† After the cold treatments, plants were returned to normal field or greenhouse conditions. All cold treatments were timed to end together for each experiment, so the subsequent conditions were identical for all treatments.

The nature of the vernalization process

Not enough is known about vernalization to define the mechanism, but there is information on its relationship to metabolism, its permanence, and its site.

Vernalization requires energy. The simplest answer to the question, what happens during vernalizing cold treatments, might be that nothing happens— that there is merely a suspension or slowing down of processes occurring faster at normal growing temperatures. That this is unlikely is indicated by a number of facts. First, vernalization is an aerobic process. Table 16.6A gives some data obtained by Gregory and Purvis (1938b) with Petkus winter rye, indicating the complete ineffectiveness of a cold treatment given under nitrogen. According to other experiments, as little as one-fifth of 1 per cent of the normal oxygen level in air will allow some vernalization to take place. Later work in the same

Table 16.6   Seed vernalization in Petkus winter rye: oxygen requirement, and reversal by drying (data from Gregory and Purvis, 1938a, b).

A.	Mean leaf number preceding inflorescence	Score†
Control*	23·3	21
9 wks at 1°C, air (vernalized)	8·1	133
9 wks at 1°C, N$_2$	23·4	25
1 wk at 20°C, N$_2$	22·3	21

B.	Mean spike length at 19 wks, mm	Mean number of tillers per plant
Control*	1·6	4·7
Vernalized, dried 1 day	34·0	2·7
Vernalized, dried 6 weeks	44·3	2·9
Vernalized, dried 8 weeks	4·9	9·7
Vernalized, dried 14 weeks	2·4	16·3
Vernalized, dried 20 weeks	1·6	13·7

* Seeds were planted under long-day field conditions without indicated pretreatment.
† An arbitrary measure of floral development rate based on a number of characteristics and frequently used by these authors.

laboratory showed that, besides oxygen, a substrate is also required. This is demonstrated by excising and culturing embryos under sterile conditions on various media. Cold treatments on media without a sugar give rise to plants whose later growth is excellent, but only with a sugar present is vernalization successfully accomplished.

Vernalization, then, is not a passive absence of activity, but requires a substantial amount of energy. The size requirement for vernalization in biennials, in fact, has been interpreted as due to the need to produce and store a sufficient amount of substrate. Though mere correlations are notoriously risky, it is noteworthy that the seeds of many common biennials are small with relatively little storage material, while the cereals, at least, and winter annuals such as peas, have abundant reserves. The evidence that vernalization requires energy also makes it easier to understand the greater effectiveness of higher rather than lower temperatures with short exposure times (Fig. 16.5A) and the shift towards lower optima as exposure time increases (Fig. 16.5B). Effects of this kind might represent a balance between the process of vernalization itself, favoured by low temperatures, and the processes supplying energy, which are presumably slowed by the same low temperatures. However, other explanations for these optimum shifts have been advanced.

Devernalization.   Vernalization is not only preventable, as by limiting the supply of oxygen or substrate, it is also reversible. In winter rye, drying vernalized seeds and holding them at room temperature for various lengths of time reverses the flower-promoting effect of the cold treatment. This is shown by the data of Gregory and Purvis (1938a) in Table 16.6B, in which the rate of floral development resulting from various treatments is expressed as spike (inflorescence) length after 19 weeks of field growth. Holding dry for as long as 6 weeks had no inhibiting effect (perhaps it even promoted), but holding dry for 8 or more weeks greatly reduced the rate of floral development, and 20 weeks returned it to the unvernalized level. However, the second column of Table 16.6B makes it clear that this effect is not a simple reversal of vernalization. The number of tillers—branches arising from the base—produced by winter rye plants is usually small whether the plants are vernalized or unvernalized. The devernalization caused by holding the seed dry, however, brings about an increased number of tillers; the increased capacity for floral development is somehow converted to branching. The mechanism of this effect is not known, and must depend on relatively early events in meristem development.

Devernalization also occurs in other winter annuals, in biennials and in perennials. The usual agent of devernalization is relatively high temperature, though light or darkness and even photoperiodism may play a role. In general, the more complete the vernalization treatment and the more time elapses before attempts at devernalization, the less successful will devernalization be. Obviously certain hardy perennials that require vernalization each year must do so because the normal environmental cycle brings about devernalization. Schwabe (1955) has shown that in *Chrysanthemum*, devernalization is caused by low light intensities combined with relatively high temperatures (23–28°C), conditions that may well occur among the growing rhizomes in late summer. It is interesting to note that even the spring strain of Petkus rye, which can in a sense be regarded as genetically vernalized, is apparently slightly devernalized if the moistened seeds are held at room temperature under anaerobic conditions for several weeks. Seed treated in this manner and then planted under long days, flowers only after producing 8 or 9 leaves rather than after 7. This effect can be reversed by the usual vernalizing cold treatments (Gregory and Purvis, 1938b).

The site of vernalization.   Different portions of biennial or perennial plants can be cooled by means of small refrigerated coils. Experiments conducted in this manner indicate that the growing tip is the site of vernalization in most plants. The exposure of this portion to cold seems sufficient to make all the tissues subsequently derived from it respond in the typical vernalized fashion. Using the culture techniques mentioned earlier, even fragments of rye embryos can be vernalized if they are sufficiently large to generate a new plant. Hence vernalization appears to induce a state that, once acquired, is passed on across many cell generations from the cells originally exposed.

Wellensiek (1964) has recently urged the view that not only apical meristems, but all dividing cells including those in roots or leaves, can be regarded as the sites or potential sites of vernalization. The requirement for the exposure of the growing tip thus represents simply a requirement for dividing cells. Some experiments on the vernalization of leaf cuttings of *Lunaria biennis* (Wellensiek, 1964) provide the major evidence.

If excised leaves of *Lunaria* are set in sand, they root, and eventually regenerate into entire plants. Plants regenerated from leaves of unvernalized plants require vernalization in order to flower, but if the cut leaves are first held at 5°C for a sufficient time, the plants derived from them will respond as though vernalized. This in itself indicates that a normal apical meristem need not be present for vernalization to occur. The involvement of dividing cells is indicated as follows. If, at the end of the cold treatment, the lower 5 mm of each leaf cutting is removed, the plants that finally regenerate are not vernalized. This is apparently because only the cells in the lower portion of the petiole near the original cut were dividing during the cold treatment. Further evidence is obtained by treating entire plants at 5°C, then making cuttings of each leaf under higher temperatures and eventually assessing the degree of vernalization of plants so regenerated. In such experiments, the younger the leaf the greater the degree of vernalization in the final plant. Older leaves appear to have lost the susceptibility to vernalization entirely. This result correlates well with the varying proportion of cells in a young dividing state in the various leaves.

Vernalization and photoperiodic responses; related temperature effects

The preceding material gives an oversimplified view of the effects of low temperature on flowering in at least two respects. First, the frequent correlation between vernalization and long-day response does not exhaust the relations between vernalization and photoperiodism. Second, many effects of low temperatures on flowering resemble vernalization to a greater or lesser degree and are difficult to classify.

In certain strains of *Spinacia* (spinach) and *Trifolium* (clover), seed vernalization or plant vernalization can modify or even abolish the long-day requirement, either reducing the critical day length or causing the plants to become day length indifferent. In at least one vernalizable plant, *Chrysanthemum*, vernalization brings about a short-day rather than a long-day response. Besides these, one of the most interesting but least explored relationships between photoperiodism and vernalization is that in *Campanula medium* and *Lolium perenne*, in which short-day treatments can replace the cold requirement either wholly or in part. Thus these plants respond either as vernalizable long-day plants or as short-long-day plants, depending on the circumstances. As in all short-long-day plants, the short-day treatment (or vernalization) must precede the long-day exposure in order for flowering to occur, the reverse order being ineffective. Even the classical Petkus winter rye apparently shows a slight substitution of short days for vernalization, though not in anything like the degree of the other plants mentioned. The possible substitution of short days for low temperature may thus have general implications for the mechanism of vernalization, but little systematic work on it has been done.

It is easy to come by examples of temperature effects whose relationships to true vernalization are problematic. Flowering and fruiting in the tomato (*Lycopersicum esculentum*) for instance, is favoured by relatively low night temperatures. This phenomenon, observed also in many other plants, has been called thermoperiodism. Similar effects can also be obtained by exposing the seedlings to low temperatures before they are put on higher temperatures, so that vernalization and thermoperiodism in this plant appear to converge. Another instance of such convergence is presented by *Chrysanthemum*, in which vernalization is easily accomplished by repeated exposure to low night temperatures separated by periods of higher temperature. Finally, there are a number of plants in which it is essentially impossible to decide whether one is dealing with vernalization or simply a very low temperature optimum for flowering. A notable example is the stock *Matthiola incana*. The growing plants need at least 3 weeks below 19°C to initiate flowers and must be held under these conditions until full differentiation of the primordia has taken place; even a brief period at high temperature will inhibit the process during this time. After full differentiation has occurred, however, the plant can be returned to warm conditions and differentiation of new primordia will now continue. Obviously the response of stocks differs from true vernalization both in the higher temperatures involved and in the relative lack of persistence (induction) of the effect until major differentiation has taken place. Just as obviously, however, there is also some induction. Thus, what we have discussed as vernalization may well represent a special case of more general, less easily defined phenomena.

Intermediate physiology of the flowering response

Photoperiodism and vernalization are the chief environmental controls in the physiology of flowering, which suggests that we should now consider the internal processes mediating such control. One might imagine *a priori* what both the physiological and genetic evidence indicates: these processes are not integral parts of either photoperiodism or vernalization. Rather, photoperiodic or vernalizable plants are plants in which some step or steps in the processes leading to flowering have come to be dependent on day length or exposure to cold. Space and the complexity of the problems—which are special aspects of the general regulation of development—forbid more than a brief introductory survey.

Perception and communication of photoperiodic effects

A most important advance in the study of photoperiodism occurred in Russia when Chailakhyan (1936) showed that the short-day plant *Chrysanthemum*, partially defoliated, flowered if the leafy portion was kept on short days and the leafless portions on long days, but not with the conditions reversed. Work with other plants and in other laboratories soon established photoperiodic perception by the leaves as a general principle and opened up many technical possibilities for further experiments. In addition, the theoretical significance of this fact was immediately recognized, notably by Chailakhyan himself, and is simply this. Although the leaves perceive the photoperiodic conditions, flower initiation itself of course takes place at the growing apex. Thus, something apparently moves between leaves and apex to bring about flowering. Chailakhyan suggested that the leaves produce a specific flower-inducing hormone which he named florigen (Greek, flower-maker) (see also chapter 2). Since then much of the research on the photoperiodic control of flowering has been directed at elucidating this general situation either in terms of florigen or in whatever alternative terms will serve. For the present we will avoid the assumption of a specific florigen and use only general terms such as the floral stimulus.

Source and translocation of the floral stimulus.   Effective photoperiodic perception leading to flowering is not a property of all leaves on a plant. The effectiveness of a leaf is generally dependent on its developmental stage. For example, Khudairi and Hamner (1954) working with *Xanthium*, studied the flowering responses of plants reduced to a single leaf before exposure to a long night and thus determined the effectiveness of leaves of various sizes. Figure 16.6, based on their data, is more or less self-explanatory and shows that neither very young

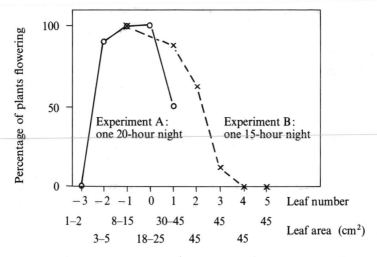

Figure 16.6   Effect of leaf size and age on ability to bring about flowering in *Xanthium*. For each experiment plants were defoliated except for a single leaf of the size and rank indicated, given one long night, and returned to the long-day conditions. (After Khudairi and Hamner, 1954.)

nor very old leaves are effective. The leaves that are most effective in *Xanthium* are apparently those in the most rapid phase of growth; however, this does not hold true for all plants. Another difference often found between species is in the effectiveness of cotyledons. In *Xanthium*, for example, the cotyledons do not bring about a response to short days, while in *Pharbitis*, they do.

Once the presumed floral stimulus has been produced by an appropriate leaf it must then reach the apex. Most studies of this phase of the process agree in indicating that the movement takes place in the phloem. Removing a ring of phloem while leaving the xylem essentially intact ('girdling') stops movement of the stimulus, as does also localized chilling, scalding with steam, or application of poison such as chloroform. Although several early reports suggested that the floral stimulus passed through a layer of cellulose interposed in a graft, it now seems clear that an actual graft union is necessary. It has since then been established that cells can grow through the interstices of a cellulose membrane and thus establish tissue contact in spite of its presence. In short, the stimulus apparently passes only from living cells to living cells. Finally, work with radioactive carbon dioxide shows that conditions affecting the transport of assimilates from the leaf to the apex also have similar effects on movement of the floral stimulus, a fact again indicative of phloem transport.

Though it moves in the phloem, the floral stimulus is not necessarily simply carried with the assimilates in a mass flow. The latter hypothesis has been widely adopted, chiefly on the basis of evidence of the kind just summarized, but it runs into difficulties in terms of the rates of movement involved. Estimates of the rate of movement of the floral stimulus give average velocities of 2–20 mm h^{-1} while estimates of the movements of carbohydrates in the phloem are often 10 times higher. Since such estimates have rarely been made for the movement of both floral stimulus and assimilates in the same plant under the same conditions particular importance attaches to essentially simultaneous studies, notably that of Evans and Wardlaw (1966).

Evans and Wardlaw worked with *Lolium temulentum* plants defoliated to a single leaf and given a single long day. Estimates of assimilate movement were made by exposing a portion of the leaf to radioactive carbon dioxide under photosynthetic conditions for a brief period, then assaying tissues at various distances from the tissue exposed to see how fast the peak of radioactivity moved. These estimates were conducted on the same schedule as the estimates of floral stimulus movement and indicated rates of 770–1050 mm h^{-1}. Stimulus movement itself was estimated as follows. The basal (proximal) 80 mm of each leaf was wrapped with aluminium foil at the end of a short day so that only the distal portion (300–340 mm) received the long-day extension. At various times during the long-day extension, leaves on various plants were cut off either at the base, 40 mm from the base, or 80 mm from the base; some remained intact throughout the long day. Obviously, cutting off the leaves inhibited flowering that would otherwise be brought about. This inhibition was a function of both the time and the place at which the leaf was cut off. As shown in Fig. 16.7, if inhibition is plotted against time of excision, three essentially parallel curves are obtained for the different positions of excision, and the time delay indicated between these curves is an estimate of the time required for the floral stimulus to pass the 40 or 80 mm distance. Thus, in Fig. 16.7, if interpolation is made at the 60 per cent inhibition level, an estimate of 80 mm in 6 hours or about 13 mm h^{-1} is obtained. In a number of such experiments Evans and Wardlaw found rates of 10–24 mm h^{-1}, completely out of accord with the roughly 1000 mm h^{-1} indicated for assimilate movement.

Other experiments in the same paper also suggest that the floral stimulus does not simply move passively with the assimilates. For example, when the effectiveness of leaves of various ages were studied, leaves roughly 20 per cent of their final area proved highly active in inducing flowering, but exported very

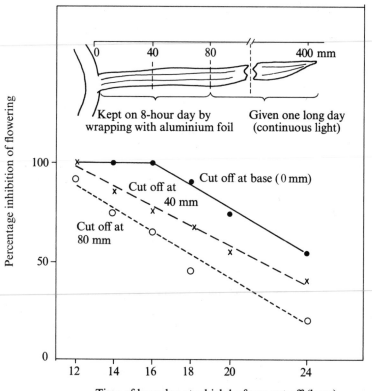

Figure 16.7 Determination of the rate of movement of the floral stimulus in *Lolium temulentum*. All plants used were defoliated except for a single leaf which was treated as indicated. All plants flowered if the leaf was not cut off at all. Inhibition of flowering was less, the later the cutting and the further from the leaf base. (After Evans and Wardlaw, 1966.)

little assimilate. Thus, though the route of floral stimulus translocation can be regarded as established, the manner cannot be, and must perhaps wait until the nature of the material is better known.

Inhibitions of flowering.   A tacit assumption in the preceding paragraphs is that the photoperiodic control of flowering involves merely the promotion of flowering by favourable photoperiods. We have ignored the possibility that non-favourable photoperiodic treatments are actively inhibitory, not simply inactive. Evidence for this has been obtained in both long-day and short-day plants. The latter have been studied in particular by Schwabe (1959). In quantitative studies with various successions of long and short days, Schwabe found that in Biloxi soy beans, *Perilla* and *Kalanchoë*, each long day annuls the effect of approximately two short days; in *Chenopodium* the inhibition is somewhat smaller. These experiments suggested that a long day acted by annulling the effect of subsequent short days, rather than by destroying previous induction. A corollary to this is that day lengths that are above the critical length when given constantly, or when given immediately following long days, may actually be inductive if they follow short-day lengths—that is, if the previous long-day inhibition has been relieved. Since these somewhat confusing statements are largely interpretations and may prove to be incorrect, perhaps it is best simply to look at some data. Table 16.7 summarizes two of Schwabe's experiments with *Kalanchoë* and illustrates the major phenomena.

In Table 16.7A, the treatments in group (i) show the effects of 20-day treatments with constant day lengths, and indicate simply that a $12\frac{1}{2}$-hour day is almost non-inductive and a 13-hour day is completely so under these conditions. However, the treatments in group (ii) show that these day lengths are inductive when preceded by shorter day lengths. For example, a comparison of the third and fourth treatments in this group shows that addition of four $12\frac{1}{2}$-hour days more than double the amount of flowering, and addition of four more 13-hour days increased it further. An interesting comparison can also be made between the last treatment in group (ii) and the last in group (iii). The average day length in both cases is the same, but the ascending treatment (group (ii)) gave twice the flowering of the descending treatment. Somewhat similar effects are shown in Table 16.7B, in which the effects on flowering of $11\frac{1}{2}$-hour day lengths given for 7, 13, or 19 days are shown to depend strongly on whether the treatment is preceded or followed by continuous light or $12\frac{1}{2}$-hour days. Clearly, non-inductive short days either following or preceding inductive short days are much more favourable to flowering than continuous light, which is relatively inhibitory. While such experiments indicate nothing concerning the details of the mechanism, they suggest that photoperiodic effects may in fact be completely continuous, ranging from strongly promotive to strongly inhibitory.

The question arises whether inhibitions of flowering by unfavourable photoperiods have a mechanism analogous to that of promotions by favourable

Table 16.7 Effect of the order of presentation of various day lengths on flowering of *Kalanchoë blossfeldiana* (data from Schwabe, 1959).

	Day length (hours) during indicated period (days)*					Flower number per plant
A.	0–4	5–8	9–12	13–16	17–20	
(i)	11	11	11	11	11	81·3
	$11\frac{1}{2}$	$11\frac{1}{2}$	$11\frac{1}{2}$	$11\frac{1}{2}$	$11\frac{1}{2}$	66·5
	12	12	12	12	12	14·2
	$12\frac{1}{2}$	$12\frac{1}{2}$	$12\frac{1}{2}$	$12\frac{1}{2}$	$12\frac{1}{2}$	0·4
	13	13	13	13	13	0·0
(ii)	11	C	C	C	C	0·0
	11	$11\frac{1}{2}$	C	C	C	2·6
	11	$11\frac{1}{2}$	12	C	C	5·9
	11	$11\frac{1}{2}$	12	$12\frac{1}{2}$	C	13·0
	11	$11\frac{1}{2}$	12	$12\frac{1}{2}$	13	17·8
(iii)	13	C	C	C	C	0·0
	13	$12\frac{1}{2}$	C	C	C	0·0
	13	$12\frac{1}{2}$	12	C	C	0·1
	13	$12\frac{1}{2}$	12	$11\frac{1}{2}$	C	2·0
	13	$12\frac{1}{2}$	12	$11\frac{1}{2}$	11	7·7
B.	2 weeks preceding	Then, either 7 days	or 13 days	or 19 days	2 weeks following	
(i)	C	$11\frac{1}{2}$	–	–	C	0·1
	C	–	$11\frac{1}{2}$	–	C	9·3
	C	–	–	$11\frac{1}{2}$	C	34·8
(ii)	C	$11\frac{1}{2}$	–	–	$12\frac{1}{2}$	2·3
	C	–	$11\frac{1}{2}$	–	$12\frac{1}{2}$	26·9
	C	–	–	$11\frac{1}{2}$	$12\frac{1}{2}$	102·9
(iii)	$12\frac{1}{2}$	$11\frac{1}{2}$	–	–	C	0·9
	$12\frac{1}{2}$	–	$11\frac{1}{2}$	–	C	20·5
	$12\frac{1}{2}$	–	–	$11\frac{1}{2}$	C	88·6
(iv)	C	–	–	$12\frac{1}{2}$	$12\frac{1}{2}$	0·0

* C: continuous light. Plants grown and kept under continuous light except as noted.

photoperiods—that is, the export of active substances by the leaves. This appears to be true in some plants, but the situation is unclear in most of those studied. One of the clearest examples of inhibition by leaves on unfavourable photoperiods may also be somewhat atypical and is to be found in annual *Hyoscyamus*. In the sort of experiment that can only be performed with plants having large storage roots, Lang and Melchers (1943) exposed completely defoliated plants to long days, short days, or total darkness. Rapid flower initiation occurred under all these conditions: leafless *Hyoscyamus* ceases to be photoperiodic. Grafting back even a single leaf, however, restores the photo-periodic dependence of flowering. The conclusion that *Hyoscyamus* leaves export an inhibitor of flowering under unfavourable photoperiodic conditions is difficult to avoid.

Most experiments on this general question of translocated inhibitors, however, are much more complicated. They are usually performed by keeping some leaves on a plant in favourable photoperiods and others on unfavourable photoperiods. The amount of flowering obtained can then be regarded as a resultant of the effects of promotive and inhibitory substances. In this type of experiment, leaves on unfavourable photoperiods generally oppose the effects of leaves on favourable photoperiods most strongly if they are situated between the latter and the apex. Rather than indicating the action of inhibitors, such results have usually been interpreted in terms of the hypothesis that the flowering stimulus is carried to the apex in the mass flow of assimilates. In these terms, leaves on unfavourable photoperiods inhibit flowering either by acting as 'sinks' for such assimilates or by producing assimilates free of the stimulus and thus diluting its presence in the total flow reaching the apex. In either case inhibiting leaves would have greater effects when situated between the promoting leaves and the apex, and the postulation of an exported inhibitor is not required. In spite of the fact that there is reason to doubt any simple version of the assimilate-flow hypothesis of floral stimulus movement, probably most experiments involving inhibition by leaves on unfavourable photoperiods can be interpreted in this way, though they need not be.

An unequivocal example of a translocatable inhibitor of flowering has been analysed by Guttridge (1959) in *Fragaria*, the strawberry plant. The evidence here is analogous to that used for the existence and movement of the floral stimulus in other plants. *Fragaria* is a short-day plant with respect to flowering —long days inhibit flowering and bring about increased vegetative development, particularly of the stolons ('runners') so characteristic of the growth habit. Guttridge made use of plantlets connected to each other by stolons so that one plant of the pair could be kept in long days and the other in short. In a number of such experiments, long-day conditions given to one plant clearly inhibited flowering and promoted vegetative growth in the other. This effect was increased by conditions that would tend to increase the flow of assimilates from the plant on long days to the other plant—such as giving more intense or earlier illumination to the plant on long days—and decreased by the opposite conditions. Thus in *Fragaria* there is no good evidence for a translocatable floral stimulus, but excellent evidence for an analogous inhibitor of flowering that also promotes vegetative growth. The extent to which such mechanisms may occur either instead of, or together with, the mechanism of a translocatable floral stimulus is completely unknown.

Substance and process in induction and flowering

Having indicated some of the physiological evidence concerning the perception and transmission of photoperiodic effects, we can now note briefly what is known of the nature of the events involved. For more detail on this extremely complex topic, see Hillman (1962) and Lang (1965).

Florigen: Physiological concept or chemical reality? The term florigen is properly restricted to mean a specific substance that causes floral initiation and is the same in various species of plants. The most striking evidence for the existence of such a substance comes from interspecific graft experiments. These have been successful between all combinations of photoperiodic response types. For example, in grafts between the long-day *Hyoscyamus* and the short-day Maryland Mammoth tobacco, the *Hyoscyamus* will flower under short days if the tobacco portion of the graft is also kept under short days, but not if the tobacco is kept under long days. Similarly, *Xanthium* can be made to flower on long days if it is grafted to certain long-day composites such as *Erigeron* or *Rudbeckia*. Flowering in photoperiodic varieties of *Glycine* (soy beans) or *Gossypium* (cotton) can be brought about under unfavourable photoperiods by grafting to a day length indifferent variety.

Though such results certainly indicate that floral stimuli produced by plants of one photoperiodic response type can be active in plants of another response type, some caution must be exercised in concluding that a single florigen exists. In the first place, successful grafts are possible only between plants within the same family or order. Thus the equivalence of the floral stimulus in distantly related plants is not demonstrable by this method. In addition, there is indirect evidence that some of the successful grafting experiments may not be due to the transfer of specific floral hormones, but to the removal of inhibitors or the provision of reserve carbohydrates. Undoubtedly the major reason for continuing scepticism concerning the existence of florigen, however, is the fact that no substance has yet been extracted or synthesized that will induce flowering in all types of plants or even in all response types within a single family. In short, florigen remains a physiological concept rather than a chemical reality.

The lack of a known substance to work with is not due to want of effort on the part of plant physiologists and biochemists. Ever since the first indications that floral stimuli existed and could be transmitted by grafting, attempts have been made to prepare extracts with florigen activity. A successful extraction of florigen must fulfil several strict criteria. The substance must be present in photoperiodically induced or day length indifferent plants, but not in photoperiodic plants under non-inductive conditions and it must be able to induce flowering in plants of different photoperiodic types and families under non-inductive conditions. The history of the subject is strewn with reports, both published and unpublished, of results that seemed promising at first but then proved unrepeatable or ambiguous. In addition to the fact that no substance has ever yet fulfilled all the criteria mentioned, ambiguity has often been provided by tests of activity on photoperiodic plants conducted under marginal conditions, with the test plants barely 'poised' at the critical day length, for example. In these circumstances many non-specific substances can affect flowering in one way or the other, but such effects generally disappear under clearly inductive or non-inductive photoperiods.

Various explanations have been advanced for the lack of success in obtaining florigen extracts. One possibility, of course, is that florigen does not exist; the phenomena attributed to it may be due to a particular combination or succession of less specific substances, or to the removal of inhibitors. Another possibility is that florigen is so labile that the techniques of extraction destroy it. Some studies of photoperiodic induction, as we shall see shortly, have suggested that florigen has the autocatalytic properties of a virus; on this basis it may resemble those viruses that can only be transferred by direct graft union.

The most recent, and so far apparently the most successful, attempt to extract florigen has been the subject of several brief reports (e.g., Lincoln *et al.*, 1962). Extracts of both flowering *Xanthium* and flowering *Helianthus* (day length indifferent) are able to bring about induction in *Xanthium* even under continuous light if prepared and applied in the proper manner. While conformity to all the criteria of florigen extraction has not yet been demonstrated, the development is encouraging.

Induction: state and process.   The fact that the flower-initiating effect of a given photoperiodic treatment may persist after the treatment ends has been the subject of many investigations. In certain plants, notably *Xanthium* and *Perilla*, the induced state is extremely long lasting; in others, such as Biloxi soy beans, it persists for only a few weeks or during the development of a few nodes and vegetative growth is soon resumed under non-inductive conditions. An understanding of the induced state thus faces several difficulties. Not only are there all gradations between near permanent and extremely transient induction, but even those plants that maintain the induced state most strongly seem to do so by at least two different mechanisms. This becomes evident from a consideration of *Perilla* and *Xanthium*, both short-day plants.

Induction in *Perilla* is entirely a property of a leaf directly exposed to an inducing photoperiod. Once such a leaf is induced it can be cut off and grafted to another plant and there cause flowering even under long-day conditions. Even detached leaves can be photoperiodically induced, as can be demonstrated by the same test. Attempts to prevent or remove leaf induction with various inhibitors have so far been unsuccessful: unless the inhibitors kill the leaf, induction is unaffected. However, only leaves directly exposed to the inducing photoperiod have this property. Leaves developing on a flowering plant subsequent to its return to long days are not themselves induced—that is, they will not bring about flowering when grafted to another plant. Thus, in *Perilla*, though leaf induction itself is permanent, the induced state of the plant as a whole is eventually lost under long days when all the induced leaves have died.

The origin and maintenance of induction in *Xanthium* differs in almost all respects from the situation described for *Perilla*. The only similarity is that induction initially requires the exposure of a leaf of a certain age (Fig. 16.6) to short days. Here again, as in *Perilla*, the test for induction is the ability of a leaf or branch to bring about flowering when grafted to a vegetative plant on long days. Unlike a *Perilla* leaf, a *Xanthium* leaf once induced loses this property as it grows older. The plant as a whole, however, does not lose the induced state. This is because, though the initial induction required photoperiodic treatment, its perpetuation does not. Younger leaves are indirectly induced even under long days by developing on a plant with active induced leaves. Thus the induced state in *Xanthium* moves from older to younger leaves much as a virus or other self-replicating translocatable infection might. In fact, it has been suggested that an autocatalytic entity of this kind may underlie flowering in *Xanthium*, but there is no direct evidence for such a notion. In this connection, however, it is striking that the only way to cause induced *Xanthium* to revert to vegetative growth seems to be to cut it back several times, removing flowers and leaves and forcing out new growth from previously dormant buds. This is remarkably similar to the manner in which tissue free of certain virus diseases can be produced in some infected plants.

Two other remarkable characteristics of the induced state in some plants are worth noting, fractional induction and steady-state induction. Fractional induction has been observed in plants in which non-inductive photoperiods do not actively oppose the effects of inductive photoperiods in the manner discussed earlier. In such plants, exemplified by the common weed *Plantago lanceolata*, a long-day plant, the requirement for a number of inductive cycles can be satisfied even when many non-inductive cycles are interpolated into the treatment. If, say, half the number of long days required for flowering are given, the plants so treated will 'remember' the treatment even after many weeks on short days, as can be demonstrated by their rapid response to further long-day treatments. Nevertheless, such plants are not otherwise detectably different from plants never previously exposed to long days. This kind of situation may be an extreme case of what can be called steady-state induction, which is easily observed by giving plants minimal or high numbers of inductive cycles and then returning them to non-inductive conditions. In plants such as *Xanthium*, in which induction is maintained indefinitely, it is nevertheless maintained and expressed at very different rates in such experiments. With a large number of inductive cycles, the plants flower profusely and start to fruit within a few weeks, while with a single inductive cycle, development is very slow and fruiting may take the better part of a year. Thus, different steady states of induction can be achieved. It is possible, using the terms employed previously, to see such a situation as due to a very mild or intense 'infection' with the induced state.

Self-maintaining or steady states are not unique to the photoperiodic control of flowering. The perpetuation of the vernalized condition in tissues descended from a vernalized meristem is the same sort of phenomenon. In fact, morphogenic studies on both plants and animals provide many other examples of analogous conditions—the review by Brink (1962) should be consulted in this regard. There is little or no direct evidence about the general mechanism of such states—or even as to whether there is a single general mechanism. Developments in recent genetic theory, especially the ideas of regulatory genes and of whole groups of genes that may be expressed or repressed together—'operons'—may eventually supply some of the basic clues. A major difficulty is that such concepts have developed almost entirely from work with bacteria and the techniques are so far lacking to demonstrate their direct relevance in higher plants. Experimental work of the highest order will be needed to develop the systems in which this can be done. However, as we shall see in the concluding section, there is some evidence already for the direct involvement of DNA and RNA in these processes.

The pharmacology of flowering.   A great many investigations have dealt with promotions or inhibitions of flowering by known chemicals, natural and synthetic. Such work is roughly divisible into two classes with some overlap between them: studies on the effects of plant growth substances and related compounds, and studies on the effect of metabolic inhibitors.

Among the substances generally classed as plant growth substances, it is clearly the gibberellins (see chapter 2) that have the greatest effects on flowering. These compounds generally cause flowering in most rosette plants under noninductive conditions, whether the plants normally require vernalization, longday treatment, or both. One of the earliest detectable events following environmental induction in such plants is a rapid increase in the level of native gibberellins. Hence these compounds are unquestionably part of the chain of internal processes leading to flowering in many plants. That gibberellins cannot be florigen, however, is indicated by their relative inactivity on short-day plants, in which the effects on flowering are at best marginal and occasionally inhibitory. It has recently been suggested by Chailakhyan, the original proponent of florigen, that two substances, gibberellin and 'anthesin', constitute florigen, with the former limiting in long-day plants and the latter limiting in short-day plants. But the evidence is tenuous at best, since anthesin has not been identified.

Plant growth substances other than gibberellins, notably the auxins and cytokinins, seem to be only indirectly related to photoperiodism and vernalization, and interact strongly with these processes only in marginal or anomalous circumstances. Naturally, since these substances are of primary importance in the general control of plant cell growth and differentiation, a subsidiary role in any plant process is difficult to exclude.

Several other substances whose actions may be related to that of plant growth substances have interesting effects on flowering, though perhaps of no general significance. Ethylene chlorohydrin can cause the flowering of *Xanthium* under completely non-inductive conditions; whether the action is related to the hormonal action of ethylene in plant cells is unknown. Several compounds active as growth retardants (perhaps through inhibition of gibberellin synthesis) have brought about flowering in species of *Rhododendron* under conditions normally unfavourable. The implication of these results is that gibberellin levels in some plants may be so high as to prevent flowering.

Many compounds known to act as metabolic inhibitors have, over the years, been tested on flowering systems. Two general difficulties accompany such investigations. First is the problem of making certain that any inhibition of flowering obtained is relatively specific, not merely a consequence of general growth inhibition. In some investigations it has even seemed possible to identify the mode of action of the inhibitor, as, for example, an effect on the synthesis rather than the action of a floral hormone, though a number of complex assumptions underlie such analyses. The second difficulty is inherent in all forms of inhibitor analysis—the question of inhibitor specificity. When inhibitors are first used their biochemical action seems to be highly specific, but the longer they are used and studied, the less certain the specificity. Nevertheless, some meaning is often conferred on such experiments by showing specific reversals of the inhibition with substances consonant with the proposed interpretation —for example, by the reversal of the action of a supposed anti-gibberellin by gibberellin. Even in such cases, however, the reversal may turn out to be unspecific.

Of the large number of such investigations, most have shown only that various stages in the progress toward flowering are more or less easily poisoned, but supply little specific information about their nature. One recent investigation, however, suggests that the floral stimulus in *Xanthium* and *Pharbitis* might be steroid. Bonner *et al.* (1963) applied to these plants several compounds known to inhibit late steps in cholesterol synthesis and found marked inhibitions of photoperiodic induction. These inhibitors were most effective when applied to leaves shortly before an inductive long night, and were ineffective when given to buds or leaves after a long night, suggesting that the action was specifically on hormone synthesis in the leaves rather than on hormone action at the meristem. Unfortunately, no material tested would reverse the inhibitions, so the specificity remains in doubt.

Much recent work with inhibitors has been devoted to the effects of inhibitors of nucleic acid synthesis for reasons touched on in the preceding section. Zeevaart (1962), for example, found that the photoinduction of *Pharbitis* was inhibited by 5-fluorouracil and 5-fluorodeoxyuridine; partial reversals were obtained with thymidine and related compounds. Thus a specific effect on DNA synthesis is indicated. Inhibitor application was effective even some time after the end of the dark period and was thus probably on the action of the floral stimulus in the meristem. Similar experiments on *Xanthium* in the same and other laboratories, and other work with vernalizable plants, suggest that the synthesis of both DNA and RNA may be necessary for either vernalization or photoperiodic induction. These results are consistent with indications obtained by other means that active buds are necessary for photoinduction to be successful in *Xanthium*, and that dividing cells may be necessary for vernalization to occur. A note of caution is advisable, however, before accepting the widespread assumption that such results represent a great advance toward understanding these processes. Inhibitor studies and direct analyses are beginning to suggest that many processes, including the action of auxins, gibberellins, and phytochrome, are at some point closely related to DNA or RNA synthesis—to gene activation, in some sense. The commoner the situation appears to be, the less it explains except as an intermediary. What will be necessary is some idea of precisely *what* further processes are thereby set in motion, and how their interactions result in the development obtained.

The writing of this chapter was carried out at Brookhaven National Laboratory under the auspices of the U.S. Atomic Energy Commission.

Bibliography

Selected further reading

Results rapidly overtake books in this field, so it is best to keep in touch with recent issues of the plant physiological journals and of the *Annual Reviews of Plant Physiology*. Nevertheless, the following works are worth noting.

Evans, L. T. (ed.), 1963. *Environmental Control of Plant Growth*. Academic Press, New York. xvii + 449 pp. Papers presented at a symposium on climatically-dependent plant processes, including photoperiodism and vernalization, as studied under controlled conditions.

Hillman, W. S., 1962. *The Physiology of Flowering*. Holt, Rinehart & Winston, New York. xii + 146 pp. A brief critical survey of the entire field with emphases substantially different from those of this chapter.

Lang, A., 1965. Physiology of Flower Initiation. *Encyclopedia of Plant Physiology*, Vol. 15(1). Springer-Verlag, Berlin-Heidelberg-New York, pp. 1380–1536. The single most outstanding scholarly treatment of the subject, though difficult reading at times. A number of the viewpoints are in marked contrast to those of Hillman and this chapter.

Salisbury, F. B., 1963. *The Flowering Process*. The Macmillan Company, New York. xii + 234 pp. Most useful as a compendium of work on *Xanthium*, its author's pet.

Went, F. W., 1957. *The Experimental Control of Plant Growth*. Chronica Botanica, Waltham, Mass. xvii + 343 pp. A summary of pioneering work on 'phytotronics'—

the large-scale study of plant growth under artificial climates—by one of the great biologists, discoverer of plant hormones.

Withrow, R. B. (ed.), 1959. *Photoperiodism and Related Phenomena in Plants and Animals.* Publ. No. 55, American Association for the Advancement of Science, Washington, D.C. xi + 903 pp. A valuable collection of articles on photoperiodism, photomorphogenesis, rhythms, etc. in organisms ranging from *Paramecium* to *Homo.*

Papers cited in text

Blaney, L. T. and K. C. Hamner, 1957. Interrelations among effects of temperature, photoperiod, and dark period on floral initiation of Biloxi Soybean. *Bot. Gaz.* **119**, 10–24.

Bonner, J., E. Heftmann and J. A. D. Zeevaart, 1963. Suppression of floral induction by inhibitors of steroid biosynthesis. *Plant Physiol.* **38**, 81–88.

Borthwick, H. A. and R. J. Downs, 1964. Roles of active phytochrome in control of flowering of *Xanthium pennsylvanicum. Bot. Gaz.* **125**, 227–231.

Brink, R. A., 1962. Phase change in higher plants and somatic cell heredity. *Quart. Rev. Biol.* **37**, 1–22.

Bünning, E., 1963. *The Physiological Clock.* Academic Press, New York.

Bünning, E. and H. Kemmler, 1954. Über den Phasenwechsel der endogenen Tagesrhythmik bei Langtagpflanzen. *Z. Bot.* **42**, 135–150.

Chailakhyan, M. H., 1936. On the mechanism of photoperiodic reaction. *C. R. (Doklady) Acad. Sci. U.R.S.S.* **10**, 89–93.

Claes, H. and A. Lang, 1947. Die Blütenbildung von *Hyoscyamus niger* in 48 Stündigen Licht-Dunkel-Zyklen und in Zyklen mit aufgeteilten Lichtphasen. *Z. Naturf.* **26**, 56–63.

Cumming, B. G., S. B. Hendricks, and H. A. Borthwick, 1965. Rhythmic flowering responses and phytochrome changes in a selection of *Chenopodium rubrum. Can. J. Bot.* **43**, 825–853.

Downs, R. J., 1956. Photoreversibility of flower initiation. *Plant Physiol.* **31**, 279–284.

Evans, L. T., 1960. Inflorescence initiation in *Lolium temulentum* L. I. Effect of plant age and leaf area on sensitivity to photoperiodic induction. *Austr. J. Biol. Sci.* **13**, 123–131.

Evans, L. T. and I. F. Wardlaw, 1966. Independent translocation 14C-labeled assimilates and of the floral stimulus in *Lolium temulentum. Planta,* **68**, 310–326.

Fredericq, H., 1964. Conditions determining effects of far-red and red irradiations on flowering response of *Pharbitis Nil. Plant Physiol.* **39**, 812–816.

Garner, W. W. and H. A. Allard, 1920. Effect of the relative length of day and night and other factors of the environment on growth and reproduction in plants. *J. Agr. Res.* **18**, 553–606, plates 64–79.

Garner, W. W. and H. A. Allard, 1923. Further studies in photoperiodism, the response of the plant to relative length of day and night. *J. Agr. Res.* **23**, 871–920, with 19 plates.

Gregory, F. G. and O. N. Purvis, 1938a. Studies in the vernalization of cereals. II. The vernalization of excised mature embryos, and of developing ears. *Ann. Bot.,* N.S., **2**, 237–251.

Gregory, F. G. and O. N. Purvis, 1938b. Studies in vernalization of cereals. III. The use of anaerobic conditions in the analysis of the vernalizing effect of low temperature during germination. *Ann. Bot.,* N.S., **2**, 753–761.

Guttridge, C. G., 1959. Further evidence for a growth-promoting and flower-inhibiting hormone in strawberry. *Ann. Bot.,* N.S., **23**, 612–621.

Hillman, W. S., 1963. Photoperiodism: An effect of darkness during the light period on critical night length. *Science,* **140**, 1397–1398.

Hillman, W. S., 1964. Endogenous circadian rhythms and the response of *Lemna perpusilla* to skeleton photoperiods. *Am. Naturalist*, **98**, 323–328.

Hillman, W. S., 1965. Red light, blue light, and copper ion in the photoperiodic control of flowering in *Lemna perpusilla* 6746. *Plant Cell Physiol.* **6**, 499–506.

Hillman, W. S., 1966. Photoperiodism in *Lemna*: Reversal of night-interruption depends on color of the main photoperiod. *Science*, **154**, 1360–1362.

Huxley, J., 1949. *Soviet Genetics and World Science. Lysenko and the meaning of heredity.* Chatto & Windus, London.

Inamura, S. and A. Takimoto, 1955. Photoperiodic responses in Japanese Morning Glory, *Pharbitis Nil* CHOIS., a sensitive short-day plant. *Bot. Mag.* (Tokyo) **68**, 235–241.

Khudairi, A. and K. C. Hamner, 1954. The relative sensitivity of *Xanthium* leaves of different ages to photoperiodic induction. *Plant Physiol.* **29**, 251–257.

Kofranek, A. M. and R. M. Sachs, 1964. Effect of far-red illumination during the photoperiod on floral initiation of *Chenopodium amaranticolor*. *Am. J. Bot.* **51**, 520–521.

Könitz, W., 1958. Blühhemmung bei Kurztagpflanzen durch Hellrot-und Dunkelrotlicht in der Photo-und Skotophilen Phase. *Planta*, **51**, 1–29.

Lang, A., 1951. Untersuchungen über das Kältebedürfnis von Zweijährigem *Hyoscyamus niger*. *Der Züchter*, **21**, 241–243.

Lang, A. and G. Melchers, 1943. Die photoperiodische Reaktion von *Hyoscyamus niger*. *Planta*, **33**, 656–702.

Lincoln, R. G., D. L. Mayfield, R. O. Hutchins, A. Cunningham, K. C. Hamner, and B. H. Carpenter, 1962. Floral initiation of *Xanthium* in response to application of an extract from a day-neutral plant. *Nature, Lond.* **195**, 918.

Murneek, A. E., 1948. History of research in photoperiodism. In *Vernalisation and Photoperiodism*, ed. A. E. Murneek and R. O. Whyte. Chronica Botanica, Waltham, Mass.

Nakayama, S., H. A. Borthwick, and S. B. Hendricks, 1960. Failure of photo-reversible control of flowering in *Pharbitis Nil*. *Bot. Gaz.* **121**, 237–243.

Oda, Y., 1962. Effect of light quality on flowering of *Lemna perpusilla* 6746. *Plant Cell Physiol.* **3**, 415–417.

Pittendrigh, C. S., 1966. The circadian oscillation in *Drosophila* pseudo-obscura pupae: a model for the photoperiodic clock. *Z. Pflanzenphysiol.* **54**, 275–307.

Salisbury, F. B., 1965. Time measurement and the light period in flowering. *Planta*, **66**, 1–26.

Schwabe, W. W., 1955. Factors controlling flowering in the *Chrysanthemum*. V. Devernalization in relation to high temperature and low light intensity treatments. *J. Exp.* **6**, 435–450.

Schwabe, W. W., 1959. Studies of long-day inhibition in short-day plants. *J. Exp. Bot.* **10**, 317–329.

Stolwijk, J. A. J. and J. A. D. Zeevaart, 1955. Wave length dependence of different light reactions governing flowering in *Hyoscyamus niger*. *Proc. Koninkl. Ned. Akad. Wetenschap.* Ser. C, **58**, 386–396.

Takimoto, A. and K. C. Hamner, 1965. Studies on red light interruption in relation to timing mechanisms involved in the photoperiodic response of *Pharbitis Nil*. *Plant Physiol.* **40**, 852–854.

Wellensiek, S. J., 1958. Vernalization and age in *Lunaria biennis*. *Proc. Koninkl. Ned. Akad. Wetenschap.* Ser. C, **61**, 561–571.

Wellensiek, S. J., 1964. Dividing cells as the prerequisite for vernalization. *Plant Physiol.* **39**, 832–835.

Whyte, R. O., 1948. History of research in vernalization. In *Vernalisation and Photoperiodism*, ed. A. E. Murneek and R. O. Whyte. Chronica Botanica, Waltham, Mass.

Withrow, R. B., 1959. A kinetic analysis of photoperiodism. In *Photoperiodism and Related Phenomena in Plants and Animals*, ed. R. B. Withrow, American Association for the Advancement of Science, Washington, D.C.

Zeevaart, J. A. D., 1962. DNA multiplication as a requirement for expression of floral stimulus in *Pharbitis Nil. Plant Physiol.* **37**, 296–304.

17. Germination and dormancy

P. F. Wareing

Introduction

Nearly all land plants pass through a phase of dormancy at some stage in the life-cycle, either as spores, in the lower plants such as bryophytes and ferns, or as seeds, in the case of higher plants. Moreover, thallophytes, especially fungi, show resting spores which constitute a dormancy phase. Apart from spores and seeds, however, perennial plants show many other types of resting organ, such as the winter resting buds of trees, tubers, rhizomes, bulbs, corms, and so on.

Usually the phase of dormancy coincides with a period of unfavourable climatic conditions, either of low temperature, or of high temperature and drought. Thus, many resting organs show adaptation to adverse climatic conditions, in the form of special structures, such as bud-scales or subterranean organs, and by the development of a physiological state which confers greater frost or heat resistance than is shown during the actively-growing phase of the plant. It is not the intention, here, to consider the physiological basis of frost and drought resistance, but it is important to recognize that dormancy is typically a phase which shows special adaptation to adverse environmental conditions.

It is not easy to give a clear-cut, all-embracing definition of dormancy, but the term may be applied in a general sense to any phase in the life-cycle of the whole plant, or of particular organs, in which active growth is temporarily suspended. This arrest of active growth may be due to unfavourable environmental conditions, such as low temperature or drought, when we may speak of *imposed dormancy* or 'quiescence'.* On the other hand, growth may be arrested even though environmental conditions appear to be favourable. Thus, many temperate tree species form resting buds in August, at a time when temperature conditions are normally very favourable for growth. Similarly, many seeds fail to germinate when freshly shed, even though placed under favourable conditions of moisture, temperature, and aeration, whereas they will readily germinate under those same conditions at a later stage. In such cases the dormancy appears to be due to conditions within the dormant organ itself and we then speak of *innate dormancy* or 'rest'. Thus, innate dormancy may be defined as a condition in which germination or growth fails to occur even though external conditions

* For more detailed discussion of the definition of various forms of dormancy, see Vegis (1965) and Romberger (1963).

appear to be favourable. The fully dormant condition of a bud or other organ is not attained suddenly, but is gradually developed over a period. During this phase the organ may be induced to resume growth by various treatments, such as high temperature (p. 631), or defoliation in the case of shoots of woody plants. The organ is then said to be in a state of *predormancy* or *early rest*. As predormancy progresses, however, it becomes increasingly difficult to induce resumption of growth by changes in environmental conditions and the organ then enters a state of full dormancy or *mid-dormancy*. This phase is followed by a gradual emergence from dormancy, during which it becomes progressively easier to induce resumption of growth and this state is referred to as *post-dormancy*, or *after-rest*.

With some organs it is possible to induce growth under rather specific conditions, for example, under long photoperiods and over a very narrow temperature range (p. 631), even at the time of their deepest dormancy. Such organs are said to show *relative dormancy*, as contrasted with *true dormancy*, in which growth cannot be induced under any set of environmental conditions. However, the conditions which will induce growth of organs showing relative dormancy are not normally met with during the rest period of these organs in nature, and so they do normally show a period of dormancy.

In many cases, organs which have not fully emerged from dormancy can be thrown back into full dormancy by certain environmental conditions, such as high temperature. In such cases, we speak of *secondary* or *induced dormancy*. Instances are also known in which dormancy may be induced in organs which do not normally become dormant; for example, the 'lower' seeds of *Xanthium pennsylvanicum* can be rendered dormant by maintaining them at 30°C under conditions which restrict their gaseous exchange (p. 620).

Most of this chapter will be concerned with various forms of innate dormancy, which poses many interesting problems. The dormancy of buds or resting organs which are modified shoots (e.g., bulbs, potato tubers) will be discussed first, and then seed dormancy and germination.

Bud dormancy

The induction of bud dormancy

Most woody plants form resting buds at some stage in the annual cycle of growth. In temperate zone species, such buds are formed typically in the late summer or autumn and the whole tree or shrub enters a dormant or resting phase. Resting buds are also formed by many tropical species, but not all the shoots of a given tree may enter the resting phase simultaneously, so that both dormant and actively-growing shoots may be found on the same tree at any given time.

The resting buds of woody plants normally contain a number of unexpanded leaf primordia whose further development has been arrested by the onset of dormancy. These leaf primordia may be surrounded by a number of bud scales which may represent either modified stipules (e.g., *Betula, Fagus*) or modified leaves (e.g., *Pinus, Malus, Ribes, Acer*). In some species, such as *Viburnum* spp., the leaf primordia are not surrounded by any special covering structure and the buds are described as 'naked'.

A number of phases in the development of a resting bud may be recognized. The following description of bud development may be taken as applicable to species which have 'foliar' bud scales, such as *Acer pseudoplatanus* (Schüepp, 1929) or *Ribes nigrum*. (For details of other species, see Romberger, 1963.) First, there is cessation of extension growth, so that no further leaves are expanded or internodes extended. However, meristematic activity at the apical meristem still continues for some time and a number of new leaf primordia are normally formed during this period of bud development. At the same time, some of the outer primordia show greater marginal growth than is the case for a primordia destined to develop into normal leaves, and they give rise to the bud scales, or 'cataphylls'. As this process continues the bud swells, but ultimately all growth and meristematic activity ceases with the formation of the fully developed bud. Similar stages may be recognized in the development of both terminal and axillary buds. In some species, such as pines, growth of the bud may continue over many months, from June to the end of September (Sacher, 1954).

Apart from these morphological changes seen during bud development, physiological changes occur. When an axillary bud is first formed it is inhibited from growing by 'correlative inhibition', since it will grow out if the main shoot apical region is removed. Later in the season, however, decapitation does not result in outgrowth of the axillary buds, in many species. Thus, originally the bud is not itself innately dormant, but it is held in check by influences arising outside the bud itself, whereas later the bud itself becomes dormant. There may be no visible difference between a dormant and a non-dormant bud, although sometimes the latter is rather larger than the former.

What causes an actively growing shoot to cease growth and form a resting-bud? Is the onset of dormancy brought about by certain environmental conditions, or is it due to causes inherent in the tree itself? The answers to these questions depend upon species and upon the age of the tree. Day length is one environmental factor which has been shown to have a marked effect on the onset of dormancy in many woody species (Wareing, 1956; Nitsch, 1957). In the majority of woody species so far investigated it has been found that short days strongly promote the formation of resting-buds and the onset of dormancy (Table 17.1). Not only those species which form terminal resting-buds and thus have a 'monopodial' growth habit, but also those species, such as *Tilia, Ulmus*, and *Robinia*, in which the main shoot apical region is shed with the onset of dormancy ('sympodial' growth), show these photoperiodic effects. These short-day responses occur not only in broad-leaved (dicotyledonous) trees, but also in conifers, such as pines and larches. Certain woody species, such as *Betula* and

Table 17.1   Woody species in which the formation of resting buds and the onset of dormancy is accelerated by short days.

Acacia melanoxylon	Morus alba
Acer spp.	Paulonia tomentosa
Aesculus hippocastanum	Phellodendron amurense
Ailanthus glandulosa	Picea abies
Alnus incarna	Pinus spp.
Berberis Thunbergii	Platanus occidentalis
Betula spp.	Populus spp.
Carpinus betulus	Prunus persica
Catalpa speciosa	Pyrus ussuriensis
Cercis canadensis	Quercus spp.
Citrus spp.	Rhus typhina
Cornus florida	Ribes nigrum
Corylus avellana	Robinia pseudacacia
Fagus sylvatica	Salix spp.
Larix decidua	Taxus baccata
Liquidambar styraciflua	Ulmus americana
Liriodendron tulipifera	Weigela florida

Robinia, are very sensitive to day length and may be kept in continuous growth for as long as 18 months if they are maintained under long-day conditions in a warm greenhouse (Kramer, 1936; Downs and Borthwick, 1956), whereas they cease growth in about 14 days, if they are transferred to short photoperiods. Other species, such as *Fraxinus excelsior* and other members of the family Oleaceae, and certain members of the Rosaceae, such as apple, show very little response to day length.

Photoperiodism in woody plants shows certain parallels with the photoperiodic phenomena found in herbaceous plants, in relation to flowering (chapter 16). Thus, the effects of long dark periods can be nullified by a relatively short 'night interruption' by light in woody plants, as in herbaceous species (Wareing, 1950; Downs and Borthwick, 1956). It appears that phytochrome is involved in these photoperiodic dormancy responses, although a clear demonstration of 'red/far red reversibility' is still awaited.

In leafy seedlings of some woody species, e.g., *Acer pseudoplatanus*, the photoperiodic responses are determined primarily by the day lengths to which the mature leaves are exposed, but in *Betula pubescens* even the very young leaves in the shoot apical region are also sensitive (Wareing, 1954).

How far do these photoperiodic responses of woody plants control the induction of dormancy under field conditions? There seems little doubt that they are important in determining the onset of dormancy in young trees of certain species, such as poplar and larch, which continue active extension growth well into September or October, when the natural day lengths are declining rapidly; this conclusion is supported by the fact that extension of the natural day length by artificial illumination in the autumn prolongs the growth of the seedlings. On the other hand, older trees of most woody species cease extension growth much earlier in the season, often in June, July, or early August, when the natural photoperiods are still long. Thus, it seems unlikely that declining day lengths play a major role in determining the duration of extension growth in mature trees of most woody species. What, then, controls the formation of resting buds in such trees? Other environmental conditions, especially soil moisture and mineral nutrient levels, undoubtedly have a marked effect on the duration of extension growth of trees. It seems unlikely, however, that these factors play an overriding role in controlling the duration of extension growth since even under favourable conditions of moisture and soil nutrients mature trees grow for only a limited period each year. It would seem more probable that internal competition for nutrients between the various branches and shoots of a tree may be important in determining the annual increment of extension growth. Thus, the determining factor may not be the depletion of nutrients in the soil, but exhaustion of nutrients, including organic metabolites, within the tree itself (see Kozlowski, 1964, 1966; Priestley, 1962). It is clear, however, that we need more precise experimental data on this problem.

It would appear that the buds formed under long days are first in a state of pre-dormancy, since they will expand prematurely in the summer if the trees are defoliated, either experimentally by hand or by insect attack. Later in the season these buds enter a state of full dormancy; although short-day conditions are not required for the initial formation of these buds, they may play an important role in the ultimate development of deep dormancy of the buds, as has been shown for birch (Wareing and Black, 1958).

Removal of bud dormancy

Buds of temperate species which have entered a state of dormancy in the autumn resume growth in the following spring. Evidently the dormant state has been removed during the winter and it is now well known that the buds of many woody species require a period of winter-chilling before growth can be resumed (Colville, 1920; see also Vegis, 1961). The most effective temperatures for overcoming dormancy appear to be in the range 1–10°C, and the length of chilling period required varies from 260 to over 1,000 hours (Samish, 1954). In regions with mild winters, such as occur in California and South Africa, the chilling requirements of woody plants may not be fully met, with the result that bud-break in the spring is delayed and irregular. This delayed foliation may present a serious problem for the cultivation of fruit crops, such as peach, in these regions.

As we have seen, short photoperiods generally promote bud dormancy and once the fully dormant condition has been attained, in most species it can only be overcome by a period of chilling. In some species, such as *Betula* spp., and

Fagus sylvatica, however, bud dormancy can be overcome by transferring seedlings from short-day to long-day conditions. The buds of these species are themselves capable of responding to long photoperiods, since even the leafless seedlings will resume growth under long days (Wareing, 1953).

In many woody plants, the chilling requirements may have been met by January or February, but bud-break may be delayed for much longer because the outdoor temperatures remain too low for growth. With such buds, innate dormancy (rest) has been replaced by imposed dormancy (quiescence). Thus, although chilling is required to overcome dormancy, the actual time of bud-break is determined by rising temperature in the spring.

Apart from the natural removal of bud dormancy by chilling temperatures, dormancy can be overcome by a wide range of substances, including ethylene chlorhydrin, thiourea, and dinitrophenol. Immersion for several hours in a warm water bath at 40°C is also effective. Special interest attaches to the observation that gibberellic acid will overcome bud dormancy in a number of woody species, since chilling treatment has been shown to lead to an increase in the levels of endogenous gibberellins in the resting buds of several species (p. 627).

Dormancy in various organs

Dormancy is shown by a wide variety of organs of perennial plants, including rhizomes, corms, bulbs, bulbils, root-tubers, stem-tubers, and winter resting-buds of certain aquatic plants. Dormancy of the corms of *Gladiolus*, the rhizomes of *Convallaria*, and the tubers of *Helianthus tuberosus* is most rapidly overcome by storage at chilling temperatures. On the other hand, the tubers of most varieties of potato emerge from dormancy more rapidly when stored at 22°C than at 10° (see Burton, 1963).

The dormancy of the winter resting buds (turions) of the aquatic plants *Utricularia*, *Stratiotes*, and *Hydrocharis* has been studied in some detail, especially by Vegis (1961, 1965). In *Stratiotes aloides*, the dormancy period of the turions was, in general, shorter at 10°C than at 20–25°C. It was also shown that short days, in association with high temperature, promote the induction of dormancy in turions of *Stratiotes*. Short days have also been found to promote the formation of resting buds (hibernaculae) in the insectivorous plant, *Pinguicula grandiflora*, and the dormancy of the buds is overcome by chilling (Heslop-Harrison, 1962). By contrast, dormancy in bulbs of onion (*Allium cepa*) is promoted by long photoperiods, and the period of dormancy is shortest when the bulbs are stored in cool temperatures (Mann and Lewis, 1956). Similarly, dormancy in *Lunularia cruciata*, a desert liverwort from Israel, is promoted by long days and overcome by short days (Schwabe and Nachmony-Bascombe, 1963). It is interesting to note that this appears to be an adaptation to the hot, dry conditions prevailing during the summer in its natural habitat.

Seed dormancy

Forms of seed dormancy

Although the seeds* of some species, including the crop plants, corn (*Zea mays*), peas (*Pisum*), and beans (*Phaseolus*) will germinate almost immediately after harvesting, when placed under suitable conditions of moisture, temperature, and aeration, innate dormancy of the seeds occurs in a large number of other species. A number of distinct forms of dormancy can be recognized in seeds, the main features of which will be summarized briefly. (For more detailed accounts see: Barton and Crocker, 1948; Crocker and Barton, 1957; Barton, 1965a, b; Evenari, 1965; Lang, 1965; Stokes, 1965. The reader is also referred to these works for supporting references, where these are not given in this section.)

Impermeable seed coats. The seeds of certain families, including the Leguminosae, Malvaceae, Chenopodiacea, Convolulaceae, and Solanaceae, have testas which are impermeable to water when freshly shed and hence they are liable to lie dormant in the soil until the impermeable layers have been broken down by the action of soil micro-organisms. Alternatively, the testas can be rendered permeable by mechanical scarification or by treatment for short periods with concentrated sulphuric acid.

Immaturity of the embryo. In a number of seeds the development of the embryo is incomplete when they are shed, and germination will not occur until further embryo development has occurred. Seeds showing this type of dormancy include *Anemone nemorosa*, *Ficaria verna*, and *Caltha palustris*. The embryos of *Fraxinus excelsior* are morphologically developed but undergo considerable growth after shedding.

Need for after-ripening in dry storage. The seeds of many species are dormant when they are harvested, but need no special treatment to overcome their dormancy, and if they are kept under dry storage conditions at normal temperatures they gradually emerge from dormancy over a period ranging from a few weeks to several months. Many of the common cereals show this type of dormancy, e.g, barley, oats, and wheat (Barton, 1965a). In some cases, however, such seeds show relative, rather than true dormancy. For example, fresh seeds of some barley varieties will not germinate at 15°C, but will do so at 10°C; again, fresh seeds of *Thlaspi arvense* will germinate only at 28–30°C (see Barton, 1965a).

* In the following discussion the term 'seed' will be used for a variety of propagules or dispersal units. Strictly speaking the seed consists of the embryo and its testa, together with the endosperm, if present. However, in many species the actual dispersal unit is a fruit, in which the seed is surrounded by additional coats derived from fruit (pericarp), and which may be so intimately associated with the testa that it is difficult to distinguish the individual layers. For convenience the term 'seed' will be applied also to these latter organs, including caryopses, achenes, schizocarps, etc.

After dry storage such seeds become capable of germinating over a much wider temperature range. In some species the after-ripening process takes place most rapidly at high temperatures, e.g., seeds of *Malva* spp. become non-dormant after 2 hours at 70°C. Dormancy of cereal seeds may be removed by storing them at 35–40°C for 2–4 days. Usually this type of dormancy can be overcome by removing the seed coats, e.g., in cereal grains and *Avena fatua*.

Seeds having a chilling requirement.   Many seeds of temperate species show a form of dormancy which is overcome by chilling (see Barton and Crocker, 1948; Stokes, 1965). For example freshly-harvested seeds of apple, rose, and peach will not germinate if they are planted under moist conditions and maintained at 20°C, but if they are first kept under moist conditions at 0–5°C for several weeks, they will germinate when transferred to warmer conditions. Under field conditions this chilling requirement is normally met by natural winter temperatures, and this type of dormancy has the result that germination does not take place in the autumn, immediately after the seeds are shed, but is delayed until the following spring. The seeds of a considerable number of woody plants, especially among the Rosaceae, show this type of dormancy, but it is by no means confined to the seeds of woody plants (Table 17.2).

At one time it was thought that seeds of this type must be subjected to freezing temperatures, in order to fracture the coats, but it is now known that temperatures just above freezing (0–5°C) are most effective (Fig. 17.1). In order for the

Table 17.2   Species requiring chilling to overcome seed dormancy.

A. Woody species

Acer saccharum	*Malus* spp.
A. pseudoplatanus	*Picea* spp.
Betula spp.	*Pinus* spp.
Cornus florida	*Prunus* spp.
Corylus avellana	*Pyrus* spp.
Crataegus spp.	*Ribes grossularia*
Cupressus macrocarpa	*Rosa* spp.
Fagus sylvatica	*Sequoia gigantea*
Fraxinus excelsior	*Tilia* spp.
Hamamelis virginiana	*Thuja occidentalis*
Juglans nigra	*Vitis* spp.
Liriodendron tulipifera	

B. Herbaceous species

Alisma plantago-aquatica	*Polygonum* spp.
Gaultheria procumbens	*Scirpus campestris*
Gentiana spp.	*Typha latifolia*
Iris versicolor	

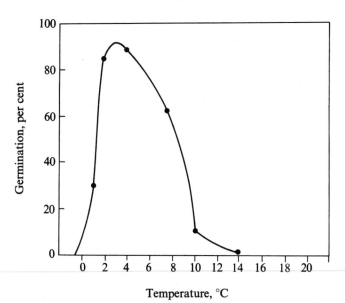

Figure 17.1 Germination of apple seed after chilling for 85 days at various temperatures. (After Schander, 1955.)

chilling treatment to be effective the seeds must be imbibed and they must have adequate aeration. The period of chilling required to give full germination varies from 1 to 6 months, according to the species.

In some species the intact seed shows a chilling requirement whereas when the seed coat is removed the isolated embryo germinates readily, e.g., in *Alisma plantago aquatica* and other aquatic plants, in many grasses and cereals (see Crocker, 1948), and in the seeds of *Betula pubescens* and *Acer pseudoplatanus*.

In other species the embryo itself shows dormancy, which requires to be overcome by chilling, e.g., *Sorbus aucuparia*, peach (*Prunus persica*), *Fraxinus* spp. The distinction between these two forms of dormancy is not sharp, since in samples of hazel (*Corylus avellana*) seed, for example, both types of dormancy may be found. Moreover, in seeds with dormant embryos, if the chilling requirement is only partially met, the embryos become non-dormant, but the intact seeds still require further chilling for germination to take place. For example, intact nuts of hazel require 12 weeks of chilling, whereas the embryo will germinate after 4 weeks of chilling if the coats are removed (Frankland and Wareing, 1966). Similar effects are observed in apple seeds (Fig. 17.2).

In some cases dormant embryos may be induced to germinate without chilling, after removal of the seed coats. The germination is sluggish and the growth of the resulting seedlings is stunted and dwarfed. Such dwarfed seedlings may continue to show this habit of growth for as long as 10 years if they are kept in a

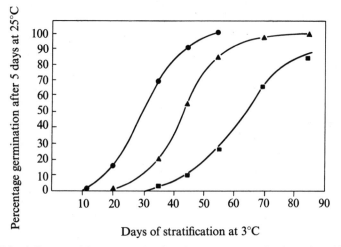

Figure 17.2 Influence of integuments and endosperm on germination of unchilled apple embryos at 25°C.
  ● Naked embryos
  ▲ Embryos with endosperm intact
  ■ Embryos with both endosperm and integuments intact. (After Visser, 1954.)

warm greenhouse, but if they are exposed to chilling temperatures they lose their stunted habit and show normal growth. The dwarfed habit may also be partially overcome by treatment with gibberellic acid in some species, e.g., peach.

Some seeds show epicotyl dormancy; for example, acorns (*Quercus* spp.) germinate and put out a radicle in the autumn, without a chilling treatment, but growth of the epicotyl is dependent upon chilling and does not normally take place until the spring. Epicotyl dormancy is also shown by *Viburnum* spp. A further variation of this type of dormancy is shown by the seeds of *Convallaria* and *Polygonatum* in which a chilling treatment is necessary for emergence of the radicle, but further growth of the epicotyl does not take place until the seeds have been exposed to a second period of chilling. In nature, seedlings of these species do not appear above ground until the second spring after shedding, and hence they are called 'two-year seeds'.

Another type of two-year seed is seen in *Crataegus* spp., in which the seeds are surrounded by impermeable endocarp, and hence they cannot imbibe water. During the first winter, chilling is ineffective and it is not until the surrounding coats have been eroded by microbial action during the following summer that the seed can absorb water and then respond to the second winter's chilling.

The dormancy of many seeds with a chilling requirement may be overcome by treatment with various chemicals, including nitrates, thiourea, ethylene chlorhydrin, gibberellic acid, and kinetin. The possible role of endogenous hormones in the regulation of this type of dormancy is discussed later.

Light-sensitive seeds.* The germination of many seeds is affected by light. Such seeds are said to be *photoblastic.* A large number of types of seed show a form of dormancy which is overcome by exposure to light, and which are said to be *positively photoblastic,* while a smaller number have their germination inhibited by light, and are described as *negatively photoblastic.* Some examples of the two types of photoblastic seeds are given in Table 17.3, but the list could be greatly extended.

Table 17.3 Photoblastic species.

A. Germination promoted by light

Amaranthus retroflexus	*Lepidium virginicum*
Betula spp.	*Lycopersicum esculentum* (tomato)
Bidens tripartitus	*Lythrum salicaria*
Capsella bursa-pastoris	*Nicotiana tabacum* (tobacco)
Digitalis purpurea	*Paeonia* spp.
Echium vulgare	*Perilla ocymoides*
Epilobium hirsutum	*Phleum pratense*
Eschololtzia spp.	*Poa* spp.
Fragaria virginiana	*Primula* spp.
Juncus spp.	*Rumex* spp.
Kalanchoë blossfeldiana	*Verbascum thapsus*
Lactuca sativa (lettuce)	

B. Germination inhibited by light

Helleborus niger	*Nigella damascena*
Lamium amplexicaule	*Phacelia tanacetifolia*
Nemophila insignis	*Silene armeria*

The light requirements of positively photoblastic seeds are often extremely low; for example, exposure to 100 m.c.s. (meter-candle-seconds) of white light gives maximum germination in one variety of tobacco seed, while 'Grand Rapids' lettuce seed requires 400–600 m.c.s. Within certain limits the 'reciprocity law' holds, i.e., germination response depends upon the *quantity* of light received (light intensity × time of exposure). At high light intensities, exposures as short as 0·1 have an appreciable affect on the germination of seeds of *Lythrum salicaria.*

The spectral requirements of light-promoted seeds have been studied in some detail and, indeed, it was the classical studies of Borthwick, Hendricks, and their co-workers which led to the discovery in 1952 of the 'red/far red' effects and hence to the discovery of phytochrome (chapter 14). It had been shown

* For a more detailed account, see Evenari (1965).

earlier by Flint and McAlister (1935; 1937) that it is the red region (650 nm) of the spectrum which is most effective in promoting the germination of lettuce seed and that infrared ('far-red') radiation in the region 730 nm *inhibits* germination. Borthwick and Hendricks showed later that the effects of these two spectral regions are mutually antagonistic and that if they are given successively, then whether germination occurs or not depends upon the type of radiation to which they were *last* exposed (Table 17.4). They postulated the presence of a

Table 17.4   Photoreversal of promotion and inhibition of germination of 'Grand Rapids' lettuce seed (from Borthwick *et al.*, 1952).

Irradiation	Percentage germination
R	70
R–F	6
R–F–R	74
R–F–R–F	6
R–F–R–F–R	76
R–F–R–F–R–F	7
R–F–R–F–R–F–R	81
R–F–R–F–R–F–R–F	7

R = Red; F = Far-red

photoreversible pigment which can exist in two alternative states, one absorbing in the red region (P_{650}) and the other in the far red (P_{730}). When seeds are exposed to red radiation the pigment is converted from P_{650} to P_{730}, and it is evidently the latter form which sets up a chain of reactions which lead ultimately to germination. Exposure of the pigment in the form P_{730} to far-red radiation brings about its re-conversion to P_{650}, which evidently is not a form capable of initiating the germination processes. Since the properties of phytochrome have been described in detail in chapter 14, further detailed treatment is not necessary here.

Since the demonstration of red/far red reversibility in lettuce seeds this phenomenon has been shown also for a considerable number of other species, and it would appear to be of universal occurrence in positively photoblastic seeds.

Studies of the spectral sensitivity of light-inhibited seeds has shown that it is the far-red region of the spectrum which is most effective with these seeds. The effect of red radiation on negatively photoblastic seeds varies considerably from species to species. In some, red counteracts the inhibitory action of far-red, but in others, such as *Nemophila insignis* and *Avena fatua*, red appears to have no effect and acts like darkness.

Apart from the effects of red and far-red radiation, blue light has an effect on many seeds. The effects of blue are complex and it may inhibit or promote, depending upon the species and the conditions under which it is applied. In lettuce seed, blue may promote germination if applied for short periods immediately after imbibition, but it inhibits if applied for longer periods. Blue light also inhibits the germination of certain negatively photoblastic seeds, e.g., *Nemophila*. In general, considerably higher energy levels are required with blue than with red or far-red radiation.

The difference between positively and negatively photoblastic seeds appears to lie in their relative sensitivities to red and far-red. Under natural conditions, seeds receive white light which includes both red and far-red; in positively photoblastic seeds the promotive effects of red evidently predominate, whereas in negatively photoblastic seeds it is the far-red effect which is over-riding.

Some seeds appear to exhibit a form of photoperiodism. Thus, seeds of *Nemophila insignis* give a higher germination percentage if exposed to short daily light periods than if exposed to longer photoperiods. Thus we may speak of this as a 'short day' seed. On the other hand, seeds of *Betula pubescens* require long daily periods of irradiation when they are maintained at 15°C, the germination percentage rising linearly with the length of the photoperiod (Fig. 17.3).

The responses of photoblastic seeds are markedly affected by temperature. Thus, seeds of 'Grand Rapids' lettuce will germinate in the dark over the temperature range 10–20°C, but become light-requiring at 20–30°C. At 35°C

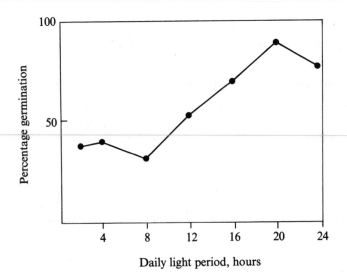

Figure 17.3   Effect of day length on germination of unchilled seeds of birch (*Betula pubescens*) at 20°C. (Black and Wareing, 1955.)

germination is inhibited in both light and dark. In many seeds alternating temperature regimes have a marked effect on germination, and in some cases alternating temperatures may replace a light-requirement. For example, at a constant temperature of 25°C, seeds of *Rumex crispus* are light-requiring, but if exposed to alternating temperatures of 15°C and 25°C daily, they will germinate in darkness.

The responses of many photoblastic seeds tend to decline with age. Thus, freshly harvested seeds of 'Grand Rapids' lettuce are light-requiring, but show increasing germination percentage in the dark after several months of dry storage. A similar decline in light-sensitivity is shown by light-inhibited seeds.

If positively photoblastic seeds, such as lettuce, are kept in darkness, in an imbibed condition, for prolonged periods, they become incapable of responding to light and are said to be 'skotodormant'. Conversely, negatively photoblastic seeds, such as *Phacelia*, become 'photodormant' ('light-hard') if maintained in the light for prolonged periods. Both types of dormancy may be overcome by low temperature.

A variety of substances will overcome the dormancy of light-requiring seeds. Thus, potassium nitrate, thiourea, and gibberellic acid will replace the light-requirements of seeds of lettuce and many other species. Kinetin does not completely replace the light requirement of lettuce seeds, but it renders them more light-sensitive, so that they will germinate in response to a very brief light exposure. By contrast, coumarin inhibits the germination of lettuce seeds, and its effect can be overcome by red light.

In many species, only the intact seed is light-requiring and the isolated embryo will germinate equally well in light or dark, e.g., *Betula pubescens*. This phenomenon will be considered further in the discussion on the role of seed coats in dormancy.

Seed coat effects in dormancy *

As we have already seen, seed coats play an important role in the dormancy of many species, not only where the seed coat is impermeable to water, but also in seeds which require after-ripening in dry storage, in those with a chilling requirement, and in light-sensitive seeds. Indeed, it is clear that in all cases where the embryo itself is not dormant, the dormancy of the intact seed depends upon the presence of the coats, which will include the testa, together with the endosperm and pericarp in some 'seeds'. Embryo dormancy is most common in seeds with a chilling requirement (p. 613 and see chapter 16), and also in some with a need for after ripening in dry storage, e.g., *Ambrosia trifida*; apparently very few light-sensitive seeds show embryo dormancy.

There is much evidence to suggest that in some species the seed coat effects

* For more detailed accounts see Barton (1965b), Stokes (1965), and Evenari (1965).

are due to interference with gaseous exchange, especially with oxygen uptake. Thus, certain dormant seeds may be induced to germinate in high concentrations of oxygen, e.g., in cereals, *Avena fatua* and *Rumex crispus*, but this treatment is not effective with many others. Moreover, it has been shown that the coats of certain species constitute a barrier to oxygen diffusion and that when the coat is removed oxygen uptake by the embryo is greatly increased, e.g., in *Xanthium pennsylvanicum* and apple (Visser, 1954). In *Cucurbita pepo* the seed coats are less permeable to oxygen than to carbon dioxide (Brown, 1940).

In many cases it is not necessary to remove the entire seed coat, and it is sufficient to remove the coat at the radicle end only. In other cases, slitting or pricking the seed-coat leads to increased germination and the effect is further enhanced if the seeds so treated are kept in an atmosphere of increased oxygen tension, e.g.. *Betula pubescens* (Table 17.5). There is thus considerable evidence

Table 17.5 The influence of oxygen tension on the percentage germination of seeds of *Betula pubescens* treated in various ways (Black and Wareing, 1959).

Treatment	Air	50% oxygen
Isolated embryos	100	100
Intact seeds	0	0
Pericarp and endosperm slit	70·1	100
Pericarp scratched	25·0	100
Pericarp pricked	9·4	57·6

that seed coats may limit oxygen supplies to the embryo in certain species. The effects of seed coats on oxygen uptake appear also to be important in secondary dormancy (p. 620).

On the other hand, there is also evidence that the effect of seed coats may be due to mechanical resistance to the growth of the radicle. Chen and Thimann (1964, 1966) have shown that the negatively photoblastic seeds of *Phacelia tanacetifolia* will germinate equally well in light or dark if the seed coat is removed in the radicle region. If seeds treated in this way are placed in 0·3 M mannitol solutions (which will reduce the ability of the root cells to take up water) their light sensitivity is restored, and they will then not germinate in the light, but they will still do so in the dark. The inhibition of germination in the light can be overcome by gibberellic acid both in intact seeds, and in treated seeds kept in mannitol. The authors concluded that in *Phacelia* the seed coat exerts a mechanical resistance to the growth of the radicle and that light reduces the ability of the radicle to overcome this resistance, in some way not yet understood. It might be suggested that the effect of gibberellic acid is to stimulate the synthesis of hydrolytic enzymes, which will increase the level of osmotically-active substances in the tissues by hydrolysis of storage macromolecules, such as starch and proteins; but the levels of amylase, protease, and lipase apparently remain unchanged, whether the seeds are in the light or dark, or treated with gibberellic acid.

Ikuma and Thimann (1963) have carried out parallel studies with the positively photoblastic seeds of lettuce. They showed that the inhibitory effect of the seed coats on germination resides in the endosperm of the radicle region, which also appears to be the photosensitive region. Germination will occur if the mechanical resistance of the endosperm is reduced, either by cutting or treatment with cellulolytic enzymes. They postulated that exposure to red light stimulates the synthesis of cellulolytic enzymes, although they were unable to demonstrate any increase in the activity of such enzymes in the seeds following exposure to red light.

It seems questionable whether the effects of seed coats can be attributed to mechanical resistance in all species. Thus, the intact seeds of *Citrullus colocynthis* are inhibited by light, but this inhibition is greatly reduced in the naked embryos. Within the thick testa is a thin membrane surrounding the embryo, and if the testa is removed so that only this membrane is left, the germination of the embryo is still inhibited in the light, although it occurs readily in the dark (Koller *et al.*, 1963). The effect of the inner membrane can be simulated by enclosing the naked embryo in moist filter paper. Since it appears that neither the natural inner membrane, nor the filter paper, can exert any appreciable mechanical resistance, it seems more likely that interference with gaseous exchange is involved.

Secondary dormancy of seeds and buds

Secondary or induced dormancy may occur in both seeds and buds. This phenomenon was first clearly shown for the seeds of *Ambrosia trifolia* and *Xanthium pennsylvanicum* (Davis, 1930a, b). The freshly harvested seeds of *Ambrosia* contain dormant embryos, but these may be after-ripened in dry-storage or by chilling in moist storage. However, some of the after-ripened seeds fail to germinate, although the embryos are no longer dormant, apparently due to limitation of oxygen uptake by the nucellar membranes. If such seeds are kept at 30°C for 4 weeks the embryos become dormant again and this secondary dormancy can again be overcome by after-ripening in dry storage or by chilling. The seeds of *Xanthium* may be similarly thrown into secondary dormancy by embedding them in clay (thus restricting gaseous exchange) and keeping them at 30°C.

Similar secondary dormancy phenomena have been shown for a range of species, including members of the Polygonaceae and Rosaceae (*Sorbus, Rhodotypos*, apple, pear). In all these cases, the secondary dormancy can be overcome by chilling treatment, and it appears that the development of secondary dormancy is the reverse of after-ripening (Abbott, 1956).

Secondary dormancy may also be induced in the winter resting-buds of *Stratiotes aloides* (Vegis, 1949). As these buds emerge from true dormancy following chilling, they will sprout within a certain temperature range, but if they are maintained above the maximum temperature for growth they are thrown into secondary dormancy and then require a further period of chilling before growth can occur. A similar phenomenon has been shown for the buds of woody plants, such as peach (*Prunus persica*), in which high temperatures during the early phases of post dormancy (after winter-chilling) can induce secondary dormancy (Vegis, 1964b).

The mechanism of dormancy

As we have seen, dormancy is shown by a wide range of plant organs, of very different morphology. In many such organs we are dealing, essentially, with dormancy of the shoot, i.e., with dormant buds, whether we are considering the shoot of a woody plant, a potato tuber, or a rhizome. On the other hand, in the germination of a seed the primary criterion of emergence from dormancy is the growth of the radicle. Moreover, the seed is surrounded by one or more coats (testa, endosperm, pericarp) which are not found in dormant buds, although bud-scales may perform some of the functions of seed coats. Because of their very different structure and morphology, it might be expected that the physiological mechanism of dormancy would be different in seeds and buds. However, a comparison of dormancy phenomena in the two types of organ shows striking parallels. Thus, dormancy phenomena in seeds with a chilling requirement show close similarities with those seen in the winter resting buds of many species; moreover, substances such as gibberellic acid, thiourea and ethylene chlorhydrin, are effective in overcoming dormancy in both types of organ. Parallels between photoblastic seeds and buds are less striking, but phytochrome is known to be involved in both types of dormancy, and there are similarities in the responses to light of both seeds and buds in the same species, e.g., *Betula pubescens*. Again there are striking similarities in the phenomena of induced or secondary dormancy in seeds and buds (p. 632).

These parallels in the dormancy phenomena of different types of organ suggest that a common mechanism may be involved, at least in certain species. Thus, it would seem reasonable to look for many common features in the mechanism of dormancy in seeds and buds; at the same time the occurrence of special structures, the coats, in seeds should put us on our guard against pressing the similarities too far.

In recent years, attention has been directed mainly towards two hypotheses as to the cause of dormancy, in both seeds and buds. The first hypothesis assumes that the regulation of dormancy is primarily hormonal, while the second stresses the importance of interference with gaseous exchange by seed coats and other structures. These two hypotheses will now be discussed in turn.

Hormonal regulation of dormancy

It is clear that in true dormancy there must be some 'internal' block to growth, which is thereby prevented, even though external conditions are favourable. In terms of growth-substances, we can envisage that growth may be prevented either (a) because some essential growth-promoting substance is deficient, or (b) by the presence of actively inhibitory substances. These two possibilities are not mutually exclusive, of course.

The idea that growth-inhibiting substances may have a regulatory function was first suggested by Molisch (1922), to explain why the seeds of succulent fruits do not normally germinate *in situ*. Since then, a considerable amount of evidence has accumulated, indicating that germination inhibitors may occur not only in succulent fruits, but also in the pericarp of non-succulent fruits (Evenari, 1949; Wareing, 1965). When the dispersal unit consists of a seed or seeds enclosed within pericarp tissue, the presence of the latter may cause delayed germination, as has been shown for *Sinapis alba* (Sroelov, 1940) and beet (Tolman and Stout, 1940). From this situation it is but a short step to envisage that inhibitors present in the testa, endosperm, or the embryo itself may cause delayed germination, and hence give rise to seed dormancy.

Bud dormancy.   The hypothesis that inhibitors are involved in bud dormancy appears to have been first suggested by Hemberg for potato tubers and resting buds of *Fraxinus excelsior* (Hemberg, 1949a, b). The hypothesis was based upon a correlation between the level of the endogenous growth inhibitors in the tissues and the state of dormancy of the buds; thus treatments, such as exposure to ethylene chlorhydrin, which remove the dormancy of potato tubers, also cause a marked reduction in the levels of the endogenous inhibitors. Similarly, it has been shown for a number of woody species that the levels of endogenous inhibitors in resting buds decline during the course of the winter, and this change is correlated with the gradual emergence of the buds from dormancy (Phillips and Wareing, 1958; von Guttenberg and Leike, 1958; Dörffling, 1963) (Fig. 17.4).

Some of the best evidence for the view that bud dormancy involves growth inhibitors comes from the study of photoperiodic responses in seedlings of woody plants. As we have seen, in many woody species short days cause the cessation of extension growth and the formation of dormant winter resting-buds. In *Acer pseudoplatanus* this photoperiodic induction of dormancy is determined by the day length conditions to which the *mature leaves* are exposed, although the response actually takes place at the shoot apex (Wareing, 1954). This observation in itself suggests that there must be the transmission of some

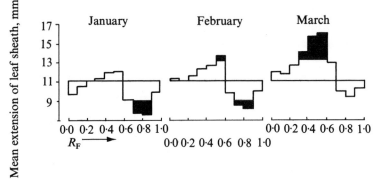

Figure 17.4 Effect of chilling on the gibberellin and inhibitor content of dormant buds of blackcurrant (*Ribes nigrum*), at different stages during the winter. The paper chromatograms of bud extracts were developed in isopropanol : ammonia : water (10 : 1 : 1) and assayed by the dwarf maize leaf section test, which is sensitive to both gibberellins and inhibitors. (El-Antably, unpublished.)

inhibitory influence from leaves maintained under short days to the apex. This conclusion is strongly supported by the experiments of Waxman (1957) with *Cornus florida rubra*, in which it was shown that if the tip of a shoot growing under long days is cut off the axillary buds grow out, but that if the uppermost pair of leaves is maintained under short days they remain inhibited (Fig. 17.5).

Similar conclusions are indicated by experiments with *Betula pubescens* plants (Wareing, 1954) which have first been induced to form resting-buds by exposure to short days. If the whole plants (leaves and buds) are then transferred to long days, the buds will swell and resume growth; they will also do so if the seedlings are defoliated, and the buds themselves exposed to long days. On the other hand, if the buds are exposed to long days and the leaves to short days, the buds remain dormant. Quite clearly, leaves maintained under short days have an inhibitory

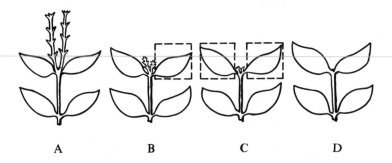

Figure 17.5 Inhibitory effect of short-day leaves on bud growth in *Cornus florida rubra*. Decapitated branches were kept either under long days (A, B, C) or under short days (D). The development of the top axillary shoots is completely prevented when the whole plant is placed under short days, and partially prevented when only one (B) or two (C) of the uppermost leaves are subjected to short days. (From Nitsch, 1963, adapted from Waxman, 1957.)

effect upon the buds, and it is reasonable to postulate the production of growth inhibitory substances by the leaves under short days, and their transport to the shoot apices. It is of interest, therefore, that it has been shown for several woody species, including *Betula*, that higher amounts of growth inhibitors can be extracted from the leaves and buds under short days than under long days (Nitsch, 1957; Phillips and Wareing, 1958, 1959) (Fig. 17.6). Moreover, in-

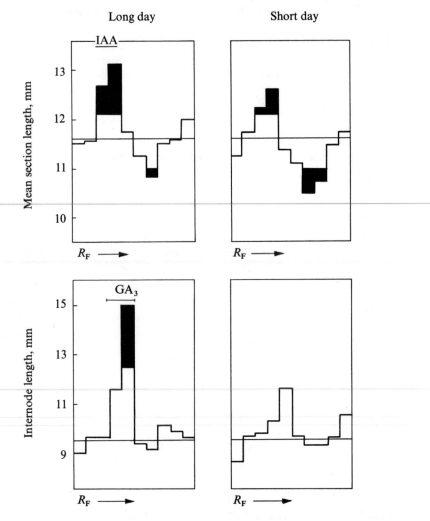

Figure 17.6   Effect of day length on auxin and inhibitor levels (above) and gibberellin levels (below) in shoots of birch (*Betula pubescens*). The paper chromatograms were assayed by the *Avena* coleoptile section test for auxins and inhibitors (above), and by the dwarf pea epicotyl test for gibberellins (below). (After Digby and Wareing, 1966.)

creased amounts of inhibitor can be detected two days after transfer from long-day to short-day conditions, at which time there are no detectable morphological changes at the shoot apex (Phillips and Wareing, loc. cit.); the changes in inhibitor level therefore *precede* the formation of resting buds. Thus the study of these inhibitor changes suggests that the inhibitory effect of short-day leaves on the buds is due to the observed increased levels of growth inhibitors under short days.

The chemical nature of the inhibitory material in the leaves of woody plants, and which varies with photoperiod, has recently been elucidated. It was shown that in *Acer pseudoplatanus* the inhibitory activity resides in a single, highly-active substance, which was called 'dormin' (Robinson *et al.*, 1963; Robinson and Wareing, 1964). In the meantime, Addicott and his co-workers were studying a substance present in young cotton fruits which accelerates their abscission and they isolated a crystalline fraction which they called "abscisin II' (Okhuma *et al.*, 1963) which they later identified as a sesquiterpenoid with the following structure (Okhuma *et al.*, 1965):

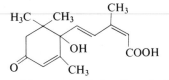

This structure was confirmed by synthesis by Cornforth *et al.* (1965b). Comparison of the infrared absorption spectra and other physical properties of dormin and abscisin II showed that they were identical (Cornforth *et al.*, 1965a). It has now been agreed that this substance shall be called abscisic acid (ABA).

The *trans, trans-isomer* of abscisic acid possesses little physiological activity. The molecule contains an asymmetric atom (C*) and therefore exhibits optical isomerism. The naturally-occurring form is the (+)-enantiomorph. The optical rotatory dispersion of (+)-abscisic acid is anomalous, showing a very intense Cotton effect in the ultraviolet. This property can be used under certain conditions to detect the presence of the compound in partially purified plant extracts. (Cornforth *et al.*, 1966). Using this property it has been shown that abscisic acid is present in a wide range of plant species, both herbaceous and woody (Table 17.6).

Pure abscisic acid is a highly active growth inhibitor (at less than 1 p.p.m.) in the *Avena* coleoptile test and other growth tests. It seems likely, therefore, that abscisic acid represents a new type of growth hormone.

A critical test for the inhibitor hypothesis is whether dormancy can be induced in seedlings of woody plants growing actively under long days, by applying abscisic acid to the leaves. This has been successfully achieved with seedlings of *Betula pubescens*, *Acer pseudoplatanus*, and *Ribes nigrum*, which ceased growing and formed terminal resting buds in response to applied abscisic acid. (Eagles and Wareing, 1963, 1964; El-Antably *et al.*, 1967) (Fig. 17.7).

Table 17.6   Species in which the presence of abscisic acid has been established (Cornforth *et al.*, 1966a).

Species	Organ
Sycamore (*Acer pseudoplatanus*)	Buds, leaves
Birch (*Betula pubescens*)	Leaves
Rose (*Rosa arvensis*)	Achenes, pseudocarps
Potato (*Solanum tuberosum*)	Tubers
Cabbage (*Brassica oleracea*)	Leaves
Avocado (*Persea grantissima*)	Seed
Lemon (*Citrus medica*)	Fruit
Plum (*Prunus domestica*)	Stems
Lupin (*Lupinus luteus*)	Young pods
Strawberry (*Fragaria*)	Leaves
Willow (*Salix viminalis*)	Aphid honeydew

Thus, both studies of variation in the levels of endogenous abscisic acid, and the results of application of synthetic abscisic acid, are consistent with the hypothesis that it plays an important role in the regulation of bud dormancy. On the other hand, the occurrence of abscisic acid in a range of herbaceous, as well as woody, plants suggests that it has a variety of roles, and is probably involved in the regulation of a number of growth processes, in addition to the control of dormancy.

Figure 17.7   Experimental induction of resting buds in birch (*Betula pubescens*) by application of abscisic acid to the leaves. *Left-hand side*: shoot apical region of control seedlings growing actively under long days. *Right-hand side*: Dormant bud induced by application of abscisic acid to leaves of similar plants maintained under long days. (From Eagles and Wareing, 1964.)

It is necessary to consider also the possible role of growth-promoting substances in the control of bud dormancy. There is little evidence that auxins are important in the regulation of bud dormancy. On the other hand, there are good grounds for thinking that gibberellins, and possibly also cytokinins, may play a key role in dormancy regulation. Firstly, during the course of winter-chilling, there is a progressive build-up of endogenous gibberellins in the buds of *Acer pseudoplatanus* and other woody plants parallel with the decline in inhibitors (Eagles and Wareing, 1964) (Fig. 17.4). Similarly, the emergence of potato buds from dormancy is accompanied by an increase in the levels of endogenous gibberellins in the tubers (Smith and Rappaport, 1961). Moreover, there is a reciprocal variation in the levels of endogenous gibberellin and inhibitor in woody plants under different day-length conditions (Fig. 17.6). Furthermore, bud dormancy in many woody plants can be overcome by the application of exogenous gibberellic acid. Gibberellic acid is also successful in delaying or preventing the onset of dormancy under short days in many woody species (see Vegis (1964b) for reference). Thus, it is possible that bud dormancy is regulated by a balance between gibberellins and endogenous inhibitors such as abscisic acid. Indeed, if gibberellic acid and abscisic acid are applied simultaneously to growing birch seedlings, the inhibitory effect of the abscisic acid is nullified and the plants remain in active growth. These observations raise the question as to whether, possibly, abscisic acid is a specific antagonist of gibberellic acid, i.e., an 'anti-gibberellin'. Studies of the interaction between abscisic acid and gibberellic acid in the synthesis of amylase by barley endosperm have given some support to this view (Thomas *et al.*, 1965; Chrispeels and Varner, 1967), but in other tests the two growth substances appear to act independently (Robinson and Wareing, 1964; Milborrow, 1967). Some evidence has been obtained for the view that abscisic acid may affect gibberellin biosynthesis in certain systems (Wareing *et al.*, 1968).

Seed dormancy. We now turn to the possible role of growth substances in seed dormancy. In considering the possible significance of endogenous inhibitors, it is necessary to maintain a critical attitude, since a range of substances present in seeds and fruits have been shown to inhibit germination when extracted and re-applied to test seeds, but it cannot be assumed that such substances necessarily play a regulatory role in nature. In order to establish such a role, it is necessary to study carefully how far the changes in levels of endogenous inhibitors during seed development, and during emergence from dormancy, are correlated with the state of dormancy of the seed.

As an example of a study of this type we will consider seed dormancy in *Xanthium pennsylvanicum*, which has been generally considered to provide a good instance of dormancy due to impermeability of the testa to oxygen (Davis, 1930b). The fruits each contain two seeds, the lower of which germinates readily after a period of dry storage, whereas the 'upper' seed remains dormant when sown under similar conditions. The view that dormancy is due to impermeability of the testa to oxygen is based upon the following facts: (a) the intact upper seeds are dormant at 20°C under normal atmospheric conditions, but can be induced to germinate in high oxygen concentrations; (b) if the testa is removed, the isolated embryos will germinate in air if planted on a moist medium; (c) oxygen uptake by the embryo is considerably increased by removal of the testa. On the other hand, the seeds contain two water-soluble inhibitors, present mainly in the embryo itself (Wareing and Foda, 1956). Being water-soluble these inhibitors are rapidly leached out of the naked embryo, but the testa is impermeable to the inhibitors and there is no leaching from the intact seeds. Earlier workers claimed that isolated embryos from upper seeds were non-dormant, but the conditions of their experiments were such that rapid leaching of inhibitors would have occurred from the naked embryo during the experiments. When the precautions are taken to ensure that there is no leaching into the medium, then the embryos themselves may be shown to be dormant. These dormant embryos can be induced to grow, however, by leaching out the inhibitors, or by placing them in an atmosphere of pure oxygen, under which conditions the inhibitors are rapidly broken down in the tissues. Thus, the high oxygen requirement of these seeds is apparently due to the presence of the inhibitors, and the effect of the testa is not primarily due to interference with oxygen uptake, but arises from the fact that it prevents leaching out of the inhibitor from the intact seed.

The hormonal aspects of dormancy in seeds with a chilling requirement have been studied in relation both to inhibitors and to growth promoters. The presence of growth inhibitors in the embryos, testa, endosperm, and pericarp has been demonstrated for a number of species (Evenari, 1949; Wareing, 1965). However, the role of inhibitors in this type of seed dormancy has been more difficult to demonstrate. In several species it has been found that the levels of the endogenous inhibitors do not decline appreciably during chilling, in contrast to the position in dormant buds (Villiers and Wareing, 1965; Frankland and Wareing, 1966). On the other hand, leaching of dormant embryos, which presumably removes or dilutes the inhibitors, leads to germination in several species, e.g., *Fraxinus excelsior* (Villiers and Wareing, 1965), *Rosa* spp. (Jackson and Blundell, 1963) as in the case of *Xanthium pennsylvanicum* referred to above. In *Rosa* the pericarp has an inhibitory effect on the germination of the seed within, which cannot be fully explained in terms of mechanical resistance since the effect is only partly removed by cutting the pericarp. Since the latter has been shown to contain abscisic acid (Jackson and Blundell, 1966) it seems likely that part of the inhibitory effect of the pericarp is due to the presence of this inhibitor.

The interaction between germination promoters and inhibitors has been studied in several seeds with a chilling requirement. It has been shown for seeds of *Fraxinus excelsior* that extracts of embryos which have been subjected to chilling contain a substance which is capable of stimulating germination of dormant embryos, whereas this stimulator is not present in extracts of unchilled embryos (Villiers and Wareing, 1965). Moreover, this germination promoter is able to counteract the effect of the endogenous inhibitor on the germination of the embryos. More recently, it has been shown that gibberellins increase in embryos of *Fraxinus excelsior* during chilling (Kentzer, 1966) and it is possible that the germination promotor observed by Villiers and Wareing is a gibberellin. Chilling leads to a similar increase in the levels of endogenous gibberellins in seeds of *Corylus avellana* (Frankland and Wareing, 1962). Thus, although it is possible that endogenous inhibitors play a role in the dormancy of seeds with a chilling requirement, one of the effects of chilling appears to be an increase in the levels of endogenous gibberellins, and the regulation of dormancy may involve an interaction between these two types of growth substances.

The role of inhibitors in light-sensitive seeds is not yet clearly established. As mentioned above, examples of embryo dormancy are uncommon in photoblastic seeds, but seed coat effects are generally very important. An example of a light-requiring seed in which inhibitors may be important is provided by *Betula pubescens*. This 'seed' (actually an achene) shows a chilling requirement, but if unchilled it will germinate in response to light (p. 617). This light response is present only in *intact* seeds, the isolated embryos being able to germinate in complete darkness. Evidently coat effects are important, and this conclusion is supported by the observation that if the seed coats are slit, greatly increased germination in the dark can be obtained by increasing the oxygen tension (Black and Wareing, 1959). The pericarp contains a germination inhibitor (which may be identical with that in the leaves, which is now known to be abscisic acid), and when isolated embryos are planted on a medium containing the inhibitor the light-requirement is restored. Pretreatment of the intact seeds by leaching (which reduces the level of inhibitor), on the other hand, reduces the light requirement. Evidence was obtained that the presence of the inhibitor increases the oxygen requirements of the intact seed, and it appears that dormancy of birch seeds is partly due to the high oxygen requirements of the embryo resulting from the presence of the inhibitor, and partly to physical interference with oxygen uptake by the pericarp.

Temperature and restricted oxygen uptake as factors in dormancy

Earlier workers, notably Davis (1930a, b), Thornton (1935; 1945), and Crocker (1948), stressed the importance of restricted oxygen uptake, especially in association with high temperatures, as a factor in the induction of dormancy in seeds and other organs. A more recent protagonist of this theory has been Vegis (1964a, b; 1965), whose views will now be considered.

Vegis has drawn attention to interesting changes in the temperature requirements for bud-break and seed germination during the rest period. These changes had been observed by earlier workers, including Askenasy (1877), but have been largely overlooked more recently. It appears to be a general rule, in all types of resting organ, that as they become progressively more dormant the range of temperatures over which they can maintain growth becomes narrower, until they reach a state of full or relative dormancy. Conversely, as organs emerge from dormancy, at first they will grow only over a narrow temperature range, but later they become capable of growing over a progressively wider temperature range. For example, the immature seeds of cereals will at first germinate over a comparatively wide temperature range, but as ripening proceeds the seeds lose their ability to germinate at high temperatures, although they will still germinate at 10°C. On the other hand, as after-ripening proceeds they recover the

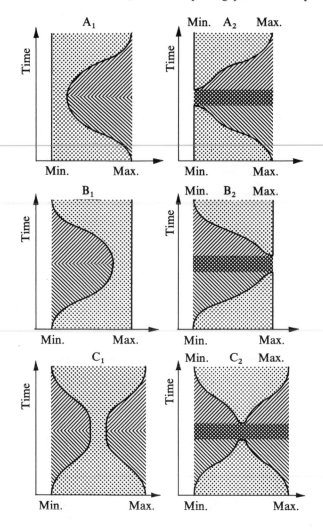

ability to germinate at high temperatures. Vegis (1964a, b; 1965) has pointed out that this type of change in temperature response, narrowing through a lowering of the maximum temperature to permit germination and subsequently widening through raising of this temperature, (Fig. 17.8A$_2$) appears to be very common in various types of dormant organ. For example, the bulbs of hyacinth, tulip, and daffodil will not grow at high temperatures during the early stages of dormancy in the summer, but they will continue to grow slowly at 5–8°C at this time. During the winter there is a gradual widening of the temperature range for growth towards higher temperatures. According to Vegis (loc. cit.) the winter resting buds of many trees also show similar phenomena.

Another kind of widening of the temperature range for seed germination and bud break is seen in species in which the dormant organs will grow at higher temperatures, but not at lower temperatures (Fig. 17.8B$_2$). Thus, the unchilled seed of *Betula* spp. will germinate only at temperatures of about 30°C, but after chilling they ultimately become able to germinate at 0°C. Similar behaviour is shown by the tubers of potato (*Solanum tuberosum*), of which the buds will at first grow only at high temperature, but after storage they will grow also at low temperatures.

In yet another group of species the narrowing of the temperature range during dormancy involves both a raising of the minimum and lowering of the maximum for growth (Fig. 17.8C$_2$). This type of behaviour is shown by the seeds of *Rumex crispus*, *Chenopodium album* and many other species, by the turions of *Stratiotes aloides*, and by the buds of many tree species.

Except in species showing dormancy of the embryo, the narrowing of the temperature range for germination is normally seen only in the intact seeds, and removal of the coats allows germination over a wide range of temperatures. Light-sensitive seeds also show these temperature effects. Some species will

---

Figure 17.8   Diagrammatic representation of the various types of narrowing and widening of the temperature range for germination and bud break at times of change in growth activity.
A. Narrowing of the temperature range through decrease of the maximum temperature at which germination and bud break can occur; and widening through increase.
B. Narrowing of the temperature range through increase of the minimum temperature at which germination and bud break can occur; and widening through decrease.
C. Narrowing of the temperature range through simultaneous decrease of the maximum and increase of the minimum temperature at which germination and bud break can occur; and widening through increase of the maximum and decrease of the minimum.
A$_1$, B$_1$, C$_1$: Predormancy is followed directly by postdormancy, with no true dormancy.
A$_2$, B$_2$, C$_2$: A period of true dormancy separates pre-and postdormancy.
Min.: Minimum temperature for germination and bud break.
Max.: Maximum temperature for germination and bud break.
Dotted areas stand for the temperature range within which germination and bud break occur.
Slant lines in lower portion of each diagram stand for the temperature range within which seeds and buds in a state of predormancy do not germinate or break.
Slant lines in upper portion of each diagram stand for the temperature range within which seeds and buds in a state of postdormancy do not germinate or break.
Crisscross lines stand for the period of true dormancy. (From Vegis, 1964a.)

germinate in darkness over a narrow temperature range but will do so over a wider range in the light. Thus, light-sensitive varieties of lettuce seed will germinate in darkness only at temperatures below 18°C, but in the light they will germinate also at high temperatures. In many dormant seeds and buds, gibberellic acid (GA_3) is able to bring about germination at temperatures which are normally inhibitory.

Vegis (1956; 1964) has put forward a general theory of dormancy in which it is postulated that high temperatures, in association with restricted oxygen uptake due to the presence of covering structures such as seed coats and bud scales, are the primary causes of dormancy in both seeds and buds. We have seen (p. 620) that there is good evidence that secondary dormancy may be brought about in seeds by maintaining them at high temperatures under conditions of restricted oxygen supply. According to Vegis (loc. cit.), buds of woody plants which are still partially dormant due to insufficient winter chilling may also be thrown into secondary dormancy by high temperatures. It is an essential part of Vegis' theory that primary and secondary dormancy are similar in nature and that primary dormancy is, therefore, probably also brought about by high temperatures and restricted oxygen uptake. Thus, he states that 'under natural conditions the cause of the formation of the resting condition is a temperature too high for growth of the young, recently formed cells, which are surrounded by structures limiting diffusion'. He points out that the embryos of developing seeds are liable to experience oxygen deficiency because of the surrounding seed coats and maternal tissues, and postulates that under such conditions of partial anaerobiosis certain metabolic changes occur, leading to the formation of fatty acids and fats, which tend to accumulate in dormant tissues.

Direct evidence for the view that high temperatures and restricted oxygen diffusion are responsible for the onset of seed and bud dormancy under natural conditions is lacking, and the theory is based mainly upon the assumption that natural primary dormancy is essentially similar to experimentally-induced secondary dormancy. Vegis points out that there is evidence that the uptake of oxygen by the meristematic tissues of *Acer* buds is limited by the scales (Pollock, 1953), but clearly until such scales have been formed they do not constitute any barrier to oxygen diffusion and it is difficult to see, therefore, how interference with oxygen uptake can be important in the initial formation of a resting bud. Further, the formation of resting buds in woody plants under short days appears to be controlled by some stimulus arising in the *leaves* (p. 608), and it is difficult to reconcile this observation with the view that dormancy is caused by conditions arising within the bud itself. To overcome this difficulty, Vegis (1964) suggests that short days only cause the cessation of growth and the formation of terminal buds and not the onset of dormancy, which is brought about by high temperature and restricted oxygen supply after the buds have formed.

The possible importance of oxygen deficiency within the seed as a factor in dormancy is suggested by the recent work of Roberts (1964a, b) on seeds of rice, which show dormancy immediately after harvesting, but gradually emerge from dormancy during dry storage. In rice, the husks play an important role in the dormancy since their removal leads to greatly increased germination; the residual dormancy appears to be due to the influence of the pericarp and testa. The inhibitory effect of the seed coverings does not appear to be due to mechanical resistance. On the other hand, storage in oxygen greatly accelerates the removal of dormancy, suggesting that the coats constitute a barrier to oxygen uptake by the embryo. It was found that the Q_{10} for the rate of loss of dormancy had a constant value of 3·38 over the range 27–47°C, suggesting that some oxidation reaction, possibly not enzymic and not connected with respiration, may be involved in the breaking of dormancy during dry storage.

In order to test this hypothesis, Roberts (loc. cit.) investigated the effect of pretreatment with various respiratory inhibitors, on the basis that if the breaking of dormancy is connected with respiration then it should be slowed down by such treatment. He obtained the unexpected result that inhibitors of cytochrome oxidase, such as carbon monoxide, cyanide, azide, hydrogen sulphide, and hydroxylamine, all showed a marked *stimulatory* effect on the breaking of dormancy. On the other hand, a wide range of other respiratory inhibitors, which do not act primarily on the terminal oxidase system, had little effect on dormancy. Nitrate and nitrite, and certain oxidizing agents such as hydrogen peroxide, were effective in overcoming dormancy, whereas ammonium ions and urea were not. Similar results were obtained with seeds of barley (Major and Roberts, unpublished), in which, in addition to the terminal oxidase inhibitors found to be effective with rice, it was found that iodoacetate, sodium fluoride (which inhibit glycolysis), and dinitrophenol (which uncouples oxidative phosphorylation) are also effective. It was shown that pretreatment with sodium azide, malonate and cyanide depresses the oxygen uptake of both dormant and non-dormant seeds of rice and barley and that, in general, the depression of oxygen uptake is associated with *increased* germination of dormant seeds. Cyanide has also been shown to be effective in overcoming bud dormancy (Samish, 1954).

Roberts has put forward a hypothesis to account for these apparently paradoxical observations. He suggests that it is necessary for some oxidation reaction to proceed to a certain level before germination can take place. The oxygen tension inside the seed is probably very low; this is indicated, for instance, by the high R.Q. and low oxidation-reduction potential which is found in rice seeds in the initial stages of germination. Cytochrome oxidase has a very high affinity for oxygen and is therefore a strong competitor for the available oxygen. Inhibition of cytochrome oxidase would remove this competition and make available more oxygen for the oxidation reaction necessary for breaking dormancy. This hypothesis is further supported by the observation that, when set to germinate in water, dormant barley seeds were found to have a *higher* oxygen uptake than non-dormant. The nature of the postulated oxidation reaction responsible for the breaking of dormancy is unknown.

So far we have considered the two main theories of the mechanism of dormancy regulation, the 'hormonal' and the 'restriction of gaseous exchange' theories, as if they were mutually exclusive alternatives. However, there is some evidence to suggest that the high oxygen requirements of certain seeds, e.g., *Xanthium pennsylvanicum* and *Betula pubescens*, arises from the presence of endogenous inhibitors (p. 628). If this interpretation can be shown to be correct and of general application, then a bridge between the two theories would be established. It is possible that the after-ripening of rice-seed involves the oxidation of a germination inhibitor, but attempts by Roberts to demonstrate the presence of an active germination inhibitor in extracts of rice seed were unsuccessful.

Metabolic aspects of dormancy

Studies on the metabolic aspects of dormancy have been carried out by numerous workers during the past 40–50 years, but it must be admitted that, in general, these studies have been singularly unrewarding. Many earlier workers studied changes in the activity of certain enzymes, such as catalase, peroxidase and lipase, but this work was based upon concepts of metabolism which now appear as over-simplified and out-of-date. Other workers have investigated differences in respiratory metabolism between dormant and non-dormant tissues. This approach is still pertinent, especially in view of the apparent importance of oxygen deficiency in the dormancy of certain seeds (p. 619). Vegis (1964) has suggested that under conditions of oxygen deficiency, such as occur in seeds and buds, especially at high temperatures, the oxidation of acetyl coenzyme A is limited so that this compound and certain glycolytic intermediates are diverted into other pathways, particularly into the formation of fatty acids and lipids.

Studies have been carried out on the respiration of a photoblastic ('Grand Rapids') and a non-photoblastic variety ('Progress') of lettuce seed, after irradiation with red light, with far-red light, and in darkness, and at different stages of after-ripening (Evenari *et al.*, 1955). The non-light-requiring variety 'Progress' was found to have a higher respiration rate in the dark than 'Grand Rapids'. Irradiation of 'Grand Rapids' with red light leads to increased respiration, while far-red lowers it. During after-ripening the respiration rate increases as the seeds become less dormant. It is not clear from these results whether dormancy is due to a respiratory block, which is removed by exposure to red light, or whether the increased respiration following red light is the *result* of the removal of dormancy.

It is possible that dormancy is controlled by the regulation of energy availability. Pollock and Olney (1959) showed that chilling of sour cherry (*Prunus cerasus*) seeds caused an increase in the levels of phosphorus in the embryo and suggested that dormancy may be due to a block in phosphate metabolism. Bradbeer and Colman (1967) have investigated the effects of chilling on the metabolism of seeds of *Corylus avelana*. They found that the tricarboxylic acid (TCA) cycle, the glycolytic pathway and the pentose phosphate pathway (PPP) are all active in the dormant seeds. They were unable to establish any clear pattern of changes in the activity of the TCA cycle during chilling, but the activity of the glycolytic pathway increased relative to that of the PPP. In experiments involving the feeding of adenine-8-^{14}C to cotyledon slices, it was found that after 24 hours of feeding at 5°C, 95 per cent of the activity was in adenine, adenosine, adenosine monophosphate (AMP), and adenosine diphosphate (ADP). During the first 8 days of chilling there was a marked fall in the labelling of adenosine and corresponding rises in AMP and ADP. These observations are consistent with the hypothesis that the breaking of dormancy may depend upon increased availability of energy to the embryo.

Another approach to the problem of dormancy is through studies on nucleic acid metabolism. Tuan and Bonner (1964) have studied RNA synthesis in dormant and non-dormant buds of potato. In dormant buds there is only very limited RNA and DNA synthesis, but the rates of synthesis are markedly increased by treatments which break the dormancy, such as exposure to ethylene chlorhydrin. RNA synthesis by dormant buds is inhibited by actinomycin D and hence is DNA-dependent. Preparations of 'chromatin' (i.e., deoxyribonucleoprotein) of dormant potato tubers are incapable of stimulating RNA synthesis *in vitro*, whereas chromatin of non-dormant buds is highly effective in supporting RNA-synthesis by added RNA-polymerase. They conclude that in dormant potato tissue the DNA is blocked in some way (in a 'repressed' state) and hence cannot support DNA-dependent synthesis. Since there is evidence that plant growth hormones may act through nucleic acid metabolism (see chapters 1, 2, and 3), the effects of gibberellin and abscisic acid on dormancy may be exerted through the regulation of RNA-synthesis.

Germination

Conditions for germination

As is well known, the most important factors commonly affecting the germination of seeds are water, aeration, temperature, and light. Since most seeds have a relatively low water-content when shed, they need to take up considerable amounts of water before germination can occur. In the initial stages, the process of water uptake is primarily by imbibition, and certain seeds, such as those of *Xanthium*, can take up water by this process even from relatively dry soil. The imbibitional properties of such seeds derive from the colloidal materials they contain, especially proteins and starch. The uptake of water may be limited by impermeable coats, however, in the case of 'hard-coated' seeds (p. 611). Such seeds are liable to lie dormant in the soil for considerable periods, until erosion of the impermeable layers by soil micro-organisms allows them to take up sufficient water for germination. The germination of certain seeds, e.g., barley, is adversely affected if they are soaked in water for 2–3 days before being planted under conditions favourable for germination. The cause of this 'soaking injury' is not known, but it may be due to the leaching out of metabolites essential for germination (Lang, 1965).

The 'cardinal' temperatures, viz., minimum, optimum, and maximum, for germination vary considerably from one species to another. In general, the minimum temperatures for germination are lower for temperate species than for tropical ones. Some species, such as beech (*Fagus sylvatica*), *Trifolium repens*, and various alpine plants, will germinate at temperatures which are little above 0°C (Crocker and Barton, 1957). As we have already seen (p. 620), many seeds will only germinate over a narrow temperature range when freshly harvested, some germinating only at lower temperatures, and others only at higher ones, but usually this temperature range widens considerably during after-ripening. Some seeds will germinate well at constant temperatures but with others germination is promoted by a daily alternation in temperature, e.g., *Rumex crispus*, *Oenothera biennis*, and *Holcus lanatus* (Crocker and Barton, 1957).

The oxygen requirements of different seeds vary considerably, but most seeds are capable of germinating at oxygen tensions below that of air. In general, however, seeds require higher oxygen tensions for germination than do growing plants of the same species. These relatively high oxygen requirements of many seeds are evidently due to the presence of the surrounding coats, which limit the rate of diffusion of oxygen into the seed (p. 619). There are marked differences between species in their ability to germinate under water, some 50 per cent of 78 species tested by Morinaga (1926) being incapable of germinating under water, presumably because of the reduced tension of dissolved oxygen as compared with that of air. Differences in ability to germinate under water evidently reflect differences in respiratory metabolism. Rice, which germinates readily under water, appears to have a particularly active anaerobic glycolysis system (Taylor, 1942), but most seeds require an active aerobic oxidation system for germination, e.g., barley.

Carbon dioxide will inhibit the germination of many species at high concentrations, which are not likely to be met with in the soil under normal conditions. On the other hand high carbon dioxide concentrations promote the germination of some species, e.g., *Xanthium pennsylvanicum*, lettuce (Thornton, 1935), and *Trifolium subterraneum* (Ballard, 1958).

The effects of light on germination have already been dealt with in some detail (p. 615) and need not be elaborated on here.

Metabolic aspects of germination*

The usual criterion of germination is the emergence of the radicle through the seed coats. In lettuce seed, both cell-division and cell-elongation are taking place at 12–14 h after sowing at 25°C, when radicle elongation can first be detected (Evenari *et al.*, 1957). In maize, the initial growth of the coleorhiza appears to involve primarily cell elongation. However, it is likely that many biochemical processes precede the first visible signs of germination, but the nature of these processes still remains obscure.

One of the most striking changes which occurs when a seed is planted under conditions favourable for germination is a rapid increase in the respiration rate. Thus, in peas an increased respiration rate can be detected 2–4 hours after soaking.

The patterns of carbon dioxide output (Q_{CO_2}), oxygen uptake (Q_{O_2}) and of the respiratory quotient (R.Q.) during germination vary considerably from one seed to another. In barley and wheat there is a fairly uniform rise of both Q_{CO_2} and Q_{O_2} during the early stages of germination. In peas, on the other hand, after an initial rise the respiration rate reaches a 'plateau' when it remains constant for several hours, and then rises again (Spragg and Yemm, 1959). This plateau appears to end at about the time the testa is broken, suggesting that in the preceding period gaseous exchange is limited by the coats. Although Q_{O_2} and Q_{CO_2} rise with time, they do so at different rates so that there are marked changes in the value of R.Q. during germination. These changes in R.Q. reflect the changes in the nature of the respiratory substrate, i.e., whether it is carbohydrate, fat or protein, but the R.Q. is also affected by the differential permeability of the seed coats to oxygen and carbon dioxide.

During the early stages of germination of peas respiration appears to be predominantly anaerobic, owing to the restriction of oxygen uptake by the testa (Spragg and Yemm, 1959). There is much evidence for the occurrence of anaerobic glycolysis in peas and other seeds, and ethyl alcohol is known to accumulate in germinating seeds under anaerobic conditions. The enzymes of glycolysis have been shown to be present in pea seeds (Hatcher and Turner, 1958), and extracts of peas will take up inorganic phosphate from the medium by glycolytic phosphorylation (Mayer and Mapson, 1962).

Under aerobic conditions the pyruvic acid produced by glycolysis undergoes oxidation to carbon dioxide and water via the tricarboxylic acid (TCA) cycle. The enzymes involved in this pathway are located in the mitochondria and the activity of the TCA cycle has been shown for mitochondrial preparations from germinating seeds (Conn and Young, 1957; Neal and Beevers, 1960). However, the mitochondria of dry seeds are apparently considerably less active than those from germinating seeds, and it is likely that the TCA cycle is not fully operative

* For a more detailed account, see Mayer and Poljakoff-Mayber (1963).

in the initial stages of germination. Moreover, mitochondria from dry seeds are apparently incapable of carrying out oxidative phosphorylation. ATP is absent from dry beans but rapidly appears following imbibition. Thus, it appears that aerobic respiration does not proceed actively in freshly harvested seeds, but increases gradually as germination progresses.

Cytochrome–oxidase has been shown to be present in several types of seed, suggesting that 'terminal oxidation' via the flavoprotein–cytochrome chain is occurring, but the existence of other oxidase systems has also been reported. Thus, the presence of an electron transport system involving pyridine nucleotide, glutathione, ascorbic acid, and ascorbic acid oxidase has been demonstrated for pea seedlings (Mapson and Moustaffe, 1957), but is inactive in the dry seeds. Phenolases are present in certain seeds. Whether these other oxidase systems are important in respiration, however, is doubtful.

The pentose phosphate provides an alternative mechanism for the aerobic metabolism of carbohydrates and has been shown to be present in mung beans (Chakravorty and Burma, 1959).

Although the activity of certain enzymes can be demonstrated in resting seeds, in general there is a marked increase in the activity of many enzymes during germination. For some enzymes this increase appears to be due to their conversion from an inactive form. However, there is also clearly a *de novo* synthesis of many enzymes, as shown by the effects of inhibitors of protein synthesis, such as chloramphenicol. The stimulation of amylase formation in barley endosperm by gibberellic acid provides a good example of the *de novo* synthesis of an enzyme during germination.

During germination of the barley grain the starch reserves in the endosperm are digested, under the action of various amylases. The enzyme, α-amylase, is not active in the dormant barley grain, but it appears during germination. Haberlandt, in 1890, concluded that the α-amylase in the endosperm of germinating barley is secreted by the aleurone layer and more recent work has fully confirmed this conclusion. In 1887, Sachs showed that digestion of the endosperm does not occur if the embryo is removed. Following the discovery of gibberellins it was found that the application of GA_3 to whole grains stimulates the development of α-amylase activity. This discovery led to the idea that possibly the effect of the embryo on the development of α-amylase activity was mediated through the endogenous gibberellins which it produces. This hypothesis was fully confirmed in experiments by Paleg (1960, 1964), who showed that application of GA_3 to half-seeds of barley, from which the embryo had been removed, stimulated the development of α-amylase activity, and that this effect of GA_3 is specifically on the aleurone layer. This system has been further studied by Varner and his co-workers. In a series of elegant experiments they have shown that the α-amylase is synthesized *de novo* by the aleurone layer, in response to applied GA_3 (Varner and Chandra, 1964). Thus, we have here a clear example of the hormonal control of enzyme synthesis. It was further shown that the formation of α-amylase is inhibited by Cycloheximide, and therefore requires protein synthesis, and by Actinomycin D, and therefore requires RNA synthesis. It is not clear, however, whether gibberellins act primarily at the 'transcription' or at the 'translation' level.

Changes in nucleic acid metabolism occurring during germination have been studied in other seeds. There is active synthesis of RNA during germination (Ingle and Hageman, 1965) and this apparently involves all fractions of RNA (Cherry et al., 1965). Detailed studies have been made on the development of ribosomal system in the endosperm of germinating castor bean seeds (Marré et al., 1965). During maturation of the seeds all RNA fractions decrease markedly and ribosomes almost completely disappear in the endosperm, but during the early stages of germination these processes are reversed and ribosomes increase markedly. Actinomycin D strongly inhibits the synthesis of RNA in the seeds, indicating its dependence on DNA. During germination mono-ribosome and poly-ribosome fractions develop rapidly. Pulse labelling experiments with ^{32}P phosphate showed that immediately after the pulse most of the radioactivity is in the poly-ribosome fractions and in what appear to be ribosomal precursors. Short-term ^{14}C-leucine labelling experiments indicated that incorporation of activity (into protein) is associated mostly with the poly-ribosome fractions. Thus it appears that in the germinating castor bean endosperm one of the early changes involves the reconstitution of poly-ribosomes (ribosome-messenger RNA complexes), a process which is DNA-dependent. In further studies with wheat embryos it was shown that during the first 30 minutes of imbibition of water by the embryos there is a rapid formation of functional poly-ribosomes with a corresponding increase in the capacity of ribosomal preparations to incorporate ^{14}C-leucine into protein (Marcus and Feeley, 1966). Ribosomes prepared from dry wheat embryos or peanut cotyledons show little or no capacity to incorporate ^{14}C amino acids into protein, whereas preparations from imbibed embryos or cotyledons are active (Marcus and Feeley, 1964). In the presence of polyuridylic acid, the ribosome system of dry seed is as active as that from imbibed seed in catalysing ^{14}C-phenylalanine incorporation. These results suggest the formation or activation of messenger-RNA during imbibition, and the observed formation of poly-ribosomes during germination is clearly consistent with this latter conclusion. By contrast with the foregoing results other workers, using germinating cotton embryos, have found that incorporation of ^{14}C-leucine into protein is not inhibited by Actinomycin D during the first 16 hours of germination, nor is there any loss of poly-ribosomes by this treatment (Dure and Waters, 1965; Waters and Dure, 1966). The authors suggest that the protein synthesis occurring during the early stages of germination is directed by 'long-lived' mRNA which exists in the mature seed and which is not rapidly degraded and resynthesized. It is possible, however, that the failure of Actinomycin D to inhibit protein synthesis is due to the inability of the antibiotic to penetrate into the cells.

It is not possible, here, to deal with the mobilization of reserves during germination and the reader is referred to the accounts of Mayer and Poljakoff-Mayber (1963) and Koller *et al.* (1962) on this subject.

Bibliography

Selected further reading

Chapters by Barton, Evenari, Lang, Stokes, Vegis, and Wareing, 1965. In *Encyclopedia of Plant Physiology*. Vol. 15.

Barton, L. and W. Crocker, 1948. *Twenty Years of Seed Research*, Faber, London.

Mayer, A. and A. Poljakoff-Mayber, 1963. *The Germination of Seeds*, Pergamon Press, Oxford.

Samish, E. M., 1954. Dormancy in woody plants. *Ann. Rev. Plant Physiol.* **5**, 183–204.

Wareing, P. F., 1956. Photoperiodism in woody plants. *Ann. Rev. Plant Physiol.* **7**, 191–214.

Papers cited in text

Abbott, D. L., 1956. Temperature and the dormancy of apple seeds. *Rep. 14th International Hort. Congr.* 743–753, The Hague-Scheveningen.

Askenasy, E., 1877. Über die jähreliche Periode der Knospen. *Bot. Z.* **35**, 793–815, 817–848.

Ballard, L. A. T., 1958. Studies of dormancy in the seeds of subterranean clover (*Trifolium subterraneum* L.) I. Breaking of dormancy by carbon dioxide and by activated carbon. *Austr. J. Biol. Sci.* **11**, 246–260.

Barton, L. V., 1965a. Seed dormancy: General survey of dormancy types in seeds, and dormancy imposed by external agents. *Encyc. Plant Physiol.* **15**(2), 699–720.

Barton, L. V., 1965b. Dormancy in seeds imposed by the seed coat. *Encyc. Plant Physiol.* **15**(2), 727–745.

Barton, L. V. and W. Crocker, 1948. *Twenty years of Seed Research.* Faber, London.

Black, M. and P. F. Wareing, 1955. Growth studies in woody species VII. Photoperiodic control of germination in *Betula pubescens* Ehr., *Physiol. Plant.*, **8**, 300–316.

Black, M. and P. F. Wareing, 1959. The role of germination inhibitors and oxygen in the dormancy of the light sensitive seed of *Betula* spp. *J. Exp. Bot.* **10**, 134–145.

Borthwick, H. A., S. B. Hendricks, M. W. Parker, E. H. Toole and V. Toole, 1952. A reversible photoreaction controlling seed germination. *Proc. Nat. Acad. Sci., U.S.*, **38**, 662–666.

Bradbeer, J. W. and B. Colman, 1967. Studies in seed dormancy. I. The metabolism of $(2-{}^{14}C)$ acetate by chilled seeds of *Corylus avellana* L. *New Phytol.* **66**, 5–15.

Brown, R., 1940. An experimental study of the permeability to gases of seed coat membranes of *Cucurbita pepo*. *Ann. Bot.* N.S. **4**, 379–395.

Burton, W. G., 1963. Concepts and mechanism of dormancy. In *The Growth of the Potato*. Butterworths, London.

Chakravorty, M. and D. P. Burma, 1959. Enzymes of the pentose phosphate pathway in the mung-bean seedling. *Biochem. J.* **73**, 48–53.

Chen, S. S. C. and K. V. Thimann, 1964. Studies on the germination of light-inhibited seed of *Phacelia tanacetifolia*. *Israel J. Bot.* **13**, 57–73.

Chen, S. S. C. and K. V. Thimann, 1966. Nature of seed dormancy in *Phacelia tanacetifolia*. *Science*, **153**, 1537–1538.

Cherry, J. H., H. Chroboczek, W. J. G. Carpenter, and A. Richmond, 1965. Nucleic acid metabolism in peanut cotyledons. *Plant Physiol.* **40**, 582–587.

Chrispeels, M. and J. E. Varner, 1967. Hormonal control of enzyme synthesis: On the mode of action of gibberellic acid and abscisin in aleurone layers of barley. *Plant Physiol.* **42**, 1008–1116.

Colville, F. V., 1920. The influence of cold in stimulating the growth of plants. *J. Agr. Res.* **20**, 151–160.

Conn, E. E. and L. C. Young, 1957. Oxidative phosphylation by lupine mitochondria. *J. Biol. Chem.* **226**, 23–32.

Cornforth, J. W., B. V. Milborrow, G. Ryback, and P. F. Wareing, 1965a. Identity of sycamore 'dormin' with abscisin II. *Nature, Lond.* **205**, 1269–1272.

Cornforth, J. W., B. V. Milborrow, and G. Ryback, 1965b. Synthesis of ($\pm$) abscisin II. *Nature, Lond.* **206**, 715.

Cornforth, J. W., B. V. Milborrow, and G. Ryback, 1966a. Identification and estimation of (+) abscisin II ('dormin') in plant extracts by spectropolarimetry. *Nature, Lond.* **210**, 627–628.

Crocker, W., 1948. *Growth of Plants.* Reinhold Pub. Corp., New York.

Crocker, W. and L. V. Barton, 1957. *Physiology of Seeds.* Chron. Bot. Waltham, Mass.

Davis, W. E., 1930. Primary dormancy, after-ripening and development of secondary dormancy in embryos of *Ambrosia trifolia. Am. J. Bot.* **17**, 58–76.

Davis, W. E., 1930b. The development of dormancy in seeds of cocklebur (*Xanthium*). *Am. J. Bot.* **17**, 77–87.

Digby, J. and P. F. Wareing. 1966. The relationship between endogenous hormone levels in the plant and seasonal aspects of cambial activity. *Ann. Bot.,* **30**, 607–622.

Dörffling, K., 1963. Über das Wachstoff-hemmstoffsystem von *Acer pseudoplatanus* I. Der Jahresgung der Wachs- und Hemmstoffe in Knospen, Blättern und im Kambium. *Planta,* **60**, 390–412.

Downs, R. J. and H. A. Borthwick, 1956. Effects of photoperiod on growth of trees. *Bot. Gaz.* **117**, 310–326.

Dure, L. and L. Waters, 1965. Long-lived messenger RNA: evidence from cotton seed germination. *Science,* **147**, 410–412.

Eagles, C. F. and P. F. Wareing, 1963. Dormancy regulators in woody plants. *Nature, Lond.* **199**, 874.

Eagles, C. F. and P. F. Wareing, 1964. The role of growth substances in the regulation of bud dormancy. *Physiol. Plant.* **17**, 697–709.

El-Antably, H. M. M., P. F. Wareing, and J. Hillman, 1967. Some physiological responses to d,l abscisin (dormin). *Planta,* **73**, 74–90.

Evenari, M., 1949. Germination inhibitors. *Bot. Rev.,* **15**, 153–194.

Evenari, M., 1965. Light and seed dormancy. *Encyc. Plant. Physiol.* **15**(2), 804–847.

Evenari, M., G. Newmann, and S. Klein, 1955. The influence of red and infra-red light on the respiration of photoblastic seeds. *Physiol. Plant.* **8**, 33–47.

Evenari, M., S. Klein, H. Anchori, and N. Feinburn, 1957. The beginning of cell-division and cell elongation in germinating lettuce seed. *Bull. Res. Council,* Israel, **D6**, 33–37.

Flint, L. H. and E. D. McAlister, 1935. Wavelength of radiation in the visible spectrum inhibiting the germination of light sensitive lettuce seed. *Smithsonian Inst. Misc. Coll.* **94**, 1–11.

Flint, L. H. and E. D. McAlister, 1937. Wavelength of radiation in the visible spectrum promoting germination of light sensitive lettuce seed. *Smithsonian Inst. Misc. Coll.* **96**, 1–8.

Frankland, B. and P. F. Wareing, 1962. Changes in endogenous gibberellins in relation to chilling of dormant seeds. *Nature, Lond.* **194**, 313–314.

Frankland, B. and P. F. Wareing, 1966. Hormonal regulation of seed dormancy in hazel (*Corylus avellana* L.) and beech (*Fagus sylvatica*). *J. Exp. Bot.* **17**, 596–611.

Güttenberg, H. von and H. Leike, 1958. Untersuchungen über den Wuchs- und Hemm-stoffgehalt ruhender und treibender Knospen von *Syringa vulgaris*, L. *Planta*, **52**, 96–120.

Hatch, M. D. and J. F. Turner, 1958. Glycolysis by an extract of pea seeds. *Biochem. J.* **69**, 495–501.

Hemberg, T., 1949a. Significance of growth-inhibiting substances and auxins for the rest-period of potatoes. *Physiol. Plant.* **2**, 24–36.

Hemberg, T., 1949b. Growth-inhibiting substances in the terminal buds of *Fraxinus*. *Physiol. Plant.* **2**, 37–44.

Heslop-Harrison, Y., 1962. Winter dormancy and vegetative propagation in Irish *Pinguicula grandiflora*. *Proc. Roy. Irish Acad.* **62B**, 23–30.

Ikuma, H. and K. V. Thimann, 1963. The role of the seed coats in germination of photo-sensitive lettuce seeds. *Plant and Cell Physiol.* **4**, 169–185.

Ingle, J. and R. H. Hageman, 1965. Metabolic changes associated with the germination of corn. II. Nucleic acid metabolism. *Plant Physiol.* **40**, 48–53.

Jackson, G. A. D. and J. B. Blundell, 1963. Germination in *Rosa*. *J. Hort. Sci.* **38**, 310–320.

Jackson, G. A. D. and J. B. Blundell, 1966. Effect of dormin on fruit-set in *Rosa*. *Nature, Lond.* **212**, 1470–1471.

Kentzer, T., 1966. Gibberellin-like substances and growth inhibitors in relation to the dormancy and after-ripening of ash seeds (*Fraxinus excelsior L.*) *Acta Soc. Bot. Pol.* **35**, 575–585.

Koller, D., A. Poljakoff-Mayber, A. Berg, and T. Diskin, 1963. Germination-regulating mechanisms in *Citrullus colocynthis*. *Am. J. Bot.* **50**, 597–603.

Kozlowski, T. T., 1964. Shoot growth in woody plants. *Bot. Rev.* **30**, 335–392.

Kozlowski, T. T., 1966. Food relations of woody plants. *Bot. Rev.* **32**, 294–382.

Kramer, P. J., 1936. The effect of variation in length of day on the growth and dormancy of trees. *Plant Phys.* **11**, 127–137.

Lang, A., 1965. Effects of some internal and external conditions on seed germination. *Encycl. Plant Physiol.* **15**(2), 848–893.

Major, W. and E. H. Roberts (in press).

Mann, L. K. and D. A. Lewis, 1956. Rest and dormancy in garlic. *Hilgardia*. **26**, 161–189.

Mapson, L. W. and E. M. Moustaffa, 1957. Ascorbic acid and glutathione as respiratory carriers in the respiration of pea seedlings. *Biochem. J.* **62**, 248–259.

Marcus, A. and J. Feeley, 1964. Activation of protein synthesis in the imbibition phase of seed germination. *Proc. Nat. Acad. Sci. U.S.* **51**, 1075–   .

Marcus, A. and J. Feeley, 1966. Protein synthesis in imbibed seeds. II. Polysome for-mation during imbibition. *J. Biol. Chem.* **240**, 1675–1680.

Marrè, E., S. Cocucci and S. Sturani, 1965. On the development of the ribosomal system in the endosperm of germinating castor bean seeds. *Plant Physiol.* **40**, 1162–1170.

Mayer, A. M. and L. W. Mapson, 1962. Esterification of inorganic phosphate by gly-colysis in extracts from normal and pentan-2-d treated pea cotyledons. *J. Exp. Bot.* **13**, 201–212.

Mayer, A. and A. Poljakoff-Mayber, 1963. *The Germination of Seeds*. Pergamon Press, Oxford.

Milborrow, B. V., 1967. The effects of synthetic d,l-dormin (Abscisin II) on the growth of the oat mesocotyl. *Planta*, **70**, 155–171.

Molisch, H., 1922. *Pflanzenphysiologie als Theorie der Gärtnerei*. 5th Ed. Jena.

Moringa, T., 1926. Germination of seeds under water. *Am. J. Bot.* **13**, 126–140.

Neal, G. E. and H. Beevers, 1960. Pyruvate utilization in castorbean endosperm and other tissues. *Biochem. J.* **74**, 409–416.

Nitsch, J. P., 1957. Growth responses of woody plants to photoperiodic stimuli. *Proc. Am. Soc. Hort. Sci.* **70**, 512–525.

Nitsch, J. P. 1963. The mediation of climatic effects through endogenous regulating substances. In, *Environmental Control of Plant Growth*, Ed. L. T. Evans, Academic Press, New York.

Okhuma, K., J. L. Lyon, F. T. Addicott, and O. E. Smith, 1963. Abscisin II, an abscission-accelerating substance from young cotton fruit. *Science*, **142**, 1592.

Okhuma, K., F. T. Addicott, O. E. Smith, and W. E. Thilssen, 1965. The structure of abscisin II. *Tetrahedron Letters*, **29**, 2529.

Paleg, L. G., 1960. Physiological effects of gibberellic acid. II. On starch hydrolyzing enzymes of barley endosperm. *Plant Physiol.* **35**, 902–906.

Paleg, L. G., 1964. Cellular localization of the gibberellin-induced response of barley endosperm. *5th Internat. Conf. Plant Growth Regulation, Paris.*

Phillips, I. D. J. and P. F. Wareing, 1958. Studies in dormancy of sycamore I. Seasonal changes in the growth-substance content of the shoot. *J. Exp. Bot.* **9**, 350–364.

Phillips, I. D. J. and P. F. Wareing, 1959. Studies on the dormancy of sycamore II. The effect of daylength on the natural growth inhibitor content of the shoot. *J. Exp. Bot.* **10**, 504–514.

Pollock, B. M., 1953. The respiration of *Acer* buds in relation to the inception and termination of the winter rest. *Physiol. Plant.* **6**, 47–77.

Pollock, B. M. and H. O. Olney, 1959. Studies of the rest period I. Growth, translocation and the respiratory changes in the embryonic organs of the after-ripening cherry seed. *Plant Physiol.* **34**, 131–142.

Priestley, C. A., 1962. Carbohydrate resources within the perennial plant. *Comm. Bur. Hort. and Plantation Crops, Tech. Comm. 27.*

Roberts, E. H., 1964a. The distribution of oxidation-reduction enzymes and the effects of respiratory inhibitors and oxidizing agents on dormancy in rice seed. *Physiol. Plant.* **17**, 14–28.

Roberts, E. H., 1964b. A survey of the effects of chemical treatments on dormancy of rice seed. *Physiol. Plant.* **17**, 30–43.

Robinson, P. M., P. F. Wareing, and T. H. Thomas, 1963. Isolation of the inhibitor varying with photoperiod in *Acer pseudoplatanus*. *Nature, Lond.* **199**, 874–876.

Robinson, P. M. and P. F. Wareing, 1964. Chemical nature and biological properties of the inhibitor varying with photoperiod in sycamore (*Acer pseudoplatanus*) *Physiol. Plant.* **17**, 314–323.

Romberger, J. A., 1963. *Meristems, growth and development in woody plants.* U.S. Department of Agriculture, Tech. Bull. No. 1293.

Sacher, J. A., 1954. Structure and seasonal activity of the shoot apices of *Pinus lambertiana* and *Pinus ponderosa*. *Am. J. Bot.* **52**, 841.

Schander, H., 1955. Keimungsphysiologische Studien auf Kernobst. II. Untersuchungen über die allgemeinen Temperaturansprüche der Kernobstsamen während der Keimung *Z. Pflanzenzücht.* **34**, 421–440.

Samish, R. M., 1954. Dormancy in woody plants. *Ann. Rev. Physiol.* **5**, 183–204.

Schüepp, O., 1929. Untersuchungen zur beschreibenden und experimentellen Enwicklungsgesichte von *Acer pseudoplatanus* L. *Jahrb. f. wiss. Bot.* **70**, 743–804.

Schwabe, W. W. and S. Nachmony-Bascombe, 1963. Growth and dormancy in *Lunularia cruciata* (L) Dum. II. The response to daylength and temperature. *J. Exp. Bot.* **14**, 353–378.

Smith, O. E. and L. Rappaport, 1961. Endogenous gibberellins in resting and sprouting potato tubers. *Adv. Chem. Sev.* **28**, 42–48.

Spragg, S. P. and E. W. Yemm, 1959. Respiratory mechanisms and the changes of glutathione and ascorbic acid in germinating peas. *J. Exp. Bot.* **10**, 409–425.

Sroelov, R., 1940. On germination inhibitors. IV. Germination inhibitors of *Sinapis alba* and other seeds when enclosed in their fruit. *Pales. J. Bot.*, Jerusalem Ser. **11**, 33–37.

Stokes, P., 1965. Temperature and seed dormancy. *Encyc. Plant Physiol.* **15**(2), 746–803.

Taylor, D. L., 1942. Influence of oxygen tension on respiration, fermentation and growth in wheat and rice. *Am. J. Bot.* **29**, 721–738.

Thomas, T. H., P. F. Wareing, and P. M. Robinson, 1965. Action of the sycamore dormin as a gibberellin antagonist. *Nature, Lond.* **205**, 1269–1272.

Thornton, N. C., 1936. Factors influencing germination and development of dormancy in cocklebur seeds. *Contr. Boyce Thompson Inst.* **7**, 477–496.

Thornton, N. C., 1945. Importance of oxygen supply in secondary dormancy and its relation to the inhibiting mechanism regulating dormancy. *Contr. Boyce Thompson Inst.* **13**, 487–500.

Tolman, B. and M. Stout, 1940. Toxic effect on germinating sugar beet seed of water soluble substances in the seed-ball. *J. Agr. Res.* **61**, 817–830.

Tuan, D. Y. H. and J. Bonner, 1964. Dormancy associated with repression of genetic activity. *Plant Phys.* **39**, 768–772.

Varner, J. E. and G. Ram Chandra, 1964. Hormonal control of enzyme synthesis in barley endosperm. *Proc. Nat. Acad. Sci., U. S.* **52**, 100–

Vegis, A., 1949. Durch hohe Temperatur bedingter Weidereintritt des Ruhezustandes bei den Winterknospen. *Svensk bot. Tidskr.* **43**, 671–714.

Vegis, A., 1956. Formation of the resting condition in plants. *Experienta*, **12**, 94–99.

Vegis, A., 1961. Kältebedurfnis in Wachstum und Entwicklung Samenkeimung und Vegetative Entwicklung der Knospen. *Encyc. Plant Physiol.* **16**, 168–298.

Vegis, A., 1964a. Climatic control of germination, bud break and dormancy. In *Environmental Control of Plant Growth*, ed. L. T. Evans. Academic Press, New York.

Vegis, A., 1964b. Dormancy in higher plants. *Ann Rev. Plant Physiol.* **15**, 185–224.

Vegis, A., 1965. A. Ruhezustände bei höhren Pflanzen. Induktion, Verlauf und Beendigung: Ubersicht, Terminologie, allgemeine Probleme. *Encycl. Plant Physiol.* **15**(2), 499–533.

Villiers, T. A. and P. F. Wareing, 1965. The growth substance content of dormant fruits of *Fraxinus excelsior*. *J. Exp. Bot.* **16**, 534–544.

Visser, T., 1954. After-ripening and germination of apple seeds in relation to seed coats. *Kon. Ned. Akad. von Wetensch.* Amsterdam, Ser. C, **57**, 175–185.

Wareing, P. F., 1950. Growth Studies in Woody Species I. Photoperiodism in First-Year Seedlings of *Pinus silvestris*. *Physiol. Plant.* **3**, 258–276.

Wareing, P. F., 1953. Growth Studies in Woody Species V. Photoperiodism in Dormant Buds of *Fagus sylvatica* L. *Physiol. Plant.* **6**, 692–706.

Wareing, P. F., 1954. Growth Studies in Woody Species VI. The Locus of Photoperiodic Perception in Relation to Dormancy. *Physiol. Plant.* **7**, 261–277.

Wareing, P. F., 1956. Photoperiodism in Woody Plants. *Ann. Rev. Plant. Phys.* **7**, 191–214.

Wareing, P. F., 1959. Photoperiodism in seeds and seedlings of woody species. In *The Physiology of Forest Trees*, ed. Thimann. Ronald Press, N.Y. pp. 539–556.

Wareing, P. F., 1965. Endogenous inhibitors in seed germination and dormancy. *Encyl. Plant Physiol.* **15**, 909–924.

Wareing, P. F. and H. Foda, 1956. Growth inhibitors and dormancy in *Xanthium* seed. *Physiol. Plant.* **10**, 266–280.

Wareing, P. F. and M. Black, 1958. Photoperiodism in seeds and seedlings of woody species. In *The Physiology of Forest Trees*. The Ronald Press Co., New York.

Wareing, P. F., J. E. G. Good, and J. Manuel, 1968. Some possible physiological roles of abscisic acid. *Proc. 6th Inter. Conf. on Plant Growth Substances*, Ottawa.

Waters, L. C. and L. S. Dure, 1966. Ribonucleic acid synthesis in germinating cotton seeds. *J. Molec. Biol.* **19**, 1–27.

Waxman, S., 1957. *The development of woody plants as affected by photoperiodic treatments*. Ph.D. thesis, Cornell University, Ithaca, N.Y. 193 pp.

18. Circadian rhythms in plants

Malcolm B. Wilkins

General introduction

Under normal circumstances plants grow in an environment in which a number of parameters fluctuate rhythmically, with a period of 24 hours. Of these parameters light intensity and temperature are the most obvious, but fluctuations also occur in such geophysical factors as the strength of the earth's magnetic field and the intensity of cosmic ray bombardment. It is perhaps not surprising, therefore, that in such an environment plants, in common with other living organisms, exhibit 24-hour rhythms in their physiological and chemical activity. Rhythmic behaviour in plants under natural conditions has been known for nearly two thousand years. The so-called sleep movements of leaves were recorded by Pliny in the first century A.D. and by Albertus Magnus in the thirteenth century.

The ease with which leaf movement could be observed and recorded led to almost all the early work on plant rhythms being carried out on this phenomenon. In 1729, de Mairan carried out a most important experiment in which he transferred plants to a darkened room and found that the leaves continued to move in a rhythmic manner for a number of days. This finding, later confirmed both by Duhamel (1758) and Zinn (1759), clearly focused attention upon the existence in plants of a rhythmic system capable of controlling the movement of leaves under uniform environmental conditions.

The concept of the endogenous rhythm, as distinct from the exogenous rhythm, thus became firmly established. Exogenous rhythms are found in many plants under natural conditions, but they do not persist in a uniform environment and appear to be due solely to the periodic stimulation of the organism by environmental parameters. For example, cultures of the fungi *Pilobolus crystallinus* (Uebelmesser, 1954) and *Sordaria fimicola* (Ingold and Dring, 1957) discharge their spores rhythmically under normal conditions but this rhythmicity is lost as soon as they are transferred to a uniform environment.

Definitions

Before proceeding to a detailed consideration of the occurrence of endogenous rhythms in plants, the technical terms most frequently used in this field must be briefly explained. In this chapter only rhythms having a period of about 24 hours are considered. In the past these have been termed endogenous, diurnal rhythms but this term is unsatisfactory since the word diurnal implies a periodicity of precisely 24 hours. In almost all cases the free-running period of plant and animal rhythms is not 24 hours but something between 21 and 28 hours. Rhythms having a period between these approximate limits are now termed *circadian rhythms* (circa = about, diem = day). The *period* of a rhythm is the time between the repetition of a definite phase of the oscillation, that is, for example, the time between successive peaks or troughs. The *free-running* or *natural* period of a rhythm is the steady-state period exhibited under uniform environmental conditions. The *phase* is the instantaneous state of an oscillation within a period, and the *phase difference* is the displacement of an oscillation along the time axis. The phase difference between two circadian rhythms is usually expressed as the time in hours between the occurrence of a particular instantaneous state in the cycle. *Entrainment* of a rhythm is the coupling of an endogenous, circadian oscillating system to rhythmicity in an environmental parameter. All circadian systems are entrained to a precise 24-hour period under natural conditions and it is also possible to entrain them to cycles having other periods. The *damping* of a rhythm is the gradual decrease in the amplitude with time, a phenomenon observed in most circadian rhythms in plants kept under uniform conditions. A more complete list of terms has been given by Aschoff (1965).

Establishment of the endogenous nature of the rhythm

Continuation of a rhythm under conditions uniform with respect to light and temperature is not critical evidence for the endogenous nature of that rhythm. In such a controlled environment a rhythm might result directly from the periodic stimulation of the organism by an uncontrolled environmental parameter such as one of the geophysical factors to which reference has already been made. In order to establish that a rhythm is endogenous, it is now generally accepted that the rhythm must conform to the following five conditions which were first proposed by Pittendrigh (1954).

(a) The rhythm should continue in an environment in which as many environmental parameters as possible are held constant.
(b) The phase of the rhythm should be able to be shifted and the new phase subsequently retained under uniform environmental conditions.
(c) The rhythm should be initiated by a single stimulus. If the rhythmicity observed in plants is endogenous, that is, reflecting the activity of an internal oscillating system, it would be expected that the oscillating system and the natural periodicity of the system would be inherited from one

generation to the next. In other words, there is no question of the period of oscillation being acquired from the environment by experience of the cyclical variation in light intensity and temperature. In plants that require an extended period of time for growth and development, it is not possible to apply this test, but a number of plants can be grown under standard conditions of temperature and in either light or darkness, and a rhythm can be induced by a single stimulus such as a change from light to darkness. Seedlings of *Avena sativa* grown entirely under red light and at a uniform temperature do not exhibit rhythmicity in the growth rate of their coleoptiles. However, if the seedlings are transferred to darkness about 40–50 hours after imbibition, a circadian rhythm is induced in the growth rate of the coleoptiles (Ball and Dyke, 1954). A rhythm of leaf movement in *Phaseolus multiflorus* can similarly be induced by transferring the seedlings from continuous light to continuous darkness or vice versa (Lörcher, 1958; Wassermann, 1959). In arhythmic cultures of *Gonyaulax polyedra* which have been maintained in continuous light for three years a single change in the light intensity suffices to initiate the circadian rhythm of luminescence (Sweeney and Hastings, 1962).

(d) The phase of the rhythm should be delayed under hypoxia. As might be expected of an endogenous mechanism, the operation of the oscillating system is arrested in a number of plants and animals when the organisms are exposed to anaerobic conditions for several hours (Ball and Dyke, 1957; Bünning *et al.*, 1965; Wilkins, 1967). Continued rhythmic oscillation is therefore dependent upon energy being made available by aerobic metabolism.

(e) The period of the rhythm should not be *exactly* 24 hours. Of the five conditions for establishing the endogenous nature of a rhythm this is the most decisive. It is possible for a rhythm to conform to the previous four conditions and still be due to an external periodic influence affecting the plant in the supposedly uniform environment. The periodic stimulation might elicit a rhythm in the plant by affecting the basic oscillating process which controls the various aspects of physiological and chemical activity which are normally monitored as the observed rhythm. If the periodic stimulus and the basic oscillating system can be in any phase relationship to each other, it would be possible to adjust the phase of the observed rhythm by a treatment involving, for example, light, temperature change, or anaerobiosis. In such a system, however, the period of the rhythm would be expected to be precisely 24 hours. If it can be demonstrated that the natural frequency of the endogenous oscillating system is significantly different from 24 hours and, in addition, that it can be altered by adjusting the uniform level of one of the environmental parameters, then the possibility of the rhythm being the result of a 24-hour periodic stimulation by an uncontrolled environmental parameter becomes extremely remote.

A large number of plant rhythms have been shown to have a free-running period that is significantly different from 24 hours: for example, those of the rhythms of leaf movement in *Phaseolus multiflorus* (Leinweber, 1956) and of carbon dioxide emission of detached *Bryophyllum* leaves (Wilkins, 1962a) are 28·0 hours and 22·4 hours respectively at 25–26°C. When these plants are in a uniform environment the peaks of the rhythms will thus occur respectively later and earlier on successive days. After several days the difference between the times of occurrence of the peaks in these plants and those remaining in a normal environment can be as much as 12 hours.

Rhythms which meet these five conditions are thus termed endogenous, that is to say they reflect the activity of a cellular system capable of sustained oscillation, at least for a limited period of time. This cellular system is often referred to as a biological 'clock'. Rhythms in plants usually continue for up to one or two weeks in a uniform environment but those in animals continue for a considerably longer time, in some cases for as much as several months. The more rapid damping of plant rhythms is probably due in part to the exhaustion of nutrients under the controlled conditions of darkness and temperature, although other factors are almost certainly involved. Unfortunately the starvation effect cannot be overcome by exposing plants to prolonged light of sufficient intensity to bring about photosynthesis because the rhythms are usually inhibited by these levels of irradiation. In *Gonyaulax polyedra*, however, while the rhythm of luminescence is inhibited by bright light, it persists for much longer in dim light than in prolonged darkness (Hastings and Sweeney, 1959).

The occurrence of rhythms in plants

Persistent circadian rhythms have been observed in members of each major plant group with the exception of the *Bryophyta*. They do not appear to occur in all members of any one group, and of two closely related species, one may possess a persistent rhythm and the other not. The persistent rhythm of spore discharge found in *Pilobolus sphaerosporus* but not in *P. crystallinus* exemplifies this point (Uebelmesser, 1954). Some examples of plant rhythms are given in Table 18.1.

Organization of circadian systems

Rhythms have been observed in many different physiological activities and in many different species of plant. Many of the characteristics of the rhythms in different species are closely similar and this has naturally led to the idea that there may be a similar basic oscillating system or 'clock' in the different

Table 18.1 Examples of circadian rhythms in plants.

Group	Organism	Rhythm	Reference
Photosynthetic Protozoans	*Gonyaulax polyedra*	luminescence, photosynthetic capacity, growth	Hastings and Sweeney (1959)
	Euglena gracilis	phototaxis	Bruce and Pittendrigh (1956)
Algae	*Hydrodictyon reticulatum*	photosynthesis, respiration	Pirson, Schön, and Döring (1954)
	Oedogonium cardiacum	sporulation	Bühnemann (1955)
	Acetabularia major	photosynthetic capacity	Sweeney and Haxo (1961)
Fungi	*Daldinia concentrica*	spore discharge	Ingold and Cox (1955)
	Pilobolus sphaerosporus	sporangium discharge	Uebelmesser (1954)
	Neurospora crassa	growth	Pittendrigh *et al.* (1959)
Pteridophyta	*Selaginella serpens*	plastid shape	Busch (1953)
Spermatophyta	*Phaseolus multiflorus*	leaf movement	Bünning (1958)
	Kalanchoë blossfeldiana	petal movement	Bünsow (1953)
	Avena sativa	growth rate of coleoptile	Ball and Dyke (1954)
	Bryophyllum fedtschenkoi	CO_2-fixation in darkness	Warren and Wilkins (1961)

organisms. This basic system is thought to control the various aspects of physiological and chemical activity which are normally monitored in the study of endogenous rhythms. While there is now some evidence that the basic oscillating systems may not be identical in all living organisms, and this point will be discussed later (p. 667), the general idea that somewhere in the cell there is a basic oscillating system controlling a number of observable rhythms seems to be correct (Hastings and Sweeney, 1959). If such is the case then the phases of all observable rhythms controlled by the basic system should bear a fixed relationship to each other. In the photosynthetic armoured marine dinoflagellate *Gonyaulax polyedra* it has been shown that the rhythms in photosynthetic capacity, luminescence, and cell division do have a fixed phase relationship to one another. In other words, if the phase of one rhythm is shifted, the phases of the other rhythms are shifted as well (Hastings and Sweeney, 1959).

The central problem in the study of circadian rhythmicity is obviously the identification of the basic oscillating system. This has not yet been achieved in any living system, plant or animal. However, much information is now available on the characteristics of the basic oscillating system in a number of plants. This information has been gained from physiological studies of the responses of the system to changes in environmental parameters, and of the dependence of oscillation upon aerobic metabolism. In these investigations the rhythm of physiological activity is monitored and changes in the phase and period are used to deduce the behaviour of the basic oscillating system.

Control of circadian oscillating systems by environmental factors

The environmental parameters with the greatest effect upon the basic oscillating systems are light, temperature, and the partial pressure of oxygen. The effects range from the induction of phase shifts to changes in the steady-state period of oscillation. In addition, cyclical variation in environmental parameters can entrain the basic systems to non-circadian frequencies. A large number of investigations have been made of the influence of environmental factors on the phase and period of circadian systems. In an introductory review such as this it is obviously impossible to treat the subject in a comprehensive manner. Discussion will therefore be confined to a few organisms which have been thoroughly investigated and which may be regarded as showing typical responses to changes in environmental factors. For a more detailed discussion reference should be made to Bünning (1967).

Entrainment

Entrainment of circadian oscillating systems to frequencies other than their natural frequency has been achieved in a large number of plants by periodic variation in light intensity and temperature (Bruce, 1960). At a uniform temperature, alternating exposures to light and darkness can entrain circadian systems to periodicities ranging from 6 hours to 36 hours (Fig. 18.1). In most organisms there appears to be a limit to the rapidity with which the circadian systems can be forced to oscillate by imposed cycles of light and darkness. The leaf movement rhythm in *Phaseolus multiflorus* for example, can easily be entrained to an 18-hour period by 9:9 hour cycles of light and darkness. When exposed to 8:8 hour cycles, however, the rhythm of leaf movement does not show a 16-hour period but reverts to its natural frequency (Bünning, 1957b). These limits of entrainment are, however, not a fixed property of the basic oscillating system but are a function of the energy input into the system. This fact has been demonstrated with the rhythm of carbon dioxide metabolism in *Bryophyllum* (Wilkins, 1962b). When a light intensity of 1,000 lux was employed

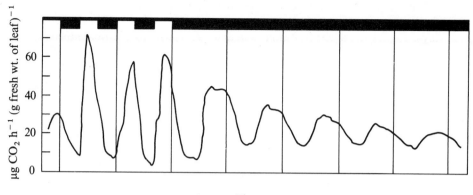

Time

Figure 18.1 Entrainment of the rhythm of carbon dioxide metabolism in leaves of *Bryophyllum* to 8:8-hour cycles of light and darkness, and the reappearance of the natural period on the restoration of continuous darkness. Temperature 25°C. Light and dark times are indicated by the bar above the figure. Vertical lines show midnight. (From Wilkins, 1962b.)

entrainment occurred with 3:3 hour, 6:6 hour, and 8:8 hour cycles of light and darkness. At 500 lux entrainment occurred with 6:6 and 8:8 cycles but not with 3:3 hour cycles, and at 100 lux entrainment occurred with 8:8 hour cycles but not with 3:3 hour and 6:6 hour cycles.

In all except two of the cases reported in the literature, the period of an entrained basic oscillating system returns to its natural value immediately the organism is returned to a uniform environment. This is illustrated in the case of *Bryophyllum* in Fig. 18.1. There is, therefore, in most organisms, no evidence that the basic oscillator can acquire a new frequency from the environment. The two organisms which appear to be exceptions to this general rule are *Pilobolus sphaerosporus* and *Hydrodictyon reticulatum*. In *Pilobolus* the period of the entrained rhythm appears to be retained for a limited time after uniform environmental conditions are restored, and then the natural period reappears (Uebelmesser, 1954). This apparent retention of the entrained frequency by the oscillating system has been discussed by Wilkins (1960a) and attributed entirely to the development and maturation time of the sporangiophore. The sporangiophores are initiated by a change from darkness to light and require approximately 28 hours to attain maturity at room temperature. The last peak to occur on the entrained frequency after a uniform environment was restored was always found approximately 28 hours after the culture last experienced a change from darkness to light.

In *Hydrodictyon* there seems to be clear evidence for the retention of the entrained frequency by the basic oscillating system after a uniform environment is restored (Pirson, Schön, and Döring, 1954). This is a unique situation and one that is rather unexpected in the light of our knowledge of physical oscillating systems. After exposure to cycles of 10·5 hours light and 7 hours darkness the rhythm persisted in subsequent darkness with a period of 17·5 hours for several days. This aspect of the rhythmic system in *Hydrodictyon* undoubtedly warrants further study.

Periodic fluctuations in temperature can also entrain circadian oscillating systems, although they have been less well investigated than cycles of light and darkness. The rhythm of sporulation in *Oedogonium cardiacum* can be entrained to temperature cycles having periods of 18 hours and 24 hours but not to those having periods of 12 hours, 30 hours, and 48 hours. A maximum temperature change of 2·5°C is sufficient to achieve entrainment (Bühnemann, 1955). In *Kalanchoë blossfeldiana* entrainment of the rhythm of petal movement has been achieved with a temperature fluctuation of only 1°C (Oltmanns, 1960).

Frequency demultiplication

Frequency demultiplication is a special case of entrainment that has been observed in a number of plants. When this phenomenon occurs the period shown by the circadian system is a multiple of the period of the entraining cycles rather than equal to it, as is the case in normal entrainment. Biale (1940) showed that the rhythm of transpiration in lemon cuttings had an exactly 24-hour period when exposed to 8-hour cycles of light and darkness. In *Phaseolus multiflorus* the rhythm of leaf movement shows periods of 28 hours and 32 hours when exposed to 14-hour and 16-hour cycles of light and darkness respectively (Flügel, 1949). Other examples of frequency demultiplication have been given by Bruce (1960).

Inhibition

Of the environmental factors which inhibit oscillation of the basic circadian systems in plants, light, high and low temperature, and anaerobic conditions have the most dramatic effects.

In the Crassulacean plant *Bryophyllum fedtschenkoi* and the marine dino-flagellate *Gonyaulax polyedra*, two living systems at widely different levels of organization, the circadian oscillators are inhibited by prolonged exposure to bright light (Wilkins, 1959, 1960a; Hastings and Sweeney, 1959). Oscillation begins again, however, with the restoration of darkness. The first peak occurs a definite time after the restoration of darkness, regardless of the time of day at which this change occurs. Similar reports have been made for the rhythm in the growth rate of the *Avena sativa* coleoptile (Ball and Dyke, 1954). The change from light to darkness thus not only initiates oscillation but also determines the phase. The behaviour observed in *Bryophyllum* and *Avena* suggests that during exposure to light the circadian oscillator is inhibited by being driven to and held at some specific phase in its cycle.

In *Bryophyllum* exposure to high temperature has an effect on the circadian system almost identical to that of bright light (Wilkins, 1962a). At 35–36°C the system is inhibited but oscillation begins again when the temperature is lowered to 25–26°C. The first peak of the rhythm occurs a definite time after the temperature is lowered and this time is almost identical to that found after a change from prolonged light to darkness at 25°C. The implication in *Bryophyllum* is, therefore, that both bright light and high temperature inhibit oscillation by forcing the basic system to, and holding it at, the same phase of the cycle. The mechanism whereby light and high temperature exert this effect has not yet been elucidated.

Since metabolism is severely reduced at temperatures near 0°C it might be expected that the circadian system would be inhibited at these temperatures. In *Avena sativa* (Ball and Dyke, 1957) and *Phaseolus multiflorus* (Bünning and Tazawa, 1957) seedlings, and in *Bryophyllum* leaves (Wilkins, 1962a), prolonged exposure to low temperature delays the phase of oscillation, the magnitude of the delay being related to the length of chilling. Reduction of the rate of metabolism by exposure to anaerobic conditions has also been shown to inhibit oscillation of the circadian systems in *Avena* (Ball and Dyke, 1957; Wilkins and Warren, 1963), *Phaseolus* (Bünning, Kurras, and Vielhaben, 1965) and *Bryophyllum* (Wilkins, 1967).

Phase control

Phase control of circadian systems has been thoroughly investigated in a number of plants. This is one of the most important aspects of research in this field because in some animals, particularly man, where there appear to be several independent circadian systems operating in one organism, phase control and hence the correlation of physiological functions under the control of the independent circadian systems, can have important consequences which are reflected in the health and proficiency of the individuals.

Light and temperature change are the major factors which determine the phase of circadian oscillators. In plants such as *Avena* and *Bryophyllum* for example, the phase of the rhythm is determined by the time that darkness begins, and similar findings have been reported for the systems in *Neurospora* (Pittendrigh *et al.*, 1959) and *Gonyaulax* (Hastings and Sweeney, 1959). In *Bryophyllum* and *Gonyaulax* a decrease in the light intensity rather than a change from light to darkness can also be used as a phase determining signal (Wilkins, 1960b; Hastings and Sweeney, 1959). In *Phaseolus* and *Oedogonium*, however, transfer from darkness to light can initiate the rhythm and determine the phase (Bühnemann, 1955).

When an organism is being maintained in prolonged darkness, the phase of the oscillating circadian system can be reset by a single, brief exposure to light.

In *Bryophyllum*, *Phaseolus*, and *Gonyaulax*, for example, a few hours illumination can reset the phase, but the magnitude of the phase shift depends upon the position in the cycle at which this light treatment is given. In one part of the cycle the phase of the rhythm is quite unaffected by the exposure to light, whereas in other parts of the cycle a large phase shift is induced. This is illustrated in Fig. 18.2 in the case of *Bryophyllum*. In this plant at least, the magnitude of the phase shift is determined by the time at which the light treatment ends, and not by its duration. This variation in the magnitude of the phase shift

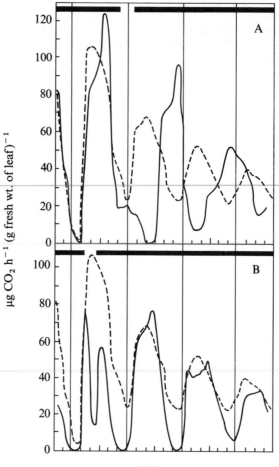

Figure 18.2   The effectiveness of a 5-hour light treatment in resetting the phase of the rhythm of carbon dioxide metabolism in *Bryophyllum* leaves otherwise maintained in darkness. Treatments were given either between the peaks (A) or at the apex of a peak (B) as shown by the bar above each figure. Vertical lines show midnight. Broken line: control leaves in continuous darkness. Continuous line: light-treated leaves. (From Wilkins, 1960b.)

induced by a light treatment given at different positions in the cycle would have been anticipated on the basis of the suggestion made in the previous section. If the circadian oscillating system is driven to and held at one phase of its cycle by prolonged illumination then during the course of normal oscillation the system will be at this phase for a specific time. If a short exposure to light is given so that it ends at the time when the oscillator would normally be at this position in its cycle then no phase shift would be expected.

The phase of circadian rhythms in organisms kept in continuous dim light can also be changed by a single exposure to darkness for a few hours. This has been observed in *Bryophyllum*, *Phaseolus* and *Gonyaulax*. In *Bryophyllum* the effectiveness of a dark treatment in resetting the phase of the circadian system again depends upon the part of the cycle in which it is administered.

In addition to light and dark perturbations, exposure of organisms kept in a uniform environment to changes in temperature can induce predictable phase shifts in circadian systems. For example, exposure of *Bryophyllum* leaves, otherwise kept at 26°C, to 36°C for a few hours can induce phase shifts which are identical with those induced by exposure to light (Wilkins, 1962a). As has been pointed out in the previous section, the effects of light and exposure to 36°C on this system appear to be identical, and this similarity extends even to the positions in the cycle at which the treatments are effective and ineffective in shifting the phase. A closely similar situation is found in *Phaseolus* where the magnitude of the phase shift induced by exposure to elevated temperatures for several hours again depends upon the position in the cycle at which the treatment is given (Moser, 1962).

Exposure of circadian systems to low temperature or to anaerobic conditions for a few hours can also induce phase shifts. With both types of treatment it is again found that their effectiveness depends to a great extent upon the position in the cycle at which the treatments are applied. For example, the circadian oscillators in both *Phaseolus* (Wagner, 1963; Bünning *et al.*, 1965) and *Bryophyllum* (Wilkins, 1962a, 1967) are arrested, and hence a phase delay induced, when they are chilled or exposed to anaerobic conditions in some parts of the cycle, and are quite unaffected by similar treatment in other parts of the cycle.

The implications of results such as these are that over some parts of its cycle the operation of the circadian system is dependent upon metabolic energy, whereas over other parts of the cycle the system can operate independently of the availability of metabolic energy. The oscillating systems therefore appear to have the characteristics of a relaxation oscillator, that is one in which energy is taken in over one portion of the cycle and given out again over the other portion. Bünning (1958) had indeed suggested that circadian oscillating systems might in fact be of the relaxation type, but recent findings have shown that this is most unlikely to be the case. In both *Bryophyllum* and *Phaseolus* it has been found that the parts of the cycle over which low temperature and anaerobic conditions can be applied without delaying the phase are quite different. Since both types of treatment might be expected to have similar effects in reducing the availability of metabolic energy to the circadian system, it might have been anticipated that they would have had similar effects in similar parts of the cycle.

The induction of phase shifts in circadian systems undergoing steady-state oscillation in uniform environmental conditions can thus be achieved with a variety of stimuli. The effectiveness of all these stimuli depends upon the position in the cycle at which they are given. At least in *Bryophyllum* for treatments given in those parts of the cycle over which they are effective, the magnitude of the phase shift induced by light and high temperature depends upon when the treatment ends. For treatment with low temperature and anaerobic conditions, however, the delay is more closely related to the duration of treatment. The latter treatments thus appear to have the effect of holding the oscillating system in the phase it had reached at the onset of the treatment. This contrasts with the proposed mechanism discussed earlier for the phase shift induced by light and high temperature.

Control of period

In an oscillating system which is to form the basis for time measurement, control and regulation of the period is of vital importance. It has already been mentioned that the natural period of circadian oscillators is within the range 22–28 hours. In *Phaseolus*, varieties having distinctly different natural periods have been found (Bünning, 1935b). Within one variety, however, the period of the circadian oscillator is not precisely fixed, but can vary slightly with changes in environmental parameters.

In animals, the uniform white light flux to which the organism is subjected can modify the free-running period of the circadian systems. Although there are exceptions, animals generally react to the level of illumination in a predictable way and follow what has become known as Aschoff's Rule. Generally, as the light intensity increases the period of circadian oscillation becomes shorter in diurnal animals and longer in nocturnal animals (Aschoff, 1958). The circadian systems in plants do not show this response as a general rule although in *Gonyaulax* the period appears to become shorter with increasing light flux (Hastings and Sweeney, 1959). However, exposure to uniform radiant fluxes in different bands of the spectrum has been shown to influence the period of the circadian system in *Phaseolus multiflorus* (Lörcher, 1958). The spectral bands between 610 and 690 nm and between 690 to 850 nm increase and decrease respectively the period in comparison with that observed in other spectral bands and in total darkness.

The environmental factor most likely to affect the periods of circadian oscillating systems is temperature. Since it seems reasonable to suppose that the basic oscillating system in the plants comprises a number of interlocked chemical reactions, it might be anticipated that the frequency of oscillation would be a function of temperature. In most organisms that have been investigated this is, in fact, so, but to a much smaller extent than would have been expected. While the rate of a normal chemical reaction doubles or trebles with a rise in temperature of 10°C, the period of most circadian oscillators in plants increases or decreases by a factor of from 0·8 to 1·03 with such a temperature change.

Some examples of the changes in the natural period of circadian oscillators with temperature are shown in Table 18.2. These values refer to the steady-state

Table 18.2   Variation of the period of circadian oscillating systems with ambient temperature.

Organism	Temperature °C	Period hours	Reference
Phaseolus *multiflorus*	15	28·3 ± 0·4	Leinweber (1956)
	20	28·0 ± 0·4	
	25	28·0 ± 1·0	
Oedogonium *cardiacum*	17·5	20	Bühnemann (1955)
	25·0	22	
	27·5	25	
Gonyaulax *polyedra*	15·9	22·5	Hastings and
	21·0	24·4	Sweeney (1957)
	26·6	26·8	
Bryophyllum *fedtschenkoi*	16	23·8 ± 0·3	Wilkins (1962a)
	26	22·4 ± 0·4	
Avena sativa	15–17	24	Ball and Dyke
	26–28	24	(1954)

period, that is the period of oscillation after the organism has been kept in the stated conditions for a sufficient time for the period to attain a regular value. In a number of plants, such as *Bryophyllum*, *Oedogonium*, and *Gonyaulax*, the period shows a slight but significant dependence upon ambient temperature. In a few plants, for example *Phaseolus* and *Avena* no significant difference can be observed in the period of oscillation between approximately 15° and 25–28°C. On the other hand, there are one or two reports of a quite marked variation in the period with temperature. In *Cestrum nocturnum* the rhythm of flower opening has a period of 27 hours at 17°C and 31 hours at 13°C (Overland, 1960).

The slight change in the period of most circadian oscillators at different ambient temperatures clearly shows that the systems must incorporate some kind of temperature compensating process. The mechanism of this process is unknown. The temperature compensating mechanism of most circadian systems is not, by any means, perfect, but it does dramatically reduce changes in the period at different temperatures. In a few organisms it appears to reduce variation in the period below the level of statistical significance.

The operation of a compensating process can be observed after organisms are subjected to a single change of temperature. When plants of *Phaseolus multiflorus* are kept at 25°C, the period of the leaf movement rhythm is about 28 hours. After the plants are transferred to 15°C the period at first increases to 33·4 hours but then gradually decreases and finally attains a steady-state value somewhat less than 28 hours (Leinweber, 1956).

Modification of the period of circadian oscillating systems has been achieved with several other types of treatment. In one strain of *Phaseolus* the period was increased from the normal value of about 24 hours to between 30 and 35 hours, when the atmospheric pressure surrounding the plant was reduced to 30–44 mm of mercury (Bünning, 1935b). In addition, lengthening of the period has been achieved by chemical treatments such as the application of colchicine, phenylurethane (Bünning, 1956a; 1957a; 1957b), theobromine and thophylline (Keller, 1960) to *Phaseolus* plants. Even 2 per cent ethanol supplied in this way increased the period of oscillation in *Phaseolus* by 5 hours (Keller, 1960).

The increase in the period resulting from lowered oxygen tension is consistent with the results of exposing organisms to anaerobic conditions as described in the previous section. A reduction in the rate of oscillation over one portion of the cycle would give rise to the increased period. The mechanism whereby the various chemical substances lengthen the period is quite unknown at the present time.

The light effects

In the study of any photobiological response it is important to identify the pigment responsible for photon capture. This is usually achieved by determining an action spectrum which is, basically, a measure of the relative effectiveness of quanta of different energies in eliciting the response. The energy of a quantum is inversely proportional to wavelength. The action spectrum is then compared with the absorption spectra of the various pigments present in the organism. The pigment with the absorption spectrum which coincides exactly with the action spectrum is the one which is responsible for absorbing the light energy which initiates the response.

Most studies of the effect of wavelength of light on plant rhythms have been made with spectral bands which are too broad to allow the construction of an action spectrum. These studies have, however, given a general indication of the most active regions of the spectrum. In higher plants the red end of the spectrum is undoubtedly the most active in controlling the circadian oscillating systems. The rhythm in the growth rate of coleoptiles of *Avena sativa* is initiated by a change from continuous red illumination, in which the seeds have been germinated, to darkness (Ball and Dyke, 1954). In *Phaseolus multiflorus* the rhythm of leaf movement can be initiated by exposure of dark-grown plants to irradiation in the spectral band between 600 and 700 nm (Lörcher, 1958). On the other hand, while irradiation of plants with the band between 700 and 800 nm does not initiate the rhythm, it does have the effect of cancelling out the inductive effect of a previous exposure to the band between 600 and 700 nm. This cancellation can itself be overcome by a further exposure to the band between 600 and 700 nm. This type of response to these spectral regions is characteristic of responses brought about by activation of the pigment phytochrome (see chapter 14). There is other evidence that these two spectral bands are the most active in controlling the circadian oscillator in *Phaseolus*. It will be recalled that the period of oscillation can be lengthened or shortened by continuously exposing the plants to light in these regions of the spectrum (Lörcher, 1958). The oscillating system in the succulent plant *Bryophyllum* seems also to be affected only by longer wavelengths of visible radiation; prolonged exposure to the red end of the spectrum inhibits oscillation, and interruption of continuous darkness by a few hours of red light induces a phase shift (Wilkins, 1960b).

While experiments described hitherto reveal prominent activity in the red end of the spectrum they do not provide concrete evidence for the identification of the pigment responsible for photon capture. This can only be achieved when an action spectrum for the physiological response has been determined and so far this has been done with only one organism. The action spectrum for shifting the phase of the circadian oscillator in *Gonyaulax polyedra* (Fig. 18.3) shows a major peak in the blue region and a minor peak in the red region (Hastings and Sweeney, 1960). Unfortunately, the peaks of the action spectrum do not coincide exactly with the absorption spectrum of any of the pigments in *Gonyaulax*, but Hastings and Sweeney (1959) think it probable that one of the chlorophylls may be the primary photoreceptor pigment. Even with this action spectrum, therefore, it is not possible to identify with certainty the pigment responsible for mediating in this phase-shifting effect of light. An action spectrum has also been determined recently for shifting the phase of the circadian oscillating system in *Bryophyllum* (Wilkins, unpublished). It shows only one peak, located between 600 and 700 nm with maximum activity at about 630–660 nm. The activity of wavelengths between 600 and 700 nm in *Bryophyllum* is similar to that found in the case of *Phaseolus* but in *Bryophyllum* it has not yet been possible to demonstrate activity in the 700–800 nm spectral band, either in inducing a phase shift or in reversing the effect of a previous exposure to the 600–700 nm band. At present it is impossible to establish whether phytochrome is the photoreceptor for light-induced phase shifts in *Bryophyllum*. In the fungi, activity appears to be confined to the blue end of the spectrum. The evidence for this is largely indirect. It is known that white light can entrain and reset the phase of the circadian system in, for example, *Pilobolus* (Uebelmesser, 1954) and, as far as the author

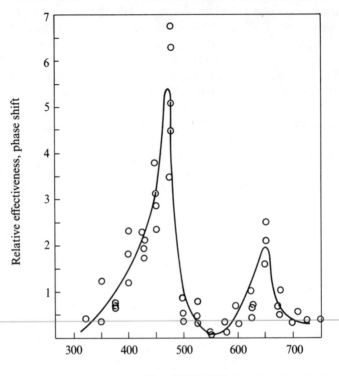

Figure 18.3 Action spectrum for shifting the phase of the rhythm of luminescence in *Gonyaulax* calculated as the relative effectiveness of equal incident quanta at each wavelength. (After Hastings and Sweeney, 1960.)

is aware, no red-light responses are known in the fungi. The implication is, therefore, that the activity must be associated with the shorter (blue-green) visible wavelengths. Sargent and Briggs (1967) have recently confirmed that inhibition of the rhythm of growth in *Neurospora crassa* by light is attributable to wavelengths below about 530 nm.

Quite different wavelengths of light are active in controlling the circadian systems in higher plants, in *Gonyaulax*, and in the fungi. This finding clearly indicates that different pigments are involved as photoreceptors for the light-induced responses in these organisms. In none of these organisms has the photoreceptor pigment been identified with certainty.

Ultraviolet radiation has been shown to induce stable phase shifts in the circadian oscillator in *Gonyaulax* (Sweeney, 1963). As with other light-induced phase shifts, the magnitude of the shift depends upon the time in the cycle at which irradiation occurs.

Location of the circadian oscillating system

In unicellular organisms there is little doubt that each cell has a circadian system controlling the rhythms manifest in its physiological and chemical activity. In multicellular organisms, however, the question arises as to whether the circadian system is confined to specialized cells, as seems to be the case in animals, for example, the cockroach, *Periplaneta americana* (Harker, 1958), or whether each cell in the organism has its own circadian system. Evidence obtained from the rhythm of carbon dioxide metabolism in leaves of *Bryophyllum* suggests that each cell has its own circadian system. Small cubes of leaf mesophyll from which the epidermis has been removed still exhibit circadian rhythms of carbon dioxide metabolism regardless of the region of the leaf from which they are taken (Wilkins, 1959). Furthermore, tissue cultures of leaf mesophyll cells of *Bryophyllum* also show circadian rhythmicity in carbon dioxide metabolism (Wilkins and Holowinsky, 1965). So in this plant at least, rhythmicity does not depend upon the organization of the leaf.

In other higher plants it is not clear whether the controlling circadian system is localized or not. In *Phaseolus*, for example, the periodic changes in the turgidity of the cortical cells of the pulvinus give rise to the rhythm of leaf movement (see chapter 8). These changes in cortical cell turgidity appear to be dependent upon the presence of the lamina (Brauner and Arslan, 1951), but it is not known whether they are due to a rhythm in the synthesis of a hormone in the lamina, to a rhythm in the transport of the hormone from the lamina to the pulvinus or to periodic changes in the sensitivity of the pulvinus cells to the hormone. All three possibilities could, of course, be involved simultaneously.

A most interesting and important problem in this general field of study is to ascertain the location of the circadian oscillating system within the cell. The solution of this problem is of fundamental importance since it demands the identification of the system's chemical mechanism.

Three organisms have been investigated with a view to ascertaining the intracellular location of the circadian oscillating system. The umbrella-shaped, giant unicellular alga *Acetabularia* has certain advantages as an experimental material in this type of investigation, since the nucleus is conveniently located at the base of the stalk and can therefore be removed easily. It has been found that the rhythm in photosynthetic capacity continues in enucleate and normal cells in a similar manner. Furthermore, the phase of the rhythm is easily shifted by light regardless of whether the nucleus is present in the cells or not (Sweeney and Haxo, 1961). This evidence suggests that the basic oscillating system is not located in the nucleus, and that the nucleus is not essential for mediating in the light-induced phase shifts. To assume that the nucleus has no role in the circadian system would, however, be unjustified. When cells of *Acetabularia* are grown in cycles of light and darkness differing in phase by 12 hours, the phases of the rhythm of photosynthetic capacity of the two stocks differ by 12 hours. If nuclei from one stock are now transplanted into cells of the other stock and the stocks then transferred to a uniform environment, the phase of the rhythm in those cells that have been subjected to nuclear transplantation corresponds to that of the light–dark cycle to which the nucleus has been subjected (Schweiger and Schweiger, 1965). This is shown in Fig. 18.4. The circadian oscillating system thus appears to be located in the cytoplasm or cytoplasmic organelles, and can operate entirely independently of the nucleus. When the nucleus is present, however, it apparently plays a dominant role in controlling the phase of the cytoplasmic oscillating systems.

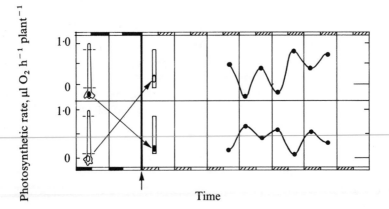

Figure 18.4   The effect on the rhythm of photosynthetic capacity in *Acetabularia* cells of nuclear transplantation between cultures which have been held on light–dark cycles differing in phase by 12 hours. Time of transplantation is shown by the arrow. The plants were afterwards kept in constant illumination but the phases of the previous cycles of light and darkness to which the cell was exposed are shown by the hatched area of bars at the top and bottom of the figure. Vertical lines are 24 hours apart. (From Schweiger and Schweiger, 1965.)

In enucleate cells of *Acetabularia*, the persistence of the rhythm, and the sensitivity of the phase to light stimuli, do not in themselves establish that nucleic acid metabolism is not involved in the circadian oscillating system. In *Acetabularia* considerable amounts of DNA have been found in the chloroplasts (Gibor and Izawa, 1963) and the synthesis of RNA has been demonstrated in cells from which the nucleus has been removed (Schweiger and Berger, 1964). In some recent experiments it has been found that the phase and period of the rhythm of photosynthetic capacity in *Acetabularia* are unaffected by the high concentrations of Actinomycin D, Puromycin, and chloramphenicol, which markedly reduce RNA and protein synthesis (Sweeney, Tuffli, and Rubin, 1967). At first sight these results point to circadian rhythmicity being independent of nucleic acid and protein synthesis. However, only a small fraction of the total nucleic acid and protein synthesis might be involved in the circadian oscillating system. Since a considerable fraction of RNA metabolism appears to be unaffected by Actinomycin D (Sweeney et al., 1967), the absence of an effect of this substance on the circadian system may be due to it having little or no inhibiting action on that fraction of nucleic acid synthesis which is involved in the circadian system. Even the experiments of Sweeney et al. (1967) do not provide unequivocal evidence which can resolve the problem of the extent to which nucleic acid metabolism and protein synthesis are involved in the circadian oscillating system of *Acetabularia*. It would be extremely valuable to know more of the chemical changes which accompany the rhythm of photosynthetic capacity in cells. For example, information on whether or not the amounts and specific activities of the enzymes responsible for photosynthetic carbon dioxide fixation vary periodically would indicate whether rhythmic enzyme synthesis is involved. If periodic enzyme synthesis were involved then obviously much interest must centre on nucleic acid metabolism being involved in the basic oscillating system.

In two other organisms attempts have been made to analyse the chemical changes occurring in cells exhibiting circadian oscillations. In *Gonyaulax* the rhythms of both luminescence and photosynthesis appear to be due to periodic changes in the level of enzymes involved in these reactions. The specific activity and total activity of luciferase has been shown to vary diurnally (Hastings and Keynan, 1965). Similarly, assays of ribulose-diphosphate carboxylase in homogenates of *Gonyaulax* cells showed a distinct cycle of activity, with approximately the same amplitude and phase as those of the rhythm in photosynthetic activity in the cell suspension from which the samples for enzyme assay were obtained (Sweeney, 1965).

These data suggest that there is a periodic synthesis of luciferase and ribulose-diphosphate carboxylase in *Gonyaulax* and one explanation for this periodic synthesis of enzyme is that there is a periodic synthesis of messenger RNAs coding for these two enzymes. If this were the case, inhibitors of RNA synthesis, such as Actinomycin D, and of protein synthesis, such as Puromycin and chloramphenicol, ought to inhibit the rhythm and induce phase shifts. The effects of these substances on the rhythm in *Gonyaulax* have been investigated and it has been found that Actinomycin D abolishes the rhythm in luminescence (Karakashian and Hastings, 1963). The inhibition is not, however, immediate; one peak of luminescence occurs after the beginning of the treatment. This finding suggests that if messenger RNA synthesis in *Gonyaulax* is totally blocked by Actinomycin D, a state of affairs which has not been checked experimentally but presumed from experiments on other tissues, synthesis of the RNA responsible for a particular peak of luminescence takes place some 24 hours earlier. In other words, RNA synthesis occurring at a particular time commits the cells to a programme of luminescence for the next 24 hours. The action of Actinomycin D supports the idea that the peaks of luminescence are due to rhythmic *de novo* synthesis of luciferase. However, the effects of inhibitors of protein synthesis only partially support this idea. Puromycin caused an almost immediate abolition of the rhythm, a finding consistent with the general scheme just outlined, but chloramphenicol had the effect of enhancing the amplitude of the rhythm which otherwise persisted normally! Substances which inhibit DNA synthesis ought to have little or no effect on the rhythm, if the rhythmicity is based on periodic RNA synthesis. Mitomycin, which inhibits DNA synthesis, had no effect on the rhythm of luminescence until about three days after application began, but then the rhythm was abolished. Karakashian and Hastings (1963) have suggested that this somewhat unexpected finding could be explained on the basis that DNA turnover occurs at a slow rate and that the delayed inhibition of the rhythm is due to the lack of newly synthesized DNA to replace that broken down.

Experiments involving the prolonged exposure of tissues to inhibitors are difficult to interpret. Abolition of the observed rhythm does not enable the inhibition of the circadian oscillating system to be established with certainty since inhibition of the observed physiological process or a breakdown of the linking mechanism between the oscillator and the observed physiological process might occur while the oscillator functions normally.

To determine whether or not a substance affects the basic oscillating system it is necessary to apply the substance for short periods of time. The induction of a phase shift in the rhythm by such a brief treatment would suggest an effect on the circadian oscillating system. Even this approach has severe disadvantages when chemical substances are being used since it is often difficult to remove the inhibitors from the tissues at the end of the treatment. Karakashian and Hastings (1963) used pulse-type treatments with inhibitors of nucleic acid and protein synthesis in different parts of the cycle, but could find no phase-shifting effects. A brief treatment with Actinomycin D inhibited the rhythm in the same way as a prolonged treatment; suggesting a strong retention of the inhibitor in the tissues. This approach has not, therefore, yielded additional evidence which can help in assessing the extent to which nucleic acid metabolism is involved in the circadian oscillating system in *Gonyaulax*.

Investigation of the rhythm of carbon dioxide emission by leaves of *Bryophyllum* has begun at the chemical level. The rhythm has been shown to be due to the periodic activity of a carbon dioxide fixation system in the leaves (Warren and Wilkins, 1961). Several enzymes have the capability of bringing about carbon dioxide fixation in these leaves in total darkness but the enzyme most likely to be principally involved is phosphoenolpyruvic carboxylase (Warren, 1964). Studies have been made of the causes underlying the rhythmic activity of the system responsible for carbon dioxide fixation in the leaf. There are large amounts of phosphoenolpyruvic carboxylase present in the leaf, even in those phases of the cycle in which the carbon dioxide fixation activity of the leaf is zero. In contrast to the findings on *Gonyaulax*, therefore, the rhythm of carbon dioxide metabolism in *Bryophyllum* does not appear to be due to rhythmic variation in the amount of enzyme in the leaves. Some other mechanism must clearly be involved—cycles in the availability of substrates for the carboxylation reaction, or in the concentration of an inhibitor of the fixation reaction are the two most obvious possibilities. When carbon dioxide concentration is non-limiting, infiltration of leaves with phosphenolpyruvic acid at times when carbon dioxide fixation was zero caused no increase in fixation. It seems un-likely, therefore, that substrate concentration is limiting when fixation by the leaves is extremely low. Possible periodic variation in the availability of co-enzymes has not yet been investigated. However, the rate of oxygen uptake by *Bryophyllum* leaves in darkness shows no rhythmic variation. The periodic appearance of an inhibitor of phosphoenolpyruvic carboxylase in the tissues has been examined. At least two inhibiting substances are present in the leaves, but they are present at all phases of the cycle of carbon dioxide fixation. One of these substances appears to be citric acid.

Interpretation of these results is rather difficult at this stage of the investigation but a tentative conclusion is that at least one of the inhibitors arises as an end product of the carbon dioxide fixation reaction. Further fixation might then be prevented until the concentration of inhibitor at the site of fixation is reduced by removal of the inhibitor to a new intracellular location (Warren, 1964).

The present primitive status of our knowledge of the chemistry of the rhythm in *Bryophyllum* obviously permits only a speculative assessment of the mechanism involved. The absence of periodic fluctuation in extractable enzyme activity does, however, point to the mechanism being somewhat different to that involved in *Gonyaulax*. At present there appears to be no evidence to suggest that nucleic acid metabolism is intimately involved in the basic oscillating system in *Bryophyllum* leaves.

Conclusions

Plants and animals at all levels of organization from the fungi to man appear to have developed the ability to measure the passage of time and to regulate their behaviour and metabolism on a temporal basis. The concept of the biological clock has arisen from the study of three biological phenomena: (a) endogenous circadian rhythms, (b) photoperiodism, and (c) orientation and direction finding in animals. Photoperiodic responses are discussed in detail in chapter 16 but they clearly indicate that both plants and animals have developed the ability to detect small changes in the length of the night and day. To achieve such a precise measurement of night length or day length a reliable reference system must be available, and the operation of this system must be largely independent of fluctuation in environmental factors such as ambient temperature. The basic oscillating system which controls circadian rhythms in plants and animals has all the characteristics that would be required in such a reference mechanism. Direction finding is achieved in many animals by means of the sun's position in the sky. In order to operate a solar compass an organism must be able to compensate for the relative movement of the sun across the sky during the day. There is now unequivocal evidence that animals use a biological clock system to achieve this compensation (Hoffmann, 1954).

Considerable survival value stems from the operation of the biological clock. Emergence of *Drosophila* flies from their pupae occurs just before dawn when the humidity is highest and they have the greatest chance of avoiding desiccation. The time of emergence is determined by the biological clocks of the pupae being set by the previous dusk. Synchronization of the times at which individuals of a species attain sexual maturity is of obvious significance. The tolerance of animals to drugs and ionizing radiation appears to vary with time of day (Aschoff, 1964), and in cockroaches serious pathological conditions arise when the physiological functions of the animals are not maintained on a circadian cycle (Harker, 1958). A comprehensive review of the survival value and adaptive significance of biological clocks in animals has recently been made by Aschoff (1964).

It is becoming increasingly obvious from studies of circadian rhythms in man that, quite apart from their intrinsic interest, biological clocks have important implications in the medical, sociological, and behavioural sciences (Wilkins, 1968). The full extent to which circadian oscillating systems regulate cellular metabolism and physiological activity in plants and animals will only be revealed by future research. At all events, elucidation of the mechanism of the biological clock will have consequences far beyond the boundaries of fundamental biology.

Bibliography

Selected further reading

Aschoff, J. (Editor), 1965. *Circadian Clocks*. North Holland Publishing Co., Amsterdam.
Bünning, E., 1967. *The Physiological Clock*. Longmans, Springer-Verlag, New York.
Chovnick, A. (ed.), 1960. 'Biological clocks', *Cold Spring Harb. Symp. quant. Biol.* **25.**
Emme, A., 1966. *The Clock of Living Nature*. Peace Publishers, Moscow.
Harker, J. E., 1964. *The Physiology of Diurnal Rhythms*. University Press, Cambridge.
Wilkins, M. B., 1968. Biological Clocks. *Advancement of Science*, **24**, 273–283.

Papers cited in text

Aschoff, J., 1958. Tierische Periodik unter dem Einfluss von Zeitgebern. *Z. Tierpsychol.* **15**, 1–30.
Aschoff, J., 1964. Survival value of diurnal rhythms. *Symp. Zool. Soc. Lond.* **13**, 79–98.
Aschoff, J., 1965. *Circadian Clocks, x–xix*. North Holland Publishing Co., Amsterdam.
Ball, N. G., and I. J. Dyke, 1954. An endogenous 24-hour rhythm in the growth rate of the *Avena* coleoptile. *J. Exp. Bot.* **5**, 421–433.

Ball, N. G., and I. J. Dyke, 1957. The effects of decapitation, lack of O_2, and low temperature on the endogenous 24-hour rhythm in the growth rate of the *Avena* coleoptile. *J. Exp. Bot.* **8**, 323–338.

Biale, J. B., 1940. Periodicity in transpiration of lemon cuttings under constant environmental conditions. *Proc. Am. Soc. Hort. Sci.* **38**, 70–74.

Brauner, L., and N. Arslan, 1951. Experiments on the auxin reactions of the pulvinus of *Phaseous multiflorus*. *Rev. Fac. Sci. Univ. Istanbul.* **16/B**, 257.

Bruce, V. G., and C. S. Pittendrigh, 1956. Temperature independence in a unicellular 'clock'. *Nat. Acad. of Sci.* **42**, 676–682.

Bruce, V. G., 1960. Environmental entrainment of circadian rhythms. *Cold Spring Harb. Symp.* **XXV**, 29–48.

Bühnemann, F., 1955. Das endodiurnale System der Oedogoniumzelle 11. *Biol. Zent.* **74**, 691–705.

Bünning, E., 1935a. Zur Kenntnis der endonomen Tagesrhythmik bei Insekten und bei Pflanzen. *Ber. d. Bot. Ges.* **35**, 594.

Bünning, E., 1935b. Zur Kenntnis der erblichen Tagesperiodizität bei den Primärblättern von *Phaseolus multiflorus*. *Jb. wiss. Bot.* **81**, 411–418.

Bünning, E., 1956a. Versuche zur Beeinflussung der endogenen Tagesrhythmik durch chemische Faktoren. *Z. f. Bot.* **44**, 515–529.

Bünning, E., 1956b. Endogenous Rhythms in Plants. *Ann Rev. Plant Physiol.* **7**, 71–90.

Bünning, E., 1957a. Über die Urethanvergiftung der endogenen Tagesrhythmik. *Planta*, **48**, 453–458.

Bünning, E., 1957b. Endogenous diurnal cycles of activity in plants. In *Rhythmic and Synthetic Processes in Growth*. ed. Rudnick. Princeton Univ. Press.

Bünning, 1958. Cellular Clocks, *Nature, Lond.* **181**, 1169–1171.

Bünning, E., 1967. *The Physiological Clock*. Longmans, Springer-Verlag, New York.

Bünning, E., and M. Tazawa, 1957. Über der Temperatureinfluss auf die endogene Tagesrhythmik bei *Phaseolus*. *Planta*, **50**, 107–121.

Bünning, E., S. Kurras, and V. Vielhaben, 1965. Phasenverschiebungen der endogenen Tagesrhythmik durch reduktion der Atmung. *Planta* (Berl.) **64**, 291–300.

Bünsow, R., 1953. Endogene Tagesrhythmik und Photoperiodismus bei *Kalanchoë blossfeldiana*. *Planta*, **42**, 220–252.

Busch, G., 1953. Über die photoperiodische Formänderung der Cloroplasten von *Selaginella serpens*. *Biol. Zbl.* **72**, 598–629.

Duhamel de Monceau, 1758. *La Physique des Arbres*. Vol. 2, Paris, pp. 158–159.

Flügel A., 1949. Die Gesetzmassigkeiten der endogenen Tagesrhythmik. *Planta*, **37**, 337–375.

Gibor, A., and M. Izawa, 1963. The DNA content of the chloroplasts of *Acetabularia*. *Proc. Nat. Acad. Sci. U.S.* **50**, 1164.

Harker, J. E., 1958. Experimental production of midgut tumours in *Periplaneta americana* L. *J. Exp. Biol.* **35**, 251–259.

Hastings, J. W., 1960. Biochemical aspects of rhythms: Phase shifting by chemicals. *Cold Spring Harb. Symp.* **XXV**, 131–143.

Hastings, J. W., and B. M. Sweeney, 1957. On the mechanism of temperature independence in a biological clock. *Proc. Nat. Acad. Sci. U.S.* **43**, 804–811.

Hastings, J. W., and B. M. Sweeney, 1959. The *Gonyaulax* Clock. In *Photoperiodism and Related Phenomena in Plants and Animals*, ed. Withrow. American Association for the Advancement of Science, Washington, D.C.

Hastings, J. W., and B. M. Sweeney, 1960. The action spectrum for shifting the phase of the rhythm of luminescence in *Gonyaulax polyedra*. *J. Gen. Physiol.* **43**, 697–706.

Hastings, J. W., and A. Keynan, 1965. Molecular aspects of circadian systems. In *Circadian Clocks*, ed. Aschoff. North Holland Publishing Co., Amsterdam.

Hoffmann, K., 1954. Versuche zu der im Richtungsfinden der Vögel enthaltenen Zeitschätzung. *Z. Tierpsychol.* **11**, 453–475.

Ingold, C. T., and V. J. Cox, 1955. Periodicity of spore discharge in *Daldinia. Ann. Bot.* **19**, 201–209.

Ingold, C. T., and V. J. Dring, 1957. An analysis of spore discharge in *Sordaria. Ann. Bot.*, N.S. **21**, 456–477.

Karakashian, M. W., and J. W. Hastings, 1963. The effects of inhibitors of macromolecular biosynthesis upon the persistent rhythm of luminescence in *Gonyaulax. J. Gen. Physiol.* **47**, 1–12.

Keller, S., 1960. Über die Wirkung chemischer Faktoren auf die tagesperiodischen Blattbewegungen von *Phaseolus multiflorus. Z. f. Bot.* **48**, 32–57.

Leinweber, F. J., 1956. Über die Temperaturabhängigkeit der Periodenlänge bei der endogenen Tagesrhythmik von Phaseolus. *Z. f. Bot.* **44**, 337–364.

Lörcher, L., 1958. Die Wirkung verschiedener Lichtqualitäten auf die endogene Tagesrhythmik von *Phaseolus. Z. f. Bot.* **46**, 209–241.

De Mairan, 1729. *Observation botaniqué. Histoire de l'Academic Royale des Sciences.* Paris, p.35.

Moser, I., 1962. Phasenverschiebungen der endogenen Tagesrhythmik bei *Phaseolus* durch Temperatur- und Lichtintensitätsänderungen. *Planta (Berl.)*, **58**, 199–219.

Oltmanns, O., 1960. Über den Einfluss der Temperatur auf die endogene Tagesrhythmik und die Blühinduktion bei der Kurztagpflanze *Kalanchoë Blossfeldiana. Planta*, **54**, 233–264.

Overland, L., 1960. Endogenous rhythm in opening and odor of flowers of *Cestrum noctornum. Am. J. Bot.* **47**, 378–382.

Pirson, A., W. H. Schön, and H. Döring, 1954. Wachstums- und Stoffwechselperiodik bei *Hydrodictyon. Z. Naturforsch*, **9b**, 350–353.

Pittendrigh, C. S., 1954. On temperature independence in the clock system controlling emergence time in *Drosophila. Proc. Nat. Acad. Sci. U.S.* **40**, 1018–1029.

Pittendrigh, C. S., V. G. Bruce, N. S. Rosenweig, and M. L. Rubin, 1959. Growth patterns in Neurospora. *Nature, Lond.* **184**, 169–171.

Sargent, M. L., and W. R. Briggs, 1967. The effects of light upon a circadian rhythm of conidiation in *Neurospora. Plant Physiol.* **42**, 1504–1510.

Schweiger, H. G., and S. Berger, 1964. DNA—dependent RNA synthesis in chloroplasts of *Acetabularia. Biochim. Biophys. Acta*, **87**, 533.

Schweiger, H. G., and E. Schweiger, 1965. The role of the nucleus in a cytoplasmic diurnal rhythm. In *Circadian Clocks*, ed. Aschoff. North Holland Publishing Co., Amsterdam.

Sweeney, B. M., 1963. Resetting the biological clock in *Gonyaulax* with ultra-violet light. *Plant Physiol.* **38**, 704–708.

Sweeney, B. M., 1965. Rhythmicity in the biochemistry of photosynthesis in *Gonyaulax.* In *Circadian Clocks*, ed. Aschoff. North Holland Publishing Co., Amsterdam.

Sweeney, B. M., and J. W. Hastings, 1962. Rhythms. In *Physiology and Biochemistry of Algae.* Academic Press Inc., New York.

Sweeney, B. M. and F. T. Haxo, 1961. Persistence of a photosynthetic rhythm in enucleated *Acetabularia. Science*, **134**, 1361–1363.

Sweeney, B. M., C. F. Tuffli, and R. H. Rubin, 1967. The circadian rhythm of photosynthesis in *Acetabularia* in the presence of Actinomycin D, Puromycin, and Chloramphenicol. *J. Gen. Physiol.* **50**, 647–659.

Uebelmesser, E. R., 1954. Über den endonomen Tagesrhythmus der Sporangientragerbildung von *Pilobolus. Archiv für Mikrobiol.* **20**, 1–33.

Wagner, R., 1963. Der Einfluss niedriger temperatur auf die Phasenlage der endogentagesperiodischen Blättbewegungen von *Phaseolus multiflorus. Z. f. B.* **51**, 179–204.

Warren, D. M., 1964. *Endogenous rhythms in the carbon dioxide fixation of plant tissues.* Ph.D. Thesis, University of London, U.K.

Warren, D. M. and M. B. Wilkins, 1961. An endogenous rhythm in the rate of dark-fixation of carbon dioxide in leaves of *Bryophyllum fedtschenkoi*. *Nature, Lond.* **191**, 686–688.

Wassermann, L., 1959. Die Auslösung endogen-tagesperiodischer Vorgänge bei Pflanzen durch einmalige Reize. *Planta*, **53**, 647–669.

Wilkins, M. B., 1959. An endogenous rhythm in the rate of carbon dioxide output of *Bryophyllum*. 1. Some preliminary experiments. *J. Exp. Bot.* **10**, 377–390.

Wilkins, M. B., 1960a. The effect of light upon plant rhythms. *Cold Spring Harb. Symp.* **XXV**, 115–129.

Wilkins, M. B., 1960b. An endogenous rhythm in the rate of carbon dioxide output of *Bryophyllum*. II. The effects of light and darkness on the phase and period of the rhythm. *J. Exp. Bot.* **11**, 269–288.

Wilkins, M. B., 1962a. An endogenous rhythm in the rate of carbon dioxide output of *Bryophyllum*. III. The effects of temperature changes on the phase and period of the rhythm. *Proc. Roy. Soc. Lond.* **B. 156**, 220–241.

Wilkins, M. B., 1962b. An endogenous rhythm in the rate of carbon dioxide output of *Bryophyllum*. IV. Effect of intensity of illumination on entrainment of the rhythm by cycles of light and darkness. *Plant Physiol.* **37**, 735–741.

Wilkins, M. B., 1967. An endogenous rhythm in the rate of carbon dioxide output of *Bryophyllum*. V. The dependence of rhythmicity upon aerobic metabolism. *Planta (Berl.)*, **72**, 66–77.

Wilkins, M. B., 1968. Biological Clocks. *Advancement of Science*, **24**, 273–283.

Wilkins, M. B. and D. M. Warren, 1963. The influence of low partial pressures of oxygen on the rhythm in the growth rate of the *Avena* coleoptile. *Planta*, **60**, 261–273.

Wilkins, M. B and A. W. Holowinsky, 1965. The occurrence of an endogenous circadian rhythm in a plant tissue culture. *Plant Physiol.* **40**, 907–909.

Zinn, J. G., 1759. Von dem Schlafe der Pflanzen. *Hamburgischen Magazin*, **22**, 40–50.

Index

THIS BOOK HAS BEEN SET IN MONOPHOTO TIMES NEW ROMAN 10 ON 12 POINT
AND PRINTED AND BOUND IN GREAT BRITAIN BY
WILLIAM CLOWES AND SONS, LIMITED, LONDON AND BECCLES